León
Tamaulipas
Gulf of Mexico
Hidalgo
D.F.
Tlaxcala
Veracruz
Puebla
Oaxaca
Tabasco
Chiapas
Campeche
Yucatán
Quintana Roo
Belize
Guatemala

Corozal
Orange Walk
Belize
Cayo
BELIZE
Stann Creek
Toledo
Petén
MÉXICO
Huehuetenango
Alta Verapaz
Quiché
Izabal
Cortés
San Marcos
Baja Verapaz
Zacapa
Santa Barbara
Totonicapán
Quezaltenango
Chimaltenango
El Progreso
Copán
Guatemala
Sololá
Chiquimula
Sacatepéquez
Jalapa
Ocotepeque
Intibuca
Retalhuleu
Suchitepéquez
Santa Rosa
Jutiapa
Lempira
Santa Ana
Chalatenango
Escuintla
La Paz
San Salvador
Cabañas
Ahuachapán
Cuscatlán
Morazán
La Libertad
San Vicente
Sonsonate
San Miguel
La Paz
Usulután
GUATEMALA
EL SALVADOR
0
100 km
Pacific Ocean

Caribbean Sea
HONDURAS
Atlantida
Colón
Yoro
Gracias a Díos
Olancho
agua
Francisco Morazán
El Paraíso
alle
Choluteca
NICARAGUA
DG

HONDURAS
Nueva Segovia
Jinotega
Atlántico Norte
Madriz
Estelí
Matagalpa
Chinandega
León
Atlántico Sur
NICARAGUA
Boaco
Masaya
Chontales
Managua
Granada
Carazo
Lago de Nicaragua
Rivas
Río San Juan
Caribbean Sea
Alajuela
Heredia
Guanacaste
Limón
Cartago
San José
Pacific Ocean
COSTA RICA
Puntarenas
Bocas del Toro
Chiriquí
0
100 km

PANAMÁ
San Blas
Colón
Panamá
Coclé
raguas
Darién
Herrera
Golfo de Panamá
Los Santos
COLOMBIA

SIDA, BOTANICAL MISCELLANY 23
BOTANICAL RESEARCH INSTITUTE OF TEXAS

TAXONOMY, DISTRIBUTION, AND ECOLOGY OF THE GENUS *PHASEOLUS* (LEGUMINOSAE-PAPILIONOIDEAE) IN NORTH AMERICA, MEXICO AND CENTRAL AMERICA

Taxonomy, Distribution, and Ecology of the Genus *Phaseolus* (Leguminosae-Papilionoideae) in North America, Mexico and Central America

George F. Freytag
&
Daniel G. Debouck

2002

ISSN 0833-1475
ISBN 1-889878-11-1

Sida, Botanical Miscellany, No. 23
Founded by
Wm. F. Mahler and Barney L. Lipscomb, 1987
Editor: Barney Lipscomb
Botanical Research Institute of Texas
509 Pecan Street
Fort Worth, Texas 76102-4060 USA
Contributing Spanish Editor: Félix Llamas
Dpt. Biología Vegetal, Universidad de León
Campus de Vegazana s/n, 24071 León, Spain
Cover Design: Linny Heagy

Taxonomy, distribution, and ecology
of the genus Phaseolus (Leguminosae–Papilionoideae)
in North America, Mexico and Central America
By George F. Freytag and Daniel G. Debouck

Front Cover Photograph:
Seeds of wild species of *Phaseolus*
Photo credit: Aracely Ospina CIAT
Back cover Photographs:
Flowers of *Phaseolus* species of extreme size:
background photo of *Phaseolus chiapasanus*
Photo credit: Arsenio Ciprian, CIAT
and bottom left hand insert *Phaseolus macvaughii*
Photo credit: Aracely Ospina CIAT

Inside cover maps:
The inside cover maps are provided for the easy reference to geographical names and locations mentioned in the text, and are not to be considered authoritative. They do not necessarily mean an endorsement neither of such names or boundaries of any kind by the authors nor by the institutions that have supported this publication or been involved in whatever way.

Distribution of copies by:
Botanical Research Institute of Texas
509 Pecan Street
Fort Worth, Texas 76102-4060 USA
Telephone: 1.817.332.4441
Fax: 1.817.332.4112
Website: www.brit.org/sida
Email: sida@brit.org

BRIT

Sida, Botanical Miscellany, No. 23
31 December 2002

Printed in the United States of America

Taxonomy, Distribution, and Ecology of the Genus *Phaseolus* (Leguminosae-Papilionoideae) in North America, Mexico and Central America

The following institutions, corporations, and individuals kindly provided financial support for the publication of this book.
The authors would like to thank them for their support.

Colorado Bean Network (CBN) is a non-profit organization which supports the dry bean industry in Colorado.

The International Center for Tropical Agriculture (CIAT) is a not-for-profit organization that conducts socially and environmentally progressive research aimed at reducing hunger and poverty and preserving natural resources in developing countries.

The International Plant Genetic Resources Institute (IPGRI) is the world's largest nonprofit agricultural research and training organization devoted solely to the study and promotion of agricultural biodiversity.

Rogers is part of Syngenta, the world's leading agribusiness company operating across all major areas of crop protection and seeds. The company is committed to sustainable agriculture through innovative research and technology.

The US Agency for International Development (USAID) is the government agency providing U.S. economic and humanitarian assistance worldwide for more than 40 years.

The United States Department of Agriculture (USDA) seeks to enhance the quality of life for the American people by supporting production of agriculture.

Mark J. Bassett

Paul Gepts

Table of Contents

New names and new combinations

The origin of a 250-years taxonomical problem:
Nature has found different ways at twisting a keel!

Photo by Juan Carlos Quintana, Centro Internacional de Agricultura Tropical (CIAT), Apartado Aéreo 6713, Cali, COLOMBIA

"Carina ... spiraliter contra solem revoluta ..." Linneaus 1754
"Phaseolus *carina spiralis* Vigna *omnia* Phaseoli
nisi carinaerostris vel rostro obliquo valde incurvo sed spiram
perfectamnon efficiente" Bentham 1865

and early descriptions were not that precise...

TAXONOMY, DISTRIBUTION, AND ECOLOGY OF THE GENUS *PHASEOLUS* (LEGUMINOSAE-PAPILIONOIDEAE) IN NORTH AMERICA, MEXICO AND CENTRAL AMERICA[1]

George F. Freytag

Research Plant Geneticist (retired)
Tropical Agriculture Research Station (TARS)
USDA/ARS, P.O. Box 70
Mayagüez, PR 00681
Present address:
Collaborator, National Seed Storage Laboratory
USDA/ARS, 1111 South Mason Street
Fort Collins, CO 80521-4500, U.S.A.

Daniel G. Debouck

Head, Genetic Resources Unit
Centro Internacional de Agricultura Tropical (CIAT)
Apartado Aéreo 6713
Cali, COLOMBIA

ABSTRACT

Review of many herbaria and recent field collections have shown the genus *Phaseolus* (sensu stricto—Verdcourt) to be more variable than previously recognized. This treatment encompases the geographical area from Panama northward, an area in which both authors have considerable field experience. We are recognizing 15 sections based on major morphological differences of previously recognized species and are describing 44 new taxa, many of which are of subspecific rank. The sections and subsections are: *Acutifolii, Bracteati, Brevilegumeni, Chiapasana, Coccinei, Coriacei, Digitati, Falcati, Minkelersia, Paniculati* subsects. *Volubili* and *Lignosi, Pedicellati, Phaseoli, Revoluti, Rugosi* and *Xanthotricha*. The new taxa are: *P. acinaciformis, P. albiflorus, P. albinervus, P. albiviolaceus, P. altimontanus, P. carteri, P. campanulatus, P. coccineus* subsp. *coccineus* vars. *argenteus, condensatus, lineatibracteolatus, parvibracteolatus, pubescens, semperbracteolatus, splendens, tridentatus*, and *zongolicensis, P. coccineus* subsp. *striatus* vars. *guatemalensis, minuticicatricatus, pringlei, purpurascens, rigidicaulis*, and *timilpanensis, P. dasycarpus, P. esquincensis, P. gladiolatus, P. leptostachyus* vars. *lobatifolius, nanus*, and *pinnatifolius* fs. *purpureus* and *albus, P. longiplacentifer, P. maculatifolius, P. magnilobatus, P. nodosus, P. parvifolius, P. persistentus, P. polymorphus* var. *albus, P. pyramidalis, P. reticulatus, P. rotundatus, P. scrobiculatifolius, P. teulensis*, and *P. trifidus*. The new combinations are: *P. coccineus* subsp. *coccineus* var. *griseus, P. coccineus* subsp. *coccineus* var. *strigillosus, P.coccineus* subsp. *striatus* var. *striatus, P. leptostachyus* var. *intonsus, P. maculatus* subsp. *ritensis, P. polystachyus* subsp. *sinuatus*, and *P. polystachyus* subsp. *smilacifolius*. Data about geographic distribution, ecology, habitats, and species affinities are presented as available.

RESUMEN

La revisión de varios herbarios y recientes colecciones de campo han mostrado que el género *Phaseolus* (sensu stricto—Verdcourt) es más variable que previamente reconocido. Este trabajo cubre la áerea geográfica desde Panamá hacia el norte, donde ambos autores tienen bastante experiencia en el campo. Quince secciones están reconocidas basandose en diferencias morfológicas principales de especies reconocidas previamente y estamos describiendo 44 taxa nuevas, muchas de las cuales son de rango subespecífico. Las secciones y subsecciones son: *Acutifolii, Bracteati, Brevilegumeni, Chiapasana, Coccinei, Coriacei, Digitati, Falcati, Minkelersia, Paniculati* subsects. *Volubili* y *Lignosi, Pedicellati, Phaseoli, Revoluti, Rugosi* y *Xanthotricha*. Los taxa nuevos son: *P. acinaciformis, P. albiflorus, P. albinervus, P. albiviolaceus, P. altimontanus, P. carteri, P. campanulatus, P. coccineus* subsp. *coccineus* vars. *argenteus, condensatus, lineatibracteolatus, parvibracteolatus, pubescens, semperbracteolatus, splendens, tridentatus*, y *zongolicensis, P. coccineus* subsp. *striatus* vars. *guatemalensis, minuticicatricatus, pringlei, purpurascens, rigidicaulis*, y *timilpanensis, P. dasycarpus, P. esquincensis, P. gladiolatus, P. longiplacentifer, P. leptostachyus* vars. *lobatifolius, nanus*, and *pinnatifolius* fs. *purpureus* and *albus, P. maculatifolius, P. magnilobatus, P. nodosus, P. parvifolius, P. persistentus, P. polymorphus* var. *albus, P. pyramidalis, P. reticulatus, P. rotundatus, P. scrobiculatifolius, P. teulensis*, y *P. trifidus*. Las nuevas combinaciones son: *P. coccineus* subsp. *coccineus* var. *griseus, P. coccineus* subsp. *coccineus* var. *strigillosus, P.coccineus* subsp. *striatus* var. *striatus, P. leptostachyus* var. *intonsus, P. maculatus* subsp. *ritensis, P. polystachyus* subsp. *sinuatus*, y *P. polystachyus* subsp. *smilacifolius*. Se presentan datos sobre la distribución geográfica, la ecología y los habitats, y las afinidades interespecíficas cuando disponibles.

INTRODUCTION

Phaseolus beans are a fascinating group! So much variability exists that perhaps over 400 species have been described since the 1700s and five distinct species have been domesticated already in pre-Columbian times; they were a staple food of many ancient American civilizations, and have since

[1]The research reported herein and its publication were funded in large part by the U.S. Government through the Agricultural Research Service, USDA and therefore are in the public domain and not copyrightable. It may be freely reprinted with customary crediting of the source. The publication costs of this article were defrayed in part by page charge payment and must therefore be hereby marked "advertisement" in accordance with 18 U.S.C. § 1734 solely to indicate this fact.

turned into a key part of the diet for millions of people throughout the world. Being nutritionally and economically important, beans have spread and cropping has intensified with increasing pressures from diseases and pests. Although the frequent object of many scientific studies (and of three important reviews in 1988, 1991, and 1999), beans are less known botanically as compared to other major crops.

How many *Phaseolus* species are there, and which are they? How do wild species relate to the cultivated ones? have been the central questions behind this study—and answers to them its main objectives—, for two co-authors who came to legume taxonomy because of their interest in bean breeding. Bean breeding with interspecific hybridizations (Fermond 1855) predates the founding of genetics by Mendel, who himself used two bean species in his experiments (Mendel 1865). While broadening the genetic basis of the crop continues to be a primary objective of bean breeders (see Kelly et al. 1999; Singh 2001, for reviews), wide crossing in spite of its long history has not raised the level of achievements as compared to tomato or potato. Among the different questions looking for explanations to this, the question whether the appropriate species were crossed seemed relevant (Debouck 2000a,b). Asking that question by the late 1970s would have pointed out to the fact that perhaps less than 200 accessions of wild *Phaseolus* were available to breeders, some of them misclassified. It was acknowledged that a review of the genus was overdue (Westphal 1974), and comprehensive and useful treatments of the genus at the generic level came shortly (Maréchal et al. 1978b; Lackey 1981,1983). The landmark world legume conference held in Kew in 1978 set a consensus view about the generic limits, but it was too early to indicate the exact content of *Phaseolus*. Delgado (1985) mostly using herbarium specimens (4,500 voucher specimens from 30 institutions) did the first exhaustive treatment applying the newly defined generic delimitations. When the co-authors separately carried out germplasm explorations from the late 1970s onwards, they anticipated that in addition to taxonomic knowledge breeders would need germplasm and insight about species relationships.

Linnaeus (1753) who defined the genus first recognized *Phaseolus* as a legume with a twisted keel and having small boat or canoe shaped pods, and described eleven species in his *Species Plantarum* only four of which are recognized as belonging to the genus as currently understood. In his prodromus De Candolle (1825) recognized three sections: *Euphaseolus* with compressed pods, *Strophostyles* with terete pods, and a third unnamed and uncharacterized section. *Euphaseolus* was divided into five subsections (*Caracallae*, *Perennes*, *Macropodii*, *Brachypodii*, and *Heterophylli*) with a total of 36 species. *Strophostyles* had terete pods and reniform or cylindrical seeds, and a total of 17 species. The third unnamed section had four species. Don (1832) in his system of gardening and botany followed this treatment and named some new species. Bentham (1837, 1840) divided the genus into seven sections some of which are now considered as genera, viz. *Macroptilium* or subgenera within *Vigna*. Bentham and Hooker (1865) recognized six sections and were generally followed by Hassler (1923) who also proposed new categories of series and subseries for some of the sections.

In a timely review of American Phaseoleae, Charles Vancouver Piper (1926) recognized eight sections for the 92 species (of which 38 were new), a system generally followed by Urban (1928), Morton (1944) and others. Urban (1928) separated *Macroptilium* as a genus to include (South American) species with three-angles keel. Verdcourt (1970) redefined *Phaseolus* and the other closely related genera of *Macroptilium* and *Ramirezella*, and separated *Vigna*. Maréchal and co-workers (1978a,b; 1981) further defined the differences between *Phaseolus* and *Vigna*, and Lackey (1983) endorsed the narrow definition of *Phaseolus* (by then often named as '*Phaseolus* sensu stricto'). Delgado (1985) further refined the classification and recognized four sections in the genus, viz. *Chiapasana*, *Phaseolus*, *Minkelersia*, and *Xanthotricha* for a total of 36 species. Recent changes about South American (mostly Brazilian) species of *Macroptilium* have been presented by Barbosa-Fevereiro (1986-87). Authors currently agree to place *Phaseolus* sensu stricto within the subtribe Phaseolinae, the tribe Phaseoleae, the subfamily Papilionoideae (see Isely & Polhill 1980), and the family Leguminosae.

Our treatment of the genus has tried to analyze in depth the variability revealed in herbarium desk study and during field work, and to assign appropriate level epithets to those distinctive taxa which can be easily recognized in accordance with the criteria mentioned below. The viable seed received from recent collections and from other scientists was especially valuable since it permitted obtaining data on germination, seedlings, roots, growth rates, daylength and temperature requirements, fertility, selfing, and crossability with other species both by ourselves and by others. In our field collecting we have made an effort to obtain viable seed of all taxa. Most of this seed and that which the senior author has been able to obtain from other sources (totaling 573 numbers) was planted out in replicated field trials at three locations in Puerto Rico (Mayagüez, Isabela and Ponce) and/or in screenhouse, greenhouse or environmental chambers for study and seed increase (see Sullivan & Freytag 1986). The original seed and that the senior author has been able to increase are on deposit at several institutions as is given in the section "List of Living Germplasm ..." near the end of this work.

We hope this revision will stimulate others to make additional field collections and pursue further studies to confirm or clarify our determinations. We also feel that many of the species and varieties in the genus are very recent and still actively evolving and so will change with time. We express our regrets that this study did not encompass all the herbaria that it might have, especially those in Mexico, some of which we could not have access to; this monograph therefore will be but a step along the way to a useful and correct classification of all the taxa within the genus.

TAXONOMIC CRITERIA

A generally held concept for a species level determination is that there must be a relatively free gene exchange among all indivuals of that entity in the wild. Those individuals are thus genetically compatible and share most morphological and genetical characteristics among the species populations in the wild. Unfortunately a taxonomist will seldom have good evidence of this from study of herbarium collections. Thus isolated populations from different portions of a species range may or may not seem to be distinct species depending on the individual taxonomist's evaluation.

In the first place this study has tried to distinguish clearly between wild and cultivated material, focusing on the former. If there has been a strong doubt whether a particular specimen came from a natural wild population, based either on morphology or on collection data, then that particular collection has been considered cultivated (escaped, introgressed or hybrid) and not used in the determination of species limits. Cultivated material has been considered previously (see Freytag 1955, 1965; Freytag et al. 1956; Westphal 1974) and in the present treatment generally receives the Latin epithet currently used.

Because of problems with precise species determination, we have tried to characterize species on the basis of three distinct criteria: 1) traditional species in the genus, that is, those species with which there has been little argument by plant taxonomists and plant breeders as to whether they are 'good' species, with or without gene exchange (introgression with other species); 2) presence of a genome containing a significant and distinctive group of morphological, anatomical, and/or physiological characteristics; and 3) a natural and distinct geographical distribution.

The subspecies category is used when there is a small number of major characters correlated with a well defined distribution and relatively little apparent gene exchange with other subspecies. The botanical variety category is used when there is a small number of relatively minor characters correlated with a less well defined distribution and often considerable gene exchange with other varieties. The botanical form category is used when there is only one or a few minor characters easily identifiable.

In order to set our taxonomic criteria, we have been using evidence from different disciplines of biology. While morphology of adult plants grown in natural original habitats as observed on herbarium voucher specimens (the authors have examined jointly or separately over 5,000 specimens from institutions listed in the 'Acknowledgments') has generated most of our data, other types of evidence have been used as well as explained hereafter. When none of us has seen a voucher specimen, and when one (or both) of the authors is sure about its location, it is indicated 'n.v.'

COMMENTS ON BEAN MORPHOLOGY

Previous taxonomic studies of the genus, especially those of Maréchal et al. (1978b), Lackey (1981, 1983) and Delgado (1985, 2000), have presented very exhaustive treatments of *Phaseolus* morphology, including pollen characters and chromosome numbers. We shall only mention a few characteristics in the following sections for the purpose of clarification of our viewpoints or to add information to these previous studies. Botanical terms are used along Hickey and King (2000), Stearn (1993), and Usher (1996).

Root

Although many herbarium sheets contain notes on the type of root, such as annual or perennial (i.e. that can resume growth after the first fruiting period), and fibrous to variously tuber-like, including size and shape, there are few complete roots mounted on the sheets, and fewer yet of photographs. The senior author's collection of *P. maculatus* in 1948 (see Color Plate III, photo 36) is the only one we have seen of the root of that species and one of the few photographs available for any species). We feel that this type of information is invaluable since we have found many variations in shape, as well as size, branching, and surface texture which seem to give precise diagnostic information for species and species groups. Wherever we have been able to obtain viable seed of taxa we have grown the plant out in the greenhouse and studied the roots for as long as we could keep them alive. This has often been quite difficult since the wild species have been notably susceptible to disease, more so as they enter maturity after a year or two. This was especially difficult with *P. maculatus* which we could seldom keep for a second year in the greenhouse in Mayagüez, though we have dug and photographed very large roots of this species in the field which indicates that they could live for many years. We have also dug and photographed many other species in the field. Most of this information has been incorporated into the drawings and descriptions. Unfortunately viable seed was not available for all taxa and so we cannot provide complete descriptions in many cases. Almost all taxa of the genus have been found to have perennial, thick and fleshy roots, often much branched and some quite woody (see Color Plate III, photos 25–36).

Stem and aerial shoot habits

Most plant habits are of some kind of indeterminate vine; however there is one species, *P. leptostachyus* (Section K. *Falcati*), which seems to be an indeterminate prostrate vine but is actually often determinate. An indeterminate vine refers to auxotelic axis development which continues vegetative growth producing lateral inflorescences only, and determinate plant type refers to anthotelic axis development with production of lateral inflorescences and finally ending stem growth in the production of a terminal inflorescence (Debouck 1983a,b; Lackey 1981, 1983).

Phaseolus leptostachyus seems to be the only case of a truly determinate plant type in wild material, and attention has not been called to this fact in any previous taxonomic treatment. Determinate plant types are well known to exist in cultivated *P. vulgaris* (Section B. *Phaseoli*) and in a few cultivated *P. coccineus* (Section C. *Coccinei*) and *P. lunatus* (Section D. *Paniculati*), but usually are not found in the wild types of these species. This fact makes one wonder if it is possible to derive the determinate plant type of the cultivars through introgression from *P. leptostachyus* as the third species involved in domestication as once proposed (Freytag, 1955) even though this latter species has only 20 chromosomes (x=10) while most of the rest for which data exist have 22 chromosomes (x=11). The senior author has observed fields of cultivated beans mixed with extremely variable wild plants of *P. leptostachyus* from small 6 inch tall plants to many medium sized plants near Telixtlahuaca just north of Cd. Oaxaca in Oaxaca, Mexico. It may be in this area where introgression has taken place that characteristics of this species have entered the common bean. As an alternate explanation, the junior author following Bassett (1988), Gepts (1999), Koinange et al. (1996), Leakey (1988) considers that all wild materials have the dominant *Fin* gene causing indeterminate growth habit, and thinks that the mutation to determinate growth habit—where trifoliolate leaves are replaced by primary bracts and axillary buds by racemes of racemes—is inherent in the functioning of terminal meristems in all species of the genus (see Debouck 1983a, 1991).

In addition to being indeterminate vines, most wild species, especially the very large ones, are vigorously climbing plants sometimes resulting in perennial lianas which climb up brush and shrubs and even small trees. The smaller plants are usually less aggressive and often found with grasses which are used as support of the flowering vines. There seem to be some vines which refuse to climb, rather preferring to grow prostate or pendant and lay over or under other vegetation in several ways such as using the stiff, extended stipules as hooks to maintain a foothold over other competitors or simply pendant down steep slopes or down eroding gullies (see Color Plate V, photo 55).

Another plant habit we have observed but not previously described is that of the erect woody shrub with vining tips, probably most pronounced in the Subsection II. *Lignosi* of Section D. *Paniculati*. These species are densely pubescent, shrubby plants from enlarged woody taproots, with long-lived perennial stems especially in the basal portions, and generally have glanduliferous stipules and stipels, producing a sweet exudate (see Color Plate II, photo 24) usually attracting ants (myrmecophily).

Leaves

Leaves of *Phaseolus* are typically trifoliolate, with petiole, petiolule (or foliar rachis), pulvini, stipules and stipels (see also Lackey 1981, 1983). Stipules are not produced below the point of insertion (Verdcourt 1970).

Although we have found that all primary leaves (eophylls) are unifoliolate, leaves formed by meristems of terminal or axillary buds after the second plastochron (Debouck 1983a) are generally trifoliolate as indicated by previous revisions. However we have found several exceptions to this. In some cases the first few basal leaves produced at the beginning of each season's growth are unifoliolate (*P. maculatus* subsp. *ritensis*) (see Color Plate V, photo 60) and in some taxa very healthy and vigorous plants will produce at least some pinnate leaves with 4 to 5 unequally sized leaflets (*P. leptostachyus* var. *pinnatifolius*) (see Fig. 77).

Uncinate or hooked hairs early noted by Doutt (1932) are often present on petioles and petiolules. Their taxonomic value was realized much later on, and they have proven to be a key character, already at vegetative stage, of Neotropical *Phaseolus* sensu stricto, *Centrosema* and *Clitoria* (Baudet 1973; Baudet & Maréchal 1976; Lackey 1978). Stipels are always present in normal leaves but are not always present on eophylls nor on unifoliolate leaves.

Previous treatments of the genus and other works (e.g. Buhrow 1983; Lackey 1983) generally indicate that leaf characteristics are unreliable for diagnostic purposes. However we feel that leaves of most species, though they may be quite different from base to apex of the plant (i.e. heteroblastic development, see Debouck 1983a,b, 1991) and are affected during development by environment, do strictly conform to species limitations and can often be used for determining precise taxa. It is by use of the total variability on a plant or within a population, and not the use of a specific leaf, that correct identification of a species can be made. A case in point is that of the lower leaves (on the first few internodes of the seedling plantlet) of *P. carteri* with greatly lobed leaflets (as an inverted heteroblastic development!) and which separates it from *P. filiformis* with lower leaflets nearly entire (see Color Plate V, photo 59).

Perhaps the most unique leaf characteristic is that shown by *P. leptophyllus* (Section H. *Revoluti*) where the leaflet has two lateral veins nearly parallel to the mid-rib on which the margins are revolute and appressed against the middle part of the leaflet during periods of drought stress (dried specimens). Almost all of the stomata are also located on the abaxial surface which certainly provides protection from water loss when the leaflet is folded under, and the mid-rib is also slightly winged below (see Figs. 62 & 63).

Another unique characteristic is that of *P. chiapasanus* whose leaves turn jet black on drying as is seen in almost all herbarium specimens, possibly due to a very high polyphenol content.

Inflorescences

The characteristics of the inflorescence have long been considered valuable for diagnostic use in the genus, especially the presence of uncinate or hooked hairs. In all taxa of *Phaseolus* at least a few hooked hairs are found on the peduncle, rachis or pedicels even though absent on all other parts of the plant (Section F. *Minkelersia*).

The basic inflorescence structure seems to be a panicle, but in most of the genus the lateral branches are extremely reduced to a flowering node of a fascicle of one, two, three or more flowers, thus the use of the term pseudoraceme (Debouck 1983b; Lackey 1981, 1983; Tucker 1987). However, with perennialism (i.e. continued growth of the plant and on the older, more basal inflorescences) under tropical conditions, sometimes the secondary lateral axes produced additional flowers (see Color Plate I, photo 2). There are also rare cases where flowers are born on cymes such as in *P. maculatus* subsp. *ritensis* (see Color Plate II, photo 22). Anthesis is generally acropetal in the pseudoracemes, and in the individual secondary racemes.

We have also made use of primary bract characteristics such as the incision of the blade margin with the formation of serrate lobes as originally observed in the species *P. oaxacanus* but found in other taxa as well (e.g. *P. grayanus*). Contrary to the feelings expressed by Delgado (1985) that this characteristic shows all degrees of variability, we have found that it is a rather constant diagnostic character for taxa and showing little evidence of variability from outcrossing. The exception occurs on the varietal level of *P. coccineus* var. *tridentatus* (Section C. *Coccinei*) where there is often poor penetrance of the gene effects and perhaps variability produced by introgression on the edges of the variety's main distribution. The bracts are mostly persistent in the genus with the exception of *P. talamancensis* (Section E. *Bracteati*) where they are usually caducous just after anthesis.

Bracteoles

Characteristics of the bracteoles have often been used (e.g. Piper 1926) as significant diagnostic characters within the genus and we have found them to be especially useful for determination of species and infraspecific variants. Bracteoles range from very small or practically non-existent to those which are very large and often of great variability. Other valuable characteristics of the bracteoles are pubescence, shape, venation, position of insertion, and duration (caducous, deciduous, or persistent).

Flower

Characteristics of the flower are certainly most valuable for diagnostic use at all levels (Delgado 1985; Lackey 1983; Maréchal et al. 1981); however they are in large part the most difficult to measure, describe and understand on dried specimens. Although we have tried to use these characteristics for differentiating taxa, we have not always been successful and for this reason have tried to provide outline drawings for the complete flower (side and frontal views), and for all the floral parts, made from fresh living material of as many of the new taxa and species as possible. Our hopes are that when this type of information is available for all taxa it can then be intelligently used for more precise classification.

Nevertheless, we would like to call attention to some characteristics that perhaps have not yet received adequate description and use. Among these are flower color and size, and of course the morphology of those structures involved in flower tripping or pollination, especially the coiled keel (see Color Plates I & II).

Flower colors are probably very important in the attraction of specific insects or birds for pollination. Collection labels are not always very useful for color determination since a single color will often be given various descriptive names by different collectors (see also Lackey 1983). Neither have we, nor others, made any use of filters or fluorescent light for determination of flower appearance as perhaps insects or birds might see them (see Sousa-Peña et al. 1996) and which we now believe might be a more scientific approach. On the other hand, the majority of species are of colors with a tinge of blue, mostly in the purple range, though some are violet, and in some groups red is predominant. As flowers age, the fresh color generally fades, often becoming greenish, yellowish or whitish and reds become purplish. In the wild material, white is very rare and is probably caused by a single epistatic gene effect and occurs in only a few species (e.g. in the Digitati group). It is much more common in cultivated material in association with other recessive traits (Leakey 1988). Red (known to attract hummingbirds; see Sousa-Peña et al. 1996) is only found in the Section C. *Coccinei*, and truly pink flowers (as well as scarlet) are only found in hybrids and cultivars of the 'Scarlet Runner'.

Flower size is certainly of great importance for determining the specific pollinator as very large and stiff flower parts usually require a larger, stronger pollinator than very small flower parts. The largest flowers not only have large standards but the wings are very large and showy and sometimes the keel is very long with several very large terminal coils (e.g. *P. chiapasanus* or Section I. *Digitati*). Some taxa of Section F. *Minkelersia* have large, nearly erect flowers (see Color Plate II, photo 13) instead of horizontal ones as found in most of the genus and are elongate, which results in only a minute force being required for tripping. We would guess that most of these are perhaps pollinated by butterflies, moths or flies, though field collection labels lack these data. Perhaps most striking of all are the very tiny flowers found in species *P. macvaughii* (see Color Plate II, photo 19). These flowers (about 7 mm long) are only a fraction of the size of others (perhaps 1/2 the size of the next smallest found in the same section) and certainly require a very small insect to pollinate them.

Elongate flowers, often hooded or nearly enclosed by the standard, are found in Section G. *Xanthotricha* (see Color Plate II, photos 14 & 23) where we find a peculiar flower structure described by Delgado (1985) as having coils in two opposing directions. We believe that this is in error as we find all coils in a counterclockwise direction with the first coil rather large and twisted somewhat laterally and bearing at its end two close, smaller coils twisted towards the center of the flower so that the tip of the stigma emerges directly forward and somewhat against the base of the first coil (see Color Plate II, photo 15). Carried to an extreme, there are species in this group which produce cleistogamous flowers which never open.

Fruit

The fruit is a legume or pod, borne on the aerial shoot; geocarpy is so far unknown in *Phaseolus* while it exists in many legumes (e.g. the groundnut), namely in some species of *Macroptilium*. Characteristics of the pods are very useful for diagnostic purposes and we have used them extensively to define the sections (see Color Plate IV, photos 44, 45 & 48 for some examples). Extreme pod types are: 1) many-seeded and linear-tubular, erect (Section F. *Minkelersia*); 2) papyraceous (Section J. *Rugosi*) and if 1-seeded rhombic (e.g. *P. microcarpus*); 3) short, broad, very hard and fibrous (sections *Paniculati* & *Falcati*; 4) very large, flattened and turning black at maturity (Section N. *Chiapasana*); and of course the many unusual pod types found in *P. vulgaris* produced by plant breeders for snap and edible green pods, though none of these are found in the wild (and thus not considered here).

Seed

We have found seed characteristics to be very useful for species and variety determinations (see Color Plate IV) but unfortunately the seed is unknown for many taxa. Wild collections of most species always have very small seed, while derived cultivars are always considerably larger. Usually population samples of seed show little variation in size and coloration (see Color Plate IV) but some collections show a great deal of variability (see Color Plate IV, photos 40–43). Extremes are found in *P. chiapasanus* seed which is large and disk-shaped, while in *P. esperanzae* it is nearly spherical and often jet black. Species whose seeds have a rough surface (rugose) are treated together in Section J. *Rugosi*.

We have tried to obtain viable seed from collectors of all of the *Phaseolus* taxa in order to study their characteristics and variability and to obtain plants to study morphology and the life cycles at three field locations in Puerto Rico and also in some cases in the greenhouse, screenhouse, and environmental chambers. Seed increase has often been produced from these plantings. A list of these collections is given near the end of the monograph in the section titled "List of Living Germplasm." Germplasm can also be obtained from CIAT by consulting its web site (http://www.ciat.cgiar.org).

Germination and Seedling

Probably the most consistently lacking of valuable diagnostic information for the wild bean species is that concerning the germination type and characteristics of the seedling plant (see Baudet 1974, for terminology and results about nine *Phaseolus* species). This information could be obtained by looking for seedlings near mature plants and then include them on the sheet with foliage and flower, or alternatively to collect seed to be germinated later. Nevertheless, we have found only a few collection sheets where this has been done. As an alternative, we have tried to obtain this type of information by germinating seeds in the greenhouse whenever seed is available; unfortunately viable seed is lacking for many species (see Color Plate V, photo 59).

As pointed out by Delgado (1985) the majority of species are hypogeal in germination, which generally correlates with thickened, fleshy, tap roots or tuberous roots, while epigeal germination is found in only a few species with mostly fibrous roots (and cultivars are mostly found of this latter type, with the exception of *P. coccineus*). We assume the former is therefore more primitive and that germination is a valuable diagnostic characteristic. In epigeal germination use could be made of early stem elongation, that is the length of hypocotyl, epicotyl and subsequent internodes, but again information is not available for many species.

The primary leaf or eophyll also is a valuable diagnostic characteristic as concerns size, shape and variegation, as well as the size and amount of reduction of the petiole and its parts, including the presence or absence of stipels. For some taxa (*P. parvifolius*, *P. ritensis*, *P. carteri*, etc) these characteristics are precise indicators for the species. As mentioned by Baudet (1974), seedling characteristics are fairly constant within a species; presence or absence of stipels on seedling petiolules in *P. lunatus* is one exception.

COMMENTS ON BREEDING SYSTEMS AND CYTOGENETICS

The genus contains both self pollinating and cross pollinating species, though the majority are of the former type. This characteristic is mostly dependent on the flower structure, especially the shape of the stigma with lateral introrse stigmas common in the former and terminal extrorse stigmas in the latter, and the amount of nectar produced at the basal disc. For most species it is not known for sure if there are self sterility factors; species such as *P. coccineus* are self-fertile (Ibrahim & Coyne 1975), but pollen needs to be brought on the stigmatic surface mostly by bees as early observed by Darwin (1858). However there do seem to be strong infertility barriers between most species, specially when they belong to other evolutionary phylums, except for those with cultivar derivatives within the same biological species (see Debouck 1999; Delgado 1985; Hucl & Scoles 1985, for reviews). Cleistogamy is shown in a few species usually by fertilization before the bud opens. Most species have 22 chromosomes (reviewed by Lackey 1979; Maréchal 1970), but several species have been reported to have a reduced number to 20 chromosomes (see Mercado-Ruaro & Delgado Salinas 1996, 1998). The chromosomes of *Phaseolus* species studied so far are extremely small, and without any peculiar morphology (with the possible exceptions of *P. chiapasanus* and *P. pauciflorus*: see Mercado-Ruaro & Delgado Salinas 1998), and thus of little use for taxonomic purposes.

COMMENTS ON RECENT RESEARCH IN MOLECULAR TAXONOMY

Recent research with molecular markers of DNA (total genomic, chloroplastic, mitochondrial, nuclear) or products of specific gene activities substantially confirm that the genus *Phaseolus* sensu stricto does represent a natural monophyletic group (Bruneau & Doyle 1990; Bruneau et al. 1995), while American species of *Vigna* of the subgenus Sigmoidotropis (i.e. *Vigna adenantha*, *Vigna gentryi*) and of related genera such as *Macroptilium* and *Misanthus* should be excluded from *Phaseolus* (Bruneau et al. 1995; Delgado Salinas et al. 1993).

Cluster analysis of electrophoretic patterns of seed storage proteins has been used to predict relationships between species (Baudoin et al. 1991; Mirkov et al. 1994; Schmit & Debouck 1990; Sullivan & Freytag 1986) with some accuracy where seed collections have been correctly identified. With improved techniques and additional seed collections, this technique can give additonal insight into the relationships among species.

There is some variation in cpDNA (Fofana et al. 1999; Llaca et al. 1994; Schmit et al. 1993) and in ribosomal genes (Jacob et al. 1995), which can be exploited for phylogenetic studies. The genus can be organized into natural groups, as indicated from ITS DNA sequencing data (Delgado et al. 1999; Gaitán et al. 2000), some of which have fairly close relationships while other groups show higher heterogeneity (see *Comments* under individual species). Unfortunately, given technical complexities and costs, most of the molecular research so far has been based on limited collections of species, thus limiting the full exploitation of exceptionally rich information (see Gepts 1993, for advantages of molecular markers over other kinds of diversity descriptors). As well noted by Moritz and Hillis (1996), it is not here to adopt 'an empty argument' in opposing the information provided by molecular markers and that coming from other disciplines; the botanist specially in the field has to find the most efficient way to give the right name to unknown material, often seeking the necessary convergences.

DISTRIBUTION AND HABITATS

Species and taxa in general have very definite and often distinctive distributions, and also colonize defined types of vegetation, in large part due to their genetic physiological makeup (see also Debouck 2000a). Distribution maps are provided for many taxa to point out areas for additional collecting.

Points on the maps where one or more collections have been made are represented by numbers or symbols and may be correlated with the data in the manuscript for each collection for which coordinates are given. The coordinates if given on the collectors labels have no parentheses; however if the senior author has determined them by using a number of maps (e.g. Dirección General de Geografía 1982) then they are given in parentheses. In the 'Specimens examined,' we have tried to give the readership a representative information about the geographic distribution; specimens are reported as per the (first) collector and his/her collection number, and accepted acronyms of herbaria where examined by direct visits or through loans.

THE GENUS *PHASEOLUS*

Phaseolus L. Sp. Pl. 723. 1753. Type Species: *Phaseolus vulgaris* L.

Minkelersia Martens & Galeotti. Bull. Acad. Roy. Sci. Bruxelles 10:200. 1843. Type Species: *Phaseolus galactioides* (Mart. & Gal.) Maréchal, Mascherpa & Stainier (= *P. pauciflorus* Sessé & Mociño ex G. Don).
Alepidocalyx Piper. Contr. U.S. Natl. Herb. 22:672. 1926. Type Species: *Phaseolus parvulus* Greene.

Aerial shoots are indeterminate, rarely determinate, usually annual, prostrate to climbing vines, counterclockwise (dextrorotary). ***Roots*** are annual and fibrous to perennial and more or less thickened, fleshy or woody and globose to elongate. ***Stems*** terete, branched near base, usually covered with reflexed-hispid or strigose and minute uncinate hairs. ***Stipules*** present, basifix. ***Leaves*** 1- (3-) to 5-pinnate, 5–20 cm long; petiole usually canaliculate and longer than the blade; petiolule (or foliar rachis) usually shorter as compared to the petiole; stipels usually present; pulvini at base of petioles and at base of leaflets; terminal leaflet rounded-ovate to narrow lanceolate, mostly symmetrical, frequently lobed, usually pubescent, rarely nearly glabrous; lateral leaflets similar, mostly inequilateral (basiscopic part of the lamina from midvein broader than acroscopic part) and usually broader. ***Inflorescence*** usually a pseudoraceme, or a panicle with few or numerous lateral branches, always with at least a few long and short uncinate hairs present; primary and secondary bracts present, usually persistent to early deciduous; no extra floral nectaries. ***Bracteoles*** present, variously inserted, usually caducous shortly after anthesis, rarely persistent through mature pod. ***Flower*** papilionaceous, zygomorphic to terminally asymmetrical, usually purple, rarely red or white; calyx usually bilobate, or campanulate to tubular, the upper 2 lobes variously connate, usually emarginate, the lower 3 lobes variously short triangular and subequal, rarely foliaceous; standard with a basal claw and auricles, the blade usually reflexed with the tip erect and emarginate; wings with basal claw and auricle variously affixed to keel, the blade variously obovate to rounded and usually spreading, the right one more revolute than the other; keel with 2 basal claws, the tubular portion with an upward bend and a terminal 1 1/2 to 3 coils; stamens diadelphous, a single vexillary stamen on the upper side with a short basal claw and a usually geniculate globose knob on the adaxial side tapering into a delicate filament, and 9 stamens united into a tube to slightly beyond the upward bend and then variously dividing into separate delicate filaments; a basal collar subtending the ovary usually denticulate and nectariferous; ovary usually nearly straight, only a few mm long, usually densely covered by white, silky pubescence, with few–many ovules; style longer than the keel, with a thickened terminal coil and bearded apex, not pronounced beyond the stigma; stigma usually apical or lateral, variously shaped and introrse to extrorse. ***Pod*** a two valved-legume, non septate, straight to falcate, somewhat flattened to rounded, often inflated, variously dehiscing at maturity by twisting lateral valves, sparsely to densely covered by hispid hairs and densely covered with minute uncinate hairs, sometimes reticulate in flattened pods, or striped, 1–20 seed/pod, a beak present, straight to recurved. ***Seed*** from a few mm to several cm long, oblongoid to spherical, rounded to angular, smooth to rugose, variously colored and patterned in one or more colors, often a black ring around hilum; hilum centric, rounded to oblong, small, with or without epihilum; lens small, divided. ***Seedling*** from hypogeal or epigeal germination; eophyll simple, ovate to ovate-lanceolate to nearly linear, entire, truncate to cordate, obscurely 3-nerved, acuminate to acute, often petiolate.

Distribution.—Native to the western hemisphere, mostly in the tropics and subtropics, from North America (mostly Mexico and scarce in Central America) to Argentina; possibly absent from Canada (although Sousa & Delgado 1993 reported it from southern Québec and Ontario) and Chile.

Color Plate I. Flowers of *Phaseolus* species

1. Wild *P. vulgaris* from the Gentry collections of Jalisco and Nayarit, Mexico in 1967 showing variation in size and coloration (grown in the field at El Zamorano, Honduras, 1967).

2. A multi-branched inflorescence of wild *P. vulgaris* (Gentry collections from Jalisco and Nayarit, Mexico) produced by 2-year's continuous growth in 1967 at El Zamorano, Honduras.

3. Wild *P. dumosus* showing flower shape and purple color, the very large, rounded wings and the long bracteoles (grown in the USDA environmental chamber at Mayagüez, PR in 1984 from seed of *Cojolún 78-Guat-6* collected near Semetabaj, Guatemala).

4. A carpenter bee (*Xylocopa*) visiting a flower of a *P. dumosus* cultivar in 1977 at the Lemani Experiment Station, University of Puerto Rico.

5. *Phaseolus glabellus* showing the flower shape and red color, the glabrous pedicel and the lack of a bracteole (grown in the USDA environmental chamber at Mayagüez, PR in 1984 from seed of *Debouck et al. 2075* collected near Cd. Maíz, San Luís Potosí, Mexico).

6. Wild *P. coccineus coccineus coccineus* showing flower shape and red color, the pubescent pedicel and the small bracteole (grown in the USDA screenhouse at Mayagüez, PR in 1984 from seed of *INIA, Mor. 662* collected near Progreso, Morelos, Mexico).

7. Wild *P. coccineus coccineus strigillosus* (collection *Freytag et al. 78-G-97*) growing over a pine tree near San Cristobal de las Casas, Chiapas, Mexico (photo courtesy N. Vakili).

8. *Phaseolus coccineus striatus purpurascens* showing flower shape and dark purple color, the large rounded wings, and the large bracteoles (grown in USDA environmental chamber at Mayagüez, PR in 1984 from seed of *Freytag et al. 81-4* collected at Tres Marias, Morelos, Mexico).

9. A flower inflorescence of *P. coccineus striatus timilpanensis* showing the very large bracteoles (photo courtesy of D.G. Debouck).

10. *Phaseolus polystachyus* showing the bright purple flower color, slight pubescence and flower shape (grown in the USDA greenhouse at Mayagüez, PR in 1982 from seed of TARS #127 collected from near the Horticulture Experiment Station, University of Florida at Gainesville).

11. *Phaseolus jaliscanus* showing the dark purple color with dark stripes on the standard, flower shape and dense pubescence on the inflorescence (grown in the USDA greenhouse at Mayagüez, PR in 1985 from seed of TARS #321 collected as *Buhrow M-25* from near Compostela, Nayarit, Mexico).

12. *Phaseolus marechalii* showing the rather stout and heavily pubescent peduncle, the flower shape, the dark purple, striped standard, the short pedicel and the minute bracteole (grown in the USDA greenhouse at Mayagüez, PR in 1984 from seed of *INIA, Mor. 623* collected near Cuernavaca, Morelos, Mexico).

Color Plate I. Flowers of *Phaseolus* species

Color Plate II. Flowers of *Phaseolus* species.

13. *Phaseolus pauciflorus* showing the erect buds and flower and the large, dark purple wings with pendant tips (grown in the USDA environmental chamber at Mayagüez, PR in 1984 from seed collected near UNAM, Pedregal, D.F., Mexico).

14. *Phaseolus hintonii* showing the nearly glabrous rachis and pedicel, the minute bracteoles, the shape of the flower, especially the hooded shape of the standard and the light purple color (grown in the USDA screenhouse at Mayagüez, PR in 1984 from seed of *Freytag et al. 81-26* collected near Los Mazos, Jalisco, Mexico).

15. Flower of *P. hintonii* showing the keel (standard removed) with a large first coil and the two greenish terminal coils (grown in the USDA screenhouse at Mayagüez, PR in 1989 from seed of *Freytag et al. 81-26* collected near Los Mazos, Jalisco, Mexico).

16. *Phaseolus albiviolaceus* showing the glabrous rachis and pedicel, the minute bracteoles, the thumb-shaped buds and the large showy, light purple flower (grown in USDA environmental chamber at Mayagüez, PR in 1988 from seed of *Debouck et al. 2063* collected near Cd. Victoria, Tamaulipas, Mexico).

17. *Phaseolus altimontanus* showing the slightly twisted standard and wings (grown in USDA environmental chamber at Mayagüez, PR in 1988 from seed of *Debouck et al. 1523* collected from Nuevo León, Mexico).

18. *Phaseolus angustissimus* showing the puberulent pedicel, minute bracteoles, flower shape and purple flower color (grown in the USDA screenhouse at Mayagüez, PR in 1982 from seed of *Buhrow SCAN* collected near Sunset Crater, Arizona, USA).

19. *Phaseolus macvaughii* showing the very small flower size and white color (grown in the USDA screenhouse at Mayagüez, PR in 1989 from seed of *Norvell M-6999*, possibly collected near Mazatlán, Sinaloa, Mexico).

20. *Phaseolus leptostachyus leptostachyus* showing the short, pubescent pedicels, minute bracteoles, flower shape and purple color (grown in the USDA screenhouse at Mayagüez, PR in 1982 from seed of *Freytag et al. 81-14* collected near Cd. Oaxaca, Oaxaca, Mexico).

21. The habitat and flower of *P. laxiflorus* which shows the very dense shrub vegetation with which the species is associated (field collection *Freytag et al. 78-Mex-29*, photo courtesy N. Vakili).

22. An atypical inflorescence (a compound dichasium or cyme) found on a plant of *P. maculatus ritensis* (grown in a USDA screenhouse at Mayagüez, Puerto Rico in July 1987 from seed of *Buhrow GVRI* from the Santa Catalina Mts., Arizona, USA).

23. Four different flower types found in the species (from l-r) *P. microcarpus*, *P. hintonii*, *P. maculatus maculatus*, and *P. leptostachyus leptostachyus* from flowering plants grown in the USDA screenhouse in Oct. 1982 at Mayagüez, Puerto Rico.

24. A leaf of *P. jaliscanus* showing droplets of nectar produced on the stipels when grown in insect free USDA environmental chambers at Mayagüez, Puerto Rico in May, 1985 (from seed of *Buhrow M21* collected near Tequila, Jalisco, Mexico).

Color Plate II. Flowers of *Phaseolus* species.

Color Plate III. Roots of *Phaseolus* species.

25. Wild *P. vulgaris* root, from the Gentry collections (USDA) made in Jalisco and Nayarit, Mexico in 1967, showing a two-year old root beginning to deteriorate from root rots but with a still vigorous stem in the upper left of the photo (grown in the field in 1967 at El Zamorano, Honduras).

26. A two-year old root of *P. glabellus* showing the shape and coloration of the multi-branched root (grown in the USDA environmental chamber at Mayagüez, PR in 1984 from seed of NI-820 = *Leroi s.n.* collected near Jalapa, Veracruz, Mexico).

27. A one-year old root of wild *P. coccineus coccineus coccineus* showing the root shape and size and the large lenticels (grown in the USDA screenhouse at Mayagüez, PR in 1984 from seed of *INIA, Mor. 663* collected near Progreso, Morelos, Mexico).

28. One-year old roots of *P. coccineus striatus purpurascens* showing size, shape and coloration and the prominent elongate lenticels (grown in the USDA screenhouse at Mayagüez, PR in 1984 from seed of *Freytag et al. 81-4* from Tres Marías, Morelos, Mexico).

29. One-year old roots of *P. polystachyus polystachyus* showing size, shape and coloration (grown in the USDA screenhouse at Mayagüez, PR in 1982 from seed of TARS#127 collected by *Bassett s.n.* at the Hort. Farm, Gainesville, Florida, USA).

30. One-year old roots of *P. pluriflorus* showing size, shape and coloration and the prominent elongate lenticels (grown in the the USDA screenhouse at Mayagüez, PR in 1982 from seed of *Freytag et al. 81-21-C* collected near La Calaverna, Jalisco, Mexico).

31. One-year old roots of *P. pauciflorus* which shows their size, shape and coloration (grown in the USDA screenhouse at Mayagüez, PR in 1982 from seed collected from near UNAM, Pedregal, Cd. México, D.F., Mexico).

32. A one-year old root of *P. magnilobatus* showing the nearly spherical shape and dark coloration (grown in the USDA screenhouse at Mayagüez, PR in 1982 from seed of TARS #132A, collected by *Freytag et al. 81-22* near Veludero, Jalisco, Mexico).

33. A one-year old root of *P. angustissimus* showing the long erect crown, narrow root and fibrous branched roots (grown in the USDA screenhouse at Mayagüez, PR in 1982 from seed of TARS #166, collected by *Buhrow SAAN* in the Organ Mts., New Mexico, USA).

34. A two-year old root of *P. leptostachyus leptostachyus* showing the shape and light coloration of the multi-branched root (grown in the USDA screenhouse at Mayagüez, PR in 1982 from seed of *Debouck 365*, from near Corrales, Zacatecas, Mexico).

35. A two-year old root of *P. oligospermus* showing the shape and coloration of the multi-branched root (grown in the USDA screenhouse at Mayagüez, PR in 1984 from seed of *Cojolun 78-Guat-1* collected near the microwave tower near San Juan Acatempa, Guatemala).

36. Probably a very old root of *P. maculatus maculatus* showing the very large size, shape and the reddish colored flesh. Photographed in the field near Cd. Cuauhtemoc at La Junta on Hwy 260, Chihuahua, Mexico in 1947 as *Freytag & Baxter 26* (photo courtesy J. Baxter).

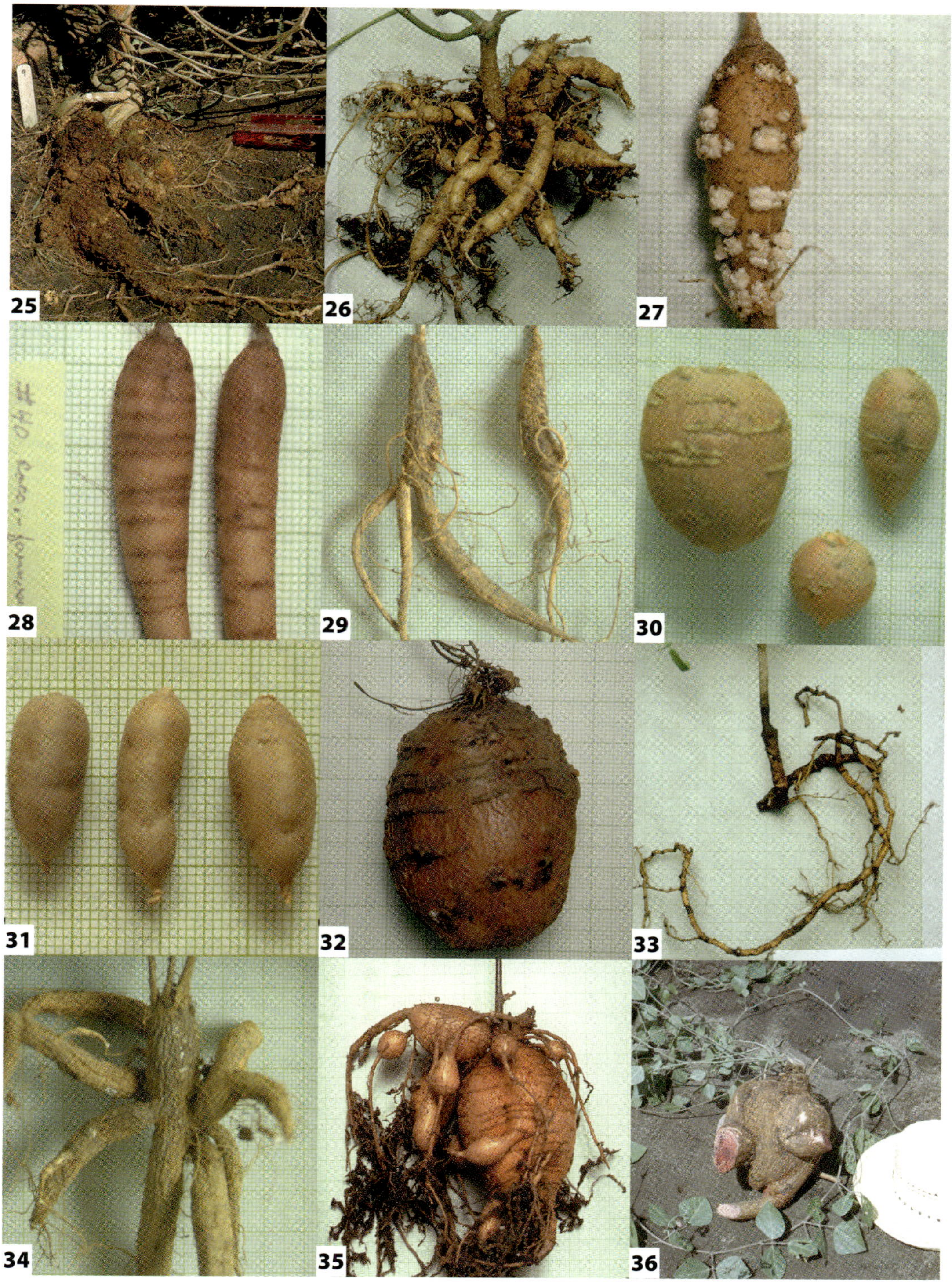

Color Plate III. Roots of *Phaseolus* species.

Color Plate IV. Seeds of *Phaseolus* species.

37. Wild *P. acutifolius acutifolius* seed showing the variation in size, shape, seed pattern and colors (grown in a field planting at the UPR Fortuna substation near Ponce, Puerto Rico in 1982).

38. Wild *P. acutifolius tenuifolius* seed showing the variation in size, shape, and seed pattern and colors (grown in a field planting at the UPR Fortuna substation near Ponce, Puerto Rico in 1982).

39. Wild *P. parvifolius* seed showing the variation in size, shape, pattern and colors from a random population sample (from the type collection of *Freytag et al. 81-13* collected at Km 35, on the road Mitla to Mixes, Oaxaca, Mexico).

40. Wild *P. vulgaris* seed showing the variation in size, shape, pattern and colors from a random population sample growing with *Zea diploperennis* (from *Freytag et al. 81-36* collected near El Chante, Jalisco, Mexico).

41. The original seed mixture showing the variation in size, shape, and seed pattern and colors of the mixed population sample (*vulgaris–coccineus*) growing in a nearly desert habitat (from *Freytag et al. 81-19* collected near Chiquilistlán, Jalisco, Mexico).

42. Wild *P. coccineus coccineus coccineus* seed showing the variation in size, shape, and seed pattern and colors of the *coccineus* portion of a mixed population sample (*vulgaris–coccineus*) growing in a nearly desert habitat (from *Freytag et al. 81-19B* collected near Chiquilistlán, Jalisco, Mexico).

43. Seed of *P. coccineus striatus purpurascens* showing the great variation in color and patterns in the field of a seed population sample (from *Freytag et al. 81-4* collected at Tres Marias, Morelos, Mexico).

44. Seed and pod of *P. hintonii* (grown in the USDA screenhouse at Mayagüez, PR in 1982 from seed of *Freytag et al. 81-26* collected at Mazos, Jalisco, Mexico).

45. Seed and pods of *P. oaxacanus* (grown in the USDA screenhouse at Mayagüez, PR in 1982 from seed of *Freytag et al. 81-11* collected at Km 15 on Hwy. Oaxaca-Tuxtepec, Oaxaca, Mexico).

46. Seed of wild *P. leptostachyus intonsus* showing the variation in size, shape, and seed pattern and colors (grown in a field planting at the UPR Fortuna substation near Ponce, Puerto Rico in 1982 from seed of NI 714a collected near Corrales, Zacatecas, Mexico as *Debouck 369*).

47. Seed of wild *P. filiformis* (grown in a field planting at the UPR Fortuna substation near Ponce, Puerto Rico in 1982).

48. Seed and pod of *P. angustissimus* (grown in the USDA screenhouse at Mayagüez, PR in 1982).

Color Plate IV. Plate IV. Seeds of *Phaseolus* species.

Color Plate V. Habitats and plants of *Phaseolus* species.

49. Joe Sullivan (l) and George F. Freytag (r) standing in a small plot of wild *Zea diploperennis* (wild corn) at bend in the road a few km before El Chante, Jalisco, Mexico and showing samples of wild beans (*P. vulgaris* of seed collection of TARS #200 = *Freytag et al. FS 81-13*).

50. Unique habitat near Chiquilistlán, Jalisco, Mexico where both *P. vulgaris* and *P. coccineus* are found growing with cacti (collection TARS #34 from *Freytag et al. FS 81-19*).

51. Plants of *P. vulgaris* from the Gentry collections (for USDA) of Jalisco and Nayarit, Mexico in 1967 showing variation in size and coloration (growing for the second year in the garden of Mrs. Freytag and daughter at El Zamorano, Honduras in 1967).

52. Habitat for *P. coccineus striatus purpurascens* on a grassy plateau in the pine forest near the mountain pass on the old road from Cd. Mexico to Cuernavaca at about Km 45 near Tres Marias, Morelos where the type collection, *Freytag et al. 78-M-4* was made.

53. The senior author talking with students at the *P. coccineus* living collection made by Prof. E. Hernández X. (deceased) at the National School of Agriculture at Chapingo in 1978.

54. Typical roadside embankment habitat, common in southern Mexico and contiguous Guatemala, where *P. leptostachyus* was found by the senior author in 1978.

55. Wild *P. leptostachyus leptostachyus* plants from two populations showing differences in their daylength sensitivity when grown under short day periods in the USDA screenhouse at Mayagüez, Puerto Rico in July, 1987 (l.–*Freytag et al. 78-M-39* from 1750 m, near Cd. Oaxaca, Oaxaca, Mexico; r.–*Debouck & Muruaga 423* from 1220 m, near El Tepozal, Nayarit, Mexico).

56. A plant of *P. micranthus* showing the prostrate habit and erect inflorescences with flowers (grown in the USDA screenhouse at Mayagüez, PR in 1982 from seed of *Buhrow M-28* collected from near Tuito, Jalisco, Mexico.

57. Russ Buhrow standing between small clumps of *P. maculatus ritensis* growing alongside the road to Mt. Lemmon, near Tucson, Arizona, USA.

58. Russ Buhrow standing in a desert thicket with *P. acutifolius tenuifolius* found in a dry canyon river bed near Tucson, Arizona, USA.

59. Plantlets of *P. carteri* (top two), *P. filiformis* (middle), and *P. macvaughii* (lower two) showing the differences in seedling eophyll and trifoliolate leaflet shapes and sizes (grown in the USDA screenhouse at Mayagüez, PR in 1982).

60. A young plantlet of *P. maculatus ritensis* showing the unifoliolate nature of the first mature leaves produced after the eophylls (grown in the USDA screenhouse at Mayagüez, PR in 1982 from seed of *Buhrow SRRI* collected in the Santa Rita Mtns. of Arizona, USA).

Color Plate V. Habitats and plants of *Phaseolus* species.

The center of main diversification is in the highlands of central (Jalisco to Mexico) and southern (Chiapas) Mexico and bordering Guatemala, with extensions towards southwestern and eastern United States, and with a secondary (perhaps later) center of diversification in the Andes. Five species have been domesticated.

SECTIONS

As seen above, very early taxonomic treatments of the genus recognized sections which included many diverse species (some of them now elevated to genera): De Candolle (1825)—three sections and four subsections; Don (1832)—three sections; Bentham (1837)—seven sections; Bentham and Hooker (1865)—six sections. Later treatments also recognized sections (some previously seen as genera): Piper (1926)—eight sections; Maréchal et al. (1978b)—three sections; Delgado (1985)—four sections; and Lackey (1983)—two sections, who also contributed to criteria for determining sections.

We have also followed the practice of recognizing sections which have been determined as a matter of convenience in order to group together species with similar morphological (and as far as possible genetical) characteristics so as to make more manageable sense of a very diverse number of species within the genus. Sections should thus reflect morphological similarities as well as phylogenetic affinities. Some might be more natural groupings than others, as the amount and kind of information available might be uneven. Results from molecular genetic research are beginning to confirm the validity of some sections as natural entities, while they also show some unexpected relationships between species of different sections which may eventually require a reassessment of the sections and species placement.

KEY TO SECTIONS

1. Leaflets 7–9 cm long, linear with revolute margins, the mid-vein laterally winged below, the two secondary veins as distinct and as long as the mid-vein giving the aspect of a leaflet with 3 parallel nerves ______ Section H. **Revoluti**
1. Leaflets not as above, usually broadly ovate, rarely lanceolate to linear, the mid-vein always more pronounced than the lateral veins.
  2. Plant a very large, more than 5 m long, climbing vine, all parts of which turn jet black on drying ______ Section N. **Chiapasana**
  2. Plant small to large, a prostrate to climbing vine or shrub, remaining greenish on drying.
    3. Plant a prostrate, non-twining, thick, straight stemmed vine; leaflets usually rhombic or nearly so, coriaceus, often waxy ______ Section O. **Coriacei**
    3. Plant usually twining, especially at distal ends of stems; leaflets mostly ovate to lanceolate, variously membranaceous.
      4. Inflorescence a panicle, sometimes with only 1–2 reduced secondary branches ______ Section D. **Paniculati**
      4. Inflorescence a pseudoraceme with secondary branches completely reduced to knobs or pads, sometimes bearing fascicles of 3 or more flowers.
        5. Primary bracts of inflorescence notably large, 8–30 mm long ______ Section E. **Bracteati**
        5. Primary bracts small to medium, 4–12 mm long.
          6. Calyx and standard somewhat tubular; calyx lobes mostly longer than tube; root usually globose, sometimes branched; seed 1–6 mm long and globose ______ Section F. **Minkelersia**
          6. Calyx variously campanulate; calyx lobes shorter than tube; root mostly thick and fleshy, seldom globose, sometimes fibrous; seed not globose, usually oblong or reniform and seldom very small.
            7. Keel structure of one large basal coil slightly twisted laterally bearing 2 small coils at the distal end ______ Section G. **Xanthotricha**
            7. Keel structure without a large basal coil, only bearing 1 1/2–2 small coils at the distal end.
              8. Seedling eophylls short petiolate, consisting mainly of a pulvinus; pod compressed with pronounced sutures; root fibrous ______ Section A. **Acutifolii**
              8. Seedling primary leaves petiolate, the petiole longer than the pulvinus.
                9. The standard in bud elongate and thumb-shaped; flowers 15–17 mm long and showy, pale violet or white, drying yellow or pale salmon ______ Section I. **Digitati**
                9. The standard in bud rounded, purse-shaped and usually slightly pointed at distal end; flowers usually 10–13 mm long, rarely much smaller, usually purple, sometimes red, rarely white.
                  10. Seed rugose or undulate knobby, sometimes pyramidal (pointed); pod small, chartaceous or papyraceous; bracteoles minute ______ Section J. **Rugosi**
                  10. Seed smooth, usually oblong or reniform; pod fibrous; bracteoles usually large, rarely small.

11. Pod usually less than 4 cm long, narrow, usually inflated and falcate; plant a prostrate or scandent small vine ______ Section K. **Falcati**
11. Pod usually more than 4 cm long, usually broad and flattened, fairly straight; plant a small or large climbing vine.
    12. Pod short, 5–6 cm long, and broad, of 2–4 ovules, covered by long, dense, yellow-brown strigose pubescence; leaflets ovate, truncate at base to somewhat basally lobed; plant a large climbing vine ______ Section L. **Brevilegumeni**
    12. Pod small to large, of 2–10 ovules, glabrous to sparsely covered by long strigose hairs; leaflets usually broadly ovate, sometimes variously lobed near base, plant a small to large climbing vine.
        13. Plant usually a small climbing vine from a thick, fleshy, elongate root; leaflets small, often lobed; bracteoles 1–3 mm long ______ Section M. **Pedicellati**
        13. Plant usually a large climbing or prostrate vine; leaflets broadly ovate, entire; bracteoles more than 3 mm long.
            14. Germination epigeal; root apparently mostly fibrous but often becoming somewhat thickened and fleshy after several years of growth; stigma usually lateral introrse to terminal; flower usually purple (rarely cerise or white) ______ Section B. **Phaseoli**
            14. Germination hypogeal; root mostly thick, fleshy or woody; stigma terminal, capitate or oblique extrorse; flower usually red, cerise or purple (rarely white) ______ Section C. **Coccinei**

Section A.—Acutifolii Freytag, sect. nov. Type Species: *Phaseolus acutifolius* A. Gray, Pl. Wright. 1:43–44. 1850.

Herbae annuae volubiles ad 4 m longae, radicibus fibrosis, lamina folioli pleurumque anguste linearis, inflorescentiae breves et pauciflorae, bracteolae minutissimae 1–3 mm longae, stigma laterale introrsum, legumen pleurumque compressum suturis manifeste et pleurumque trichomatibus uncinatis, folia plantulas petiolis brevia minus quam 1 cm longis.

Delicate annual vines to 4 m long, with fibrous roots; leaflet blades mostly narrowly elongate; inflorescences short, few-flowered; bracteoles very small, 1–3 mm long; stigma lateral introrse; pod mostly compressed, the carpels with pronounced sutures and mostly covered with distinctly uncinate hairs; seedling eophylls short petiolate (less than 1 cm long; usually reduced to the pulvinus).

Comments.—This small section consists of only two closely related species and their variants. As indicated by Buhrow (1983), four forms of *P. acutifolius* are known: domesticated variety *latifolius* (Freeman 1912, 1913; Baudet 1977), wild variety *latifolius*, wild variety *acutifolius*, and wild variety *tenuifolius*. Commonly known as tepary beans by the native peoples of the southwestern United States, they are found growing wild in dry gullies, stream beds and the foothills in and around the deserts or drier areas of the southwestern United States and northern Mexico and sparsely scattered south through the dry central highland areas to central Chiapas and Guatemala. All are small, slender, climbing annual vines with fibrous roots and in the wild populations have long acuminate to nearly linear leaflets. The cultivated teparies, distributed in arid Mesoamerica (Debouck & Smartt 1995; Nabhan & Felger 1978), generally have more broadly ovate leaflets and larger seed (domestication syndrome!) which are usually pintos variously speckled or flecked but also solid colors of white, tan or black. The seeds of some small white-seeded teparies are almost indistinguishable from seed of similar cultivars of the common bean (*P. vulgaris*) known as panamitos and some navy beans (Voysest & Dessert 1991).

The type description is based on a wild form, contrary to the case of the other *Phaseolus* cultivars (as noted by Debouck 1991; Pratt & Nabhan 1988). Pratt and Nabhan (1988) have provided a detailed study of the nomenclature of tepary bean and its wild variants. On the one hand, the collection record of wild populations is surprisingly sparse, and on the other hand ecological variation from biotaxonomic studies in the field has little been reflected in taxonomic treatments. In addition, molecular studies are just beginning to provide insight. The few collections of species *P. parvifolius* found in the herbaria have generally been confused with *P. acutifolius* var. *tenuifolius* which it closely resembles in having very narrow leaflets; however, in var. *tenuifolius* the base of at least some leaflets will be more or less lobed (more apparent in the inequilateral lateral leaflets) while leaflets of species *parvifolius* are always uniformly and equally tapered or rounded at the base. Additionally the pods of the later species are more rounded and less pubescent and the seed are smaller and smoothly oblongoid (see color Plate IV, photo 39).

KEY TO THE SPECIES AND VARIETIES

1. Lateral leaflets always distinctly inequilateral, narrowly lanceolate to broadly ovate, sometimes truncate to lobed at base, pods distinctly asperous by minute hooked hairs, seed small to medium (mostly more than 4 mm long), usually somewhat angular to smoothly rounded ________ A.1. **P. acutifolius**
   2. Lateral leaflets somewhat ovate to broadly ovate and rounded at base, seed 3.5–7 mm long, mostly rounded and thick but sometimes flattened with somewhat angular edges.
      3. Leaflets lanceolate to ovate-lanceolate, seed usually about 5 mm long, angular; very scarce in SW U.S. and NW Mexico; 500–1000 m ________ A.1.1. **P. acutifolius** var. **acutifolius**
      3. Leaflets ovate to broadly ovate and usually somewhat lanceolate, seed usually more than 5 mm long, thick, usually rounded on the ends, flattened and angular in the few wild populations (cultivated at lower elevations, in the drier, hotter areas along the Pacific coast from California to Panama, and sometimes in the drier, upland valleys and as far east as Campeche, Mexico); 335–1930 m ________ A.1.2. **P. acutifolius** var. **latifolius**
   2. Lateral leaflets narrowly lanceolate and truncate to sublobate at base, seed 3–5 mm long and distinctly angular; SW U.S. and adjacent NW Mexico and S to Jalisco; 500–1000 m ________ A.1.3. **P. acutifolius** var. **tenuifolius**
1. Lateral leaflets rounded or acute at base and scarcely inequilateral, very narrow and almost linear, 4–7 cm long, all parts of plant including pod covered by minute pubescence not noticeably uncinate, seed very small (mostly less than 4.5 mm long), oblongoid and smoothly rounded; scarce, through the drier highlands of Central Mexico from Durango, S to Chiapas and Guatemala; 1000–2000 m ________ A.2. **P. parvifolius**

A.1.1.—Phaseolus acutifolius A. Gray var. **acutifolius**, Pl. Wright. 1:43–44. 1850. (**Figs. 1–4**). TYPE: UNITED STATES. TEXAS: Mountain valley, 30 mi E of El Paso, (31°50'N, 106°4'W), Sep 1849, *Wright s.n.* (= *1311*, see Shaw 1987) (HOLOTYPE: GH, a single specimen?; ISOTYPE: K n.v. (according to Delgado, 1985, but not found in K).

Plant of delicate, climbing, indeterminate vines, 1–2 m long. ***Root*** annual, fibrous. ***Stems*** terete, slender, much branched near base, reflexed-hispid and covered with minute uncinate hairs, the internodes 10–15 cm long. ***Stipules*** triangular to lance-linear, 2–3 mm long, 1 mm wide, strongly 3-nerved, acute, puberulent, ciliate. ***Leaves*** 4.5–7.5 cm long; petiole delicate, 1–3 cm long, somewhat striate, puberulent; petiolule 5–12 mm long, uncinate hairs; pulvini, the lower 3 mm long, the upper 2 mm long, covered by strigose and long hirsute hairs adaxially; stipels linear, 1–2 mm long, 1–4 mm wide, 1-nerved, ascending; terminal leaflet lanceolate to ovate-lanceolate, 3–4 cm long, 0.75–1 cm wide near base, base deltoid, obscurely veined, acuminate to acute, apiculate, puberulent to nearly glabrous adaxially, puberulent to densely covered with small hispid and minute uncinate hairs abaxially, ciliate on margins, dull or infrequently shiny on adaxial surface; lateral leaflets similar, always distinctly inequilateral and usually broader. ***Inflorescence*** a small, few-flowered pseudoraceme; peduncle 0.5–5 cm long, shortest on upper portions of plant; rachis 0–8 cm long, long and covered by short, uncinate hairs; primary bract ovate-lanceolate to linear, 2–3 mm long, 1- to 3-veined, persistent to early deciduous; pedicel short and stout, 3–7 mm long, covered by minute uncinate hairs; pedicellar bracts linear 1 mm long 1-nerved covered with a few white hairs. ***Bracteoles*** narrowly lanceolate, 1–1.5 mm long, heavily 1-nerved, puberulent and ciliate, persistent often through mature pod. ***Flower*** purple to pink, 9–10 mm long; calyx campanulate, 3.5 mm long, the upper 2 lobes united into one rounded, slightly apiculate, 0.5–1 mm long, 2.5 mm wide, the lower 3 lobes triangular, acuminate, subequal, 1 mm long, 0.75 mm wide, the center one somewhat longer and with sparse white hispid and uncinate pubescence, ciliate on margins; standard purple, reflexed at 3 mm from base and 3.5 mm more to apex, 9 mm wide, broadly rounded, the tip erect, emarginate, the claw 1 mm long, the auricles pronounced, 1.5 mm long, 0.75 mm deep; wings purple to pink, the blade obovate to rounded, 7 mm long, 5–6 mm wide, the claw 2 mm long, the auricle 1.5 mm long, 1 mm wide, firmly affixed to keel, the right one more revolute than the other; keel, the claws 1.75 mm long, 1.5 mm more to bend and 2.5 mm more to base of the terminal 1 3/4 coils, the 1st portion of coil 2.5 mm in diam. and the terminal coil 2 mm in diam., green, the auricle 1 mm in diam.; vexillary stamen, the claw 1.25 mm long, the geniculate knob globose, 1 mm long, 0.75 mm wide, the thickened portion 1.5 mm more to filament; stamen tube 3.5 mm to bend and 3.5 mm more to divided filaments, 1.5 mm wide, the ridges scarcely developed; basal collar 0.5 mm long, denticulate; ovary straight, 5 mm long, 1 mm wide, densely covered by white, silky pubescence, 6–10 ovules; stigma lateral introrse, 0.8 mm long. ***Pod*** straight to falcate, somewhat flattened, 4–5 cm long, 5–7 mm wide, 4–6 mm thick; pronounced sutures and chartaceous valves early dehiscent, twisting tightly 2–3 times at maturity, the base somewhat stipi-

Fig. 1. Illustrations of plants of species of Section A. *Acutifolii*.–**a.** Seedlings several weeks after germination of seeds, except eophylls only of **1.a.** and **2.a.**–**b.** Roots after a couple of weeks of growth.–**c.** Terminal portions of stems with mature leaves, full-sized inflorescences with flowers and pods, except **2.c.** Mature leaf only of *Nabhan 610* (TARS #307) from Pima Co., Arizona.–**f.** Mature dehisced pods. All drawings made from living plants grown at Mayagüez from seed as indicated: **1.** *P. acutifolius* Gray var. *acutifolius* from PI 319439 (TARS #302); **2.** *P. acutifolius* Gray var. *latifolius* Freeman from PI 321637 (TARS #377); **3.** *P. acutifolius* Gray var. *tenuifolius* Gray from *Debouck 388* (TARS #158) from Mezquital, Durango, México; **4.** *P. parvifolius* Freytag from *Freytag & Sullivan 81-13* (TARS #52) from Zacatepec Mixes, Oaxaca, México.

tate, sparsely covered by white hispid hairs and densely covered by minute uncinate hairs, sometimes reticulate in flattened pods or striped in rounded pods, 2–9 seeds/pod; beak short and straight to recurved, 1–4 mm long. ***Seed*** oblongoid to rounded, angular, 4–6 mm long, 3.5–4.5 mm wide, 2–3 mm thick, speckled brown, gray, black on light tan or brown, a black ring around hilum; hilum rounded-oblong, 0.6–0.75 mm long, 0.4 mm wide; lens small, pronounced, not clearly divided. ***Seedling*** from epigeal germination; hypocotyl 1–2.5 cm long; epicotyl 1–3 cm long; eophylls opposite, simple, the blade ovate to ovate-lanceolate, truncate to subcordate, obscurely 3-nerved, acute, nearly glabrous adaxially, puberulent of minute uncinate and hispid hairs on abaxial surface; stipule at eophyll node, bifid; petiole much reduced to 1–2 mm; pulvinus 1–2 mm long; stipels absent.

Specimens examined: **MÉXICO. Baja California:** Agua del Palmillar, Cañon La Burrera, Sra. de La Laguna, 23°28'N, 109°59'W, 950 m, 14 Sep 1989, *León 3975* (US). **Chihuahua:** Villa Matamoros, approxim.20 km después Matamoros, carr. Parral–Durango, 1 km de vía, 26°40'N, 105°30'W, 1770 m, 2 Oct 1978, *Debouck et al. 286* (CHAPA, COL, K, MICH, US); ca 20 (air) km ENE of Cd. Jiménez, NW of summit of Sierra Chupaderos, 27°12'N, 104°43'W, 1524 m, 2 Oct 1973, *Henrickson 13753b* (TEX-LL); Sierra en Media, 28 Sep 1899, *Nelson 6482* (in part) (GH, K); Km 50 carr. Chihuahua–Cuauhtémoc, 1800 m, 28 Aug 1995, *Yen et al. 4646* (BRIT). **Durango:** Peñon Blanco, grown in screenhouse at Mayagüez, PR from seed of NI-843 (= TARS #4, collected in Peñon Blanco, 24°45'N, 104°10'W, 1750 m, 5 Oct 1978, by *Debouck et al. 300*), Sep.-Dec. 1985 as Study Collection *Freytag, G.F. #SC-4* (US, MO, CSU, ARIZ). **Jalisco:** Mpio. Autlán, entre Mezquitán y El Corcovado, 5 km NE of Autlán, 19°50'N, 104°17'W, 1000 m, 6 Oct 1995, *Santana et al. 7563* (BRIT). **Nayarit:** Mpio. Nayar, Las Cabezas, camino Jesús María–Cañaveral, 22°16'N, 104°35'W, 800 m, 29 Sep 1989, *Tenorio 16584* (MEXU). **Sinaloa:** Culiacán, grown in screenhouse at Mayagüez, PR from seed of NI 720 (= TARS #2, collected near Culiacán, Sin., Mex., 24°49'N, 107°11'W, 210 m, 2 Nov 1978, by *Debouck et al. 420*), Sep.-Dec. 1985 as Study Collection *Freytag, G.F. #SC-2* (US). **Sonora:** near San Carlos Bay, 24 Oct 1939, *Gentry 4728* (MO); 5 mi S of Mazatán, between Colorado and Mazatán, 7 Sep 1941, *Wiggins et al. 380* (MO); 2.5 mi N of Matape on road to Batuc, 710 m, 11 Sep 1941, *Wiggins et al. 464* (MO).

UNITED STATES. Arizona: Cochise Co.: 1 mi S of Paradise in Chiricahua National Area, (31°53'N, 109°9'W), 20 Sep 1958, *Fearing 2067* (TEX). **Pima Co.:** N of Guest House, Baboquivari Mts., (31°57'N, 111°35'W), 1311 m, 8 Oct 1944, *Gould et al. 2807* (in part) (CAS, US); Bowie, (32°16'N, 109°31'W), Sep 1884, *Jones 4252* (in part) (CAS-DS(3)). **Santa Cruz Co.:** Hank and Yank Spring, Sycamore Canyon National Area, (31°22'N, 111°8'W), 4 Sep 1977, *Nabhan GN701* (ARIZ). **New Mexico: Doña Ana Co.:** Organ Mts., (32°24'N, 106°52'W), 1372 m, 3 Sep 1897, *Wooten 528* (ARIZ). **Luna Co.:** Florida Mts., 0.5 mi NW of top of Baldy Peak, (32°3'N, 107°40'W), 168 m, 25 Aug 1990, *Worthington 18606* (UCR). **Texas: El Paso Co.:** Franklin Mts., 2 mi W of Farm Road 3255 on Transmountain Hwy, 1539 m, 4 Oct 1982, *Sanders et al. 3104* (UCR); Hueco Mts., E edge of the county N of Hwy 180, 140 m, 5 Oct 1982, *Sanders et al. 3127* (UCR); North Mt., Hueco Tanks State Park, 11 Sep 1988, *Worthington 17306* (UCR). **Jeff Davis Co.:** Buffalo Trail Scout Camp, Aguja Canyon, Davis Mts., 7 Aug 1966, *Correll 33340* (TEX-LL); W branch Goat Canyon, Mt. Livermore, (30°28'N, 104°6'W)?, 14 Sep 1935, *Hinckley 403* (F, TEX); Point of Rocks, 11.8 mi WSW of Fort Davis, 1664 m, 26 Aug 1990, *Worthington 18653* (UCR). **Presidio Co.:** near San Esteban Lake, Trans-Pecos, 1300 m, 31 Aug 1940, *Hinckley s.n.* (ARIZ, GH).

Distribution.—Very scarce in SW USA and NW Mexico, including Baja California, at elevations of 500–1,700 m.

Habitat.—Usually found growing along stream beds and dry washes in pine-oak forest, scattered among small bushes and cacti in somewhat drier places in comparison to var. *latifolius*.

Genetics.—According to Maréchal et al. (1978b), this species is midway between the *vulgaris-coccineus* complex and other species of the genus, with a strong relationship (87%) to *P. filiformis*. They recognized only two varieties within the species, but the senior author believes the type variety is also still recognizable at least from some collections though most show a great deal of intergradation. They also recognized the species characteristic of very short petiolate eophylls on the seedling plant.

Both species of this section are highly daylength sensitive, growing under natural field conditions during the long days of summer and beginning to bloom late August and throughout September. The type variety apparently has not been used in species intercrossing work. The six collections of this variety grown in Mayagüez have been completely fertile in the screenhouse and have produced copious pollen, setting most pods by selfing without requiring tripping. The seed are quite similar in appearance to var. *tenuifolius* but are considerably larger (see Color Plate IV, photo 37). The senior author has found with interspecific crosses in the greenhouse in Mayagüez that *P. parvifolius* will produce seed using varieties of *acutifolius* as the female parent, even though most of these crosses abort in 1–2 weeks, but the reciprocals are infertile, perhaps because of the warm night temperatures under these conditions (see discussion of this subject under Comments for species *parvifolius*). Recent work in the greenhouse at Colorado State University with a complete diallel cross between all the taxa of the section made by the senior author during the winter when greenhouse conditions are quite cool have shown fairly good reciprocal fertility between these species and among all varieties of the section. Such hybridization does not seem to take place in nature between these two species, perhaps due to spatial and habitat isolation as well as the relatively warm temperatures at the time of blooming in the areas where the species are found.

Comments.—True *P. acutifolius* of the type variety is rarely found in the herbaria and in the field. According to Buhrow (1983), it can still be found near El Paso (type locality), as well as in Sycamore Canyon of the Atacosa Mountains and in the Coyote Mountains of Arizona. Most collections seen of type variety *acutifolius* seem to be growing mixed with var. *tenuifolius*.

A.1.2.—Phaseolus acutifolius A. Gray var. **latifolius** Freeman, Ariz. Agric. Exp. Sta. Bull. 68:589–592. 1912. (**Figs. 1–4**). TYPE: UNITED STATES. ARIZONA: [Cochise Co.] Valleys of Sonora, 14 Sep 1851, *Wright 517* (= Gray's 949, see Shaw 1987) (HOLOTYPE: GH; ISOTYPES: GH, K, MO, US).

Phaseolus pauper Standl., Publ. Field Mus. Nat. Hist., Bot. Ser. 22:30. 1940. TYPE: MÉXICO. SONORA: San Bernardo, Río Mayo, 7 Sep 1935, *Gentry 1674* (HOLOTYPE: F; ISOTYPE: DES).

All characteristics similar to type species except for the following (data in parentheses refer to cultivated materials): ***Aerial shoot*** a small climbing vine 1–4 m long. ***Stipules*** lanceolate, heavily 4-nerved, appressed. ***Leaves*** 6.5–20 cm long; petiole slender, 2–6.8(–10) cm long; pulvini striate, heavily uncinate; terminal leaflet broadly ovate to ovate, 4–6.4(–9.5) cm long, 2.2–3.7(–5.2) cm wide, average ratio of length to width 1.74, base acute; lateral leaflets are generally the same as the terminal leaflet but slightly smaller and inequilateral, 3.8–5.6(–7.8) cm long, 2.0–3.3(–5.0) cm wide. ***Inflorescence*** peduncle to 2–5 cm long, densely uncinate; rachis 1–3 cm long bearing 2–5 flowers, densely uncinate; primary bract triangular to lanceolate, strongly 1-nerved, 2–2.5 mm long, 0.5 mm wide, puberulent and ciliate, persistent; pedicel stout, 5–6 mm long; pedicellar bracts linear 1 mm long sparsely ciliate. ***Flower*** purple (to pale violet or white), 15–20 mm long; calyx, broadly campanulate, 3–4 mm long, the lower teeth 1–1.25 mm long, 0.25–1.5 mm wide, pubescence on teeth and calyx scattered and fragile, mostly minute uncinate; standard purple (to white), squarish-rounded, broad, emarginate, 8–10 mm long, 10 mm wide, reflexed at slightly below the middle, thickened at flexure, terminal portion erect, biauriculate at the base, 0.5 mm long; wings purple (to white), the blade rounded to obovate to spatulate, somewhat cupped, 10–15 mm long, the claw narrow, 4 mm long, 0.3 mm wide, the auricle well-developed, 1.2 mm diam.; keel, the claws divided portion 2 mm long, 2 mm more to bend and 2.5 mm more to the terminal 1 3/4 coils of 2 mm diam.; vexillary stamen, the claw 0.8–1.2 mm long, the geniculate knob, 0.25 mm high; stamen tube 4 mm to bend and 3 mm more to divided filaments, the lateral ridges not developed; basal collar less than 0.5 mm long; ovary 4.5 mm long, 1.25 mm wide, 2–8 ovules; stigma 0.75 mm long. ***Pod*** straight or slightly curved and mostly rounded, somewhat flattened, 5(–7.3–10) cm long, 6(–10.5–13) mm wide, 6(–8) mm thick; valves deshiscing explosively (only weakly twisting at dehiscence), when young ciliate-puberulent of mostly uncinate hairs and a very few weak white hirsute hairs, when mature slightly uncinate or reticulate (smooth); beak straight to recurved, thin, 6–17 mm long; (2–)5–7 seeds. ***Seed*** angular, smaller, flattened and dark colored (rounded oblongoid to nearly spherical or sometimes strongly flattened), 5–7(–10) mm long, 3.5(–7.5) mm wide, 2.5(–5.5) mm thick, variously speckled brown (or bluish black to deep violet, seldom white or yellow), on a tan or gray background (very rarely red pinto flecked, often solid color in cultivars), surface usually smooth but sometimes wrinkled or pitted at maturity; hilum 0.75–1.25 mm long, 0.4–0.6 mm wide; placental tissue sometimes broken. ***Seedling*** hypocotyl 4.5–5.5(–7.5) cm long; epicotyl 5.7–8.7(–10.5) cm long; eophyll blades broadly ovate to ovate-lanceolate, mostly subcordate, acute to acuminate; petioles reduced to pulvini.

Specimens examined: **MÉXICO. Chihuahua:** Sierra en Media, 28 Sep 1899, *Nelson 6482* (in part) (US). **Coahuila:** Soledad, 25 mi SW from Monclova, (26°45'N, 101°37'W), 9–19 Sep 1880, *Palmer 261* (in part) (US); Cañon del Indio Felipe, Sierra Hechiceros, close to the Chihuahuan boundary, 28°33'N, 103°30'W, 27–29 Sep 1940, *Stewart 63* (GH). **Durango:** Durango, Castillo Najera, 1.5 km N 30 km E pueblo, 24°20'N, 104°28'W, 1930 m, 29 Sep 1978, *Debouck et al. 268* (CHAPA, COL, K, MICH, US); 21 mi NE of Cd. Durango along Rt. 45, 1828 m, 24 Oct 1966, *Gentry 22043* (US). **Nayarit:** 2 km S de San Juan Peyotán a Rancho Viejo, Mpio. Nayar, 22°27'N, 104°26'W, 700 m, 22 Sep 1989, *Flores et al. 1249* (MO); 2 km al N de Rancho Viejo, brecha a San Juan Peyotán, Mpio. Nayar, 22°20'N, 104°21'W, 710 m, no date, *Tenorio et al. 16325* (MO). **Queretaro:** Lomas Alamos, 16 Aug 1975, *Arguelles 264* (MEXU). **Sinaloa:** Mpio. Sinaloa de Leyva, Ejido Cuitanaca, 35 km N de Agua Caliente de Zevada, (25°49'N, 108°13'W)?, 14 Oct 1976, *Pérez 74* (in part) (CAS). **Sonora:** Río Bavispe just S of Bavispe, (30°29'N, 108°55'W), 20 Aug 1976, *Nabhan et al. X484* (in part) (DES); Agua

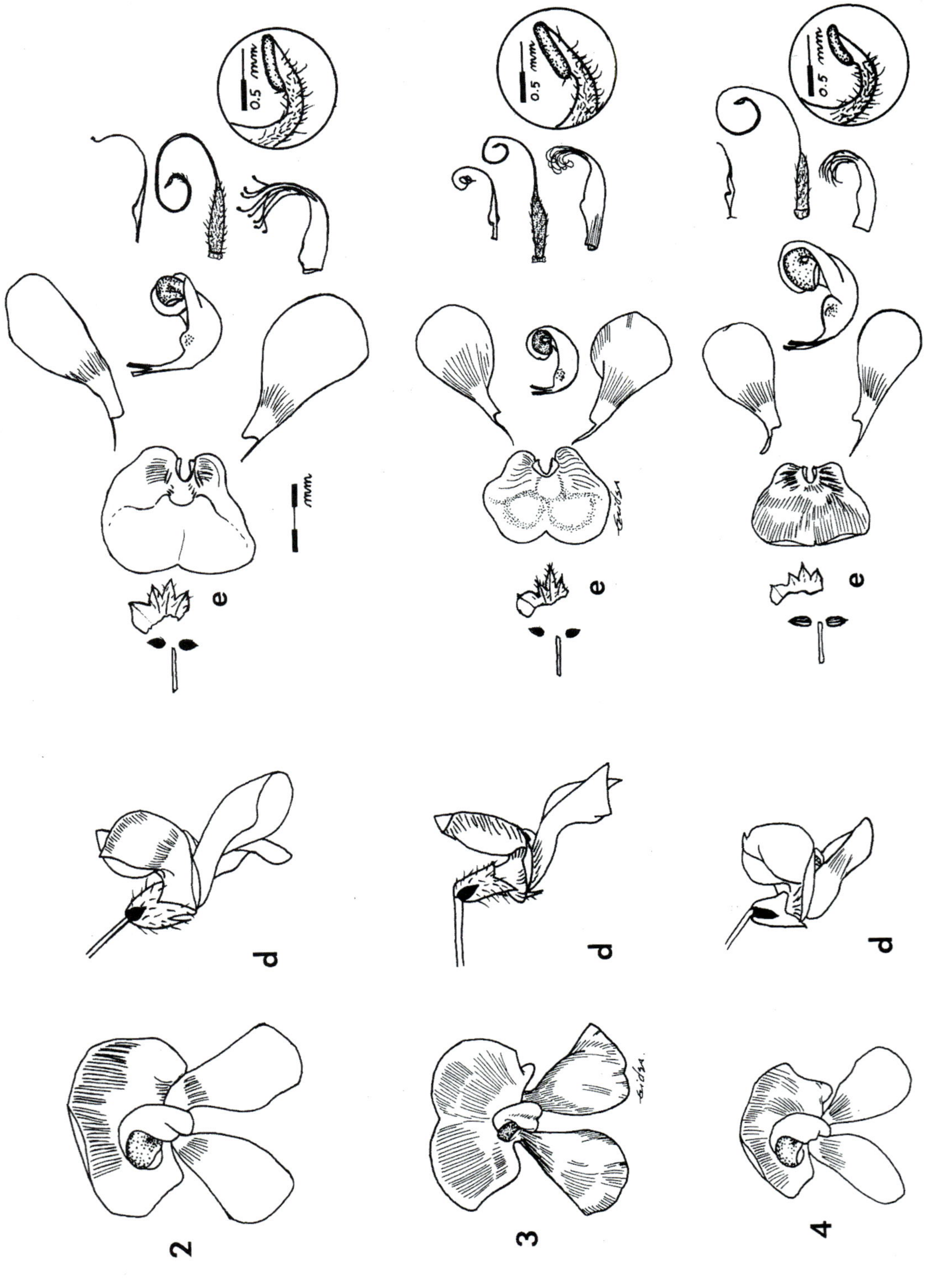

FIG. 2. Illustrations of flowers of species of Section A. *Acutifolii*.—**d.** Flowers, showing front and lateral views.—**e.** Exploded view showing all parts of the flower, including style tip and stigma as seen under the microscope. All drawings made from living plants grown at Mayagüez from seeds as indicated: **2.** *P. acutifolius* Gray var. *latifolius* Freeman from PI 321637 (TARS #377), except **2.d.** from *Nabhan 610* (TARS #307) from Pima Co., Arizona; **3.** *P. acutifolius* Gray var. *tenuifolius* Gray from *Debouck 388* (TARS #158) from Mezquital, Durango, México; **4.** *P. parvifolius* Freytag from *Freytag & Sullivan 81-13* (TARS #52) from Zacatepec Mixes, Oaxaca, México.

Nueva on dirt road to Soyopa from Tecoripa, (28°41'N, 109°59'W), 23 Aug 1976, *Nabhan et al. X505* (DES); 3 mi N of Arizpe, (30°23'N, 110°10'W), 26 Aug 1976, *Nabhan et al. X515* (DES); 1 mi S of Bacanuchi in Río Sonora drainage, 26 Aug 1976, *Nabhan et al. X516* (DES); microwave station, 15 km E of Navojoa above the road to Alamos, Río Mayo drainage, near 27°15'N, 109°17'W, 500 m, *Sanders et al. 9293* (in part) (DES); grown at UCR from seed (collected near San Bernardo, (27°30'N, 108°45'W), 305 m, Oct–Dec 1966, by *Gentry 22292*), 9 Nov 1981, *Waines et al. s.n.* (UCR); 10 mi from Cumpas on road from Baviacora, (29°57'N, 109°54'W), 23 Sep 1934, *Wiggins 7424a* (CAS-DS).

UNITED STATES. Arizona: Cochise Co.: Muleshoe Ranch Preserve, Hot Springs Canyon, 30 mi NW of Wilcox, (32°25'N, 110°15'W), 1219 m, 20 Sep 1990, *Rogers et al. 113* (UCR). **New Mexico: County Unknown:** banks of the Gila, 29 Aug 1880, *Greene 268* (K). **Texas: Brewster Co.:** Iron Mt., 29 Sep 1938, *Cory 30602* (GH). **El Paso Co.:** 1.2 air mi W Jct. Trans-Mt. Rd. with Gateway South, Franklin Mts., 31°53'48"N, 106°27'3"W, 1463 m, 20 Nov 1982, *Worthington 9220* (ASU). **Jeff Davis Co.:** 22 mi N of Ft. Davis on Rt. 17, 30°45'N, 101°45'W, 19 Sep 1966, *Correll 33663* (in part) (GH).

Distribution.—Wild types are usually found in mixed collections and are scarce in the drier areas of NW Mexico, SW US and Texas at elevations ranging from 335–1930 m, but most commonly at around 975 m (Buhrow 1983).

Habitat.—As wild it has been found growing to 4 meters length in the understory of open forests of spiny shrubs and cactus with sparse grasses in canyon bottoms and flood plains (Buhrow 1983). Nabhan (1979) also has found it to colonize sides of trails, abandoned fields and fence rows, and in arid subtropical short-tree forests, and believes it to be an opportunistic mesophyte growing during late summer rainy periods which permits it to "escape" most droughts. The cultigen is well known for its tolerance to heat and drought (Freeman 1913). Review of field collections shows it may also grow associated with wild *P. vulgaris* and var. *tenuifolius* (see Debouck 2000b).

Diseases and pests.—Cultivated tepary beans have long been known for their resistance to common bacterial blight (*Xanthomonas campestris* pv *phaseoli*) (Honma 1956); however, it is less common knowledge that the resistance in some cases is strain specific (Zapata et al. 1985) and due to a series of dominant (Drijfhout & Blok 1987) linked genes (Freytag 1989), and that some lines such as *Freytag & Vakili 78-Mex-114* (TARS #379) are highly susceptible (Freytag, unpublished). Perhaps one of the most serious diseases of teparies under irrigation is powdery mildew (*Erysiphe polygoni*) which causes defoliation and is prevalent during the cool, dry fall and winter months but is seldom seen in the tropics during the rainy seasons. Cultivated lines PI 485595, and PI 502217 (TARS #419) are highly resistant to mildew in Puerto Rico (Freytag, unpublished). Other diseases reported on tepary are reviewed by Kaiser (1981) and include: root rots—*Fusarium solani* f.sp. *phaseoli*; white mold—*Sclerotinia sclerotiorum*; rust—*Uromyces appendiculatus*; and virus—alfalfa mosaic, bean yellow mosaic, bean common mosaic, bean golden mosaic, curly top, pod mottle and four white fly-transmitted viruses.

Additional diseases seen on teparies in Puerto Rico are: ashy stem blight—*Macrophomina phaseoli*, root rot (field resistance noted by Federici, pers. comm.)—*Sclerotium rolfsii*, and soybean rust—*Phakopsora pachyrhizi* (Vakili, unpublished).

Plant pests on tepary are mostly those found on common bean, viz. leaf-eating beetles, leaf miners, leaf hoppers, flea beetles and leaf-eating caterpillars. Good sources of resistance to the leafhopper, *Empoasca kraemeri* Ross & Moore, have been found (see Debouck 1999).

Common names.—'Cuck bavi' and 'aap' are names given by the Papago and Seri native cultures, respectively for wild teparies (see Nabhan & Felger 1978).

Ethnobotany.—According to Kaplan (1956) teparies appear first in the Sacaton phase of the Hohokam at about 1,000 A.D. at Snaketown (white-seeded *P. vulgaris* appears first at about 300 B.C. at Tularosa Cave) and the oldest archaeological record of cultivated teparies (about 2,300 years before present) is from the Tehuacán Valley, Puebla, Mexico (Kaplan & Lynch 1999). Nabhan (1979) has observed that selection seems to have resulted in a reduction of seeds/pod (5.5 seeds/pod in wild vs. 4.5 seeds/pod for cultivated) as the selection for larger seed was made. There was also a selection from the black speckled seed of the wild collections for bright and solid colored seeds. Additionally primitive collections dehisce explosively and the seed (hard-seed) germinate sporadically, while domesticates dehisce slowly or not at all and germinate readily.

Genetics.—The var. *latifolius* is daylength sensitive, growing during the long days of summer in cultivated fields in the U.S. and beginning to flower during mid-August and throughout September.

Apparently it begins flowering a week or so later than the type variety and some of the cultivars continue to flower for a longer period than the wild collections.

Many collections of this taxon have been grown in the field at several locations in Puerto Rico and in the screenhouse in Mayagüez (see "List of Germplasm ..."), and all have been highly self-fertile without requiring tripping. It has been used in crosses with cultivars of *P. vulgaris* with which there is a degree of fertility, though embryo rescue is generally utilized to prevent later abortion (Alvarez et al., 1981; Pratt, 1983) and the production of allopolyploids has been used to reestablish fertility in the normally sterile F_1 (Prendota et al., 1982), and with var. *tenuifolius* with which it has been completely fertile reciprocally (Federici & Waines 1988; Waines, Stockinger, pers. comm.). In Puerto Rico it has only been receptive as a maternal parent with *P. parvifolius*, and that with difficulty as most of these crosses abort in 1–2 weeks (see discussion under *Comments* for species *parvifolius*). There is only one report of compatibility with *P. coccineus* (Coyne 1964) and that by use of *P. vulgaris* as a bridging species using *acutifolius* var. *latifolius* as the female parent with a backcross on *vulgaris* with a *vulgaris* × *coccineus* F_1.

From results of phaseolin (Schinkel & Gepts 1988) and isozyme (Garvin & Weeden 1994; Manshardt & Waines 1983; Schinkel & Gepts 1989) variability, it seems that the tepary might have been domesticated in only a very few places of arid Mesoamerica, and thus has limited probabilities of further genetic improvement. The potential for future tepary breeding might thus rely mostly on wild forms (Debouck 1992; Garvin & Weeden 1994).

Comments.—We have only listed above in exsiccatae those collections which seem to have been made from wild populations. Additionally we have seen 73 field collections of obviously cultivated materials.

The first publication by Freeman (1912) of his studies with tepary beans in which he proposed the new variety *latifolius* was in the Experiment Station Bulletin No. 68 of the University of Arizona. Apparently feeling this had only a limited circulation, he published extracts of the article, including the same illustrations, a year later in the Botanical Gazette. It is this latter publication which is generally cited in most revisions including Piper (1926), Maréchal et al. (1978b), and Delgado (1985). Unfortunately there is little agreement by most taxonomists on the validity of varietal status for *latifolius* (see Pratt & Nabhan 1988, for a discussion of the pros and cons). The senior author feels there is a basis for maintaining the variety, not only for the morphological differences but because he has found that there are wild populations still existing in nature today with the characteristics of larger, flattened seed and broader leaflets than on the type variety for the species. The junior author considers after Pratt and Nabhan (1988) that var. *latifolius* is a 'nomen confusum' because it has been assigned to two materials (an impossible treatment according to the Code of Botanical Nomenclature), viz. a wild material and the cultivated tepary (see Debouck 1991). From the molecular works done so far (Garvin & Weeden 1994; Muñoz et al. 2002; Schinkel & Gepts 1989), it is not certain that the cultivated tepary has originated from a wild population of var. *latifolius*; it could have originated from any of the other two botanical varieties and gained its larger leaflets, pods and seeds as a result of the domestication process.

Freeman based his work on his wide experience in collecting and studying cultivated teparies in the Southwest and felt that these materials were sufficiently different from the type species to merit variety recognition. Through correspondence with the Gray Herbarium he came to the conclusion that Gray had intended to name the cultivated types with a varietal name but failed to do so, therefore Freeman picked Wright's specimen at GH as his type (a tracing of which he received from GH and published as a line drawing in his original publication). Thus the type of var. *latifolius* becomes *Wright 949* (= *517*, see Shaw 1987) in the Gray Herbarium (see Delgado 1985). Somewhat later Freeman apparently visited GH and placed a handwritten note on the type specimen as follows: "This may be taken as the type specimen of *P. acutifolius* var. *latifolius*. See Bull. 68 Ariz. Agric. Exp. Sta. (signed) Geo. F. Freeman." Curiously enough this specimen also has written on it in Gray's hand "*Phaseolus acutifolius*, Gray var. *latioribus*." There is also a duplicate at GH, acquired from Herb. John A. Lowell, on which Gray had written "*Phaseolus acutifolius* Gr. var. fol. maj. Pl. Wr. 2:33."

Freeman carefully studied seed size and shape, as well as color and markings, and distinguished

47 distinct cultivars (from 216 collections). He considered the best cultivars to be whites, yellows, and brown speckled, but did not note any truly reds nor blacks (though some nearly black were in his collection) and recommended them for their earliness, heat and drought resistance. He clearly noted the differences between *P. acutifolius* and other species in eophylls, flower, pod, and seed.

Although teparies are grown in the Southwestern U.S., mostly by indigenous peoples, and to some degree in the dryer bean production areas on the west coast of Mexico and Central America, certainly there is a greater potential for extensive use of this crop throughout the world where drought is a major production constraint (National Research Council 1979; Debouck 1992). Teparies evaluated by African/Middle Eastern panels as used in typical dishes were found by Tinsley et al. (1985) to be highly acceptable compared with traditional legumes such as cowpea (which they most closely resembled), chickpea and faba beans. Protein contents in cultivated teparies is generally lower than in cultivated common bean (Waines 1978). Though higher in lectin activity, cooking resulted in lower levels of anti-nutritional factors (similar to Middle East legumes) and were considered safe for consumption (Scheerens et al. 1983; Tinsley et al. 1985). The protein of teparies was also deficient in sulphur-amino acids, as are most legumes, confirming earlier observations (Nabhan & Felger, 1978; Scheerens et al. 1983).

Cultivated forms of the tepary bean have been shown to be distant from both the common and the Lima bean (Jaaska 1996; Jacob et al. 1995; Skroch et al. 1993), and may have a secondary gene pool with *P. parvifolius* (Zink & Nagl 1998). The most recent molecular work using ITS DNA sequencing (Delgado et al. 1999; Gaitán et al. 2000) show the tepary bean remotely related to the common bean clade; there would thus be justification for a separate section '*Acutifolii*' but linked to the clade of *P. vulgaris*. This would in turn explain the fertility problems in wide crossing the two cultivated species (reviewed by Debouck 1999).

A.1.3.—Phaseolus acutifolius var. **tenuifolius** A. Gray, Pl. Wright. 2:33. 1853. (**Figs. 1–4**). TYPE: UNITED STATES. NEW MEXICO: Mountain-sides near the copper mines, New Mexico, and in Guadalupe Pass, (5–13?) Oct 1851, *Wright 716 & 768* (= Gray 950, see Shaw 1987) (HOLOTYPE: GH; ISOTYPES: G, GH, K, MO, UC, US). *Phaseolus tenuifolius* (A. Gray) Wooton & Standley, Contr. U.S. Natl. Herb. 16:140. 1913.

Phaseolus montanus Brandegee, Proc. Calif. Acad. Sci. Ser. 2, 3:130. 1893. TYPE: México. Baja California: Sierra de San Francisquito, (27°45'N, 113°15'W), 18 Oct 1890, *Brandegee 162* (LECTOTYPE: UC n.v., designated by Delgado (1985)).

Aerial shoot a scandent and climbing, indeterminate vine, 1–4 m long. ***Root*** an annual, long and fibrous. ***Leaves*** 9.5–18.5 cm long; petiole 3.5–4 cm long, striate, covered by uncinate and hispid hairs; petiolule 0.6–1 cm long, sparsely covered by uncinate hairs; pulvini 1.5 mm long, densely uncinate and hispid on adaxial surface, less so abaxially; terminal leaflet lanceolate to linear-lanceolate, 4–6–12 cm long, 0.3–5 cm wide, base hastate to somewhat sublobate, with a strong mid-vein, lateral veins not developed, tapering to an acute apex, often with large glands at base of hispid hairs on adaxial surface; lateral leaflets similar, at least the older and larger leaves inequilateral and frequently distinctly lobed. ***Inflorescence*** a short to long pseudoraceme; peduncle 1–20 cm long, densely covered by uncinate hairs; rachis densely uncinate; primary bract linear, 1 mm long, ciliate faintly 1-nerved; pedicel 4–5 mm long, moderately covered by minute and short uncinate hairs, slightly purplish; pedicellar bracts linear 1 mm long sparsely covered with minute whitish hairs. ***Bracteoles*** ovate-acuminate, 2–3 mm long. ***Flower*** purple to light purple, 8–9 mm long; calyx narrow campanulate, 2–2.5 mm long, upper 2 lips united, 3 mm wide, the lower 3 lobes acute, 0.75 mm long, 1 mm wide; standard reflexed at 2 mm from base and 4 more to apex, broadly flattened and rounded, the claw 0.5 mm long, the auricles pronounced, the thickened flexure white, the base green; wings purple to light purple, the blade nearly round, 6 mm long, 5 mm wide, the claw narrow, 2.5 mm long, the auricle greenish, loosely affixed to keel; keel white with green tip, the claws divided portion narrow, 2.5 mm long, 1 mm more to bend and 3 mm more to base of the terminal 1 1/2 coils; vexillary stamen, the claw 1 mm long, the geniculate knob 1 mm long, 0.5 mm wide, 0.5 mm high, the thickened portion of stamen 2 mm more to filament; stamen tube 3 mm long to bend and 3 mm more to divided filaments, 1 mm wide; ovary 4 mm long, 0.8 mm wide, white hirsute on sutures and long, white strigose on valves, 6–8 ovules. ***Pod*** small and narrow, 3–6 cm long, 5–6 mm wide, distinctly pubescent. ***Seed***

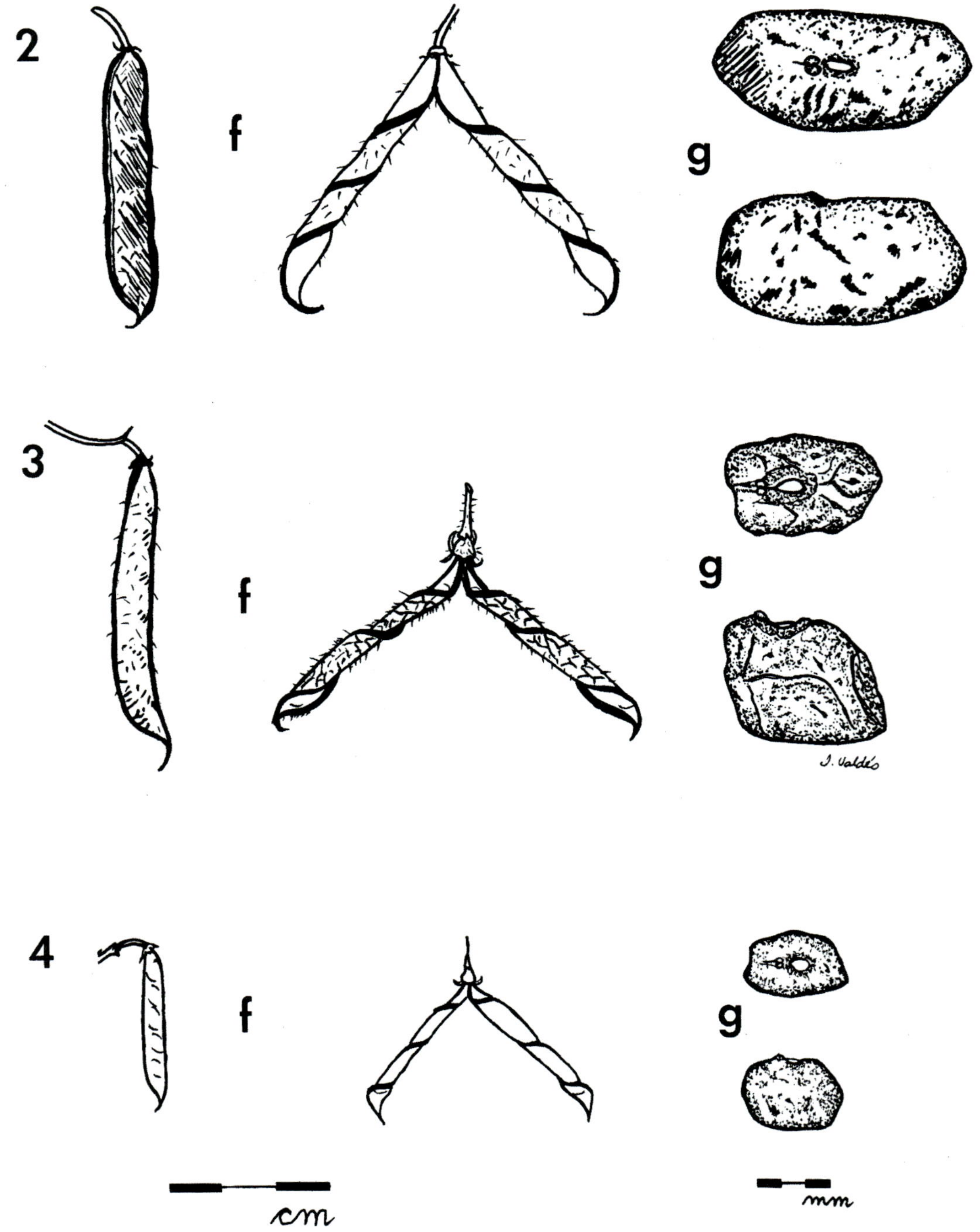

Fig. 3. Illustrations of pods and seeds of species of Section A. *Acutifolii*.—**f.** Mature pods, side view and dehisced.—**g.** Seeds, side view and view from hilum. All drawings made from living plants grown at Mayagüez from seeds as indicated: **2.** *P. acutifolius* Gray var. *latifolius* Freeman from PI 321637 (TARS #377); **3.** *P. acutifolius* Gray var. *tenuifolius* Gray from *Debouck 388* (TARS #158) from Mezquital, Durango, México; **4.** *P. parvifolius* Freytag from *Freytag & Sullivan 81-13* (TARS #52) from Zacatepec Mixes, Oaxaca, México.

squarish or trapezoidal, angular, 4–5 mm long, 4–5 mm wide, 2–3 mm thick, undulate and ridged sides, variously mottled and speckled black or brown on a brown or tan background. ***Seedling*** from epigeal germination; hypocotyl 4–6–7 cm long; epicotyl 4–6.5–7.2 cm long; eophylls, the blade broadly to narrowly lanceolate, 3–8 cm long, 0.3–1 cm wide, truncate to slightly lobed at base, acute to acuminate, petioles reduced to pulvini 1.5–2 mm long.

Specimens examined: **MÉXICO. Baja California:** Summit of Cerro Azufre, 27°30'N, 112°36'W, 1650 m, 20 Oct 1971, *Moran 18725* (ARIZ, UC). **Chihuahua:** Mpio. Batopilas, Km 27.3 Batopilas road, (27°N, 107°50'W), 1750 m, 21 Oct 1980, *Bye 10059* (UCR); near Chihuahua, 29 Sep 1886, *Pringle 728* (BM, F, G, GH, K, MEXU, MO, MSC, NA, UC, US(2)). **Coahuila:** Soledad, 25 mi SW from Monclova, (26°45'N, 101°37'W), Sep 1880, *Palmer 261* (GH, K). **Colima:** Tuxpan Canyon, (19°20'N, 103°40'W), 3 Nov 1910, *Orcutt 6525* (US). **Durango:** Villa Ocampo, Rcho. La Pastoria, Cerro abajo del Rancho, 26°20'N, 105°40'W, 2080 m, 3 Oct 1978, *Debouck DG et al. 287* (CHAPA, COL, K, MICH, US); Fco. I. Madero, Jeronimo Hernández, Lomas Charco del Indio, 24°27'N, 104°14'W, 2080 m, 6 Oct 1978, *Debouck et al. 308* (CHAPA, COL, K, MICH, US); Fco. I. Madero (?), Agua Blanca, cerro El Mezquite, 24°22'N, 104°18'W, 2070 m, 28 Oct 1978, *Debouck et al. 401* (CHAPA, MICH, US). **Jalisco:** La Barranca, Guadalajara, (20°45'N, 103°20'W), 23 Nov 1930, *Jones 27191* (CAS-DS). **Sinaloa:** Calmos, Río Fuerte, 21 Nov 1933, *Gentry 908M* (CAS-DS, MICH); Quebrado de Mansana, Sierra Surotato, (25°50'N, 107°37'W)?, 1320–1650 m, 10–14 Sep 1941, *Gentry 6472* (ARIZ, MICH, MO); Mazatlán, Colonia Esperanza, 20 m, Nov 1934, *González Ortega 7437* (G); Rancho Tasajeras, Choix, (26°47'N, 108°16'W), 490 m, 24 Sep 1919, *Narváez et al. 870* (US); Mpio. Sinaloa de Leyva, Ejido Cuitanaca, 35 km N de Agua Caliente de Zevada, (25°49'N, 108°13'W)?, 14 Oct 1976, *Pérez 74* (in part) (BM, MEXU, MO). **Sonora:** Sierra Charuco, Río Mayo, 10 Sep 1935, *Gentry 1711* (in part) (ARIZ, K, MEXU); Guasaremos, Río Mayo, (27°38'N, 108°42'W), 28 Sep 1936, *Gentry 2893* (ARIZ); Sonora-Sinaloa border on cerro with microondas station, (26°22'N, 108°57'W), 152 m, 10–20 Oct 1966, *Gentry et al. 22020* (US); Cruz del Diablo, above Huasabas in lower Río Bavispe watershed, (29°55'N, 109°14'W), 21 Aug 1976, *Nabhan et al. X495* (ARIZ); 11.8 mi NE of Imuris on Mex. Hwy 2, 3 Sep 1990, *Van Devender et al. 90-496* (ARIZ); 2 km N of Rancho la Brisca on Río Sto. Domingo (tributary of Río Sarachi and Río San Miguel), 7 km N of Rancho Agua Fría, 30°25'30"N, 110°33'W, 10 Sep 1978, *Van Devender et al. s.n.* (ARIZ); along Río de los Alisos, 31 mi S of Nogales, 8 Sep 1934, *Wiggins 7024A* (CAS-DS).

UNITED STATES. Arizona: Cochise Co.: Ramsey Canyon Nature Preserve, Miller Peak Quad., 1737 m, (31°25'N, 110°15'W), 3 Sep 1983, *Adams 209-83* (ARIZ); Stronghold Canyon E along USFS Trail 279, Dragoon Mts., (31°55'N, 109°55'W), 1524 m, 7 Sep 1983, *Daniel 3019* (ASU); Clifton, 1 Sep 1900, *Davidson 136a* (GH). **Coconino Co.:** Flagstaff, (31°10'N, 111°20'W), Oct 198?, *Rusby 583* (MICH). **Graham Co.:** Graham Mts., Gillespie Wash, N of Stockton Pass near Gillespie Well, 32°34'N, 109°45'W, 1311 m, 18 Sep 1988, *McLaughlin et al. 5364* (ARIZ); 3 mi S of confluence of Turkey Creek and Aravaipa Creek, Oak Grove Canyon, (32°55'N, 110°30'W), 1036 m, 30 Aug 1981, *Warren et al. 385* (ARIZ). **Pima Co.:** Toro Canyon, Baboquivari Mts., 2 Sep 1931, *Gilman B95* (CAS, GH); Greenlee Co., Big Lue Range, Hwy 78 at the state line, (33°10'N, 109°5'W), 1798 m, 21 Oct 1946, *Gould et al. 4031* (ARIZ); Baboquivari Canyon, W side of Baboquivari Mts., Papago Reservation, (31°49'N, 111°40'W), 18 Sep 1976, *Nabhan et al. X521* (ARIZ). **Santa Cruz Co.:** Santa Rita Mts, Madera Canyon, 1320 m, 21 Aug 1969, *Isely 10869* (BRIT); Rucker valley, Sep 1881, *Lemmon 597* (GH, UC); Pandora Ranch, N foothills, Sta. Catalina Mts., Apr 1881, *Lemmon s.n.* (UC); Santa Rita Mts., Stone Cabin Canyon, 14 Sep 1903, *Thornber s.n.* (BRIT). **New Mexico: Doña Ana Co.:** E side of Organ Mts., (32°28'N, 106°30'W), 1524 m, 16 Aug 1895, *Wooton s.n.* (US); mouth of Filmore Canyon at old mine, Organ Mts., 1768 m, 11 Oct 1980, *Worthington 6749* (ARIZ). **Hidalgo Co.:** Peloncillo Mts., upper Clanton Draw, 2.6 mi above the forest boundary along the road to Douglas, 1707 m, 3 Oct 1982, *Sanders et al. 3002* (UCR); Granite Pass, Little Hatchet Mts., 1524 m, 9 Sep 1984, *Worthington 12646* (TEX). **County Unknown:** Jornada Range Reserve, NE slope Doña Ana Mts., 2042 m, 24 Aug 1930, *Campbell 719* (US); Coronado National Forest S of Animas, (31°32'N, 108°55'W), 18 Sep 1937, *Goodding 9157* (ARIZ); (Arizona) boundary line, Pedregosa Mts., 12 Sep 1892, *Mearns 851* (US); Wood Canyon, Mangas, (34°5'N, 108°19'W), 19 Aug 1897, *Smith s.n.* (US(2)). **Texas: Davis Co.:** Point of Rocks rest area on Tx 166, Ft. Davis, 19 Aug 1977, *Nabhan 689* (ARIZ); Limpia Cañon, (30°37'N, 104°W), 1889, *Neally 648* (104?) (F, US). **El Paso Co.:** Franklin Mts., 1.2 mi W of junction Trans-Mountain Rd. with Gateway South, (31°55'N, 106°5'W), 1463 m, 20 Nov 1982, *Worthington 9220* (UCR).

Distribution.—This taxon seems to be the most common wild tepary in SW United States and NW Mexico, S through Durango to Colima, at elevations ranging from 900–2000 m, but most commonly at around 1500 m (Buhrow 1983). In contrast to Delgado's statement (1985, p. 178), it seems that there are a few records of this variant from Texas.

Habitat.—According to Nabhan (1979) and Buhrow (1983) this taxon is found growing to 2 m length most commonly in canyon bottoms and rocky slopes in open (oak-acacia-cactus) forests or shrubby chaparral of foothills (see Color Plate V, photo 58).

Diseases and pests.—Bean rust is common and leaf spots are sometimes present. It has been reported as eaten by grazing animals.

Ethnobotany.—Reported by collectors with common names such as "frejolillo" or "ejotillo."

Genetics.—This variety is highly daylength sensitive growing during the long days of summer and flowering in mid-August through mid-October. This variety has been used in inter- and intraspecific crossing, most notably at the University of California-Riverside (Waines et al. 1988; Thomas et al. 1983), and has been found to give some fertility problems in crosses with var. *latifolius*. It has been most fertile as a maternal parent. It is possible to obtain crosses with *P. vulgaris* by the use of embryo rescue (Andrade Aguilar & Jackson 1988; Mejía Jiménez et al. 1994), but the F_1 is highly sterile. Several backcrosses to *P. vulgaris*, the maternal parent, are required to establish some fertility.

Comments.—Collections of this variety are the most common representatives of wild tepary beans in the major herbaria and show it to be the most widely distributed in the field. Variety *tenuifolius* leaflets vary from very narrow and indistinctly lobed, to very large leaflets with very large lobes. Some specimens are hardly distinguishable from the type variety, but in these the seed of var. *tenuifolius* is always smaller and much more angular than for the type variety (see Color Plate IV, photo 38). Some of the leaflet types are very narrow and only slightly lobed at base, including the type specimen of var. *tenuifolius* so that it is difficult to distinguish these from species *parvifolius*. However, in those of var. *tenuifolius* the pubescence is more distinctly uncinate, the pods are larger and flatter and the seed are always somewhat larger and much more angular than in the latter species. The senior author suggests that perhaps the variety *tenuifolius* is the original and most primitive form from which the others are derived and so has the most variability and widest distribution. Schinkel and Gepts (1989) found no clear-cut isozyme differentiation between the variants var. *acutifolius* and var. *tenuifolius*. A similar conclusion was also raised by Muñoz et al. (2002) after a study of a large collection of cultivated and wild teparies with AFLPs on total genomic DNA.

A.2.—Phaseolus parvifolius Freytag, sp. nov. (**Figs. 1–4**). TYPE: MÉXICO. OAXACA: Km 35, Hwy Oaxaca-Zacatepec Mixes, 17°5'N, 96°W, 1680 m, 27 Nov 1981, (seed = TARS 52), *Freytag & Sullivan 81-13* (HOLOTYPE: US; ISOTYPES: BR, EAP, MEXU, MO).

Phaseolo acutifolio var. *tenuifolio* similis, sed volubili tenuis brevis folioli lamina anguste elongatis quasi linearibus, base acuta rotundata, bracteolis minutis oblongis vel linearibus, legumine parva et recta aliquantum compressa valvis fere glabris, semene parvo 4.5 mm longo, minusve oblongoideo vel rotundato atromaculato vel raro omnino atro, plantuli folia petiolis breviter minus quam 2–3 mm longis et laminis brevibus lanceolatis ad anguste linearibus.

Similar to *P. acutifolius* var. *tenuifolius* but: ***Plant*** a small, slender, scrambling and climbing, indeterminate vine, 0.5–1 m long. ***Root*** an annual (although it may survive for several years, longer than other taxa of this section), fine fibrous. ***Stems*** terete, 1.5–2 mm thick, mostly much less, much branched from near base, young stems covered moderately with minute apiculate or reflexed-hispid hairs, a few hairs to 0.25 mm long; internodes to 12–13 cm long. ***Stipules*** triangular, acute, 1.25 mm long, 0.75 mm wide, strongly 3-nerved, slightly covered by hispid hairs, ciliate. ***Leaves*** 7–12.5 cm long; petiole about 2.5–4.5 cm long; petiolules 5–10 mm long, covered by minute hispid and uncinate hairs; stipels linear, 1–1.25 mm long, ciliate; pulvini 1–1.5 mm long, heavily covered by hispid hairs especially on adaxial surface; terminal leaflet very narrow, almost linear to somewhat broader, acutely rounded at base, 4–5.5–7 cm long, 4–6 mm wide, some to 1 cm wide at about midpoint, obscurely veined with mostly a single mid-vein, the lateral veins arching forward and joining near the leaf edge, usually reticulate veined below, acute, covered with minute hispid, hooked and glandular hairs; lateral leaflets mostly the same but sometimes very slightly inequilateral. ***Inflorescence*** mostly a small, erect pseudoraceme to 4–5–17 cm long, mostly 1-flowered at each node with a single, often reduced secondary rachis at base; peduncle 1.3–3.5–6.5 cm long; rachis 2–2.2–12 cm long, moderately covered with minute hooked hairs, with 1–4–8 nodes, most nodes setting pods; primary bract linear, 1–2 mm long, 0.5 mm wide, heavily 3-nerved, ciliate, the secondary (pedicellar) bracts minute 0.5 mm long scale-like 1-nerved hyaline; the rachis and bracts dark purplish; pedicel 4–4.5 mm long, moderately covered with minute hooked hairs, purple. ***Bracteoles*** oblong to linear, 1.25–2 mm long, 0.5–0.75 mm wide, heavily 3-nerved, acute, green to purple, ciliate, persistent until mid-pod. ***Flower*** light purple to pink, 7–8 mm long; calyx campanulate 3–3.5 mm long, the upper 2 teeth united into a single lobe, rounded, acute, 2 mm wide, 0.5 mm long, the lower central lobe 1.25 mm long, 1 mm wide, acuminate, the lateral 0.9 mm long, 0.75 mm wide, slightly acuminate, the whole calyx hispid abaxially and minutely uncinate on adaxial side, ciliate, purplish adaxially; standard light purple to pink, distinctly purple-nerved and greenish at base, 5.5–6 mm long, sharply reflexed at 3 mm from base, thickened at flexure, the blade 4 mm long, 5 mm wide, not laterally enrolled, slightly emarginate at tip and glabrous or with a few minute hispid hairs, the claw 1 mm long, the auricles 0.5 mm deep; wings purple to pink, the blade obovate, 5 mm long, 3–4 mm wide, cupped and somewhat enrolled

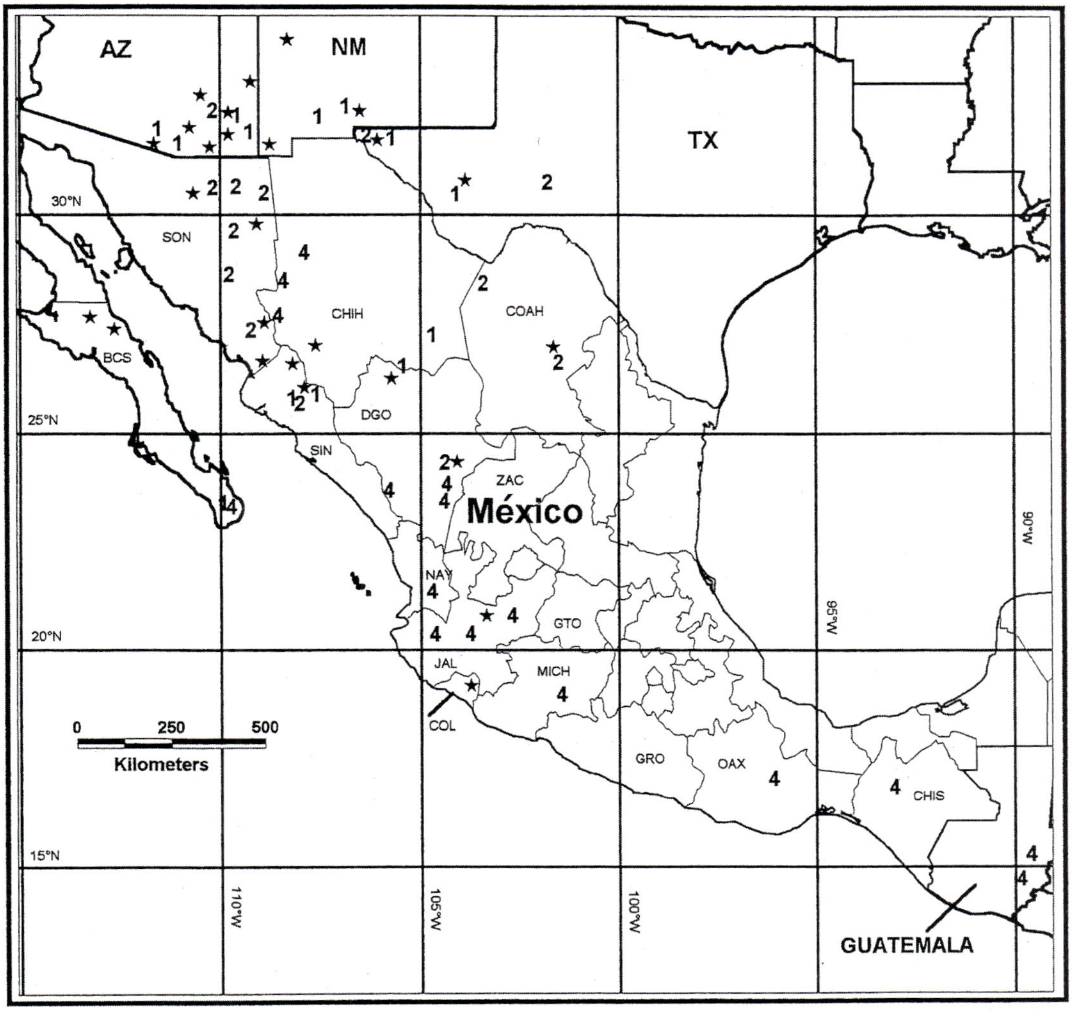

FIG. 4. **Distribution of species and varieties of Section A.** ***Acutifolii*****, as follows: 1 =** ***P. acutifolius*** **var.** ***acutifolius*****; 2 =** ***P. acutifolius*** **var.** ***latifolius*** **(wild); * =** ***P acutifolius*** **var.** ***tenuifolius*****; 4 =** ***P. parvifolius*****.**

lengthwise, the claw 2–2.5 mm long, 0.25 mm wide, the auricle green, 0.5–0.75 mm wide, firmly attached to keel; keel about 4 mm long to base of the terminal 1 1/2 coils of 1.75 mm diam., the apical coil yellowish to green and slightly smaller, the claws 2 mm long, the auricles well-developed; vexillary stamen the claw about 0.75 mm long, the geniculate knob globose and somewhat cup shaped, about 0.5 mm in diameter; stamen tube 4.75 mm long, 3 mm to bend, the ridges not developed; anthers nearly round, 0.25 mm long, flattened, yellow; basal collar 0.25 mm long, denticulate; ovary straight, 3 mm long, 0.7 mm wide, covered by very short, white hirsute pubescence, 8–9 ovules; style 3.5 mm long to the terminal thickened 1 1/4 coil of 1.25 mm diam.; stigma lateral, introrse somewhat twisted laterally, about 0.75 mm long, stylar pubescence extending beyond base of stigma. ***Pod*** straight to falcate, somewhat compressed, 4–5 cm long, 5–6 mm wide, 1.5 mm thick; pronounced sutures; valves tightly twisting 2–3 times at dehiscence, obscurely reticulate, appearing nearly glabrous but densely covered with minute hispid and hooked hairs mostly less than 0.1 mm long; beak straight, weak, 2–4 mm long; highly fertile and setting several pods per inflorescence and most filled with 6–9 seeds. ***Seed*** mostly oblongoid and smoothly rounded, sometimes with flattened ends, 3.4–4.5 mm long, 3–3.5 mm wide, 1.3–2.5 mm thick, smooth, more or less shiny, black-speckled on light tan, sometimes almost completely solid black, black ring around hilum; hilum oblong, 0.6–0.7 mm long, 0.3–0.5 mm wide; lens pronounced, not divided. ***Seedling*** from epigeal germination; hypocotyl 3–3.5–4 mm long;

epicotyl 4–6.5–7 mm long; eophylls opposite, simple 20–25 mm long 2–3 mm wide, very short petiolate less than 2–3 mm long, the blade short, narrowly lanceolate to nearly linear, rounded at base, stipules lanceolate sometimes bifid, 1-nerved, the lateral veins much reduced, acute, nearly glabrous adaxially, puberulent abaxially.

PARATYPES. **GUATEMALA. Jalapa:** 5 km NW de San Luís Jilotepeque, 14°39'N, 89°45'W, 1010 m, 5 Dec 1987, *Debouck et al. 2427* (MO, US, USCG, WIS). **Zacapa:** Sierra de las Minas, along trail above Río Hondo, (15°5'N, 89°35'W), 250–900 m, 11 Oct 1939, *Steyermark 29543* (F).

MÉXICO. Baja California: 'El Piedrón', Cañon de la Zorra, NW Santiago, 23°30'N, 109°52'W, 900 m, 1 Nov 1986, *León 2166* (US). **Chiapas:** Km 13.9, Hwy Tuxtla Gutiérrez-Sumidero, 16°49'N, 93°06'W, 1080 m, 26 Sep 1978, *Freytag et al. 78-Mex-77* (US, BR, EAP, F, GH, K, MEXU, MO, UC). **Chihuahua:** Guasaremos, Río Mayo, (27°38'N, 108°42'W), 28 Sep 1936, *Gentry 2893* (F); Cañon Huahuatán, 10 mi SE of Madera, Mpio. Temósachic, (29°8'N, 107°58'W), 23 Sep 1939, *Muller 3443* (NA, UC(2)). **Durango:** Minerva, Sierra El Registro, 23°58'N, 104°25'W, 2090 m, 10 Oct 1978, *Debouck 328-A* (CHAPA, COL, K, MO, US); Mezquitál, El Troncón, camino a San Juan, 23°30'N, 104°26'W, 1920 m, 28 Oct 1978, *Debouck 390* (CHAPA, COL, K, MICH, MO, US); El Salto, Revolcaderos, 23°37'N, 105°50'W, 1960 m, 2 Nov 1978, *Debouck 413* (CHAPA, K, MICH, MO, US). **Guerrero:** Tixtla, 9 Km E of Chilpancingo de los Bravos, 17°34'N, 99°29'W, 1620 m, 31 Oct 1987, *Debouck et al. 2356* (CHAPA, COL, G, MICH, MO). **Jalisco:** Tecolotlán-Guadalajara, grown in screenhouse at Mayagüez, PR from seed of TARS #53 (collected at Km 65, along Hwy 80 Tecolotlán-Guadalajara, Jalisco, Mex., 20°25'N 103°40'W, 1420 m, 6 Dec. 1981 by *Freytag GF et al. 81-40*), Sep.-Dec., 1985 as Study Collection *Freytag, GF #SC-53* (ARIZ, CSU, F, MO, US, WIS); 1–3 mi above (S of) Río Verde, near Hwy. from Tepatitlán to Yahualica, (20°59'N, 102°49'W), 1500–1700 m, 28 Aug 1958, *McVaugh 17441* (MICH); 4 mi NNE of Talpa de Allende, (20°25'N, 104°48'W), 1450–1500 m, 12–13 Oct 1960, *McVaugh 20130* (CAS, MICH, MO). **Michoacán:** Mpio. Carácuaro, 12 km al NW de Paso de Núñez, (18°55'N, 101°3'W), 8 Oct 1978, *Soto et al. 1178* (CAS, MO). **Nayarit:** Cerro Alto, 3.5 km W de El Tepozal, 21°23'N, 105°W, 1220 m, 6 Nov 1978, *Debouck et al. 422* (CHAPA, COL, K, MICH, MO, US); Mpio. Santa María del Oro, 3 Km E de Santa María del Oro, 21°20'N, 104°33'W, 18 Oct 1986, *Téllez 9789* (MEXU); Mpio. Ixtlán, Km 7–10 terraceria a Cacalotan, 21°07'N, 104°17'W, 8 Oct 1987, *Téllez 11072* (MEXU); 15 km by road S of Tepic to Compostela, ca 914 m, 18 Oct 1970, *Webster et al. 15700* (MICH(2)). **Oaxaca:** Dist. Yautepec, El Laurel, est. de microondas a 13 Km al E-SE de El Camarón, 1400 m, 18 Sep 1978, *Sousa et al. 9455* (BRIT, MEXU). **Sonora:** La Mesa Colorada, Sonora-Chihuahua boundary, 17 Oct 1933, *Gentry 567Mb* (MICH); Sierra Charuco, Río Mayo, (28°25'N, 108°45'W)?, 10 Sep 1935, *Gentry 1711* (in part) (F, GH, K, MO, UC, US); 3 km al N de La Lobera, Ejido Zahuarivo, Mpio. Alamos, 27°12'N, 108°53'W, 1500 m, 25 Aug 1986, *Tenorio et al. 11945* (MO).

Distribution.—Found scattered throughout western Mexico from Sonora down to Chiapas, and in central Guatemala. Growing at elevations of 250–2090 m and more common at the higher elevations.

Habitat.—Rare, found growing on rocky soils (schist, basalt, granitic, limestone) in dry grassland parks and along stream beds in open oak or pine-oak forest with *Agave, Acacia, Calliandra.*

Diseases and pests.—Collectors report chrysomelid beetles and some powdery mildew and rust, and some grazing by animals.

Genetics.—This species is highly daylength sensitive and did not flower in Mayagüez, Puerto Rico under a 12 hour daylength with normal greenhouse temperatures, which is adequate for flowering of all the other species and varieties of this section. However, it did bloom at the same time as these other taxa in a cool greenhouse during the winter at Colorado State University (CSU), Fort Collins, Colorado. In its native habitat this species grows vegetatively during the long days of summer, and is the latest to flower of the wild species of the section, beginning near the end of September and continuing through November into December. This species is highly self-fertile under greenhouse conditions in Mayagüez, producing copious pollen and setting a full complement of seeds per pod. However, at this location, it is only possible to intercross it with other varieties of the section using it as a pollen parent, and this with difficulty since most pods and ovules abort. These results were obtained at Mayagüez and confirmed at both the University of California-Riverside (Federici, pers. comm.) and at CIAT by the junior author. However, several years later the senior author made the same crosses at CSU in Colorado, under cool greenhouse conditions, and found this species to be reciprocally fertile with all the other taxa of this section both wild and cultivated; though at present writing we do not know yet whether there will be sterility in the F_1 generation.

Comments.—Although this species has been collected in isolated spots throughout highland Mexico, perhaps the senior author would not have recognized it from the herbarium specimens if he had not first collected it in the field in 1978 from a small, open pasture or park about half-way up the mountain road to the "Mirador Tepehuane" of El Sumidero near the city of Tuxtla Gutiérrez in Chiapas, Mexico. On a later trip to Oaxaca in 1981, he again collected this species in a dry wash in the mountain on the road to Zacatepec Mixes, east of the ruins at Mitla (southeast of Cd. Oaxaca). No other teparies, cultivated or wild, have come to his attention from these areas. It is also apparently found in SE Guatemala, first collected in 1939 by Julian Steyermark. It is possible that the specimens collected from northern Mexico (cited above) may not all be of this species in a pure form but rather highly introgressed into var. *tenuifolius* cytoplasm. They most closely resemble *P. parvifolius* morphologically and are therefore placed within this species. A couple of collections by the junior author in 1978 were previously identified as of var. *tenuifolius*. A recent study of a large collection of cultivated and wild teparies with AFLPs on total genomic DNA by Muñoz *et al.* (2002) has shown that all accessions available of *P. parvifolius* form a distinct group and separate from other wild teparies at about the same level that gene pools separate within *P. vulgaris* or *P. lunatus*. In a microsatellite-primed PCR analysis, Dorothea Zink and Walter Nagl (1998) showed the very close relationship between *P. parvifolius* and *P. acutifolius*. This result lead the junior author (Debouck 2000a,b) to think about a phylum of tepary bean.

Section B.—Phaseoli DC, Prodr. 2:390. 1825. emend. Freytag. TYPE SPECIES: *Phaseolus vulgaris* L., Sp. Pl. 723. 1753.

(In the following the characteristics of cultivars are given in parentheses.) Herbae perennes breviter viventes (annuae) magnae scandentes (vel prostratae vel fruticosae ad anguste erectarum) indeterminatae 3–10 m longae (vel raro determinatae 0.5–2 m longae) volubiles, radix plerumque fibrosa et aliquanto carnosa in accesionibus pristinissimis et species, folia satis parva 8–40 cm longa (ad maximam) foliolis ovatis ad ovatiacuminatis anguste integra sed aliquando basaliter lobata vel truncata, inflorescentia pseudoracemos vel aliquando paniculatorum multiflorum com crecit, bracteolae pleurumque calycem longiorem et latae vel angustissimae, flores grandes 17–28 mm longi, purpurei rare albi (albi), calycis bilabiatis sed campanulatis rare, stigma laterale introrsque ad aliquantum terminale capitatumque cum lateralisque, legumen strictum ad leviter curvatum anguste ad modice latum (ad latissimum) breve vel aliquantum latum (longissime) fibrosum (vel non fibrosum) eruptione dehiscentium (indehiscentium), semina 4–8 globosa vel oblongoidea 6–13 mm longa (ad 15 mm longa et crassa) compressa in plantas sylvestres, saepe compressa, plantula germinate epigea.

Short-lived perennial (annual in most cultivars), climbing (prostrate or bush to narrowly erect), indeterminate 3–10 m long (determinate 0.5–2 m long) vines; root mostly fibrous, but somewhat thick and somewhat fleshy in some wild collections and species (or thin, fibrous); leaves 8–40 cm long, leaflets ovate to ovate-acuminate, entire but sometimes slightly basally lobed or truncate; inflorescence a pseudoraceme sometimes becoming a large, many-flowered panicle after continued growth; bracteoles usually as long as or longer than calyx and broad to very narrow; flowers large, 17–28 mm long, purple to lilac or rarely white (often white), the calyx usually bilabiate but may be campanulate, the stigma lateral and introrse, to somewhat terminal and capitate yet with a decurrent introrse portion in wild collections; pod straight to slightly curved, narrow to medium broad (to very broad), short to medium (very long), fibrous (to non fibrous), explosively dehiscent (indehiscent); 4–8 seed, nearly spherical to oblongoid, 6–13 mm long (to 15 mm long and thick), wild types often flattened; seedling from epigeal germination.

Comments.—This section has a revised definition based on the combined original species concepts of *P. vulgaris* and *P. dumosus* (= *polyanthus*), since it is well known that these two species are closely related and may be hybridized easily using *P. vulgaris* as the female parent but somewhat more difficult with the reciprocal cross (Maréchal 1971; Camarena & Baudoin 1987). They are certainly more closely related to each other than to any other cultivated species. Perhaps their most distinctive common characteristics are their epigeal germination and fairly large showy flowers. Species *P. costaricensis* closely resembles *P. dumosus*, and some natural hybrids between the two have been observed in Costa Rica (Debouck et al. 1989a; Schmit 1992); however, it also resembles *P. coccineus* and *P. tuerckheimii* with which it has often been confused. Recent ITS DNA sequencing data (Delgado et al. 1999; Gaitán et al. 2000) show clearly that the four species claimed here within the *Phaseoli* are closely related (and to *P. coccineus*, but germination!). As we indicate below, the species of the Section B. *Phaseoli* would represent the true secondary gene pool of the common bean, and thus practical possibilities for effective widecrossing perhaps without the use of special techniques such as embryo rescue, etc (see also Debouck 1999). According to the system of nomenclature proposed in this treatment, the section should be named *Phaseoli*; therefore the senior author has changed the name proposed by DeCandolle in his Prodromus to coincide with the International Rules of Nomenclature.

KEY TO THE SPECIES

1. Stigma lateral introrse; bracteoles large, usually as long as or longer than calyx (for Mesoamerican materials), 4–6 mm long; usually small annuals (perennial in some wild types); flower purple or white, the calyx bilobate, the wings extended forward, the ovary usually of 6–8 ovules; pod long and narrow, considerably twisted at dehiscence; common in dry, oak-pine forests and grasslands throughout upland Mexico and S through Central America, and through Andean South America to Argentina; 800–2750 m ______ B.1. **P. vulgaris**
1. Stigma terminal, capitate or nearly so.
   2. Bracteoles much longer than the calyx, 5–9 mm long; pedicel stout, more than 1.5 mm thick, 6–7 mm long at anthesis.
      3. Bracteoles usually rounded ovate to linear, somewhat longer than calyx; flower dark pink or lilac to purple, the wings 23 mm long, clasping, the ovary of 6 ovules; pod 10 cm long, yellowish-brown tomentose; scarce, in humid pine forests and cloud forests in central and SE Costa Rica and NW Panama; 1200–2080 m ______ B.2. **P. costaricensis**
      3. Bracteoles linear to narrowly oblong, nearly twice as long as calyx; flower purple or white, the wings 19 mm long, nearly round, cupped, widely spreading, the ovary of 4 ovules; pod 7 cm long, nearly smooth at dehiscence; wild populations are rare (weedy or cultivated forms are common) in humid locations along streams in broadleaf forests of western and central Guatemala; 1300–1940 m ______ B.3. **P. dumosus**
   2. Bracteoles about as long as calyx, 3–4 mm long; pedicel delicate, less than 1.5 mm thick, 15–20 mm long at anthesis; flower pale purple, the wings large and spreading; rare, in humid locations along streams in pine forests in mountains of W Michoacán, Mexico; 1600–1900 m ______ B.4. **P. albescens**

B.1.-Phaseolus vulgaris L., Sp. Pl. 723. 1753. (**Figs. 5, 9, 63**). TYPE: Uppsala, microfiche No. 899.1. (lectotype chosen by Verdcourt (1971) and deposited in LINN, n.v., "cultivated at Uppsala, specimen 899/1 (LINN, lecto.!)). *Phaseolus vulgaris* L. var. *vulgaris* Maréchal, Mascherpa and Stainier, Boissiera 28:136–137. 1978.

Phaseolus esculentus Salisb., Prodr. 335. 1796.
Phaseolus communis L. ex Pritzel, Ic. Bot. Ind. 832. 1855.
Phaseolus aborigineus Burk., and *Phaseolus aborigineus* Burk. var. *hondurensis* Burk., Legum. Argent. 545. 1952.
Phaseolus vulgaris L. subsp. *aborigineus* Burk. ex Burk., In: A. Burkart & H. Brücher. Der Züchter 23(3):71. 1953. *Phaseolus vulgaris* L. var. *aborigineus* (Burk.) Baudet, Bull. Soc. Roy. Bot. Belg. 110:74. 1977.

Aerial shoot a perennial climbing indeterminate vine, tending to be annual in many wild populations (and in almost all cultivated types). ***Root*** fibrous. The main ***Stems*** mostly yearly dying back nearly to the ground but surviving a couple years under favorable tropical climates, 1-6 m long, at ground level to 2-3 cm thick, the smaller stems terete, to 2-3 mm in diameter, strongly twining erect, often branching, a few branches from base but mostly branched in upper portion, the internodes mostly 12-15 cm long but often longer usually on upper portions of plant which may be a dark red to brownish color particularly in sun-exposed parts, usually thin, nearly glabrous, a few reflexed-strigose hairs to 1.25 mm long. ***Stipules*** to 3 mm long, 1 mm wide, strongly 3-nerved, somewhat reflexed, membranous, glabrous. ***Leaves*** 8-16 cm long; petiole 3-5 cm long; petiolule 1.5-2.5 cm long, somewhat striate-nerved, moderately covered with strigose and shorter uncinate hairs; pulvini 3-5 mm long, moderately covered with strigose hairs; terminal leaflet ovate to broadly ovate, 4-8 cm long, 3-7 cm wide at just below midpoint, the blade with 3 main veins at the base with 2 much smaller lateral veins, all somewhat webbed at their basal union, the tip acuminate and minutely apiculate, membranous, the abaxial surface more or less moderately strigose mostly on the veins with many smaller uncinate hairs, nearly glabrous on the adaxial surface; lateral leaflets similar, inequilateral, 3.7-7.5 cm long, 2.6-6 cm wide at about 1/3 from the base. ***Inflorescence*** a pseudoraceme, often of 2-4 flowers, sometimes many-flowered, vertical to extending horizontally; peduncle 5-8 cm long; rachis 3-9 cm long, usually of 4-6 flowering nodes but may have as many as 8 or more, with 2 flowers per node, most flowers setting pods, heavily to moderately covered with short uncinate hairs; primary bracts broadly ovate, 4 mm long, 3.5 mm wide, heavily 8- to 12-nerved, glabrous to pubescent, minutely ciliate near tip, the secondary bracts much smaller and triangular; pedicel 5-12 mm long, delicate, nearly glabrous. ***Bracteoles*** usually sessile, oval to broadly ovate, 4-6 mm long, 3-5 mm wide,

FIG. 5. Illustrations of *Phaseolus vulgaris* L.—**a.** Seedling a few days after germination.—**b.** Stem and root of mature plant.—**c.** Vine tip with mature leaves and pods.—**d.** Flowers, side view and front view.—**e.** Exploded view of the flower showing all parts, including—**s.** Style tip and stigma as seen under the microscope.—**f.** Pods, side view and dehisced carpels.—**g.** Seed types, side view and view from the hilum; primitive types (1–4) and the South American type (5), all seed produced at Mayagüez from the following sources: **g.1.** NI 406A, collected by INIA in Guerrero, México. **g.2.** NI 401, collected as Mor. 662 by INIA in Morelos, México. **g.3.** *Freytag GF & Sullivan 81-31*, collected at El Chante, Jalisco, México. **g.4.** NI 578 from Miranda Oax-910 (Cambridge V1449), from Oaxaca, México. **g.5.** NI 621 from CIAT 10025, from Argentina. All other drawings (**a–f**) from living material grown at Mayagüez of NI 406A (TARS #104) collected by INIA in Guerrero, México.

strongly 8- to 12-nerved, nearly glabrous, ciliate near tip, caducous shortly after pollination. ***Flower*** purple, white, or bicolor, infrequently veined purple; calyx bilabiate, 4 mm long, the upper two teeth united, scarcely elongated, emarginate, 0.75 mm long, 4.5 mm wide, the lower 3 teeth about equal, 0.75 mm long, 1.5 mm wide, minutely hooked pubescent, a few scattered strigillose hairs near the tips of the lower teeth; standard mostly purple, rarely white, sometimes veined purple, often greenish on abaxial surface, 10 mm long, 10 mm wide, reflexed at about midpoint, the terminal portion erect, the stipe 3 mm long; wings purple, rarely white, sometimes veined purple, the blade 10–19 mm long, 8 mm wide, spreading, the basal claw 5.75 mm long, 0.5 mm wide, the spur about 2 mm wide adhering to keel; keel with 1 1/2 to 2 coils of 2–3 mm in diam., tip white or greenish, the basal claw 4–5 mm long, about 0.6 mm wide, the lateral knobs well-developed, rounded, about 1.75 mm in diameter, firmly adhering to the wings; vexillary stamen 10–12 mm long, the basal knob about 1.25 mm from the base, developed for about 1.25 mm into a nearly erect sheath; united stamen tube about 10.5 mm long, the basal spurs poorly developed; anthers 0.7 mm long, 0.25 mm wide, yellow, the basal collar about 1 mm long, well-developed; ovary about 6.5 mm long, 1–1.25 mm wide, glabrous or a few minute hairs near tip, 4–8 ovules, rarely 10, the style 6.5 mm long to the 1 1/2 thickened coils of about 2 mm in diameter, the stigma 1 mm long, lateral introrse in some wild specimens but tending to be longer and extending around the tip of the style about 0.5 mm on the extrorse side in the cultivated types, with stylar hairs subtending and almost to the tip. ***Pod*** nearly linear to slightly curved, 5–6.5 cm long, 5–7 mm wide, 3–5 mm thick, inflated but compressed, the beak curved, 3–6 mm long, stout, when immature the pod a green color, at maturity tan, yellow or whitish and often marked by purple flecks or streaks, when dry tan or faded white, if flecked often indistinct, twisting tightly at dehiscence, glabrous and covered by minute glandular lenticels. ***Seed*** very small in some wild specimens, usually oblong-rhomboid to spherical (spherical, reniform, flattened or oblongoid in cultivated and weedy forms), 6 mm long, 4.6 mm wide, 2.8 mm deep (larger in the cultivated types), somewhat compressed, the ends flattened except for seed not in contact with others in which case they are rounded, the colors mostly black flecks on a tan or brown background, or sometimes solid colors of tans, yellows, browns, black or nearly white, rarely reds (these colors in case of introgression with cultivated forms), a black ring around the hilum, shiny, the hilum oval, 1.25 mm long, 0.8 mm wide, the funicular scar tissue complete, the lens pronounced. **Seedling** from epigeal germination; hypocotyl 2.2–3.1–3.9 cm long; epicotyl 6.5–7.2–8.5 cm long; primary leaves opposite, simple, the petiole 3.5–4.7 cm long, the blade triangular-ovate, 4–6 cm long, 3.5–5 cm wide at base, the base auriculate, the tip slightly acuminate and not apiculate, membranous, sometimes variegated along main nerve, glabrous except sparsely hispidulose on main veins.

Specimens examined: **COSTA RICA. Alajuela:** Zarcero, 2 km SE de Zarcero hacia San Juanillo y Naranjo, valle Río Seco, 10°10'N, 84°23'W, 1610 m, 8 Feb 1998, *Debouck et al. 3121* (CR). **San José:** 1 km SW de Tabarca, 9°49'N, 84°07'W, 1750 m, 8 Jan 1987, *Debouck et al. 2097* (CR); al pie de la Piedra de Aserrí, 1.5 km E de Aserrí, 9°52'N, 84°7'W, 1560 m, 11 Jan 1987, *Debouck et al. 2111* (BR, CR, US).

EL SALVADOR. Morazán: Montes de Cacaguatique, 13°46'N, 88°13'W, 1440 m, 30 Dec 1941, *Tucker 667* (K).

GUATEMALA. Chimaltenango: along road from Chimaltenango to San Martín Jilotepeque, (14°40'N, 90°48'W), 1500 m, 25 Nov 1936, *Standley 57988* (F). **Chiquimula:** Ipala, SW de Laguna de Ipala, orillas de Laguna, 14°33'N, 89°38'W, 1390 m, 5 Dec 1987, *Debouck et al. 2423* (COL, G, SI, USCG); S of Lempa river between Esquipulas and Atulapa, (14°30'N, 89°20'W), 800 m, 11 Dec 1969, *Molina et al. 25349* (EAP, F, MO, US); Volcán Quezaltepeque, 3–4 mi NE of Quezaltepeque, 1500–2000 m, 8 Nov 1939, *Steyermark 31441* (F). **El Progreso:** 0.2 km NE Albores, San Agustín Acasaguastlán, 1250 m, 15°4'N, 89°56'W, 3 Dec 1987, *Debouck 2417* (COL, USCG). **Guatemala:** 1 km NE de Calderas, 12 km E de San Vicente Pacaya, 14°25'N, 90°35'W, 1740 m, 14 Dec 1987, *Debouck et al. 2459* (USCG); 1 km E de Sta Helena Barrillas, cerro de La Torre, 14°26'N, 90°31'W, 1740 m, 15 Dec 1987, *Debouck et al. 2461* (USCG). **Huehuetenango:** along road to Jacaltenango, 1.5–2.5 km ENE of San Antonio Huista, 15°40'N, 91°45'W, 1300–1400 m, 8–9 Jan 1976, *Iltis et al. G124* (WIS); idem, 8–9 Jan 1976, *Iltis et al. G128* (WIS). **Jalapa:** Km 124, Hwy 18 Jalapa–San Pedro Pinula, 14°40'N, 89°56'W, 1180 m, 6 Oct 1978, *Freytag et al. 78-Guat-30* (EAP, GH, F, MO, UC, US). **Sacatepéquez:** 4 km S de Alotenango, 14°27'N, 90°49'W, 1200 m, 5 Dec 1985, *Debouck et al. 1611* (K, US, USCG); 1.5 km O de San Miguel Dueñas, 1550 m, 14°32'N, 91°50'W, 6 Dec 1985, *Debouck et al. 1616* (COL, SI, US, USCG). **Santa Rosa:** Laguna de Ayarza, San Rafael Las Flores, 14°26'N, 90°8'W, 1430 m, 7 Dec 1987, *Debouck et al. 2440* (USCG); El Pino, 1320 m, Nov 1893, *Heyde et al. 6124* (K). **Suchitepéquez:** between Finca El Naranjo and upper slopes Volcán Santa Clara, (14°35'N, 91°25'W), 1250–2650 m, 23 May 1942, *Steyermark 46685* (F).

HONDURAS. Comayagua: Siguatepeque, (14°30'N, 87°50'W), 1050 m, 1 Jul 1936, *Yuncker et al. 5576* (GH). **El Paraíso:** Montaña Monserrat, (14°N, 86°45'W), 1600 m, 25 Nov 1958, *Molina 8690* (EAP, F, GH, US). **Fco. Morazán:** carr. Tatumbla a Tegucigalpa, 1500 m, 17 Oct 1993, *Linares et al. 795* (MO); Cuesta La Morolica, camino entre El Llano de Las Gallinas y Santa Lucía, (13°30'N, 86°50'W),

1300 m, 15 Jan 1951, *Molina 3897* (EAP, F); San Juan del Rancho, N of Cerro de Uyuca, (13°50'N, 87°5'W), 1500 m, Nov-Dec 1948, *Standley 522* (EAP).

MÉXICO. Chiapas: Mpio. Ixtapa, Hwy 190 in the Zinacantán paraje of Muctajoc, 1067 m, 26 Oct 1965, *Breedlove 13814* (MEXU, MICH, US); San Cristobal Las Casas, 2103 m, 3 Aug 1964, *Breedlove s.n.* (US); Km 41, N of Tuxtla Gutiérrez on dirt branch road to Gabriel Esquinca from main Hwy Tuxtla Gutiérrez to Chicoasén, 15°50'N, 93°15'W, 1080 m, 25 Sep 1978, *Freytag et al. 78-Mex-76* (EAP, GH, F, MO, UC, US); Km 38 on Hwy 190 from Tuxtla Gutiérrez to San Cristobal, 16°45'N, 92°45'W, 1120 m, 28 Sep 1978, *Freytag et al. 78-Mex-86* (EAP, GH, F, MO, UC, US); Mpio. El Zapotal, along trail from Zinacantán paraje of Paste' to San Lucas, 1067 m, 15 Dec 1966, *Laughlin 2972* (ENCB, US). **Durango:** Fco I. Madero, Jeronimo Hernández, Lomas Charco del Indio, 24°27'N, 104°14'W, 2030 m, 6 Oct 1978, *Debouck et al. 306* (CHAPA, COL, G, K); Cerro Las Manzanillas, Gamon Sur, Fco I Madero, 24°30'N, 104°16'W, 2270 m, 7 Oct 1978, *Debouck et al. 317* (CHAPA, COL, K); 21 mi SE of Cd. Durango along Rt 45 by watermarked cliffs, (23°59'N, 104°30'W), 1829 m, 24 Oct 1966, *Gentry et al. 22042* (GH(2), NA, US); City of Durango, Apr-Nov 1896, *Palmer 866* (F, GH, K, MO, UC, US). **Guanajuato:** Mpio. Valle de Santiago, cañada 0.5 km W del Volcán Batea, 20°19'N, 101°12'W, 1850 m, 16 Nov 1987, *Debouck et al. 2401* (CHAPA). **Guerrero:** Taxco, 12 Oct 1937, *Abbott 486* (GH); Mpio. Iguala y Buenavista, Cañón de La Mano, entre Los Amates y El Naranjo, 10 km N of Iguala, 900-1000 m, 19 Oct 1986, *Catalán et al. 343* (MO); Tixtla, 9 km E de Chilpancingo, 17°34'N, 99°29'W, 1620 m, 31 Oct 1987, *Debouck et al. 2355* (CHAPA, COL, MICH); along road to Arcelia, 17 mi W of Teloloapan, (18°25'N, 100°W), 1463 m, 13 Dec 1967, *Gentry et al. 22544*, (GH, MEXU, MICH, NA); Vallecitos, Montes de Oca, 19 Nov 1937, *Hinton et al. 11567* (F, GH, K, MEXU, MICH, US); along Hwy 51, 6 km NW of Teloloapan at Km 70.5 (7 km NW of Teloloapan by air), 1,600 m, 20 Nov 1971, *Iltis et al. 2* (WIS); Mpio. Chilpancingo, 22 km al S de Chilpancingo por la carr. a Acapulco, (17°25'N, 99°30'W), 1150 m, 16 Oct 1982, *Koch et al. 8269* (F, MICH). **Jalisco:** Mpio. Pihuamo, 7 km S of San José de Tule, 19°24'N, 103°24'W, 1140 m, 9 Nov 1978, *Debouck et al. 430* (CHAPA, COL, G, M); Mpio. de Tecalitlán, Barranca de San Juan de Dios, about 15 km E of Pihuamo, E slope of Sierra de los Corales, (19°15'N, 101°15'W), 1200-1300 m, 23 Oct 1963, *Dieterle 3011* (MICH, MO); Grown in screenhouse at Mayagüez, PR (= TARS 108A, from seed of *Freytag et al. 81-21A* collected at Km 17, road to Chiquilistlán, 1630 m, 1 Dec 1981), Sep.-Dec. 1985, as Study Collection *Freytag, G.F. #SC-108A* (ARIZ, CSU, F, MO, US, WIS); Canyon in W footslope of Volcán Colima, 2 mi N of Cofradía, 3 mi S of Cd. Guzmán along road to Colima, (19°40'N, 103°45'W), 1372 m, 21 Nov 1966, *Gentry 22189* (GH, NA, US); 6 mi S of Tecalitlán along road to Colima, 1219 m, 23 Nov 1966, *Gentry 22199* (NA, US); 3 mi S of Cd. Guzmán along road to Colima, 1524 m, 23 Nov 1966, *Gentry 22202* (NA, US); 2 mi NE of La Garrita, 10 mi E of Tamazula, (19°40'N, 103°5'W), 1341 m, 17 Nov 1966, *Gentry et al. 22180* (GH, NA, US); Colima-Jalisco border by Tonila, 1128 m, 20 Nov 1966, *Gentry et al. 22184* (GH, NA, US); 2 mi N of Tonila, 1219 m, 20 Nov 1966, *Gentry et al. 22186* (NA, US); 3 mi W of San Marcos on lower slope of Volcán Colima, 1402 m, 20 Nov 1966, *Gentry et al. 22187* (US); (0.3 km) S of Manantlán, 11 km S of El Chante, 1450 m, 4 Jan 1979, *Iltis et al. 1190a* (MICH); near Mascota, Jal. (20°38'N, 104°58'W), (grown from seed in greenhouse at Glenn Dale, MD, 11 Oct 1967 as *Gentry et al. s.n.), Norvell M7016* (GH); Mpio. de Ayutla, 9 km al N de Ayutla, camino a Talpa de Allende, (20°10'N, 104°25'W), 7 Sep 1979, *Solís et al. 1909* (MICH). **México:** Dist. Temascáltepec, Rincón, 1960 m, 26 Feb 1934, *Hinton et al. 5605* (F, GH, US); Mpio Temascaltepec, 1 km SW de Temascaltepec, 19°02'N, 100°03'W, 1630 m, 1 Nov 1987, *Debouck et al. 2369* (CHAPA, COL, G, M, MICH); Mpio. de Ocuilán, 4 km NE de Chalma, 18°57'N, 99°26'W, 1890 m, 4 Nov 1987, *Debouck et al. 2371* (CHAPA, MICH). **Michoacán:** Morelia, à l'ouest lieux innondés, 1860 m, 7 Oct 1909, *Arsène 3161* (GH, MO, US); 8 mi S of Tzitzio, 914 m, 15 Dec 1967, *Gentry 22551* (MEXU, NA); 10 mi S of Uruapan along road to Apatzingán, (19°10'N, 102°2'W), 1341 m, 9 Nov 1966, *Gentry et al. 22153* (GH(2), NA, US(2)); 1 km N of San Felipe, camino Zitácuaro-Angangueo, Mpio. Zitácuaro, 1900 m, 16 Oct 1982, *Martínez et al. 2280* (MO); Puerto Arúmbaro, carr. Temazcal-Huetamo, 1820 m, 13 Oct 1983, *Soto et al. 5565* (MO); S slope Mt. Punguato, Morelia, 2200 m, 9 Sep 1962, *Ugent et al. 1800* (WIS). **Morelos:** Cuernavaca, Iturbide, 14 Nov 1865, *Bourgeau 1375* (GH, K); Mpio. Tepoztlán, 4.5 km NW of Yautepec, 15 Nov 1986, *Cabrera et al. 12250* (MO); Tepoztlán, (19°N, 99°5'W), 1676-1829 m, 6 Nov 1967, *Gentry 22404* (GH, MEXU, MICH, NA). **Nayarit:** Mpio. Amatlán de Cañas, 9 km SE of Ahuacatlán, 21°08'N, 104°28'W, 1335 m, 7 Nov 1993, *Flores-Franco et al. 3127* (MO); Mpio. Xalisco, 17 km al SW de Xalisco, 21°23'N, 104°57'W, 1100 m, 17 Oct 1994, *Flores-Franco et al. 4173* (MO); 9-10 mi W of road fork on Hwy 15 along road to Jalcocotán, (21°30'N, 105°10'W), 914 m, 25 Nov 1966, *Gentry 22237* (NA, US); 2-3 mi SE of Tetitlán, 823 m, 6 Nov 1967, *Gentry et al. 22234* (GH, SI); Mt. 9 mi N of Compostela, (21°20'N, 105°W), 1000-1200 m, 13 Nov 1959, *McVaugh et al. 563* (MICH); Mpio. Tepic, 6.2 km of El Roble, brecha a Huajimic, 21°31'N, 104°28'W, 1500 m, 27 Oct 1989, *Tenorio et al. 16820* (MO). **Oaxaca:** Pochutla, Candelaria Loxicha, 40 km S de San José Pacífico, 15°59'N, 96°31'W, 1700 m, 27 Oct 1987, *Debouck et al. 2348* (CHAPA, US); 4 mi NE of Hwy junction on road to Ixtlán de Juárez, 1676 m, 2 Nov 1967, *Gentry 22395* (GH, MEXU, MICH, NA, SI); Mpio. de Ixtlán de Juárez, Dist. Ixtlán, Vivero Rancho Teja, 5 km E of Ixtlán de Juárez on the road to Capulalpán, (17°20'N, 96°30'W), 2200 m, 13 Oct 1980, *Martin 227* (MO, US); Sierra de Clavellines, 2743 m, Oct 1894, *Smith 566* (UC); Valley of Etla, (17°10'N, 96°55'W), 1615 m, 23 Oct 1895, *Smith 853* (GH). **Puebla:** Cholula, Barranco Puente Los Molinos, 4 km NE de Atlixco, 18°58'N, 98°23'W, 1880 m, 22 Oct 1987, *Debouck et al. 2325* (CHAPA, MICH (2)). **Queretaro:** Pinal de Amoles, 3 km S de Escanelilla en la carr. Mex-120 a Cadereyta, 21°8'N, 99°37'W, 1190 m, 14 Nov 1986, *Debouck 2077* (CHAPA, COL, MICH, MO, P, UC, US). **Sinaloa:** 3.5 mi E of Copala, along road to Durango, (23°25'N, 105°55'W), 823-975 m, 7 Dec 1966, *Gentry et al. 22277* (GH). **Tamaulipas:** Sierra between Jaumave and Victoria, (22°35'N, 99°25'W), 14 Oct 1931, *von Rozynski 139* (in part) (MICH). **Veracruz:** 19 km antes de Orizaba (autopista Puebla-Orizaba), 1 km de la desviación a Maltrata, (18°50'N, 97°15'W), 1550 m, 18 Oct 1969, *Lot 510* (F).

NICARAGUA. Estelí: Llano Almaciguera, 8.4 km S of Hwy 1 on road to Estanzuela, 2.8 km S of Río Estanzuela bridge, 13°00'N, 86°21'W, 1260-1280 m, 13 Nov 1978, *Stevens 10738* (MO). **Jinotega:** La Bastilla, 10 Km NE of Jinotega, 11 Jan 1969, *Zelaya 2143* (BRIT).

Other specimens seen.—An additional 250 collection numbers were examined and determined to be of cultivated types because of information by the collector (label: "cultivated," "escaped," "landrace" or some similar notation) or because of large leaf, pod, or seed size. Specimens of this kind were not used in forming the species description nor in defining the distribution range and are not listed here since our subject matter deals with wild species and forms only.

Habitat.—Often found in dryer localities of desert to thorny shrub (see Color Plate V, photo 50) on hillsides or on steep slopes in open pine-oak forests and usually at altitudes of 800–2000 m, growing in and over shrubs and brush, often at edges of thickets, trails, and over stone walls but seldom in mature forests or other climax vegetation. Soils of these areas are usually shallow mineral or volcanic types derived from limestone, schist, granite, lava or tuffa, usually friable but sometimes rather heavy clay, often sandy with good drainage and moisture (for a complete description of habitats see Gentry 1969; Delgado et al. 1988).

Diseases and pests.—The species is highly susceptible to diseases (root rots, *Xanthomonas* blight, downy mildew, rust, angular leafspot, BCMV and other virus) and insects including red spiders, mealy bug, white fly, leafhoppers, weevils, many types of caterpillars, flea beetles, *Diabrotica*, and Mexican bean beetles (additional information is also provided by Miranda Colín, 1967). Some bruchid resistance has been reported due to a globulin storage protein named arcelin causing antibiosis (Romero-Andreas 1984; Osborn et al. 1986) and documented further in Colombia (Cardona et al. 1989; Kornegay et al. 1993).

Common names.—Generally given the names of "frijol de ratón," "frijol de rata," or "frijol de coyote" in Mexico (Delgado et al. 1988; Gentry 1969), or "frijol de venado" en Costa Rica (Araya et al. 2001). Called "chaneca" (meaning the ancient one) in the Zapotec part of Oaxaca, Mexico (Gentry 1969). Called "frijol miscaite" in Honduras according to Molina and "matz" or "cumatz" according to McBryde (1947) in Huehuetenango, Guatemala.

Ethnobotany.—Normally not eaten by indigenous peoples though they admit to using the seed in periods of scarcity. Reported as growing with different teosinte species *Zea mexicana* and *Z. diploperennis* in several Mexican states, Jalisco, México, Michoacán (Benz et al. 1990; Delgado et al. 1988; Miranda Colín 1967, respectively; and Iltis (pers. comm.), and in Huehuetenango, Guatemala (McBryde 1947). The presence of wild bean near El Chante, Jalisco and growing with teosinte was confirmed by the senior author during the 1981 field collecting trip to Mexico and Guatemala and was also found to be extremely variable (see Color Plate V, photo 49 and Color Plate IV, photo 40). In January 1995, the junior author found wild *P. vulgaris* (# 3074) growing together with *Zea mays* subsp. *huehuetenangensis* near Bujxup in the Huista region of Huehuetenango, Guatemala. Others have found it growing in corn and sorghum fields and grazed by cows and goats.

Genetics.—*Phaseolus vulgaris* is fairly easily hybridized when used as the female parent with *P. dumosus* (=*polyanthus*) (see Maréchal 1971; Camarena & Baudoin 1987). Though the reciprocal cross with the cytoplasm of *P. dumosus* seems much more difficult, at least when crossing cultivated forms (Camarena & Baudoin 1987; Schmit et al. 1992), the use of heterozygous cultivars may overcome this obstacle (Kedar & Bemis 1960).

Phaseolus vulgaris is also fertile with pollen of *P. coccineus* ssp. *coccineus* and *P. coccineus* ssp. *striatus* (Alvarez et al. 1981; Al-Yasiri & Coyne 1966; Baggett 1956; Cheng et al. 1981; Fermond 1855; Ibrahim & Coyne 1975; Manshardt & Bassett 1984; Shii et al. 1982; Smartt 1970), and production of polyploids has been used in some of these crosses to increase fertility (Smartt & Haq 1972). However these two species are rarely found growing together in a wild situation (see Debouck 2000b, for a list of places) with one notable exception found in central Jalisco where both *P. coccineus* and *P. vulgaris* were found to be highly variable but with no signs of hybridization (see Plate IV, photos 41 & 42). As domestication has advanced it seems isolating mechanisms have increased over the wild progenitors (Miranda-Colín & Evans 1973). Although many variants of these two species have been crossed on multiple occasions, little fertile and useful progeny have yet been achieved (see reviews by Hucl & Scoles 1985, and Smartt 1990). One could thus conclude that these two species are not very close to each other and can be located in different sections of the genus; a confirmation of this statement has been recently obtained by analysis of cpDNA polymorphisms with two different techniques (Fofana et al. 1999; Schmit et al. 1993).

Phaseolus vulgaris is considerably less compatible with *P. acutifolius* (Andrade-Aguilar & Jackson

1988; Mejía-Jiménez et al. 1994; Mok et al. 1978; Pratt 1983; Rabakoarihanta et al. 1980; Smartt 1970; Thomas et al. 1983; Thomas & Waines 1984; Waines et al. 1988). This information from interspecific crossing indicates that tepary bean has separated early from the phylum of common bean (justifying thus to have them in separate sections), and this is consistent with data of RFLPs of rRNA genes (Jacob et al. 1995) and ITS DNA sequencing (Delgado et al. 1999; Gaitán et al. 2000).

It has not been possible to get fertile hybrids from any other species crosses: *P. angustissimus*—Belivanis & Doré 1986; Petzoldt & Dickson 1987 (F_1 hybrids were obtained by embryo rescue, but did not produce mature seed); *P. filiformis*—Maréchal & Baudoin 1978; *P. lunatus*—Honma & Heeckt 1959 (this report seems doubtful); Kuboyama et al. 1991; Leonard et al. 1987; Mok et al. 1978; *P. maculatus*—Braak & Kooistra 1975); and *P. ritensis*—Petzoldt & Dickson 1987. Production of polyploids of the sterile F_1 has also been used to restore fertility in some of these wide crosses (Braak & Kooistra 1975; Weilenmann de Tau et al. 1986).

Abnormals (lethals and semilethals), influenced in part by high temperatures, have been observed in wide intraspecific crosses (Gepts & Bliss 1985; Koinange & Gepts 1992; Shii et al. 1980), indicating incipient divergence within wild *P. vulgaris* itself, a possibility which has been supported by studies of mtDNA (for reviews see: Khairallah et al. 1992; Bannerot & Debouck 1992; Gepts 1993).

Comments.—An adaptable species found in many different habitats, most on the dry side, sometimes found associated with either teosinte (wild corn) or with other bean species (see Debouck 2000b). Usually locations have fairly constant rainfall for early growth and then the rains abruptly cease and a dry season initiates during or shortly after flowering which usually takes place from August to November in Central America. Seed are violently thrown by the twisting dehiscent pods and then remain inactive on the dry soil surface or in the leafmold for months until the rains begin again. Some seed may remain viable for a couple of years before germinating. Although generally thought to be a fibrous-rooted annual, as is the case with most all cultivated collections seen by us, it appears that at least some of the wild types are definitely perennials (as suggested by Gentry 1969) and as observed by the senior author on planting out some of these Gentry collections on separate occasions in Honduras and Puerto Rico where they survived for several years and produced nearly perennial roots (see Color Plate III, photo 25 and Color Plate V, photo 51). These perennial types are mostly found in the states of Nayarit, Jalisco, Michoacán and Guerrero in western Mexico, and in Quetzaltenango and Sololá, Guatemala and are quite variable (see Color Plate I, photo 1; Color Plate IV, photo 41).

Early collectors found no wild types at all, even in Mexico and Guatemala, thus leading to erroneous conclusions about the probable origin of cultivated beans (see Bukasov 1930; Ivanov 1928; and Vavilov 1931). It is interesting to note that there are few wild collections from Central America (i.e. El Salvador, Nicaragua) (see Delgado 2001; Toro et al. 1990), in fact none from Panama (although Brücher 1988 claimed it was found in the Chiriquí mountains) were found in the herbaria studied and only a few from Honduras and Costa Rica (for a distribution in this country, see Araya et al. 2001). This would seem to indicate that the Central American area is rather on the edge of the distribution of the wild species or that the areas suitable for it are restricted in that part of the Neotropics. Its northernmost habitat seems to be in Chihuahua (Nabhan 1985).

Delgado (1985) recognizes three taxa of varietal rank: a) *vulgaris* for the cultivated types, b) *aborigineus* Burk. for the wild types of South America, and c) *mexicanus* Delgado for the wild types of Mexico. We could perhaps recognize a var. *mexicanus* for the northern hemisphere wild variants, a var. *aborigineus* for the South American wild variants with larger seed as observed by Burkart in Argentina, and an additional var. *ecuadorensis* for a small-seeded wild type with apparently some self sterility, found in Ecuador (Debouck et al. 1993; Kami et al. 1995). A fourth group found in the eastern Colombian highlands might perhaps be recognized too (Tohme et al. 1996).

After a numerical taxonomic study, Maréchal and co-workers (1978b) concluded that the common bean and the Lima bean represent the two ends of an orthogenetic sequence, with many *Phaseolus* species distributed between them (Le Marchand & Maréchal 1977b; Maréchal et al. 1978b) and confirmed by research with PCR-RFLPs of cpDNA where at least two branches appeared in the genus, one including *P. lunatus* and another one including *P. vulgaris* (Fofana et al. 1999). Significant differences between northern and southern gene pools of common bean have been confirmed with the help of molecular markers (allozymes: Koenig & Gepts 1989; phaseolin: Gepts et al. 1986; Koenig et al. 1990; RFLPs of mtDNA: Khairallah et al. 1992; photosynthesis physiological parameters: Lynch et al. 1992). Research on cpDNA polymorphisms indicates that *P. vulgaris* probably shares a common progenitor with other species of Section B. *Phaseoli* (Llaca et al. 1994; Schmit et al. 1993) as well as with Section C. *Coccinei* (Hervieu et al. 1994). Additional evidence was obtained with RFLP markers of the rRNA genes (Jacob et al. 1995) and by microsatellite fingerprinting (Hamann et al. 1995). Other biochemical differences in seed storage proteins have been found with globulins in common bean (Osborn 1988); lectins are more important as storage proteins in Lima bean (Gutiérrez Salgado et al. 1995) and in the case of *P. glabellus* the major fraction is different from either of the former mentioned species (Schmit & Debouck 1990). For more information see Becerra & Gepts 1994; Delgado et al. 1988; Pratt & Nabhan, 1988; and Waines et al. 1988. The senior author believes cultivated material should all carry the identification of the type var. *vulgaris*.

B.2.—Phaseolus costaricensis Freytag & Debouck, Novon 6:157–163. 1996. (**Figs. 6, 9**). TYPE: COSTA RICA. PROVINCIA DE SAN JOSÉ: San Isidro El General, 3 km SE de Copey, orillas del Río Pedregosa, 9°37'N, 83°55'W, 1800–2080 m, 14 Jan 1987, *Debouck et al. 2135* (HOLOTYPE: US; ISOTYPES: BR, CR, UC).

Aerial shoot a large perennial, woody, shrubby indeterminate vine, 4–8 m long. ***Root*** extensive and fibrous but not penetrating deeply into soil. ***Stem*** stout, terete, striate-angled, at base of plant to 3–4 mm in diam., corky and knobby, younger stems with densely short white reflexed-strigose hairs; internodes 12–16 cm long, covered by reflexed, short and long, fine, strigose, hispid and uncinate, whitish hairs. ***Stipules*** triangular-lanceolate, 6 mm long, 3 mm wide, 5- to 7-nerved, acute, thin, sparsely covered with brownish strigose hairs. ***Leaves*** 13–34 cm long; petioles stout, rounded canaliculate, 2.5–8–16 cm long, covered with tawny-villous to hirtellous hairs; petiolules 1.6–5.5 cm long; stipels 6 mm long, 2.5 mm wide, the lower triangular-ovate, 8- to 10-nerved, acute, the upper oblong-linear, slightly falcate, 5 mm long, 1–1.5 mm wide; pulvini 5–6 mm long, covered by densely yellowish to tawny brown and long strigose hairs; terminal leaflets broadly ovate to oblong-ovate, 4–7–11 cm long, 4.5–12 cm wide at about 1/3 from base, nearly obtuse to short acute, apiculate, the main veins prominently whitish below, secondary veins inconspicuous, adaxially sparsely pubescent, below sparsely to densely covered with yellow strigose to nearly tomentose hairs; lateral leaflets similar and slightly inequilateral. ***Inflorescence*** a very long raceme, 10–38–87 cm long, much exceeding leaves; peduncle 11–18–41 cm long; rachis stout, 3–15–46 cm long, with hirtellous, tawny hairs, glabrate below, with many flowering nodes to 30 or more spaced 2–5–30 mm apart on axis; primary bracts broadly rounded to ovate-lanceolate, 5–12 mm long, 1.5–8 mm wide, 3- to 12-nerved, acute, thin to hyaline, abaxially heavily covered with white strigose hairs, glabrous within, somewhat persistent; pedicel stout, 6–7 mm long becoming 16–18 mm long at mature pod, sparsely covered with minute white uncinate hairs; pedicellar bracts lanceolate 3 mm long faintly 3-nerved purplish sparsely covered with white hairs hyaline margins. ***Bracteoles*** often rounded ovate to elongate-acuminate to linear, 5–8–9 mm long, 2.5–4–8 mm wide, acuminate, sparsely covered by white hirsute to strigose hairs abaxially, glabrous adaxially, purplish. ***Flower*** dark pink or lilac to purple, 23 mm long, very numerous in fascicles of 2–5; calyx campanulate, 5–6 mm long, with a small knob at base, pubescent, the two upper lobes united into one emarginate scarcely developed, 1 mm long, 6 mm wide, the 3 lower lobes subequal, dentate-acute, 2 mm long, 2.5 mm wide, the center one slightly longer, sparsely covered with weak, white, short strigose hairs, the mouth of the calyx oblique; standard dark pink or lilac to purple, broadly rounded-obovate, cupped, 18 mm long, 14–15 mm wide, emarginate, reflexed, 8 mm to bend, somewhat thickened at flexure, lateral edges recurved with a spur at lower edge 1 mm long, the claw poorly developed, 0.5 mm long, the auricles 3 mm long, 2 mm wide and nectariferous, inserted 3–4 mm from base; wings lilac to purple, broadly rounded, cupped and clasping, unequal, spreading laterally, 23 mm long, the claw 6–8 mm long, 0.5 mm wide, the blade 13–16 mm long, 10–11 mm wide, the spur well-developed, nearly round, 1.5–2.5 mm long; keel tubular, straight, the claws divided 4 mm, 5 mm more to flexure and 6 mm more to the terminal 1 3/4 tight coils of 3.5 mm diam.; vexillary stamen, the claw 1.5 mm long to geniculate knob developed into a

Fig. 6. Illustrations of *Phaseolus costaricensis* Freytag & Debouck.—**a.** Seedling a few days after germination.—**b.** Stem and root of mature plant.—**c.** Vine tip with mature leaves and pods.—**d.** Flowers, side view and front view.—**e.** Exploded view of the flower showing all parts, including—**s.** Style tip and stigma as seen under the microscope.—**f.** Pods, side view and dehisced carpels.—**g.** Seeds, side view and view from the hilum. All drawings from plants grown in the greenhouse at Mayagüez of *Debouck et al. 2135* (TARS #437) from Costa Rica, except **f. & g.** of the field collection *Debouck et al. 2128* (TARS #420) from Costa Rica.

thin sheath 2 mm long, 1.5 mm wide, the thickened portion 7 mm long to filament; stamen tube straight, ridges not developed, the basal collar 1.5–2 mm long, denticulate; ovary straight, 10 mm long covered by fine white pubescence, 6 or less ovules, the style 10–12 mm long with a terminal thickened coil 3 mm in diam., the stigma terminal, capitate, 0.75 mm long. ***Pod*** when young, straight, flat, broad, about 3 times longer than broad, covered with yellowish brown tomentose hairs to sparsely covered by strigose hairs, mature pods straight, 10 cm long, 13 mm wide, 7–10 mm thick; the beak strong, 7 mm long, recurved. ***Seed*** squarish ovate, rounded and flattened, 10–11.6 mm long, 8.3–9.4 mm wide, 3.8–5.1 mm thick, brown and black speckled and streaked on brown and tan, a black ring around hilum; the hilum oblong oval, 4.2 mm long, 1.8 mm wide; the lens medium in size and raised. ***Seedling*** large from epigeal germination, the hypocotyl 1.7 cm long, the epicotyl 2.2 cm long; stipules united; the next internode very long to 12 cm or more; primary leaves opposite, simple, the petiole 3 cm long, stipels present, the blade broadly rounded-ovate, 8 cm long, 7.5 cm wide near the base, the base auriculate, the tip acuminate and not apiculate, membranous, nearly glabrous.

Specimens examined: **COSTA RICA. Alajuela:** Tres Ríos and Ochomogo, vicinity San Ramón, (9°55'N, 84°25'W), Dec 1912, *Brenes 18* (GH); Zarcero, 2 km SE de Zarcero hacia San Juanillo y Naranjo, valle Río Seco, 10°10'N, 84°23'W, 1610 m, 8 Feb 1998, *Debouck et al. 3122* (CR, MO). **Cartago:** Cuesta Chinchilla, 1.2 km SW de Cot (también en barranco del Río Páez cerca de Cot), 9°54'N, 83°53'W, 1650 m, 7 Jan 1987, *Debouck et al. 2093*, (BR, CR, US); Cerro de la Carpintera, 2 km S de San Vicente, 3 km SE de Tres Ríos, 9°53'N, 83°58'W, 1600 m, 11 Jan 1987, *Debouck et al. 2118*, (CR, US); cerca de la planta eléctrica María del Rosario, 3 km NE de Dulce Nombre, Mpio. Tres Ríos, 9°57'N, 83°57'W, 1750 m, 11 Jan 1987, *Debouck et al. 2119* (BR, CR, MICH, US, WIS); Pacayas, 1.5 km SE de Pacayas hacia Cervantes, 9°54'N, 83°47'W, 1570 m, 7 Feb 1998, *Debouck et al. 3118* (CR, MO); N of Cartago, Río Reventado, 1950 m, 11 Dec 1984, *Khan et al. 957* (CR); Cartago, San Ramón de Tres Ríos, 20 Feb 1959, *Rodríguez 595* (CR). **San José:** San Isidro de Coronado, 1400–1600 m, 1 Dec 1937–1 Jan 1938, *Allen 555* (F, GH, MO); Río Herradura, tributary of Río Chirripó del Pacífico, NW of Canaán, General Valley, 1600 m, 29 Dec. 1969, *Burger et al. 7097* (F, MO); 2 km SW de Aserrí, 0.2 km después de Km 13 de Ruta 4, 1470 m, 9°51'N, 84°06'W, 8 Jan 1987, *Debouck et al. 2095*, (BR, CR, UC, US); Alajuelita, potreros arriba de San Antonio, al pie del cerro de San Miguel, 9°52'N, 84°07'W, 1620 m, 8 Jan 1987, *Debouck et al. 2102* (BR, CR, US); 1.5 km E de Aserrí, al pie de la Piedra, 9°52'N, 84°07'W, 1590 m, 11 Jan 1987, *Debouck et al. 2116* (BR, CR, COL, US); 1.8 km E de Sta. María a Copey, frente Quebrada Salta, orilla del Río Pirris, 9°40'N, 83°57'W, 1660 m, 12 Jan 1987, *Debouck et al. 2122* (BR, CR, COL, MICH, US); San Isidro El General, 0.8 km S de Pueblo Nuevo, orillas del Río Buenavista, 9°26'N, 83°40'W, 1550 m, 13 Jan 1987, *Debouck et al. 2126* (CR, UC, US); San Isidro El General, al pie de la Fila Villarevia, orillas del Río Blanco, 2 km N de Herradura, 9°30'N, 83°37'W, 1690 m, 13 Jan 1987, *Debouck et al. 2128* (BR, CR, MICH, SI, US, WIS); San Isidro El General, 1 km N de Providencia, orillas del Río Brujo, 9°34'N, 83°51'W, 1990 m, 14 Jan 1987, *Debouck et al. 2132* (BR, CR, US); San Isidro El General, 3 km SE de Copey, orillas del Río Pedregosa, 9°37'N, 83°55'W, 2080 m, 14 Jan 1987, *Debouck et al. 2135* (BR, CR, UC, US); Herradura, 1.5 km N de Herradura, 9°29'N, 83°37'W, 1550 m, 5 Feb 1998, *Debouck et al. 3111* (CR); Buena Vista, 2 km N de La Piedra, 9°31'N, 83°40'W, 1500 m, 5 Feb 1998, *Debouck et al. 3112* (CR); barranco entre Aserrí y Tarbaca, 1150 m, 25 Nov 1964, *Jiménez 2581* (F); below Cerro Pico Blanco, 4 km S of Escazú, 9°52'N, 84°8'W, 1900 m, 12 Dec 1976, *Lent 3966* (F); vicinity of El General, 1560 m, Dec 1936, *Skutch 2990* (GH, K, MO); La Cruz de Alajuelita, 1810 m, 22 Dec 1935, *Solis 392* (F, MO); between Aserrí and Tarbaca, 1200–1700 m, 6 Dec 1925, *Standley 41362* (F); Llanos de Alajuelita, 1 Dec 1889, *Tonduz 1475* (G, US).

PANAMÁ. Boquete: Bajo Mono, (8°35'N, 82°27'W), 1500 m, 3 Apr 1938, *Davidson 496*, (F, GH, MO). **Chiriquí:** valley of the Río Chiriquí Viejo, near New Switzerland, 1800–2000 m, 6–14 Jan 1939, *Allen 1397* (EAP, MO); Methodist Camp near Nueva Suiza, 19 Feb 1971, *Croat 13508* (F, MO); top of peak between Baru and Respinga, aproxim. 3000 m, 27 Nov 1975, *D'Arcy 10109* (MO); Río Chiquero, 22 Mar 1977, *D'Arcy 11005* (MEXU, MO); N of Volcán City, valley of R. Chiriquí Viejo, 1700–1800 m, 9 Dec 1966, *Duke 9019* (MO); Nueva Suiza near Audubon Society Cabin, 1750 m, 12 Sep 1972, *Gentry 5986* (F, MO); Finca Lérida to Peña Blanca, 1750–2000 m, 9 Jul 1940, *Woodson et al. 313* (MO, US); Casita Alta, Volcán de Chiriquí, 1500–2000 m, 28 Jun–2 Jul 1938, *Woodson et al. 918* (GH, MO, NA, US).

Habitat.—The large vines from 3–8 meters long are often found climbing and sprawling over shrubs (*Dahlia*, *Coffea*, Compositae, Solanaceae) and trees, and are found in sunny openings often in disturbed cut-over or burned over areas, coffee plantations or pastures in humid montane forest (*Alnus, Erythrina*) with bamboo and numerous epiphytes (ferns, orchids, aralias and bromeliads), and are most frequent and abundant on steep slopes and in inaccessible areas at 1400–2100 m, most often at 1600–1800 m. It is found growing on dark brown soils derived from lava or volcanic ash, or metamorphic schists, with high organic content and rocky, very moist areas, often near streams or riversides.

Common names.—Known by local inhabitants by the names of "frijol de montaña" around Cot, Provincia de Cartago, Costa Rica and "cubá de venado" in the area south of Alajuelita, Costa Rica.

Ethnobotany.—Not eaten.

Diseases and pests.—It is reported to be damaged by anthracnose, angular leaf spot, lacebugs, thrips and weevils.

Genetics.—It has been reported as visited by hummingbirds and carpenter bees. The following specimen was determined by the collectors to be an obvious hybrid, probably with *P. dumosus* found at the same location: **COSTA RICA. San José:** al pie de la Piedra de Aserrí, 1.5 km E de Aserrí, 9°52'N, 84°7'W, 1560 m, 11 Jan 1987, *Debouck et al. 2114* (BR, US). It has been observed to cross naturally with *P. dumosus* in the field (D.G. Debouck, pers. obs., 1987), though very rarely, since *P. dumosus* is an introduced plant in Costa Rica, and can also be crossed with *P. vulgaris* (H. Bannerot, pers. comm., 1989; Singh et al. 1997), in both cases as pollen parent.

Comments.—This species is named in honor of the country, Costa Rica, in which it is most widely distributed (see also Araya et al. 2001). This species has been confused with *P. coccineus* from which it differs by the dark pink or fuschia flower color, large and long primary bracts and bracteoles, epigeal germination, and fibrous roots. *P. costaricensis* often seems to produce a panicle with several to many lower lateral branches, especially on the larger inflorescences, much as in some populations of *P. lunatus* L., *P. maculatus* Scheele, *P. polystachyus* (L.) Britton, Sterns & Poggenberg and other related, more or less woody *Phaseolus* species. It has also been confused with *P. tuerckheimii* from which it differs by having much less tawny pubescence, having long pods with 4–6 seed, and details of flower structure which are very different, especially the large, long and broad bracts and bracteoles, and the considerably larger standard and wings.

Since *P. costaricensis* shares several cpDNA polymorphisms with *P. vulgaris* and *P. dumosus* (Schmit 1992; Schmit et al. 1993), the placement of this species in the same section as the later two taxa seems appropriate, and perhaps is ancestral to them or to the latter. Results from comparative cpDNA and nuclear DNA analysis of *P. dumosus* (Llaca et al. 1994) have shown a possible reticulate origin for this taxon, its cytoplasm donor parent having been crossed several times to *P. coccineus*. It is still premature to conclude that *P. costaricensis* is the cytoplasm donor parent to *P. dumosus*, but there is little doubt that the present novel species should be included in any further phylogenetic study including the three bean cultigens *P. vulgaris*, *P. dumosus* and *P. coccineus*.

B.3.—Phaseolus dumosus Macfady. Fl. Jamaica 1:279. 1837. (**Figs. 7, 9**). TYPE: (none from Macfadyen's work so far located; however the very complete and precise description of MacFadyen leaves no doubt as to the characteristics of his species, in all probability an escaped or feral variant (since these forms are common in the Caribbean islands). LECTOTYPE: GUATEMALA. SOLOLÁ: on very steep hill of volcanic rim, in and along ravine of small stream, not in forest, at Km 110 Hwy 11, Patzicia-Panajachel, 14°43'N, 91°5'W, 1750 m, 6 Oct 1978, *Freytag & Vakili 78-Guat-47* (NEOTYPE, here designated, US; ISONEOTYPES: BR, EAP, MEXU, MO, UC).

Phaseolus polyanthus Greenm., Field Mus. Bot. Ser. 2:253. 1907. TYPE: MÉXICO. VERACRUZ: On railroad banks near Jalapa, 10 Sep 1906, *Barnes et al. 20* (HOLOTYPE: F; ISOTYPE: US). *Phaseolus coccineus* subsp. *polyanthus* (Greenm.) Maréchal, Mascherpa & Stainier, Taxon 27:199. 1978a.

Phaseolus flavescens Piper, Contr. U.S. Natl. Herb. 22:686. 1926. TYPE: COLOMBIA. Caldas, at Río San Rafael, below Cerro Tatamá, 2400–2600 m, 7–11 Sep 1922, *Pennell 10334* (HOLOTYPE: US 1,143,511).

Phaseolus leucanthus Piper, Contr. U.S. Natl. Herb. 22:686. 1926. TYPE: MÉXICO. VERACRUZ: San Cristóbal, 3 Sep 1857, *Mohr 123* (HOLOTYPE: US 773,886).

Phaseolus harmsianus Diels, Biblioth. Bot. Heft 116, 29:98. 1937. TYPE: ECUADOR. TUNGURAGUA: bei der Müdung des R. Topo, im Wald, 1270 m, 14 Sep 1933, *Diels 1028* (HOLOTYPE: location unknown).

Phaseolus coccineus subsp. *darwinianus* Hernández. & Miranda, Revista Soc. Mex. Hist. Nat. 20:114–116. 1959. TYPE: MÉXICO. PUEBLA: Huauchinango, 2200 m, *Hernández & Tapia s.n.* (HOLOTYPE: CHAPA).

Aerial shoot a very large, indeterminate vine, to 8–10 m or more. ***Root*** an annual tending to be perennial, fibrous but somewhat thickened, to 2–3 cm thick near ground level. ***Stems*** terete, younger stems 3 mm in diameter or more, ribbed, sparsely covered with strigose pubescence, older stems near ground level robust, angular-striate, to 2 cm thick, tending to be perennial. ***Stipules*** 5–6 mm long, 2.5–3 mm wide, foliaceous, extended to slightly reflexed, nearly glabrous except for a few strigose hairs on adaxial surface and margins. ***Leaves*** 30–39 cm long; petiole 15–20 cm long; petiolule 4–5 cm long, stout to 3 mm in diameter, heavily ribbed, sparsely covered with forward appressed-strigose hairs; pulvini large, 4–5 mm long, densely covered with strigose pubescence; terminal leaflet very broadly ovate, to 13 cm long, 12.5 cm wide at 1/3 from base, acuminate, apiculate; lateral leaflets inequilateral, 13 cm long, 11.5 cm wide near base, moderately strigose pubescent adaxially mostly on veins, abaxial surface nearly glabrous, 3 main nerves with slight webbing between with many long hairs at base be-

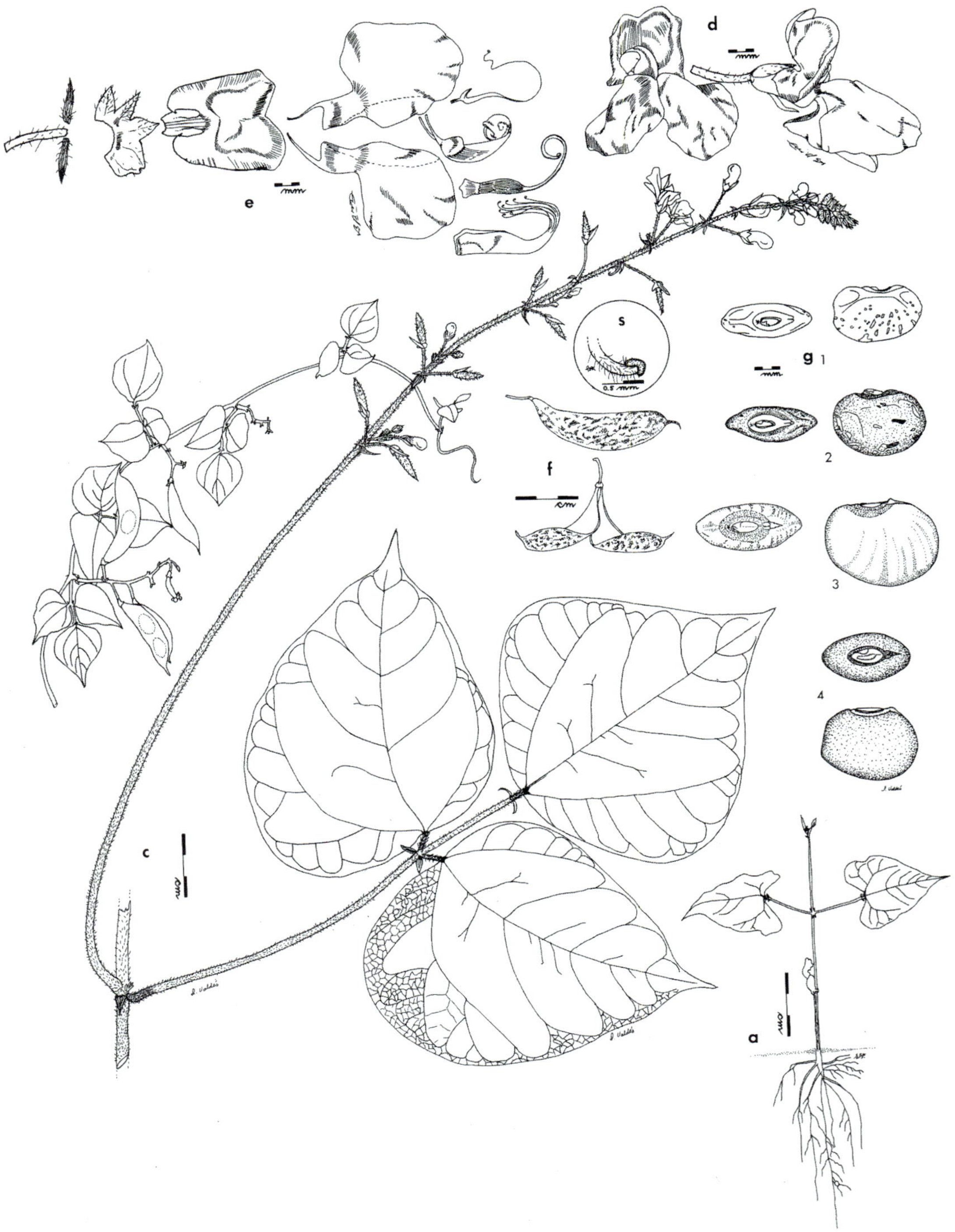

FIG. 7. Illustrations of *Phaseolus dumosus* Macfady.–**a.** Seedling a few weeks after germination.–**b.** Stem of mature plant.–**c.** Vine tip with mature leaves, inflorescence and pods.–**d.** Flowers, side view and front view.–**e.** Exploded view of the flower showing all parts, including–**s.** Style tip and stigma as seen under the microscope.–**f.** Pods, side view and dehisced carpels.–**g.** Seed types, side view and view from the hilum. Drawings **a,d,e,f,g.2**, & **s** of plants grown in greenhouse at Mayagüez of *Cojulún 78-Guat-6* (TARS #309) from Semetabaj-Panajachel, Guatemala; drawings **c.** & **g.1** of field collection *Freytag et al. 78-Guat-47* (TARS #71) from Patzicia-Panajachel, Guatemala; drawings of cultivated seed **g.3** of *Debouck 1601* from Chimachoy, Chimaltenango, Guatemala, and **g.4** of "Piloy" (market seed) from Antigua, Guatemala.

tween veins, often 2 marginal smaller veins, very thin almost membranous. ***Inflorescence*** a panicle with 2 early flowers per node and a secondary branch axis produced between them at about anthesis, later flowers produced on the branch, the very long main peduncle to 20 cm or more long, stout, erect, sparsely covered with strigose hairs, the rachis as long or longer than peduncle, densely covered with hirsute hairs with scattered smaller hooked pubescence; primary bracts lanceolate, 6–7 mm long, 1.5 mm wide, densely covered with strigose pubescence; pedicel rather stout, 6–7 mm long, covered with moderately yellowish-brown strigose and hooked pubescence, most hairs 0.5–0.75 mm long; pedicellar bracts narrowly lanceolate 2–4 mm long ciliate 1-nerved often caducous hyaline borders. ***Bracteoles*** linear to narrowly oblong, 6–8 mm long, 1–1.5 mm wide, mostly 3- to 5-nerved, acute, hyaline margins, sparsely to heavily covered with strigose hairs, caducous shortly after pollination. ***Flower*** purple rarely white; calyx tube 3–4 mm long, covered with yellowish-brown strigose hairs especially on adaxial and abaxial parts and nearly glabrous laterally, upper two teeth united into a single lobe, scarcely elongated, 5.75 mm wide, slightly emarginate, the lower 3 lobes slightly rounded dentate, subequal, central lobe 2.5 mm long, 2 mm wide, the lateral 2 mm long, 2 mm wide, centrally nerved, acute, slightly covered with strigose pubescence on both surfaces, ciliate; standard purple rarely white, the blade broadly rounded, erect, 5 mm from base to flexure and 7 mm more to emarginate tip, 12 mm wide, laterally enrolled, a few minute hairs at apex, the claw very broad and not well defined, 3 mm long, 2 mm wide at base and 6 mm wide at upper end, the auricles little developed, 0.5–1 mm long; wings purple rarely white, round, widely spreading and cupped, unequal, the claws 5 mm long, 0.75 mm wide, the spur well-developed 1.5–1.75 mm long, adhering to keel, the blade 14 mm long, 12 mm wide; keel, the claws 5 mm long, 3 mm more to bend and 4 mm more to base of the terminal 1 1/2 coils of 4 mm diam., tip whitish; vexillary stamen, the claw 1 mm long, the geniculate flap well-developed, about 1.75 mm long, 1.5 mm wide extending forward as a thin sheath, 4 mm more to end of thickened portion; stamen tube 8.5 mm to bend and about 4 mm more to end of united portion, the ridges scarcely developed; anthers 0.75 mm long, 0.4 mm wide; basal collar denticulate, 1.5 mm long; ovary straight, 9 mm long, 1.8 mm wide, densely covered with long white cannose hairs, 4 ovules; style 7 mm long to the thickened 1 1/2 coils of 2.5–3 mm in diam.; stigma terminal, capitate and introrse, about 0.6 mm long, oblique at about a 45° angle, the stylar hairs to within about 0.5 mm of the stigma. ***Pod*** straight, tapered near stem, 7 cm long, 12–13 mm wide, 7–10 mm thick; sutures heavy, the lower suture rounded; valves smooth, farinose-pustulate, puberulent of short, white strigose hairs, fibrous; beak nearly straight, 7–8 mm long. ***Seed*** oblongoid, flattened, 11.5 mm long, 8.6 mm wide, 3.9 mm thick, smooth, black streaked and striped on a tan backbround, black ring around hilum; hilum ovate-oblong, 4 mm long, 1.75 mm wide, scar continuous; lens prominent, divided. ***Seedling*** from epigeal germination; hypocotyl 3–6.5 cm long; epicotyl 3.5–4 cm long; stipules united; next internode as long as or longer than epicotyl, to 5 cm or more; eophylls opposite, simple, the petiole 2.2 cm long, stipels present, the blade triangular ovate, 4 cm long, 3–3.5 cm wide near the base, the base somewhat angular to broadly auriculate, the tip acuminate and not apiculate, membranous, nearly glabrous.

Specimens examined: **GUATEMALA. Alta Verapaz:** Km 197 Hwy CA-14, Route Cobán–Guatemala City, 15°10'N, 90°23'W, 1330 m, 12 Oct 1978, *Freytag et al. 78-Guat-132* (BR, EAP, F, GH, K, MEXU, MO, UC, US). **Guatemala:** Amatitlán, en el Cerro Hoja de Queso, 2–3 km NE de Calderas, 14°26'N, 90°35'W, 1800 m, 15 Dec 1987, *Debouck et al. 2460* (BR, COL, MICH, MO, SI, UC, US, USCG). **Quetzaltenango:** E side of dam on Samalá River, 0.5 km E Santa María de Jesús, 1460 m, 14°42'N, 91°32'W, 25 Jan 1995, *Debouck et al. 3093* (MICH, USCG). **Sacatepéquez:** 8 km S de Santa María de Jesús, camino a Palín Km 16.5 desde Antigua, 1550 m, 14°29'N, 90°42'W, 4 Dec 1985, *Debouck et al. 1608* (COL, K, MICH, UC, US, USCG, WIS); 8 km W de San Miguel Dueñas, 1940 m, 14°33'N, 90°51'W, 6 Dec 1986, *Debouck et al. 1622* (BR, COL, G, K, M, MICH, SI, UC, US, USCG). **San Marcos:** Finca El Porvenir along Río Chopal, S-facing slopes of Volcán Tajumulco, (15°N, 91°59'W), 1300–1500 m, 11 Mar 1940, *Steyermark 37495* (F(2)). **Sololá:** 2.2 km N de Panajachel, terracería hacia Sololá, 1680 m, 14°46'N, 91°10'W, 10 Dec 1985, *Debouck et al. 1631* (COL, K, US, USCG).

MÉXICO. Chiapas: Mpio. Motozintla de Mendoza, 45–50 km NE Huixtla to Motozintla, 1900 m, 17 Nov 1971, *Breedlove 22691* (MEXU).

Other specimens seen.—A total of 70 additional collection numbers were examined which are all believed to be of cultivated or feral types (present in islands of the Caribbean, Costa Rica and abundant in the northwestern Andes; see also Berglund-Brücher & Brücher 1974) and were not used in forming the species description nor distribution.

Habitat.—Found growing in humid pine-oak or humid montane forests with bamboo, *Wimmeria*, *Podocarpus*, and caña brava and with epiphytes (Bromeliads) and many vines (*Ipomea*, *Cucurbita*, *Dioscorea*) and with abundant herbaceous plants (Labiateae, Geraniaceae, Comelinaceae, Araceae, Dioscoreaceae, and *Dahlia*). Sometimes growing very abundantly on very steep hills of volcanic rim, in and along ravines of small streams, mostly in open places and not in the forest and mostly climbing over shrubs, weeds, and small trees, to 8–10 m long or more. This species prefers very deep, moist, well-drained, humus soil, derived from volcanic ash, which is often sandy and rocky.

Diseases and pests.—It is reported to be damaged infrequently by thrips, leaf miners, and weevils.

Common names.—Names given on collectors labels are "piligüe de monte" in Guatemala for wild types and "cubá" in Costa Rica for feral forms. For vernacular names in the Andes see Schmit & Debouck (1991). In Jamaica, for white-flowered feral forms, it is known as "year-bean" (specimen *Maxon 1136*, at NY).

Ethnobotany.—It is reported to be grazed by animals such as cows and horses.

Comments.—There seems to be no good reason for maintaining this species with the name of *P. polyanthus* Greenman since Macfadyen's work antidates that of Greenman by a full 70 years, and in both cases the species were based on collections of feral material. The lack of a Macfadyen type for *P. dumosus* is unfortunate and has lead to confusion (for instance Smartt 1973, p. 425, considered it as a synonym of *P. lunatus*). The description is however more than adequate to leave no doubt as to the species in hand (see also Debouck 1991). Perhaps one might argue that since the epithet *P. polyanthus* has been in use for a long time that this name should be maintained, but the species is little known by most plant scientists in either the United States or Europe and little used by scientists in Latin America, and has only recently been of interest in Mexico, Costa Rica, Colombia or the Caribbean Islands (all areas where the cultivated and feral types occur). The senior author following Lackey (1983) therefore argues that the *a priori* name of *P. dumosus* now replaces that of *P. polyanthus* as the valid name for this taxon, and the species description has therefore been revised and is based on the neotype herein designated from the first field collection in Guatemala of a wild population. We now know that the wild form is found at about 1500 m on south facing slopes of the central-western Cordillera Volcánica of Guatemala or nearby and can best be recognized by the relatively small inflorescence, light purple flower color (see Color Plate I, photo 3) and the small (11 mm × 8 mm), distinctly flattened seed of a black speckled and striped pattern on tan background and with an oblong hilum some 4 mm long. Cultivar seed is generally much larger and thicker (nearly spherical).

Piper (1926) created several new species (*P. leucanthus*, *P. flavescens*), obviously based on cultivated or feral variants of this species, and which have key characteristics the same as those of other cultivated *P. dumosus*. They were keyed out in his work on the basis of differences in bracteole length—not even a good varietal difference within this species. Nevertheless, all do have very long, narrow bracteoles, a key difference from other *Phaseolus* species as noted by Maréchal et al. (1978b), who also noted similarities of *P. dumosus* with both *P. vulgaris* and *P. coccineus* with which it is interfertile. Data from mtDNA (Hervieu et al. 1993; Hervieu et al. 1994) and cpDNA (Schmit 1992; Schmit et al. 1993) analysis also seem to confirm the placement of this species together with *P. vulgaris*, but not with *P. coccineus*, a distinction that native Indian farmers who do grow *P. dumosus* and *P. coccineus* together have always made (Breedlove & Laughlin 1993; Schmit & Debouck 1991).

The junior author has been puzzled during his explorations in southern Nariño and western Putumayo in Colombia, and in northern Cotopaxi in Ecuador by the local abundance of natural hybrids between *P. coccineus* and *P. dumosus* (both mostly as feral forms), but curiously not yet collected from central Guatemala (particularly as wild forms). Pollination occurs from visits by large carpenter bees (see Color Plate I, photo 4). Natural crosses have been demonstrated to occur with *P. coccineus* as the cytoplasm donor through analysis of cpDNA polymorphisms (Schmit et al. 1993). Experimental crosses (Maréchal 1971; Schmit et al. 1992) have confirmed the close relationship between these two taxa. One could thus think that the hypothesis of *P. dumosus* (*polyanthus*) being a variant of *P. coccineus* as claimed elsewhere (Hernández X. et al. 1959; Piñero & Eguiarte 1988) would be valid. Reciprocal crosses using *P. dumosus* as the cytoplasm donor show high but not total fertility with *P. coccineus* and reduced fertility with *P. vulgaris* (Maréchal et al. 1978b). A hypothesis to account for these different levels of fertility would be that *P. dumosus*, although with a chloroplast genome close to the one of *P. vulgaris*, would have its nuclear DNA 'contaminated' by *P. coccineus* genes during its genesis. The scenario of a reticulate origin for *P. dumosus* presented by Llaca et al. (1994) is compatible with this hypothesis. Whether *P. costaricensis*, or another taxon of the *Phaseoli* section yet to be discovered or already extinct, might be the original cytoplasm donor to *P. dumosus* still needs to be demonstrated.

As explained elsewhere (Debouck 1992; Debouck & Smartt 1995; Schmit & Debouck 1991), *P. dumosus* is the fifth case of plant domestication within the genus *Phaseolus*, and has led to the existence of about 350 cultivated populations (collected and maintained at CIAT, being possibly one third of all cultivated types available in traditional Latin America). The presence of wild *P. dumosus* in central western Guatemala sympatrically with cultivated forms (and their free hybridization) points to this region as a possible center of origin of this bean crop. Whether this is the only center or whether there is another one in Mexico (possibly Chiapas and/or Oaxaca) remains to be demonstrated.

B.4.—Phaseolus albescens McVaugh ex R. Ramírez & A. Delgado. Sida 18:637–646. 1999. (**Figs. 8, 9**).

TYPE: MÉXICO. JALISCO: Mpio. Ciudad Guzmán, 2 km sobre la desviación que va al Fresnito de la carretera Cd. Guzmán–El Grullo, 1640 m, 19 Nov 1996, *Delgado et al. 1705* (HOLOTYPE: MEXU n.v.; ISOTYPE: IBUG n.v.).

Aerial shoot a large climbing, indeterminate vine, to 5 m long. ***Root*** unknown. ***Stems*** terete, striate, puberulent of minute uncinate hairs and scattered hirsute and reflexed-pilose hairs to 5 mm long; internodes 10–15 cm long. ***Stipules*** broad triangular, 4–5 mm long, 3–3.5 mm wide at base, acute to obtuse, 6- to 8-nerved, glabrous, ciliate margins. ***Leaves*** 28–30 cm long; petioles 11–13 cm long, strongly striate ridged, puberulent of minute uncinate and scattered pilose hairs; petiolules 4 cm long, similar to petioles; pulvini 5–6 mm long, glabrous abaxially, covered with white strigose hairs adaxially; stipels ovate, lower 5–6 mm long, upper 3–4 mm long, 1–2 mm wide, acute, 3- to 6-nerved, glabrous; terminal leaflet broadly ovate to deltoid, 12 cm long, 10 cm wide at near midpoint, short acuminate, apiculate, indistinctly veined, ultimate veinlets dark and distinct, thin membranous, sparsely covered with appressed-white pilose hairs adaxially, sparsely covered with white pilose and strigose hairs abaxially, especially on veins; lateral leaflets similar but inequilateral. ***Inflorescence*** long multi-flowered panicle with 1 or 2 short branches at base of rachis; peduncle 15 cm long, sparsely pubescent; rachis 20–25 cm long of 15 or more nodes, sparsely covered with uncinate and strigose hairs; bracts lanceolate to aciculate, 5–6 mm long, 1.25 mm wide, indistinctly 3-nerved, acute, glabrous; pedicel very delicate, 15–20 mm long, about 0.25 mm thick (dried specimens), glabrous. ***Bracteoles*** about as long as calyx, 3–4 mm long, oblong-ovate, 1–1.25 mm wide, acute, glabrous, indistinctly 3-veined, caducous. ***Flower*** 28 mm long, showy pale purple fading yellowish, in bud purplish; calyx campanulate, the tube 3–4 mm long, glabrous, the upper 2 lobes joined into 1 scarcely elongated, slightly emarginate, the lower 3 lobes subequal, rounded dentate, short, 0.75 mm long, 2 mm wide, hyaline margins, somewhat ciliate; standard pale purple, the claw stout, short and broad, 1.5 mm long, 2.5 mm wide, the blade broadly rounded, 7 mm long to flexure and 11 mm more to deeply emarginate apex, 14 mm wide, lateral edges enrolled, the auricles small, 1 mm diam., rounded and bulbous; wings pale purple, spreading and cupped, the claw 5 mm long, 0.75 mm wide, the spur well-developed, 2 mm long, the blade round, 23 mm more to apex, 16–17 mm wide; keel, the claws 5 mm long, the ridges well-developed, 2.5 mm diam, firmly affixed to wings, 5 mm more to bend and 5 mm more to base of the terminal 1 3/4 coils of 3.5 mm diam.; vexillary stamen, the claw 1 mm long, the geniculate flap well-developed, 1.5 mm long, 1.5 mm wide, 3–4 mm more to end of thickened portion; stamen tube 3 mm broad, 7 mm long to bend and 5 mm more to end of united portion, the ridges scarcely developed; basal collar dentate 1 mm long; ovary straight, 4.5 mm long, 1.3 mm wide, covered with short, white, silky pubescence; style 10 mm long to the terminal thickened coil of 2.5 mm diam.; stigma terminal, capitate, 0.5–1 mm diam. ***Pod*** (not seen by us). ***Seed*** not seen by us). ***Seedling*** (not seen by us).

Specimens examined: **MÉXICO. Jalisco:** Sierra de Manantlán, 15–20 mi SE of Autlán, bajada S and W of divide between Aserradero San Miguel Uno and Durazno, 2000–2350 m, 6 Nov 1952, *McVaugh 13943* (NY). **Michoacán:** 12 km SE of Aserradero Dos Aguas and nearly W of Aguililla,(18°50'N, 102°45'W), 1600–1900 m, 26–27 Nov 1970, *McVaugh 24754* (CAS, MICH, NY).

Habitat.—Known from few collections, this species is an herbaceous vine to 10 m long climbing on trees in wet ravines and on steep slopes between volcanic and calcareous rocks in cloud forests of firs

Fig. 8. Illustrations of *Phaseolus albescens* McVaugh ex Ramírez & Delgado.—c. Portion of stem with mature leaves and inflorescence.—e. Exploded view of the flower showing all parts, including—s. Style tip and stigma as seen under the microscope. All drawings from specimen *McVaugh 24754* from Michoacán, México.

and broad-leaved trees. It is locally abundant but apparently restricted to montane cloud forests of western Mexico (note the ecological vicariance with wild *P. dumosus* in the same kind of forest in central-western Guatemala: Schmit & Debouck 1991), presently under threat because of excessive logging and habitat alteration.

Comments.—Although this species looks much like *P. dumosus* the bracteoles are only as long as the calyx, very different from those very elongate bracteoles, often 2 to 3 times longer than the calyx, of the latter species, and has long, delicate pedicels in contrast to the rather stout pedicels of species *dumosus*. In addition, Ramírez-Delgadillo and Delgado-Salinas (1999) reported up to five seeds per pod (one less in wild *P. dumosus*); these authors have reported the epigeal germination, key trait for locating this taxon into this section. *P. albescens* seems to have been noticed first by McVaugh (1987, p. 655) who wrote "whether or not *P. polyanthus* and *P. coccineus* subsp. *darwinianus* are the same, our plant is apparently not identical with either one." As early as 1992 the junior author (Debouck 1992) raises the question of its relationship with wild *P. dumosus*, and only recently Delgado et al. (1999) have shown the two taxa to be closely related but distinct.

Section C.—Coccinei (L.) Freytag, emend. & sect. nov. Type Species: *Phaseolus coccineus* L., Sp. Pl. 724. 1753.

Herbae volubiles scandentes caulibus indeterminatis 1-10 m longis, radice perennibus crassis et carnosis, flos grandis 11-30 mm longus coccineus vel purpureus (albus), nectariis pleurumque hyperagilibus, stigma terminale capitate vel obliquum extrorsum, plantula germinatibus hipogea.

Climbing or prostrate vines 1-10 m long. Root perennial, thick, fleshy. Flower large, 11-30 mm long, red or purple (white only in cultivars); floral nectaries mostly hyperactive; stigma terminal, capitate or oblique, extrorse. Germination hypogeal.

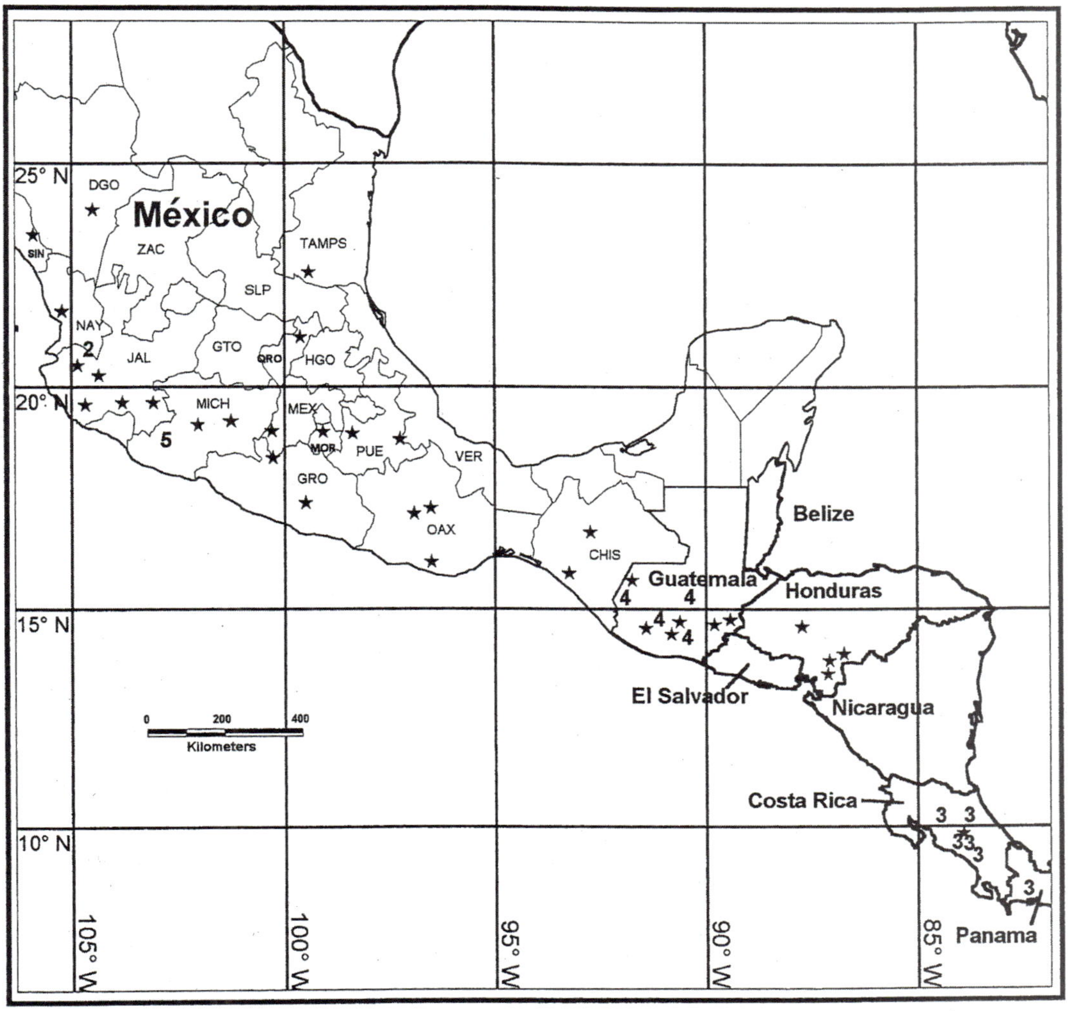

Fig. 9. Distribution of species of Section B. *Phaseoli* as follows: * = *Phaseolus vulgaris* L.; 3 = *P. costaricensis*; 4 = *P. dumosus*; 5 = *P. albescens*. (2 = *P. campanulatus*, see text).

Comments.—This section consists of one species *P. coccineus* which is extremely variable with two closely related series (subspecies) of taxa one with red flowers and one with purple flowers. All taxa of this section are perennials with roots which are large, narrow or thickened, fleshy to woody, globular to clavate to fusiform, often much branched. They are also found in pine, oak, hackberry or highland tropical forests, usually at elevations over 1,500 m and are distinguished by having a large, extrorse or terminal, capitate or oblique stigma and are normally cross-pollinated by carpenter bees (*Xylocopa*).

The senior author has made an effort to provide an overall view in the form of a morphological "family tree" of the *P. coccineus* complex based on some of the principal differentiating characters (see Fig. 10.-Family tree for the *P. coccineus* complex). This shows a family tree divided into two main species branches with the more variable of the two (*P. coccineus*) also subdivided into two main subspecies branches based on two morphologic characteristics traditionally used for determining taxa in this group, namely flower color and the shape and size of the bracts and bracteoles (arranged on the tree with the larger sizes, round shape and dense pubescence at the top).

Thus the senior author places the species *P. glabellus*, with minute glabrous bracts and bracteoles and a standard somewhat erect and fleshy, at the bottom of the tree (see Color Plate I, photo 5). Evidently this species is infertile with other members of the complex since intermediates with other

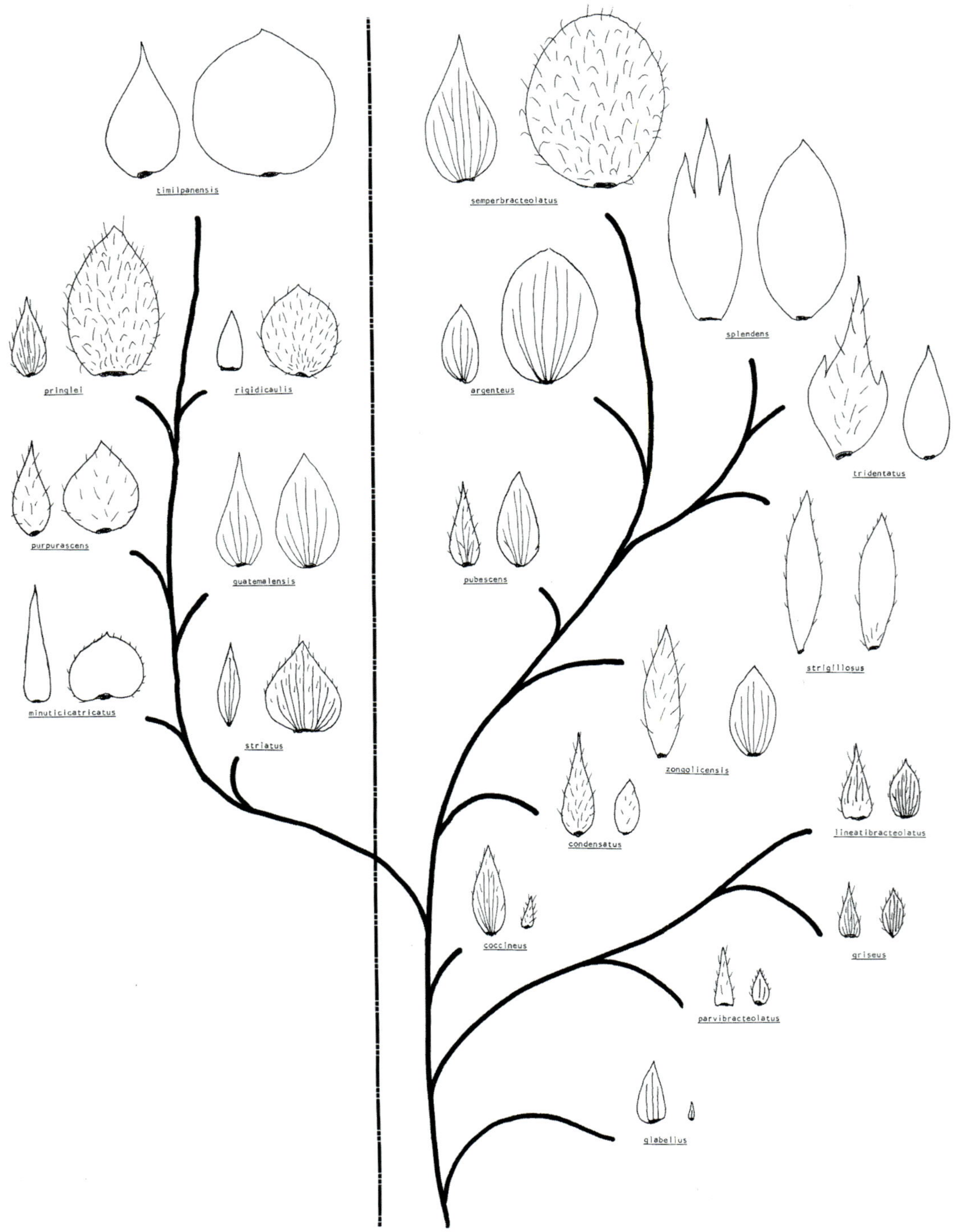

Fig. 10. Morphological family tree for species of Section C. *Coccinei*. See discussion in text.

taxa of this section are not found in the field, nor has the senior author been able to obtain fertile crosses with subsp. *coccineus* var. *condensatus* and subsp. *striatus* var. *purpurascens*, which the senior author has had in environmental chambers all at the same time and producing numerous selfed seed. Similar results have been obtained by Dr. R. Maréchal (pers. comm.). Evidence from cpDNA and other molecular genetic analyses seems to indicate that *P. glabellus* is considerably different (the senior author considers it as an archaic species) from other species or subspecies, etc, of this section than might be apparent from the great similarities in morphological characteristics (Delgado et al. 1999; Gaitán et al. 2000; Jacob et al. 1995; Llaca et al. 1994; Schmit et al. 1993).

On the main subspecies branches of *P. coccineus*, the largest series (*P. coccineus* L. subspecies *coccineus*) is composed of twelve varieties of mostly climbing or scrambling, short to long vines with large, showy, brick-red or scarlet flowers, usually a hooded and fleshy standard and small to large bracteoles (see Color Plate I, photo 6). Here one finds several branches from the main trend which goes from the minute, slightly pubescent bracts and bracteoles through ever larger size and more pubescence to arrive at the largest, rounded and tomentose bracteoles of var. *semperbracteolatus*. One of the distinct bracteole types near the base shows small, scale-like, strongly nerved bracteoles ending at var. *lineatibracteolatus*. Another trend is that shown by elongate primary bracts beginning with var. *strigillosus*, often with a tendency to climb in and over pine trees (see Color Plate I, photo 7), and with a distinctly southern distribution in Mexico, Guatemala, and more rarely in Honduras, and passing into the tridentate types of var. *tridentatus* found in W Mexico and var. *splendens* found at a single location in Nuevo Laredo in NE Mexico.

The other series (*P. coccineus* L. subspecies *striatus*) has seven varieties, all with purple flowers with a hooded and fleshy standard (see Color Plate I, photo 8), most of which are nearly prostrate or scandent vines except for var. *guatemalensis*. On the left of the main stream of red-flowered ssp. *coccineus* types, the senior author has placed several unusual variants—directly opposite a corresponding counterpart from the purple-flowered branch of ssp. *striatus*. Thus one finds a red-flowered type with a peculiar type leaflet of dark olive green adaxial surface and silvery abaxial surface as shown by var. *pubescens* with a similar purple-flowered counterpart in var. *guatemalensis*. Both are found on the southern edge of distribution of the total complex in Guatemala and Nicaragua. Although there is a tendency for most of subspecies *striatus* to be prostrate vines (except for the long climbing vine of var. *guatemalensis*), this prostrate habit is carried to an extreme in var. *rigidicaulis* found near Mexico City and which closely resembles *P. maculatus* (the "Metcalf" bean), and has a counterpart in red-flowered var. *argenteus*, also a small prostrate vine but with silvery leaves. At the top of the subspecies *striatus* branch, the senior author has placed var. *pringlei* with its very large, densely pubescent bracts and var. *timilpanensis* which is nearly glabrous but has the largest, roundest bracteoles of the purple-flowered series (see Color Plate I, photo 9), both of which are similar to a counterpart var. *semperbracteolatus* in the red-flowered group.

Delgado (1985) has recognized five subspecies in *P. coccineus*: *coccineus*, *formosus*, *darwinianus*, *glabellus* and *griseus*. The senior author believes two of these are untenable: *formosus* with large bracts and red flowers which would make it synonymous with *coccineus*, and *darwinianus* which both authors take to be *P. dumosus*. The subspecies *coccineus*, *glabellus*, and *griseus*, which he recognizes, will be discussed in more detail below.

KEY TO THE SPECIES, SUBSPECIES AND VARIETIES

1. Flowers brick-red or scarlet (cultivars often white-flowered); mostly long (over 5 m) climbing vines.
   2. Plant variously pubescent; leaves variously ovate, membranous; bracteoles usually more than half the length of the calyx tube, 2 mm or longer, variously nerved; young pod variously pubescent ____ C.1. **P. coccineus** subsp. **coccineus**
   2. Plant completely glabrous; leaflets broadly ovate, somewhat fleshy and lustrous, often drying thin, nearly transparent membranous; bracteoles minute, less than 1 mm long, 1-nerved; young pod glabrous; scarce, with a rather limited distribution in Mexico along the E escarpment of the Sierra Madre Oriental; from S Tamaulipas to Jalapa, Veracruz and bordering Puebla and Hidalgo; 800–2600 m ____ **P. glabellus** (see at the end of section)
      3. Bracteoles 3.5 mm long or less, usually less than half as long as the calyx, strongly nerved, pubescent; leaves ovate to oblong-ovate, densely whitish pilose or hispid especially on the lower surface; young pod heavily white pubescent.
         4. Bracteoles 1- to 2-nerved, minute and scale-like; rare, in W Mexico; Durango; 2100–2200 m ____ C.1.2. **P. coccineus** subsp. **coccineus** var. **parvibracteolatus**
         4. Bracteoles multi-nerved.
            5. Bracteoles strongly 3- to 6-nerved; scarce, in W Mexico; Jalisco, Michoacán, Guerrero and Oaxaca; 1100–2980 m ____ C.1.3. **P. coccineus** subsp. **coccineus** var. **griseus**

5. Bracteoles strongly 6- to 9-nerved; scarce, in central Mexico; Oaxaca to San Luís Potosí; 1800–2200m ______ C.1.4. **P. coccineus** subsp. **coccineus** var. **lineatibracteolatus**
3. Bracteoles medium to large (more than 3.5 mm long), more than half as long to twice as long as the calyx.
6. Primary bracts mostly tridentate, the 2 lateral teeth obscure to about 1 mm long.
7. Primary bracts 8–11 mm long, about 4 mm broad, moderately pubescent; scarce, in W Mexico; mostly in Michoacán; 1859–2700 m ______ C.1.5. **P. coccineus** subsp. **coccineus** var. **tridentatus**
7. Primary bracts 11–14 mm long, about 5 mm broad, densely pubescent; rare, in NE Mexico; Nuevo León & Tamaulipas; 1500–1900 m ______ C.1.6. **P. coccineus** subsp. **coccineus** var. **splendens**
6. Primary bracts mostly entire, ovate to nearly linear.
8. Bracteoles nearly linear, 5–10 mm long, about twice as long as the calyx, 1–2 mm wide; common, in S and W Mexico, Guatemala and Honduras; 1000–2580 m ______ C.1.7. **P. coccineus** subsp. **coccineus** var. **strigillosus**
8. Bracteoles broadly lanceolate to ovate.
9. Bracteoles orbicular to round, 7–9.5 mm long, 5–8 mm wide, glabrous or tomentose, seldom intermediate (plant sparsely pubescent), usually persistent through young pod; scarce, in central Mexico, mostly Oaxaca; 2430–3040 m ______ C.1.8. **P. coccineus** subsp. **coccineus** var. **semperbracteolatus**
9. Bracteoles variously ovate, oblong, or lanceolate, mostly 1.5–7 mm long (to 11 mm), pubescent, caducous before or shortly after anthesis.
10. Inflorescence with many, very short internodes of 1.5–4 mm long, with a total length usually 10–15 cm long; scarce, in N & central Mexico; 1350–2500 m ______ C.1.9. **P. coccineus** subsp. **coccineus** var. **condensatus**
10. Inflorescence with few to many, medium to long (10–25 mm) internodes.
11. Seed small, 5–7 mm long (much larger in cultivars), streaked or speckled with black, browns, and tans, rarely a solid color.
12. Leaves dark green adaxially, lighter green abaxially, extremely variable especially in cultivated varieties which have larger plant, flower and seed and are more widely distributed; common, throughout upland Mexico and S through the highlands of Central America; 1170–3250 m ______ C.1.1. **P. coccineus** subsp. **coccineus** var. **coccineus**
12. Leaves dark olive green adaxially, silvery or grayish abaxially.
13. Terminal leaflets broadly ovate to ovate; young pod whitish or yellowish strigose, mostly on the valves; rare, in high mountains in W central Mexico and N Guatemala; 1700–3100 m ______ C.1.10. **P. coccineus** subsp. **coccineus** var. **pubescens**
13. Terminal leaflets nearly triangular, truncate at base; rare, in dry mountains NE of Teotihuacán, Oaxaca; about 2100 m ______ C.1.11. **P. coccineus** subsp. **coccineus** var. **argenteus**
11. Seed medium, 9.6–11 mm long, nearly spherical, solid black; rare, limited to dense hackberry forests of the eastern escarpment of central Mexico; near Zongolica, Veracruz; about 1800 m ______ C.1.12. **P. coccineus** subsp. **coccineus** var. **zongolicensis**
1. Flowers purple to violet; scandent or climbing vines on grass and brush ______ C.2. **P. coccineus** subsp. **striatus**
14. Bracteoles distinctly 5- to 11-striate, somewhat longer than the calyx, sparsely pubescent to glabrous.
15. Plant scandent; leaves 7–15 cm long.
16. Bracteoles persistent through anthesis; rare, in central Mexico; near Puebla–Oaxaca border N of Huahuapan de León, in dry pine-oak forests; 2340–2740 m ______ C.2.1. **P. coccineus** subsp. **striatus** var. **striatus**
16. Bracteoles very early caducous usually absent in late bud with a small white scar visible at base of calyx; rare, in the eastern portion of the valley of Mexico City, in rocky, old lava flows; about 2,200 m ______ C.2.2. **P. coccineus** subsp. **striatus** var. **minuticicatricatus**
15. Plant climbing; leaves 23–27.5 cm long; rare, in N central Mts. of Guatemala and S to central Nicaragua, in mountain forests; 2000–2500 m ______ C.2.3. **P. coccineus** subsp. **striatus** var. **guatemalensis**
14. Bracteoles not distinctly striate, pubescent.
17. Bracteoles 3–6 mm long, as long as the calyx to somewhat longer.
18. Pedicels 11–21 mm long; leaf petioles 5–7 cm long, the leaflets broadly ovate-acuminate to somewhat rhomboid, acute; scarce, in the volcanic mountains around Mexico City; Hidalgo, D.F., Morelos, Tlaxcala, Puebla, in bunch-grasslands and open pine forests; 2500–3000 m ______ C.2.4. **P. coccineus** subsp. **striatus** var. **purpurascens**
18. Pedicels 7–8 mm long; leaf petioles 3–3.5 cm long, the leaflets rhomboidal to ovate, obtuse; rare, in pine forests just W of Mexico City, D.F.; near Contreras; about 2500 m ______ C.2.5. **P. coccineus** subsp. **striatus** var. **rigidicaulis**
17. Bracteoles 7–10 mm long, to twice as long as the calyx.
19. Bracteoles very densely white or yellowish strigose, nearly vellose (also inflorescence and most of

plant); rare, in central Mexico; near Mexico City, D.F. and N, to San Luís Potosí; about 2370 m ___ C.2.6. **P. coccineus** subsp. **striatus** var. **pringlei**

19. Bracteoles glabrous to only slightly pubescent; rare, in central Mexico; north of Toluca, Mexico; about 2620 m ___ C.2.7. **P. coccineus** subsp. **striatus** var. **timilpanensis**

C.1.1.—Phaseolus coccineus L. subsp. **coccineus** var. **coccineus**, Sp. Pl. 724. 1753. (**Figs. 11, 12; Color Plate 1–6**). TYPE: Cultivated at Uppsala, LINN, microfiche no. 899.2 (LECTOTYPE: LINN, designated by Delgado (1985) n.v.).

Phaseolus multiflorus Lam., Ency. Bot. 3:70. 1789.

Phaseolus multiflorus Willd., Sp. Pl. 3:1030. 1810.

Phaseolus formosus HBK, Nov. Gen. & Sp. 6:449. 1823. *Phaseolus coccineus* subsp. *formosus* (HBK) Maréchal, Mascherpa & Stainier, Taxon 27:199. 1978. *Lipusa formosa* (Kunth) Alefeld, Landwirth. Fl. 26. 1866. TYPE: MÉXICO. MÉXICO: prope Tolucam, 2760 m, Sep 1803, *Humboldt & Bonpland s.n.* (LECTOTYPE: P, designated by Delgado (1985) n.v.).

Phaseolus sylvestris HBK, Nov. Gen. & Sp. 6:450. 1823. TYPE: MÉXICO. MICHOACÁN: Crescit in sylvis inter Valladolid et Ario, prope Patzcuaro, 2260 m, Sep 1803, *Humboldt & Bonpland s.n.* (LECTOTYPE: P, designated by Delgado (1985) n.v.).

Phaseolus superbus DC, Mém. Soc. Phys. Genève 7:310. 1836. TYPE: unknown.

Phaseolus obvallatus Schlecht., Linnaea 12:328. 1838. *Phaseolus coccineus* subsp. *obvallatus* (Schlecht.) Maréchal, Mascherpa & Stainier, Taxon 27:199. 1978a. TYPE: MÉXICO. HIDALGO: Mineral del Monte, 1831–1837, *Ehrenberg 299* (LECTOTYPE: HAL, designated by Piper (1926) n.v.).

Phaseolus proriferus Jones, Contr. West. Bot. 12:14. 1908. TYPE: MÉXICO. CHIHUAHUA: Sierra Madre, Guayanapa Canyon, 3,600 ft., Sep 1903, *Jones s.n.* (HOLOTYPE: POM(?) n.v.), syn. nov.

Phaseolus leiosepalus Piper, Contr. U.S. Natl. Herb. 22:685. 1926. TYPE: MÉXICO. OAXACA: 18 mi SW of the City of Oaxaca, 2286–2896 m, 10–20 Sep 1894, *Nelson 1358* (HOLOTYPE: US 43734).

Plant a vigorous climbing or rarely prostrate indeterminate vine, 1–5 m long, over bushes and small trees, branches often pendent. ***Root*** fleshy to woody, globose to fusiform, frequently branched, pluriannual, growing vertically to horizontally to 1 m long, 2–5 cm thick. ***Stems*** terete, at base to 1 cm thick, 1–2 mm thick above; internodes short to 11 cm long, strigose, the hairs to 1 mm long, mostly appressed, reflexed, with many hooked hairs mostly at nodes, green or purplish. ***Stipules*** foliaceous, nearly triangular, 3.5–5 mm long, 4- to 8-nerved, moderately pubescent to nearly glabrous. ***Leaves*** 6.5–16 cm long; petiole 2.3–7.5 mm long; petiolule 0.9–2 cm long; pulvini 2.5–4 mm long, heavily pubescent; terminal leaflet broadly ovate to rhomboid to lanceolate, 3–6 cm long, 2.5–5 cm wide at about 1/4 of length from base, acuminate, apiculate, lightly veined, often thickened foliaceous to nearly coriaceous, glabrescent to moderately covered by strigose hairs mostly with glandular bases and on veins, very minute mostly hooked hairs between veins, somewhat ciliate on margins, often variegated along main veins; lateral leaflets similar but inequilateral. ***Inflorescence*** a short, erect raceme to 20 cm long, the peduncle 7–14 cm long, the rachis 1–8 cm long, of mostly 2–6–(14) flowering nodes, nearly glabrous to heavily hooked pubescent; primary bracts ovate to lanceolate, 4.5–10 mm long, 2–2.5 mm wide, moderately 3- to 15-nerved, nearly glabrous to heavily pubescent; pedicel delicate, 12–20 mm long, slightly to moderately hooked pubescent, mostly on adaxial surface. ***Bracteoles*** small to very large, ovate to nearly round to oblong, 1.5–11 mm long, 0.75–8 mm wide, 3- to 16-nerved, nearly glabrous to densely brown hooked pubescent, caducous somewhat after fertilization to nearly persistent. ***Flower*** brick red (scarlet), very brittle especially the standard and wings, mostly 2 per node but often 3, relatively few opening at any one time, about half set pods, maturing 2–3 or more per inflorescence; calyx 5 mm long, upper 2 teeth joined into a single round tooth 1 mm long, 6 mm wide, the lower central tooth 1–1.5 mm long, 1.5–2.5 mm wide, the lateral teeth 0.5–1 mm long, 1–2.75 mm wide, acute, glabrous to somewhat covered by strigose hairs mostly on the lower central tooth; standard light to dark brick red, very large, thickened, nearly round, 14–18 mm long, 14–19 mm wide, reflexed at near middle, the tip erect, glabrous or a few hairs near the apex; wings light to dark brick red, delicate, 15–21.5 mm long, the blade rounded, 11–16 mm long, 10 mm wide, cupped, clasping, the claw 4–5.5 mm long, 0.5 mm wide, the spur about 4 mm long, 3 mm wide; keel of 1 1/2 forward close coils, the claws 2.5–4 mm long, 0.75 mm wide, the basal knobs not developed or to 2 mm wide; vexillary stamen 12–14 mm long, 0.75 mm wide, the basal knob 2 mm wide, projected forward into a membranous sheath 1–2 mm long; stamen tube 12–13 mm long, 3.5 mm wide; anthers 0.75–0.9 mm long, 0.4 mm wide; basal collar about 1.5–2 mm long; ovary straight, 5.5–6.5 mm long, 1–1.25 mm wide, moderately to densely covered with long strigose hairs, 3–4 ovules; style 9 mm long

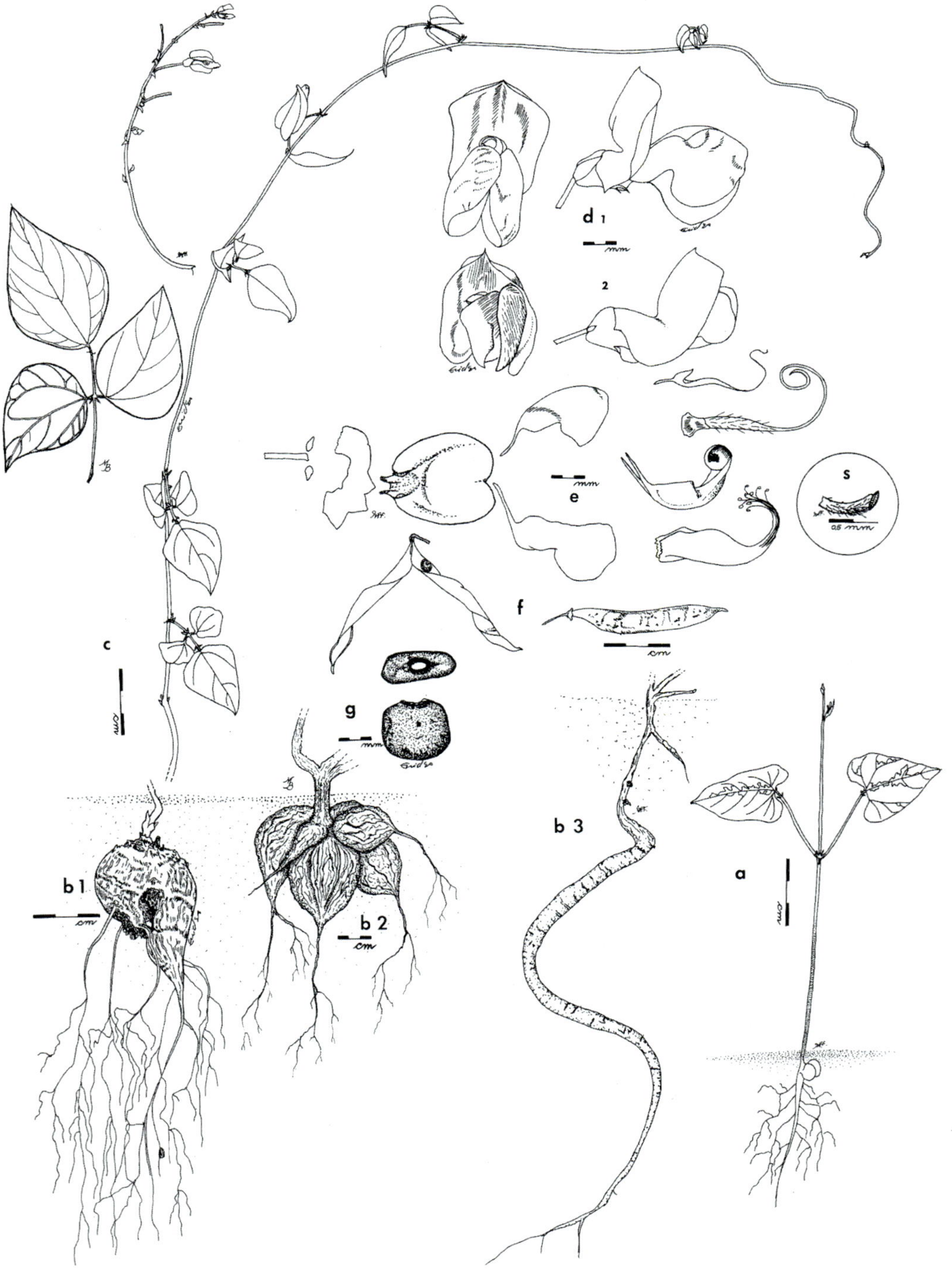

FIG. 11. Illustrations of *Phaseolus coccineus* L. subsp. *coccineus* var. *coccineus*.–**a.** Seedling a few weeks after germination of seed.–**b.1.** 1 year root.–**b.2.** 2 year root.–**b.3.** Unusual root type.–**c.** Portion of plant with mature leaves, large mature leaf, inflorescence, and vine tip.–**d.1.** Cultivated type flowers showing lateral and front views; notice erect open standard.–**d.2.** Wild type flowers showing lateral and front views; notice hooded and clasping standard.–**e.** Exploded view of the wild type flower showing all parts including–**s.** tip of style with stigma as seen under microscope; notice short, stout pedicel and clasping, deeply notched standard.–**f.** Pods, lateral view and dehisced.–**g.** Seeds, lateral view and view from hilum. All drawings made from living plants grown in environmental chambers and screenhouses at Mayagüez, PR as follows: 1 yr root, vine, pod and seed of TARS #287 (INIA, Mor 663 from Progreso, Morelos, México); 2 yr root of TARS #286 (INIA, Mor 664 from Progreso, Morelos, México); unusual root type from field collection *Freytag et al 78-Mex-97* from near Teopisca, Chiapas, México; wild type flowers and exploded views are from TARS #17 & #288 (INIA, Mor 662 from Progreso, Morelos, México); cultivated type flowers from TARS #46 (*Freytag et al. 81-9* from near Zongolica, Veracruz, México).

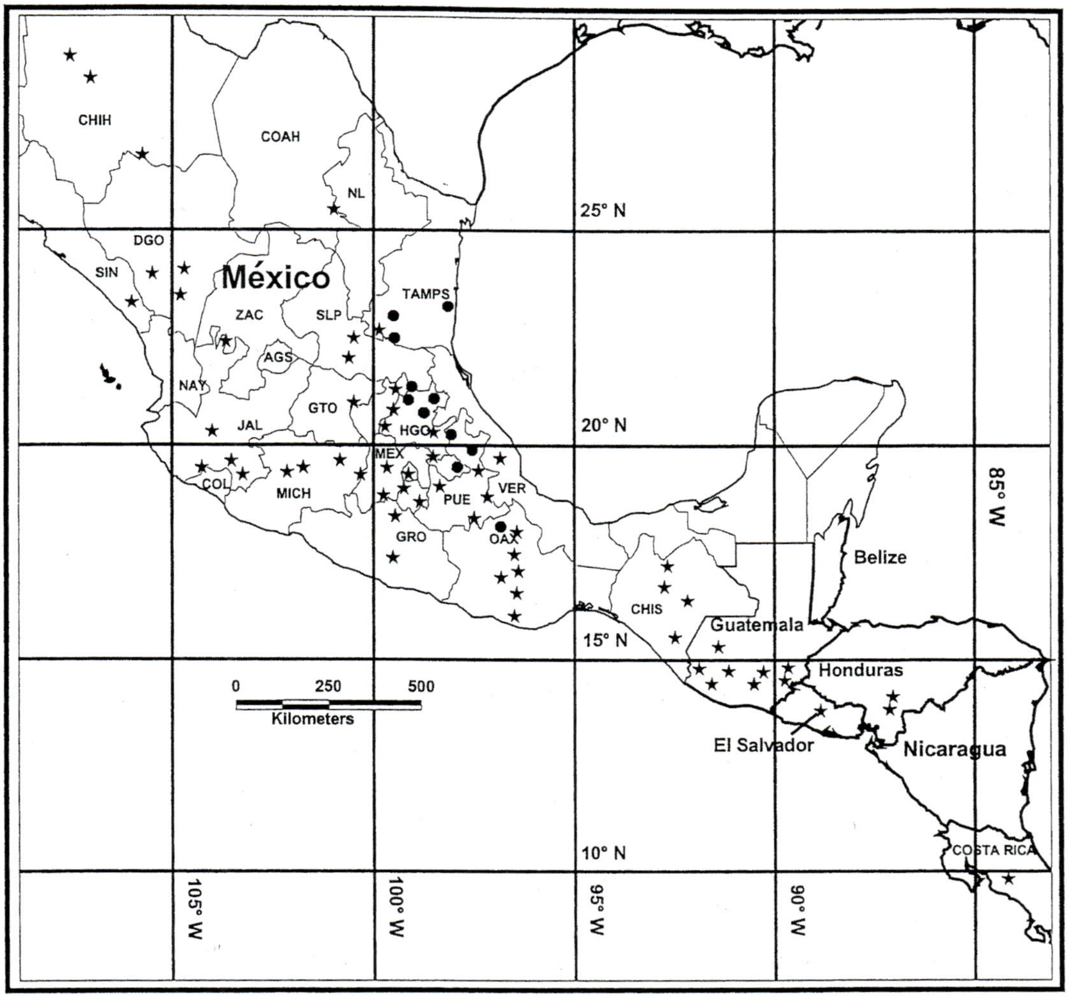

Fig. 12. Distribution of major red-flowered species, as follows: o = *Phaseolus glabellus* Piper, and * = *Phaseolus coccineus* L. subsp. *coccineus* var. *coccineus*.

up to the terminal thickened 1 1/2 coils of 2.5 mm in diam., the stylar hairs to about 0.5–1.5 mm from the tip; stigma terminal, capitate to nearly flat and extrorsely oblique at about a 45 degree angle. ***Pod*** small, nearly straight, 6 cm long, 1 cm wide, 0.5–0.8 cm deep, inflated, the valves fibrous, with somewhat thickened sutures, pubescent, green or tan with purple stripes, early dehiscent by one complete twist. ***Seed*** oblong squarish, flattened, 7 mm long, 6 mm wide, 3 mm thick, (much larger to 25 mm long in cultivars), shiny, variously solid color or black or dark brown spotted and striped on a lighter brown, tan, red, or purple background (raarely white in cultivars), a black ring around the hilum; hilum oval 1.5 mm long; lens prominent. ***Seedling*** from hypogeal germination; hypocotyl not elongated; epicotyl 9 m long, the next internode 7 cm long; stipules at eophyll bearing node, united; primary leaves opposite, simple, the petiole 33 m long, stipels present, the blade triangular ovate, 4 cm long, 3.5 cm wide near the base, the base somewhat auriculate, apex acute, slightly apiculate, pubescent and often variegated dark and light green.

Specimens examined: **COSTA RICA. Alajuela:** Guadalupe, Alfaro Ruíz, (9°50'N, 84°10'W), 1550 m, 28 Dec 1939, *Smith P2207* (GH).

EL SALVADOR. Chalatenango: bosque de El Pital, (13°55'N, 88°55'W), 2000–2700 m, 25 Aug 1976, *Montalvo 4797* (MO).

GUATEMALA. Chimaltenango: San Andrés Itzapa, Aldea Chimachoy, 12 km S Parramos, 14°35'N, 90°48'W, 2360 m, 3 Dec 1985, *Debouck et al. 1602* (COL, K, US, USCG); Patzicia, 6 km NW de Patzicia, Aldea El Sitio, Finca La Sierra, 14°40'N, 90°57'W, 2030 m, 11 Dec 1987, *Debouck et al. 2445* (US, USCG); Chichavac, 2400–2700 m, Nov–Dec 1930, *Skutch 5* (US). **Escuintla:** no location, 1942, *Aguilar 1594* (F). **Guatemala:** Amatitlán, cerro de la Torre, aproxim.1 km E de Santa Helena Barrillas, 14°26'N,

90°31'W, 1830 m, 15 Dec 1987, *Debouck et al. 2462* (COL, US, USCG); Santa Catarina Pinula, 4 km SE de Piedra Parada, 14°33'N, 90°30'W, 1750 m, 17 Dec 1987, *Debouck et al. 2475* (BR, MICH, US, USCG). **Huehuetenango:** arriba de San Pedro Necta, (15°25'N, 91°45'W), 1800 m, Nov 1923, *Salas 478* (US); Mts. W of Aguacatán, on the road to Huehuetenango, (15°20'N, 91°25'W), 1950 m, 27 Dec 1940, *Standley 81315* (F). **Jalapa:** 3 km SW de El Paraíso, 14°36'N, 90°04'W, 1870 m, 6 Dec 1987, *Debouck et al. 2436* (MICH, US, USCG); 0.5 km NE de El Mirador, approxim.11 km E de Miramundo, 14°35'N, 90°05'W, 2010 m, 6 Dec 1987, *Debouck et al. 2437* (MICH, SI, US, USCG); 9 km SW de Mataquescuintla, 14°30'N, 90°15'W, 1800 m, 7 Dec 1987, *Debouck et al. 2442* (BR, US, USCG); Km 92 and 88 Hwy 18 Guatemala-Jalapa, 2390 m, 5 Oct 1978, *Freytag et al. 78-Guat-41-8* (MEXU, MO, UC, US); between Jalapa and Montaña Miramundo, (14°25'N, 90°10'W), 1500–2000 m, 7 Dec 1939, *Steyermark 32891* (F). **Jutiapa:** Volcán Suchitán, NW of Asunción Mita, (14°15'N, 89°45'W), 600–2050 m, 18 Nov 1939, *Steyermark 31898* (F). **Quetzaltenango:** lower N slopes of Volcán de Santa María, above Palojunoj, (14°40'N, 91°35'W), 2500–2700 m, 15 Jan 1941, *Standley 83475* (F); region of Las Nubes, S of San Martín Chile Verde, 2250 m, 16 Jan 1941, *Standley 83604* (F); Mts. above Río Samala, Sierra Madre, 2 km W of Zuñil, 2300 m, 14 Dec 1962, *Williams et al. 22996* (EAP, F, GH). **Sacatepéquez:** Santiago, 1981 m, 1891, *Gómez 1017* (G, US(2)). **San Marcos:** 5 km W of Palestina, San Antonio Sacatepéquez, 14°57'N, 91°43'W, 2600 m, 11 Dec 1986, *Debouck et al. 1635* (COL, K, US, USCG); Aldea San Andrés Chapí, 8 km N of San Pedro Sacatepéquez, 15°0'N, 91°47'W, 2720 m, 12 Nov 1985, *Debouck et al. 1637* (COL, M, US, USCG); Barranco Eminencia, road between San Marcos and San Rafael Pie de la Cuesta, between Finca La Lucha and Buena Vista, 2500–2700 m, 6 Feb 1941, *Standley 86298* (F). **Sololá:** San Andrés Semetabaj, 1 km N of Godínez, 14°44'N, 91°07'W, 2170 m, 9 Dec 1985, *Debouck et al. 1628* (BR, COL, US, USCG); Nahuala, 6 km E of Sololá, 14°47'N, 91°19'W, 2440 m, 10 Dec 1985, *Debouck et al. 1633* (BR, COL, K, US, USCG); Km 163 Hwy CA-1, Quetzaltenango-Nahuala, 14°52'N, 91°25'W, 2570 m, 7 Oct 1978, *Freytag et al. 78-Guat-62* (BR, EAP, MEXU, MO, US); Mpio. Panajachel, finca Media Cuesta, 1820 m, 30 Oct 1974, *Mathewson 39* (WIS); Volcán Santa Clara, (14°40'N, 91°25'W), 2100–3000 m, 5 Jun 1942, *Steyermark 46907* (F); 4 km E of Godínez, 2100 m, 5 Dec 1963, *Williams et al 25187* (F). **Suchitepéquez:** Volcán Atitlán, 2438 m, 24 Oct 1934, *Skutch 1533* (F, GH, US). **Zacapa:** Monte Virgen, Sierra de Las Minas, (15°N, 89°50'W), 2200 m, 12–13 Jan 1942, *Steyermark 42616* (F).

HONDURAS. Fco. Morazán: Mt. Uyuca, cerca de Las Flores, drainage of the Río Yeguare at about 13°50'N, 87°W, 2000 m, 4 Sep 1948, *Molina 1116* (EAP, F, GH); between El Eden and El Rancho, Mt. Canta Gallo, 1600 m, 1 Nov 1966, *Molina 18525* (EAP, F, GH, US(2)); San Juancito, (14°8'N, 87°2'W), 2000 m, 6 Nov 1947, *Williams et al. 13329* (EAP, F, GH, US).

MÉXICO. Chiapas: Km 19 del periférico Sur de San Cristobal Las Casas, 2140 m, 20 Sep 1977, *Arias 627* (CAS-DS, F); Mpio. de Comitán, El Durazno, 1960 m, 22 Sep 1977, *Arias 647* (MO); Mpio. de San Juan Chamula, Km 13 del camino San Cristobal Las Casas-Tenejapa, 2200 m, 29 Sep 1977, *Arias 715* (WIS); Mpio. Tenejapa, above Tenejapa Center, 2164 m, 5 Aug 1964, *Breedlove 6899* (CAS-DS, F, US); Mpio. Tenejapa, at Moel Ch'en above Tenejapa Center along trail to Kulak'tik, 2286 m, 13 Jul 1965, *Breedlove 10904* (CAS-DS, TEX-LL, US); Mpio. of Amatenango del Valle, Amatenango, 1798 m, 23 Aug 1965, *Breedlove 12185* (CAS-DS); near a lake called Ik'al Nab, on the boundary between Chamula and Zinacantán along the road to Zinacantán Center, 2377 m, 21 Sep 1965, *Breedlove 12375* (F, MICH, TEX-LL, US); Mpio. of Huistán, 3 mi W of Huistán, 2134 m, 23 Sep 1965, *Breedlove 12401* (CAS-DS, US); Mpio. Zinacantán, 1 mi W of Navenchauc along Hwy 190, (16°45'N, 92°45'W), 1829 m, 26 Oct 1965, *Breedlove 13754* (CAS-DS, US, WIS); Mpio. of Motozintla de Mendoza, Cerro Mozotal below the microwave tower along the road from Huixtla to El Porvenir and Siltepec, 3000 m, 19 Sep 1976, *Breedlove 40298* (CAS-DS); Mpio. of Motozintla de Mendoza, Niquivil at the junction w/a small side ridge to Cerro Boquerón, (15°20'N, 92°15'W), 2600 m, 16 Dec 1976, *Breedlove 42744* (CAS, CAS-DS); 30 mi W of Ocosingo, Hwy Palenque-San Cristobal de las Casas, 2130 m, 7 Jun 1987, *Croat 66165* (UCR); Mpio. de San Cristobal Las Casas, 18 km sobre la carr. a Ocosingo, El Ciprés, 2340 m, 22 Sep 1977, *Delgado et al. 527* (CAS, F); Km 89 near Comitán, Hwy 190 San Cristobal-Teopisca, 16°25'N, 92°10'W, 2000 m, 29 Sep 1978, *Freytag et al. 78-Mex-97* (BR, EAP, GH, MEXU, MO, UC, US); Mpio. of Zinacantán, along Hwy 190 at paraje Sekemtik, 1585 m, 4 Sep 1966, *Laughlin 1797* (CAS-DS, US); Navenchauc, a 5 km al W de Nachij, o sea a 16 km al W de San Cristobal de Las Casas, 2300 m, 5 Dec 1980, *Sousa et al. 11328* (MO); Mpio. Pueblo Nuevo Solistahuacan, 3 km NW of Pueblo Nuevo Solistahuacan, (17°14'N, 92°55'W), 1768 m, 15 Jul 1970, *Zuill 94* (CAS-DS). **Chihuahua:** Mpio. de Guachochic, arroyo N of church at Cusarare, 2130 m, 28 Jul 1977, *Bye 7717* (UCR); 4 mi SW of Villa Matamoros, Sierra de Sta. Barbara, (26°45'N, 105°45'W), 1829–1920 m, 5 Oct 1959, *Correll et al. 22828* (TEX-LL); 12 mi W of Cuauhtémoc, 78 mi W of Chihuahua, (28°30'N, 107°W), 1981 m, 3 Jul 1959, *Gould 8745* (TEX); Soldier Canyon, Sierra Madre Mts., 2012 m, 16 Sep 1903, *Jones s.n.* (CAS-DS, MO, US); road to Namiquipa, (29°N, 107°30'W), 10–19 Oct 1935, *LeSueur 411* (F, GH, TEX); Majalca, Cañón Encantado, 26 Jul 1936, *LeSueur 686*, (F, GH, TEX); 10 km W of Cuauhtémoc, 1981–2286 m, 29 Aug 1967, *Oliver et al. 612*, (MO); Calera, E of San Isidro, 2100–2300 m, 15 Sep 1934, *Pennell 18793* (US); Mpio. de Casas Grandes, El Caracol, 25 km al S de Colonia Juárez, 2008 m, 23 Sep 1982, *Tenorio et al. 1665* (CAS). COAHUILA: Saltillo, (25°30'N, 101°0'W), 1904, *Palmer 444* (GH, US). **Distrito Federal:** Santa Fé City, 18 Oct 1903, *Holway 5170* (US); Dción. de Tlalpan, 4 km al E del Ajusco, 2750 m, 23 Jul 1967, *Jiménez 112* (WIS); SW of Contreras, (19°15'N, 99°15'W), 14 Sep 1940, *Langman 2823* (NA); Barrancas de Mixcoac, 6 Nov 1937, *Lyonnet et al. 1954* (US); Valle de México, San Jerónimo, 2400 m, 10 Sep 1950, *Matuda 19442* (CAS, US); Churubusco, 4 Oct 1910, *Orcutt 4304* (F); Bosque de Chapultepec, 3a Sección, 29 Sep 1979, *Ramos 14Ch* (US,WIS); Tacuba, Valley of México, 12 Jun 1899, *Rose et al. 4553* (US); El Mirador, 2652 m, 1 Sep 1930, *Russell et al. 99* (US); Dción. Tlalpan, Topilejo, 2600 m, 15 Sep 1976, *Ventura 2133* (MO). **Durango:** 3 mi NE of Otinapa on plateau near Río Chico, (24°3'N, 105°W), 2300–2400 m, 8 Jul 1950, *Maysilles 7330* (MICH); Quebrada de San Juan, 26 mi N of Railroad at Coyotes on road to San Luís, (24°N, 105°30'W)?, 1900–2100 m, 10 Aug 1955, *Maysilles 8346* (MICH); 19 mi SW of Durango, 10 Aug 1957, *Waterfall et al. 13497* (F). **Guanajuato:** Km 23, 19.4 mi W of Dolores Hidalgo, NE slopes of Mts. NE of Guanajuato, (21°15'N, 99°20'W)?, 2380 m, 7 Aug 1968, *Anderson et al. 5063* (CAS, MICH); Km 17–18 from Guanajuato to Dolores Hidalgo between Valencia and Sta. Rosa, 2438 m, 29 Aug 1948, *Moore et al. 4781* (GH, MICH). **Guerrero:** ca 10.5 km on microondas road up Cerro Alquitrán, marked "El Tejocote" on Hwy 95 W of Mazatlán, (17°25'N, 99°31'W), 2190–2220 m, 29–30 Sep 1983, *Anderson 12890* (MICH); Cerro Tuxpán, 13 km NE de Iguala, 18°23'N, 99°29'W, 1630 m, 31 Oct 1987, *Debouck et al. 2359* (BR, CHAPA, MICH, UC, US); Dist. Mina, Armenia, 2400 m, 23 Oct 1936, *Hinton 9755* (GH, K, MO, US); Dist. Mina, second ridge W of Petlacala, Sierra Madre del Sur, 1912 m, 1 Jan 1938, *Mexía 9054* (F, G, GH, K, MO, US); 19 km al SW de Filo de Caballo, camino a Puerto del Gallo, 2000 m, 19 Oct 1983, *Soto et al. 5764* (CAS); Mpio. de Alcozauca, Alcozauca, ladera NW Barranca del Agua Salada, 1400 m, 29 Oct 1983, *Viveros et al. 284* (CAS). **Hidalgo:** 6 km NE of Tepeapulco, cerro al NE del pueblo, 19°48'N, 98°32'W, 2650 m, 29 Oct 1986, *Debouck et al. 2026* (BR, CHAPA, US); Mineral del Monte, 0.5 km S de Omitlán de Juárez, Km 16 Mex 105 a Tampico, 20°09'N, 98°38'W, 2420 m, 19 Nov 1986, *Debouck et al. 2087* (CHAPA, US); Mpio. Zimapán, 2 km al NE de Zimapán, rumbo a Jacala, (20°50'N, 99°30'W), 2060 m, 6 Sep 1979, *Delgado et al. 1116* (MO); Mpio. Zempoala, 500 m al N de Sto. Tomás, en el entronque de la brecha a Rancho Mazatepéc, hacia Singuilucan, 2500 m, 5 Sep 1980, *Hernández 4901* (CAS); Mpio. Singuilucan, La Grava, 7 kms al Sur de Singuilucan, 2280 m, 11 Aug 1980, *Hernández et al. 4798* (US); Cerro de Santa Ana, 4 km al ESE de Tepeapulco, 2650 m, 19 Jul 1953, *Rzedowski 16944* (US); Cerro Santa Ana, 10 km NE of Apán, NE–SE slopes, 2950–3025 m, 19 Jun 1966, *West B-5* (WIS); Cerro Coronillas, W of Hacienda Huehuechoca, 10 km ENE of Apán, 19°4'N, 98°22'W, 2750–3075 m, 28 Jun 1966, *West G-8* (WIS); E of Cerro Jaguacillo, 10 km NE of Apán, 2650–2800 m, 18 Jun 1966, *West A-11* (WIS). **Jalisco:** Km 12 on road from Tapalpa–Chiquilistlán, 20°2'N, 103°46'W, 2100 m, 1 Dec 1981, *Freytag et al. 81-18* (MEXU, MO, US); Mpio. de Cuautitlán, 3 km al O de Las Joyas, entre Rincón de Manantlán y Las Joyas, (19°30'N, 104°15'W), 2110 m, 22 Oct 1983, *Guzmán et al. 6381* (WIS); V. Colima, 1.5 km E of El Depósito, ca 10 km WSW of Cd. Guzmán, 19°39'N, 103°32'W, 1750–1800 m, 24 Sep 1978, *Iltis et al. 586* (MICH, WIS); near a lumber road leaving Colima Hwy, 7 mi SSW of Tecalitlán and extending SE toward Sn. Isidro, Sierra del Halo, (19°20'N, 103°15'W), 2000–2200 m, 28–30 Nov 1959, *McVaugh et al. 1190* (MICH); 14 km al W de Bolaños, sobre la brecha a Tuxpán de los Huicholes, 1950 m, 19 Oct 1983, *Rico et al. 638* (MO); Mpio. Talpa, 20 km al S de Talpa, camino a la Cuesta, (20°10'N, 104°0'W), 8 Sep 1979, *Solís et al. 1957* (MICH, TEX). **México:** Mpio. Tlalmanalco, 3 km al E de San Rafael, 2800 m, 10 Aug 1969, *Aguirre 84* (CAS-DS, MICH, TEX, WIS); Chapingo, 13 Dec 1924, *Contrersis s.n.* (US); Ixtapaluca, 42 km E de Ixtapaluca, 3 km E del guardabosque "El Jaguey," 19°20'N, 98°46'W, 2630 m, 22 Oct 1987, *Debouck et al. 2320* (CHAPA, COL, MICH); Chíltepec, 4 km N de Chíltepec, 18°56'N, 99°52'W, 2470 m, 1 Nov 1987, *Debouck et al. 2367* (CHAPA, MICH); 3 km al SE de San Francisco Chimalpa, 28 Oct 1971, *Espinosa 1032* (CAS, EAP, US); 19 mi N of Toluca, (19°30'N, 99°40'W), 2134 m, 29 Jul 1956, *Fearing et al. 118* (TEX(2), US); Polotitlán, 6 Oct 1962, *Gómez 853* (GH, US); Dist. Temascáltepec, Bejucos, 610 m, 6 Oct 1932, *Hinton 2006* (GH, K, US); Dist. Temascáltepec, Tequesquipán, 2480 m, 26 Oct 1932, *Hinton 2310* (G, GH, K); Cerro del Pinal, Otzoloapan, 2300–2500 m, 18–22 Oct 1954, *Matuda et al. 31815* (CAS, US); Cañada de Nanchititla, 1800 m, 4–12 Dec 1954, *Matuda et al. 32023* (US); Tultenango, 13 Jul 1901, *Rose et al. 5439* (US); vertiente E del Cerro del Pino, cerca de Ayotla, 2600 m, 13 Jul 1967, *Rzedowski 23940* (CAS-DS); 1 km al NE de San Antonio, base de cabeza del Ixtaccíhuatl, 2500 m, 28 Sep 1980, *Sánchez 424* (WIS); La Gavia, 35 km from Toluca, Km 105 on Toluca–Morelia Hwy., 2682 m, 9 Aug 1944, *Sharp 44304* (MO); Cascada de los Diamantes, San Rafael, 28 Aug 1978, *Trejo 136* (CAS, TEX); Mpio. Texcoco, San Nicolás Tlamincas, 2500 m, 28 Apr 1983, *Ventura 788* (GH). **Michoacán:** Cerro Azul, vicinity of Morelia, 2200 m, 1910, *Arsène 6577* (US); Km 52 de la carr. Uruapan–Pátzcuaro, 1800 m, 5 Nov 1973, *Banda 1490* (CAS, MO); Mpio. Pátzcuaro, 2–3 km al SO de Pátzcuaro, (19°30'N, 101°45'W), 2380 m, 13 Sep 1979, *Caballero et al. 1045* (CAS); 4 mi S of Carapán along road to Uruapan, 2134 m, 6 Nov 1966, *Gentry 22142* (US); on road to Volcan Paricutín, near Capácuaro, (19°30'N, 102°2'W), 4 Nov 1958, *Jones 22880* (WIS); 35 mi E of Morelia, (19°40'N, 100°50'W), 14 Oct 1962, *Weber et al. 11877* (GH, MICH). **Morelos:** Cuautla, (18°49'N, 98°58'W), 1981 m, 2 Aug 1953, *Downing 164* (MICH); Sierra Morelos, Cuernavaca, 1950 m, 25 Oct 1969, *Hinton et al. 17425* (K, US); toll road to Mexico, 8 mi N of Cuernavaca, (19°N, 99°8'W), 23 Nov 1958, *Jones 22878* (WIS); Km 56–57 road to Cuernavaca, 13 Oct 1940, *Moore 112* (GH); Sierra de Tepoztlán, 21 Sep 1903, *Rose et al. 7241* (GH, US); Huitzilac, 2350 m, 13 Oct 1975, *Sousa et al. 5119* (CAS, US). **Oaxaca:** Sierra de Juárez, along Hwy 175 between Valle Nacional and Oaxaca, 26 mi above (W of) Valle Nacional, 1900 m, 30 Jun. 1977, *Croat 39875* (MO); idem, 36 mi above (W of) Valle Nacional, 2400 m, 30 Jun 1977, *Croat 39914* (MO); Miahuatlán, Tamazulapan, 13 km SE de Miahuatlán, Km 117 Mex 175, 16°19'N, 96°34'W, 2060 m, 26 Oct 1987, *Debouck et al. 2343* (CHAPA, MICH, US); Pochutla, Candelaria Loxicha, 1 km S de Portillo Sta. Ana, 27 km S de San José Pacífico, 16°N, 96°30'W, 2240 m, 27 Oct 1987, *Debouck et al. 2345* (BR, CHAPA, COL, MICH, SI, US); Dist. Ixtlán, ridge above Río San Cristobal, Sierra Madre del Sur, 2800 m, 16 Oct 1962, *Denham et al. 243/62* (CAS); 1.6 km NW of Hwy 175 (between Oaxaca and Túxtepec), on dirt road that intersects Hwy 175 2 km NE of Machín, 2750 m, 4 Jul 1979, *Diggs et al. 2333* (WIS); Dist. Pochutla, 24 mi N of Candelaria, 1796 m, 29 Oct 1967, *Gentry 22387* (GH, MICH, NA); on Hwy 175 from Oaxaca to Túxtepec; 43.6 mi from Jct. 190 & 175, 2515 m, 15 Aug 1975, *Le Doux et al. 2248* (MO); camino a Guelatao, 2500 m, 27 Oct 1971, *Matuda 38394* (CAS, US); Mpio. Miahuatlán, 19°43'N, 96°51'W, 1750 m, 21 Nov 1979, *Rodríguez 53* (F); 9 km N of Ixtlán, 2200 m, 17 Sep 1965, *Roe et al. 1996* (MICH); Mts. of San Juan del Estado, 2286 m, 13 Aug 1894, *Smith 95* (US); above Teotitlán del Camino on the road to Huautla, (18°8'N, 96°40'W), 2000–3250 m, 3 Aug 1961, *Smith et al. 4194* (US); Dist. Zaachilá, Km 14 camino Zaachilá–Sta. Inés del Monte, (16°58'N, 96°50'W), 2130 m, 4 Oct 1978, *Solano et al. 171* (MO); 1 km al N de la Cumbre, carr. Oaxaca–Guelatao, 2600 m, 24 Aug 1976, *Sousa et al. 6100* (CAS, K, MO(2)); 5 km al N–NE de Díaz Ordáz, o sea a 7 km al S de Cuajimoloya, 2600 m, 7 Aug 1977, *Sousa et al. 7815* (CAS); Mpio. Macuiltianguís, Dist. Ixtlán, El Machín, 2700 m, 11 Aug 1977, *Sousa et al. 7933* (mixed coll.) (CAS, MO, UC); Dist. Zaachilá, 7 km al W de "Santiago Clavellina," hacia San Antonio Huitepec, 18 Sep 1982, *Torres et al. 1308* (CAS, MO); 13.8 mi E of Rt 135 on road to Huautla, 17 Oct 1981, *Warnock 2512* (TEX); 5 mi from 'Llano Verde', Pass NE of Oaxtlán de Juárez, dividing drainages of Río Grande de Ixtlán and Río Cajones, watershed of Río Grande de Ixtlán, 27 Aug 1975, *Webster et al. 20204* (MO). **Puebla:** Puebla, Hda. Sta. Barbara (Alseseca), 2150 m, 16 Aug 1907, *Arsène 1381* (MO, US); Puebla, Barranca près Hda. Alamos, 2170 m, 5 Nov 1907, *Arsène 2093* (MO, US); El Mayorazgo, Puebla, 2130 m, 16 Aug 1906, *Arsène s.n.* (US); 30 km al SW de Tehuacán, por la carr. a Esperanza, (18°20'N, 97°30'W), 31 Aug 1979, *Chiang et al F-235* (CAS, MO); Tlalhuapán, 6 km E de Río Frío, 19°19'N, 98°41'W, 2730 m, 22 Oct 1987, *Debouck et al. 2321* (CHAPA, COL, M, MICH); Teziutlán, 27 Aug 1977, *Hernández et al. X-191* (F(2), MO); Chinantla, 2134 m, Jul 1841, *Liebmann 5314* (F); Amozoc, Sep 1942, *Martínez s.n.* (US). **Queretaro:** Pinal de Amoles, 3 km S de Escanelilla, en la carr. Mex 120 a Cadereyta, 21°8'N, 99°37'W, 1170 m, 14 Nov 1986, *Debouck et al. 2078* (BR, CHAPA, US); Fca. Reséndiz, Cadereyta, (Tepepán de Zochimilco), (20°40'N, 99°50'W), 27 Jul 1952, *Kelly 706* (UC); 2 mi W of Pinal de Amoles, Km 136 on Rt 120, 12 Oct 1981, *Warnock 2461* (TEX). **San Luís Potosí:** Sierra Alvarez, Hwy 70, 21.1 mi along Hwy 70, E of San Luís

Potosí, 2380 m, 5 Aug. 1983, *Buhrow et al. M13* (UCR); 5.6 mi W of San Francisco, 3020 m, 25 Jul 1978, *Daniel 344* (MICH); Zaragoza, 0.2 km al N de Alvárez, 22°02'N, 100°38'W, 2400 m, 3 Nov 1986, *Debouck et al. 2050* (BR, CHAPA, US); Zaragoza, cerro 1 km S del rancho 'El Cincuenta y Ocho', 2440 m, 4 Nov 1986, *Debouck et al. 2053* (BR, CHAPA, US); 25 mi E of San Luis Potosí and 2.5 mi up road to microwave tower, 2804 m, 29 May 1974, *McPherson 921* (CAS, MICH); Km 42 (E slope), FC Potosí and Río Verde, Sierra de Alvarez, Sierra Madre Occidental, (22°30'N, 100°30'W), 2300-2400 m, 30 Jul. 1934, *Pennell 17790* (US); Hwy 70, 27.4 mi E of San Luis Potosí, 9 Aug 1984, *Randolph 233* (UCR); Mpio. Zaragoza, Sierra de Alvarez, cerca de Puerto Huerta, 2450 m, 3 Sep 1954, *Rzedowski 4051* (TEX); in montibus San Miguelito, Aug 1876, *Schaffner 824* (GH, K). **Sinaloa:** Ocuragüí, Sierra Surotato, 1829-2134 m, 10 Sep 1941, *Gentry 6232* (in part) (GH); 4.4 mi SW of El Paraíso on road between Villa Unión and El Salto, (23°30'N, 104°55'W), 2042 m, 27 Sep 1953, *Ownbey et al. 1911* (F, GH, MICH, UC, US); 4.9 mi E of Potrerillos on Rt 40, 6 Sep 1962, *Ozment 57* (WIS); Mpio. Concordia, 2 km al NE de El Carrizo, (23°20'N, 106°0'W), 2000 m, 26 Jun 1982, *Tenorio et al. 618* (MO). **Tamaulipas:** 23 km E de Tula en el camino hacia Ocampo, 22°54'N, 99°37'W, 1520 m, 12 Nov 1986, *Debouck et al. 2071* (CHAPA, US); Rancho del Cielo, Nov 1964, *Webster et al. 93* (TEX). **Tlaxcala:** Calpulalpan, Km 44 de la carr. Mex 136 a Tlaxcala, 15 km antes de Calpulalpan, , 19°34'N, 98°41'W, 2700 m, 28 Oct 1986, *Debouck et al. 2025* (CHAPA, US); San Bernabé, Amaxac de Guerrero, 26 Aug 1945, *Hernández et al. 2184* (F). **Veracruz:** Mpio. Acultzingo, 3.4 mi SW of Acultzingo on Rt 150, 1.8 mi NE of Puebla-Veracruz border, 2120 m, 24 Sep 1985, *Luckow et al. 2963* (TEX); Maltrata, May 1937, *Matuda 1301* (MICH, US); along road to Puebla just W of Orizaba, 29 Jul 1967, *Mosquín et al. 6661* (GH); Mt. Orizaba, 2743 m, 8 Aug 1891, *Seaton 519* (GH); W of Orizaba, (18°50'N, 97°10'W), 1524 m, 15 Sep 1944, *Sharp 44866* (MO); Mpio. de Perote, Tatalco, (19°40'N, 97°10'W), 2300 m, 7 Jul 1970, *Ventura 1538* (MICH, MO). **Zacatecas:** Monte Escobedo, (22°25'N, 103°40'W), 26 Aug 1897, *Rose 2631* (US).

An additional 192 collection numbers were examined and determined to be of cultivated types. The majority of specimens in the herbaria are of cultivated types as indicated on the collector's label by "cultivated, escaped, landrace" or some similar notation, or of similar morphological type characteristic of cultivars and generally recognized by having large leaves, pods, or seed. All specimens of this type were not used in forming the species description nor in defining the distribution range and are not listed here since they were quite numerous.

Habitat.—This taxon is found growing in or around mostly mixed forests of pine-oak, or with hawthorn, liquidambar, or juniper, often on steep slopes or along stream beds, or in open pine-oak grassy areas, mesquite grasslands or sometimes in or on edges of cloud forest. The forests are often second growth and the understory is composed of shrubs and herbs including junipers, *Rhus*, *Cassia*, cacti, mesquite, magueys, *Yucca*, *Dahlia*, *Sophora* and *Dasylirion*. Soils are often rocky, from very sandy granitic with mica to rocky limestone with lots of humus, and may be yellow, red, to black, and usually in very moist locations. This taxon is usually found at elevations of 1500-2500 m.

Truly wild *P. coccineus* probably does not spread further south-east of Lake Nicaragua which is a natural barrier to many Neoboreal floristic elements (Rzedowski 1978); so-called wild *P. coccineus* of Panamá and Costa Rica as reported elsewhere (Lackey & D'Arcy 1980; Standley 1937, respectively) are in our view probably escaped from cultivation or weedy or hybrid forms (some were also *P. costaricensis*, as explained by Debouck et al. 1989a).

Diseases and pests.—The principal disease of this species seems to be rust; however, other pathogens are reported in wild plants and include: angular leaf spot, anthracnose, and Ascochyta leaf spot. The principal insect pest mentioned is *Apion* pod weevil, followed by: chrysomelid beetles, thrips, flea beetle and *Empoasca* leafhoppers.

Ethnobotany.—Known under several names in central and southern Mexico and Guatemala (see Table). Though apparently generally not eaten there are early collectors who report that the root is edible. Later collectors note that the plant is more abundant in inaccesible places and that cattle do graze it.

Genetics.—The large, black carpenter bees (*Xylocopa*) are usually responsible for pollination and are frequently seen visiting the red flowers. There are a few reports of hummingbirds visiting this species and Delgado (1985) gives *Hylocharis leucotis* and *Amazilia beryllina* as visiting wild and cultivated populations. From the works of Mendel (1865) and Fermond (1855), it was well known that this species could cross easily with *P. vulgaris*, but segregation was towards the original species (the cytoplasm donor species; see also Wall & York 1957) and not intermediate hybrids. The species is also known to hybridize easily with *P. dumosus* though few useful hybrids have been produced. The reader is referred to Delgado (1988) for a thorough treatise on the crossability with this species.

Comments.—All type variety wild *P. coccineus* collections should be identifiable by the hypogeal germination, fleshy and enlarged roots (see Color Plate III, photo 27), red flower color, fibrous, diagonally ridged or nerved carpels, and very small, nearly globose seed. Some of the outstanding major

Common names	Place	Collector
Ayocote	Central Mexico	(various, mostly cultivated)
Frijolillo	Hidalgo	West
Frijol de risa	Tlaxcala	Martínez
Yexixima		"
Engorda-caballo	various	various
Frijol de abono	various	
Frijol del Ratón	México	"
Frijolón	Queretaro	Debouck
Frijol cubinche	San Luís Potosí	"
Shut chenek	Chiapas	Delgado
Monte chenek		"
Shashan	Veracruz	Rodríguez
Kumat	Guatemala	Freytag
Frijol culebra		"

variations in one or several characters are separated from the type variety and given individual varietal names in large part to call attention to them and to encourage additional collecting.

Wild collections can best be distinguished by field notes and the small size of all organs compared to cultivated materials, especially as regards seed size (see Color Plate IV, photo 42). Frequently cultivated by the indigenous populations in highland Mexico and Guatemala (Breedlove & Laughlin 1993; Hernández Xolocotzi et al. 1979; McBryde 1947; Standley & Steyermark 1946), most herbarium specimens are of this cultivated material which can be recognized by the large size of all plant, flowering and fruiting parts–much larger than in the wild types. Cultivated material is often annual but may survive in fields several years because of perennial, thickened roots or if found in locations favorable to continued growth. In highly disturbed areas all types of intermediates may be found. These cultivars are often highly variable, with many seed colors and patterns, while the wild types from a single location are generally less variable.

Piper (1926) created two new species (*P. leiosepalus* and *P. strigillosus*) based on minor differences of bracteole shape and pubescence. Nevertheless, these are variable characteristics at best for species though perhaps useful for varietal distinction. Maréchal et al. (1978b) recognized that these two species were rather intermediates in these characteristics instead of extremes. The description of *P. leiosepalus* follows more closely the truly wild form of type var. *coccineus* as found in Chiapas and Guatemala and which is easily recognized by the generally small leaflets and hooded and fleshy standard of the flower (see Color Plate I, photo 6). There may be some truly distinct subspecific taxa presently being lumped into this group which could be separated with more complete collections from this area. Maréchal et al. (1978b) also realized the problem of placing some purple flowered "*formosus*" types, which the senior author has mostly placed in the subsp. *striatus*, in this species due to the problem of flower color change to somewhat purple while drying.

C.1.2.—Phaseolus coccineus L. subsp. **coccineus** var. **parvibracteolatus** Freytag, var. nov. (**Figs. 13, 24**). TYPE: MÉXICO. DURANGO: barranca of Río Jaral, 15 mi NW of Coyotes, ca. 80 km W of Cd. Durango, (24°N, 105°25'W)?, 2100–2200 m, 1 Oct 1962, *McVaugh 21723* (HOLOTYPE: MICH).

A *Phaseolo coccineo* var. *coccineo* in pedicellis longissimis tenellis bracteolis minutis squamiformibus 1-nervibus pubescentibus et leguminibus ubi juvenibus dense albo-pubescentibus differt.

Plant an annual, climbing indeterminate vine to 2 m long. ***Root*** a perennial, thickened, woody, elongate, to 2 cm in diam. ***Stems*** terete, 1.5–2 mm diam., somewhat striate, covered by appressed-reflexed long, white hispid and minute uncinate hairs, purplish. ***Stipules*** broad triangular, obtuse, 3 mm long, 1.5 mm wide, 5- to 6-veined, covered by minute uncinate hairs. ***Leaves*** 15–17 cm long; petioles 6–7 cm long, delicate, slightly striate, covered with reflexed hispid and minutely uncinate hairs, purplish; petiolules 2 cm long; pulvini, the lower 6–7 mm long, the upper 4 mm long, densely covered by white hispid and uncinate hairs; stipels spatulate, acute, 1-nerved, 2–2.5 mm long, 0.75 mm wide, ciliate;

Fig. 13. Illustrations of *Phaseolus coccineus* L. subsp. *coccineus* var. *parvibracteolatus* Freytag.–c. Portion of plant with mature leaf and inflorescence; notice long, erect peduncle and very long, delicate pedicels.–e. Exploded view showing all parts of the flower including–s. stigma as seen under microscope; notice minute, 1-nerved bracteoles and long sheath on vexillary stamen. All drawings made from the type, *McVaugh 21723* from Chihuahua, México.

terminal leaflet broadly ovate, acute to acuminate, minutely apiculate, 6–7 cm long, 6 cm wide at 1/3 from base, upper surface covered with minutely white hispid and uncinate hairs, somewhat farinaceous, abaxial surface more densely pubescent and hairs tending to strigose, with white veins. ***Inflorescence*** long and delicate, the peduncle 20–22 cm long, with scattered white uncinate hairs, purplish, the rachis 4–5 cm long, many-flowered; bracts triangular-aciculate, 3 mm long, 1 mm wide, indistinctly 4- to 5-nerved, covered with short hispid hairs, purplish. ***Bracteoles*** scale-like, hyaline, ovate, 1.5–2 mm long, 0.75–1 mm wide, acute, 1- to 2-nerved, puberulent of minute white uncinate hairs, ciliate margins. ***Flower*** large, scarlet; calyx campanulate, tube 3–4 mm long, puberulent of minute uncinate hairs, purplish, upper 2 lobes scarcely elongate, rounded, united into 1 emarginate, 0.5 mm long, 12 mm wide, the lower 3 lobes subequal, dentate, 2–2.5 mm long, 2 mm wide, the center one narrower and longer, slightly covered with white strigose, hispid and minute uncinate hairs; standard scarlet, the blade broadly rounded, 6–7 mm from base to flexure and 8–9 mm more to emarginate tip, 13 mm wide, the claw 5 mm long, 3 mm wide at base and 5–6 mm wide above, auricles lacking; wings scarlet, unequal, cupped and spreading, the blade rounded, 14 mm long, 10 mm wide, the claw 6 mm long, 0.75 mm wide, the spur poorly developed; keel 8 mm long to bend and 5 mm more to base of the terminal 1 1/2 coils of 2.5 mm diam., apex red, the pronounced ridges 2 mm in diam. at 3 mm from base; stamen tube 4 mm from base to slight bend downward, 4 mm more to upward bend and 4.5 mm more to end of united portion; vexillary stamen, the claw 2.25 mm long, the geniculate flap well-developed, 1.25 mm long, 2 mm wide, 4 mm more to end of thickened portion; basal collar 1 mm long, dentate; ovary straight, 5–6 mm long, 1.25 mm wide, densely covered with long white strigose hairs, 6–7 ovules; style 7 mm long to terminal thickened coil of 2 mm in diam.; stigma terminal, capitate and oblique, extrorse, 0.75 mm long. ***Pod*** immature densely covered by white strigose hairs. ***Seed*** unknown. ***Seedling*** unknown.

Habitat.—The collector indicates that this taxon is abundant on rocky rhyolitic soil and talus under W-facing bluffs in open pine-oak woodlands in a deep canyon of Río Jaral and at an elevation of 2100–2200 m.

Comments.—This variety is named for its very small but broadly ovate bracteoles. Additionally the long, almost glabrous pedicels set it distinctly apart from all other varieties of Section C. *Coccinei*. Apparently very rare since only a single collection of this variety has been examined.

C.1.3.—Phaseolus coccineus L. subsp. **coccineus** var. **griseus** (Piper) Freytag, comb. & stat. nov. (**Figs. 14, 24**). *Phaseolus griseus* Piper Contr. U.S. Natl. Herb. 22:683. 1926. TYPE: MÉXICO. JALISCO: near Guadalajara, 28 Sep 1903, *Rose & Painter 7369* (HOLOTYPE: US).

Plant an herbaceous climbing indeterminate vine, 3–4 m long, with 2–3 secondary branches, dark grayish. ***Root*** a perennial, globose, 3–5 cm long, 3–4 cm in diam., with tapering taproot to 2 dm long, the crown at soil surface about 1 cm long. ***Stems*** slender, terete, the whole herbage covered with fine grayish pubescence of retrorsely appressed hairs, at times nearly glabrous. ***Stipules*** triangular, 3 mm long, acute, striate, reflexed, somewhat pubescent. ***Leaves*** 7.3–23.5 cm long; petiole puberulent, shorter than the leaves, 3–5.5 cm long; petiolule 1–2.5 cm long; pulvini 3–5 mm long, very densely covered by strigose hairs; stipels triangular-lanceolate to ovate-lanceolate, 2 mm long, strongly 1- to 3-veined, ciliate; terminal leaflet broadly ovate to ovate-lanceolate, often somewhat hastate, 3–15 cm long, 2.5–6 cm wide at base, rounded to truncate at base, 3-nerved, inconspicuously veined, the ultimate veinlets raised on abaxial surface, acuminate, apiculate, membranous, minutely puberulent adaxially, paler and densely covered with white strigose pubescence to puberulent on abaxial surface. ***Inflorescence*** to 30 cm long with up to 24 nodes, the peduncles 11 cm long, the rachis 9–20 cm long; primary bract lanceolate, broadest at base, 3 mm long, persistent; pedicel delicate, 15–18 mm long, densely covered with minute uncinate hairs. ***Bracteoles*** oblong, strongly 3- to 6-nerved (rarely to 8-), 2.5–3 mm long, less than half as long as the calyx, 1.5–1.75 mm wide, acute to obtuse, ciliate. ***Flower*** scarlet; calyx campanulate, somewhat oblique, minutely puberulent, the hairs longer ventrally, the 2 upper teeth united into a single rounded entire lip, 8 mm long, the lower with 3 broad triangular teeth, 1 mm long, 2 mm wide; standard scarlet, round, 12–13 mm long, reflexed at 6 mm from base and 7 mm more to apex, not emarginate, the claw not developed, the auricles poorly defined, less than 0.5 mm long; wings, the blade broadly obovate, spreading, cupped, 14 mm long, 7 mm wide, the claw 4.5 mm long, the spur 1 mm high and firmly attached to keel; keel claws 4 mm long, 4 mm more to bend and 2 mm more to base of terminal 1 1/2 coils of 3 mm diam.; vexillary stamen, the claw 1 mm long, the knob developed into a cordate shaped sheath 1.5 mm long, 1.5 mm wide, the thickened portion 1.6 mm broad; stamen tube doubly reflexed, the first bend abaxially at 4 mm from base, 3 mm more to the second bend adaxially, 3 mm more to base of terminal 1 3/4 coils of 3 mm diam., the ridges developed into knobs 0.5 mm diam. at 2 mm from base; basal collar 1.25 mm long, somewhat oblique at apex; ovary linear, straight, 5 mm long, 1 m wide, densely covered with long white pubescence, the terminal thickened coil 2 mm in diam.; stigma, terminal extrorse, oblique, 0.75 mm long. ***Pod*** (immature?) nearly straight, 5–6 cm long, 7–10 mm wide, 3 mm thick; valves smooth to scabrous, fibrous twisting tightly at dehiscence, covered by short hirsute hairs; beak nearly straight to recurved, stout, 4–6 mm long. ***Seed*** oblong, compressed, orange-brown, shiny, 8 mm long, 5 mm wide, 3 mm thick; hilum lanceolate, 2 mm long. ***Seedling*** from hypogeal germination; hypocotyl not elongated; epicotyl 6 cm long, the next internode 11 cm long; stipules at the eophyll node, united; primary leaves opposite, simple, the petiole 16 mm long, stipels present, the blade triangular, 2.8 cm long, 1.7 cm wide near the base, the base truncate, apex acute, slightly apiculate, pubescent.

Specimens examined: **MÉXICO. Guerrero:** Omiltemí, W of Chilpancingo, (17°40'N, 99°50'W)?, 1900 m, 2 Aug 1967, *Sousa 3131* (GH, UC). **Jalisco:** Ixtlahuacán de Los Membrillos, 1738 m, 28 Oct 1961, *Detling 8763* (MICH); 6 mi E of Atotonilco El Alto, (20°30'N, 102°25'W), 1800 m, 23 Aug 1958, *McVaugh 17238* (MICH); 8 mi SE of Jalostotitlán, road to San Miguel El Alto, (21°N, 102°30'W),

Fig. 14. Illustrations of *Phaseolus coccineus* L. subsp. *coccineus* var. *griseus* (Piper) Freytag.–**a.** Plantlet several weeks after germination.–**c.** Portion of plant with mature leaf and inflorescences.–**e.** Exploded view of the flower showing all parts including–**s.** Stigma as seen under microscope; notice long pedicel, small, strongly nerved bracteoles and snall, rounded standard.–**h.** Bracteole showing strong parallel nervation. All drawings made from the field collection of *Freytag et al. 78-Mex-68* from Oaxaca, Mexico.

1900 m, 30 Aug 1958, *McVaugh 17539* (CAS, MICH); Sierra del Halo, 5 km toward San Isidro on a lumber-road leaving Colima, Hwy 11, 12 km SW of Tecalitlán, (19°25'N, 103°25'W), 1600–1650 m, 19 Nov 1970, *McVaugh 24493* (MICH); 10 mi W of Guadalajara on road to Tepic, (20°40'N, 103°30'W), 1524 m, 15 Sep 1959, *Soderstrom 617* (MICH, US); V. Tequila, along road to microwave station, 20°47'N, 103°50'W, 2987 m, 25 Oct 1970, *Webster et al. 15961* (MICH). **Michoacán:** Pto. Zarzamora, Coalcomán, (18°45'N, 103°10'W), 1460 m, 27 Sep 1938, *Hinton et al. 12261* (GH, K, US). **Oaxaca:** Km 145 Hwy 190 Oaxaca–Tehuantepec, near El Laurel, 16°32'N, 95°54'W, 1100 m, 23 Sep 1978, *Freytag et al. 78-Mex-68*, (BR, CSU, EAP, F, G, GH, IBUG, K, MEXU, MICH, MO, TEX, UC, US, WIS); 7–8 mi SE of Miahuatlán along new road over Sierra Madre del Sur, (16°15'N, 96°30'W), 1981–2134 m, 2 Sep 1952, *Gentry 12086* (MICH, NA, TEX); 6 km al S de la desviación por la carr. a Tlaxiaco, (17°30'N, 97°30'W), 7 Jul 1976, *Magallanes et al. 105* (MICH); Dist. Teposcolula, 4 km al NW de Tamazulapan, 1900 m, 18 Oct 1977, *Sousa et al. 8233* (CAS); Dist. Ejutla, El Vado, (16°30'N, 96°45'W), 1450 m, 19 Oct 1977, *Sousa et al. 8274* (CAS); Dist. Juquila, 19 km al SW de San Pedro Juchatengo, (16°20'N, 97°15'W), 1550 m, 20 Oct 1977, *Sousa et al. 8372* (CAS); 2 km al N de Tamazulapan, (17°45'N, 97°40'W), 1950–2210 m, 16 Sep 1976, *Téllez et al. 27* (CAS, MO(2)). **Sinaloa:** Ocuragüí, Sierra Surotato, 1829–2134 m, 1–10 Sep 1941, *Gentry 6232* (MICH, MO).

Additional collections examined which closely resemble this variety but not used for determining distribution since they may be hybrids of some type are as follows: **MÉXICO. Chiapas:** from seed purchased in market of Tenejapa, *Breedlove 14289* (CAS-DS). **Oaxaca:** cultivated in Teposcolula, *Sousa et al. 7711* (CAS). **HONDURAS.** in open forest near Ocotepeque, *Molina 22149* (CAS-DS, EAP, F).

Habitat.—Found on rolling foothills of open valleys with clay soils derived from limestone or on very steep (to 50°) slopes in very wet, thick leafmold under oak, pine or oak-pine forests, at elevations of 1100–2987 m.

Comments.—This taxon was described by Piper (1926) as a species from the collection *Rose & Painter 7369* from near Guadalajara, Jalisco. The dense, white pubescence of the type, giving a grayish cast to the foliage, suggests that the taxon is adapted to a dry desert habitat. Indeed, the only locality where the senior author has collected this taxon, in the foothills of the southern coastal zone just NW of Tehuantepec in southern Oaxaca, is a scrub oak-mesquite association bordering the dry coastal region.

In addition to the grayish cast of the foliage, the most distinctive characteristic of var. *griseus* is the small, usually heavily 3- to 5-nerved bracteoles. This taxon also often has a somewhat elongate terminal leaflet. Three specimens from the Distrito Federal, mentioned by Piper (1926) as being of this variety, are of variety *coccineus*, while *Bourgeau 581* is of var. *pringlei*. Maréchal et al. (1978b) did not think var. *griseus* was different from *P. glabellus* which he considered a subspecies, apparently missing the striking differences in pubescence and bracteole size, shape and nervation. In addition, they did not mention the distinctly separate distributions of these two taxa. On the other hand, Delgado (1985) clearly distinguished between these taxa as subspecies, with *P. glabellus* as a glabrous subspecies with a distribution limited to eastern Mexico and var. *griseus* as a pubescent subspecies with distribution limited to western Mexico.

C.1.4.—Phaseolus coccineus L. subsp. **coccineus** var. **lineatibracteolatus** Freytag, var. nov. (**Figs. 15, 24**). Type: MÉXICO. San Luís Potosí: 30 mi E of SLP, Rt 86 (near Alvárez? possibly Rt 80 or Rt 70?), (22°5'N, 100°45'W), 22 Aug 1968, *Turner 10* (holotype: WIS; isotypes: F, TEX, UC).

Persimilis *Phaseolo coccineo* var. *griseo* in bracteolis conspicue nervatis, sed plerumque grandioribus et 6–9 nervibus differt. Crecit in montibus orientalis Nova Hispania.

Plant a large (stout) trailing, indeterminate vine. ***Root*** unknown. ***Stems*** very slightly striate, densely covered by retrorse-hispid and glandular hairs, the internodes 7–10 cm long. ***Leaves*** 8–15 cm long; petioles 3.5–6 cm long, covered by delicate, minute hispid hairs, purplish; petiolules 1.5 cm long; stipels aciculate, 2.5–3 mm long, heavily 1- to 3-nerved; pulvini 3–3.5 mm long, densely covered with yellowish hispid and uncinate hairs; terminal leaflet broadly ovate-elongate, 3–4–7 cm long, 2.5–3–4.5 cm wide, densely covered by minute, white hispid hairs especially on abaxial surface of nerves, sparsely so adaxially, minute-reticulately veined. ***Inflorescence*** erect, very long (20–30 cm or more), somewhat ridged, densely covered with minute hooked hairs, the peduncle 14–20 cm long, the rachis 7–15 cm long; primary bract small, ovate-lanceolate, 3.5–4 mm long, 1–1.5 mm wide, 3- to 5-nerved, lateral nerves more pronounced, acuminate, covered with scattered hispid hairs, purplish, the secondary bracts 1.5 mm long, heavily 1-nerved; pedicels very slender, 15–20 mm long, sparsely covered by hooked pubescence. ***Bracteoles*** oblong-ovate, heavily 6- to 9-nerved, the nerves white, 2.5–3.5 mm long (1/2–3/4 to as long as calyx), 1.5–2.5 mm wide, sparsely covered by hispid hairs

Fig. 15. Illustrations of *Phaseolus coccineus* L. subsp. *coccineus* var. *lineatibracteolatus* Freytag.–c. Portion of plant with mature leaves and inflorescences; notice prostrate stem, long, erect inflorescence and long pedicels.–e. Exploded view of the flower showing all parts including–s. Stigma as seen under microscope; notice long pedicel and small, strongly nerved bracteoles. All drawings made from the type, *Turner 10*, a field collection from E of Cd. San Luís Potosí, México.

especially on edges, greenish to purplish. ***Flower*** scarlet; calyx large flaring campanulate, 3.5 mm long, upper 2 lobes joined into one rounded scarcely defined, 6 mm wide, less than 0.5 mm long, the lower 3 lobes rounded-triangular, acute, 1.5 mm wide, the lateral 0.75 mm long, the central 1 mm long, slightly covered by hooked pubescence, a few longer strigose hairs on lower lobes, purplish; standard scarlet, rounded, very slightly emarginate, 10 mm wide, reflexed at 4–5 mm from base, the terminal portion 6–8 mm long, the claw scarcely defined, the auricles at margins, 3 mm from base, 0.5 mm long; wings scarlet, rounded, cupped, 14 mm long, 7–8 mm wide, the claw 4 mm long, the spur well-developed, 1 mm long, firmly affixed to keel; keel, the claws 3.5 mm long to the ridge, 2 mm in diam., 3 mm to bend and 3 mm more to base of terminal 1 3/4 coils of 3 mm diam.; vexillary stamen with a very well-developed geniculate knob and sheath, the claw 1 mm long, the knob 1 mm diam., the sheath 2 mm long, 1 mm wide, the thickened portion about 4 mm more to filament; stamen tube short and thick at base, 3 mm to the first bend abaxially, 3 mm more to second bend adaxially and 3 mm more to filaments, the ridges scarcely developed; basal collar 2 mm long; ovary straight, 6 mm long, 0.75 mm wide, densely covered by white hirsute hairs; style with 1 terminal thickened coil of 1.5 mm diam.; stigma, terminal extrorse, oblique, 0.8 mm long. ***Pod*** unknown. ***Seed*** unknown. ***Seedling*** unknown.

PARATYPES. **MÉXICO. Durango:** 59.65 mi E of Villa Unión on Mex. 40, 2100 m, 3 Aug 1983, *Buhrow et al. M6* (UCR). **Hidalgo:** Mpio. de Zimapán, Maguey Verde, 25 kms al NE de Zimapán, (20°50'N, 99°15'W), 2100 m, 7 Sep 1979, *Hernández 3735* (CAS); Hwy 85 about 6 mi N of Jacala, (21°2'N, 99°8'W), Sep 1970, *Norris 17295* (CAS). **San Luís Potosí:** Mpio. de Guadalcazar, 3 km después de Guadalcazar, approx. 10 km de la carr. central, (22°50'N, 100°20'W), 1500 m, 24 Sep 1977, *Garcia 486* (CAS); 22 km W of Sta. Catarina on Hwy 86 (Hwy 70?) at Km 49, (22°5'N, 100°40'W), 2200 m, 29 Sep 1965, *Roe et al. 2189* (MICH, WIS). **Tamaulipas:** Rancho del Cielo area, La Perra, 2073 m, 29 Aug 1968, *Richardson 856* (TEX).

Habitat.—This taxon is apparently rare, found on rocky outcrops in open oak forests (*Quercus*) with Mimosoideae (*Tepanne*), composites and ferns, in partially shady to shady locations. Soils are reddish brown, clay, from rocky igneus zones.

Comments.—This variety is named for the 6–9 well-defined nerves of the bracteole, thus it is similar to var. *griseus* in having bracteoles heavily-nerved, though more numerous. In addition to the more numerous bracteole nerves, it differs in that the pedicel is 15–20 mm long, delicate and only slightly pubescent, the foliage is not as grayish pubescent and the leaflets are mostly ovate.

C.1.5.—Phaseolus coccineus L. subsp. **coccineus** var. **tridentatus** Freytag, var. nov. (**Figs. 16, 24**). TYPE: MÉXICO. MICHOACÁN: lava flow E of San Juan Nuevo, 8 km S of Uruapan, (19°22'N, 102°4'W), 1859 m, 11–15 Oct 1961, *King & Soderstrom 4718* (HOLOTYPE: US 2366903; ISOTYPES: MICH, TEX, UC).

Similis *Phaseolo coccineus* var. *coccineo* sed saltem aliquot bracteis conspicue tridentatis et 8–10 mm longis differt. Crecit in montibus occidentalis Nova Hispania.

Plant a small, climbing, indeterminate vine. ***Root*** unknown. ***Stems*** with internodes 8–10 cm long. ***Stipules*** broadly triangular, 4 mm long, 2.5 mm wide, 5- to 6-nerved, abaxial surface sparsely covered with hispid hairs, ciliate margins. ***Leaves*** 11–14 cm long; petiole to 5–7 cm long; petiolule 1.5–1.8 cm long; stipels 3 mm long, 1 mm wide, 3- to 5-nerved; pulvini densely covered with yellowish-brown strigose and hispid hairs; terminal leaflet ovate-rhomboid to ovate, 4.5–5 cm long, 3.5–4.6 cm wide at near mid-point, sparsely covered with glandular hispid hairs adaxially, densely covered by white hispid hairs abaxially. ***Inflorescence*** an erect raceme, the peduncle 5–12 cm long, the rachis 12–13 cm long; primary bracts oblong-lanceolate, 8–11 mm long, 4 mm wide, tridentate, the 2 lateral teeth about midway to tip and scarcely visible, to 1 mm long, covered by moderately long, white hispid to pilose hairs on abaxial surface. ***Bracteoles*** broadly ovate, 5–6–20 mm long, 3.5–4 mm wide, weakly 6- to 7-nerved, sparsely covered by white strigose hairs especially on margins, purple tinged. ***Flower*** scarlet; standard nearly round, emarginate, 12–13 mm long, 12 mm wide, reflexed at 5 mm from the base, a few hispid hairs at apex, claw 1 mm long, the auricles well-developed, 0.5 mm long; wings scarlet, the blade oblong, 17 mm long, 6–7 mm wide, cupped, spreading at nearly 90°, the claw 4.5–5 mm long, the spur well-developed, 1.5–2 mm long, firmly attached to keel; keel, the claw 4 mm long, the ridges well-developed, 4 mm more to bend and 4 mm more to base of the terminal 1 1/2 coils of 3.5 mm diam.; vexillary stamen, the claw 1.25 mm long, the knob developed into sheath 2 mm long, 2 mm wide and thickened, 5 mm more to filament; stamen tube 8 mm to bend and 5 mm more to divided filaments, the ridges scarcely developed; basal collar well-developed, 3 mm long, the longer abaxial tip cut obliquely to shorter adaxial side; ovary straight, 6 mm long, 1 mm wide, densely covered with long, white pubescence; style 6 mm long to the terminal thickened coil of 2.25 mm diam.; stigma terminal, extrorse oblique, 0.8 mm long. ***Pod*** the immature ones, nearly straight, 4 cm long, 0.7 cm wide, the beak stout, recurved, 3 mm long, sparsely covered by strigose hairs. ***Seed*** unknown. ***Seedling*** unknown.

PARATYPES. **MÉXICO. Chiapas:** beside old road, Km 14–4 Hwy 190 San Cristobal-Comitán-Teopisca, 16°31'N, 92°30'W, 1980 m, 29 Sep 1978, *Freytag et al. 78-Mex-100* (BR, EAP, MEXU, MO, US). **Jalisco:** S of El Chante on road to Manantlán, 1400 m, 20 Sep 1983, *Anderson 12726* (MICH); 12.5 mi from El Chante through foothills to the Cerro del Muñeco, or Sierra de Manantlán (30–35 km SE of Autlán), 1600 m, 30 Sep 1966, *Anderson et al. 3829* (CAS, MICH); along road El Chante-Manantlán, about 1/2 km above Rincón de Manantlán, 20°38'N, 104°14'W, 1500 m, 27 Nov 1981, *Freytag et al. 81-37* (BR, EAP, MEXU, MO, US); above Manantlán, near

FIG. 16. Illustrations of *Phaseolus coccineus* L. subsp. *coccineus* var. *tridentatus* Freytag.—**c.** Portion of plant with mature leaf and inflorescence; note long petioles and inflorescence with large, tridentate, primary bracts and many flowers grouped near apex.—**e.** Exploded view showing all parts of the flower including—**s.** Stigma as seen under the microscope; note large, striate bracteole, short, deeply notched standard, long knob on vexillary stamen, short, broad stamen tube, and long basal collar of ovary.—**f.** Primary tridentate bract. All drawings made from the type, *King & Soderstrom 4718*, a field collection from 8 km S of Uruapan, Michoacán, México.

Zarza Mora and Las Joyas, 1700–1800 m, 5 Dec 1981, *Freytag et al. 81-38* (BR, MEXU, MO, US); Amacueca, near the summit of the plateau, road to Tapalpa, (19°50'N, 103°40'W), 2100–2250 m, 2 Nov 1960, *McVaugh 20609* (CAS, MICH); 12–15 mi SSE of Autlán, on lumber road to Corralitos, 4–10 mi above (SE of) Ahuacapán, (19°40'N, 104°20'W), 1500–2200 m, 22–23 Nov 1959, *McVaugh et al. 901* (CAS, MICH). **México:** San Miguel, Temascáltepec, (19°2'N, 100°2'W), 2740 m, 12 May 1932, *Hinton 623* (F, GH, K, TEX-LL, US); Hornos, Temascáltepec, 2 Sep 1935, *Hinton et al. 8245* (F, G, GH, K, US); Cumbre-Cimientos, Temascáltepec, 15 Sep 1935, *Hinton et al. 8279* (F, G, GH, K, US). **Michoacán:** Punguato, Morelia, (19°45'N, 101°20'W)?, 2100 m, 9 Aug 1909, *Arsène 2871* (US); Zitácuaro-Cacique, Zitácuaro, (19°20'N, 100°20'W)?, 2450 m, 27 Sep 1938, *Hinton et al. 13276* (CAS-DS, F, GH, K, LIL, MICH(2), US); Tancitaro, Uruapan, 2070 m, 11 Oct 1940, *Hinton et al. 15509* (GH, TEX, TEX-LL, US); Mpio. Tancitaro, 2 mi S of Tancitaro, 1981 m, 14 Aug 1940, *Leavenworth 550* (ARIZ, F, GH, MICH, NA); 24 km al NW de Pátzcuaro, carr. Pátzcuaro-Uruapan, (19°25'N, 101°55'W), 1700 m, 9 Sep 1978, *Soto et al. 1008* (CAS); El Angangueo, salida a Villa Victoria, 2550 m, (19°38'N, 100°16'W), 10 Sep 1979, *Soto et al. 1489* (K, WIS). **Morelos:** El Parque, 31 Aug 1910, *Orcutt 4420* (F, GH, MO). **Oaxaca:** Mpio. Macuiltianguís, Dist. Ixtlán, El Machín, (17°20'N, 96°30'W), 2700 m, 11 Aug 1977, *Sousa et al. 7933* (UC). **Sinaloa:** 19 mi E of Sta. Lucía along Hwy 40, (23°30'N, 104°55'W), 2000 m, 13 Sep 1971, *Keil et al. 8864* (TEX).

Habitat.—Found growing in sparsely to densely vegetated slopes of lava flow ("malpaís") with vegetation mainly of *Agave*, *Pinus*, and low shrubs and herbs, or in clearings in pine forest or pine-juniper-oak, and in rich soil at edges of cultivated fields.

Common name.—Frijol de ratón.

Comments.—The varietal epithet is based on the very distinctive and rather large, trifid bracts. However, on some specimens not all the bracts are toothed and sometimes these may be very indistinctly so, which perhaps may be caused by introgression with variety *coccineus* which seems to occur in the mountains surrounding the volcanic area near Uruapan, Michoacán and in the Perote-Xalapa area (as shown by an intermediate specimen *Freytag et al. 78-Mex-17* not cited here).

C.1.6.—Phaseolus coccineus L. subsp. **coccineus** var. **splendens** Freytag, var. nov. (**Figs. 17, 24**). TYPE: MÉXICO. NUEVO LEÓN: dense woods, Dulces Nombres and just E of border into Tamaulipas, 24°N 99.5–100°W, 1690 m, 12 Jul 1948, *Meyer & Rogers 2742* (HOLOTYPE: MO; ISOTYPES: G, GH).

Similis *Phaseolo coccineo* var. *tridentato* in bracteam tridentatem, sed longioribus et omnibus partibus plantibus majoribus differt. Crecit montibus meridionalis provinciae Novoleonensis et Tamaulipense rarus.

Plant a long, climbing indeterminate vine to 5 m long. ***Root*** unknown. ***Stems*** terete, densely covered by retrorse-strigose hairs, the internodes to 14 cm long. ***Leaves*** 17–19.5 cm long; petiole slender, 8 cm long; petiolule 2.5 cm long; stipels aciculate, 6–7 mm long, 1-nerved, sparsely ciliate; terminal leaflet broadly ovate, short acuminate, 7–9 cm long, 6.5–9 cm wide, widest at about 1/2 from base to tip, when young very densely covered by white strigose hairs on both surfaces, at maturity becoming less dense abaxially and sparsely pubescent adaxially; lateral leaflets similar but markedly inequilateral. ***Inflorescence*** many-flowered, very long and greatly exceeding foliage, the peduncle 15–27 cm long, the rachis 10–20 cm long; primary bract lanceolate, 11–14 mm long, 4–7 mm wide, distinctly tridentate, the two lateral teeth to 1.5 mm long, distinctly multi-nerved and very densely covered with white strigose hairs; pedicels 13–25 mm long, densely hooked pubescent. ***Bracteoles*** broadly oblong-ovate, 7–14 mm long (more than twice as long as calyx), 4–6 mm wide, strongly multi-nerved and densely covered with white strigose hairs, early deciduous. ***Flower*** scarlet to flame orange, drying pink to purple; calyx broadly campanulate, tube about 3.5 mm long, the two upper teeth joined into one emarginate not at all developed, 2 lower lateral teeth scarcely defined, broadly acute, 1 mm long, the central tooth long acute, 4 mm long, 3 mm wide, all teeth hyaline at margin, ciliate, densely covered by long, white strigose hairs especially on lower teeth; standard scarlet, broad, round, reflexed at 6–7 mm from base to bend and 11–12 mm from bend to apex, 15 mm broad, a few minute hairs at the apex, the claw scarcely defined, the auricles 1 mm long; wings scarlet, the blade broadly obovate, 23 mm long, 10 mm wide, the claw 7 mm long, the spur well-developed, 3 mm in diam.; keel greenish, the claw 4 mm long, 11 mm to bend and 5 mm to base of terminal 1 1/2 coils of 4 mm diam.; vexillary stamen, the claw 1.5 mm long, the geniculate sheath flat, well-developed, 2.75 mm long, 2 mm wide, the thickened portion 5 mm more to filament; stamen tube broad at base, 9 mm long to bend and 4 mm more to filaments, the ridges developed into a thick roundish knob about 0.5 mm in diam.; basal collar 2 mm long, somewhat oblique at tip, denticulate; ovary straight, 8 mm long, 1.5 mm wide, densely covered by long, white pubescence; style with terminal thickened coil 2 mm in diam.; stigma terminal, extrorse oblique, 0.75 mm long. ***Pod*** when young very densely

Fig. 17. Illustrations of *Phaseolus coccineus* L. subsp. *coccineus* var. *splendens* Freytag.–c. Portion of plant with mature leaf and inflorescence; note very large, ovate leaf, very large inflorescence with large, tridentate primary bracts and many large flowers distributed along axis.–e. Exploded view showing all parts of the flower including–s. Stigma as seen under the microscope; note long, densely uncinate pedicel, large bracteoles, large, rounded standard, long wings, broad, nerved stamen tube and long keel base. All drawings made from a field collection, *Stanford et al. 656*, from 4 km W of Miquihuana, Tamaulipas, México.

covered with white tomentose hairs, becoming less pubescent at maturity, 7 cm long, 1 cm wide, the tip recurved, 9 mm long. ***Seed*** unknown. ***Seedling*** unknown.

Paratypes. **MÉXICO. Nuevo León:** E side Cerro Linadero, Dulces Nombres, Nuevo León and just E of border into Tamaulipas, 1900 m, 9 Aug 1948, *Meyer et al. 2898* (BM, BR, G, K, MO). **Tamaulipas:** 4 km W of Miquihuana, 23°42'N, 99°45'W, 3110 m, 4 Aug 1941, *Stanford et al. 656* (CAS-DS, GH, MO).

Habitat.–This taxon was found growing in dense woods, along stream banks in canyons, and on limestone ridges in open pine forest. However, it must be quite rare since only a few collections have been made.

Comments.—This variety is named for the very large and showy inflorescences, bracts and flowers which surpass any of the other taxa in the species, viz. the tridentate primary bracts of 11–14 mm long, bracteoles of 7–14 mm long, wing 23 mm long. All of these characteristics are exceptional for the subspecies and are typical of this variety. The foliage is also nearly glabrous in contrast to the fairly heavy pubescence on most other varieties of the subspecies.

The variety with which it might be confused is var. *tridentatus* because both have tridentate bracts, but in addition to the distinct distributions, the leaflets of var. *splendens* are larger and broader making the base less cuneate, the middle lower lobe of the calyx is much longer than the two lateral lobes while those of var. *tridentatus* are subequal, dentate, and acute, the stigma tip is more terminal and thicker on var. *splendens*, and the keel is narrower and more elongate.

C.1.7.—Phaseolus coccineus L. subsp. **coccineus** var. **strigillosus** (Piper) Freytag, comb. & stat. nov. (**Figs. 18, 24**). *Phaseolus strigillosus* Piper Contr. U.S. Natl. Herb. 22:685. 1926. TYPE: MÉXICO. NAYARIT: Sierra Madre near Santa Teresa, Territorio de Tepic, 12 Aug 1897, *Rose 2218* (HOLOTYPE: US 301132).

Plant a climbing, indeterminate vine to 5 m long. ***Root*** unknown. ***Stems*** terete, covered with reflexed-white hispid and uncinnate hairs, to nearly glabrous, slightly ribbed; internodes 8–15 cm long. ***Stipules*** broad triangular, obtuse, 2–3–6 mm long, 1.5–2 mm wide, 4- to 5-nerved, glabrous, ciliate. ***Leaves*** 10.8–21 cm long; petiole delicate, 4–5.5–9.5 cm long, glabrous or a few scattered hairs; petiolule 1–2 cm long; stipels lanceoate, the lower ones 3 mm long, 0.75 mm wide, 1- to 3-nerved, nearly glabrous, the upper ones 1.75–2 mm long, 0.5 mm wide, sparsely covered with hispid hairs; pulvini at base of petiole 5 mm long, covered by minutely uncinate hairs, those of leaflets 3 mm long, densely covered with uncinate hairs adaxially; terminal leaflet broadly ovate, 5–9 cm long, 4–9 cm wide at 1/4 to 1/3 from the base, minutely apiculate, nearly glabrous, only a few long hispid hairs on adaxial surface, covered by hispid and strigose hairs abaxially, membranous, light green adaxial surface, paler abaxially. ***Inflorescence*** delicate, 12–20–35 cm long, the peduncle 10–16–25 cm long, the short rachis 4–7–10 cm long, many nodes and flowers; primary bract aciculate to lanceolate, 5–8 mm long, 0.75–2 mm wide, 4- to 5-nerved, nearly glabrous, ciliate; pedicel delicate, striate, 12–27 mm long, glabrous to puberulent of a few hispid and minute uncinate hairs. ***Bracteoles*** nearly linear to orbicular-lanceolate, 5–6–10 mm long, 1–1.5–2 mm wide, 3- to 5-nerved, nearly glabrous with a few hispid hairs at apex. ***Flower*** scarlet; calyx campanulate, tube 3 mm long, upper 2 lobes joined into one 5 mm long, scarcely elongate, emarginate, covered by hispid hairs, the lower 3 lobes dentate, subequal, 1.75 mm long, 2 mm wide, acute, covered by scattered hispid and strigose hairs; standard scarlet, reflexed at 3–6 mm from base, terminal portion 5–9 mm long, apex with a few hispid hairs especially in bud, the large flap-like auricles 3 mm long, 1 mm wide; wings scarlet, the blade obovate, cupped, 8–9 mm long, 10–12 mm wide, the claw 3.5–4 mm long, 0.5 mm wide, spur not well-developed; keel claws 3 mm long, 2 mm more to bend and 2 mm more to base of the terminal coil of 2–3 mm diam.; vexillary stamen 0.5 mm from base to the geniculate rounded flap, 1 mm long, 2 mm more to end of thickened portion; stamen tube 6 mm long from base to the bend and 4 mm more to the end of united portion; ovary straight, 5 mm long, 0.75 mm wide; style 8 mm long to base of the terminal thickened coil of 1.5 mm diam.; stigma oblique, extrorse, straight 0.75 mm long. ***Pod*** unknown. ***Seed*** unknown. ***Seedling*** unknown.

Specimens examined: **GUATEMALA. Chimaltenango:** Aldea Pampaya, Km 30 Hwy Chimaltenango-Yepocapa, 1980 m, 3 Oct 1978, *Freytag et al. 78-Guat-15* (BR, EAP, F, GH, K, MEXU, MO, UC, US); Km 13, near Chimachoy, crossroads to Volcán Acatenango, Hwy 10 Chimaltenango-Yepocapa, 14°34'N, 90°51'W, 2230 m, 3 Oct 1978, *Freytag et al. 78-Guat-16B* (US); Km 78 Hwy 11 Patzicia-Panajachel, 1930 m, 6 Oct 1978, *Freytag et al. 78-Guat-42* (BR, EAP, F, GH, MEXU, MO, UC, US); Chipeacul, Km 90 Hwy Patzicia-Panajachel, 14°48'N, 91°15'W, 2310 m, 6 Oct 1978, *Freytag et al. 78-Guat-42-1B* (EAP, MEXU, MO, US). **Huehuetenango:** Km 321.5 Hwy 9N, Huehuetenango-Soloma, at San Juan Ixcoy, 15°35'N, 91°26'W, 2222 m, 10 Oct 1978, *Freytag et al. 78-Guat-109* (BR, EAP, F, GH, K, MEXU, MO, UC, US); Km 339, 11 km from Huehuetenango, Hwy 7W Huehuetenango-Cobán, 15°21'N, 91°25'W, 2000 m, 11 Oct 1978, *Freytag et al. 78-Guat-113* (EAP, MEXU, MO, US). **Jalapa:** 1 km from La Toma, Km 97.5 Hwy 18 Guatemala-Jalapa, 14°35'N, 90°10'W, 1960 m, 5 Oct 1978, *Freytag et al. 78-Guat-41-6* (BR, EAP, MEXU, MO, US); Km 87.3 Hwy 18 Guatemala-Jalapa, 2430 m, 5 Oct 1978, *Freytag et al. 78-Guat-41-9* (MEXU, MO, US); Km 35.5, between Mataquescuintla and Cd. Guatemala, Hwy 18 Guatemala-Jalapa, 14°35'N, 90°15'W, 1950 m, 5 Oct 1978, *Freytag et al. 78-Guat-41-10* (BR, EAP, MEXU, MO, UC, US). **Quetzaltenango:** Km 9 Hwy Palestina-Río Blanco, portion Palestina de los Altos to Sibilia, near Patzacán, 2580 m, 9 Oct 1978, *Freytag et al. 78-Guat-95* (BR, EAP, F, GH, K, MEXU, MO, UC, US). **San Marcos:** Km 3, near San Cristobal Cucho, Hwy 6 San Marcos to La Reforma, 2050 m, 8 Oct 1978, *Freytag et al. 78-Guat-88* (EAP, MEXU, MO, US); Km 19 Hwy 125 San Marcos-La Reforma, 1775 m, 8 Oct 1978,

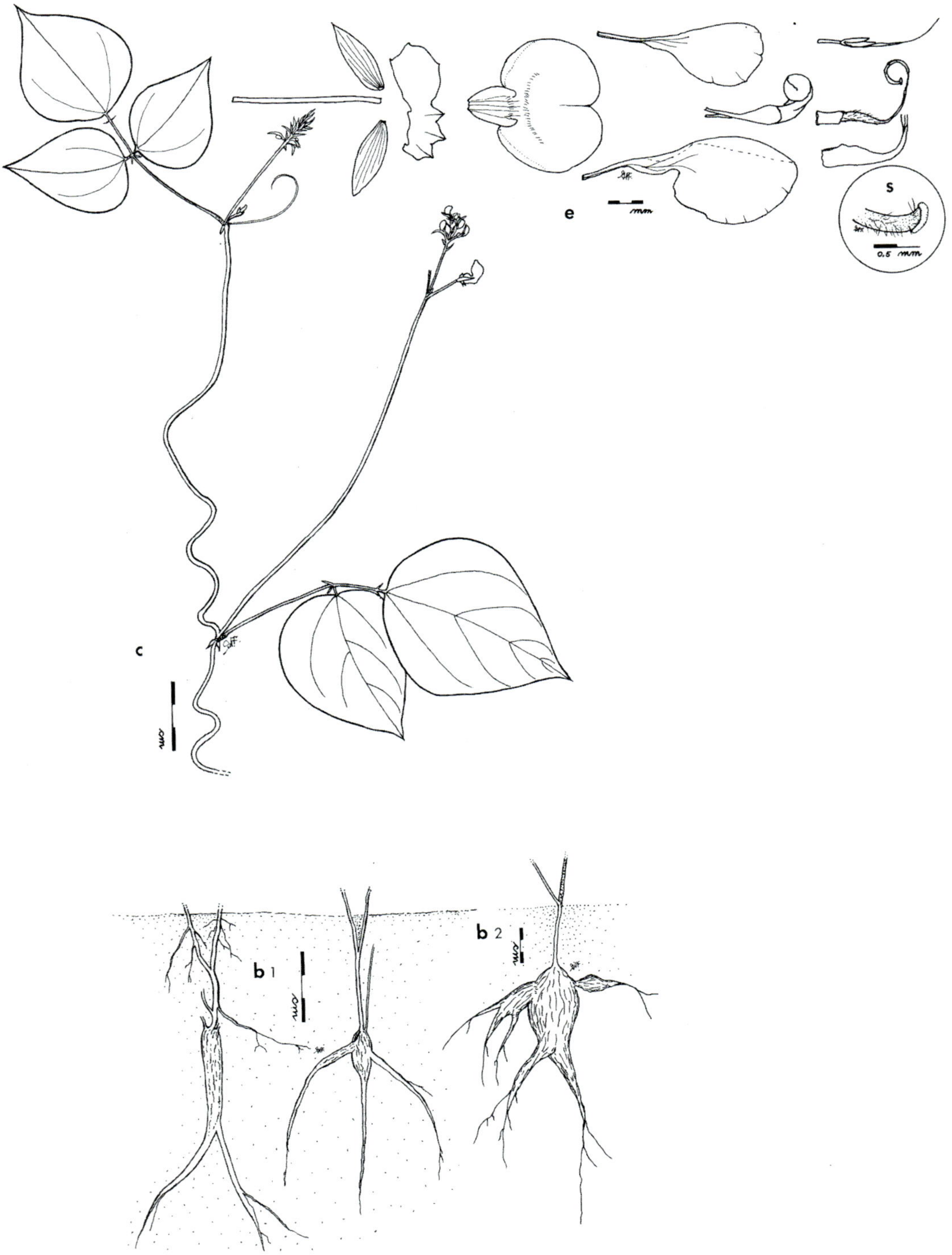

FIG. 18. Illustrations of *Phaseolus coccineus* L. subsp. *coccineus* var. *strigillosus* (Piper) Freytag.–**b.1.** Two young roots from field.–**b.2.** Mature root.–**c.** Portion of plant with mature leaf and inflorescence; note inflorescence with large primary bracts and long, narrow bracteoles, especially apparent in buds at apex.–**e.** Exploded view of flower showing all parts including–**s.** Stigma as seen under the microscope; note glabrous pedicel, large elongate bracteoles, small, rounded standard, rounded wings, short stamen tube and short keel. Drawings made from following: plant with leaf and inflorescence made from the field collection *Mejía 792* from SW of Ixtlán, Oaxaca, México; flower made from living material grown in the greenhouse at Mayagüez of seed of *Freytag et al. 78-Guat-42* from Km 78, Hwy 11 Patzicia–Panajachel, Guatemala and mature root drawn from photograph of the same field collection.

Freytag et al. 78-Guat-89 (BR, EAP, MEXU, MO, US). **Sololá:** Km 153 Hwy Sta. Lucía Utatlán–Sta. Clara La Laguna, 2330 m, 7 Oct 1978, *Freytag et al. 78-Guat-59* (EAP, MEXU, MO, US).

HONDURAS. Fco. Morazán: Mt. Uyuca cloud forest, (13°50'N, 87°5'W), Aug 1960, *Pfeifer 1822* (US).

MÉXICO. Chiapas: Mpio. of Venustiano Carranza, 3 mi S of Aguacatenango along road to Pinola Las Rosas, (15°25'N, 92°27'W), 1707 m, 13 Oct 1965, *Breedlove et al. 13125* (CAS, US); Mpio. Pueblo Nuevo Solistahuacan, 3 mi N of Pueblo N. Solistahuacan, 1615 m, 3 Aug 1967, *Clarke 709* (UCR). **Jalisco:** Nevado de Colima, 2 mi E of Jazmín, (19°35'N, 103°30'W), 1600 m, 31 Oct 1962, *McVaugh 21981* (MICH); 3–5 mi NW of San Miguel de la Sierra (ca 40 km airline, W of Ayutla), (20°20'N, 104°30'W), 2000 m, 3 Nov 1962, *McVaugh 22027* (MICH). **Nayarit:** Mpio. Tepic, Cerro San Juan, 1.5–3 km S of the road from Tepic–Sta. Cruz, 1200–1300 m, 4 Oct 1985, *Bartholomew et al. 2614* (CAS); 9.5 mi W of Tepic, (21°30'N, 105°10'W), 1000–1100 m, 25 Sep 1960, *McVaugh 19383* (CAS, MICH); Canyon near 'El Ocote', SW of Ixtlán, (20°59'N, 104°22'W), 1480 m, 29 Sep 1926, *Mexía 792* (F, UC). **Oaxaca:** Hwy 190 Km 129, Huajuapan–Oaxaca, 17°4'N, 97°2'W, 2120 m, 21 Sep 1978, *Freytag et al. 78-Mex-43* (ARIZ, BR, EAP, F, GH, K, MEXU, MO, UC, US); 46 mi S Miahuatlán on road to Puerto Angel, (16°10'N, 96°30'W), 18 Oct 1981, *Warnock 2523* (TEX).

Habitat.—In Guatemala typical variety *strigillosus* is mostly found in openings in undisturbed oak or pine forest habitats on the high mountains at elevations from 1000–2580 m with a predominance of the populations at the higher altitudes; where there is milpa cultivation and second growth there seems to be much introgression with cultivars of variety *coccineus*.

Diseases.—This taxon is reported to be heavily attacked by rust and infrequently with angular leaf spot and anthracnose.

Common names.—"Chamborote (chomborote) silvestre" and "Frijolillo" are the only two names given by collectors.

Genetics.—It has been observed to be pollinated by the large black carpenter bee, *Xylocopa*.

Comments.—Variety *strigillosus* seems to be principally a Guatemalan taxon. Very long, nearly linear bracteoles in wild beans are only typically found in the species *P. dumosus* Macfady. which only has purple or white flowers. As far as the senior author knows, the combination of red flowers (*coccineus*) and long, linear bracteoles (*dumosus*) is never found in the type variety *coccineus* nor in the advanced generation hybrids between this species and *dumosus*. Nevertheless, long, narrow bracteoles are the main distinguishing character of var. *strigillosus*, as well as the typical pyramidal cluster of buds and flowers on the developing inflorescences (see Color Plate I, photo 7). The long, narrow bracts and bracteoles are often dark, usually purplish, and nearly glabrous.

Much intergradation is found, presumably introgression with variety *coccineus*, which causes broader, shorter bracteoles, and in these intermediate plants there is more pubescence and less clustering of flowers at the apex of the inflorescence than in typical var. *strigillosus*. Some collections showing these intermediates (not given in specimens cited above nor used for variety description and distribution) are: **GUATEMALA**. Top of ridge, Hwy 10 Chimaltenango–Yepocapa, 2025 m, 3 Oct 1978, *Freytag et al. 78-Guat-16* (EAP, MEXU, MO, US); Hwy 10 Chimaltenango–Yepocapa, 2025 m, 3 Oct 1978, *Freytag et al. 78-Guat-16A* (EAP, MEXU, MO, US); Km 7.5–7 on dirt road to Totonicapán from road Nahualá–Quetzaltenango, 2490 m, 7 Oct 1978, *Freytag et al. 78-Guat-63* (MEXU, MO, US); Slopes of volcano at the wood bridge, Km 9, road from Antigua–Volcán de Agua, 1940 m, 3 Oct 1978, *Freytag et al. 78-Guat-17* (BR, EAP, MEXU, MO, US).

The apparently disjunct distribution (from Nayarit in the north to Chiapas, Guatemala and Honduras in the south) may be due to the spotty occurrence of this variety in the higher mountains and/or to insufficient collecting in similar habitats in S Mexico (Michoacán, Guerrero, Oaxaca and Chiapas). From the senior author own collecting, he concludes that the center of distribution of var. *strigillosus* is in northern Guatemala and not in western Mexico where the type was collected. In Honduras there are only very few good collections of var. *strigillosus*, while most are mixed or intergrade with variable cultivars of variety *coccineus*.

C.1.8.—Phaseolus coccineus L. subsp. **coccineus** var. **semperbracteolatus** Freytag, var. nov. (**Figs. 19, 24**). TYPE: MÉXICO. OAXACA: Mt. Oaxaca, NE of Cd. Oaxaca, Hwy 175, 2438 m, (17°9'N, 97°41'W), Oct 1894, *Smith 562* (HOLOTYPE: EAP).

Similis *Phaseolo coccineo* var. *coccineo* sed tenellus volubilisque in umbram et bracteolis grandissimis late ovatis vel rotundatis glabris vel tomentosis raro intermediis insigniter persistentibus usque ad legumen juvene differt. Crecit montibus septentrionale Oaxacae rarus.

Plant a weakly climbing, indeterminate vine, 2–3 m long. ***Root*** thick, fleshy, ovoid to globose, the

FIG. 19. Illustrations of *Phaseolus coccineus* L. subsp. *coccineus* var. *semperbracteolatus* Freytag.—**c.** Portion of plant with mature leaves and inflorescences; note climbing stem, somewhat undulate leaf margins, medium few-flowered inflorescence with large primary bracts and very large bracteoles (tomentose or glabrous) and persistent in young pods.—**e.** Exploded view of flower showing all parts including—**s.** Stigma as seen under the microscope; note large bracteoles, elongate standard, and wide knob or sheath on vexillary stamen. Drawings made of: **c.** from type field collection *Smith 562*; **e.** and **s.** from living material grown in screenhouse at Mayagüez, PR of seed from *Freytag et al. 78-Mex-60A* from Km 15, Hwy 175 Oaxaca–Tuxtepec, Oaxaca, México.

young root about 2 cm long, 1 cm diam. (mature roots unknown). ***Stems*** with internodes 4–12 cm long, nearly glabrous. ***Leaves*** 7–11 cm long; petiole 2–4 cm long; petiolule 1.2–2 cm long; terminal leaflet ovate, hastate to obscurely lobed at base, 4–5 cm long, 2–3 cm wide, acuminate to acute, distinctly white veined, nearly glabrous but sparsely covered by appressed white hispid hairs adaxially especially on veins, sparsely covered by glandular and minute hispid hairs abaxially. ***Inflorescence*** erect, few-flowered, rather stout, 8–9–13 cm long; peduncle 6–10 cm long, 2 mm thick, nearly glabrous; rachis 2–6 cm long of 6–9 nodes; primary bract large, ovate, acuminate, 4–8 mm long, 2–4 mm wide, obscurely 3- to 6-nerved, acuminate, nearly glabrous, whitish; pedicels 15–20 mm long, nearly glabrous. ***Bracteoles*** very large, orbicular to round, 7–9.5 mm long, 5–8 mm wide, distinctly 10- to 14-nerved, whitish or greenish to purple tinged, glabrous and hyaline or nearly tomentose (seldom intermediate), usually persistent through young pod. ***Flower*** scarlet, calyx 6 mm long, campanulate, the upper 2 lobes united into one 9 mm wide, not elongated nor emarginate, a few strigose hairs at line of union, the three lower teeth triangular, acute, 2.5 mm long, 2.5 mm wide, the center one somewhat narrower and densely covered by long strigose hairs; standard scarlet, round to oblong, reflexed at 4–6 mm from base, the terminal portion erect, not emarginate 10–12 mm long, 12–17 mm wide, the claw not defined, the auricles minute; wings scarlet, the blade obovate-rounded, cupped, spreading, 17 mm long, 9–10 mm wide, the claw 4 mm long, the spur 1.5 mm long, firmly fastened to keel; keel 6–7 mm to bend and 4–6 mm more to base of the terminal 1 1/2 coils of 3.5–3.75 mm diam., the claws 3–4.5 mm long, the ridges pronounced, 2.5 mm diam.; vexillary stamen, the claw 1 mm long, the geniculate knob prolonged into a flat sheath 1.5 mm long, 2 mm wide, the thickened portion 2–3 mm long to the filament; stamen tube broad at base, nearly straight, 7 mm to bend and 5 more to filaments; basal collar 2.5 mm long, dentate at margin; ovary straight, 6 mm long, 1 mm wide, pronounced sutures, densely covered by long hirsute hairs, 6–8 ovules; style 5 mm long to the terminal thickened coil of 3 mm diam.; stigma terminal, extrorse oblique, 1 mm long. ***Pod*** slightly curved and flattened, 4–6 cm long, 1.1 cm wide, 5–7 mm thick; valves fibrous, somewhat asperous, mottled brown or purplish, when young densely covered by yellowish, long strigose hairs, twisted 1–2 times at dehiscence; beak nearly straight, narrow, 5–6 mm long; 7 seed. ***Seed*** orbicular to globose, 6–7 mm long, 4–6.4 mm wide, 3–4 mm (immature?) thick, solid black or veined and mottled black on brown or red, black ring around hilum; hilum oblong, 1 mm long, 0.5 mm wide. ***Seedling*** unknown.

PARATYPES. **MÉXICO. Jalisco:** Km 20 Hwy from Cd. Guzmán–El Grullo, near Piedra Ancha, 19°38'N, 103°35'W, 2000 m, 3 Dec 1981, *Freytag et al. 81-24* (BR, EAP, F, GH, MEXU, MO, US); 3 mi S of Mazamitla, Sierra del Tigre, (19°52'N, 103°5'W), 2100–2200 m, 16 Sep 1952, *McVaugh 12935* (BRIT, MICH, US). **Michoacán:** 6 km al SE de Villa Madero, camino a Nocupetaro, (19°15'N, 101°20'W), 2400 m, 30 Sep 1982, *Soto 4730* (CAS). **Morelos:** 10 km above Cuernavaca on Hwy 95, (19°5'N, 99°10'W), 22 Nov 1981, *Freytag et al 81-2* (BR, EAP, MEXU, MO, US). **Oaxaca:** 25 km by road NE of Hwy 190 on road to Guelatáo, Hwy 175, alt 2440 m, 14 Oct 1983, *Anderson 13109* (MICH); Dist. del Centro, Cerro San Felipe, 2000 m, 20 Sep 1908, *Conzatti 2305* (F, GH); Cañada de San Gabriel, 3000 m, 8 Aug 1897, *Conzatti et al. 378* (GH, K); Km 15 Hwy 175 Oaxaca–Tuxtepec, 2250 m, 22 Sep 1978, *Freytag et al. 78-Mex-60A* (pubescent) &-*60B* (glabrous) (BR, EAP, F, GH, K, MEXU, MO, NA, UC, US); Km 15 Hwy 175 Oaxaca–Tuxtepec, about 15 km from turn off towards Tuxtepec on Hwy 190 Oaxaca–Tehuantepec, 2250 m, 27 Nov 1981, *Freytag et al. 81-10* (US); SW slopes of Cerro San Felipe, above San Felipe, ca 12–14 km NNW of Oaxaca, 17°9'N, 96°18'W, 2700–3000 m, 22 Aug 1960, *Iltis et al. 1253* (MICH, WIS); Mts. about Yalalag, (17°20'N, 96°15'W), 2134 m, 1 Aug 1894, *Nelson 975* (US); Sierra de Clavellinas, 2743 m, Oct. 1894, *Smith 565* (EAP, F); 1 km al N de La Cumbre, carr. Oaxaca–Guelatáo, 2600 m, 24 Aug. 1976, *Sousa et al. 6100* (MICH); Dist. Coixtlahuaca, 7 km al SW de San Cristobal Suchixtlahuaca, (17°45'N, 97°30'W), 2400 m, 24 Sep 1978, *Sousa et al. 9778* (CAS); 6 km E de Ayutla, 2400 m, 18 Sep 1976, *Téllez et al. 86* (CAS, MO(2)); 4–5 km up logging road W from La Cumbre, ca 30 km NNE of Oaxaca on Hwy 175 between Oaxaca and Ixtlán de Juárez, 2800 m, 4–7 Jul 1963, *Tillett 637121* (CAS-DS, US); S of road Km 140, Rt 190, ca 12 mi NW Jct. with Rt. 131, (17°10'N, 97°5'W), 20 Oct 1981, *Warnock 2545* (TEX). **Puebla:** above Teotitlán del Camino on the road to Huautla, Tehuacan area, (18°8'N, 96°40'W), 2000–3250 m, 3 Aug 1961, *Smith et al. 4194* (F). **Queretaro:** Mpio. Pinal de Amoles, 13 km al NW de Pinal de Amoles, carr. Pinal de Amoles–Jalpán, (21°8'N, 99°37'W), 1620 m, 22 Oct 1982, *Tenorio et al 2330* (CAS, MO). **Zacatecas:** Sierra de los Morones, near Plateado, 1 Sep 1897, *Rose 2724* (US).

Habitat.—This variety is found climbing on and over shrubs on steep N-facing slopes, from the lower edge of dense oak-pine forest to montane rain forests above, usually more abundant at the edges of the forests. These forests have many ferns, bromeliads and lichens, with bunch grasses below, and associated with *Baccha* and *Crataegus*. Soils are rocky and sandy with lots of humus and very moist, often in heavy red to black clay loam from largely sedimentary or igneous rock.

Diseases.—There is one report of susceptibility to bean rust.

Comments.—This varietal epithet refers to the very striking persistence of the large, nearly round bracteoles, seldom falling before the pod is at least half developed in length and almost always persistent until after the flower has fallen. In most varieties of *P. coccineus* the bracteoles typically are present only through anthesis and fall off shortly after fertilization. In this variety they tend to persist during the elongation of young pods until they reach nearly full size. Thus in this variety the young pods look as though they have two earmuffs attached at the receptacle end. *P. oaxacanus* is also present in the same habitat. *Smith CL 562* and *563* were apparently collected at the same location, just north and east of Cd. Oaxaca; *562* has glabrous bracteoles and is from Mt. Oaxaca at 2438 m altitude, while *563* has pubescent bracteoles and is from Sierra San Felipe at 3048 m altitude. The senior author also collected these two variants together on Hwy 175 in Oaxaca-Tuxtepec at 2250 m altitude. Thus it seems likely that the pubescence of the bracteole is not an important diagnostic character in this variety, though it is curious that intermediates (moderate pubescence) are seldom found.

C.1.9.—Phaseolus coccineus L. subsp. **coccineus** var. **condensatus** Freytag, var. nov. (**Figs. 20, 24**). TYPE: MÉXICO. MORELOS: Tepoztlán, (19°N, 99°5'W), 1829 m, 6 Nov 1967, *Gentry 22405* (HOLOTYPE: US; ISOTYPES: GH, MICH, NA).

Similis *Phaseolo coccineo* var. *coccineo* sed inflorescentia internodiis multis brevissimis et bracteis primariis longis angustisque differt.

Plant a scandent, indeterminate vine, to 2–3 m long. ***Root*** subglobose, perennial, some secondary branches near the crown, to 12–14 cm long, 6 cm in diam., corky and ringed by lenticular areas. ***Stems*** obscurely striate, nearly glabrous but with a few scattered glandular and short hispid hairs; internodes 5–11 cm long. ***Stipules*** triangular, 3–4 mm long, 5- to 7-nerved, acute, hispidulose on adaxial surface, glabrous abaxially. ***Leaves*** 10–13 cm long; petioles 4–6 cm long, ridged, sparsely covered by hispid hairs; petiolules 1–1.75 cm long; stipels lanceolate, 1-nerved, nearly glabrous; terminal leaflet rhombic-ovate, 5–5.5 cm long, 3–4 cm wide, slightly acuminate, minutely apiculate, distinctly whitish or fine yellowish-nerved abaxially, nearly glabrous but sparsely covered by glandular and appressed white hispid hairs on abaxial surface, mostly on veins, less so adaxially. ***Inflorescence*** many-flowered, short; peduncle 3–8 cm long; rachis 3–10–30 cm long, with very short internodes of 1.5–4 mm long, purplish, covered with many minute hooked hairs; primary bract lanceolate to nearly linear, extended, 3.5–5 mm long, about 1–2 mm wide at base, 3- to 4-nerved, slightly covered by white hispid hairs on abaxial surface; pedicel 5 mm long at anthesis to 15 mm long in young pod. ***Bracteoles*** ovate, 2–4 mm long, 1.75–2 mm wide, less than 1/2 to as long as calyx, acute, puberulent. ***Flower*** scarlet, small; calyx campanulate, glabrous; standard 3 mm long, 4 mm wide; stigma terminal, oblique extrorse. ***Pod*** nearly straight, 7–8 cm long, 1 cm wide, 5.5–6.4 mm deep, beak nearly straight, 3 mm long, the carpels often striped purple, when young densely covered with yellowish strigose and minute white, hooked hairs. ***Seed*** round, thick, 5.7–9.1 mm long, 5–7 mm wide, 2.8–3.6 mm thick, speckled black (gray) on brown, shiny, black ring around hilum. ***Seedling*** from hypogeal germination; hypocotyl not elongated; epicotyl 65–75 mm long, next 3 internodes successively increasing in size from 40–53 mm long to 64–120 mm long; stipules at eophyll node, united; primary leaves opposite, simple, entire; petiole 27 mm long, pulvinus 3 mm long; stipels present; blade ovate, 5.2 cm long, 3.3 cm wide at base, the base truncate, tip acuminate, slightly apiculate.

PARATYPES. **MÉXICO. Durango:** barranca of Río Jaral, 15 mi W of Coyotes RR station about 80 km W of Cd. Durango, (23°55'N, 105°20'W), 2100–2200 m, 1 Oct 1952, *McVaugh 21723* (CAS, MICH); 5 mi W of Buenos Aires, Hwy Durango-Mazatlán, 11 Sep 1951, *Muller 510* (UCR). **Hidalgo:** Nicolás Flores, Rancho de Jaguey, 9 km E entronque Mex 85 a Nicolás Flores, 20°48'N, 99°14'W, 2370 m, 30 Oct 1986, *Debouck et al. 2031* (BR, CHAPA, COL, M, MICH, US); Zimapán, 4 km NW de entronque Mex 85 (Hda. La Estancia) en camino a San Francisco del Oro, 20°48'N, 99°20'W, 2150 m, 31 Oct 1986, *Debouck et al. 2034* (BR, CHAPA, US); Zimapán, El Salto, 32 km NE de Trancas en la Mex 85 a Tamazunchale, 20°53'N, 99°14'W, 2030 m, 1 Nov 1986, *Debouck et al. 2037* (BR, CHAPA, COL, US(2)). **Jalisco:** 2 km S of San Agustín, Unión de Tula, 20°5'N, 104°9'W, 1350 m, 26 Nov 1978, *Debouck et al. 471* (CHAPA, COL, US); Zarza Mora, 2 km E of Las Joyas, 6 km WSW of Rincón de Manantlán, 19°35'N, 104°16'W, 1900 m, 6 Jan 1979, *Iltis et al. 1313*, (MICH). **México:** 9 km NW de Coatepec, 18°55'N, 99°46'W 1980 m, 1 Nov 1987, *Debouck et al. 2365* (CHAPA, MICH); Ocuilán, 4 km NE de Chalma, 18°57'N, 99°26'W, 1890 m, 4 Nov 1987, *Debouck et al. 2372* (CHAPA, COL, MICH). **Morelos:** Km 64 along toll road NE of Cuernavaca, 1981 m, 30 Nov 1967, *Gentry 22490* (GH, MICH, NA). **Oaxaca:** 10 mi SE of Miahuatlán, (16°15'N, 96°35'W), 1829 m, 29 Oct 1967, *Gentry 22388* (GH, MICH, NA). **Puebla:** San Juan, 2180 m, Aug 1906, *Arsène 42* (US); Moulin de Huexotitla, Puebla, 2155

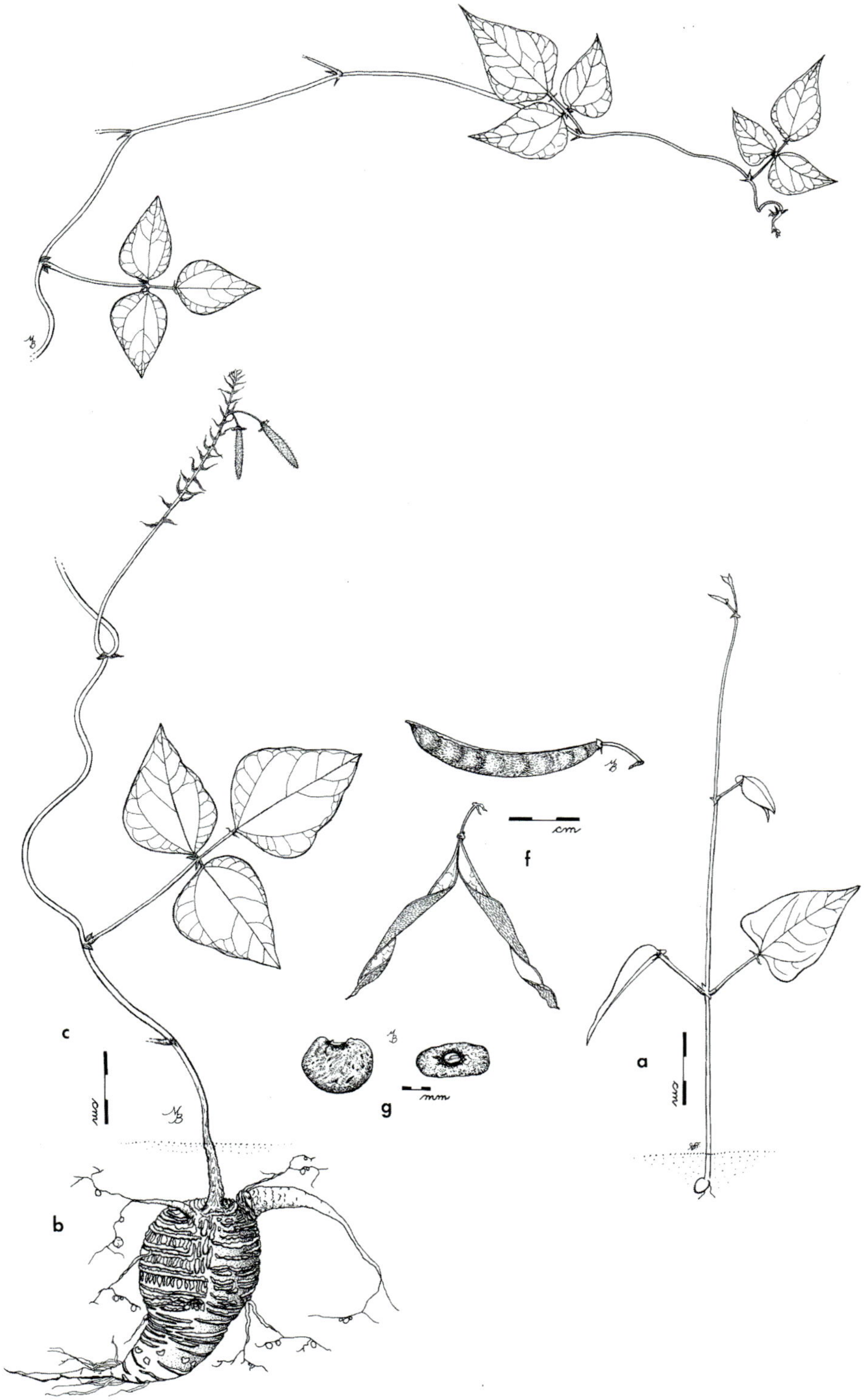

Fig. 20. Illustrations of *Phaseolus coccineus* L. subsp. *coccineus* var. *condensatus* Freytag.—**a.** Plantlet a couple of weeks after germination of seed; note triangular primary leaf.—**b.** Root after several years of growth; note short, clavate to turbinate, slightly branched form with distinct horizontal lenticels.—**c.** Portions of plant with mature leaves and inflorescences and vine tip; notice climbing stem, short inflorescence with very short internodes and long bracts.—**f.** Pods, side view and dehisced.—**g.** Seeds, side view and view from hilum; note rounded and flattened shape. Drawings made of: **c.** from field collection of *Gentry 22405* from Morelos, México; all others from living material grown in environmental chambers and greenhouse at Mayagüez, PR from seed of TARS #411 (*Debouck et al. 2037* from near Zimapán, Hidalgo, México).

m, 23 Sep 1906, *Arsène 320* (US); Rancho Posada E of Puebla, 2194 m, 25 Sep 1906, *Arsène 358* (US); au Mayoraz sur l'Atoyac, 2120 m, 9 Oct 1907, *Arsène 1270* (US); Santa Isabel Cholula, barranco del Puente Los Molinos, 4 km NE de Atlixco, 18°58'N, 98°23'W, 1880 m, 22 Oct 1987, *Debouck et al. 2323* (CHAPA, COL, MICH(2)). **Tlaxcala:** Mpio. San Luís Teolocholco, Huaxinca, 2500 m, 3 Aug 1977, *Martínez et al. 1714* (CAS, MO). **Veracruz:** Azoul, région d'Orizaba, 2 Oct 1865-66, *Bourgeau 3173* (in part) (K, US).

Habitat.—This variety is found growing on shrubs in the semi-shade of pine-oak or mixed forests of oak, *Cupressus, Alnus, Juniper, Carpinus, Tilia* and *Magnolia*, with an understory of grass, composites or maguey, at altitudes of 1800–2400 m. The soils are rocky and sandy, usually red and may be derived from schists, limestone, basalt, rhyolite or lava.

Diseases and pests.—Rust and anthracnose are most frequently mentioned and *Apion* weevil may attack the pods.

Common names.—Some collectors have given "frijol de la rata" and "frijol de risa" as names used by Spanish peoples and "yeximima" as the name used by the native people.

Comments.—The variety *condensatus* in its extreme form is easily recognized by the long, lanceolate to linear bracts produced closely together on inflorescences of strikingly short internodes (mostly less than 4 mm long). However, throughout its range which overlaps type var. *coccineus*, it seems to introgress with all types of intermediates being observed. For this reason, it is very difficult to place all specimens accurately and only the extreme types are given in the list of exsiccatae.

C.1.10.—Phaseolus coccineus L. subsp. **coccineus** var. **pubescens** Freytag, var. nov. (**Figs. 21, 24**). TYPE: GUATEMALA. SAN MARCOS: forests 6 km N of San Marcos, Sierra Madre Mts., (15°N, 91°45'W), 2700 m, 13 Dec 1963, *Williams et al. 25901* (2 sheets, in flower) (HOLOTYPE: F(2); ISOTYPE: EAP).

Similis *Phaseolo coccineo* var. *coccineo* sed brevis tenellis volubilisque, foliolis in superficibus atro-olivaceis infra et argenteis inflorescentia paucifloris, pedicellis pilis luteis longis conspicua differt.

Plant a climbing, indeterminate vine, to 5 m long. ***Root*** unknown. ***Stems*** terete, slender, striate, sparsely covered by yellow, retrorse pubescence; internodes 6–14 cm long, mostly glabrous except for a few strigose hairs at nodes, young stems densely covered with yellow strigose hairs. ***Stipules*** oblong, acute, 4 mm long, 1.5 mm wide, 4- to 7-nerved, glabrous above, covered with long strigose hairs abaxially, ciliate on margin, extended. ***Leaves*** 9.6–15.6 cm long, dark gray-green adaxially, silvery and reticulately veined abaxially; petioles, delicate, 1.5–5.5 cm long, sparsely covered with long, yellowish strigose hairs; petiolule 1.5–2.5 cm long, sparsely covered with long, yellowish strigose hairs; stipels, the lower ones aciculate, 4 mm long, dark, 1-nerved, the upper linear, 2 mm long, dark purple, glabrous; pulvini dark, 3 mm long, densely covered by hispid and long, yellow strigose hairs; terminal leaflet broadly ovate to ovate, 6–7 cm long, 4.5–5 cm wide at 1/3 from base, whitish veined, dark green adaxially and silvery and sparsely covered with long hispid hairs abaxially, some hairs at leaf apex with large glandular bases. ***Inflorescence*** a short to long pseudoraceme; peduncle 5–19–27 cm long, glabrous; rachis 1–5 cm long of 2–14 flowers, sparsely covered by long, yellowish strigose hairs; primary bract broadly ovate to lanceolate, acute, 4.5–5 mm long, 2–3.5 mm wide, 3- to 11-nerved, purplish, glabrous adaxially, with long, yellowish strigose hairs on abaxial surface; pedicel 10–12 mm long in flower to 18 mm long at young pod, glabrous to densely covered by long strigose hairs, dark purplish. ***Bracteoles*** broad, oblong, 5 mm long, 2.3–3 mm wide, acute, dark purple, nearly glabrous, a few hispid hairs, 4- to 8-nerved. ***Flower*** large and delicate, red or scarlet to salmon, drying purple, thin membranous; calyx flaring campanulate, 4 mm long, the upper 2 lobes united into one entire, 1 mm long, 3.5 mm wide, the lower 3 dentate, subequal, 1 mm long, 1.5 mm wide, the central one slightly longer and bearded by long, white strigose hairs, the lateral ones nearly glabrous, purple; standard red, sharply reflexed, 5 mm long to flexure and 6–8 mm more to tip, rounded, 12 mm wide, the claw indefinite, 1 mm long, the auricles missing; wings red, the blade rounded, cupped, spreading, 17 mm long, 8 mm wide, the claw 5 mm long, 0.5 mm wide, the spur well-developed 0.5–1 mm long, firmly affixed to keel, the blade 10 mm diam.; keel claws 5 mm long, 3 mm more to bend and 4 mm more to base of the terminal 1 1/2 coil of 3 mm diam.; vexillary stamen, the claw 1.5 mm long, the knob a rounded sheath 0.5–1 mm long, the thickened portion 3 mm long; stamen tube nearly straight; style with terminal thickened coil of 2–2.5 mm diam.; stigma extrorse oblique, 0.75 mm long; ovary densely covered with long, yellow strigose hairs except on sutures. ***Pod*** the young pod long, covered by white or yellowish strigose hairs, mostly on valves, the mature pods 3–5 cm long, 0.8–1 cm wide, of 2–4

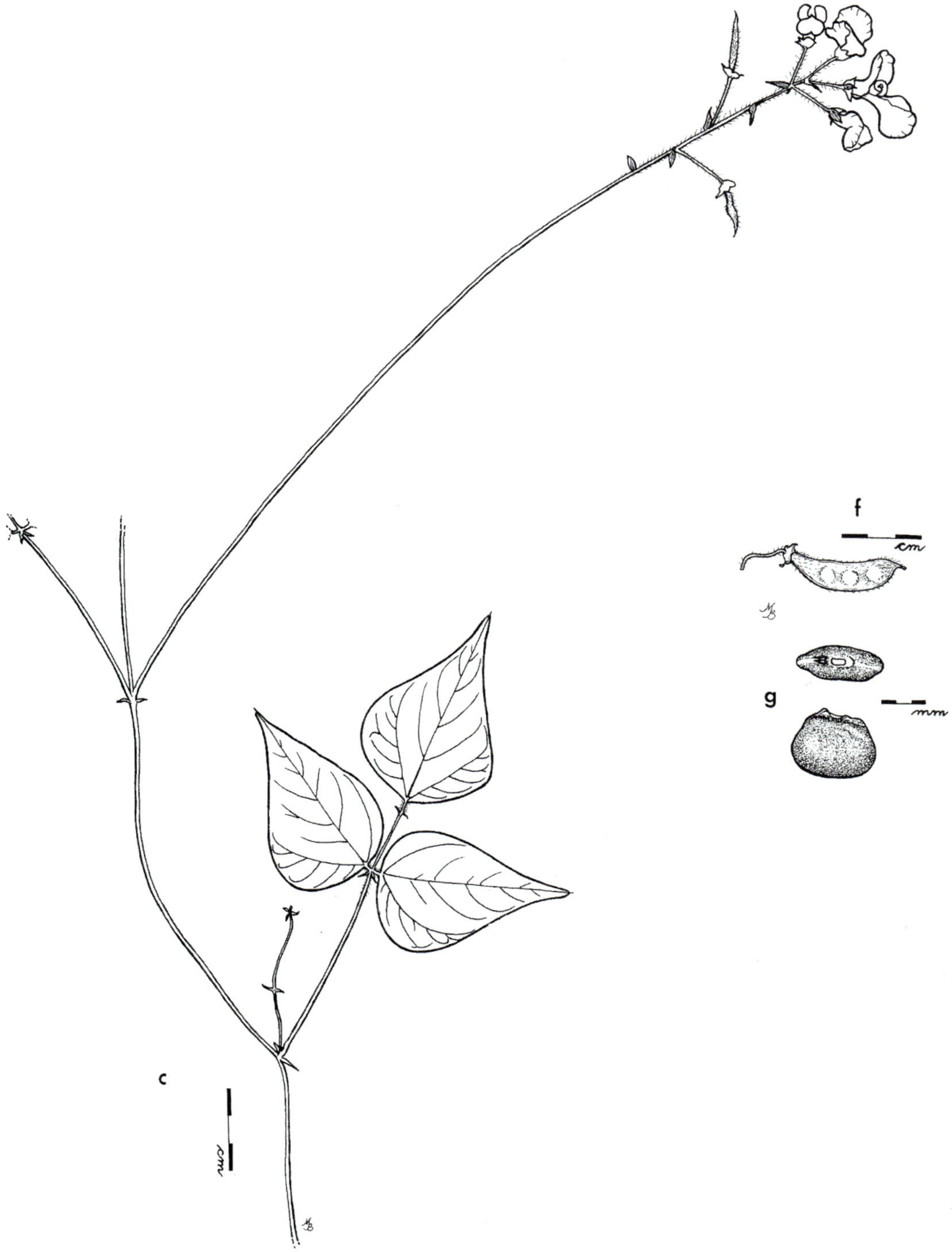

FIG. 21. Illustrations of *Phaseolus coccineus* L. subsp. *coccineus* var. *pubescens* Freytag.—**c.** Portion of plant with mature leaf and inflorescence; notice delicate climbing vine and inflorescence with distinct long pubescence on raceme and pedicels. Drawings from type field collections: **c.** from type *Williams et al. 25901* collected near San Marcos, Guatemala, and **f.** & **g.** from *Standley 86104* collected from Los Vajos near Quetzaltenango, Guatemala.

seed, broad and blunt, the base somewhat stipitate, smooth, brittle, glabrous, dehiscing 1 turn; beak short, straight, 1 mm long; suture somewhat thickened. ***Seed*** unknown. ***Seedling*** unknown.

PARATYPES. **GUATEMALA. Chimaltenango:** along road to Panajachel between Los Idolos and Chocoyos, 2400 m, 12-23 Jan 1966, *Molina et al. 16205* (F). **Guatemala:** San Andrecillo, (14°30'N, 90°30'W)?, 1700 m, 26 Sep 1972, *Molina et al. 27546* (EAP, F, MICH). **Quetzaltenango:** Los Vajos, Cerro Quemado, (14°45'N, 91°30'W)?, 42900-3000 m, 5 Feb 1941, *Standley 86104* (F); Finca La Arabia, Quetzaltenango, 2438 m, 30 Oct 1944, *White 5268* (MICH); Cuesta de El Caracol, Sierra Madre Mts., about 5-8 km N of San Juan Ostuncalco, 3000-3100 m, 11 Dec 1962, *Williams et al. 22733* (EAP, F, GH, US, WIS). **San Marcos:** Sierra Madre Mts., about 6 km (air-line) N of San Marcos, 2700 m, 13 Dec 1963, *Williams et al. 25872* (EAP, F, US).

MÉXICO. Michoacán: Cerro Santa María, 8-10 km SW of Jiquilpán and 5 km NE of Quitupán, (20°N, 102°5'W), 2000 m, 8-9 Aug 1959, *Feddema 197* (CAS, MICH(2)); between Río del Salto and La Polvilla, 18 mi E of Morelia, (19°40'N, 101°W), 2195-2438 m, 9 Nov 1981, *King et al. 5048* (BRIT, MICH, TEX, UC, US); 18 mi S of Pátzcuaro, (19°15'N, 101°45'W), 2713-2743 m, 20-25 Nov 1961, *King et al. 5186* (MICH, TEX, US); 3-6 km SE of Aserradero Dos Aguas, W of Aguililla, (18°45'N, 102°45'W), 2000-2100 m, 25 Nov 1970, *McVaugh 24685* (MICH). **Oaxaca:** 14-18 mi NE of Cd. Oaxaca, along road to Ixtlán, (17°15'N, 98°30'W), 2286-2743 m, 29 Aug 1952, *Gentry 12063* (TEX(2)); S facing slopes along Hwy Oaxaca-Tuxtepec, 12.3 mi N of Jct. 190 and 175 on 175, 2591 m, 15 Aug 1975, *Le Doux et al. 2238* (MO, TEX-LL).

Habitat.—This taxon is found growing over shrubs in moist thickets on steep slopes in pine-oak forest, sometimes with spruce and with an understory of composites, *Lamiaceae*, and lupines. Soils are deep organic loams derived from volcanic or limestone rock.

Diseases and pests.—Rust has been reported and *Apion* weevils, thrips and *Diabrotica* beetles are the main insect pests.

Common names.—"Frijol de venado" and "frijolillo" are given by collectors.

Comments.—The varietal epithet refers to the long strigose pubescence on the inflorescence and pedicels. Usually the inflorescence in this variety is relatively short, few-flowered and delicate in comparison to most varieties of this species. There are relatively few collections in the herbaria, and they are often variously and incorrectly identified because of resemblances of the short, glabrous pod (*P. lunatus*), foliage (*P. macrolepis*) and flower (var. *coccineus*) of these taxa.

C.1.11.—Phaseolus coccineus L. subsp. **coccineus** var. **argenteus** Freytag, var. nov. (**Figs. 22, 24**). TYPE: MÉXICO. OAXACA: 14.1 km N of Teotitlán del Valle on the road to Benito Juárez, (18°10'N, 97°3'W), 2100 m, 20 Oct 1985, *Bartholomew et al. 3213* (HOLOTYPE: CAS 755360).

Similis *Phaseolo coccineo* var. *coccineo* sed brevis prostratis volubilisque, foliolis ad marginibus leviter undulatis et aspectu argenteo differt. Crecit unica orientalis Tehuacanensis Oaxacanae.

Plant apparently prostrate, indeterminate vines, to 3 m long. ***Root*** unknown. ***Stems*** 1 mm thick, nearly glabrous, sparsely covered by reflexed, white pilose and minute uncinate hairs; internodes 8-11 cm long; stipules broadly oblong, 4 mm long, 2 mm wide, strongly 6- to 8-nerved, acute, glabrous. ***Leaves*** medium; petiole 2-4 cm long, delicate, canaliculate, sparsely covered by appressed-pilose and minute uncinate hairs; petiolule 1-2 cm long; pulvini 3 mm long, densely covered by pilose and strigose hairs; stipels linear, the lower 3 mm long, the upper 1.5 mm long, acute, strongly 1-nerved, glabrous; terminal leaflet nearly triangular, elongate, with undulate margins, often truncate at base, 5-7 cm long, 2.5-3 cm wide near base, obscurely veined, apiculate to 1 mm long, nearly glabrous but puberulent along veins, silvery abaxially and dark green adaxially, with glandular-pustulate hairs on both surfaces, ciliate on margins. ***Inflorescence*** short, erect; peduncle 5-10 cm long, striate, puberulent; rachis short, 1-2 cm long, covered with white uncinate hairs; primary bracts elliptic, 4 mm long, 2 mm wide, 3- to 5-nerved, acute, glabrous; pedicel 18-20 mm long, delicate, puberulent. ***Bracteoles*** broadly oblong, 7 mm long, 5-6 mm wide, multi-veined, obtuse to acute, glabrous, persistent to young pod. ***Flower*** reddish-orange; calyx flaring campanulate, the tube 3 mm long, the upper 2 lobes united into one scarcely elongate, emarginate, lower 3 subequal, dentate, 1.5 mm long, 3 mm wide, center lobe strigose; standard reddish, the blade broadly rounded, 6 mm from base to flexure and 9-10 mm more to scarcely emarginate tip, 15-16 mm wide, not enrolled laterally, veined, the claw scarcely developed, 0.5 mm long, the auricles less than 0.5 mm long; wings reddish, the blade spreading, round and cupped, 11-12 mm long, 10-11 mm wide, nerved, the claw 5 mm long, 0.5 mm wide, the spur 1.5 mm diam.; keel, the claws 3 mm long, 4 mm more to bend and 5 mm more to base of the terminal 1 3/4 coils of 3.5 mm diam., the ridges well-developed, 2-3 mm diam. and firmly affixed to wings; vexillary stamen, the claw 1.5 mm long with a well-developed geniculate flap, 1.5 mm long, 1.5 mm wide, and 4

FIG. 22. Illustrations of *Phaseolus coccineus* L. subsp. *coccineus* var. *argenteus* Freytag.—c. Portion of plant with mature leaves and inflorescences; note prostrate stem, wavy undulate leaf margins of elongate leaves, short, erect inflorescences with capitate group of flowers at apex, large bracteoles and heavily veined flowers. Drawings from type field collection *Bartholomew et al. 3213* from near Teotitlán del Valle, Oaxaca, México.

mm more to end of thickened portion; stamen tube 9 mm long to bend and 4–5 mm more to end of the united portion, 3 mm broad, the ridges not developed; basal collar denticulate, 2 mm long, 2 mm wide; ovary straight, 6 mm long, 1 mm wide, densely covered with white canescent hairs, 5–6 ovules; style 11 mm to the terminal thickened coil of 2.5 mm diam; stigma terminal, oblique, extrorse, 1 mm long. ***Pod*** unknown. ***Seed*** unknown. ***Seedling*** unknown.

Habitat.—This taxon is a wiry vine growing in pine-oak forest at elevations around 2100 m.

Comments.—The varietal name refers to the lustrous silvery appearance of the under surfaces of the foliage. The prostrate habit of the vine is also quite unusual and perhaps indicative of a habitat lacking shrubs or weeds on which to climb. These two characteristics suggest a very dry and perhaps grassland habitat just east of Tehuacán, Puebla. This taxon must be very limited in distribution since only one specimen has been examined.

C.1.12.—Phaseolus coccineus L. subsp. **coccineus** var. **zongolicensis** Freytag, var. nov. (**Figs. 23, 24**).

TYPE: MÉXICO. VERACRUZ: 5 km NW of Zongolica, 20 km S of Orizaba, 1800 m, on very steep escarpments covered by almost pure stands of hackberry (probably *Celtis monoica* Hemsl., "capulín macho"), 18°41'N, 97°5'W, 26 Nov 1981, *Freytag & Sullivan 81-8* (HOLOTYPE: US; ISOTYPES: EAP, GH, MEXU, MO, WIS).

Similis *Phaseolo. coccineo* var. *coccineo* sensu Maréchal et al., sed foliolis terminalibus mediocribus vel majoribus, seminibus crassis et rotundatis differt. Crescit unica var. in silvis densis *Celtidis monoicae* Hemsl. (hackberry, capulín macho) in saltibus valde praeruptis provincinae Veracrucis circa 1800 m.

Plant a climbing, indeterminate vine, to 3 m long. ***Root*** perennial, fusiform, woody, to 0.5 m long, 3–5 cm thick. ***Stems*** terete, 2.5–3–15 mm diameter, moderately striate to ribbed. ***Stipules*** foliaceous, triangular, 4 mm long, 3 mm wide, moderately covered by strigose hairs on abaxial surface. ***Extrafloral nectaries*** on stipules and stipels only slightly active. ***Leaves*** to 30 cm long; petiole 5–13 cm long; petiolule 2.5–3.5 cm long; basal pulvinus 7 mm long; upper pulvini 3.5 mm long, densely covered by hispid hairs; terminal leaflet stipels lanceolate, 3–3.5 mm long, 0.5 mm wide, 1-nerved, glabrous; lateral leaflet stipels lanceolate, 4–5 mm long, 1 mm wide, 2-nerved, glabrous; terminal leaflet broadly ovate to nearly round, to 12 cm long, 8.5 cm wide, widest at 1/4 of length from base, short acuminate,

Fig. 23. Illustrations of *Phaseolus coccineus* L. subsp. *coccineus* var. *zongolicensis* Freytag.–**a.** Plantlet several weeks after germination of seed.–**b.** Root after several years of growth.–**c.** Portion of plant with mature leaf and inflorescence.–**d.** Flowers, showing lateral and front views.–**e.** Exploded view of flower showing all parts including–**s.** Stigma as seen under the microscope.–**f.** Pods, side view and dehisced.–**g.** Seeds, side view and view from hilum. All drawings made from living plants grown in environmental chambers and greenhouses at Mayagüez, PR from seed of type collection *Freytag et al. 81-8* (TARS #32) from near Zongolica, Veracruz, México except b. drawn from a photograph taken at type locality, and f. & g. which were drawn from the type field collection.

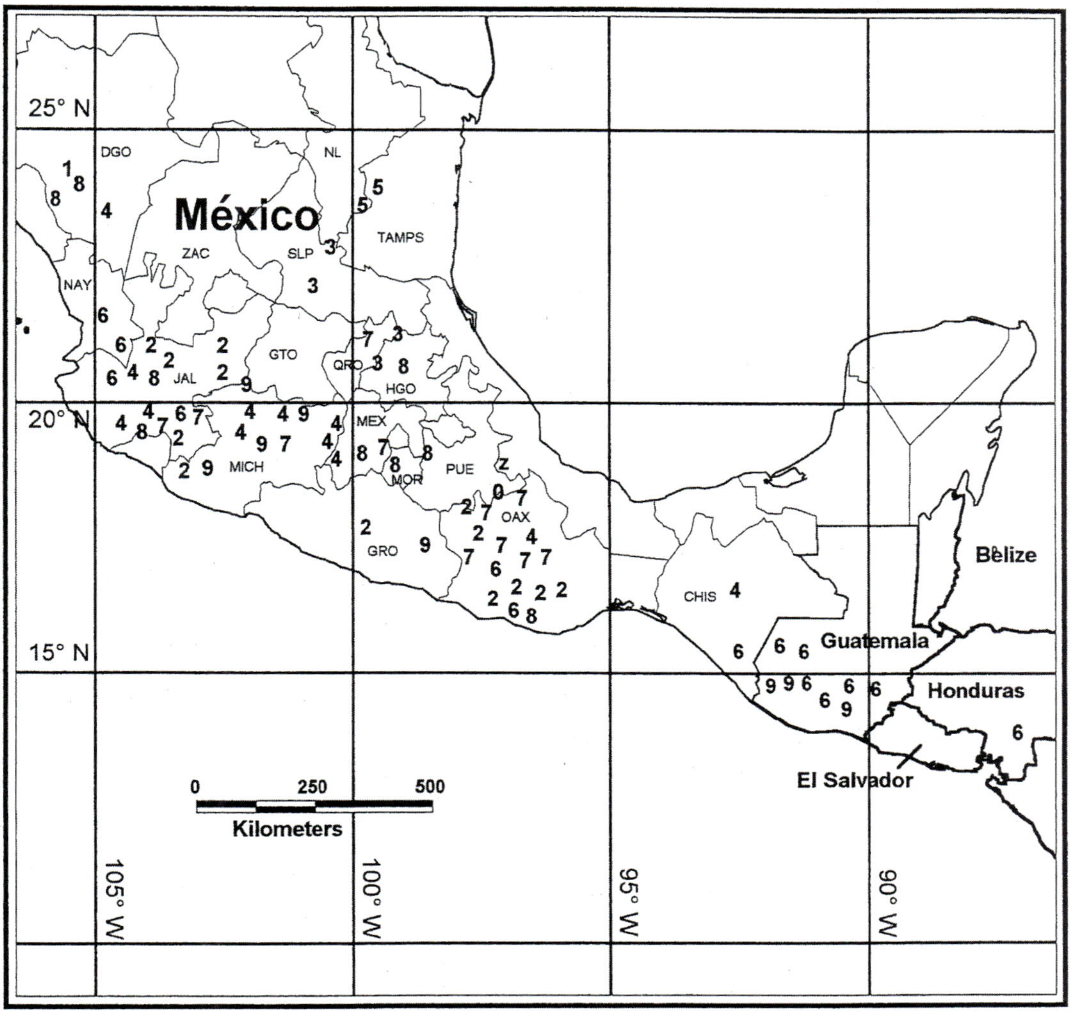

FIG. 24. Distribution of minor varieties of Section C. *Coccinei* species *Phaseolus coccineus* L. subsp. *coccineus*, as follows: 1 = *parvibracteolatus*; 2 = *griseus*; 3 = *lineatibracteolatus*; 4 = *tridentatus*; 5 = *splendens*; 6 = *strigillosus*; 7 = *semperbracteolatus*; 8 = *condensatus*; 9 = *pubescens*; 0 = *argenteus*; z = *zongolicensis*.

apiculate, thin membranous, the abaxial surface slightly covered by white strigose hairs mostly on veins, becoming glabrescent except on the veins at maturity, slightly webbed between main veins at base; yellowish-green on both surfaces, the abaxoal surface a lighter color; lateral leaflets similar except inequilateral and to 10.5 cm long, 7.5 cm wide, widest at 1/5 of length from base. ***Inflorescence*** a slightly curved, erect raceme, exserted above foliage, the peduncle 10–20 cm long, the rachis 17–20 cm long, slightly covered by strigose hairs and densely covered by hooked pubescence, strongly ribbed, with usually 7–15 flowering nodes; primary bracts 5–7 mm long, 2–2.5 mm wide, lanceolate, 8- to 16-nerved; pedicels 11–15 mm long at pod stage, slender, densely covered with strigose and long hooked hairs. ***Bracteoles*** oblongate, 5 mm long, 2.5 mm wide, faintly 6- to 8-nerved, nearly glabrous, early caducous. ***Flower*** large, scarlet, drying purplish; calyx campanulate, 10 mm long, densely covered with glandulose and strigose hairs, with ciliate margins, the lobes acute, the lower about 1.5 mm long; standard scarlet, nearly round, large and broad, 12–13 mm long, 10 mm wide, the basal claw lacking, reflexed near base, the basal portion 5 mm to bend, the upper portion 8 mm long, erect; wings scarlet, the blade rounded-spatulate, 15–18 mm long, spreading, the basal claw 4 mm long, 0.5 mm wide, the spur indistinct, 2 mm wide, the blade 9 mm long, 6–7 mm wide; keel 9 mm long to the base of the terminal 1 1/2 coils, reddish, the basal claws 1.5 mm long, the lateral knobs distinct, the tip dark scarlet; vexillary stamen with a basal lip 1 mm long; stamen tube slightly inflated, 6 mm long to the bend and 3 mm more to the divided filaments; anthers 0.7 mm long; ovary 5 mm long, 0.75 mm wide, very densely covered with white strigose hairs, soon turning yellowish or brownish, 6 or less ovules; basal collar 1.5–1.75 mm long, denticulate; style 8.75 mm long to base of the terminal, thickened, single complete coil of 2.25 mm diameter; stigma terminal, extrorsely oblique, 0.75 mm long. ***Pod*** straight and broad, 6.5–10 cm long, 1–1.7 cm wide, 0.55 cm thick, with heavy sutures and glandular knobs on sutures and valves, moderately covered by small strigose and smaller hooked hairs, the beak large, 8–10 mm long, strong and slightly curved; plants generally very fertile, setting 1–4 pods/inflorescence, usually 1 per node. ***Seed*** nearly spherical, 9.6–11.3 mm long, 7.2–9.9 mm wide, 4.7 mm thick, smooth and shiny, solid black; hilum oblong, 2.25–2.5 mm long, 1.15 mm wide; lens pronounced. ***Seedling*** from hypogeal germination; epicotyl 6–8.2 cm long; primary leaf petiole 4–5 cm long; eophyll blade broadly ovate, 5–9 cm long, 4.7–8.5 cm wide, widest at 1/7–1/5 of length from base, the base auriculate with lobes 1–1.5 cm long, acuminate, obtusely rounded but not apiculate, membranous, the adaxial surface nearly glabrous, minutely glandular-puberulent, also on veins, faintly variegated light green along midvein.

Habitat.—This variety was found on the forest floor in the talus at the base of cliffs on very steep escarpments covered by almost pure stands of hackberry (*Celtis*) forest at an elevation of 1800 m.

Comments.—This variety has been grown in the greenhouse at Mayagüez and has grown and flowered better than most varieties of *P. coccineus* under these warm, humid conditions. The large, solid black seed are unusual for wild varieties of this section; however, in the apparent hybrids mentioned below most seed are spotted or striped and variously colored. Along the eastern escarpment there are apparent zones of introgression of this taxon with cultivated types of var. *coccineus* giving rise to escapes and intermediates of all types but with a nearly round seed and predominance of dark colors including black seed. Among these collections are: **MÉXICO. Veracruz:** Paso del Viento, Hwy 150 Km 30 Orizaba-Tehuacán, 18°45'N, 97°10'W, 2120 m, 19 Sep 1978, *Freytag et al. 78-Mex-28*, (BR, EAP, MEXU, MO, UC, US); idem, *Freytag et al. 81-7* (BR, EAP, MEXU, MO, US). No other collections from the Zongolica area of Veracruz have been found in the major herbaria.

C.2.1.—Phaseolus coccineus L. subsp. **striatus** (Brandeg.) Freytag var. **striatus**, comb. & stat. nov. (**Fig. 31**). *Phaseolus striatus* Brandeg., Zoe 5:248. 1908. TYPE: MÉXICO. PUEBLA: Cerro de la Yerba, rocky soil, Jul 1907, *Purpus 2678* (lectotype, designated by Delgado: UC 101647; ISOTYPES: G, UC).

Plant probably scandent, indeterminate vines, to 5 m long. ***Root*** fleshy, perennial. ***Stem*** sparsely pubescent. ***Stipules*** ovate-acuminate, 5- to 6-nerved. ***Leaves*** 7–15 cm long; petiole 3–7.5 cm long, the stipels lanceolate, 1-nerved, covered with strigose hairs; terminal leaflet rhombic to triangular, 4–6 cm long, 2.5–5 cm wide, the base deltoid, long acuminate, apiculate, sparsely covered by strigose hairs adaxially, more densely so abaxially, with minute ciliate hairs mostly on veins; lateral leaflets obliquely inequilateral, deltoid-acuminate-mucronate. ***Inflorescence*** a pseudoraceme slightly surpassing foliage, to 30 cm long; peduncle 4.5–18 cm long; rachis 4–15 cm long of 6–14 flowering nodes, covered by small white hooked pubescence; primary bract long lanceolate, 4.5 mm long, 1 mm wide, 3-nerved, puberulent; pedicel 10–20 mm long, densely covered by short hooked hairs. ***Bracteoles*** broadly elliptical, slightly longer than calyx, 4.5–5 mm long, 3–4.5 mm wide, heavily 8- to 11-nerved, puberulent of minute ciliate hairs especially on margins. ***Flower*** purple; calyx campanulate, upper 2 teeth rounded, the lower 3 teeth subequal, broadly rounded, acute, apiculate, sparsely pubescent; standard purple, 12 mm long, wings purple, the blade orbicular, spreading, about 8 mm long, the claw about 8 mm long; keel of about 1 1/2 close coils; stigma terminal, extrorse. ***Pod*** when young very densely covered with yellowish or brownish strigose or villous pubescence, slightly curved, 6.5 cm long, 1 cm wide, the beak 3 mm long. ***Seed*** unknown. ***Seedling*** unknown.

Specimens examined: **MÉXICO. Michoacán:** 4 mi S of Carapán, on road to Uruapan, (19°50'N, 102°2'W), 27 Aug 1957, *Irwin 1315* (TEX). **Oaxaca**: Cerro del Corral del Piedra (SE of Huahuapan), (17°45'N, 97°45'W), 2438–3048 m, Aug 1909, *Purpus 3890* (UC). **Puebla:** San Luís Tultitlanapa (near San Luís Atolotitlán ?, 18°10'N, 97°25'W, 2000 m)?, Jul 1908, *Purpus 2378a* (BM, F, GH, MO, UC, US); San Luís Tultitlanapa, near Cerro Verde, (18°5'N, 97°15'W)?, Jul 1908, *Purpus 2378b* (UC).

Habitat.—Most of the area of Oaxaca and Puebla, where this variety is native, is covered with mature forests of pine, rather on the dry side, and with a rather open nature with grass and shrubs in the clearings. The area is quite high with accompanying cold and frosts through most of the winter months.

Comments.—No recent collections have been seen, which would indicate that this taxon is rather scarce and thus data is lacking on the exact habitat and soil preference (probably recent volcanic). Nevertheless the distinctly nerved bracteoles and location are greatly different from the only other taxon, var. *purpurascens*, with which it might be confused. It would be valuable to have more collections and living germplasm to confirm this distinction.

C.2.2.—Phaseolus coccineus L. subsp. **striatus** (Brandeg.) Freytag var. **minuticicatricatus** Freytag, var. nov. (**Figs. 25, 31**). TYPE: MÉXICO. DISTRITO FEDERAL: Santa Marta Ixtapalapa, (19°20'N, 99°W), 2340 m, 13 Oct 1975, *Sousa & Delgado 5112* (HOLOTYPE: CAS 615045; ISOTYPE: US).

Similis *Phaseolo coccineo* var. *striato* sed bracteolis caducis precosissimis et cicatrice conspcue parvula alba ad basim calycem differt. Crecit valli orientalis proximis civitat capital Nova Hispania.

Plant a prostrate, indeterminate vine, to 5 m long. ***Root*** unknown. ***Stems*** terete, 1–2 mm thick, striate, farinaceous, covered by appressed-reflexed-pilose hairs; internodes 6–8 cm long. ***Stipules*** spatulate, 4–6 mm long, 1.25–1.5 mm wide, strongly 5- to 8-nerved, acute to obtuse, covered with hispid hairs. ***Leaves*** 9–15 cm long; petioles 4–6 cm long, delicate, canaliculate, striate, sparsely covered with white strigose hairs; petiolule 1–2–2.5 cm long, sparsely covered with white strigose hairs; pulvini, the lower 4–5 mm long, nearly glabrous, upper 3–4 mm long, covered with white pilose and strigose hairs; stipels, the lower lanceolate, 3–4 mm long, strongly 1-nerved, the upper aciculate to linear, 2–2.5 mm long, strongly 1-nerved, glabrous or with a few pilose hairs; terminal leaflet broadly ovate, 3.3–5.9 cm long, 2.7–4.4 cm wide, acuminate, apiculate, green on both surfaces but covered by white strigose and uncinate pubescence; lateral leaflet similar but slightly inequilateral. ***Inflorescence*** an erect pseudoraceme; peduncle 9–11 cm long; rachis 3–10 cm long, covered with reflexed-pilose and strigose hairs; primary bract ovate-acuminate, 4–6 mm long, 2–3 mm wide, 4- to 5-nerved, acuminate, sparsely covered by pilose hairs, somewhat ciliate; pedicel delicate, 10 mm long, purplish, sparsely covered by pilose hairs. ***Bracteoles*** nearly round, 4–5 mm diam., acute, heavily multi-nerved, nearly glabrous, very early caducous leaving a highly visible whitish scar at base of calyx. ***Flower*** purple (lilac); calyx purplish, campanulate, tube 3 mm long, farinaceous, with bracteole scars at base, upper 2 lobes rounded, 1.5 mm long, 3 mm wide, lower 3 lobes subequal, dentate, 1.5 mm long, 2 mm wide, acute, a few pilose hairs on central lobe, ciliate; standard purple, striate, the blade broadly rounded and cupped, 4 mm from base to flexure and 6 mm more to emarginate apex, lateral edges not enrolled, the claw not developed, the auricles lacking; wings purple, the claw 4 mm long, 0.5 mm wide, the spur well-developed, 1.25 mm high, the blade broadly rounded and cupped, 11 mm long, 8 mm wide; keel, the claws 3.5 mm long, 4 mm more to bend and 3 mm more to base of the terminal 1 1/2 coils of 3.5 mm diam., the ridges well-developed, 1 mm diam. and firmly affixed to wings; vexillary stamen stout, the claw 0.75 mm long, the geniculate flap well-developed 1.25 mm long, 1.75 mm wide, 4 mm more to end of the thickened portion; stamen tube very broad and short, 7 mm to bend and 3 mm more to end of united portion, 2.5 mm wide; basal collar denticulate, 1.5 mm long; ovary straight and broad, 5 mm long, 1.5 mm wide, densely covered by long white cannose hairs; style 6 mm long to the terminal thickened coil of 2.5 mm diam.; stigma terminal, oblique, extrorse, 0.75 mm long. ***Pod*** unknown. ***Seed*** unknown. ***Seedling*** unknown.

Habitat.—This variety is found on the eastern edges of the Valley of Mexico, in what is now Mexico city in the area around Santa Marta Ixtapalapa, where the senior author found wild plants still growing in 1981. These plants, with fairly large roots, were surviving in high voltage powerline right-of-ways and other vacant plots in very sandy and dry soils. Most of this area is now covered with small houses packed one next to the other, almost completely covering the land. The only open space is now in the right-of-way under the powerlines which cross near the women's penitentiary and where the land in this area is used as a garbage dump and pasturage for goats and burros.

The senior author had seen this area in 1953 when it was occupied by small farmers who farmed the level spots between rocks and rock outcroppings. There were many stone walls throughout. The

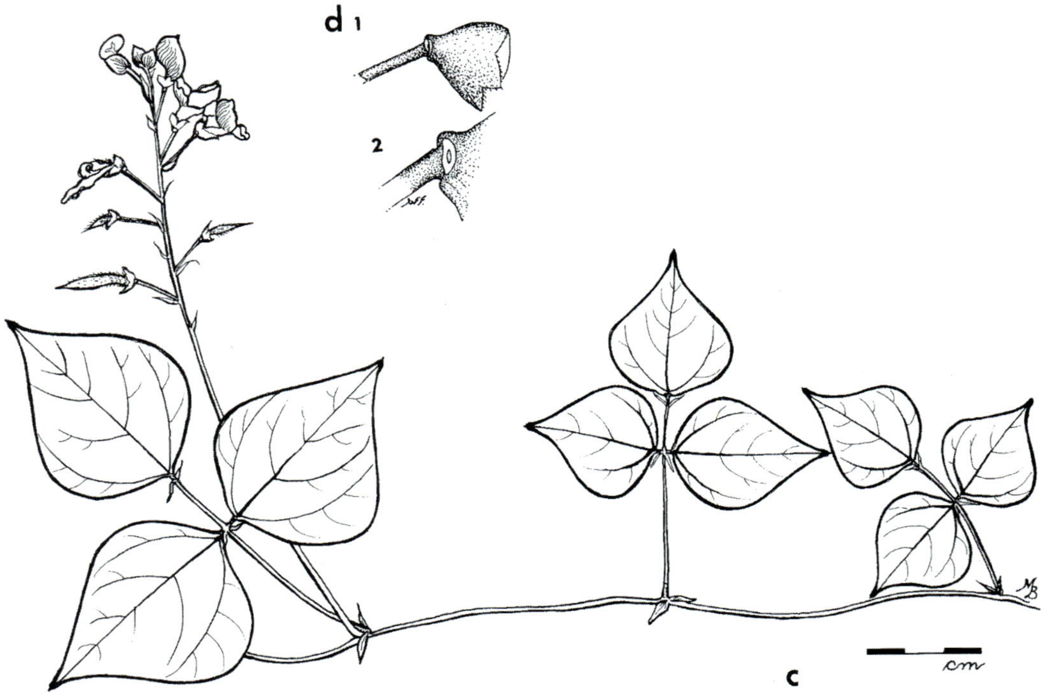

FIG. 25. Illustrations of *Phaseolus coccineus* L. subsp. *striatus* (Brandeg.) Freytag var. *minuticicatricatus* Freytag.–**c.** Portion of plant with mature leaves and inflorescence; note prostrate stem with short internodes, small, rounded-oval, acuminate leaves, short, erect inflorescence, large primary bracts, and large, early dehiscent bracteoles which leave noticeable scars.–**d.1.** Lateral view of the calyx of the flower and pedicel.–**d.2.** Close-up of bracteole scar at base of calyx. All drawings made from the type field collection *Sousa and Delgado 5112* from Santa Marta Ixtapalapa on the eastern city limit of Cd. México, D.F., Mexico.

native vegetation was all but gone with a few scraggley trees and grasses. Wild bean vines covered some of the piles of rock outcroppings and rock fences and there were prostrate vining beans of the *P. coccineus* complex, (presumably of this variety), blooming during the summer and fall months. This area was extremely dry and the soil very sandy with scarcely enough nutrients and moisture for corn to reach flowering. Beans grown in association with corn were generally mixtures of common beans and ayocotes (cultivated *P. coccineus*).

Comments.—This taxon is named for the very prominent scars which remain on the base of the calyx after the caducous bracteoles are shed very early in flowering.

C.2.3.—Phaseolus coccineus L. subsp. **striatus** (Brandeg.) Freytag var. **guatemalensis** Freytag, var. nov. (**Figs. 26, 31**). TYPE: GUATEMALA. HUEHUETENANGO: Los Alisos bridge on way to Huehuetenango, (15°10'N, 91°30'W), 2010 m, 11 Jan 1974, *Molina et al. 30271* (seed and pod) (HOLOTYPE: MO; ISOTYPES: EAP, F).

Certe ad *Phaseolo coccineo* subsp. *striato* pertinenti, sed amplis volubilisque, foliolis grandibus rotundatis tenuibus membranaceis plerumque in superficie olivaceis et infra argenteis differt, et leguminibus parvis illis *Phaseolo lunato* facile confusis. Crecit in montibus altus Centrali Americanae.

Plant a large climbing, indeterminate vine, to 10 m long. ***Root*** unknown. ***Stems*** terete, 2 mm thick, nearly glabrous; internodes to 15 cm long. ***Stipules*** ligulate, 5 mm long, 2 mm wide, 5-nerved, sparsely ciliate on margins. ***Leaves*** 23–27.5 cm long; petiole 9–12 cm long, delicate, sparsely covered with yellowish-strigose hairs; petiolule 3.5–4 cm long; stipels linear, 4–5 mm long, 1-nerved, glabrous; pulvini 4 mm long, densely covered by long, white hispid hairs; terminal leaflet broadly rounded ovate, 10–10.5 cm long, 8–9 cm wide, widest at midpoint, acuminate, apiculate, thin membranous, nearly glabrous, sparsely covered with long, white hispid hairs adaxially especially along veins, somewhat farinaceous-glandular and with a whitish cast abaxially contrasting sharply with the ultimate dark reticulate veinlets; lateral leaflets similar but slightly inequilateral. ***Inflorescence*** a long, erect, few-

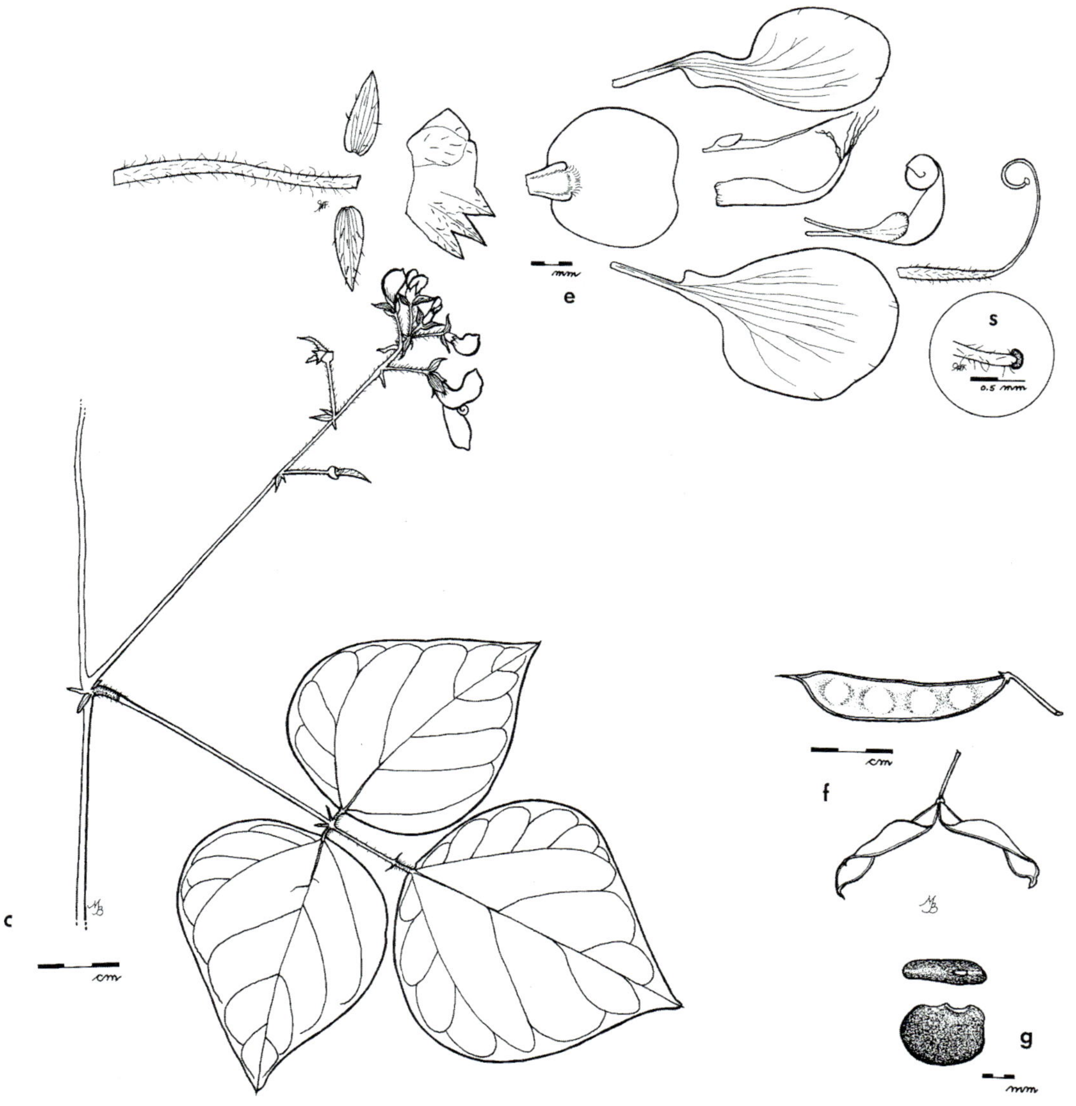

Fig. 26. Illustrations of *Phaseolus coccineus* L. subsp. *striatus* (Brandeg.) Freytag var. *guatemalensis* Freytag.—c. Portion of plant with mature leaf and inflorescence; note the very broad leaf shape and the inflorescence which is quite long but few-flowered.—e. Exploded view of the flower showing all the parts including—s. Stigma as seen under the microscope; note the very large bracteoles and large, rounded standard and wings.—f. Pods, side view and dehisced.—g. Seeds, note rounded and flat shape. All drawings from the field collection *Williams et al. 29221* from near Matagalpa, Nicaragua.

flowered pseudoraceme; peduncle to 24 cm long; rachis 7–9 cm long of 5–7 nodes, sparsely covered by strigose pubescence; primary bract ovate-acuminate, 6–8 mm long, 2.5 mm wide, 3-nerved, sparsely covered with strigose hairs; pedicel 20–22 mm long, stout, covered with strigose hairs. ***Bracteoles*** oblong-lanceolate, 6–8 mm long, 3.5 mm wide, distinctly 6- to 7-nerved, acute, sparsely covered with strigose hairs to glabrous. ***Flower*** large, purple to lilac, drying and fading to cream or yellow; calyx flaring oblique campanulate, the 2 upper lobes somewhat rounded-dentate, acute, the lower 3 subequal, 2 mm long, 1.25 mm wide, acute, covered by strigose hairs; standard broadly ovate, 11 mm long, 10 mm wide, reflexed at 4 mm from base, enrolled at margin, the claw obscure, the auricles poorly developed, 0.5 mm long; wings, the blade broadly oblong, 23 mm long, 15 mm wide, cupped, spreading, the claw 5 mm long, the spur well-developed, 0.5 mm long, firmly affixed to keel; keel short, claws 5 mm long, 3 mm to bend and 5 mm more to the terminal 1 1/2 coil of 3 mm diam., the ridges well-developed, 1.5 mm high; vexillary stamen, the claw 1 mm long, the geniculate knob prolonged into a flap 1.5 mm long, 2 mm wide, the thickened portion 6 mm more to filament; stamen tube broad only at base, 8 mm long to bend and 3 mm more to filaments, the ridges scarcely developed; basal collar, 1.5 mm long; ovary straight, 7 mm long, 1 mm wide, densely long, white pubescent, 4–6 ovules; style with thickened terminal coil of 2.5 mm diam.; stigma terminal capitate. ***Pod*** falcate to nearly straight, 4.5–6.5 cm long, 1–1.2 cm wide, farinaceous when immature, when mature compressed, fibrous brittle, very sparsely covered by strigose and minute uncinate hairs, loosely twisted 1–2 turns at dehiscence; beak short to long, 3–8 mm long, straight to recurved, stout; 2–5 seed. ***Seed*** orbicular to rounded, flattened, 8–9 mm long, 6.8–7.4 mm wide, 3–3.4 mm thick, pinto black on reddish brown background, shiny, darkened ring around hilum; hilum orbicular 1.5 mm long, 0.75 mm wide; lens pronounced, raised and divided. ***Seedling*** unknown.

PARATYPES. **GUATEMALA. Chimaltenango:** 3 km E of Chocoyos, (14°35'N, 90°45'W), 2100 m, 5 Dec 1963, *Williams et al. 25107* (EAP, F, GH, US). **Huehuetenango:** Aguacatán road E of Huehuetenango at Km 13–14, (15°20'N, 91°25'W), 1950 m, 2 Jan 1941, *Standley 81974* (F); Río Chixoy near Malacatancito about 20 km SW of Huehuetenango, 1600 m, 1 Dec 1962, *Williams et al. 22155* (EAP, F, US).

NICARAGUA. Jinotega: Ocotillo near Sta. Lastemia, Cordillera Central, (12°55'N, 86°W), 1550 m, 17 Jan 1965, *Williams et al. 28714* (F). **Matagalpa:** Sta. María de Ostuma, Cordillera Central, (12°45'N, 86°W), 1300–1500 m, 8–15 Jan 1963, *Williams et al. 23420* (CAS-DS, F, UC); Forest on Mt. near Xelajú, Cordillera Central, 13°2'N, 85°55'W, 1500 m, 12 Feb 1965, *Williams et al. 29221* (flowers) (EAP, F, GH).

Habitat.—This variety is found growing as vines in thickets in high mountain cloud-forests, or along streams in dense pine-oak forests at high altitudes.

Comments.—The varietal epithet gives honor to the country of Guatemala where the first herbarium specimens seen by the senior author were collected. The apparent disjunct distribution of this taxon may simply be the result of spotty collecting since it most certainly is found in the high mountain cloud forests of Honduras as well. It must be very scarce as few specimens have been collected and it has never been seen in the field by the authors even though we have collected in Santa María de Ostuma and in the cloud forests of Honduras.

C.2.4.—Phaseolus coccineus L. subsp. **striatus** (Brandeg.) Freytag var. **purpurascens** Freytag, var. nov. (**Figs. 27, 31**). TYPE: MÉXICO. MORELOS: Km 44.9 of Hwy 95 México to Cuernavaca, near Tres Cumbres (Tres Marías), in open pine savannahs, 2700 m, 19°4'N, 99°14'W, 15 Sep 1978, *Freytag & Vakili 78-Mex-5* (HOLOTYPE: US(2); ISOTYPES: ARIZ, BR, CSU, EAP, F, GH, IBUG, K, MEXU, MICH, MO, NA, TEX, UC, UC-R, WIS, WY).

Similis *Phaseolo coccineo* var. *striato* sed statura prostratis, radice magna claviformibus, bracteolis non conspicue nervatis, floribusque purpurascentibus pallidis differt. Crecit montibus altus proximis civitat capital Nova Hispania.

Plant a perennial, prostrate, indeterminate vine, 2–5 m long. ***Root*** perennial, fleshy, claviform, to 30 cm long, often with a crown to 10 cm diameter. ***Stems*** terete, about 2 mm thick, branched several times mostly near base; internodes generally short, 6–7 cm long, the upper to 12 cm long, densely covered by reflexed-strigose hairs on the adaxial surface, sparsely so on older stems, the hairs to 0.75 mm long. ***Stipules*** foliaceous, lanceolate, acute, 4 mm long, 1.5 mm wide, obscurely 6-nerved, reflexed, purplish, moderately pubescent on abaxial surface. ***Extrafloral nectaries*** on stipules and stipels, inactive. ***Leaves*** 11.6–18.5 cm long; petiole 5–7 cm long; petiolules 1.5–3 cm long; basal pulvinus 4 mm long, lateral pulvini 4–5 mm long, terminal pulvinus 3 mm long; terminal leaflet stipels, 1.5 mm long, 0.5 mm wide, 1-nerved, nearly glabrous; lateral leaflet stipels, 2.75 mm long, 0.5 mm wide, 1-nerved, pubescent; terminal leaflet broadly ovate-acuminate to somewhat rhomboid, mostly 4.6–(7.5) cm long, 3.4 cm wide with widest portion at 1/3 of length from base, the base nearly truncate, the apex acute to slightly acuminate, apiculate, the venation moderate; adaxial surface generally densely covered by strigose hairs, the abaxial surface with more and many smaller scattered hooked hairs and minute glandular scales especially near veins, the older leaves sometimes nearly glabrous, somewhat coriaceous, dark green on both surfaces; lateral leaflets similar, but inequilateral. ***Inflorescence*** an erect raceme exserted above foliage, with sometimes 2 at a node, the peduncle 12–15–30 cm long, the rachis 4–5–15 cm long, usually 6- to 12-flowered, densely reflexed-strigose and hooked pubescent; primary bracts lanceolate, 5 mm long, 1–1.5 mm wide, 6-nerved, densely covered by strigose hairs, purple; secondary bracts broadly ovate, 3.5 mm long, 1–1.5 mm wide; pedicels generally 11–15–

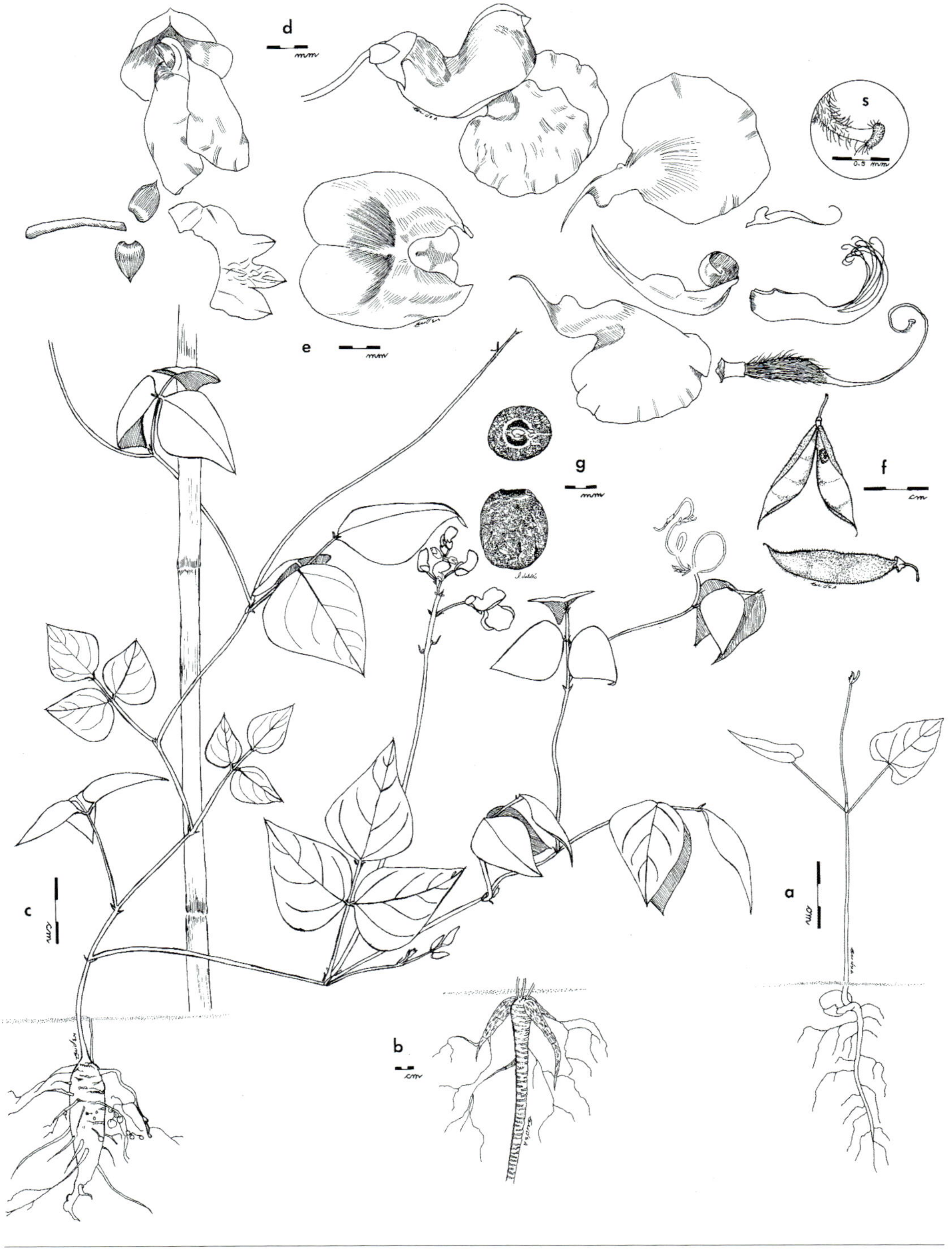

FIG. 27. Illustrations of *Phaseolus coccineus* L. subsp. *striatus* (Brandeg.) Freytag var. *purpurascens* Freytag.—**a.** Plantlet several days after germination of seed.—**b.** Root after 2 years of growth.—**c.** Portion of plant with root after 1 year of growth, mature leaves, inflorescence, and small lateral branch; note the prostrate nature of growth habit.—**d.** Flowers, showing lateral and front views; note the large, showy, rounded wings.—**e.** Exploded view of flower showing all parts including—**s.** Style tip and stigma as seen under the microscope; note the capitate shape of the stigma.—**f.** Pods, lateral view and dehisced.—**g.** Seeds, lateral view and view from hilum; note the thick and trapezoidal shape. All drawings made from living plants grown in environmental chambers at Mayagüez, PR from seed of the type field collection *Freytag et al. 78-Mex-5* from near Tres Cumbres, Morelos, México.

21 mm long, densely pubescent almost entirely of hooked hairs. ***Bracteoles*** broadly ovate, 3–5 mm long, 2–3.5 mm wide, moderately 6-nerved, acute, moderately to densely covered by strigose hairs, caducous after anthesis, often purple. ***Flower*** light to dark purple when fresh; calyx campanulate, 4–5 mm long, the 2 upper lobes rounded, emarginate, 0.75 mm long, 2.5 mm wide, the lower central lobe acute, 1.5 mm long, 3 mm wide, the lateral lobes 1 mm long, 1.25 mm wide, moderately covered by strigose hairs abaxially; standard nearly round, light to dark purple, 12 mm long, 10–12 mm wide, reflexed at middle, the upper portion hooded; basal nectaries copiously secreting; wings, the blade nearly round, light to dark purple, somewhat clasping, 22 mm long, 9–10 mm wide, cupped, the basal claw about 8 mm long, the spur pronounced, about 2.5 mm long and wide; keel 13 mm long to base of the terminal 1 1/2 coils, purplish, the tip whitish, the basal claws 2.5 mm long, the lateral knobs indistinct; vexillary stamen about 30 mm long, the basal knob 2 mm long, 1.5 mm wide; stamen tube 15 mm long, the base somewhat enlarged; anthers 0.75 mm long; 4–15 cm long; ovary 7 mm long, 1.5 mm wide, densely covered with strigose hairs, ovules 6–8; basal collar 1.25 mm long, denticulate; style 8 mm long to base of the terminal, thickened, single complete coil of 2.5 mm diameter; stigma terminal, capitate. ***Pod*** straight to slightly curved, 5.2–5.8–7.5 cm long, 0.8–1 cm wide, 0.5–0.7 cm thick; carpels often wrinkled or ridged, fleshy when green, very densely covered by long strigose and bristly, minute uncinate hairs, many minute glandular scales, especially on young and maturing pods, very fibrous at maturity, slightly striped or purplish turning brown and fading tan, with fairly pronounced sutures, moderately twisted at dehiscence, the beak, stiff, slightly to strongly curved, 3–4 mm long; plants generally very fertile, setting 2 pods per node at times and 3–4 pods per inflorescence, usually with 4–5 seeds per pod. ***Seed*** oblongoid to ovoid, 5–7.8 mm long, 5.1–6.4 mm wide, 3.2–4.3 mm thick, with slightly flattened or squarish ends, smooth and shiny, the sides speckled black and brown on lighter brown, turning darker with age, a pronounced black ring around hilum; hilum broadly orbicular to roundish, 1 mm long, 0.75 mm wide. ***Seedling*** from hypogeal germination; epicotyl 6–10.2 cm long; primary leaf petiole 2.6–3 cm long; blade ovate-triangular, 2.6–3.5 cm long, 1.7–3 cm wide, widest at 1/5 of length from base, the base nearly truncate to slightly auriculate with lobes 2–3 mm long, the apex acuminate, obtusely rounded but not apiculate, coriaceous, the adaxial surface of blade strongly hispid, hispid-ciliate on the margins, faintly light-green variegated along midvein, the abaxial surface nearly glabrous.

PARATYPES. **MÉXICO. Distrito Federal:** Cerro Pelado, 2 km N and NW of La Cima RR Station, 3050–3300 m, 13 Jul 1960, *Iltis et al. 189* (MICH, WIS); Desierto de los Leones, (19°19'N, 99°18'W), 10 Jul 1938, *Kenoyer A73* (F); 4 km E of Ajusco, Tlalpan, 2750 m, 27 Jul 1961, *Salinas 45* (WIS). **Hidalgo:** 1.5 km SSW of Tezoantla, Real del Monte, 2950 m, *Medina 477* (IBUG); Mesas near Métepec Station, (20°10'N, 98°10'W), 2530 m, 27 Jun 1904, *Pringle 11968* (BRIT, CAS, F, GH, K, MICH, MO, SI, US). **México:** Mpio. Chalco, Colonia Avila Camacho, (19°15'N, 98°55'W), 2750 m, 11 Nov 1966, *Chávez 47* (CAS-DS(2), F, IBUG, MO, TEX, TEX-LL, US, WIS); Km 45, Hwy 136, Texcoco–Apitzaco–Tlaxcala, at the "summit" of pass going E from Texcoco, 19°33'N, 98°39'W, 2670 m, 16 Sep 1978, *Freytag et al. 78-Mex-8* (BR, EAP, F, GH, K, MEXU, MO, UC, US); Mpio. Amecameca, 1 km al E de San Antonio, 2500 m, 27 Sep 1981, *García 144* (CAS); Chapingo, (19°25'N, 99°55'W), 2240 m, Oct 1924, *Gómez s.n.* (US(3); Rancho San Luís Aculco, Tenango del Aire, (19°10'N, 99°35'W), 2300 m, 14 Nov 1980, *Hinton et al. 18026* (TEX); Amecameca, 20 Oct 1903, *Holway 5188* (US); Amecameca, 2590 m, 10 Oct 1906, *Johnson s.n.* (US); Cerros de Toluca, (19°0'N, 99°45'W), May 1945, *Martínez s.n.* (US); Ixtaccíhuatl, Mar–Jul 1903, *Purpus 311* (US); Mpio. Iztapaluca, Cañada Zaltomatla, SE of Chapingo, 2600 m, 11 Aug 1984, *Randolph 281* (UCR); on Popocatépetl, (19°5'N, 98°40'W), 22 Aug 1901, *Rose et al. 6258* (US); Toluca, 4 Sep 1903, *Rose et al. 6789* (US); Tultenango, 13 Oct 1903, *Rose et al. 7834* (GH, US); 1 km E of San Antonio, Amecameca, 2500 m, *Solís 2V-38* (IBUG); Mpio. de Chalco, Tlacoyote, 2800 m, 21 Aug 1977, *Ventura 3008* (CAS(2), MICH). **Morelos:** 45 km S of Cd. México, 3048 m, 10 Jul 1940, *Hitchcock et al. 7046* (CAS-DS, NA, UC, US); Km 44.5 de la carr. Cuernavaca libre, a 6.8 km de Tres Cumbres, 2860 m, 13 Oct 1975, *Sousa et al. 5113* (CAS, US). **Puebla:** Río Otlati, 72 km SE of Mexico City, 2652 m, 13 Aug 1942, *Weaver 953* (GH, NA, TEX); 5 km al E de Río Frío, orilla de la super-Carr. a Puebla, (19°15'N, 98°40'W), 2800 m, 5 Jul 1970, *Weber 435* (CAS-DS). **Tlaxcala:** Hills at Contadero Station, (19°28'N, 98°30'W), 2590 m, 30 Aug 1901, *Pringle 8606* (F, G, GH, K, MO, UC, US(2)).

Habitat.—The senior author has collected this taxon in the mountains S of Mexico City, along the side of the old road to Cuernavaca near Tres Cumbres (Tres Marias) where it is found scattered in open rolling hills of pine-oak interspersed with grasslands (*Festuca–Muhlenbergia–Stipa ichu*, tussock grass, 'zacatonal') and some Maguey, mostly on S-facing slopes, often under old, large pines and among bunch grasses (see Color Plate V, photo 52). A population of some 50–100 plants about 2–5 m apart was also found on almost flat land E of Mexico City near the summit on the road to Tlaxcala and the seed population sample shows a fair amount of variability (see Color Plate IV, photo 43).

Most are small scrambling vines to 2 m long, often growing in and on low shrubs (*Eleagnus*?) and form large parsnip shaped roots (see Color Plate III, photo 28). Other collectors have found it along stream banks in the mountains E of Mexico City on the road to Puebla at elevations from 2200–3300 m. The soils are relatively recent volcanic lava rock mixed with deep sandy loam soil.

Diseases and pests.—Angular leafspot, bacterial blight and anthracnose disease, though spotty, are reported. Black aphids are found sometimes on raceme tips.

Genetics.—Apparent hybrids between the type var. of *coccineus* and this variety were collected by: MÉXICO. Mexico: above Texcoco at 2600 m, *Freytag et al. 78-Mex-7*; and on the SE slope of Cerro Pelado, 2 km NNW of La Cima RR station at 3050–3300 m, *Iltis et al. 189*. This elevation is normal for the variety. Hybrids have also been collected at a quarry S of Cd. México on the old Highway to Cuernavaca, 2570 m, *Freytag et al. 78-Mex-4*. None of these specimens were used in forming the description and distribution.

Comments.—This variety is named for the purple flower color which is also typical of this subspecies group (see Color Plate I, photo 8). The plant resembles the type var. *coccineus* in many ways with the most striking difference being the flower color. Additionally, around Cd. México these two varieties are separated by altitude preferences with the purple-flowered varieties at the highest elevations. It is surprising that there are so few collections in the herbaria since this taxon is apparently found on all the high mountain passes around Mexico City where it would be easily collected. However, it is often obscured by the bunch grass and is found in spots rather than uniformly distributed. USDA collectors (H. S. Gentry) had not been able to locate it in the Tres Cumbres area until the senior author personally showed them the precise location on the old highway to Cuernavaca where it had been collected.

C.2.5.—Phaseolus coccineus L. subsp. **striatus** (Brandeg.) Freytag var. **rigidicaulis** Freytag, var. nov. (**Figs. 28, 31**). TYPE: MÉXICO. DISTRITO FEDERAL: Cañada de Contreras, (19°18'N, 99°15'W), 2700–2900 m, May 1954, *Gold 565* (HOLOTYPE: US 2791162; ISOTYPES: CAS, NA).

Similis *Phaseolo coccineo* var. *striato* in florem structuram, sed volubilis prostratis, strictis, et foliolis parvis ovatis obtusis illis *Phaseolo. maculato* similis et inflorescentis longis erectis in quoque nodo floribus brevipedicellatis ad apicem aggregatum et florentem in abril differt. Crecit in montibus proximis civitat capital Nova Hispania.

Plant large, prostrate, indeterminate vines, to 5 m long. ***Root*** unknown. ***Stems*** terete, 2–2.5 mm diam.; internodes straight, short and stout, striate, densely covered by short reflexed-hispid and minute uncinate hairs, 6–9 cm long. ***Stipules*** ligulate to triangular-lanceolate, 4–4.5 mm long, 1.25–1.5 mm wide, obtuse, obscurely 7- to 9-nerved, glabrous on adaxial surface, densely pubescent of uncinate hairs on abaxial surface. ***Leaves*** 7.5–10 cm long; petioles about same length as leaflets, 3–3.5 cm long, delicate, sparsely covered with white strigose hairs and densely covered with minute uncinate pubescence; petiolules 1 cm long; pulvini about 3 mm long, very densely covered with white hispid and minute uncinate hairs, especially on adaxial surface; stipels narrow spatulate, 2–2.5 mm long, 1- to 3-nerved, puberulent; terminal leaflet ovate to rhomboidal, 3–4 cm long, 3 cm wide at about 1/3 from base, obtuse to retuse, minutely apiculate, covered with appressed white hispid and strigose hairs adaxially, densely so abaxially, veinlets rusty brown, adaxial surface an olive green, lighter abaxially; lateral leaflets inequilateral, otherwise similar. ***Inflorescence*** stout, at every leaf node, erect and exserted much above foliage; peduncle 17–20 cm long, strongly striate, covered with minute uncinate hairs; rachis 4–7 cm long, densely covered with uncinate hairs; primary bracts ovate-lanceolate, 3–4 mm long, 1–1.3 mm wide, obscurely veined, covered by strigose hairs on abaxial surface, glabrous adaxially; secondary bracts about same size but much broader, caducous; pedicel 7–8 mm long, densely covered by white uncinate hairs. ***Bracteoles*** nearly round, somewhat larger than calyx, 5–6 mm long, 4–5 mm wide, retuse, obscurely 10- to 11-striate, densely covered by yellowish strigose and uncinate hairs on abaxial surface, glabrous adaxially, persistent through anthesis. ***Flower*** purple; calyx campanulate, tube 2.5 mm long, densely covered by white strigose hairs on both surfaces, puberulent on sides, upper lobes rounded, 0.25 mm long, united into 1 emarginate, ciliate, the lower 3 lobes dentate, acute, subequal, 1.5 mm long, 1.4–2 mm wide, ciliate, the center lobe long strigose bearded; standard purple, the claw large, not well defined, 2 mm wide at base, 6 mm wide at top, the

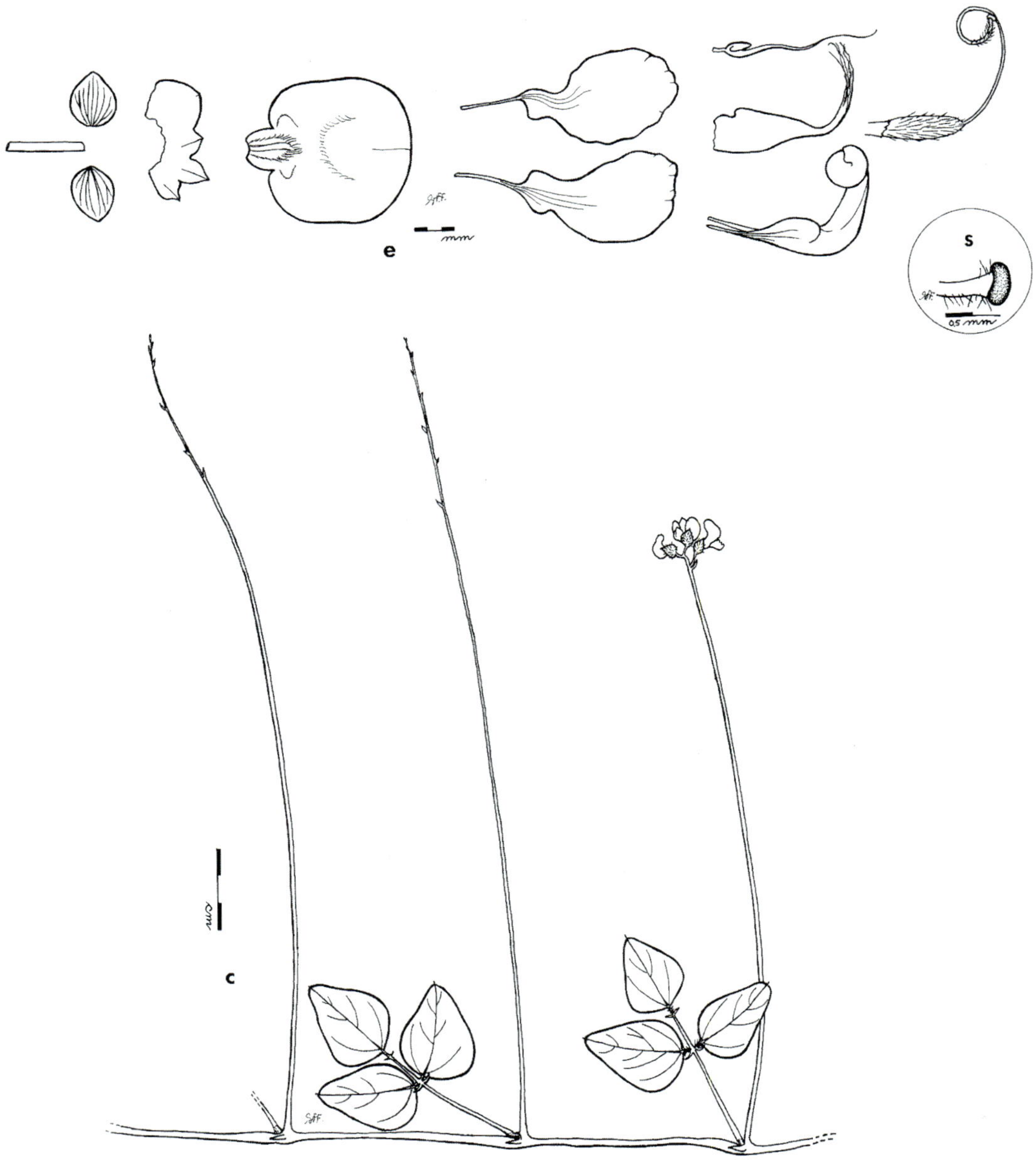

Fig. 28. Illustrations of *Phaseolus coccineus* L. subsp. *striatus* (Brandeg.) Freytag var. *rigidicaulis* Freytag.–c. Portion of plant with mature leaves and inflorescences; note straight, stout, prostrate stem with short internodes, small, triangular-rounded leaves, and long, erect inflorescences with long peduncles and short racemes.–e. Exploded view of the flower showing all parts including–s. Stigma as seen under microscope; notice short pedicel, rounded-ovate bracteoles, short, squarish standard, and short, broad stamen tube. All drawings made from the type field collection *Gold 565* from Cañada de Contreras, D.F., México.

blade rounded-oblong, 5 mm long from base to flexure and 9 mm more to apex, 10 mm wide, edges enrolled, the auricles small flaps 1 mm wide, 0.5 mm long; wings purple, the blade rounded and cupped, unequal, 15 mm long, 9 mm wide, the claws 4 mm long, 0.5 mm wide, the spur well-developed, 1.5 mm high, firmly attached to keel; keel the claws 3.5 mm long, the auricles 2 mm diam., 8 mm long from base to bend and 6 mm more to base of the terminal 1 1/2 coil of 2.5–3.5 mm in diam.; vexillary stamen, the claw 0.5 mm long, the geniculate flap elongated into a large sheath parallel to stamen, 1.5 mm long, 1.25 mm wide, with lower edges elongated into short horn-like points, 5.5 mm more to end of the thickened portion; stamen tube 6.5 mm to bend and 4 mm more to end of the united portion, the small knob-like ridges at 1 mm from base, 0.5 mm high; basal collar white, apex oblique, denticulate, 1.5–2 mm long; ovary straight, 5 mm long, 1 mm wide, densely covered with long, fine, white silky pubescence; style 7.5 mm to the terminal thickened coil of 2.5 mm diam.; stigma terminal, capitate, slightly oblique, 0.6 mm long. ***Pod*** unknown. ***Seed*** unknown. ***Seedling*** unknown.

Habitat.—Although the precise habitat for var. *rigidicaulis* is not given on the collector's label, the town of Contreras (dínamos) is close by (SW) Mexico City and probably is similar to a near-by and well-known recreational area of Desierto de los Leones which is an old pine forest, quite dense and of heavy shade with a few rock outcroppings.

Comments.—The varietal epithet is based on the stiff, straight, prostrate type of vine which is most characteristic for this taxon. In addition the small and short petiolate leaves, reminiscent of *P. maculatus*, and the fairly tall, erect inflorescences with flowers somewhat bunched at the tip are most striking and different than all other taxa of the species. It must be very scarce since it is found so close to Mexico City, yet only one specimen has been collected. The flowering period, apparently in April–May, is very unusual for none of the other taxa in the species *coccineus* are anywhere near this early (or late), most flowering in August–September.

C.2.6.—Phaseolus coccineus L. subsp. **striatus** (Brandeg.) Freytag, var. **pringlei** Rose ex Freytag, var. nov. (**Figs. 29, 31**). TYPE: MÉXICO. DISTRITO FEDERAL: Pedregal (lava beds), Valley of Mexico, (19°17'N, 99°11'W), 7,800 ft (2377 m), 25 Aug 1896, *Pringle 6430* (HOLOTYPE: US 942134; ISOTYPES: G, GH, K, MO, MSC, UC(2), US(2)).

Similis *Phaseolo coccineo* var. *striato* sed volubilis prostratis omnibus partibus dense albo- vel flavidi-pubescentibus vel tomentosis, inflorescentia et bracteolis grandibus et late ovatis usque ad calycis duplo longioribus differt. Crecit vallibus citati capitali Nova Hispania rarus.

Plant a prostrate, indeterminate vine, to 3 m long. ***Root*** unknown. ***Stems*** strongly to obscurely ribbed, striate or nerved, covered with farinaceous, glandular pubescence and sparsely covered with short, white, retrorse-hispid hairs; internodes straight, 6–8–14 cm long. ***Stipules*** prominent, triangular-spatulate, 4–5 mm long, 1.5–2 mm wide, 6- to 8-nerved, covered with strigose hairs. ***Leaves*** 11.5–15.6 cm long; petioles 3–6 cm long, distinctly striate, sparsely covered with white strigose and minute uncinate hairs; petiolules 1–2 cm long; stipels narrow ligulate, the upper 2.5 mm long, the lower 3.5–5 mm long, 0.5 mm wide, distinctly 1- to 4-nerved, acute, sparsely covered with hispid hairs; pulvini 3 mm long, densely covered by white uncinate hairs; terminal leaflets rhombic-ovate, 7 cm long, 5 cm wide at 1/4 from base, cuneate at base, acuminate, apiculate, distinctly whitish or yellowish, densely covered with white strigose hairs abaxially, especially along veins, less so adaxially, somewhat farinaceous; lateral leaflets similar but more rounded at base. ***Inflorescence*** erect, exserted, twice as long as foliage; peduncle 12–17–30 cm long; rachis 3–10 cm long, densely covered with white strigose and uncinate hairs; primary bracts ovate-acuminate, 5–10 mm long, 2–3 mm wide, 10- to 11-nerved and moderately covered with white strigose and uncinate hairs especially on margins, the secondary bracts wider (to 3 mm broad) and more pubescent on abaxial surface, nearly glabrous on adaxial surface; pedicels 12–18 mm long, heavily covered by white or yellowish strigose and uncinate hairs. ***Bracteoles*** large, ovate-oblong to nearly round, 8–10 mm long, 4.5–5 wide, indistinctly many-nerved, very densely covered with white or yellowish strigose hairs, nearly vellose. ***Flower*** purple; standard purple, nearly round, 10–12 mm long, 13 mm wide, sharply reflexed at 4 mm from base and 5–6 more to apex, emarginate, veined, 10–12 mm wide, the claw scarcely developed, 0.5 mm long, the auricles lacking; wings purple, the blade deeply cupped, obovate, 17 mm long, 10 mm wide, the claw 4 mm long, the spur 2 mm in diam., firmly attached to keel; keel long stipitate, the claws 4 mm long, the center portion broad and short, 4 mm to bend and 3 more to the terminal 1 3/4 coils of 2.5–3.75 mm diam.; vexillary stamen, the claw 1 mm to the knob developed into a triangular shaped, large, flat sheath, 1 1/2 mm long, 1 1/2 mm wide, the thickened portion 3–4 mm long to filaments; stamen tube short and broad, 3.5 mm long from base to bend abaxially and 3 mm more to the main adaxial bend, and 3 mm more to end of the united portion; basal collar 1–1.5 mm long, minutely denticulate; ovary straight, 6.5 mm long, 1.75 mm wide, densely covered by long, white pubescence, 2–3 ovules; style with a heavily thickened coil of 2.5 mm diam.; stigma terminal, extrorse oblique, 0.75 mm long. ***Pod*** when young silvery pubescent, covered by dense whitish, long, strigose hairs, when mature 5.5 cm

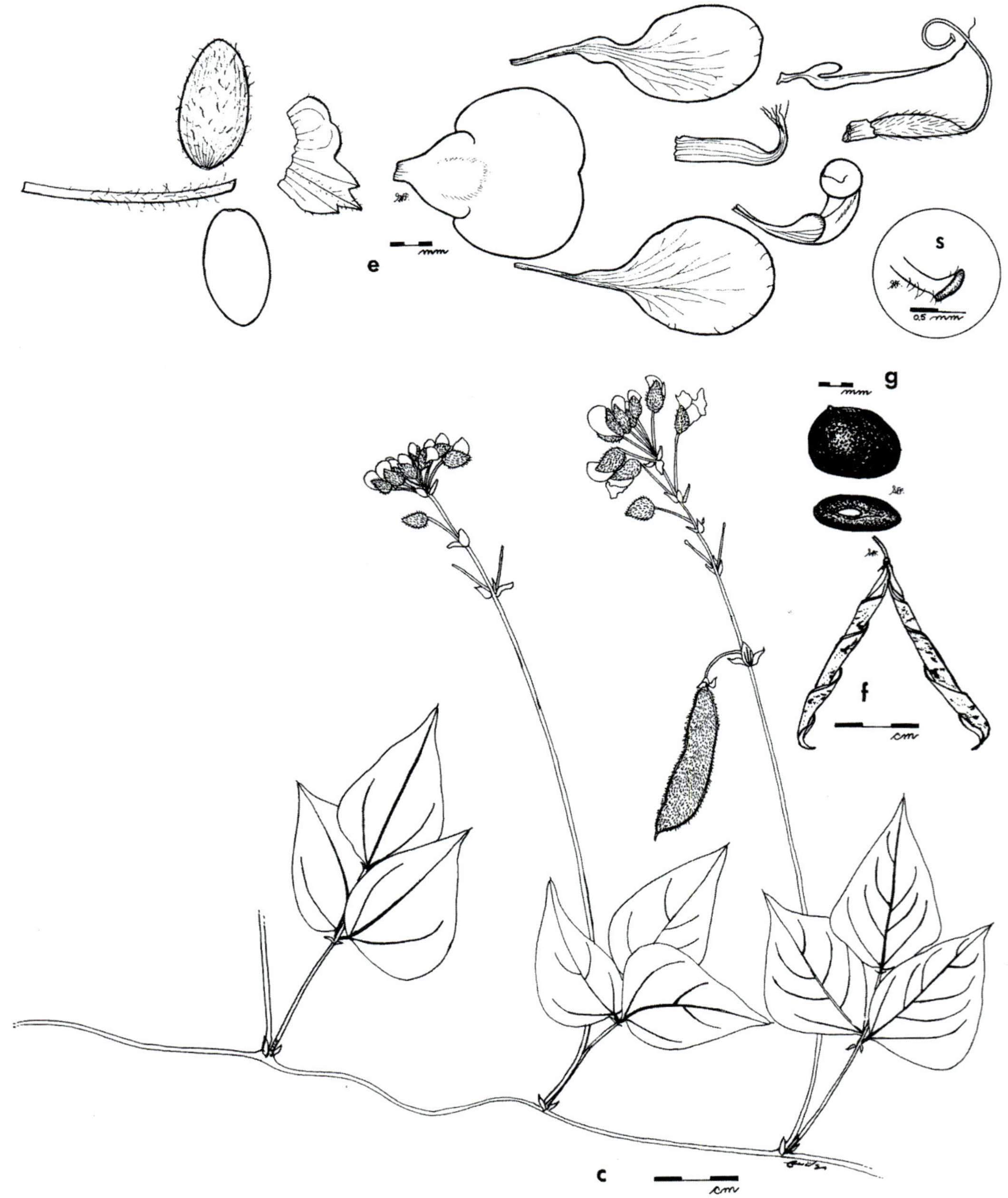

Fig. 29. Illustrations of *Phaseolus coccineus* L. subsp. *striatus* (Brandeg.) Freytag, var. *pringlei* Rose ex Freytag.–**c.** Portion of plant with mature leaves and inflorescences, and immature pod; note the prostrate stem with short internodes, somewhat short petioled and deltoid leaves, and a whitish pubescence covering most parts of the plant especially dense on the long, erect inflorescences with long peduncles and capped by a group of flowers.–**e.** Exploded view of the flower showing all parts including–**s.** Style tip and stigma as seen under microscope; notice densely pubescent pedicel and tomentose bracteoles which are very large and rounded-ovate.–**f.** Pod, dehisced.–**g.** Seeds, side view and view from hilum. All drawings made from the type field collection *Pringle 6430* from the lava beds of the Pedregal in the southern part of the Valley of México, D.F., México.

long, 1.3 cm wide; sutures fairly strong; valves twisted tightly at maturity, wrinkled and somewhat farinaceous, sparsely covered by long, white strigose hairs; beak short, 2–3 mm long, slightly curved. ***Seed*** rounded oblong, speckled black and brown on tan or solid black, 7–8.5 mm long, 7–8 mm wide, 4–5 mm thick; lens pronounced; hilum oval, 1.9–2 mm long, 0.8–1 mm wide. ***Seedling*** unknown.

PARATYPES. **MÉXICO. Distrito Federal:** Vallée de Mexique, Tacubaya, Aug 1865–66, *Bourgeau 580* (G, GH, K, US); Vallée de Mexique, Santa Fé, (19°23'N, 99°15'W), 17 Aug 1865, *Bourgeau 734* (G, GH, K, US); Lomas, Aug 1937, *Lyonnet 1585*, (CAS, MO, US); lava fields near Eslaba, 2438 m, 22 Sep 1903, *Pringle 11431* (BRIT, CAS, F, GH, K, MICH, SI, TEX, US). **México:** Cerro Sacromonte, Amecameca, (19°10'N, 98°45'W), 6 Aug 1967, *Castillo s.n.* (CAS-DS, TEX); Chalco region, (19°15'N, 98°50'W), 4 Oct 1921, *Kempton et al. s.n.* (GH, US); Ixtaccíhuatl, 2100–2400 m, Mar–Jul 1903, *Purpus 311* (MO, UC); Salto de Agua, (2200 m), (19°58'N, 99°16'W), Oct 1905, *Purpus 1748* (F(2), GH, MO, UC, US); Cascada de los Diamantes, San Rafael, 28 Aug 1978, *Trejo 136* (CAS). **San Luís Potosí:** Alvárez, (2700 m), (22°2'N, 100°37'W), 5–10 Sep 1902, *Palmer 63* (in part) (CAS, F, GH, K, MO(2), UC, US).

Habitat.—According to the collection labels this variety is found in the "pedregal" just S of Cd. México, which are open, recent lava flows with a minimum of grasses and other annuals and almost no trees. It is also found on the eastern slopes of the mountains W of Mexico City and to the north, around the Valley of Mexico and into Hidalgo and San Luís Potosí. These areas are certainly very drought prone and the whitish tomentose covering of vegetative parts of the plant would seem to be an adaptation to this environment. All of these localities are rather lower than for other varieties of subspecies *striatus*, that is, about the same elevation as Cd. México. Curiously enough there are few recent collections of this taxon, perhaps indicating that it is quite rare now. However, the variety is also found along brooks at somewhat higher altitudes on the western slopes of the volcanoes E of Mexico City, and down the slopes into the Valley of Mexico to Chalco and south to Amecameca. All of this area of the volcanoes and neighboring slopes is covered with pine forest, open in parts, and probably quite dry, thus the variety is probably found along the streams where more moisture is available, though detailed data are lacking.

Comments.—The varietal epithet was originally proposed by Rose to honor Pringle who had collected the specimens for him, though it apparently was never published. Piper, in dealing with the *coccineus* complex, recognized as unique the heavy pubescence, especially of the bracteoles, of a number of collections from Mexico and placed them all in the species *obvallatus*, separated from those with puberulent bracteoles which he placed in the species *formosus*. Nevertheless he had doubts as to whether this was a valid distinction since he did mention that several collections were noted as having purple flowers which had been recognized by two unpublished names (as given on herbarium specimens), one by Rose and one by Piper. Perhaps some of his problems were caused by so many cultivated forms within the collections from the same locations, accompanied by sparse field notes, which did not allow him to appreciate that they were not all wild types from undisturbed habitats. Maréchal et al. (1978b) noted the similarity of *Pringle 6430* with the purple-flowered "*formosus*" taxon of Piper with the exception of the dense pubescence of the bracteoles, but was not sure of the constancy of the purple flower color. He has indicated (pers. comm.), however, that pollen grain morphology could be a good way of distinguishing species differences in the group, but Delgado (1985) concluded that he was not yet able to do so (the senior author's reason for not using pollen characters).

The senior author now believes the purple flower color is a primary distinguishing characteristic within the *coccineus* complex, occurring in several taxa around Mexico City and south to the border between the states of Puebla and Oaxaca, S through Chiapas and Guatemala, and as far south as Nicaragua. One of the first of these purple flowered collections was *Pringle 6430* from the lava beds (pedregal) just south of the National University, and recognized by Rose as a distinct species, *P. pringlei*, but not published. This taxon (as well as var. *minuticicatricatus*) may also be found east of the city at Santa Marta Ixtapalapa (or at least was until the early 1950's, since the area is now covered by dense, low-cost housing), and seems to occur sporadically as far east as mid-Hidalgo, apparently as far north as El Salto, and west to Santa Fé in the Sierra de Ajusco. With the growth of Mexico City in the last few decades, the variety *pringlei* has perhaps become practically extinct, possibly existing in public parks, rare vacant lots, under high voltage transmission lines and other right-of-ways, and possibly in Hidalgo. These areas need to be carefully surveyed and collected!

C.2.7.—Phaseolus coccineus L. subsp. **striatus** (Brandeg.) Freytag var. **timilpanensis** Freytag, var. nov. (**Figs. 30, 31**). TYPE: MÉXICO. MÉXICO: 5 km SE de Timilpán, 19°51'N, 99°41'W, 2620 m, 13 Nov 1987, *Debouck & Muruaga 2392* (HOLOTYPE: US; ISOTYPES: BR, CHAPA, MICH).

Similis *Phaseolo. coccineo* var. *striato* sed volubilis fere glabris, foliolis latis ovatis acuminatis et petiolatis et petiolis brevibus, pedicellis longis tenellis et bracteolis majoribus fere persistentibus differt. Crecit varietatus septentrionalis mayoribus in montibus proximis civitat capital Nova Hispania rarus.

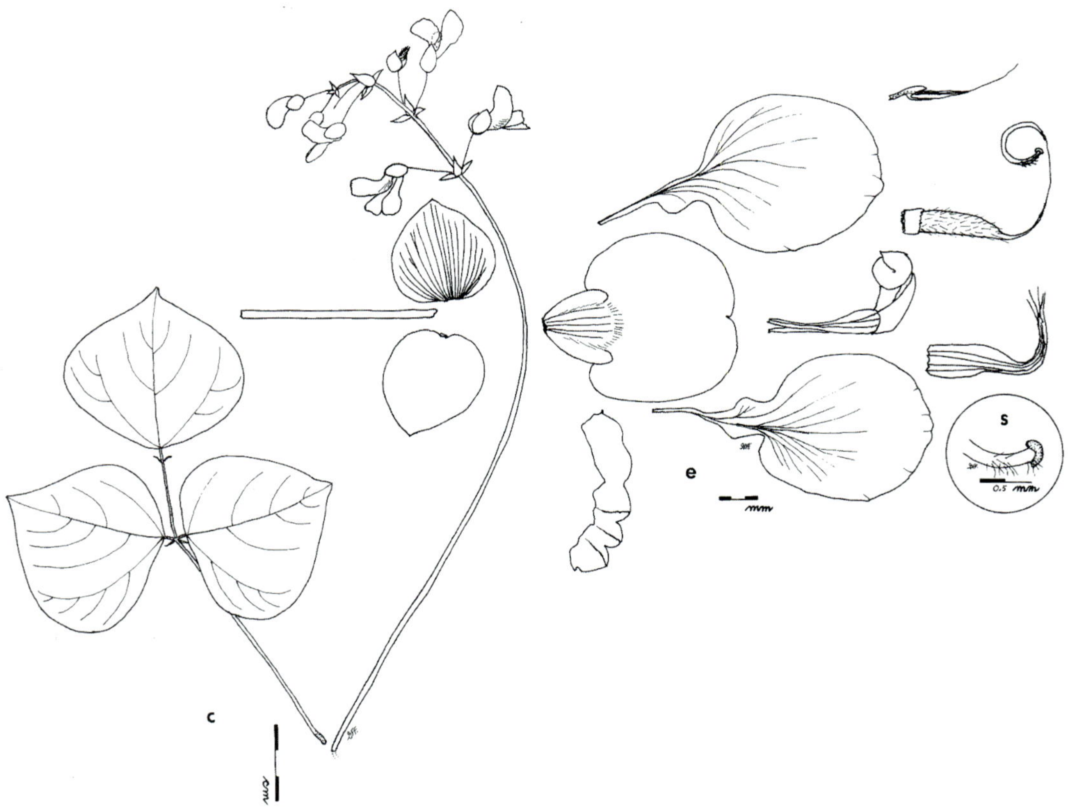

Fig. 30. Illustrations of *Phaseolus coccineus* L. subsp. *striatus* (Brandeg.) Freytag var. *timilpanensis* Freytag.–c. Mature leaf and inflorescence; note the somewhat long petioled and very broad ovate leaflets.–e. Exploded view of the flower showing all parts including–s. Style tip and stigma as seen under microscope; notice nearly glabrous pedicel, very large and glabrous bracteoles which are very large and rounded-ovate, and the very large, rounded standard and wings. All drawings made from the type field collection *Debouck & Muruaga 2392* from near Timilpán, in the NW part of the state of México, México.

Plant a prostrate and climbing, indeterminate vine, to 2–3 m long. ***Root*** unknown. ***Stems*** terete. ***Leaves*** 19–20.5 cm long; petiole 9.5 cm long, puberulent; petiolule 3 cm long, sparsely covered by white strigose and uncinate hairs; pulvini, the lower 6–7 m long, the upper 3–4 mm long, covered with white strigose and uncinate hairs; stipels narrow lanceolate, 3 mm long, glabrous to glabrescent; terminal leaflet nearly round, broader than long, 6–7 cm long, 6.5–7.5 cm wide, scarcely nerved, short acute and apiculate, membranous, bright green, glabrescent adaxially, sparsely covered with long, white strigose hairs abaxially; lateral leaflets similar but inequilateral. ***Inflorescence*** a long, nearly capitate pseudoraceme to 30 cm long; peduncle 25 cm long, fairly stout, purplish, glabrescent; rachis 6 cm long of 3–6 flowering nodes, covered by white uncinate hairs, purplish; primary and secondary bracts similar, broad ovate, acuminate, 6–7 mm long, 2–3 mm wide, sparsely pubescent, purplish; pedicels delicate, 15–16 mm long, covered by minute, white uncinate hairs. ***Bracteoles*** very large, rounded or broad orbicular, 7–9 mm long, 6–8 mm wide, glabrous to glabrescent with a few scattered, white strigose hairs, purplish, indistinctly multi-nerved. ***Flower*** purple, large; calyx campanulate, 4–5 mm long, upper lobes united into one rounded scarcely emarginate, 7 mm wide, lower 3 lobes subequal, rounded dentate, 1.5 mm long, 2.5 mm wide, glabrescent of minute white uncinate hairs; standard dark purple, the claw very broad, 4 mm long, 6 mm wide at uppper end, the blade rounded, 6 mm from the base to flexure and 8 mm more to the emarginate apex, 12–14 mm broad, distinctly purple-nerved, a few small, white strigose hairs at apex, the auricles about 1 mm long; wings purple, the blade nearly round, 17.5–20 mm long, cupped and spreading, subequal, the claw 5 mm long, 0.5 mm wide, the spur 1.5 mm diam. and firmly attached to keel, the blade rounded ovate, 15 mm long, 10 mm

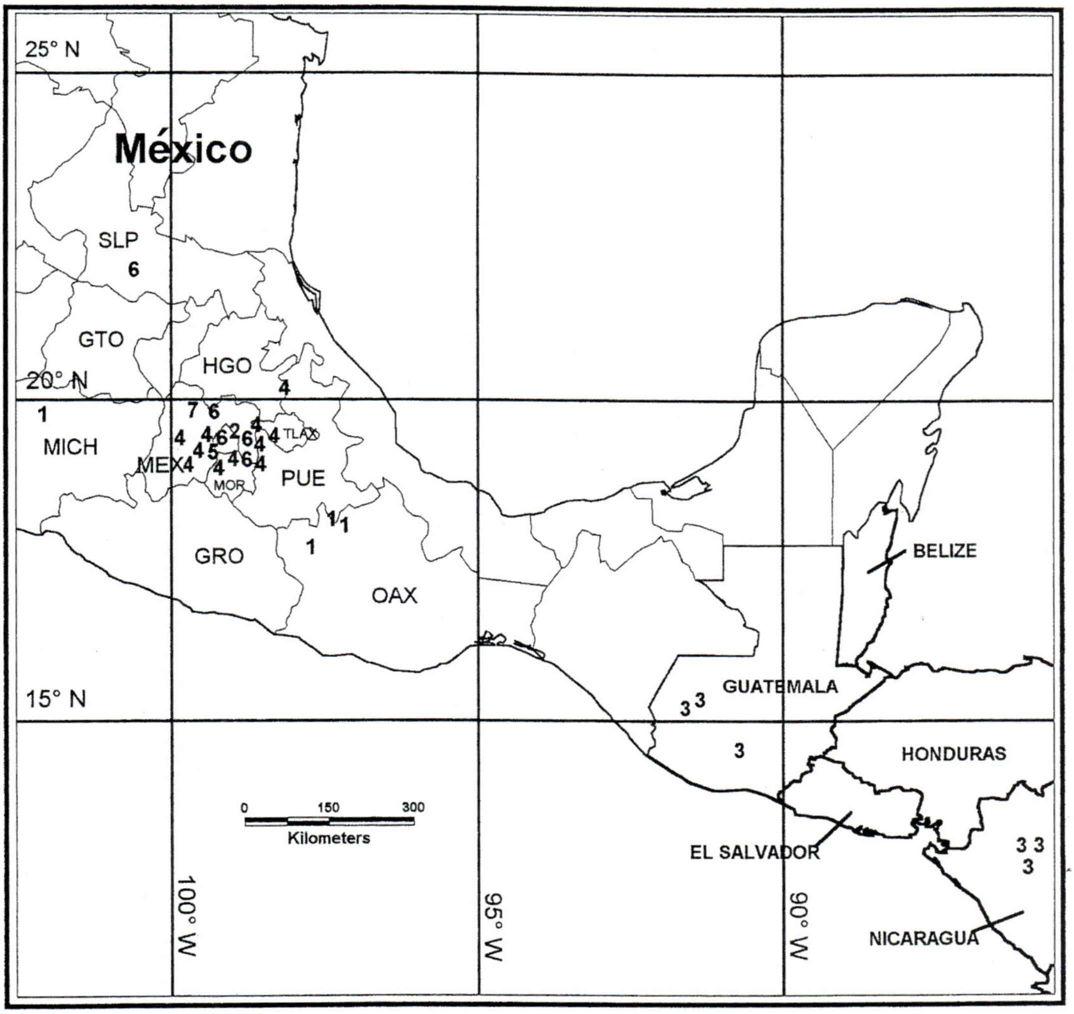

Fig. 31. Distribution of varieties of Section C. *Coccinei* species *Phaseolus coccineus* L. subsp. *striatus* (Brandeg.) Freytag, as follows: 1 = *striatus*; 2 = *minuticicatricatus*; 3 = *guatemalensis*; 4 = *purpurascens*; 5 = *rigidicaulis*; 6 = *pringlei*; 7 = *timilpanensis*.

wide, distinctly purple-nerved; keel, the claws 3.5–4 mm long, 2 mm more to curve and 6 more to base of the terminal 1 1/2 coils of 3 mm diam.; vexillary stamen very small, the claw 1 mm long to the rounded geniculate sheath and 1 1/2 mm long, 3 mm more to end of the thickened portion; stamen tube nearly straight, 2–3 mm broad at base and tapering rapidly, 8 mm to the bend and about 4 mm more to separated filaments; basal collar 1 mm long, denticulate; ovary 6 mm long, 1 mm wide, densely covered with long, white tomentose pubescence; style 8 mm to base of the terminal thickened coil of 2 mm diam.; stigma terminal, capitate, broadly rounded, about 0.4 mm diam., with several stout hairs at base; ovules 4. ***Pod*** unknown. ***Seed*** oblongoid, nearly globose, 7.5–8.5 mm long, 5.7–6.6 mm wide, 3.9–4.9 mm thick, mottled and finely flecked black on a light tan background giving the appearance of a gray seedcoat, black ring around hilum; hilum orbicular 2 mm long, 0.8 mm wide; lens large and raised, slightly divided. ***Seedling*** unknown.

Habitat.—Like most of subspecies *striatus* this variety is found on the mountains around (northwest of) Mexico City at very high altitudes.

Comments.—The varietal epithet is based on the name of the town, Timilpán, closest to the type collection location. This taxon is found the farthest north of the varieties of the purple flowered subspecies *striatus*, and can be distinguished from the other varieties by the following characters: rounded leaflets (broader than long), nearly glabrous foliage, large primary and secondary bracts, very delicate pedicels, large rounded and nearly glabrous bracteoles (see Color Plate I, photo 9), and rounded to oblongoid (rounded ends) grayish seed (somewhat like jelly beans).

The case of *Phaseolus glabellus* Piper

Phaseolus glabellus Piper. Contr. U.S. Natl. Herb. 22:683. 1926. (**Figs. 12, 32**). *Phaseolus glaber* Schlecht. Linnaea 12:327. 1838. (Not *P. glaber* Roxb. 1832.) TYPE: MÉXICO. VERACRUZ: Jalapa: in dumentis prope Jalapam, Aug 1828, *Schiede s.n.* (LECTOTYPE, HAL designated by Delgado (1985) n.v.; ISOTYPE: G; Piper (1926) gives P, n.v.).

Aerial shoot a long trailing and climbing indeterminate vine to 3–5 m long. ***Root*** perennial, fleshy, narrow, much branched, large to 0.75 m long. ***Stem*** the young ones terete, 2–3 mm thick, the old stems perennial, about 1 cm thick, somewhat striate by hyaline veins or ridges, glabrous, of two types, the prostrate stems are straight and non-vining, spreading, rooting at nodes, the twisting and climbing stems are more slender. ***Stipules*** triangular, 2.5 mm long, 1.5 mm wide, 5-nerved, glabrous. ***Leaves*** 8.5–11.5 cm long; petiole 2.5–3.5 cm long, glabrous; petiolule 1–1.5 cm long; pulvini 2.5–5 mm long sparsely covered with few hooked hairs; terminal leaflet broadly ovate, 5–6 cm long, 5 cm wide, acuminate, apiculate, glabrous, often lustrous and fleshy when green but drying very thin membranous to nearly hyaline; lateral leaflets very similar slightly inequilateral. ***Inflorescence*** usually a short pseudoraceme to 40 cm long, the peduncle 10–27 cm long, the rachis 1–15 cm long of 8–15 flowering nodes close together but rather irregularly spaced, glabrous with only a few hooked hairs; primary bracts oblong-ovate to ovate-lanceolate, 3–5 mm long, 1–1.5 mm wide, acute, heavily 3- to 5-nerved, hyaline, glabrous; pedicel 10–12 mm long, very delicate, glabrous; pedicellar bracts lanceolate 1–1.5 mm long 1-nerved glabrous hyaline. ***Bracteoles*** minute, scale-like, 0.75–1 mm long, 0.25–0.3 mm wide, 1-nerved, hyaline, glabrous. ***Flower*** brick red or scarlet; calyx campanulate, 4–5 mm long, the 2 upper lobes united into a single rounded, emarginate, scarcely developed, the lower lobes subequal, very broad, 1.5 mm long, 3 mm wide, acute, glabrous, glandular, the margins minutely ciliate; standard red, broadly rounded, reflexed at 6 mm from base, 13–15 mm long, 12 mm wide, the claw 1.5 mm long, the auricles poorly developed, 0.5 mm long; wings red, spatulate to obovate, the blade 12 mm long, 6 mm wide, spreading, the claw 4 mm long, the spur poorly developed, firmly attached to keel; keel light red to white, the claws 4 mm long, 4 mm more to bend and 4.5 mm more to base of the terminal 1 1/2 coils of 3.25 mm diam.; vexillary stamen, the claw 1.5 mm long to the geniculate knob developed into a circular hooded sheath about 2 mm long; stamen tube nearly straight, broad at base, 8 mm long to bend and 2 mm more to filaments, the ridges rounded knobs 0.75 mm high at about 1.5 mm from base; basal collar 1–1.25 mm long; ovary straight, 7 mm long, 1 mm wide, glabrous except for a few ciliate hairs near base, 6–8 ovules; the style about 7 mm long to terminal thickened coil of 2 mm diam.; the stigma terminal extrorse, oblique. ***Pod*** nearly straight, 4.5–5 cm long, 0.7–1 cm wide; valves when young dark purplish turning green, often striped red or purple, somewhat reticulate veined and chartaceous, heavy sutures, the beak short, 1–2 mm long, straight, acute, glabrous. ***Seed*** oblong to rounded, slightly curved with rounded ends or angular, 4–6.3 mm long, 3–4.8 mm wide, 2.2–4 mm thick, heavily mottled and speckled with black on brown, a black ring around hilum; hilum sunken, orbicular, whitish transparent, 1.75–2 mm long, 0.9–1 mm wide, the placenta sometimes split open down center; lens prominent. ***Seedling*** from hypogeal germination; hypocotyl not elongated; epicotyl 29 mm long, the next internode short 7–8 mm long and the next two internodes consecutively longer, 12–16 mm long; stipules at eophyll bearing node, united; primary leaves opposite, simple, the petiole 27 mm long, stipels absent, the blade triangular ovate, 4 cm long, 2.7 cm wide near the base, the base somewhat truncate, apex acuminate, slightly apiculate, glabrous.

Specimens examined: **MÉXICO. Chiapas:** Mpio. Ocosingo, 2 km de Cushul-ja, a 17 km al O de Ocosingo, (16°40'N, 92°30'W), 6 Dec 1980, *Sousa et al. 11354* (BM, MEXU, MO). **Hidalgo:** La Misión, Km 199 Mex 85 a Tamazunchale, 17 km NE de Jacala, 0.2 km antes de Rancho Viejo, 21°3'N, 99°8'W, 1610 m, 1 Nov 1986, *Debouck et al. 2041* (CHAPA, COL, MICH, MO, UC, US); Chapulhuacán, Km 228 Mex 85 a Cd. Victoria, 1 km NE de Puerto del Gavilán, 1 km W de Sta. Ana, 21°7'N, 99°W, 1410 m, 2 Nov 1986, *Debouck et al. 2042* (CHAPA, UC, US); Mpio. Molango, 1 km al N de Molango, (20°45'N, 98°45'W), 1500 m, 13 Nov 1975, *Delgado et al. 67* (CAS, MEXU); Mpio. Tlanchinol, Tlanchinol, a 40 km al NE de Molango, 1510 m, 14 Nov 1975, *Delgado et al. 90* (CAS); 2 mi S of El Alamo, 33 mi S of the northern boundary of Hidalgo beside Hwy 85, 2430 m, 12 Aug 1978, *Dunn et al. 23087* (MSC); Rancheria de San Bartolo Tutotepec, El Zopilote, 1100 m, 18 Mar. 1973, *Gimate 906* (ENCB, MEXU); Mpio. Tianguistengo, 4 km al N de

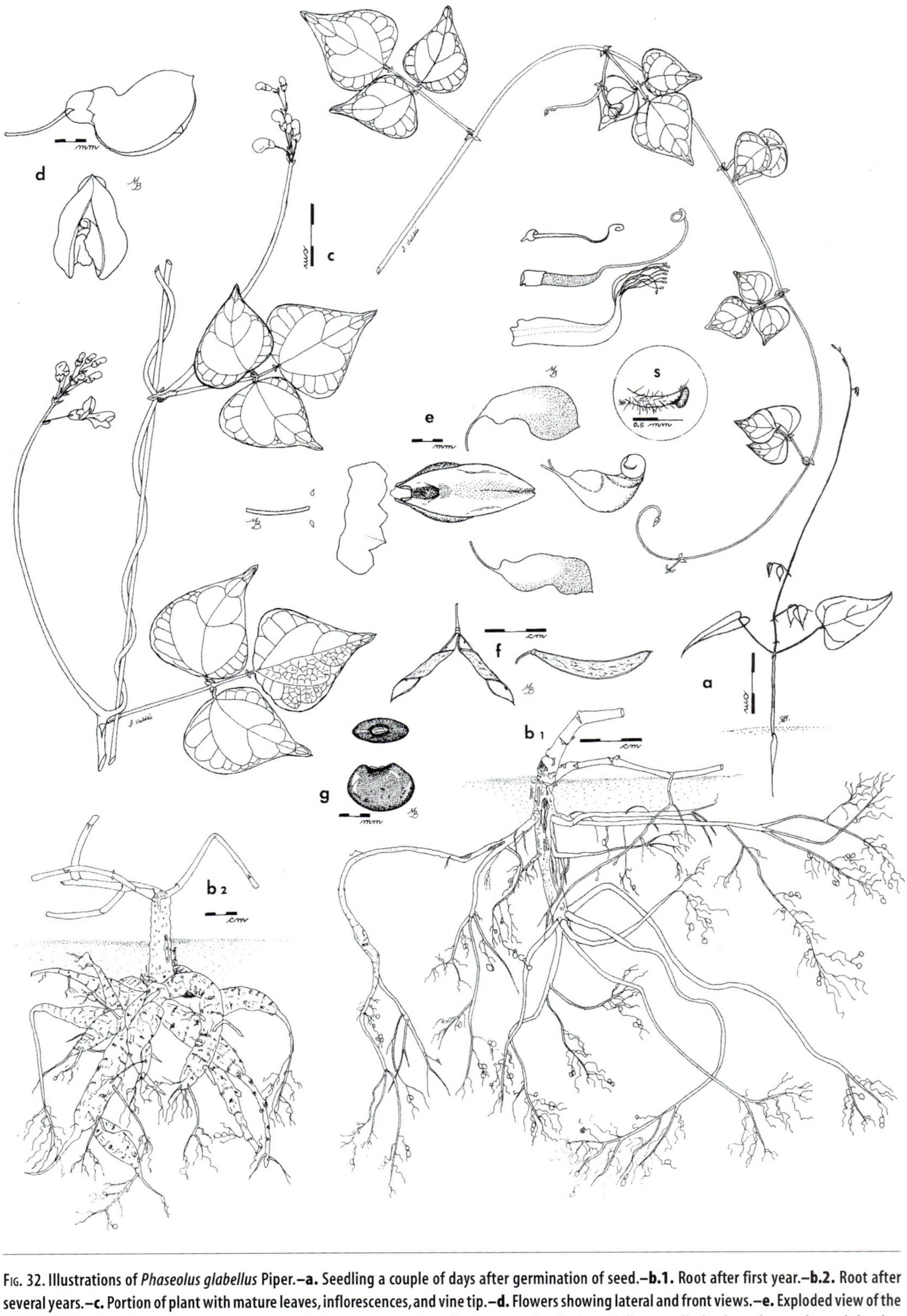

FIG. 32. Illustrations of *Phaseolus glabellus* Piper.—**a.** Seedling a couple of days after germination of seed.—**b.1.** Root after first year.—**b.2.** Root after several years.—**c.** Portion of plant with mature leaves, inflorescences, and vine tip.—**d.** Flowers showing lateral and front views.—**e.** Exploded view of the flower showing all parts including—**s.** style tip and stigma as seen under microscope; notice the long, delicate pedicel, minute bracteoles and clasping standard.—**f.** Pods, lateral view and dehisced.—**g.** Seeds, lateral view and view from hilum. All drawings made from living plants grown in environmental chambers and screenhouses at Mayagüez, PR of TARS #56 (*Leroi s.n.*, from Jalapa, México.) except the drawings of plantlet, flowers, seed and pod which are from TARS #413 (*Debouck 2075* from near Ciudad del Maíz, San Luís Potosí, México).

Tianguistengo hacia Xochicoatlán, 1600 m, 26 Mar 1981, *Hernández et al. 5671* (CAS); Mpio. San Bartolo Tutotepec, Tuto, 8 km N of San Bartolo Tutotepec, 1800 m, 22 May 1982, *Hernández 7352* (MO); between Puerto de Piedra and Santa Ana, 30 Oct 1966, *Rudd 1055* (CAS); El Rayo, 14 km NE of Pisaflores, 920 m, 27 Oct 1982, *Tenorio et al. 2410* (WIS). **Oaxaca:** Dist. Teotitlán, Puente de Fierro, 7 km al N de Huautla de Jiménez, (18°8'N, 96°50'W), 1250 m, 16 Sep 1977, *Sousa et al. 8132* (BM, CAS, CR, SI); Dist. Teotitlán, 4 km al SE de San Jerónimo Tecoatl, 1550 m, 22 Nov 1977, *Sousa et al. 8883* (BM, CAS). **Puebla:** 8 km al NW de Huauchinango, (20°15'N, 98°5'W), 6 Oct 1979, *Basurto et al. 441* (MO); Puente Ajajalpan, Jilotzingo, Zacatlán, 1200 m, 15 Sep 1988, *Basurto 566* (MEXU); La Cumbre, entronque Zacapoaxtla-Cuetzalán con Xochitlán, (19°58'N, 97°33'W)?, (2600 m?), 27 Aug 1977, *Hernández et al. X-183* (CAS, F, MO, WIS); subiendo la cuesta de Apulco rumbo a Zacapoaxtla, 1 km antes de San Carlos, 27 Aug 1977, *Hernández et al. s.n.* (CAS, F(2)); Mpio. Villa Juárez Necaxa, Presa del Necaxa, 1500 m, 18 Aug 1966, *Pascoe 50* (MEXU); Mpio. Villa Juárez Tlaxcalantongo, 750 m, Aug. 1959, *Riba 438* (MEXU). **San Luis Potosí:** Xilitla, 1 km N de Ahuacatlán, 16 km de Xilitla hacia Jalpán en la carr. Mex 120, 21°21'N, 99°3'W, 1030 m, 2 Nov 1986, *Debouck et al. 2043* (CHAPA, UC, US); Ciudad del Maíz, 4 km W de Platanito, Km 139 de Mex 80 a Ciudad Mante, 22°29'N, 99°29'W, 1150 m, 13 Nov 1986, *Debouck et al. 2075* (CHAPA, MO, SI, UC, US, WIS); Los Canos, 15-21 Oct 1902, *Palmer 216* (BM(mixed!), F, GH, MO, NA, US(2)); Tamasopo Canyon, Espinazo del Diablo, 800-900 m, 7 Aug 1934, *Pennell 17966* (US). **Tamaulipas:** Mpio. Aldama, Sierra la Borrada (Sierra de San Juan, W of Rancho Las Yucas, ca highest point in Sierra, 40 km NNW of Aldama, 23°14'N, 98°10'W, 23 Jul 1957, *Dressler 1993* (GH, MEXU, MICH, MO); Julillo to Río Sabinas, N of Mante, 4 Apr 1960, *Duke M3625* (MO); road from Rancho "Las Yucas" to Sta. María de los Nogales, through "El Columpio" from "Los Cerritos" to the Cerro de San Juan, 900-1500 m, 22 Sep 1956, *Martínez et al. F-1925* (TEX); Mpio Gómez Farias, ridge N of Rancho del Cielo above Gómez Farias, 1270 m, 26 Aug 1950, *Sharp 50235* (MEXU); 10 km NW of El Progreso, 18 km NW of Ocampo, 23°N, 99°30'W, 1450 m, 22 Aug 1941, *Stanford et al. 1016* (ARIZ, CAS-DS, GH, MO). **Veracruz:** Mpio. Jalapa, Rancho Guadalupe, 3 km W of Jalapa, carr. vieja Jalapa-Coatepec, 1450 m, 6 Sep 1975, *Calzada 1961* (MEXU); Jardín Botánico Clavijero, 3 km W of Jalapa, 1380 m, 29 Mar 1978, *Calzada 4298* (WIS); Jardín Botánico INIREB, 4 km SSW of Xalapa, (19°30'N, 97°56'W), 1300 m, 10 Oct 1978, *Iltis et al. 943* (MICH, WIS, US); base of Cerro de San Cristobal, W of Orizaba, 1 Nov 1948, *Langman 3579* (US); Jalapa, May 1838, *Linden 676*, (MICH, K); Mpio. Huatusco, cerca de Coscontla, 1265 m, 7 Jul 1972, *Ventura 5683* (ENCB, MEXU); Mpio. Atzalán, Estatino, Río Salado. 1500 m, 13 Nov 1969, *Ventura 46* (ENCB, MEXU); Mpio. Tlalnelhuayocán, 1300 m, 13 Aug 1969, Mpio. Tlalnelhuayocán, 1300 m, 13 Aug 1976, *Zola 649* (F); Mpio. Xalapa, Cerro de Macuiltépetl, 1480 m, 7 Sep 1976, *Zola 702* (F).

Also examined but not used for species description nor distribution: **Veracruz:** Jalapa, (Cultivé en serre chaude à Gembloux, sous le numéro d'introduction: NI 820. Graine provenant de: Jardín Botánico of Jalapa, Leroi, Univ. Tours, France (1982), 12 Nov. 1981), *Le Marchand et al. 4162* (BR).

Habitat.—*P. glabellus* is found mostly on grass covered steep northern slopes in moist (foggy) broadleaved evergreen forest (oak, pine, with species of *Acer, Alnus, Aliso, Carpinus, Cornus, Fraxinus, Liquidambar, Rhus*) at altitudes of 800-1200-2600 m, often cut over and with secondary growth of shrubs (Compositae, Lamiaceae, Solanaceae, and small palms) and numerous vines (*Ipomoea*), ferns and several epiphytes; more abundant in inaccessible sunny spots. Usually growing in moist, rocky, deep, light or dark brown, organic soils derived from limestone, sometimes on steep hills of yellow, red or black clay derived from volcanic rocks.

From currently available herbarium specimens, its habitat seems to be restricted to the montane rain forest of the eastern slope of the Sierra Madre Oriental bordering the Gulf of Mexico (Schmit et al. 1996), in 'Bosque mesofilo de montaña' (Hernández Xolocotzi et al. 1951; Rzedowski 1978), from southwestern Tamaulipas to the northern highlands of Chiapas (Breedlove 1973). This unique vegetation zone, benefitting from humid winds and mists from the Gulf, stretches as an arc bordering the highlands of central Mexico and Chiapas (Schmit et al. 1996), and covers only 1% of Mexico and is presently threatened as particularly suitable for coffee growing and cattle ranching (Rzedowski 1993). So far this species has not been reported from Guatemala.

The relatively uniform ecological niche of *P. glabellus* contrasts with the variety of environments in which wild *P. coccineus* are found.

Diseases and pests.—It is reported as sometimes heavily damaged by weevils.

Common names.—It is generally known as "frijol de ratón" or "frijolillo."

Ethnobotany.—It is known to be wild by local inhabitants, and reported both edible (flowers) and not edible.

Genetics.—In preliminary work in environmental chambers at Mayagüez where this taxon has flowered and produced seed (TARS #413 = *Debouck et al. 2075*), it has been impossible to obtain seed or pods when crossing it with *P. coccineus* subsp. *striatus* var. *purpurascens* (TARS #38 = *Freytag et al. 81-4*) or with *P. coccineus* subsp. *coccineus* var. *condensatus* (TARS #411 = *Debouck et al. 2037*). Neither does it seem to cross with these forms (*DGD 2042* and an escape from cultivation of *P. coccineus*) when they are found together in nature as in the humid southern Huasteca of Hidalgo (personal observations by the junior author in the Huasteca in 1987). This situation has also been acknowledged by Delgado Salinas (1988) and confirmed by Schmit et al. (1991) and Sousa-Peña et al. (1996). Attempts made at Gembloux to cross it with other taxa of *P. coccineus* have also failed so far (Schmit et al. 1992). However the spatial isolation of *P. glabellus* in a humid tropical habitat, at a lower elevation and in a dense, deciduous tropical forest generally below the range of the other subspecies, could be a reason for the genetic isolation of this taxon.

Comments.—This taxon, originally described as *P. glaber* by von Schlechtendal (1838) from material collected by Schiede at Xalapa (Jalapa, Veracruz, Mexico) was given a new name by Piper (1926). Piper clearly recognized the key distinguishing characteristics including the minute bracteoles and overall plant glabrousity, thus it is surprising that he considered *Palmer 216* from SLP (a mixed collection) as the same since this collection is mostly *P. pedicellatus* which has more acuminate, often somewhat basally lobed leaflets and considerably more pubescence. In living material, *P. glabellus* leaves are glabrous, somewhat thick and succulent, even lustrous, and when dried become extremely thin membranous, almost hyaline. Unfortunately, Piper also included in *P. glabellus* the long narrow root and short pods with 6–8 seed of the Palmer specimen, which are very different from true *P. glabellus* which has a much branched, smooth and fleshy root (see Color Plate III, photo 26), and pods which are slightly curved and fairly thin with few (3–4) seed. Although Maréchal et al. (1978b) maintained *P. glabellus* as a separate species, he apparently had some doubts about this treatment due to the sparse material at hand.

However, he was in error in believing this subspecies to come from the higher altitudes of Mexico when in reality it is more common from the mid elevations of eastern Mexico. Additionally he did not recognize the striking differences between this taxon and var. *griseus* in shape and pubescence of leaf and bracteoles, and of course, distribution and habitat. However, he subsequently obtained seed from Dr. B. Leroi, University of Tours, France and was able to bring plants to flower and fruit and recognized (pers. comm.) the close relationship of this taxon to the rest of the *coccineus* complex. The senior author is deeply indebted to Dr. Maréchal for duplicates of his material, and viable seed from which he has grown out this species in screenhouse and environmental chambers for several years, though this particular collection has never bloomed for him. On the other hand, plants grown from seed of *Debouck 2075* (= TARS #413) from near Cd. Maíz, SLP have flowered profusely under the same conditions (see Color Plate I, photo 5) and have been highly fertile though it does require tripping of flowers. These two collections have provided the material for the revised description given above. Delgado (1985), in his recent revision of *Phaseolus*, changed the status of *P. glabellus* to that of a subspecies of *P. coccineus*. However, in recent works (Llaca et al. 1994; Delgado Salinas et al. 1999; Sousa-Peña et al. (1996), he considered it as a separate species due to the sterility with other red-flowered species of *coccineus*, and molecular evidence.

The placement of *P. glabellus* outside the section C. *Coccinei* is also confirmed by SDS-PAGE electrophoresis of seed storage proteins (Schmit & Debouck 1990; Schmit et al. 1992, 1996) which shows that *P. glabellus* has seed storage proteins of lower molecular weight (28–38 kD) instead of forming globulins (45–52 kD) as do the other forms of *P. coccineus* and the metabolic pathway for their synthesis is different (recently confirmed by Pueyo & Delgado Salinas, 1997). Further evidence for considering *P. glabellus* as a separate species has come from cpDNA research (Gepts 1996; Llaca et al. 1994; Schmit et al. 1993) which have shown that *P. glabellus* may be far more distant from the group of taxa *P. coccineus*, *P. dumosus* and *P. vulgaris*, than is *P. lunatus*. These results have also been confirmed using other molecular techniques (Hamann et al. 1995; Jacob et al. 1995; Schumann & Nagl 1995; Vekemans et al. 1997). Recently, after ITS DNA sequencing, Delgado Salinas and co-workers (1999) and Gaitán and co-workers (2000) indicate a close relationship with *P. pedicellatus*, *P. polymorphus* and *P. grayanus*. This is unexpected, given the striking morphological differences between *P. glabellus* and the relatively uniform section of *P. pedicellatus*. So, including *P. glabellus* into the section M. *Pedicellati* would somehow break the morphological unity of that section; on the other hand, including it in the *Coccinei* would in turn break the uniformity of this section, for long known to be close to *P. vulgaris*, while on the basis of all experimental data cited above surely *P. glabellus* is not close to the common bean nor the scarlet runner. The junior author (Debouck 2000a) has proposed to have *P. glabellus* as an isolated taxon outside currently recognized phylums. Therefore, and pending on additional data (if needed), *P. glabellus* is tentatively maintained here but outside the Section C. *Coccinei* for convenience to the field collector due to the morphological similarities of flower color, shape of stigma and other flower parts.

Section D.—Paniculati Freytag, sect. nov. TYPE SPECIES: *Phaseolus lunatus* L., Sp. Pl. 724. 1753.

(In the following the characteristics of cultivars are given in parentheses). Herba pleurumque perennis (annua), caulis ad basim crassem plus minusve lignosus, volubilis magnascens prostratus ad scandens vel frutex cum apices volubiles, radix pleurumque perennis crassa carnosa (fibrosa annua breviter vivens), foliola pleurumque parva (grandia) coriacea nec herbacea, pubescentis densissimis abaxallare, nectaria foliolarum stipelatorum et bracteolatorum interdum agiles, panicula linearis cum ramis lateralibus brevissimi vel gongylodibus reduit, bracteolae parvissimae, vexillum ad apicem pubescentem, stigma laterale introrsum, legumen pleurumque parvum falcatum inflatum et suturae crassae pubescentissima uncinatis, valvae fibrosae durae et glaucae vel glabrae.

Plant mostly perennial (annual), the basal stem thick, more or less woody, becoming a large prostrate to climbing vine or shrubby bush with vining tips; root usually a thick, fleshy perennial (fibrous, short-lived annual); leaflets usually small (large), usually coriaceous but sometimes membranous, mostly densely pubescent abaxially; foliar nectaries present, usually found on stipels and sometimes on bracteoles and often hyperactive; inflorescence a long narrow panicle with the lateral branches very short or reduced to knobs; bracteoles very small; standard of flower usually pubescent at tip; stigma lateral introrse; pod usually short, falcate, inflated; sutures heavy, mostly densely covered with uncinate hairs, valves tough fibrous and often glaucous or glabrous.

Comments.—This is quite a natural grouping of species many of which seem to be bushy or shrubby (somewhat woody), upright plants with vining tips and adapted to xerophytic or subhumid habitats. But there are exceptions to the shrubby plant type, all of which the senior author has placed in Subsection I. *Volubili*, and of which the most notable is the species *P. lunatus*, found throughout the American tropics mostly at lower elevations and in moist habitats, often along stream banks. All the species of this section are characterized by having a panicle, not always well defined since some of these species have the lateral branches much reduced to 1–2 short basal branches. In others the branches are reduced to stubby knobs, or are not present until somewhat after the first blooming period if the habitat is favorable to growth and the inflorescence continues to grow, eventually producing lateral branches and additional flowers. In addition, the species of this section are characterized by having short, falcate, inflated, and tough fibrous pods, and often having minute pubescence on the bud tips. Many of the species are poorly known, perhaps by a single specimen, and seem to be poorly competitive and with localized distribution, being found in only widely separated and isolated spots in central Mexico. Much additional collecting is needed to obtain more complete collections and to clarify if the species are not more prevalent and have a wider distribution.

There is some evidence through the production of experimental hybrids (i.e. *P. lunatus* × *P. maculatus*; see Katanga & Baudoin 1987a; Le Marchand et al. 1976; Le Marchand & Maréchal 1977a; *P. lunatus* × *P. ritensis*; see Katanga & Baudoin 1990) that there is a close relationship between this section and Section O. *Coriacei*, which with numerous species would thus represent the secondary and tertiary gene pools of the Lima bean (Baudoin 1990; Debouck & Smartt 1995). On that basis, the junior author (Debouck 2000a) wrote about a Lima bean phylum. Recent ITS DNA sequencing data (Delgado et al. 1999; Gaitán et al. 2000) confirm this relationship, while they also support to maintain the Coriacei as a separate section.

KEY TO SUBSECTIONS

1. Plant a somewhat woody, climbing vine with mostly nearly puberulent or glabrous leaves ______ Subsection I. **Volubili**
1. Plant with a stout, woody basal stem developing into a small bush with vining branch tips with most parts covered with white or yellowish hispid or strigose to tomentose hairs ______ Subsection II. **Lignosi**

Subsection I.—Volubili Freytag, subsect. nov. TYPE SPECIES: *Phaseolus polystachyus* (L.) Britt. Sterns & Pogg.

Herba pleurumque volubilis grandis scandens (vel prostrata), radix pleurumque perennis crassa carnosa (radix annua fibrosa), folia plerumque pubescentae paucae vel puberulentae, bracteolae minutissimae, stigma laterale introrsum, legumen compressum suturis manifestis et pleurumque multitrichomatis uncinatis.

Plant mostly large climbing (or prostrate) vines; root mostly thick, fleshy and perennial (fibrous rooted annuals); leaves mostly lightly pubescent or puberulent; bracteoles very small; stigma lateral introrse; pod compressed with heavy sutures, mostly densely covered by uncinate hairs.

KEY TO SPECIES AND SUBSPECIES

1. Root fibrous to somewhat fleshy, often surviving a number of years and the stem becoming quite thickened; leaflets usually somewhat triangular-ovate, nearly glabrous; very common throughout the American tropics, usually along stream beds; sea level to 2,000 m, rarely at higher elevations in Guatemala to 2440 m ____ D.I.1. **P. lunatus**
1. Root usually quite thickened and fleshy or woody; leaflets variously rounded-ovate, mostly densely pubescent.
    2. Plant distribution in U.S., from Texas to east coast.
        3. Leaflets mostly large ovate and never lobate, a large climbing vine; once common throughout eastern US, from Florida N to Ohio and W to Arkansas; 30–910 m __________ D.I.2.1. **P. polystachyus** subsp. **polystachyus**
        3. Leaflets distinctly lobate.
            4. Plant a medium sized prostrate vine with small variously undulate or lobed leaflets; scarce, on sandy savannas throughout SE coastal US, near sea level ________________ D.I.2.2. **P. polystachyus** subsp. **sinuatus**
            4. Plant a large prostrate or climbing vine with large basally lobed leaflets; rare, in a disjunct distribution: hummock limestone areas of central Florida and perhaps cliff edges in Kerr Co., Texas; 60–200 m ______________________________ D.I.2.3. **P. polystachyus** subsp. **smilacifolius**
    2. Plant distribution in Mexico.
        5. Plant a prostrate vine with fairly small, coriaceous leaflets rare, near Zaragosa, Nuevo León; 1960 m ______________________________ D.I.4. **P. maculatifolius**
        5. Plant a climbing vine.
            6. Leaflets usually small, petioles less than 3 cm long, elongate ovate to ovate-acuminate, and quite pubescent; scarce, in dryer pine forests of W Mexico: Sinaloa, Jalisco and Michoacán; 975–2370 m ______________________________ D.I.3. **P. salicifolius**
            6. Leaflets usually medium sized, petioles more than 3.5 cm long, ovate and glabrous; in very moist deciduous forests N of Orizaba, Veracruz.
                7. Bracts about 5 mm long; bracteoles 1 mm long, hyaline and not nerved; pods distinctly long strigose pubescent; rare, found only near Huayacocotla, Veracruz; 1950 m _____________ D.I.5. **P. dasycarpus**
                7. Bracts about 2 mm long; bracteoles l mm long, 1-nerved; pods glabous; rare, found only near Santa Ana Atzacan, Veracruz; 1200 m ______________________________ D.I.6. **P. longiplacentifer**

D.I.1.—Phaseolus lunatus L., Sp. Pl. 724. 1753. (**Figs. 33, 34**). TYPE: Bengal. *Phaseolus benghalensis scandens striato* Bergen, Cat. stirp. Hort. Acad. Viadr. compl. 99. 1744. (NEOTYPES (cult.) designated by Westphal (1974): *Westphal 8622*, HOLOTYPE: WAG n.v.; ISOTYPES: K, P n.v.; for cultivated small-seeded materials; *Westphal 8652*, HOLOTYPE: WAG n.v.; ISOTYPES: K, P n.v.; for cultivated large-seeded materials).

Phaseolus inamoenus L., Sp. Pl. 724. 1753.
Phaseolus bipunctatus Jacq., Hort. Vindobom. 1:13. 1770.
Phaseolus macrocarpus Moench., Meth. Pl. 155. 1794.
Phaseolus puberulus HBK, Nov. Gen. Sp. 6:451. 1823.
Phaseolus xuarezii Zucc. ex DC, Prodr. 2:393. 1825.
Phaseolus falcatus Benth. ex Hemsl., Biol. Centr. Am. Bot. 1:304. 1830?. syn. nov. TYPE: MÉXICO. NAYARIT: San Blas to Tepic, 1830, *Coulter 576* (HOLOTYPE: Ex Herb Collegii SS, Trin, Dublin. n.v.; ISOTYPES: GH, K).
Phaseolus limensis Macfady., Fl. Jam. 1:279. 1837. TYPE: none given.
Phaseolus saccharatus Macfady., Fl. Jam. 1:282. 1837. TYPE: none given.
Phaseolus lunatus var. *macrocarpus* (Moench.) Benth., Mart. Fl. Bras. 15:181. 1859.
Phaseolus rosei Piper, Contr. U.S. Natl. Herb. 22:694. 1926. syn. nov. TYPE: ECUADOR. Hacienda Licay, Huigra, 6 Sep 1918, (grown in greenhouse, Washington, D.C. by P.L. Ricker, 1924 from seed of) *Rose et al. 24067* (HOLOTYPE: US 1241201; ISOTYPE: GH).
Phaseolus viridis Piper, Contr. U.S. Natl. Herb. 22:693. 1926. TYPE: GUATEMALA. ALTA VERAPAZ: Cubilquitz in Gebnsehen windend, 350 m, Feb 1904, *von Tuerckheim II.897* (HOLOTYPE: US 576636; ISOTYPES: F, GH, US). *Phaseolus lunatus* var. *viridis* (Piper) Delgado, comb. nov. (1985).
Phaseolus lunatus var. *silvester* Baudet, Bull. Soc. Roy. Bot. Belg. 110:71. 1977. TYPE: plant cultivated in Gembloux, Belgium, place of origin unknown, *Le Marchand & Maréchal 516* (HOLOTYPE: BR).

Aerial shoot an annual tending to perennial, climbing or scandent, much branched and very large, indeterminate vine, 2–8 m long. ***Root*** perennial tending to be annual, fibrous and sometimes somewhat fleshy, much branched, 0.5–1 m long, 0.5–1 cm thick. ***Stems*** terete, to 3–4 mm thick, at base of

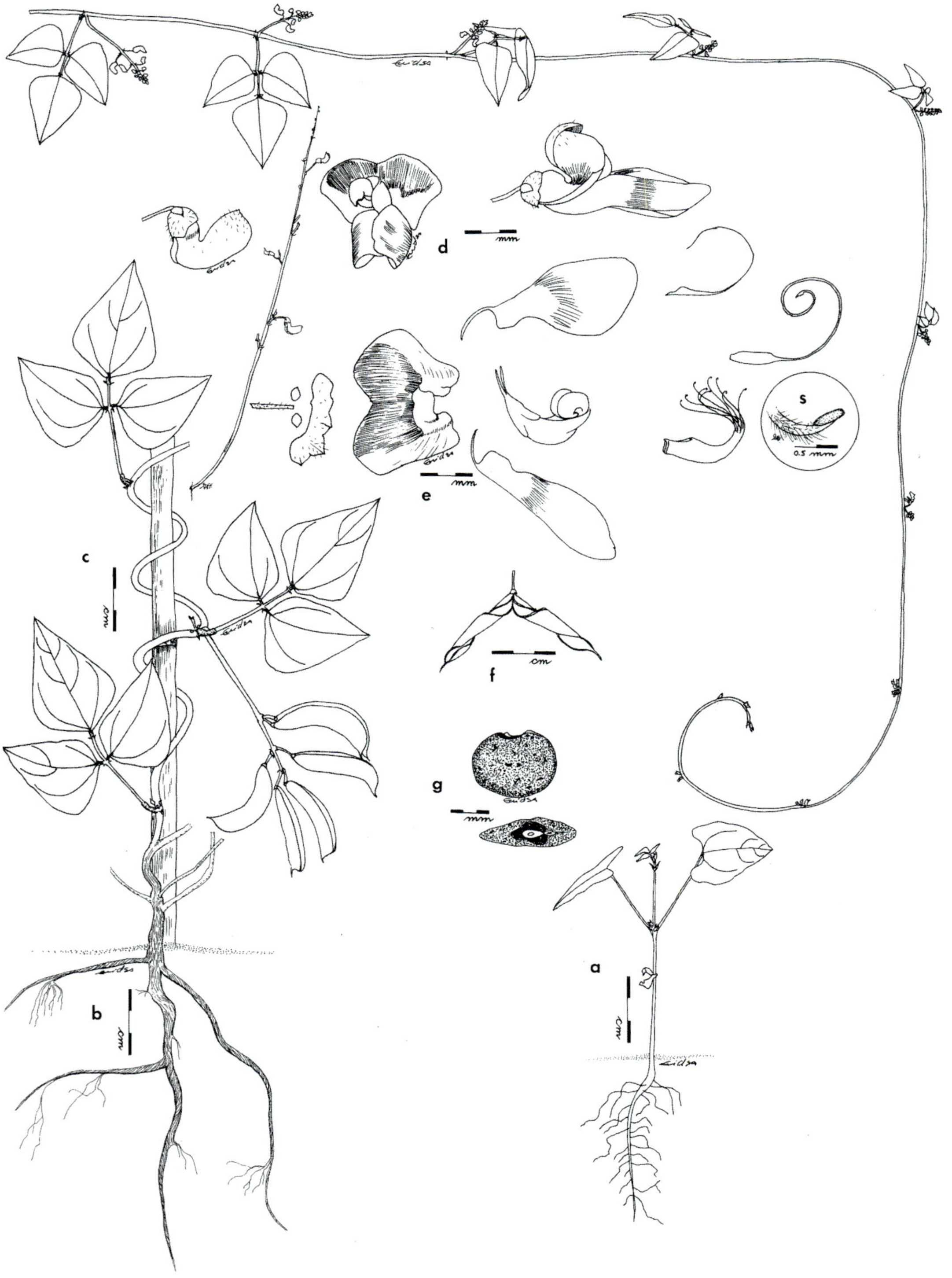

FIG. 33. Illustrations of *Phaseolus lunatus* L.—**a.** Seedling several days after germination.—**b.** Stem and root of mature plant.—**c.** Vine with mature leaves and pods, and, separately, a mature inflorescence and vine tip with young inflorescences; note the pyramidal shape of the young inflorescences and the short, lateral secondary branches of the mature inflorescence.—**d.** Bud showing pubescence, and flowers, side view and front view; note minute pubescence at tip of standard.—**e.** Exploded view of the flower showing all parts, including—**s.** Style tip and stigma as seen under the microscope.—**f.** Pod with twisted carpels.—**g.** Seed (primitive type), side view and view from the hilum. All drawings from living material grown in greenhouse at Mayagüez of TARS #64 (*Evans 16*) from Veracruz, México.

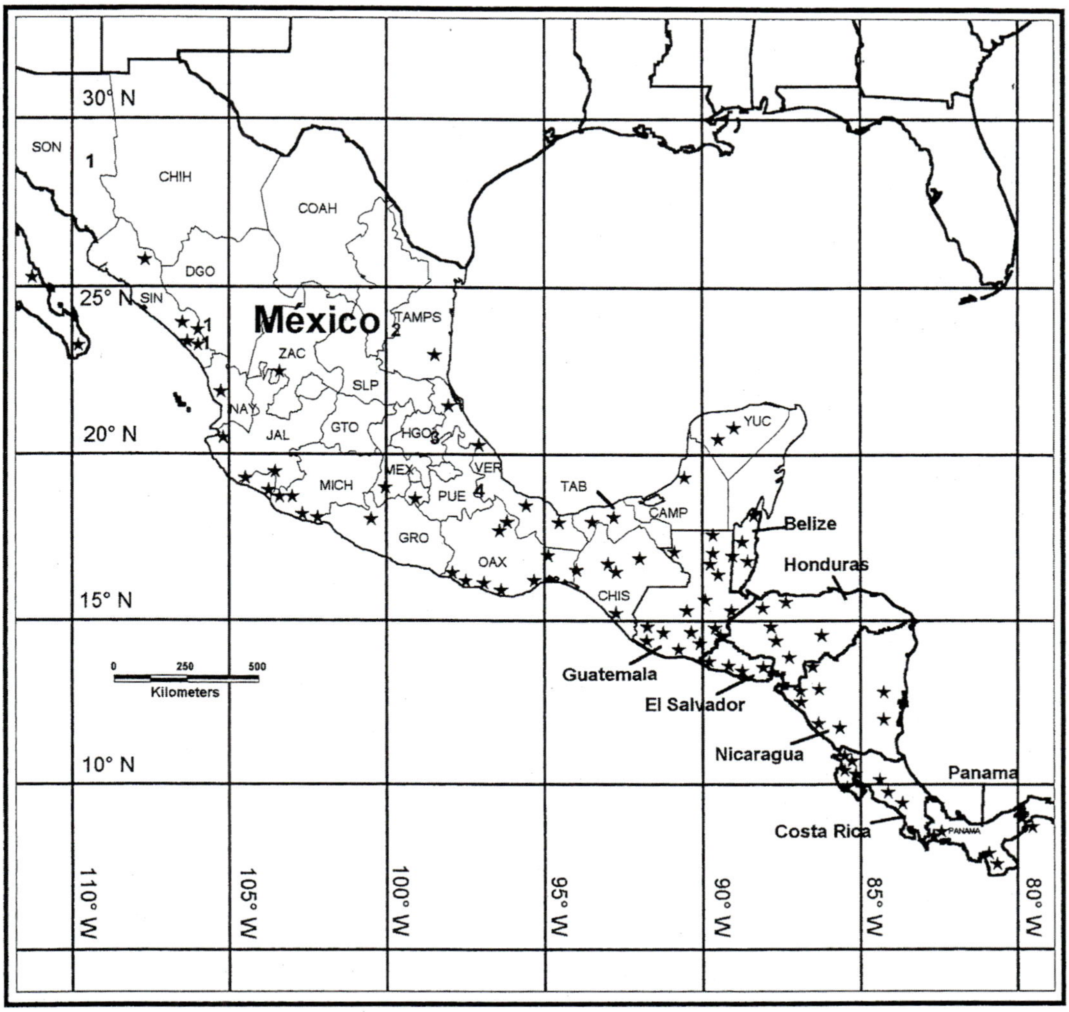

Fig. 34. Distribution of *Phaseolus lunatus* L. and minor species of section D. *Paniculati*, subsection I. *Volubili*, as follows: * = *P. lunatus*; 1 = *P. salicifolius*; 2 = *P. maculatifolius*; 3 = *P. dasycarpus*; 4 = *P. longiplacentifer*.

plant becoming woody to 1 cm or more thick, the internodes to 25–30 cm long, somewhat glabrate to copiously covered with reflexed-strigose hairs. ***Stipules*** ovate to lanceolate, 2 mm long, acute, hispidulose on abaxial surface. ***Leaves*** 11–17 cm long; petiole longer than terminal leaflet, striate, hispidulose; stipels sometimes nectariferous; terminal leaflet variously shaped, usually ovate to lanceolate, 5–6–8 cm long, sometimes very narrow, rounded to hastate base, acute to acuminate, apiculate, membranous, nearly glabrous, green; lateral leaflets similar but more or less inequilateral. ***Inflorescence*** a horizontal to erect panicle 15–30–40 cm long, usually with several small secondary branches, the peduncle 5–9–14 cm long, striate, the rachis 10–23 cm long, sparsely covered with strigose hairs; primary bract ovate-acuminate, 1 mm long, 1-nerved, sparsely covered with strigose hairs; pedicel 5–8 mm long, delicate, sparsely covered with strigose and uncinate hairs; pedicellar bracts scale-like hyaline less than 1 mm long sparsely covered with short white strigose hairs. ***Bracteoles*** ovate to oblong, acute to obtuse, about half as long as calyx, 1–1.5 mm long, 0.5 mm wide, weakly 3-nerved, glabrous to very slightly covered by strigose, sometime nectariferous hairs. ***Flower*** purple (whitish) and/or greenish, very numerous especially toward tips of inflorescences, 2–3 or more produced at each node; calyx campanulate, the tube 1.5–2.5 mm long, the upper lobes united, rounded, emarginate, lower 3 lobes subequal, broadly acute, 0.5 mm long, 1.5–1.75 mm wide, pilose especially on lower lobes, ciliate; standard purple (or white) and greenish especially on adaxial surface, in bud more or less covered by hispid or hispidulose hairs at the apex, broadly rounded-squarish, erect, 5–6 mm long, recurved and hooded forward over the keel, lateral edges enrolled, thickened at tip and point of flexure, the auricles well-developed, 0.5–0.75 mm long; wings purple (or white), the blade spatulate to obovate, 7–10 mm long, 4.5 mm wide, the basal claw 2–3 mm long, the broader blade spreading, cupped and enrolled lengthwise, the spur well-developed, 1 mm diam., slightly adhering to keel; keel short, 4 mm to bend and 3 mm more to terminal 1 3/4 coils of 1.75–2 mm in diam., greenish or whitish tip; vexillary stamen, the claw 0.75 mm long, the geniculate knob 0.75 mm wide, 0.25 mm high; stamen tube 3.5 mm to bend and 2.5 mm more to divided filaments, the ridges scarcely developed; basal collar 0.5 mm long; ovary straight, 3–3.5 mm long, 1 mm wide, covered with minute white hispid hairs, 4–6 ovules; style 5–5.5 mm to the terminal coil of 1.75 mm diam.; stigma lateral introrse, linear, 0.75 mm long. ***Pod*** when young densely covered by minute hispid hairs, becoming nearly glabrous at maturity, falcate, inflated, short, 3.5–5 cm long, 1–1.3 cm wide, brittle, with thickened sutures, dehiscent and moderately twisted at dehiscence, the beak straight, stiff, 4–5 mm long. ***Seed*** obovoid, flattened to globose, 8–11 mm long, 6–7 mm wide, 2.5–4 mm deep, speckled black and brown or solid red, tan or black, shiny, often with a black halo around the hilum and with raised lines or nerves radiating from the hilum; hilum oblong, 1.5 mm long, 0.8 mm wide; lens small. ***Seedling*** from epigeal germination; hypocotyl 2 cm long; epicotyl 5 cm long, glabrous; stipules entire; petioles 5 cm long, with basal and apical pulvini; stipels absent; eophylls opposite, simple, the blade, ovate, 3 cm long, 3 cm wide, cordate-sagittate at base, acute at tip.

Specimens examined: **BELIZE. Cayo:** El Cayo, (17°0'N, 89°5'W), Mar–Jun 1933, *Chanek 88* (F, MICH); E of Agr. & Education Guest house at Central Farm, 16 Mar 1967, *Dwyer et al. 93* (MO); Spanish lookout Crossing E of Belize River, 17 Mar 1967, *Dwyer et al. 161* (MO(2)); 25 mi S of Belmopan, Hummingbird Hwy., 22 Jan 1974, *Dwyer et al. 12109* (MO(2)); El Cayo Road, 16 Feb 1938, *Gentle 2186* (F, MICH); Little Cocquericot, Belize River, (17°25'N, 88°45'W), 25 Apr 1933, *Lundell 4380* (F, MICH, US); Central Expt. Farm, Baking Pot, 150 m, 17 Feb 1964, *McKee 11383* (US). **Corozal:** Cerros Maya Ruins, Lowry's Bight, (18°15'N, 88°30'W), 3 Feb 1981, *Crane 17* (TEX-LL). **Stann Creek:** Long Bank, near John Jose Creek, 5 Jan 1939, *Gentle 2680* (F, GH, MICH); Carib Switch, Middlesex Road, (16°50'N, 88°35'W), 13 Jan 1953, *Gentle 7840* (CAS, F, TEX-LL, US). **Toledo:** Maya mountains, canyon along Bladen Branch from Richardson Creek to Quebrada de Oro, 16°31–33'N, 88°46–49'W, 100–200 m, 12 Mar 1987, *Davidse et al. 32358* (MO).

COSTA RICA. Alajuela: valley of San Ramón, 950–1000 m, 30 Nov 1925, *Brenes 562* (F); Alajuela, camino detrás del aeropuerto Juan Santamaria, 10°01'N, 84°12'W, 930 m, 9 Jan 1987, *Debouck et al. 2103* (CR, US); Alajuela, camino a San Juan Norte, 10°07'N, 84°14'W, 1390 m, 9 Jan 1987, *Debouck et al. 2104* (CR, US); 1 km N de Naranjo, 10°7'N, 84°23'W, 1170 m, 9 Jan 1987, *Debouck et al. 2106* (CR, MICH, US); Río Maravilla, San José de Alajuela, 850 m, 31 Dec 1950, *León 2943* (EAP); Palmarés, 900 m, 23 Feb 1940, *Smith P2436* (GH, UC); camino Hatillo-Alajuelita, 1150 m, 5 Jan 1935, *Solíz 90* (CR, F); finca Fco Rojas, Hatillo, 1100 m, 26 Dec 1936, *Solíz 504* (CR, F, MO); Broussailles du Tajo, près San Ramon, 1200 m, 8 Apr 1913, *Tonduz 17866* (CR, US). **Cartago:** barranco del Río Reventado, 0.3 km W de Turbina al N de la Ciudad de Cartago, 9°53'N, 83°55'W, 1550 m, 1 Jul 1987, *Debouck et al. 2092* (CR, US); Tres Ríos, en el Rancho San Juan de Tres Ríos, 3 km W de Tres Ríos, 9°55'N, 84°00'W, 1320 m, 11 Jan 1987, *Debouck et al. 2117* (CR, US); San Isidro, laderas del cerro Guatuso, 2 km S de San Isidro, 9°49'N, 83°57'W, 1470 m, 12 Jan 1987, *Debouck et al. 2120* (CR, US); bordes del Río Segundo, Barba, 1150 m, 4 Dec 1940, *León 404* (CR, F); Dulce Nombre, 1400 m, 27 Feb 1924, *Standley 35853* (US). **Guanacaste:** The Raphael Lucas Rodríguez National Wildlife Refuge, Palo Verde, valley of the Río Tempisque, 10–100 m, 29–31 Jan 1982, *Burger et al. 11342* (CR, F); Finca Taboga, Cañas, (10°20'N, 85°10'W), 8 Dec 1969, *Daubenmire 484* (CR, F); Finca La Pacífica, 16 Jan 1969, *Gentry 311* (F, MO); San Pablo a Nicoya, (9°58'N, 85°30'W), 50 m, 27 Jan 1942, *León 927* (CR, F); Playa Naranjo, Santa Rosa National Park, 10°48'N, 85°41'W, 23 Jan 1978, *Liesner 4396* (CR, MO); between Liberia and Park at Santa María, 10°45'N, 85°17'W, 500 m, 26 Jan 1978, *Liesner 4575* (CR, MO); Santa Rosa National Park, 10°53'N, 85°37'W, 200–300 m, 6 Feb 1978, *Liesner 5043* (CR, MO); La Pacífica, 4 km NW of Cañas, 50 m, 7 Jan 1974, *Opler 2024* (CR, F, MO, UC); 3 km al E de Cuajinicuil y a 11 del parque Murciélago, 160 m, 26 Jan 1983, *Sousa et al. 12717* (MO); Líbano, 260–360 m, 15 Jan 1926, *Standley et al. 44881* (US); 15 mi NW of Liberia near Río Tempisque, (10°30'N, 85°35'W), 150 m, 12 Feb 1963, *Williams et al. 24548* (EAP, F). **Heredia:** Santo Domingo, Santo Tomás, puente Río Tibás, 9°58'N, 84°04'W, 1160 m, 18 Dec 1992, *Hammel 18640* (CR, MO); Barba, bordes del Río Segundo, 1150 m, 4 Dec 1940, *León 404* (CR). **Puntarenas:** Sabalito, 5 km E de La Lucha hacia Las Mellizas, 8°53'N, 82°47'W, 1240 m, 6 Feb 1998, *Debouck et al. 3114* (CR, MO); Monteverde, 0.1 km S de Los Cerros, 10°16'N, 84°51'W, 9 Feb 1998, *Debouck et al. 3123* (CR); alrededores de Río Potrero 20 km SE de Liberia, 10 Feb 1981, *Téllez et al. 4375* (CR, MEXU). **San José:** Aserrí, 1 km SW de Aserrí, cerca de Saurez, 0.2 km antes Km 13 de Ruta 4, 9°51'N, 84°06'W, 1420 m, 8 Jan 1987, *Debouck et al. 2094* (CR, US); 1 km N de Vuelta de Jorco, 9°48'N, 84°07'W, 1450 m, 8 Jan 1987, *Debouck et al. 2099* (CR, US); en El Llano de Alajuelita, 9°53'N, 84°07'W, 1520 m, 8 Jan 1987, *Debouck et al. 2100* (CR, US); 1 km SE de Buenavista, valle del Río Grande de Candelaria, 9°46'N, 84°7'W, 1080 m, 10 Jan 1987, *Debouck et al. 2108* (BR, CR, US); 4.0 km E de Santa María a Copey, 1800 m, 12 Jan 1987, *Debouck et al. 2123* (CR, US); San Isidro El General, 0.8 km S de Pueblo Nuevo, 9°26'N, 83°40'W, 970 m, 13 Jan 1987, *Debouck et al. 2124* (CR, UC, US); San Isidro El General, 1.5 km NE de Chirimol a Canaán, antes de Zapotal, 9°27'N, 83°37'W, 1160 m, 13 Jan 1987, *Debouck et al. 2127* (BR, CR, US); Parque Bolívar, 8 May 1949, *Echeverría 1615* (EAP); Alto de Las Palomas, paso ventoso a 11 km de San José al valle de Santa Ana, 1000 m,

22 Dec 1965, *Jiménez 3495* (CR, F, US); Piedades de Santa Ana, 9°56'N, 84°13'W, 880 m, 11 Dec 1976, *Lent 3948* (CR, F); La Verbena, 1200 m, 29 Jan 1924, *Standley 32292* (US); San José, 1150 m, 29 Feb 1924, *Standley 33263* (US); Las Pavas, 1070 m, 29 Feb 1924, *Standley 36073* (US); Buenos Aires, *Tonduz 4992* (CR, US); Buissons au bord du Río María Aguilar, 1135 m, 27 Nov 1894, *Tonduz 8964* (CR, US).

EL SALVADOR. Ahuachapán: El Refugio, 13°49'N, 89°56'W, 450 m, 9 Feb 1998, *Sandoval 1753* (MO); Apaneca, (13°45'N, 89°50'W), 1400–1600 m, 24 Jan 1947, *Standley et al. 3005* (F). **La Libertad:** Plan de la Laguna, 13°40'N, 89°14'W, 840 m, 15 Feb 1993, *Candelario s.n.* (MO); Volcán de San Salvador, 1680–1860 m, 30 Jan 1946, *Carlson 376* (F); forest El Boquerón, Volcán San Salvador, (13°30'N, 89°20'W), 24 m, Feb 1968, *Molina et al. 21671* (EAP, F, US); Atéos, 17 Apr 1922, *Standley 23388* (GH, US). **San Miguel:** San Miguel, 110 m, 24–27 Feb 1922, *Standley 21125* (GH, US); Montecristo, 15 km NE of San Miguel, 13°36'N, 88°4'W, 200 m, 6 Dec 1941, *Tucker 471* (F, K, MICH, SI, UC, US); Hacienda Potrero Santo, S side of Lake Olomega, 13°17'N, 88°04'W, 60 m, 27 Jan 1942, *Tucker 842* (F, G, K, MICH, SI, TEX, UC, US). **San Salvador:** Santo Tomás, 1922, *Calderón 1301* (US); San Salvador, 650–850 m, Jan 1923, *Calderón 1435* (US(4)); San Martín, Jan 1924, *Calderón 1906* (US(2)); San Andrés, 7 Feb 1972, *Mayorga et al. 30* (NA). **San Vicente:** San Vicente, (13°30'N, 88°45'W), 400–500 m, 7–14 Feb 1947, *Standley et al. 3646* (EAP, F). **Sonsonate:** Izalco, 110 m, 19–24 Mar 1922, *Standley 21860* (GH, US); Santa Emilia, 135 m, 22–25 Mar 1922, *Standley 22120* (GH, US). **Unión:** La Unión, 150 m, 13–21 Feb 1922, *Standley 20805* (US).

GUATEMALA. Alta Verapaz: Pancajché, (15°40'N, 89°55'W), 360 m, 5 Apr 1939, *Standley 70646* (F); between San Cristobal Verapaz and Chixoy, (15°20'N, 90°30'W), 1200–1300 m, 19 Feb 1942, *Steyermark 43900* (F); San Juan Chamelco, 1500–1600 m, 1968–69, *Wilson 41002* (F). **Chimaltenango:** Párramos, 1650–1800 m, 14 Dec 1938, *Standley 59876* (F). **Chiquimula:** Esquipulas creek, 2 km from Esquipulas, (14°30'N, 89°20'W), 1000 m, 6 Dec 1969, *Molina et al. 15183* (EAP); Atulapa, 800 m, 11 Dec 1969, *Molina et al. 25384* (EAP, F, MO, US). **El Progreso:** San Agustín Acasaguastlán, 3 Km N of Magdalena, Río Hato, 14°56'N, 89°57'W, 270 m, 2 Dec 1987, *Debouck et al. 2413* (COL, G, M, USCG); San Agustín Acasaguastlán, 2 km N de Chanrayo, 15°02'N, 89°56'W, 750 m, 3 Dec 1987, *Debouck et al. 2416* (COL, MICH, US, USCG). **Escuintla:** 2.0 km W El Rodeo, Escuintla, 14°24'N, 90°51'W, 820 m, 5 Dec 1985, *Debouck et al. 1614* (BR, SI, US, USCG); 3 km W de San Vicente Pacaya, 14°24'N, 90°38'W, 1430 m, 14 Dec 1987, *Debouck et al. 2456* (COL, G, MICH, US, USCG); along Río Guacalate, 500–550 m, 28 Nov 1938, *Standley 58173* (F); San José, 0 m, 30–31 Jan 1939, *Standley 64001* (F); Río Michatoya, SE of Escuintla, (14°10'N, 90°45'W), 250–300 m, 12 Mar 1941, *Standley 89146* (F). **Guatemala:** Finca La Aurora, 1500 m, 1938–1939, *Aguilar 209* (F); 0.3 km E de Palencia en camino hacia Los Cubos, 14°41'N, 90°22'W, 1300 m, *Debouck et al. 2411* (COL, US, USCG); Mixco, 1 km N de Ciudad Quetzal (antiguo camino de Guatemala a San Raimundo), 14°39'N, 90°34'W, 1480 m, 18 Dec 1987, *Debouck et al. 2479* (BR, MICH, US, USCG); River Villalobos, 1300 m, 12–23 Jan 1966, *Molina et al. 16074* (EAP, F, GH, US); Cuesta de Villalobos, 1100 m, Jan 1923, *Salas 199* (US). **Izabal:** Quirigua, (15°20'N, 89°5'W), 75–225 m, 15–31 May 1922, *Standley 24479* (GH, US). **Jalapa:** El Límite, 9 km NW San Luís Jilotepeque, 14°40'N, 89°47'W, 1280 m, 5 Dec 1987, *Debouck et al. 2429* (COL, MICH, USCG); Laguna de Ayarza, (14°20'N, 90°5'W), 2438 m, Oct 1892, *Heyde et Lux 4171* (in part) (G, GH, US(2)). **Petén:** El Caribal trail, Lacandón, (17°10'N, 91°10'W), 2 Feb 1962, *Contreras 3282* (F, TEX- LL); Tikal National Park, Tikal, (17°5'N, 89°40'W), 11 Mar 1966, *Contreras 5545* (EAP, ENCB, F, LIL, MO); Lake Petén Itzá, Sta. Elena, S of old Santa Ana Road, (16°50'N, 89°50'W), 11 May 1966, *Contreras 5726* (CAS-DS, F, TEX, TEX-LL, US); San Miguel, on El Boquerón, 5 Feb 1968, *Contreras 7538* (CAS, F, TEX-LL); Isla Grande, Lake Petén Itzá, NE of Flores, 6 Feb 1968, *Contreras 7550* (CAS, F, TEX-LL); Dolores, on Río Mopán, (16°25'N, 89°30'W), 21 Feb 1971, *Contreras 10578* (CAS, F, TEX-LL); 2 km N de San Miguel, (17°45'N, 89°40'W), 21 Jan 1969, *Tun 128* (EAP, F, MICH, US); Km 69 camino para el Remate, Parque Nacional, Tikal, Feb 1970, *Tun 719* (EAP, F, MICH, US); Río Mopán, 5 km de Dolores, 21 Feb 1971, *Tun 1653* (BM, EAP, F, MICH, US). **Quetzaltenango:** 2 km W of Zunil, Sierra Madre above Río Samala, 2300 m, 14 Dec 1942, *Williams et al. 22996* (EAP). **Retalhuleu:** Río Coyote, 4 km W of Retalhuleu, 300 m, 17 Feb 1941, *Standley 87414* (F); plains between Nueva Linda and Champerico, (14°15'N, 91°45'W), 120 m, 18 Feb 1941, *Standley 87624* (F). **Sacatepéquez:** 5 km S Santa María de Jesús, 14°29'N, 90°43'W, 1740 m, 4 Dec 1985, *Debouck et al. 1607* (BR, COL, SI, US, USCG); 12 km S Alotenango, cerca de Finca San José Las Lajas, 14°25'N, 90°49'W, 1000 m, 5 Dec 1985, *Debouck et al. 1613* (BR, SI, US); Volcán de Agua, S of Santa María de Jesús, 1800–2100 m, 10 Dec 1938, *Standley 59452* (F); Antigua, 1500–1600 m, Nov 1938–Feb 1939, *Standley 61716* (F); Dueñas, 1590–1800 m, 21 Jan 1939, *Standley 63218* (F); Río Guacalate, between Antigua and Chimaltenango, 1660 m, 23 Dec 1940, *Standley 80970* (F). **San Marcos:** 1 km S San Rafael Pie de la Cuesta, Cuenca del Río Cabuz, 14°55'N, 91°56'W, 900 m, 11 Dec 1985, *Debouck et al. 1636* (BR, COL, K, SI, US, USCG, WIS); Ayutla, 45 m, 14–15 Mar 1939, *Standley 68831* (F); between Ocos and Ayutla, 5–20 m, 25 Mar 1940, *Steyermark 37911* (F); S of San Marcos toward Castalia, 1600–2200 m, 16 Dec 1963, *Williams et al. 26159* (EAP, F). **Santa Rosa:** San Rafael de las Flores, 1 Km N of Laguna de Ayarza, 14°27'N, 90°08'W, 1380 m, 7 Dec 1987, *Debouck et al. 2441* (COL, G, USCG); SE of Barberena, 1100–1180 m, 21 Nov 1940, *Standley 77721* (F); Cuilapilla, 900 m, 23 Nov 1940, *Standley 78088* (F); El Ahumado, S of Los Cerritos, Region of Capulín, 75 m, 7 Dec 1940, *Standley 79630* (F). **Sololá:** Panajachel, (14°40'N, 91°15'W), 30 Jan 1915, *Holway 159* (US); 3–5 km W of Panajachel, 2100 m, 6–7 Dec 1963, *Williams et al. 25313* (EAP, F(2), GH, US, WIS). **Suchitepéquez:** Cocales, 215 m, 5 Jan 1939, *Standley 62060* (F). **Zacapa:** Gualán, (15°N, 89°25'W), 128 m, 18 Jan 1905, *Deam 306* (GH, MICH, MO, US); above Teculután, Río Teculután, 250–275 m, 7 Jan. 1942, *Steyermark 42137* (F, G).

HONDURAS. Atlantida: Tela, (15°40'N, 87°30'W), 0 m, 14 Dec 1927–15 Mar. 1928, *Standley 55163* (F, US). **Colón:** base of cerro Piedra Blanca, 5 km NE of Bonito Oriental, 15°46'N, 85°41'W, 40 m, 5 Feb 1993, *Evans 1081* (MO). **Comayagua:** Comayagua, (14°25'N, 87°40'W), 600 m, 12–23 Mar 1947, *Standley et al. 5321* (F); El Banco, 640 m, 13 Mar 1945, *Valerio 2316* (EAP, F, LIL). **Cortés:** 5 km de San Pedro Sula, Río Arenales, El Sauce, (15°25'N, 88°5'W), 40 m, 18 Dec 1950, *Molina 3835* (EAP, F, GH, SI); Pulhapanzak waterfall of Río Lindo, (15°N, 87°55'W), 500 m, 12 Dec 1982, *Molina et al. 31963* (EAP). **El Paraíso:** Danlí-El Paraíso, 22 Feb 1952, *Carlson 2514* (EAP); 10 km de El Paraíso, (13°45'N, 86°35'W), 700 m, 23 Feb 1952, *Molina 5074* (EAP, F, GH); Montaña Agua Fría, 1300 m, 14 Mar 1956, *Molina 7427* (EAP); between Danlí and La Emilia, NE of Danlí, 740–900 m, 22 Feb 1949, *Standley 16158* (EAP, F); Quebrada el Ingenio de Los Angeles, 1 km de Yuscarán, 1050 m, 28 Dec 1962, *Williams et al. 11191* (EAP, F, G, GH, US). **Fco. Morazán:** Quebrada UNAH, Tegucigalpa, (13°55'N, 87°15'W), 1000 m, 15 Feb 1977, *Alduvin 84* (MO); Playas del Río Yeguare, N del Zamorano,

850 m, 19 May 1948, *Molina 880* (EAP, F, GH); Monte Redondo, El Zamorano, 800 m, 9 Dec 1971, *Molina 27170* (EAP, F, MICH, US); Uyuca, 1238 m, Nov 1943, *Valerio 1596* (EAP, F); San Francisco, Río Yeguare Valley, 800 m, 4 Jan 1947, *Williams et al. 11477* (EAP); Río de La Orilla, near Cerro Majicarán, Yeguare Valley, 900 m, 14 Dec 1948, *Williams et al. 14845* (EAP, F, GH). **Olancho:** Juticalpa, (14°35'N, 86°15'W), 380–480 m, 5–16 Mar 1949, *Standley 17115* (EAP, F). **Yoro:** Guaymas River and Hwy., 22 Feb 1962, *Dickson 560* (EAP).

MÉXICO. Baja California: San José del Cabo, (23°N, 109°45'W), Feb 1899, *Grabendorfer s.n.* (UC); María Magdalena Is., Tres Marias Group, (25°N, 112°15'W), 27 May 1897, *Maltby 166* (US); Cerro de los Dientes, S of Mt. Evermann, Socorro Is., (25°10'N, 110°57 3/4'W, 11 Mar 1957, *Moran 5787* (CAS-DS, UC). **Campeche:** Mpio. Champotón, Cd. del Carmen, 10 km al SW de Champotón, (19°25'N, 90°50'W), 2 m, 9 Feb 1983, *Martínez et al. 3037* (MO (2)). **Chiapas:** Mpio. Villa Corzo, E base of Cerro Tres Picos near Cerro Bola along a logging road SW of Colonia Agrónomos Mexicanos, (16°45'N, 93°W), 1500 m, 9 Feb 1972, *Breedlove 23982* (CAS-DS, TEX); Mpio. Chiapilla, Río Grijalva near Chiapilla on the road from Acala, (16°30'N, 92°45'W), 549 m, 25 Feb 1966, *Laughlin 237* (CAS-DS, MEXU, US); Mpio. Venustiano Carranza, Rancho Carmen along the road from Acala–Venustiano Carranza, 762 m, 30 Nov 1966, *Laughlin 2880* (CAS-DS, MEXU, US); Escuintla, (15°15'N, 92°45'W), Sep 1937, *Matuda 9* (MEXU, US); Cascada, Siltepec, 1600 m, Feb 1945, *Matuda 5071* (MEXU, US); Jilguero, Escuintla, 27 Nov 1947, *Matuda 17288* (EAP); Hacienda Monserrate, 2 km from Monserrate, 16°33'N, 94°W, *Purpus 10279* (US); Cintalapa, 17 Feb 1896, *Seler 2112* (CAS, GH, US); 8 km al NE de San Fernando a Chicoasén, 850 m, 16 Dec 1980, *Sousa et al. 11611* (CAS, CR, MO); Cascada El Salvador, 5 km al N de Tapilula, 550 m, 8 Mar 1981, *Sousa et al. 11795* (MO); Mpio. de Venustiano Carranza, Soyatitlán, Las Rosas to Pugiltik, 1036 m, 25 Oct 1967, *Ton 3153* (ENCB, US); Mpio. de Ocosingo, Rancho Mumuntik, Ocosingo, (16°55'N, 92°W), 1280 m, 10 Jan 1968, *Ton 3434* (WIS). **Colima:** Mpio. Ixtlahuacan, Las Conchas (junto al Río Coahuayana), camino Ixtlahuacan–Tecomán, (18°45'N, 103°45'W), 24 Nov 1981, *Arturo et al. 3315* (MO); Mpio. Colima, Hacienda Albarradita, 1 km W of Colima, 13 Mar 1943, *Gilly et al. 55* (MICH); Mpio. Manzanillo, 12 km N de Manzanillo, carr. Cihuatlán, (19°10'N, 104°30'W), 9 Mar 1981, *Magallanes 2833* (MICH); puente Cihuatlán, carr. Cihuatlán–Manzanillo, 21 Mar 1980, *Magallanes et al. 2201* (CAS); 9–10 km E (or SE) by winding road from Minatitlán, 1100–1350 m, 10–11 Feb 1975, *McVaugh 26213* (MICH). **Guerrero:** Pungarabato, Coyuca, 9 Jan 1934, *Hinton et al. 5439* (G, GH, K); Vallecitos, Montes de Oca, 12 Jan 1938, *Hinton et al. 11742* (CAS-DS, GH, MEXU, MICH, US); Delta de la Balsa, 0 m, Jun 1898, *Langlassé 958* (G); Mpio. La Unión, 91 km al NE de Zihuatanejo, camino a Cd. Altamirano, (18°5'N, 100°30'W), 1890 m, 20 Nov 1983, *Martínez et al. 5491* (CAS). **Jalisco:** Mpio. La Huerta, Rancho Cuixmala, 19°23'N, 104°59'W, 2 Feb 1991, *Ayala 91-2* (UCR); Casimiro Castillo, valley of Río Purificación, El Arado, 19°36'N, 104°33'W, 440 m, 28 Nov 1978, *Debouck 473* (CHAPA, G, K) ; Pto. Vallarta, (20°40'N, 105°25'W), 20 m, 25 Jan 1977, *Delgado et al. 472* (CAS, MEXU); Km 176 Hwy 80 Autlán-Barra Navidad, 875 m, 4 Dec 1981, *Freytag et al. 81-29* (US); 6 mi S of Tecalitlán, 1219 m, 23 Nov 1966, *Gentry 22200* (US); Tonila, Colima–Jalisco border, (19°30'N, 103°30'W), 1128 m, 20 Nov 1966, *Gentry et al. 22183* (GH, NA, US); UNAM Biological Field Station, 7–8 km SE of Chamela, 122 m, 12 Feb 1975, *Gentry et al. 23538* (US); 4 mi E of Cihuatlán, 8 Mar 1973, *Johnson 422-73* (MO, US); Mpio. La Huerta, Rancho Cuixmala, W of the Puerto Vallarta-Barra de Navidad (Hwy. Mex 200), along the Río Cuixmala, 12 Jan 1991, *Lott et al. 3169* (UCR); Arroyo "El Colorado," 24 Feb. 1977, *Magallanes 532* (CAS); Mpio. Comala, 20 km NE de Comala, carr. a San Antonio, 7 Mar 1981, *Magallanes 2811* (MICH); Quimixto, Trail to San Pedro El Tuito, 60 m, 2 Dec 1926, *Mexía 1218* (CAS, CAS-DS, F, G, GH, MO, NA, UC); San Sebastian, Arroyo Seco, 1500 m, 19 Jan 1927, *Mexía 1525* (CAS, MICH); Mpio. La Huerta, Hotel El Tecuán, just N of the mouth of the Río Purificación, 1–5 m, 22 Jan 1989, *Sanders 9007* (UCR). **México:** Dist. of Temascaltepec, Nanchititla, (19°2'N, 100°2'W), Jan 1933, *Hinton 3100* (G, K, US); idem, 10 Jan 1933, *Hinton 3121* (K, MA, MEXU, MO); Dist. of Temascaltepec, Guayabal, 23 Dec 1933, 790 m, *Hinton 5383* (F, G, K, US). **Michoacán:** delta de Las Balsas, 5 Jun 1898, *Langlassé 204* (GH, K, US); Metzaltepec (?), delta de Las Balsas, (18°N, 102°12'W), 0 m, Jun 1898, *Langlassé 958* (GH, K, US); 45–48 km S of Arteaga, 12–14 km N of Playa Azul, (18°10'N, 102°40'W), 150–200 m, 25 Feb 1965, *McVaugh 22593* (CAS, ENCB, MICH); 14 km al N de Playa Azul, carr. a Nueva Italia, 250 m, 24 Mar 1980, *Soto et al. 2093* (TEX); 10 km al NE de Coalcomán, carr. a Aguililla, (18°45'N, 103°0'W), 1400 m, 28 Mar 1980, *Soto et al. 2147* (CAS); Mpio. Coahuayana, 26 km al NE de El Ranchito, en Atenquique, (18°45'N, 103°25'W), 27 Mar 1981, *Soto et al. 2807* (CAS). **Morelos:** Cuautla, (18°47'N, 99°0'W), 1500 m, 29 Oct 1900, *Pringle 8353* (ENCB, F, G, GH, K, MEXU, MICH, MO, UC, US). **Nayarit:** 3 mi SE of Tetitlán, 823 m, 24 Nov 1966, *Gentry 22236* (US); 43 mi SW of Compostela, 28 Feb 1973, *Johnson 262-73* (MO); 3 mi SE of Sayulita, 1 Mar 1973, *Johnson 285-73* (MO); Santiago, (21°50'N, 105°25'W), Feb 1895, *Lamb 547* (GH); Acaponeta, (22°30'N, 105°25'W), 23–30 Jun 1897, *Rose 3132* (MO, US). **Oaxaca:** Tehuantepec, 15 Aug 1907, *Conzatti 1972* (F); Mpio. Pochutla, 8 km al E de Chacalapa, (15°45'N, 96°30'W), 370 m, 24 Nov 1977, *Delgado et al. 691* (CAS); Dist. of Tuxtepec, Chiltepec, (18°0'N, 96°10'W), 20 m, 15 Feb 1941, *Martínez-Calderón 540* (GH, MEXU, TEX-LL, UC, US); 1 km del Valle Nacional hacia Arroyo Colorado, (17°45'N, 96°25'W), 27 Feb 1976, *Shapiro 399* (CAS, ENCB, MEXU, US, WIS); Mpio. La Ciénega, SE de La Ciénega, 1500 m, 11 Oct 1978, *Solano et al. 177* (CAS, F, MO); 10 km al SE de Jamiltepec, camino a Puerto Escondido, 150 m, 11 Feb 1976, *Sousa et al. 5257* (CAS, MEXU, US, WIS); Mpio. Tututepec, 27 km al E de Río Verde, (16°10'N, 97°30'W), 25 m, 11 Feb 1976, *Sousa et al. 5268* (CAS, MEXU, US, WIS(2)); El Llano a 20 km al E de Pinotepa Nacional, (16°20'N, 97°55'W), 220 m, 19 Apr 1976, *Sousa et al. 5540* (ARIZ, MEXU, MICH, MO); 24 km al SE de Puerto Escondido, (15°57'N, 96°55'W), 80 m, 21 Apr 1976, *Sousa et al. 5574* (ARIZ, ASU, MEXU, MICH, MO); Dist. Juquila, 7 km NW de Río Grande, (16°15'N, 95°20'W), 20 m, 5 Feb 1977, *Sousa et al. 7077* (CAS, MEXU, MO); Dist. Mixe, Ejido Díaz Ordaz, 12 km al NW de la Mixtequita, carr. a Tuxtepec, 40 m, 23 Feb 1978, *Sousa et al. 9247* (CAS); Dist. Juchitán, 11 km al NE de Almoloya, carr. Chivala–Matías Romero, 150 m, 18 Dec 1978, *Sousa et al. 10218* (ASU, CAS); Dist. Juchitán, Yerba Santa, 11 km al N de Matías Romero, (17°N, 95°W), 150 m, 19 Dec 1978, *Sousa et al. 10253* (ASU, CAS); Ubero, 30–90 m, Apr 1937, *Williams 9172* (F). **Sinaloa:** Mpio. de Concordia, El Magistral, 15 km al E de Concordia, carr. Mazatlán–Durango, 230 m, 7 Dec 1982, *Aguilar et al. 120* (CAS, MO, TEX-LL); below the settlement of Tarahumarés, Cañon Tarahumarés, Sierra Surotato, (25°50'N, 107°40'W)?, 1067 m, 20 Feb 1969, *Breedlove 16437* (MICH); 30 km W of Concordia, 1219 m, Dec 1977, *Mankel s.n.* SU); 5 mi SW of Chupaderos, 274 m, 26 Mar 1976, *Marín M76-111* (ARIZ); Coyotitán a San Miguel, San Ignacio, (23°58'N, 106°30'W), 10 Oct 1918, *Narváez et al. 655* (US); Balboa, 1923, *Ortega 5147* (US); Mazatlán, (23°15'N, 106°30'W), 3 Mar 1910, *Rose et al. 13763* (US); Villa

Unión, (23°45'N, 106°W), 2 Apr 1910, *Rose et al. 13962* (GH, US); vicinity of Guadelupe, 18 Apr 1910, *Rose et al. 14800* (NY, US); canyon below Cerro de Elefante, 5.3 mi NE of Concordia along Hwy 40, 23°18'N, 106°W, 198 m, 27 Dec 1983, *Sanders et al. 4290* (TEX, UCR); Mpio. de Rosario, camino de herradura El Alamo-Cerro Yauco, 6 km al NE del Ejido de Lo de Ponce, 158 m, 2 Dec 1982, *Tenorio et al. 2861* (CAS, MO); Mpio. San Ignacio, 2 km al N de San Jerónimo, brecha a El Guayabo, 370 m, 19 Mar 1985, *Tenorio et al. 8469* (CAS). **Tabasco:** Cárdenas, (18°N, 93°10'W), 1972, *González-Marín 6845-I* (US); Mpio. Paraíso, Rancho el Chinal en la calle Aquiles Serdán, (18°30'N, 93°10'W), 18 Mar 1980, *Magaña 41* (CAS). **Tamaulipas:** 1 mi N of Cd. Madero, 1 Mar 1961, *King 4000* (F, MICH, NY, TEX, US); Mpio. Aldama, cerro de torre de microondas, Las Palmas, Ejido Cuauhtémoc, (23°N, 98°15'W), 220 m, 27 Dec 1971, *Martínez-Ojeda 336* (MEXU, MO(2)). **Veracruz:** Vallée de Cordova, 15 Dec 1865, *Bourgeau 1335* (GH); Sihuapán, 9 Jan 1973, *Calzada 922* (F); Mpio. Emiliano Zapata, desviación para Los Baños de Carrizal, carr. Xalapa, 400 m, 14 Feb 1976, *Calzada 2189* (CR, F); Mpio. de Catemaco, 1 km al SE de Sontecomapán, (18°30'N, 95°35'W), 60 m, 15 Mar 1978, *Delgado et al. 998* (CAS); Papantla, (20°30'N, 97°W), 154 m, 14 Mar 1898, *Goldman 84* (GH, US); Catemaco, San Andrés Tuxtla, 350 m, 9 Dec 1963, *McKee 10991* (US); Sanborn, 18 Apr 1910, *Orcutt 3450* (F); Island of Juana Ramírez, about 56 km S of Tampico, (21°45'N, 97°45'W), 0 m, 8-9 Mar 1910, *Palmer 468* (GH, MO, US); Mpio. San Andrés Tuxtla, Cebollal, 15 Feb 1974, *Ponce 127* (F, MEXU); Zacuapán, Nov 1908, *Purpus 3679* (UC); idem, Oct 1917, *Purpus 8002* (GH, MO(2), UC, US); Isthmus of Tehuantepec, Coatzacoalcos, (18°10'N, 94°30'W), 8 Mar 1895, *Smith 996* (ENCB, F, GH, MEXU, MO(2), SI, UC, US); Mpio. Actopan, Guajillo, 150 m, 3 Feb 1973, *Ventura 7793* (CAS, MEXU); Mpio. Puente Nacional, Mata de Caña, 100 m, 13 Mar 1973, *Ventura 7991* (ASU, MICH); Mpio. Boca del Río, Mocambo, 5 m, 15 Oct 1973, *Ventura 9116* (MEXU, MICH, MO); Mpio. Jilotepec, San Martín, 900 m, 22 Jan 1977, *Ventura 13772* (CAS); Mpio. Puente Nacional, Río Escondido, 150 m, 4 Dec 1981, *Ventura 19186* (ASU). **Yucatán:** ruins of Mayapán, (20°30'N, 89°30'W), 3 Jan 1982, *Darwin et al. 2132* (CR, F, MO); Izamal, (20°50'N, 89°W), 1895, *Gaumer 564* (CAS, CAS-DS, F, G, K, MICH, MO, UC, WIS); Suitun woods, Nov 1916, *Gaumer 23459* (F, G, GH, MO, US); El Palmer, 15 Mar 1944, *Martin 1* (NA); Mpio. Ignacio Llave, Ejido Palmas Cuata, 6 m, 18 Dec 1966, *Martínez-Calderón 1220* (CAS, F(2), GH, MICH, MO, US).

NICARAGUA. Chinandega: Chinandega, (12°25'N, 87°10'W), 21 Jan 1903, *Baker 695* (US). **Estelí:** 5-7 km E of Estelí, (12°55'N, 86°20'W), 900 m, 26 Nov 1973, *Williams et al. 42470* (EAP, F, MICH, US). **Granada:** Lake Nicaragua, Island Ometepe, (11°25'N, 85°40'W), Jan 1893, *Smith 6* (EAP, F). **Jinotega:** La Bastilla, 10 km NE of Jinotega, (13°N, 87°W), 11 Jan 1969, *Zelaya 2143* (F, UC). **Managua:** Masaya, 27 Jan 1903, *Baker 117* (CAS, CAS-DS, GH(2), LIL, MICH, SI, UC); Sierras de Managua, 900 m, 1 Jun 1941, *Grant 941* (F, GH); S of Managua, 3 Mar 192?, *Greenman et al. 5709* (GH); Casa Colorada, (11°45'N, 86°25'W), 914 m, 1 Nov 1957, *Salas et al. 2048* (EAP). **Zelaya:** Rama, (12°N, 84°15'W), 1912, *Berger s.n.* (NA); Río Grande, (12°50'N, 84°15'W), 0-15 m, 21 Apr 1949, *Molina 2278* (EAP, F, GH); ca. 26 km E of Río Blanco along new road from Río Blanco to Río Copalar, 12°55'N, 85°05'W, 200-400 m, 14 Feb 1979, *Stevens 12251* (SI).

PANAMÁ. Canal Zone: Isla Perico (near Fort Amador causeway), (9°N, 79°50'W)?, 1 Jan 1970, *McDaniel et al. 12690* (MO); Sosa Hill, Balboa, (8°45'N, 79°35'W), 27 Nov-10 Dec 1923, *Standley 25243* (US). **Chiriquí:** Quebrada Guanabanito beyond La Represa 2 mi SW of Puerto Armuelles, (8°5'N, 82°50'W), 0-200 m, 20 Feb 1973, *Croat 22076* (CAS, F, MO, NA); from Boquete to 3 mi N, (8°35'N, 82°25'W), 1006-1280 m, 12-13 Dec 1966, *Lewis et al. 318* (F, K, MEXU, MO, UC); Boquete, Feb 1849, *Seemann 1675* (GH). **Herrera:** Los Pozos, (7°35'N, 80°40'W), 20 m, 10 Mar 1971, *Burt et al. 95* (MO, TEX-LL). **Panamá:** Bella Vista, 28 Nov 1923, *Standley 25304* (US); Las Sabanas, 4 Dec 1923, *Standley 25932* (US); Tumba Muerto Road, near Panamá, 6 Jan 1924, *Standley 29792* (US). **Veraguas:** Atalaya-Santiago, (7°55'N, 81°0'W), 100-200 m, 18 Feb 1971, *Koster 117* (MO).

Habitat.—Growing along stream banks and in moist areas in deciduous tropical forests of pine-oak, or of Bombacaceae, Cesalpinoideae or Mimosoideae at 0–1600 m. It is usually found in thickets and climbing over shrubs (*Acacia, Acalypha, Annona, Brosimum, Bursera, Celtis, Cordia, Caesalpinia, Dahlia, Desmodium, Dodonaea, Erythrina, Euphorbia, Gliricidia, Lepechinia, Liquidambar, Sida, Stizolobium*, Compositae, palms) or climbing small trees and often invading cut-over lands cleared for coffee, sugar cane, cotton or corn and along fences and paths. It is more common where inaccessible to grazing animals, usually well drained or quite steep slopes. Soils are variable from black clay to brown friable to rocky and sandy (even on sand dunes near beaches) and variously derived from basalt, volcanic ash, limestone, metamorphic schists or volcanic rock.

Diseases and pests.—Small, much branched vine tips may indicate presence of mycoplasmas. Damage is caused by Chrysomelid beetles, *Heliothis*, leafminers, aphids and thrips.

Common names.—Chilipuca colorada, Comba cimarrona, Cubacillo, Frijol de culebra, Frijol de media luna, Frijol silvestre, Frijolillo, Frijolillo de bejuco, Habas, Ib Choh, Iztagapa, Ixtapacal, Petipoa, Rat weed, and Tapakal.

Genetics.—*Phaseolus lunatus* is generally considered distinctly isolated from the gene pools of the four other cultivated species (Debouck 1991; Maréchal et al. 1978b; Schmit et al. 1993). Nevertheless, its secondary and tertiary gene pools are numerous (Debouck & Smartt 1995) and successful crosses were made by Lorz (1952) using a cultivated *P. lunatus* with a wild *P. polystachyus*, but the F_1 was sterile. Doubling of the chromosomes to form an amphidiploid (Fozdar 1963) reestablished a fair level of fertility, producing a nearly perennial hybrid with intermediate characteristics between the two parental species but more like the *P. lunatus* parent with fibrous, falcate pods and large black, flat seed (the senior author has only seen a single plant of the amphidiploid with white flowers and seed). Such hybrids with *P. polystachyus* were also obtained by Le Marchand *et al.* (1976). Successful crosses have also been reported with *P. ritensis* (possibly *P. maculatus maculatus*) though fertility is very low (Le Marchand & Maréchal 1977b), and more recently with *P. jaliscanus*, *P. maculatus ritensis* and *P.* sp. (possibly *P. marechalii*) (see Katanga and Baudoin 1990).

Comments.—Delgado in his thesis of 1985 (following Baudet 1977) recognized two varieties of *P. lunatus*: a) *lunatus* (cultivated), b) *silvester* (wild), and added c) *viridis* (Piper) Delgado, a wild type characterized by glabrous foliage and pods with larger and more numerous seeds than the common wild type. In a recent work, Delgado and co-workers (Delgado et al. 1999) consider the later taxon as a separate species. This seems excessive varietal definition even though Linnaeus apparently did base the original description on cultivated material. The senior author finds there is no great difference between cultivated and wild collections other than larger seed and some other organs and some apparently determinate plant types in cultivars, and a great deal of variability and intergradation in all aspects. The taxa *viridis* and *rosei*, as species or varietal distinctions, are even less tenable as they are based on the amount of vegetative pubescence.

The name *silvester* referred initially to Central American (likely Mexican) wild forms (Baudet 1977) of wide distribution (Gutiérrez Salgado et al. 1995). Indeed, although the distribution detailed above refers to mainland Central America, the tropical form of wild Lima bean extends into the Caribbean Islands and tropical South America east of the Andes (Maquet & Baudoin 1996). Recently, another family of wild forms have been discovered at higher elevations on the western slope of the Andes in southern Ecuador and northern Peru (Debouck et al. 1989b; Debouck et al. 1987). This group of wild Lima beans has been demonstrated to be a precursor of the large-seeded Lima beans on the basis of similarities in lectin electrophoretic patterns (Debouck et al. 1989c; Gutiérrez Salgado et al. 1995; Maquet et al. 1990). On the basis of presently available evidence (Toro Ch. et al. 1993), it is possible that *P. rosei* is the wild ancestral form of the large-seeded Lima beans.

D.I.2.1.—Phaseolus polystachyus (L.) Britt., Sterns & Pogg. subsp. **polystachyus**, Prelim. Catal. N. Y. Pl. 15:281. 1888. (**Figs. 35, 36**). *Dolichos polystachios* L., Sp. Pl. 1022. 1753. TYPE: UNITED STATES. VIRGINIA: "Virginia," *Clayton 568.* (LECTOTYPE: BM designated by Delgado (1985) n.v.; photo GH).

Phaseolus perennis Walt., Fl. Carol. 182. 1788. TYPE: UNITED STATES. "CAROLINA": no specimen designated.

Phaseolus paniculatus Michaux, Fl. Bor. Amer. 2:60. 1802. TYPE: UNITED STATES. ILLINOIS: In regione Illinoensi, *Michaux s.n.* (LECTOTYPE: P, designated by Delgado (1985) n.v.; photo GH; ISOTYPE: GH).

Phaseolus polystachios var. *aquilonius* Fernald, Rhodora 44:418–420. 1942. TYPE: UNITED STATES. CONNECTICUT: Franklin, 29 Sep 1906 *Woodward s.n.* (HOLOTYPE: GH; ISOTYPE: GH).

Aerial shoot usually large, annual, climbing, indeterminate vines, to 3–4 m long. ***Root*** a large, perennial, thick, fleshy and much branched, to 20 cm long, individual branches to 2.5 cm thick, with a corky cortex. ***Stems*** terete, very long and scandent, glabrous. ***Stipules*** small triangular, often reflexed, glabrous. ***Leaves*** 16–17 cm long; petiole about 3 cm long, glabrous; petiolule 1.5 cm long, glabrous; pulvini 2–3 mm long; stipels 2–3 mm long, ligulate; terminal leaflet 12 cm long, 9.7 cm wide at 1/3 from base, broadly ovate, acuminate, glabrous; lateral leaflet similar but slightly inequilateral, 13 cm long, 8 cm wide. ***Inflorescence*** a much branched panicle, to 30 cm long; primary bracts ovate 1.5–3 mm long 3-nerved ciliate; pedicel 5 mm long, very sparsely covered with minute hooked hairs; pedicellar bracts lanceolate 1.5 mm long caducous. ***Bracteoles*** oval, minute, hyaline, scale-like, 0.5–0.75 mm long, 0.7 mm wide, sparsely covered by hooked hairs, ciliate, early deciduous. ***Flower*** purple; calyx campanulate, upper lobes united into 1 rounded, deeply emarginate, 1.25 mm long, 4.5 mm wide, lower lobes rounded and acute, 1.25 mm long, 2.25 mm wide, glabrous to puberulent; standard purple (rarely white), somewhat striate, greenish on adaxial surface, broadly rounded, 7–9 mm long, 11 mm wide, sharply reflexed, 3–4 mm to thickened bend, sides enrolled at base, the claw 0.5 mm long, the auricles slightly developed, 0.5 mm long; wings purple (rarely white), the blade broadly orbicular, 13 mm long, 7 mm wide, sharply enrolled lengthwise, spreading, the claw 3–3.5 mm long, the spur pronounced, short, 1 mm broad, loosely adhering to keel; keel purplish (rarely white), apex greenish to white, 11 mm long, reflexed at 5 mm from base and 2.5 mm more to the terminal 1 3/4 coils of 2 mm diam., the claws 2.75 mm long, the auricles well-developed, 1 mm long; vexillary stamen,

FIG. 35. *Phaseolus polystachyus* (L.) Britt., Sterns & Pogg. subsp *polystachyus*.—**a.-1.** Seedling several days after germination from a pinto seed.—**a.-2.** Seedling several days after germination from a black seed.—**b.** Root of mature plant.—**c.** Basal stem and vine tip with mature leaves and inflorescences; note short, lateral secondary branches of inflorescences.—**d.** Flowers, side view and front view.—**e.** Exploded view of the flower showing all parts, including—**s.** Style tip and stigma as seen under the microscope; note minute bracteoles and striped standard.—**f.** Pods, side view and dehisced carpels.—**g.** Seeds, side view and view from the hilum. Drawings from living material grown in greenhouse at Mayagüez as follows: all drawings from TARS #127 (Basset) from Univ. of Florida Hort. Farm, Gainesville, Florida except **a-2.** seedling from TARS #374 (black seed) from Coal Co., Oklahoma.

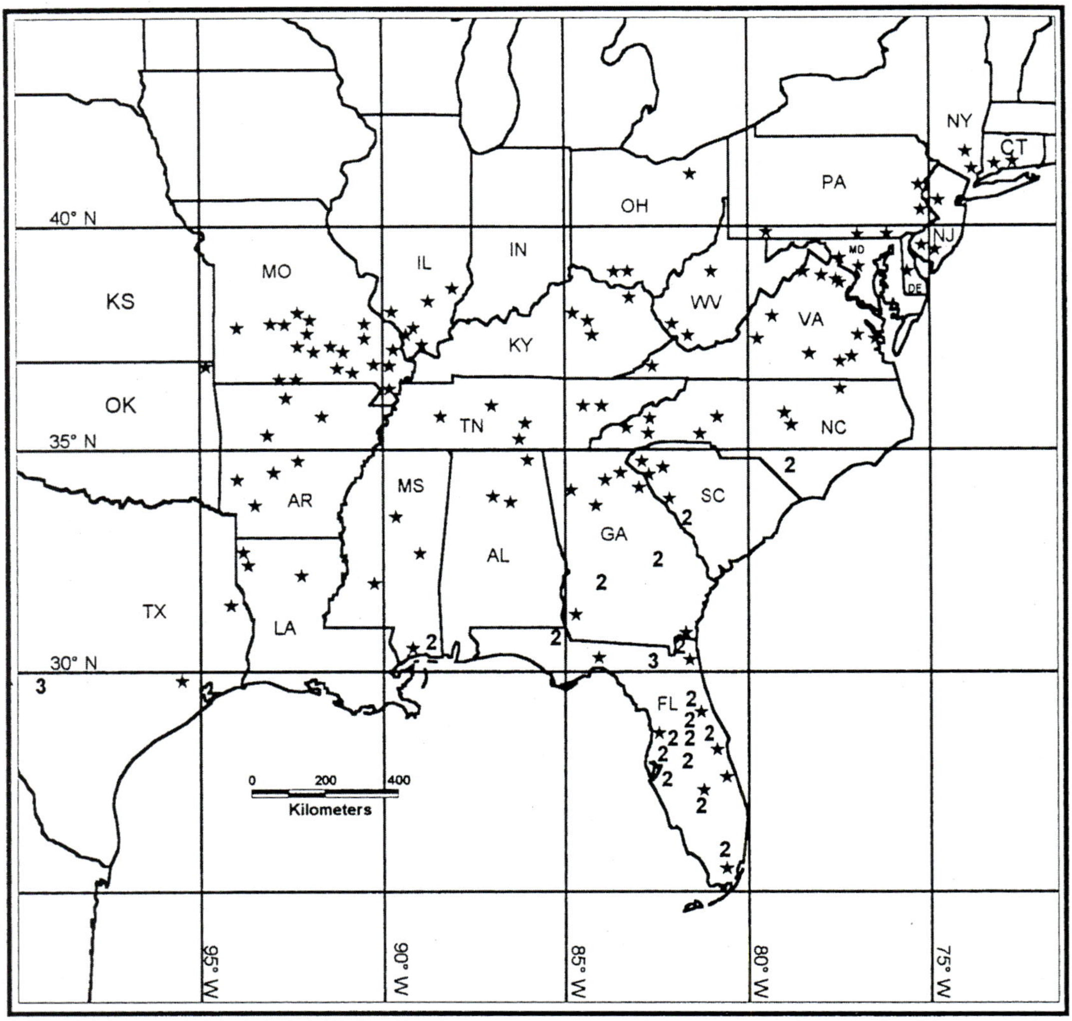

Fig. 36. Distribution of subspecies of *Phaseolus polystachyus* (L.) Britt., Sterns & Pogg. of Section D. *Paniculati*, Subsection I. *Volubili*, as follows: * = subsp. *polystachyus*; 2 = subsp. *sinuatus*; 3 = subsp. *smilacifolius*.

the claw 0.5–1 mm long, the geniculate knob 1 mm wide, 0.5 mm high; stamen tube 9 mm long, 5 mm to bend, the ridges scarcely developed; basal collar 0.75–1.25 mm long, entire to minutely denticulate; ovary straight, 3.5 mm long, 1 mm wide, glabrous, 3–4 ovules; style 10 mm long to the terminal, slightly thickened coil of 2 mm diam.; stigma lateral introrse, linear, 0.75 mm long. ***Pod*** elongate, fibrous, inflated, the immature pod glabrous, the older ones hirsute, 7 cm long, 1.2 cm wide; sutures pronounced; carpels fibrous, brittle, dehiscing to 1–2 turns; beak short to 2 mm long, stout. ***Seed*** nearly squarish or cubic in shape, 6 mm long, 6–7 mm wide, 4–4.5 mm deep, commonly black flecked, to solid black, black ring around hilum; hilum narrowly elongate, 2 mm long, 1–1.5 mm wide; lens pronounced. ***Seedling*** from hypogeal germination; epicotyl 6–7 cm long; stipules at eophyll node, united, oblong, acute; eophyll petiole 3.5 cm long, pulvini present, 4–5 mm long, stipels present or absent, the blade broadly ovate, 3–4.5 cm long, 2.5–4 cm wide, strongly 3-nerved, the base cordate, tip acute, nearly glabrous.

Specimens examined: **UNITED STATES. Alabama: Blount Co.:** Blount Springs, (33°55'N, 86°50'W), 29 Aug 1884, *Smith 464* (US). **Choctaw Co.:** 9 mi N of Toxey, 24 Sep 1967, *McDaniel 9822* (BRIT); Renfroe, 15 Oct 1893, *Mohr s.n.* (US). **County Unknown:** Sand Mountain, (33°50'N, 86°30'W), 22 Jun 1899, *Vanderbilt 4329b* (US). **Arkansas: Baxter Co.:** Cotter, 31 Aug 1915, *Palmer 8396* (CAS,

K, MO, US); bluffs of White River, near Cotter, 19 Sep 1924, *Palmer 26219* (MO). **Hampstead Co.:** Washington, (33°45'N, 93°30'W), 4 Sep 1916, *Palmer 10593* (MO). **Hot Spring Co.:** Magnet Cove, (34°30'N, 93°W), 168 m, 8 Aug 1937, *Demarée 15569* (BRIT, GH, MO). **Independence Co.:** at junction of roads from Lock #3 to Cushman and to Batesville, (35°45'N, 91°40'W), 23 Aug 1975, *Thomas et al. 46045* (BRIT, DES). **Marion Co.:** along Hwy 14, Yellville, (36°10'N, 92°40'W), 27 Jul 1935, *Moore 350127* (BRIT, TEX, WIS). **Polk Co.:** Shady Creek Camp, CCC, Ouachita National Forest, (34°20'N, 94°W), 366 m, 19 Aug 1937, *Demarée 15863* (BRIT, MO); Albert Pike Recreation Area, 213–457 m, 31 Aug 1949, *Moore 490557* (WIS). **Pope Co.:** PO Dover, Ozark National Forest, (35°20'N, 93°10'W), 183 m, 11 Aug 1939, *DeMarée 19833* (CAS, GH, MO, UC);. **Pulaski Co.:** Hwy 10, Riverside Golf course, (34°45'N, 92°20'W), 10 Oct 1938, *Merrill 1148* (WIS). **Unknown County:** Corand Prairie, Jul 1885, *Harvey s.n.* (GH); Elmo, Jul 1900, *Hindman s.n.* (US). **Connecticut:** near seashore, Norwalk, (41°5'N, 73°30'W), 23 Aug 1901, *Bissell s.n.* (GH(2), MO); near shore of Housatouie River, Huntington, (41°15'N, 73°5'W), 18 Aug 1903, *Harger 4148* (GH); East Rock Park, New Haven, (41°15'N, 72°50'W), 1 Aug 1942, *Neale et al. s.n.* (MO); Huntington, 17 Sep 1914, *Weatherby s.n.* (US). **Delaware:** Centreville, 24 Aug 1874, *Commons s.n.* (GH, MICH); Brandywine, Jun 1887, *Tatnali s.n.* (GH); Banks of Delaware River above Milford, (39°N, 75°30'W), 29 Aug 1906, *Williamson 2774* (MICH). **District of Columbia:** Washington, 14 Sep 1878, *Chickering s.n.* (MO); District line, W near reservoir, (39°N, 77°15'W), 16 Aug 1908, *Dowell 5446* (GH). **Florida: Brevard Co.:** East Coast, Cape Canaveral, (28°30'N, 80°40'W), 2–5 Apr 1904, *Burgess 679* (F). **Dade Co.:** Biscayne Bay, 1874, *Palmer 122* (F, GH, MO). **Duval Co.:** Jacksonville, (30°20'N, 81°30'W), Sep 1878, *Curtiss 663* (BRIT, F, GH, K, MO, US(2)). **Hernando Co.:** Brooksville, (28°30'N, 82°20'W), 22 Oct 1919, *Jones 145* (US). **Highlands Co.:** Parker Islands, 5 mi S of Lake Istokpoga, (27°25'N, 81°15'W), 14 Nov 1945, *Brass 15696* (US). **Indian River Co.:** S of Vero Beach, near Surf n' Sand Motel, (27°45'N, 80°25'W), 10 Oct 1963, *Lakela 26560* (GH). **Leon Co.:** 4 mi E of Tallahassee, (30°20'N, 84°5'W), 6 Sep 1955, *Godfrey 53743* (BRIT, GH, UC). **Volusia Co.:** Daytona, (29°10'N, 81°10'W), 9 Oct 1917, *Francis 90* (US). **Unknown County:** Alexander Springs, 12 Sep 1929, *O'Neill s.n.* (CAS, US); Wekiwa Springs, 5 Oct 1929, *O'Neill s.n.* (US). **Georgia: Bartow Co.:** Allatoona Creek about 1/4 mi from Etowah River, (34°10'N, 85°W), 244, 24 Aug 1948, *Duncan 8824* (GH, MO, US). **Early Co.:** Chattahoochee River, (31°15'N, 85°W), 34 m, 14 Aug 1901, *Harper 1226* (MO, US); Colomokee Creek, 8 mi S of Fort Gaines, 18 Aug 1947, *Thorne 6116* (F, US); Gulf Coastal Plain, Saffold Station, 25 Aug 1948, *Thorne et al. 8539* (MO, UC). **Elbert Co.:** along Deep Creek, 8.9 mi W 2°N of Elberton, (34°10'N, 83°W), 152 m, 9 Aug 1950, *Duncan 11678* (BRIT, GH, US). **De Kalb Co.:** base of Little Stone Mt., (33°45'N, 84°10'W), 305 m, 25 Jul 1893, *Small s.n.* (F, US). **Habersham Co.:** Toccoa Falls, 305 m, 30 Aug–3 Sep 1894, *Small s.n.* (F, MO, MSC). **Hall Co.:** along Chattahoochee River, 5.6 mi W of Flowery Branch, Piedmont Province, (34°20'N, 83°55'W), 299 m, *Adams et al. 19126* (MICH, MSC, UC). **Rabun Co.:** Estatoah Falls, on Mud Creek, (34°50'N, 83°30'W), 914 m, 12 Aug 1893, *Small s.n.* (F). **Illinois: Jackson Co.:** "Fountain B," Grand Tower, (37°35'N, 89°30'W), 18 Aug 1900, *Gleason 1768* (GH); Crab Orchard Creek, N of Carbondale, (37°45'N, 89°10'W), 1 Jul 1941, *McCree 891* (MO). **Pulaski Co.:** Karnak, (37°15'N, 89°W), 23 Sep 1919, *Palmer 16537* (MO). **Richland Co.:** SW of Calhoum, (38°30'N, 88°W), 12 Aug 1928, *Ridgway 3198* (F). **Wabash Co.:** Mt. Carmel, (32°20'N, 88°45'W), Sep 1877, *Patterson s.n.* (F). **Williamson Co.:** Big Muddy River, Bush, 11 Aug 1941, *McCree 1005* (MO). **Unknown County:** Makauda, 22 Jul 1902, *Gleason 2435* (GH); Zimmerman's Bluff, Lower Wabash Valley, 1 Sep 1915, *Shearer s.n.* (CAS). **Indiana: Crawford Co.:** 1 mi NE of West Fork, 31 Aug 1945, *Deam 63690* (BRIT). **Lawrence Co.:** 2 mi SW of Bryantsville, 16 Aug 1934, *Kriebel 2484* (BRIT). **Kentucky: Anderson Co.:** Gilbert Creek, (38°N, 85°W)?, 17 Aug 1955, *Wharton 9408* (GH). **Calloway Co.:** old Ky 94 off Ky 94, E of Mayfield, 27 Jul 1973, *Athey 2437* (BRIT). **Fayette Co.:** Kentucky River near Rogers', at the Richmond Road, Lexington, 37°54'N, 84°22'W, Jul 1834, *Peter 799r* (MICH); Lexington, 1840, *Short s.n.* (US). **Letcher Co.:** Pine Mt., (37°N, 82°45'W), 9 Jun 1933, *Braun 504* (US). **Whitley Co.:** Cumberland Falls, (37°45'N, 84°20'W), 5 Sep 1939, *Braun 2562* (US). **Louisiana: Caddo Parish:** Red River, Shreveport, (32°30'N, 93°45'W), 6 Sep 1916, *Palmer 10612* (CAS-DS, MO(2), US). **Caldwell Parish:** La 846, 1/8 mi S of Ouachita Parish, (32°10'N, 92°15'W), 12 Aug 1967, *Thomas 4314* (ASU). **Catahoula Parish:** 3 mi W of Harrisonburg, 8 Aug 1967, *Thomas et al. 3944* (BRIT). **West Feliciana Parish:** 1.5 mi E of Bains, (30°45'N, 81°15'W), 20 Aug 1938, *Correll et al. 10416* (GH); West Feliciana Parish, 24 Aug 1912, *Pennell 4335* (MICH(2)). **Maryland:** Doncaster, 4 Oct 1942, *Dimmick 2235* (BRIT); Cabin John, 21 Aug 1904, *Dowell 3238* (GH). **Mississippi: Clay Co.:** Tibbee Creek, 4.5 mi S of West Point, (33°30'N, 89°40'W), 30 Aug 1955, *Ray 5856* (GH). **Madison Co.:** Natchez Trace Parkway, (32°40'N, 89°W)?, 10 Jul 1947, *McDougall 1389* (US). **Wilkinson Co.:** 13 mi NW of Woodville, 11 Aug 1955, *Ray 5491* (GH). **Unknown County:** Crystal Springs, (32°N, 90°15'W), 3 Jun 1925, *Cook s.n.* (US); between Long Beach and Gulf Port, (30°20'N, 89°5'W), 19 Sep 1891, *Joor s.n.* (MO). **Missouri: Barry Co.:** King's River, N of Allen Ford, 4 mi SW of Viola, (36°35'N, 92°35'W), 21 Jun 1937, *Steyermark 22563* (F, MO). **Butler Co.:** 2 mi NW of Rombauer, 7 Jul 1936, *Steyermark 11372* (MO). **Carter Co.:** Current river, W of Elm Spring Hol, SW of Chilton, 2 mi E of boundary of Big Spring State Park, (37°N, 91°W)?, 19 Jun 1949, *Steyermark 68249* (F); Current River, 2 1/2–4 mi SE of Van Buren, 14 May 1952, *Steyermark 73242* (F). **Cedar Co.:** Brush Creek, 4 1/2 mi NW of Arnica, 25 Jun 1939, *Steyermark 27334* (F). **Christian Co.:** East Fork of Bull Creek, 3 mi SW of Chadwick, (37°50'N, 93°5'W), 8 Jul 1937, *Steyermark 23133* (F, MO). **Dallas Co.:** Niangua river near the 'Swinging Bridge' between Buffalo and Long's Lane, 18 Jul 1934, *Steyermark 13688* (MO). **Douglas Co.:** 1 mi NE of Bryant Creek, SE of Sweden, 20 Aug 1934, *Steyermark 14734* (MO); Spring Creek, near Roosevelt and Big Spring, 17 Jul 1937, *Steyermark 23299* (F, MO); N Fork of White River, 2 1/2 mi E of Holt, (37°55'N, 92°5'W), 24 Jul 1937, *Steyermark 23606* (F, MO); Spring Creek, below junction with Noblett creek, 4–5 mi NE of Hebron, 23 Sep 1947, *Steyermark 65101* (F). **Dunklin Co.:** Campbell, (36°30'N, 90°W), 15 Aug 1895, *Bush 91* (K, MO). **Iron Co.:** Ironton, (37°30'N, 90°30'W), Jul 1897, *Russell s.n.* (MO). **Jasper Co.:** 2 mi N of Webb City, (37°N, 94°50'W), 25 Sep 1910, *Palmer 3218* (MO); 'Castle Rock', Joplin, 29 Sep 1924, *Palmer 26309* (MO). **Laclede Co.:** 10 km N of Bennett Springs, 37°48'N, 92°50'W, 260 m, 8 Oct 1974, *Conrad 3486* (MO). **McDonald Co.:** 1 Sep 1893, *Bush s.n.* (MO). **Miller Co.:** Tavern Creek, 2 1/2 mi W of St. Anthony, (38°5'N, 92°20'W), 20 Sep 1938, *Steyermark 6883* (F, MO); Little Saline Creek, 6 1/2 mi NW of Tuscumbia, 6 Jul 1952, *Steyermark 73714* (F). **Oregon Co.:** White Creek, between Fidler Spring and Eleven Points River, Irish Wilderness, 3 1/2–4 mi S of Wilderness, (36°50'N, 91°15'W), 24 May 1938, *Steyermark 5391* (F, MO). **Ozark Co.:** N Fork of White River near Ark. line between Pumphry and White Ferry, 1 1/2–2 1/2 mi S of Udall, 29 Apr 1938, *Steyermark 5259* (F, MO). **Phelps Co.:** Jerome, (37°55'N, 92°W), 9 Aug 1929, *Kellogg 15220* (MO(2)). **Pulaski Co.:** Gasconade river, about 2 mi NE of Waynesville, (37°50'N, 82°5'W), 28 Jul 1935, *Steyermark 19332*

(MO); Big Piney river NW of Ousley Spring, 11 Aug 1935, *Steyermark 19462* (F, MO). **Ripley Co.:** Current River, 2 mi N of mouth of Buffalo, 5 mi E of Bennett, (36°50'N, 90°55'W)?, 6 Aug 1934, *Steyermark 14281* (MO). **Scott Co.:** Lost Hill, 4 mi NE of Chaffee, (37°15'N, 89°35'W), 23 Jul 1951, *Steyermark 72260* (F, MO); Mississippi River and ravines tributary to Cape La Croix Creek, W of Grays Point, 2–2 1/2 mi N and NE of Illmo, 21 Aug 1954, *Steyermark 76888* (GH). **Shannon Co.:** Current River, 7 1/2 mi SE of Rector, (37°20'N, 91°25'W), 1 Oct 1950, *Steyermark 70999* (F). **St. Louis Co.:** Bee Tree County Park, 4 Aug 1?77, *Christ s.n.* (MO); Steinige Hugel, 19 Jul 1886, *Eggert s.n.* (MO); Cliff Cave, 29 Jul 1886, *Eggert s.n.* (F(2), TEX, UC, US); Stony hills, St. Louis, 29 Jul 1886, *Eggert s.n.* (MO); Jefferson Barracks, 26 Jul 1890, *Hitchcock s.n.* (MO). **Ste. Genevieve Co.:** Mississippi River, 1/2 mi SE of Brickeys, between Brickeys and Clement, (37°55'N, 90°W)?, 29 Aug 1954, *Steyermark 76941* (MO). **Stoddard Co.:** N of Ardeola, (36°55'N, 90°W)?, 25 Apr 1938, *Steyermark 5093* (F). **Taney Co.:** White River just W of Big Cedar Hollow, 4 mi S of Ocie, 6 mi SE of Protem, (36°35'N, 92°50'W), 30 Apr 1938, *Steyermark 5271* (F, MO); White River, N and E of Brown Ferry, downstream to Cedar Hol., 5 1/2–6 mi SE of Protem, 29 Sep 1949, *Steyermark 69451* (F). **Texas Co.:** Little Paddy Creek, 3 mi SW of Slabtown, (37°30'N, 92°0'W)?, 25 Aug 1937, *Steyermark 25297* (F, MO); Jack's Fork of Current River, from 1/2 mi of Shannon Co. line to near Shannon Co. line, 5 1/2 mi SE of Arroll, (37°N, 91°40'W)?, 23 Jun 1939, *Steyermark 27164* (F). **Washington Co.:** Indian Creek, 1 mi S of North-cut P.O. (Birdie), (37°50'N, 90°30'W)?, 12 Jun 1941, *Steyermark 41155* (F, MO). **Wayne Co.:** St. Francis River just SE of mouth of Redman Creek, T26N, R7E, 3/4 mi SW of Wappapello, (36°55'N, 90°15'W), 3 Sep 1938, *Steyermark 6491* (F, MO). **Wright Co.:** Beaver Cr., near Jarrett's Ford, 3 mi N of Manes, (37°20'N, 92°20'W), 21 Aug 1937, *Steyermark 25040* (F, MO). **New Jersey: Burlington Co.:** Wading River, sea level, 16 Aug 1948, *Dimmick 2234* (BRIT). **Cape May Co.:** Dennisville, (39°10'N, 74°50'W), 1 Aug 1941, *Brown s.n.* (GH); S of Peermont, 22 Aug 1935, *Fender 867* (US); Peermont, 1–2 Sep 1929, *Pennell 14795* (MICH). **Gloucester Co.:** S Branch Timber Creek, Blackwood, (39°45'N, 75°5'W), 31 Jul 1917, *Long 17034* (ARIZ, GH); Mantua Creek, 6 Jul 1922, *Long 25946* (UC). **Hunterdon Co.:** Byram Station, (40°25'N, 75°W), 27 Aug 1897, *Fisher 2234* (GH); Delaware River road above Milford, (40°35'N, 75°W), 29 Aug 1906, *Van Pelt et al. s.n.* (GH, UC). **Passaic Co.:** E base of cliffs of trap rock ridge, First Watchung Mt., just N of Great Notch, Passaic River Basin, 122 m, 24 Sep 1944, *Clausen et al. 6458* (GH, K, MICH, MO, UC). **New York:** copse at Brooklyn, (41°40'N, 74°W), 1 Sep 1842, *Carey s.n.* (GH, K); N Staten Is., Richmond, 8 Oct 1904, *Dowell 3549* (GH). **North Carolina: Buncombe Co.:** Asheville, (35°30'N, 82°45'W), 21 Aug 1876, *Engelmann s.n.* (MO). **Iredell Co.:** Statesville, (35°45'N, 80°50'W), *Hyanis s.n.* (CAS-DS). **Macon Co.:** Wayah Bala, 18 Jul 1888, *Johnson 663* (F); King's Mt., (35°15'N, 81°20'W), 183–396 m, 27–30 Aug 1894, *Small s.n.* (F). **Orange Co.:** Chapel Hill, (35°50'N, 79°W), *Ashe et al. s.n.* (US). **Swain Co.:** Great Smoky Mts., (35°30'N, 83°20'W), 1 Aug 1891, *Beardslee et al. s.n.* (GH, MO, US). **Wake Co.:** Lake Johnson, 4 mi SW of Raleigh, (35°40'N, 78°50'W), 13 Jul 1938, *Godfrey 5002* (GH). **Unknown County:** Hot Springs, Appalachian Mts., (35°50'N, 82°55'W), 14 Aug 1924, *Wehmeyer 637* (MICH); Asbenth, Aug., *Williamson s.n.* (F). **Ohio: Adams Co.:** Turkey Creek, Cincinnati, (38°25'N, 83°15'W), 5 Aug 1926, *Braun s.n.* (US); Mineral Springs, Cincinnati, 7 Aug 1928, *Braun s.n.* (US(2)). **Highland Co.:** Fort Hill, (39°N, 83°25'W), 24 Aug 1955, *Braun s.n.* (US). **Pike Co.:** Mifflin Twp., roadside park on Rt 124, Cincinnati, W of Byington, (39°N, 83°40'W), 25 Sep 1957, *Braun s.n.* (US). **Summit Co.:** Cuyahoga River, Akron, (41°10'N, 81°35'W), 18 Aug 1889, *Foltz s.n.* (UC). **Oklahoma: Delaware Co.:** Cloud Creek 1 mi W of Colcord on State 106, 6 Sep 1958, *Wallis 7782* (BRIT). **McCurtain Co.:** upper Mountain Fork River, 8 mi NE of Smithville, 24 Jul 1978, *Taylor et al. 26884* (BRIT). **Pennsylvania: Berks Co.:** Hay Creek, 1/2 mi NE of Trap Rock station, 19 Aug 1938, *Wilkens 5646* (GH). **Chester Co.:** Mawa, 28 Jul 1889, *Brinton s.n.* (F); 1 mi E of Black Horse, 5 Aug 1933, *Fogg 5796* (GH). **Fayette Co.:** Youghiogheny River, Ohio Pyle, (39°50'N, 79°30'W), 5 Aug 1934, *Bright 9726* (CAS-DS). **Lancaster Co.:** near Giles, Lancaster City, (40°N, 76°15'W), 21 Aug 1894, *Eddy s.n.*, (ARIZ, MO); Little Conestoga, 1 mi S Millersville, Lancaster, 12 Aug 1889, *Small s.n.* (F). **Northampton Co.:** Easton, Lafayette College, (40°45'N, 75°5'W), 14 Aug 1869, *Porter s.n.* (F); Lehigh River, 3/4 mi E of Treichlers, 30 Aug 1939, *Schaeffer 5190* (GH). **Philadelphia Co.:** Philadelphia, Schuylkill, 6 Sep 1949, *Prior s.n.* (K, US). **York Co.:** Reservoir Park in Pigeon Hills, 1 3/4 mi N of Hanover, (39°50'N, 79°W), 25 Aug 1938, *Tanger s.n.* (GH). **Unknown County:** Mts. about Cold Spring, 7 Aug 1889, *Small s.n.* (F, GH, MO). **South Carolina: Anderson Co.:** Brown's Park, Anderson, (34°30'N, 82°40'W), 27 Aug 1919, *Davis s.n.* (MO, TEX, UC, US). **Greenville Co.:** Saluda Falls, (34°35'N, 82°30'W), 12 Aug 1881, *Smith 112* (US). **McCormick Co.:** Hard Labor Creek, near US 378, 4 mi E of McCormick, (33°55'N, 82°10'W), 30 Jun 1957, *Radford 26287* (GH). **Oconee Co.:** Clemson College, (34°45'N, 82°55'W), 25 Sep 1906, *House 2903* (US). **Tennessee: Blount Co.:** Chilhowee Mtn., parkway 4 mi from US 129, 17 Sep 1964, *Thomas 33217* (BRIT). **Cheatham Co.:** Kingston Springs, Turnbull Springs S of I40 bridge, 3 Jul 1972, *Churchill s.n.* (MSC); Pegram, (36°N, 87°W), 22 Aug 1940, *Svenson 10542* (UC). **Coffee Co.:** Tullahoma, (35°15'N, 86°15'W), 12 Aug 1899, *Vanderbilt 4329d* (US). **Decatur Co.:** (35°25'N, 86°10'W)?, 1855, *Ames s.n.* (MICH). **Franklin Co.:** Cumberland Mt. at Canten, (35°5'N, 86°W)?, 12 Sep 1898, *Eggert s.n.* (MO); Cumberland Mt. at Cheeroen, 8 Sep 1898, *Eggert s.n.* (MO). **Henderson Co.:** (35°45'N, 88°25'W)?, 16 Aug 1900, *Vanderbilt 4329c* (US). **Knox Co.:** 6 Aug 1894, *Kearney 1071* (US); Copses, Knoxville, (36°N, 84°W), Aug 1896, *Ruth 2475* (MO). **Polk Co.:** Cherokee Natl. Forest Rd. 27 N of Hiwassee River, 14 Aug 1970, *Somers et al. 44495* (BRIT). **Roane Co.:** Harriman, (36°N, 84°30'W), 20 Aug 1903, *Vanderbilt 4329e* (US). **Texas: Chickasaw Co.:** Houlka Creek, Pontotoc Ridge? ca 3 mi N of Houston, (29°55'N, 95°30'W), 2 Sep 1971, *McDaniel 15536* (MO). **Harrison Co.:** Karnack and Caddo Lake State Park, (32°35'N, 94°W), 11 Jul 1965, *Correll 31230* (TEX-LL). **Nacogdoches Co.:** Steven F. Austin Experimental Forest, (31°30'N, 94°10'W), 14 Jul 1981, *Johnston s.n.* (TEX). **Virginia: Augusta Co.:** Browns Ridge, vicinity of Fordwick and Craigsville, (38°N, 79°20'W), 500 m, 22 Aug 1913, *Steele 27* (US(2)). **Bedford Co.:** (37°30'N, 79°45'W)?, fl Aug, fr 15 Sep 1868, *Curtiss s.n.* (G, GH). **Dinwiddie Co.:** S of Burgess Station, (37°N, 77°30'W)?, 15 Aug 1938, *Fernald et al. 9080* (GH). **Fairfax Co.:** Great Falls, (38°45'N, 77°30'W)?, 20 Jul 1913, *Blake 5259* (TEX-LL). **Fauquier Co.:** NW of Belts, 1 mi N of Hopewell Gap, Western slope of Bull Run Mts., (38°55'N, 78°W)?, 9 Aug 1936, *Allard 2079* (F, GH); Bull Run Mts., 2 1/4 mi NE Plains, 5 Oct 1938, *Hermann 9898* (F, MICH). **Gloucester Co.:** N End Branch above James Store, 1 Sep 1925, *Wherry et al. 12631* (MO). **Greensville Co.:** N of Orion, (36°30'N, 77°30'W)?, 14 Sep 1941, *Fernald et al. 13660* (GH, US). **Loudoun Co.:** Blue Ridge—near Harpers Ferry, (39°15'N, 77°30'W), 6 Sep 1914, *Smith 2983* (US). **New Kent Co.:** Chickahominy River, Lanexa, 12 Oct 1941, *Fernald et al. 13963* (GH); Chickahominy Lake, Walker's Stat., 24 Aug 1966, *Seymour et al. 24810b* (MO). **Prince Edward Co.:** (37°10'N, 78°20'W)?, 29 Aug 1880, *Smith s.n.* (US). **Prince George Co.:** Powell's Creek, Garysville, (37°15'N, 77°10'W), 14 Jun 1938,

Fernald et al. 8325 (GH). **Roanoke Co.:** Ft. Lewis Mtn. ca. 1.5 mi NW of Salem, 5 Aug 1961, *Kral 13582* (BRIT). **Southampton Co.:** 6-7 mi S of Franklin, (37°35'N, 77°W), 19 Jun 1938, *Fernald et al. 8326* (F, GH, MO, SI, US). **Unknown County:** Dead Run, 14 Aug 1921, *Leonard et al. 897* (GH). **West Virginia: Hardy Co.:** Wardensville, (39°N, 78°30'W), 9 Aug 1932, *Core 4351* (CAS-DS). **Mingo Co.:** Varney School, (37°40'N, 82°W), 8 Jul 1930, *Berkley 993* (MO). **Ritchie Co.:** 3 mi W of Auburn, (39°N, 81°0'W), 21 Aug 1922, *Randolph et al. 1342* (GH).

Habitat.—This wide ranging species has been found growing in moist thickets and climbing on shrubs and young trees in clearings and edges of deciduous forests and in woods behind sand dunes, in deep swamps and in mixed deciduous forests of shrubs, pine and oak, *Catalpa, Sabal, Eupatorium,* and seagrapes. Also reported on: dry sandy, dry shale and rocky hillsides, in low hummocks, prairie bluffs, limestone, cherty and sandstone bluffs, burnt-over slopes of ridges, in a shell mound, on a clay embankment in gumbo soil, and in an igneous intrusive area.

Diseases and pests.—None are indicated by collectors.

Common names.—Wild Bean, Wild Kidney bean (see also Fernald & Kinsey 1943).

Genetics.—Successful crosses were made by Lorz (1952) with cultivated *P. lunatus*, but the F_1 was sterile (see explanation of results above under this latter species). These hybrids look very much like ssp. *polystachyus* and it is in these that white flowers rarely appear.

Comments.—Originally this taxon was placed in the genus *Dolichos* by Linnaeus, which led to the description of several species now in synonomy. The correct placement to *Phaseolus* took place over 100 years later. Maréchal et al. (1978b) believed this taxon occupies a position somewhere between the *vulgaris-coccineus* complex and *P. lunatus*. The senior author finds little resemblance between *P. lunatus* or *P. polystachyus* and these other species. The species *P. polystachyus* has large lateral branches on the inflorescence, a key characteristic which unites all of the species of this section, with very little pubescence and a unique flower type (see Color Plate I, photo 10). The root is also thick, fleshy and quite deeply penetrating the soil (see Color Plate III, photo 29).

There are reports in the literature (but not confirmed by the specimens examined by us) that the species may extend north as far as Michigan (Voss 1985) and Minnesota (MacMillan 1892), and perhaps even to Canada (Small 1903; Sousa & Delgado 1993). It has indeed been reported from Québec (Rydberg 1965), though this may be an error in identification of species; Lackey (1983) reports it absent from Canada. To our knowledge, this is the most extreme extent in latitude (45º latitude North) in the genus, and would let us infer particular attributes in terms of reaction to photoperiod and cool temperatures.

Delgado (1985) mentions collections of this species from Kerr Co., Texas and Dulces Nombres, Nuevo León, Mexico, the latter of which seems doubtful. The senior author has collected the location in Texas and found a single population which resembles subspecies *smilacifolius* with the same large, lobed leaflets but differs in having purple flowers rather than white ones (see *Comments* under that subspecies). The population from Dulces Nombres is probably a species of Section I. *Digitati* all of which have large basally lobed leaflets, or possibly another population of *P. maculatifolius* described below.

D.I.2.2.—Phaseolus polystachyus subsp. **sinuatus** (Nutt. ex Torrey & Gray) Freytag, comb. & stat. nov. (**Fig. 36**). *Phaseolus sinuatus* Nutt. ex Torr. & Gray, Fl. N. Amer. 1:279. 1838. *Phaseolus polystachyus* var. *sinuatus* (Nutt.) Maréchal, Mascherpa & Stainier, Taxon 27:199. 1978a. TYPE: UNITED STATES. FLORIDA: Prope Aspalaga, *Nuttall s.n.* (HOLOTYPE: BM; ISOTYPES: K, MO (photo BM). However Maréchal et al. (1978b) gave the type as *Nuttall 1829* at K, and Delgado (1985) designated East Florida, *Ware s.n.* as a lectotype (PH n.v.).

Aerial shoot an annual or perennial, prostrate, indeterminate vine. ***Root*** unknown (but probably a much branched, thick and fleshy type). ***Leaves*** 7–8 cm long; petiole 3–4 cm long, striate, glabrous; petiolule 1 cm long; terminal leaflet rounded to diamond shaped, 3 cm long, 3 cm wide at 1/3 from base, with 3 more or less well defined undulate or rounded lobes about equal in size, obtuse, minutely apiculate, reticulate veined, puberulent hispid adaxially, glabrous abaxially; lateral leaflet the same but slightly inequilateral. ***Inflorescence*** a delicate panicle; peduncle 9–13 cm long; rachis 10–23–30 cm long, paniculate with very short branches at flower nodes, 10–20 nodes, glabrous to puberulent of minute pubescence; primary bract ovate- triangular, 1.5–2 mm long, 1 mm wide, heavily 3-nerved, glabrate; pedicel delicate, 5 mm long, glabrate to sparsely pubescent of hooked hairs. ***Bracteoles*** minute, aciculate, 0.75–1 mm long, 0.5 mm wide, faintly 1-nerved, ciliate. ***Flower*** small, purple; calyx campanulate, pubescent within and margin ciliate, 3 mm long, lobes rounded acute,

0.75 mm long, 1.5 mm wide, reticulately veined, glabrous; standard purple, broad, 4 mm long to bend and 4 more to emarginate tip, 9–10 mm wide; wings purple, obovate, enrolled lengthwise, 11 mm long, the claw 3.5 mm long, the spur well-developed, 0.5 mm diam.; keel 5 mm to bend and 2.5 mm more to base of the terminal 1 3/4 coils of 2.25 mm diam.; stigma lateral introrse, linear, 0.75 mm long. ***Pod*** 3–5 cm long, 1 cm wide, somewhat stipitate, sutures thickened, glabrous, the beak curved, 2 mm long. ***Seed*** unknown. ***Seedling*** unknown.

Specimens examined: **UNITED STATES. Florida: Alachua Co.:** 12 mi N of Gainesville, 0.7 mi off Devil's Millhopper Rd in San Felasco area, 12 Aug 1960, *Gillis 3912* (MSC). **Dade Co.:** Miami, (25°45'N, 80°30'W)?, Jun 1877, *Garber s.n.* (GH); Miami, 29 Feb 1892, *Simpson 377* (GH). **Duval Co.:** Jacksonville, (30°20'N, 81°30'W)?, 14 Oct 1896, *Curtiss 5796* (BRIT, CAS, G, GH, K, MSC, MO, UC, US). **Glades Co.:** (27°N, 81°20'W)?, 1 May 1941, *Schallert 1015* (US). **Hernando Co.:** S of Robbin's House, Brooksville, (28°30'N, 82°20'W), 9 May 1939, *Murrill 587* (US). **Hillsborough Co.:** Tampa, (28°N, 82°30'W)?, 1898, *Ferguson s.n.* (MO). **Lake Co.:** Eustis, (28°55'N, 81°40'W)?, 8 Sep 1900, *Vanderbilt(?) 6063a* (US); Clermont, (28°30'N, 81°40'W), Jul, *Williamson s.n.* (F). **Manatee Co.:** Manatee, (27°35'N, 82°20'W)?, Aug 1845, *Rugel 132* (G, GH, K(2)). **Orange Co.:** Clarcona, (28°30'N, 81°W)?, 25 Sep 1899, *Meislahn 65* (US); Clarcona, 1900, *Pieters s.n.* (MICH). **Polk Co.:** Winter Haven, (28°N, 81°45'W), 26 Jun 1931, *McFarlin 5968* (CAS). **Putnam Co.:** Palatka, 3 Aug 1900, *Vanderbilt(?) 6063c* (US);. **Volusia Co.:** Beresford, 22 Sep 1910, *Hood s.n.* (MICH, MO); Seville, (29°25'N, 81°30'W), 6 Oct 1892, *Leeds s.n.* (F). **Unknown County:** Mosquito Inlet, Indian River, Sep 1845, *Curtiss 664* (BRIT, CAS, F, G, GH, K(2), MO, US(3), WIS); Lake Minneola, 26 Sep 1929, *O'Neill s.n.* (US); Mosquito Inlet, *Patterson s.n.* (F); Apalachicola et Mariana, (30°45'N, 85°15'W), Aug 1843, *Rugel s.n.* (G, K, MO(2)); S of Brickell Hammock, 26 Nov–20 Dec 1913, *Small et al. 4821* (GH, US). **Georgia: Baker Co.:** Ichauway, 28 mi SW of Albany, 29 Sep 1995, *Anderson 15917* (BRIT). **Candler Co.:** Excelsior, 0.6 mi N of Ten Mile Creek off Rt 1127, 6 Oct 1980, *Churchill 80196* (MSC). **Jenkins Co.:** Scarboro, Sep 1884, *Rusby s.n.* (MICH). **Sumter Co.:** Leslie, (30°N, 84°W), 5 Sep 1900, *Harper 592* (ARIZ, F, GH, MO, US). **Taylor Co.:** *Niesler(?) s.n.* (GH(2)). **Mississippi: Jackson Co.:** Ocean Springs, (30°30'N, 88°45'W), 25 Jul 1896, *Pollard 1017* (F, GH). **North Carolina: Cumberland Co.:** U.S. Rt 301, 5 mi S of Fayetteville, (34°55'N, 78°55'W), 8 Jul 1949, *Fox et al. 2612* (GH, MICH). **Moore Co.:** Pinehurst, (35°N, 75°25'W), Aug–Sep 1897, *Katzenstein s.n.* (GH). **South Carolina: Aiken Co.:** Aiken, (33°30'N, 81°45'W), 10–12 Sep 1866, *Ravenel s.n.* (MO).

Habitat.—Reported as having a trailing habit in open ground or on bushes in dry pine barrens or turkey-oak sandhills in sandy soil of Middle Eocene overlaid by Lafayette.

Diseases and pests.—None indicated by collectors.

Comments.—As indicated by Delgado (1985), this taxon apparently has an overlaping range with the type species, yet it is very distinct and there seems to be no intergradation (thus certainly not a variety). According to Lackey (1983), it has an annual root system. On the other hand, the similarities are so great that the senior author sees no reason at this time to maintain the specific rank. Certainly additional collecting is required since no recent collections have been seen by us and no viable seed are available for study at this time.

D.I.2.3.—Phaseolus polystachyus subsp. **smilacifolius** (Pollard) Freytag, comb. & stat. nov. (**Fig. 36**).
Phaseolus smilacifolius Pollard, Bot. Gaz. 21:233. 1896. TYPE: UNITED STATES. FLORIDA: Columbia Co.: on hummock land near Lake City, (30°10'N, 82°45'W), 29–31 Aug 1895, *Nash 2505* (HOLOTYPE: US 252673; ISOTYPES: G, GH, K, MICH, MSC, MO).

Aerial shoot a climbing, indeterminate vine, possibly perennial. ***Root*** unknown. ***Stipules*** lanceolate, 2 mm long, not nerved, acute, pubescent. ***Leaves*** 10–11 cm long, membranous; petiole 4–4.5 cm long, covered by minute white pubescence; terminal leaflet broadly ovate, 6 cm long, 5.5 cm wide, 3-lobed, the lobes large, rounded, nearly 1 cm long, apiculate; lateral leaflet inequilaterally ovate, broad to tip, 5 cm long, 4 cm wide, margin undulate lobed, acute, apiculate, puberulent adaxially, slightly pubescent abaxially on veins. ***Inflorescence*** a raceme 12–13 cm long, the peduncle 4–5 cm long, the rachis 8–9 cm long with the many flowers distributed somewhat irregularly throughout; primary bract ovate-acuminate, 2.5 mm long, 4-nerved, pubescent; pedicel very slender, about 12 mm long, glabrous. ***Bracteoles*** minute, oblong, 0.75 mm long, 0.5 mm wide, strongly 1-nerved, acute, white ciliate. ***Flower*** small, white or purple; calyx campanulate, small, 3 mm long, the lobes mostly rounded, acute, glabrous, ciliate; standard strongly reflexed, 5 mm from base to angle and 4 mm more to tip; wings white, spreading; keel with 2 coils; stigma lateral, introrse, 0.75 mm long. ***Pod*** unknown. ***Seed*** unknown. ***Seedling*** unknown.

Specimens examined: **UNITED STATES. Florida: Columbia Co.:** Lake City, Jun–Jul 1898, *Hitchcock 483* (F). **Texas: Kerr Co.:** Lamb Creek, 13 mi SW of Kerrville, 2 Oct 1946, *Cory 52444* (BRIT, US); Lamb Creek, 13 mi SW of Kerrville, on edges of cliffs along banks of creek, 29°55'N, 99°20'W, Nov 1990, *Freytag et al. 90-1* (ARIZ, CSU, F, GH, MO, TEX, UC, US).

Habitat.—"Hummocks" of Florida are sink-holes in limestone areas with very special habitats, similar to the "karst" country of Europe and the "mogotes" of Puerto Rico. Locations are very humid, shaded and protected, and there are many indigenous species in these areas. It is probable that subspecies *smilacifolius* is particularly well adapted to these special habitats, but it must now be rare since there are no recent collections from Florida and the area is now overly developed by suburban homes and clipped lawns. Perhaps the species may have been eradicated at this location by this development. The only other location (*Freytag et al. 90-1*), presumably of this subspecies because of the similar lobed leaf type and in all other respects except the flower is purple and the seed has dark colorations, is found on small, limestone cliff ledges just south of Kerrville, Texas. The most striking difference between these two disjunct locations is the white flower of the Florida collection and the purple flower of the Texas collections. The junior author has seen recent and complete collections of the Kerrville population (*O'Kennon 15106*, BRIT), and does not discard the possibility of it being another taxon.

Comments.—Delgado (1985) believes subspecies *smilacifolius* is conspecific with subspecies *polystachyus*, perhaps because of the similarity in pollen structure pointed out by Stainier (1974). However, there are many morphological differences, the most striking of which are the large lobed leaves of *smilacifolius* with no sign of intermediates, in addition to the different ranges which argue strongly for distinguishing them as separate subspecies.

Britton in his 'Flora of Bermuda' (1918) reports about the endemic *Phaseolus lignosus* Britton described therein for the first time. There is no type specimen reported. From the drawing and the comment by Britton (1918), one could tentatively conclude it to be a form of *P. polystachyus* (L.) B.S.P. subsp. *polystachyus*. Liogier (1988) has indeed reported the presence of *P. polystachyus* in Rincón, Puerto Rico. In their study using ITS sequencing, Delgado and co-workers (1999) group *P. lignosus* with *P. viridis* (considered as *P. lunatus* var. *viridis* (Piper) Delgado in Delgado, 1985) apart from *P. polystachyus* but linked to the Lima bean branch. Additional studies starting with germplasm collection are obviously needed to clarify this point.

D.I.3.—Phaseolus salicifolius Piper, Contr. U.S. Natl. Herb. 22:692. 1926. (**Figs. 34, 37**). TYPE: MÉXICO. SINALOA: Sierra de Chabarría, 1921, *Ortega 4065* (HOLOTYPE: US 1083789; ISOTYPE: MEXU).

Aerial shoot an herbaceous, climbing, indeterminate vine. ***Root*** a woody perennial with long narrow branches, the tips of which are enlarged, fleshy, obovoid tubers, 4–9 cm long, 3–4 cm thick, and covered with narrow lenticels. ***Stems*** terete, striate, slender, densely covered with yellow pilose and strigose hairs throughout; internodes rather long to 11–12 cm. ***Stipules*** triangular-ovate, 3 mm long, 1.5 mm broad at base, striate, 4- to 6-nerved, acute, densely covered with white strigose hairs. ***Leaves*** 8–13 cm long; petioles densely covered by strigose hairs, much shorter than the leaflets, 3–6 cm long; petiolules 7–12 mm long; pulvini 2.5–3 mm long, very densely covered by white strigose hairs; stipels minute, ovate, 1.25 mm long, 0.75 mm wide, 1-nerved, glabrous; leaflets all similar, lanceolate, 3–5 cm long, broadest at the nearly truncate base, 1-nerved, acute, apiculate, hispidulous and variegated adaxially and veins distinctly hirsute, abaxially paler and densely covered by white strigose hairs especially on veins. ***Inflorescence*** a pseudoraceme exceeding the leaves; peduncle 5–6 cm long; rachis early rather closely 8- to 15-flowered becoming 5–6 cm long of 4–7 flowering nodes; primary bract ovate lanceolate, about 4 mm long, 5–9-nerved, sparsely covered with white hispid hairs, promptly deciduous; pedicel 4–5 mm long, densely covered by white strigose pubescence; pedicellar bract linear 2 mm long hyaline. ***Bracteoles*** lanceolate, wider near the end, 3.5–4 mm long, indistinctly 3-nerved, acute, densely covered with white strigose hairs, early caducous (leaving white scars). ***Flower*** probably purple (given by collectors as reddish-purple to bright pink or lilac); calyx campanulate, 3 mm long, the upper 2 lobes united into a scarcely elongate lip, emarginate, ciliate, the lower 3 lobes subequal, broadly dentate, 1 mm long, 1.25 mm wide, the lateral nearly glabrous and the central lobe heavily covered with strigose hairs and ciliate; standard purple, the claw 0.75 mm long, the blade broadly rounded, reflexed at 3.5 mm from base, 5 mm more to emarginate tip, 10 mm wide; wings purple, oblong, the claw 3.5 mm long, the blade 10 mm long; keel 5 mm from base to bend and 3 mm more to base of the terminal 2 coils of 2.75 mm diam.; vexillary stamen 10 m long, the cup-shaped knob at base about 0.5 mm long; stamen tube straight, 1 mm from base to the ridges, 0.5 mm high, 6

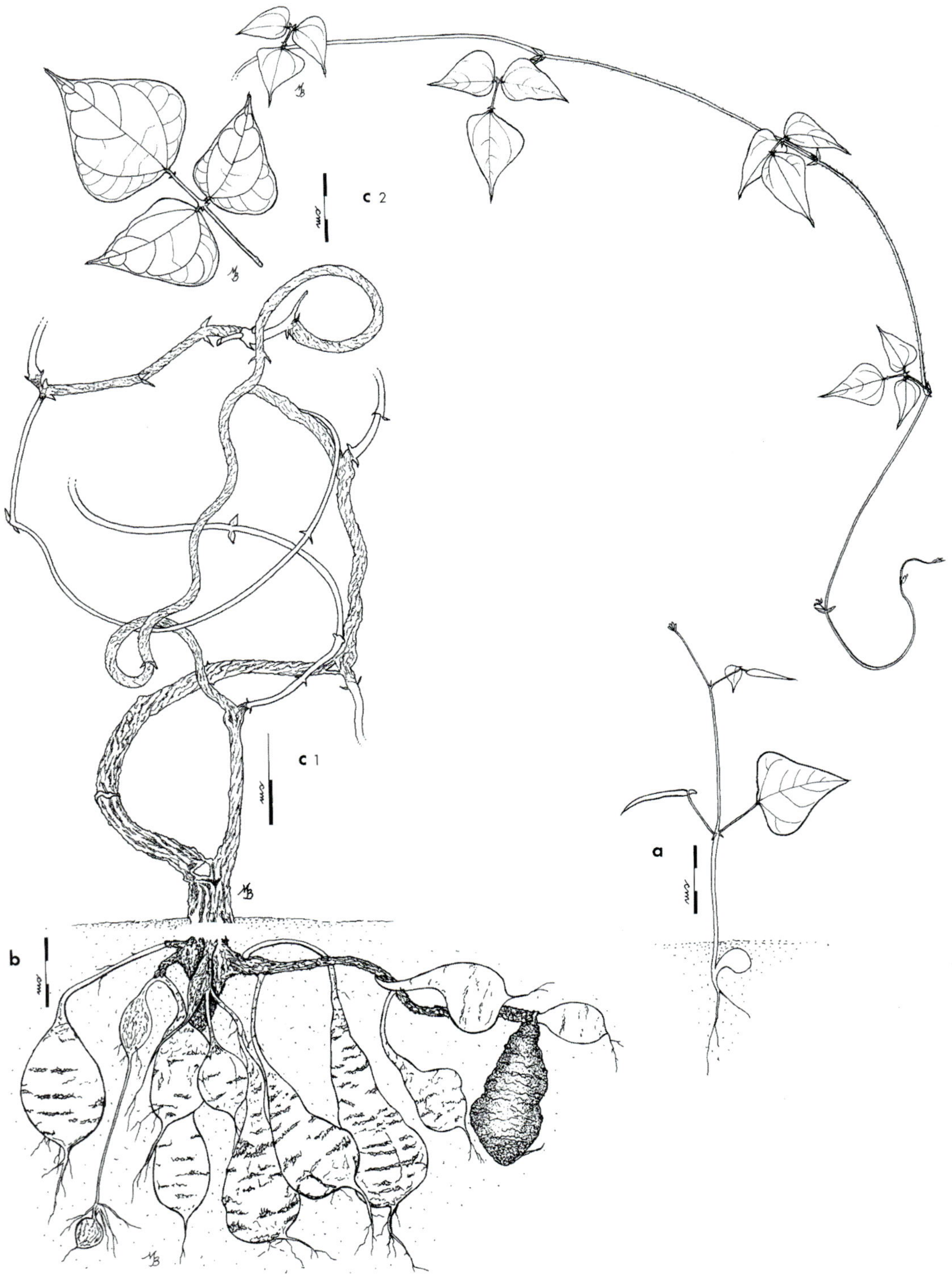

FIG. 37. Illustrations of *Phaseolus salicifolius* Piper.–**a.** Seedling several days after germination.–**b.** Root of mature plant.–**c.-1.** Old, mature lower portion of vine and detached lower leaf.–**c.-2.** Vine tip and leaves. All drawings from living material grown in greenhouse at Mayagüez from seed of TARS #316 (*Buhrow 11-2*) from near Villa Unión, Sinaloa, México.

mm more to end of the united portion; basal collar 1 mm long; ovary straight, 4.5 mm long, 0.75 mm wide, glabrous, 6 ovules; stigma lateral, introrse, linear, 1 mm long, pointed. ***Pod*** immature slightly falcate; valves glabrous; beak straight, long. ***Seed*** oblong, 4–6 mm long, 4–4.5 mm wide, brown streaked or mottled, black ring around the hilum; hilum oblong, 1 mm long. ***Seedling*** from hypogeal germination; epicotyl 5 cm long; stipules acicular, 3 mm long; eophylls triangular-ovate, 4 cm long, 4 cm wide, the base cuneate, acute; petiole 2 cm long; pulvini present; stipels absent.

Specimens examined: **MÉXICO. Durango:** 59.2 mi E of Villa Unión, 1.2 mi into Durango from Sinaloa on Hwy 40, 2000 m, 3 Aug 1983, *Buhrow et al. M3* (ARIZ); Sinaloa-Durango Border in W Durango near the Río San Lorenzo, Mar 1982, *Ferguson s.n.* (ARIZ); Mpio. El Salto, along Hwy 40 at the E end of El Espinazo del Diablo, 11.3 mi SW of La Ciudad, 16 mi E of Revolcaderos, Sierra Madre Occidental, 23°40'N, 105°45'W, 2377 m, 26 Mar 1984, *Sanders et al. 4886* (UCR); 1.3 mi into Durango from Sinaloa on Hwy 40, 1 mi past state line, 2000 m, 17 Mar 1983, *Starr 450* (ARIZ). **Sinaloa:** halfway between Mazatlán and Barranca Liebre, Durango Hwy., 18 Mar 1977, *Ames 77-120* (ARIZ); 50 mi E of Villa Unión, Durango-Mazatlán Hwy., 2000 m, Aug 1983, (grown from root at Experiment Station, Tucson, AZ), *Buhrow et al. M2* (ARIZ); 56.8 mi E of Villa Unión on Hwy 40, 2000 m, 3 Aug 1983, *Buhrow et al. M5* (ARIZ); Puerto a Tamiapa, 1524 m, 6–8 Mar 1940, *Gentry 5858* (MICH); Sierra Madre Mts., Trail "El Aguila" to "El Batel" on peak of El Aguila, 1340 m, 14 Nov 1925, *Mexía 435* (CAS, F, MO, UC); Rt 40, Forestry Station just 1 mi W of Mazatlán-side of El Palmito and Durango State line, 1890 m, 19 Mar 1975, *Pinkava et al. P12891* (ASU); Sierra Madre Occidental, 3 mi (airline) SW of El Palmito, 2.8 mi W of Hwy 40 by a dirt road that joins the Hwy. opposite a restaurant 1 1/2 mi S of El Palmito, 2134 m, 29 Dec 1983, *Sanders et al. 4419* (UCR); Sierra Madre Occidental, along Hwy 40, 1 1/2–3 mi below El Palmito, 45 mi NE of Concordia, 3 mi S of the Durango state line, 1951 m, 30 Dec 1983, *Sanders et al. 4481* (ARIZ, UCR); 55.7 mi E of Villa Unión, 1893 m, 18 Mar 1955, *Wiggins 13184* (CAS-DS, US); 1 mi SW of the Tropic of Cancer marker on Hwy 40, 6.2 mi NE of Santa Lucía, Sierra Madre Occidental, 23°26'N, 105°53'W, 1600 m, 7 Jan 1983, *Worthington et al. 9328* (UCR). **Sonora:** 18.3 mi E of the Río Yaqui bridge near Tonichi, on the road to Carrizal and Santa Rosa, Sierra Madre Occidental, 28°31'N, 109°21'W, 975 m, 27 Mar 1983, *Sanders 3710* (TEX).

Habitat.—This species is found as an uncommon vine twining up larger plants and on steep shady rocky slopes and steep N-facing slopes and on exposed ridges, near base of cliffs and talus slopes and usually in cloud forests and tropical humid deciduous forests of pine-oak with many palms and *Acacia cymbispina, Ceiba acuminata, Clethra, Jatropha, Lemaireocereus, Lippia pringlei, Lysiloma watsoni, Nolina, Pachycereus, Quercus tuberculata, Salvia, Tillandsia*, etc.

Diseases and pests.—Collectors have reported black aphids and the pollinator as a leafcutter bee.

Comments.—The collector of the type was Jesús González Ortega and the label on the type specimen is printed "Jesús G. Ortega, collector," thus Ortega is being used as the collector's last name.

This species is a long vine (more than 2 m long) with lanceolate leaflets, thus easily distinguished from the short vine with ovate leaflets in the lower portions of *P. jaliscanus* with which it might be confused. Lackey (1983) also has noted the early caducous characteristic of the primary bracts. *P. salicifolius* has been found already distant from *P. lunatus* in widecrossing experiments (Katanga & Baudoin 1990). Indeed, studying polymorphisms of cpDNA with PCR-RFLPs, Fofana and co-workers (Fofana et al. 1999) have found it related to *P. marechalii, P. maculatus, P. ritensis* and *P. rotundatus*. These results have been confirmed by ITS DNA sequencing (Delgado et al. 1999; Gaitán et al. 2000).

D.I.4.—Phaseolus maculatifolius Freytag & Debouck, sp. nov. (**Fig. 34**). Type: MÉXICO. Nuevo León: 1 km antes de llegar en Palo Bola, 9 km S de General Ignacio Zaragoza, 23°57'N, 99°47'W, 1960 m, 8 Sep 1985, *Debouck et al. 1509* (holotype: US (2 sheets); isotypes: BR, COL, K, M, UNL, UC).

Phaseolo maculato herba prostrata grandi persimili, sed foliolis minoribus ovatis, stipulis parvis, panicula ramulis secundariis manifestis differt. Crescit modo prope Zaragozam Novoleonensis rarus.

Aerial shoot is a large trailing vine. ***Root*** a perennial, thick, fleshy, and elongate. ***Stems*** terete, striate, 2–3 cm thick, densely covered by reflexed, yellow, short hirsute pubescence; internodes 9–10 cm long. ***Stipules*** lanceolate, 2.5 mm long, 0.5 mm wide, acuminate, densely covered by white hirsute hairs. ***Leaves*** 9–17 cm long; petiole 3–4–7 cm long; petiolule 1.5–2.5 cm long, densely hirsute; pulvini 2–3 mm long, very densely yellow hirsute; stipels spatulate, acute, 1–1.5 mm long, mostly ciliate; terminal leaflet ovate, acute to acuminate in older leaves, 4–5–7 cm long, apiculate, coriaceous, yellowish green becoming darker above and lighter green below, densely yellowish hirsute, almost velvety when young; lateral leaflets similar but somewhat inequilateral. ***Inflorescence*** a short to long panicle, lateral branches mostly only a few cm long; peduncle 10–12 cm long; rachis 10–12 cm long, with 5–6 lateral branches; bracts acicular, 2.5–3 mm long, hirsute, early caducous; pedicel 5 mm long, puberulent.

Bracteoles linear to acicular, 1 mm long, 0.25 mm wide, hyaline, hirsute, early caducous. ***Flower*** purple (lilac); calyx campanulate, tube 2 mm long, hirsute, upper 2 lobes rounded, 0.5 mm long, 1.5 mm wide, ciliate, lower 3 lobes subequal, rounded dentate, 0.75 mm long, 1.5 mm wide, ciliate; standard purple, the claw 1 mm long, the blade rounded, sharply reflexed at 3 mm from base and 4 mm more to emarginate apex, 10 mm wide, edges enrolled, the auricle flaps large, 2.5 mm long; wings purple, the claw 3.5 mm long, 0.5 mm wide, the spur well-developed, 1.5 mm diam., firmly attached to keel, the blade 8–9 mm long, enrolled lengthwise; vexillary stamen, the claw 1 mm long, the geniculate lobe 0.5 mm high, thickened portion 3 mm more; stamen 0.5 mm long, 0.3 mm wide; stamen tube, 1.5–2 mm from base to the lateral ridges, 0.5 mm high, 4 mm more to bend and 2.5 m more to end of the united filaments; keel, the divided claws 2.5 mm long, 3 mm more to bend and 3.5 mm more to base of the terminal 1 3/4 coils of 3 mm diam.; collar 0.75 mm long, denticulate; ovary straight, 4.5 mm long, 1 mm wide, glabrous, waxy; style 5 mm long to base of the thickened terminal coil of 2 mm diam; stigma terminal, introrse, oblique, linear, 0.7 mm long. ***Pod*** nearly straight, narrow above, broad near apex, 4 cm long, 8 mm wide; valves smooth, reticulate, fibrous; sutures thickened; beak straight, 5 mm long, stout; 3 seed. ***Seed*** unknown. ***Seedling*** from hypogeal germination, the cotyledons at ground level; epicotyl 6 cm long; stipules at eophyll node, lanceolate, entire, 2.5 mm long, 1-nerved, glabrous; eophyll petiole 1.5 cm long; stipels 1.5 mm long, lanceolate, acute, glabrous; eophyll blade ovate, acuminate, 3.5 cm long, 2 cm wide, not apiculate, sparsely hirsute; next internodes 3 cm long.

Habitat.—This species was found to be scarce on organic, rocky soils derived from schists in fairly dense forests of *Pinus, Quercus, Juglans, Cupressus, Arbutus.*

Comments.—Known only from a single collection, this species looks like *P. maculatus* in growth habit and vegetative characteristics, but the inflorescence is a panicle with definite secondary branches and the flower type resembles that of *P. polystachyus*. Using ITS DNA sequencing, Gaitán et al. (2000) have shown *P. maculatifolius* to be close to *P. jaliscanus*, *P. rotundatus* and *P. acinaciformis*.

D.I.5.—Phaseolus dasycarpus Freytag & Debouck, sp. nov. (**Figs. 34, 38**). TYPE: MÉXICO. VERACRUZ: Mpio. Huayacocotla, El Paraje, Huayacocotla, (20°30'N, 98°30'W), 1950 m, 24 Oct 1970, *Hernández & Trigos 872* (HOLOTYPE: F 1845525; ISOTYPE: MEXU).

Herba valida scandens, foliolis subtriangularis mediocria fere glabris, paniculae rigidae rectae, bracteis primariis circa 5 mm longis, bracteolis hyalinis 1 mm longis enerviis, leguminibus manifeste longestrigosis. Habitat solum in montibus prope Huayacocotlam in parte septentrionali Orizabae Veracrucis rarus.

Aerial shoot a strong climbing, indeterminate vine, 1–1.5 m long. ***Root*** unknown. ***Stems*** terete, 1.5–2 mm diam., purple striate; internodes 11–12 cm long, sparsely covered by appressed, reflexed-strigose hairs; stipules triangular, 4–5 mm long, 1.5–2 mm wide, obtuse, heavily 3- to 4-nerved, covered by short, white hispid pubescence on adaxial surface. ***Leaves*** 9–14 cm long; petiole 4.5–5 cm long, rounded and narrowly canaliculate, puberulent to glabrous; petiolule 1–2 cm long; pulvini, the lower 6 mm long, puberulent, the upper 2.5–3 mm long, dark, densely covered by yellow strigose hairs adaxially, glabrous abaxially; stipels obovate to oblong, 2–2.5 mm long, adaxial surface densely covered by short yellowish hirsute hairs, abaxial surface glabrous; terminal leaflet triangular to ovate, 3–6 cm long, 3–4.5 cm wide, the base truncate to rhomboidal, apex acute to somewhat acuminate, minutely apiculate, puberulent abaxially, a few scattered strigose hairs on adaxial surface, ciliate margins; lateral leaflets similar and distinctly inequilateral. ***Inflorescence*** a stiff, straight panicle; peduncle straight, 1–1.5 mm diam, striate, 8–15 cm long, nearly glabrous; rachis 6–10 cm long; lateral secondary branches delicate, 5–12 mm long bearing 2–4 flowers, glabrous; primary bracts lanceolate, 4–5 mm long, acute, 3- to 5-nerved, puberulent, purple; pedicel delicate, 5 mm long in flower, becoming 12 mm long in pod, glabrous, purple. ***Bracteoles*** oval, acute, 1 mm long, 0.5 mm wide, indistinctly 1-nerved, ciliate, purple. ***Flower*** purple; calyx flaring campanulate, the base restricted and angulate, the tube 2.5 mm long, the upper 2 lobes united into 1 rounded scarcely elongated, emarginate, 8–9 mm wide; the 3 lower subequal, dentate, acute, 1.25 mm long, 1.75 mm wide, glabrous to puberulous; standard purple, rounded, 5 mm to flexure and 5 mm more to emarginate tip, 10 mm wide; ovary with 6 ovules. ***Pod*** nearly straight, 4 cm long, 7 mm wide (immature); valves densely covered with long yellow strigose hairs, becoming sparsely pubescent at maturity; sutures pronounced; beak straight, 4–5 mm long. ***Seed*** unknown. ***Seedling*** unknown.

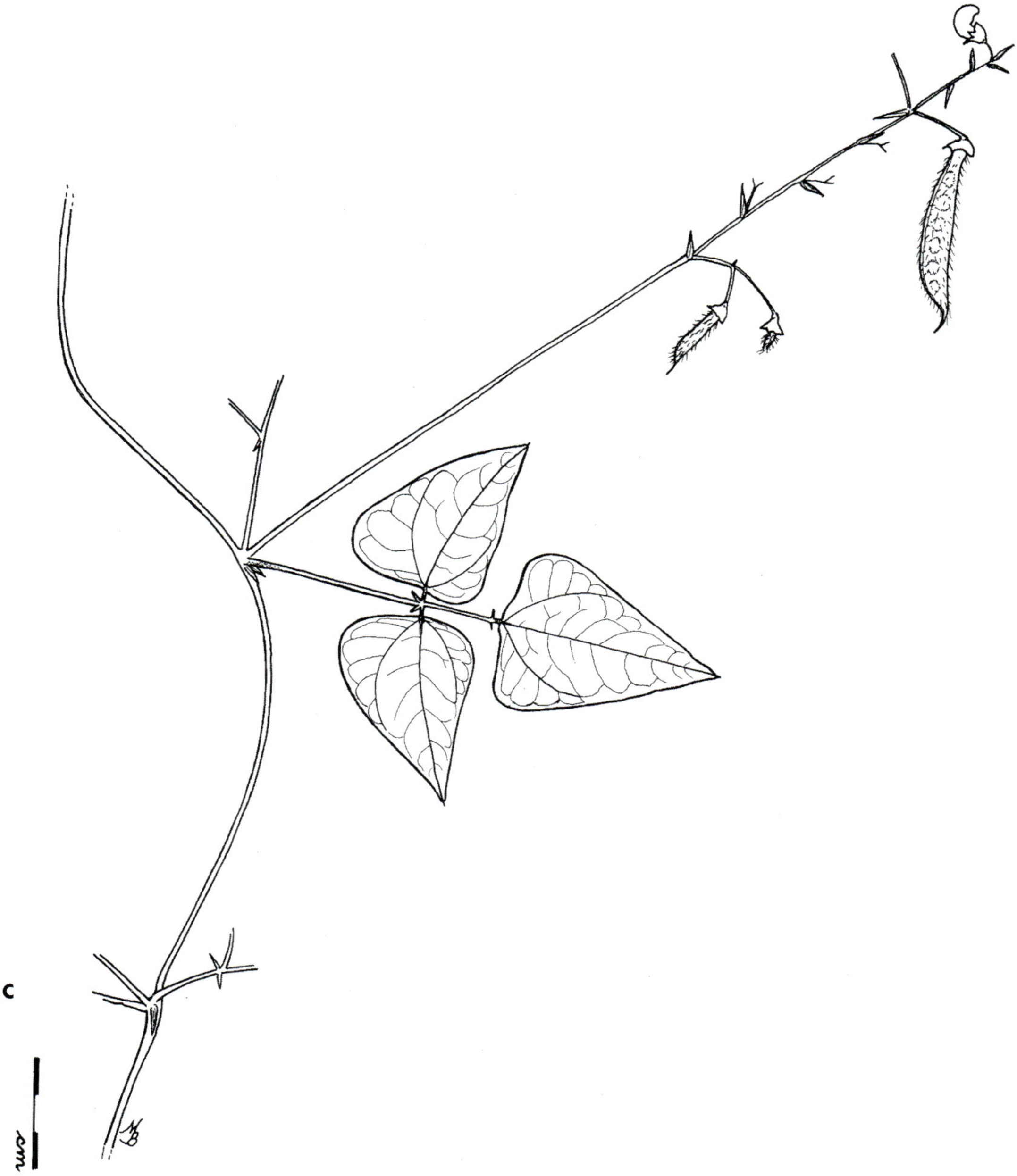

c

Fig. 38. Illustrations of *Phaseolus dasycarpus* Freytag & Debouck.—c. Small portion of vine with a mature leaf and inflorescence with bud and immature pod; note long bracts and long pubescence on pods. All drawings from type specimen *Hernández et al. 872* from El Paraje, Huayacocotla, Veracruz, México.

Habitat.—This taxon is reported to be fairly abundant on reddish soils in secondary vegetation.

Comments.—The species is known from only a single collection and though it is very much like *P. polystachyus* in most characters, the pods are quite different in being densely covered by long pubescence, and the location in Veracruz is far from the area of distribution of the former species in the SE United States.

D.I.6.—Phaseolus longiplacentifer Freytag, sp. nov. (**Figs. 34, 39**). TYPE: MÉXICO. VERACRUZ: Santa Ana Atzacán, N de Orizaba, (18°55'N, 97°5'W), 1200 m, 3 Dec 1967, *Rosas 899* (HOLOTYPE: CAS 604829; ISOTYPE: GH).

Fig. 39. Illustrations of *Phaseolus longiplacentifer* Freytag.–c. Small portion of vine with a mature leaf and, inflorescence with long, lateral secondary branches and immature pods.–f. Mature pod, side view with dehisced carpels; note long placental suspensors of immature and mature pods.–g. Seed, side view and view from the hilum. All drawings from the type specimen *Rosas 899* from Santa Ana Atzacán, N of Orizaba, Veracruz, México.

Herba valida scandens, foliolis ovatis mediocribus fere glabris, paniculae longae angustae, gongylodibus, bracteis primariis circa 2 mm longis, bracteolis uninervibus 1 mm longis, leguminibus rectis glabris, placentiis 2.5 mm longis gerentibus. Crescit solum in montibus prope Santa Ana Atzacanum in parte septentrionali Orizabae Veracrucis rarus.

Aerial shoot a large climbing, indeterminate vine. ***Root*** unknown. ***Stems*** terete, striate, 2.5–3 mm thick, glabrous; internodes 9–10 cm long. ***Stipules*** small, triangular, 3 mm long, 2 mm wide, acute, 3- to 5-nerved, glabrous. ***Leaves*** 11–14.3 cm long, dark green; petioles 3.5–5 cm long, delicate, striate, glabrous; petiolule 1–1.5 cm long, glabrous; pulvini 4 mm long, short pilose or hirsute adaxially, glabrous abaxially; stipels oblong, 1.5–2 mm long, 0.75 mm wide, acute, strongly 1- to 2-nerved, glabrous; terminal leaflet, broadly ovate, 6–7 cm long, 5 cm wide at near mid-point, acuminate, minutely apiculate, obscurely nerved, dark green on both surfaces, minutely farinose, a few scattered pilose hairs on veins on abaxial surface; lateral leaflets similar but inequilateral. ***Inflorescence*** a long narrow, 'knobby' panicle; peduncle 6–8 cm long, striate, puberulent; rachis 10–18 cm long of 10–15 nodes with lateral branches much reduced to knobs; bract ovate, acute, 1.5–1.75 mm long, 0.5–0.75 mm wide, strongly 1- to 3-nerved, glabrous; pedicel delicate, 6–8 mm long, becoming 1 cm long in pod, purplish, glabrous. ***Bracteoles*** lanceolate, 2 mm long, 0.75 mm wide, strongly 1-nerved, acute, glabrous. ***Flower*** purple (pink), (only 1 bud on type specimen—not dissected), small; calyx campanulate, tube 2.5 mm long, scabrous or farinose, upper 2 lobes united into one scarcely developed, emarginate, ciliate; lower 3 lobes subequal, dentate, acute, 0.175 mm long, 1.25 mm wide, scarcely ciliate; standard purple, tip not pubescent in bud. ***Pod*** when immature with restricted base, straight to falcate, mature pods straight, 5–6 cm long, 1.2–1.5 cm wide, widest near apex; beak straight, stout, 5–6 mm long; sutures strong, green; valves smooth, fibrous, faintly reticulate; ovules with long suspensors (placenta), 2.5 mm long becoming 1.5 mm long and brownish at maturity; 4 seed. ***Seed*** rounded, oblongoid, flattened, 9.5 mm long, 6.5 mm wide, 3 mm deep, yellowish brown, veined dark brown, dark brown ring around hilum, shiny; hilum oblong, 3 mm long, 1.5 mm wide; lens prominent, divided. ***Seedling*** unknown.

Habitat.—This taxon is reported to be abundant in secondary vegetation and near corn and sugar cane fields, yet only a single specimen has been collected.

Common name.—Frijolillo.

Comments.—Known from only a single specimen, this species has the unique characteristics found only in this species of seed borne on long placentas in the immature pods and the lateral branches of the inflorescence are nearly as long as the main axis. In other respects it seems to resemble *P. polystachyus.*

Subsection II.—Lignosi Freytag, subsect. nov. TYPE SPECIES: *Phaseolus marechalii* Delgado.

Herba perennis vel annua, rami basali crassa vel plus minusve lignosus, volubilis magnis vel frutex ad apicem volubilem, radix perennis crassa carnosa vel annotina fibrosa, foliolae plerumque parvae ad grande et coriacea pubescentae adaxile, nectaria foliolarum agiles, paniculae reduit et ramis lateralibus brevibus vel reduit gongyladibus, vexillo ad apicem pubescentem, legumen plerumque breve falcato aspero fibrose et saepe glaucum plus minusve glabrum.

Plant perennial or annual, the basal stem thick and more or less woody, becoming a large climbing vine or bush with vining tips; root perennial, thick, fleshy, or short-lived annual and fibrous; leaflets usually small to large, usually coriaceous, mostly densely pubescent below; foliar nectaries hyperactive; inflorescence usually a reduced panicle with lateral branches short or reduced to knobs; standard of flower often pubescent at tip; pod usually short, falcate, tough fibrous and often glaucous or glabrous.

KEY TO SPECIES AND SUBSPECIES

1. Flower standard pubescent at apex.
    2. Leaflets small, green and covered with whitish strigose or hirsute pubescence below, glabrous to puberulent above; rare, slopes of Volcan Colima, Jalisco; 2250–2550 m ____ D.II.3. **P. nodosus**
    2. Leaflets medium sized, dark olive green above, lighter below, both surfaces puberulent; rare, along streams in pine forests of Oaxaca; 1940–2140 m ____ D.II.7. **P. acinaciformis**
1. Flower standard glabrous at apex.
    3. Buds and flowers often bunched together at apex of inflorescence, upper leaflets often quite narrow, lanceolate to oblong-lanceolate; scarce, dry forests in W Mexico: Sinaloa to Michoacán; 450–2100 m ____ D.II.1. **P. jaliscanus**
    3. Buds and flowers distributed along inflorescence, upper leaflets about the same shape as lower ones.
        4. Upper surface of leaflets puberulent and silvery reflective from microscopic whitish pits; rare, found only near Coalcomán, Michoacán; 1330 m ____ D.II.2. **P. scrobiculatifolius**
        4. Upper surface of leaflets puberulent non-reflective or whitish strigose or hispid.
            5. Terminal leaflets roundish obtuse; rare, found near Tepatitlán, Jalisco; 1740 m ____ D.II.6. **P. rotundatus**
            5. Terminal leaflets ovate acuminate.
                6. Flower large with about 3 mm diam. coil at keel tip, leaflets whitish tomentose; rare, in Sonora; 1500–1800 m ____ D.II.9. **P. sonorensis**
                6. Flower not large and with normal size coils of about 2 mm diam. at keel tip, leaflets puberulent above or whitish strigose or hispid.
                    7. Leaflets dark olive green and puberulent above, lateral leaf veins nearly parallel; rare, in S coastal range near Pochutla, Oaxaca; 1780 m ____ D.II.8. **P. xolocotzii**
                    7. Leaflets green and whitish strigose or hispid above and below, lateral veins somewhat curved and divergent.
                        8. Inflorescence long, definitely multi-branched and 'knobby', leaflets small to medium and whitish strigose or hispid, veins below not exceptionally whitish.

9. Standard longitudinally purple-striped and strigillose on the outer surface; calyx purplish, with yellowish appressed hairs; pedicels 4–6 mm long; rare, scattered in dry pine-oak forests in C Mexico: Mexico, Morelos, Puebla and Veracruz; 1900–2500 m _______ D.II.5. **P. marechalii**
9. Standard not longitudinally purple-striped and strigillose on the outer surface; calyx purplish, with yellowish appressed hairs; pedicels 6–9 mm long _______________ D.II.10. **P. juquilensis**
8. Inflorescence short, normally not branched, leaflets quite small and whitish strigose or hispid, veins below whitish; rare, near Maguarichi, Chihuahua _____________________ D.II.4. **P. albinervus**

D.II.1.—Phaseolus jaliscanus Piper, Contr. U.S. Natl. Herb. 22:697. 1926. (**Figs. 40, 46**). TYPE: MÉXICO. JALISCO: Mts. near Talpa, (20°12'N, 105°W), 1500–1700 m, 7 Mar 1897, *Nelson 4030* (HOLOTYPE: US 763708).

Phaseolus sempervirens Piper, Contr. U.S. Natl. Herb. 22:697. 1926. TYPE: MÉXICO (possibly NAYARIT): Sierra Madre, 1849(?), *Seemann s.n.* (HOLOTYPE: BM).

Aerial shoot an annual to perennial, short to long, slender, scandent, indeterminate vine, with tips climbing erect. ***Root*** a perennial, thick, fleshy and woody globose, to 60 cm long and with several somewhat smaller branch roots. ***Stems*** terete, striate, covered by white waxy material, sparsely covered with white hispid and strigose hairs. ***Stipules*** lanceolate, 3 mm long, 1.25 mm wide, 3- to 4-nerved, acute, sparsely covered by hirsute hairs, ciliate margins. ***Leaves*** 5–10.5 cm long; petiole puberulent, much shorter than the leaflets, about 4–5 cm long on lower leaves and about 1 cm long on leaves near apex of vines; petiolule 1–1.5 cm long; stipels linear, 2 mm long, 0- to 1-nerved; leaflets, ovate on leaves at base of plants becoming oblong to oblong-lanceolate on mid portions of plants, and the terminal ones lanceolate to nearly linear, all about the same size of 3–4 cm long, 1–1.3 cm wide, rounded at base and obtuse at apex, minutely apiculate, coriaceous, adaxial surface green but with a distinctly pale, variegated band along the central rib and minutely scrobiculate, the abaxial surface heavily nerved, finely and sparsely puberulent and farinaceous and grayish. ***Inflorescence*** an erect panicle with branches to 2 cm long normally not developed but increasing in size with prolonged (biannual?) growth and flowering; peduncle 6–20 cm long, sparsely covered with minute uncinate hairs; rachis short, 1–3 cm long, covered with minute uncinate hairs, 6- to 14-flowered; primary bract lanceolate, 2–3 mm long, strongly 3- to 4-nerved, puberulent; pedicellar bracts narrowly lanceolate 1 mm long 1-nerved puberulent often purplish; pedicel as long as the calyx, to 6 mm long in pod. ***Bracteoles*** minute, linear, 0.5–0.75 mm long, ciliate, early caducous. ***Flower*** purple; calyx flaring campanulate purplish, 3 mm long, the short rounded upper lobes 0.5 mm long, 1.5 mm wide, united into one, emarginate, ciliate, the 3 lower lobes subequal, dentate, acute 1 mm long, 1.25 mm wide, ciliate, glabrous except the central lobe sparsely covered by minute hirsute hairs; standard dark purple, the claw 1.5 mm long, the blade 1 mm more to flexure and 4 mm more to emarginate apex, 7 mm wide, an intramarginal auricle on each side near the base; wings lighter purple, darkly veined at base, the slender claw 2–2.5 mm long, the blade broadly obovate, cupped, 1 cm long, orbicular, right-angled at base, the spur pronounced, 0.25–0.5 mm diam.; keel, the claws 2 mm long from base and 2 mm more to bend and 3 mm more to the terminal 1 1/2 coils of 2 mm in diam., keel tip whitish or light rose, the auricles at 2 mm from base are 1.5 mm in diam.; vexillary stamen geniculate, the claw 0.5 mm long, about 2 mm more to end of thickened portion; stamen tube 1 mm from base to the small ridges 0.25 mm high, and 5 mm more to end of the united portion; ovary 3.5 mm long, 1.3 mm wide, glabrous, 4 ovules; style 10 mm long to the terminal thickened coil of 2 mm diam.; stigma lateral, introrse, about 1 mm long. ***Pod*** falcate, 3 cm long, 1 cm wide, compressed, fibrous, somewhat brittle, glabrous; beak minute, about 1 mm long. ***Seed*** rounded, compressed, 6 mm long, 6 mm wide, 3 mm deep, heavily black flecked and streaked; hilum elongate 1.2 mm long, 0.5 mm wide; lens not pronounced. ***Seedling*** from hypogeal germination; epicotyl 5 cm long; stipules at eophyll node united acicular, 1–2 mm long; eophylls ovate-oblong, 2.5 mm long, 1.5 cm wide, cordate base, acute tip; petioles 1 cm long; pulvini present; stipels about 1 mm long.

Specimens examined: **MÉXICO. Durango:** El Salto, Revolcaderos, Km 156 carr. a Mazatlán, 23°37'N, 105°50'W, 1960 m, 2 Nov 1978, *Debouck et al. 411* (CHAPA, K, US). **Jalisco:** 35 km E of Cabo Corrientes on road from Puerto Vallarta to El Tuito, 450 m, 7 Mar 1970, *Anderson et al. 6100* (MICH); Mpio. of Atenguillo, 15 km NW of Los Volcanes on road to Talpa de Allende, 27 Nov 1983, *Breedlove et al. 60652* (CAS, MO); microondas road, 5.3 mi from Tequila, (20°45'N, 105°45'W), 1950 m, grown in screenhouse at Mayagüez, PR from seed of TARS #320 (= seed collected 11 Aug 1983 by *Buhrow M21* (ARIZ), and grown in Campbell Ave. Farm Greenhouse, Tucson, AZ,) Sep.-Dec. 1985 as Study Collection *Freytag GF #SC-320* (CSU, WIS, US, MO, ARIZ, F, TEX, UC); 5.5 km

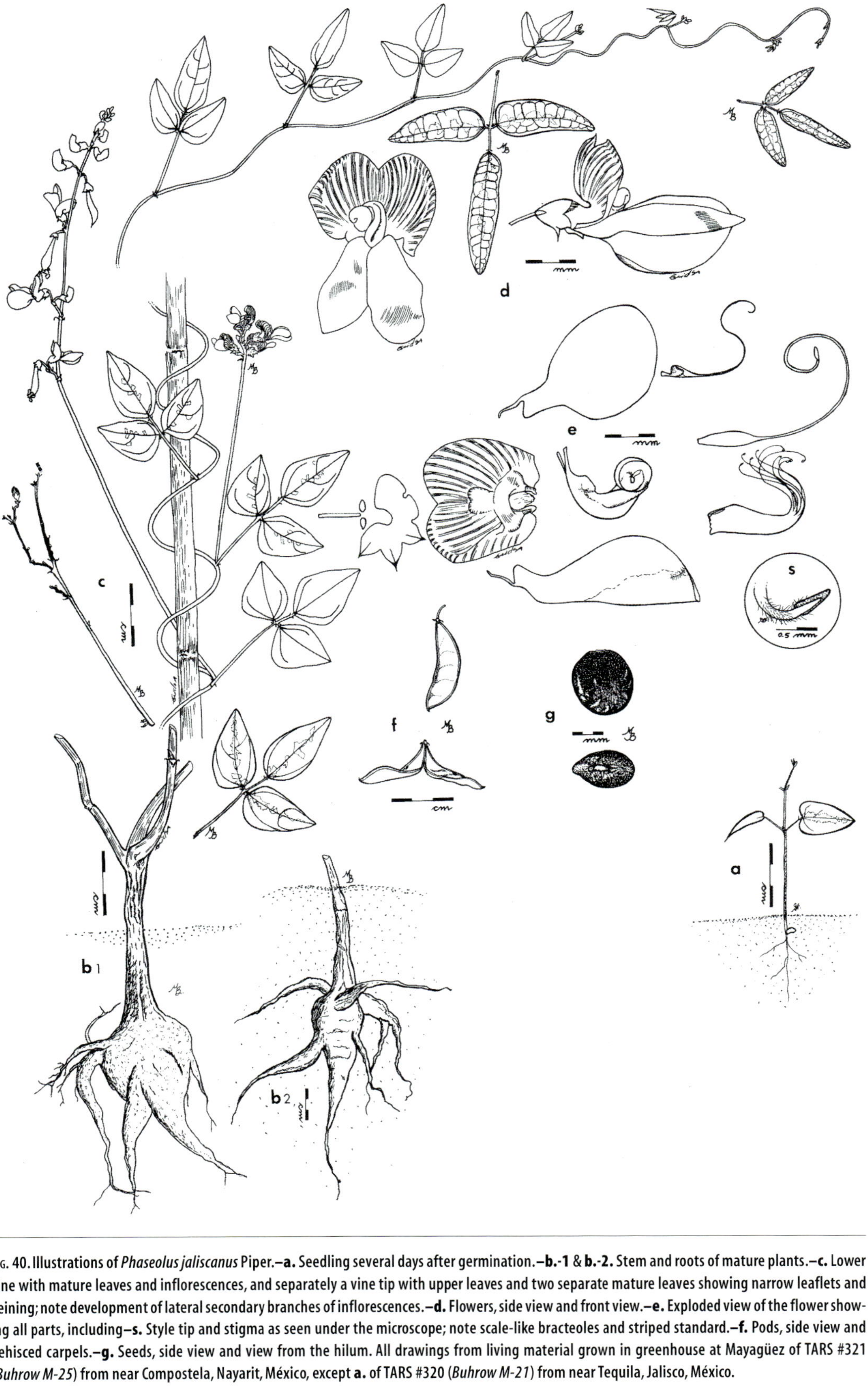

Fig. 40. Illustrations of *Phaseolus jaliscanus* Piper.—**a.** Seedling several days after germination.—**b.-1** & **b.-2.** Stem and roots of mature plants.—**c.** Lower vine with mature leaves and inflorescences, and separately a vine tip with upper leaves and two separate mature leaves showing narrow leaflets and veining; note development of lateral secondary branches of inflorescences.—**d.** Flowers, side view and front view.—**e.** Exploded view of the flower showing all parts, including—**s.** Style tip and stigma as seen under the microscope; note scale-like bracteoles and striped standard.—**f.** Pods, side view and dehisced carpels.—**g.** Seeds, side view and view from the hilum. All drawings from living material grown in greenhouse at Mayagüez of TARS #321 (*Buhrow M-25*) from near Compostela, Nayarit, México, except **a.** of TARS #320 (*Buhrow M-21*) from near Tequila, Jalisco, México.

SSE de Mascota, 20°29'N, 104°45'W, 1460 m, 4 Dec 1978, *Debouck 488* (CHAPA, K, MICH, US(2)); Atenguillo, 1 km SE Puerto La Campana Cara N, 20°16'N, 104°31'W, 1900 m, 6 Dec 1978, *Debouck 491-J* (CHAPA, G, K, M, US(2)); cara NE cerro, aproxim. 3.5 km SE Cuautla, 20°12'N, 104°27'W, 1730 m, 2 Dec 1978, *Debouck et al. 484* (CHAPA, K, US(2)); Km 20 Hwy from Cd. Guzmán-Concepción de Buenos Aires, at Jaramillo, 1 km de Dos Pozos, 19°50'N, 103°30'W, 2100 m, 2 Dec 1981, *Freytag et al. 81-20* (BR, CSU, EAP, F, G, GH, IBUG, K, MEXU, MICH, MO, NA, TEX, UC, US, WIS); 6-7 mi NW of San Miguel de la Sierra, 40-50 km airline W of Ayutla, 1900-2000 m, 4 Nov 1962, *McVaugh 22074* (CAS, MICH, MO). **Nayarit:** 10 mi S of Compostela, (21°05'N, 105°W), 12 Aug 1984, grown in Campbell Ave. Farm Greenhouse, Tucson, AZ, *Buhrow M25* (ARIZ); km 14-15 along route 28 between Tepic and Jalcacatlán, 1000 m, 7 Jan 1979, *Croat 45236* (MO); carr. Tepic-Miramar, Km 9.5 terraceria al Cuarenteno, 21°29'N, 104°55'W, 17 June 1987, *Téllez 10424* (MEXU). **Sinaloa:** Mpio. Concordia, 1-2 km N of the Mazatlán-Durango road at Liberas, (22°20'N, 106°W), 1850 m, 2 Oct 1985, *Bartholomew et al. 2549* (CAS).

Habitat.—This species is found growing on steep slopes and in dry ravines in mixed open forest of pine, oak, palm, *Arbutus* and with broad leaved shrubs and Mimosoideae, many grasses, ferns, and composites. Soils are shallow, rocky, organic and clay of brownish-red color, or limestone or basalt.

Diseases and pests.—Collectors report round and silvery leafspots and some leaf-eating worms, red spider and weevils.

Comments.—We follow Delgado (1985) in placing *P. sempervirens* as a synonym of this species, noting that McVaugh (1987) has excluded it as 'doubtful' but without much explanations. Allozyme essays (Maquet et al. 1999) have shown that *P. jaliscanus* is close to *P. salicifolius* and *P. ritensis*, as some sort of tertiary gene pool of the Lima bean, in agreement with the widecrossing experiments (Katanga & Baudoin 1990) and ITS DNA sequencing data (Delgado et al. 1999; Gaitán et al. 2000).

The species is best recognized by a rather short, upright vine and coriaceous leaflets, the lower ones ovate and the upper ones quite small and nearly linear. The flower has a distinctly striate standard (see Color Plate I, photo 11).

D.II.2.—Phaseolus scrobiculatifolius Freytag, sp. nov. (**Fig. 46**). TYPE: MÉXICO. MICHOACÁN: 16 km (10 mi) O de Coalcomán hacia Coahuayana, (18°35'N, 103°15'W), 1330 m, *Cowan 4902* (HOLOTYPE: TEX; ISOTYPE: MEXU n.v.).

Herba parvula scandens, foliolis parvulis ovatis subra argenteas ob foveolas albidas minutas infra aliquanto farinoso-cinerascenti-viridas, paniculae breves curvae ramulos secundarios perdeminutos 1-2 mm longus tantum gerentibus. Habitat solum in montibus prope Coalcomanum Michoacanae rarus.

Aerial shoot a small scrambling or climbing, indeterminate vine, to 2 m long. ***Root*** unknown. ***Stems*** terete, 1.5 mm in diam., somewhat striate, younger stems greenish, sparsely covered by reflexed-hispid and strigose hairs, older stems turning purple to jet black from an apparently algal or fungal layer; internodes 5 cm long in lower portions of plants to 10-15 cm long above. ***Stipules*** triangular, acute 1-1.5 mm long, 1 mm wide, obscurely 3-nerved, covered by a few scattered hispid hairs. ***Leaves*** 5-8.7 cm long; petiole 1.5-3 cm long, covered by uncinate hairs; petiolule 6-11 mm long; stipels obovate, acute, 1 mm long, 0.5 mm wide, covered by hispid pubescence; pulvini 3 mm long, densely covered by uncinate hairs; terminal leaflet ovate, acute to obtuse, not apiculate, 2.5-4 cm long, 1.5-3.5 cm wide, adaxial surface minutely white pitted giving off a grayish-silvery sheen, abaxial surface somewhat grayish-green and farinaceous, lightly covered by minute uncinate and hispid hairs especially on veinlets. ***Inflorescence*** a short, curved panicle; peduncle 1.2-4 cm long, covered by minute uncinate hairs; rachis 3-6 cm long of 3-10-16 nodes; lateral branches much reduced to 1-2 mm long; primary bracts ovate, 0.75-1 mm long, 0.5 mm wide, acute, covered with short hispid hairs; pedicel 2 mm long in bud to 3.5 mm long in pod, covered by minute uncinate hairs. ***Bracteoles*** orbicular, minute, 0.6-0.75 mm long, 0.3-0.4 mm wide, covered with short hirsute hairs, ciliate, purple, early deciduous. ***Flower*** purple, striate; calyx campanulate, tube 2.25 mm long, upper lobes rounded and joined into one emarginate, 0.5 mm long, 1.5 mm wide (each lobe), puberulent and ciliate, purple, the lower 3 dentate, subequal, acute, 0.75 mm long, 1.25 mm wide, a few scattered strigose hairs on middle lobe, ciliate, purple; standard purple, striate, 3.5 mm to flexure and 4 mm more to apex, 7 mm wide, a few hispid hairs at tip. ***Pod*** strongly falcate, narrow at top, (immature) 3 cm long, 8 mm wide; strong sutures; valves wrinkled and glabrous, dark; beak strong, curved abaxially, 4-5 mm long; 3-4 ovules. ***Seed*** unknown. ***Seedling*** unknown.

PARATYPE. **MÉXICO. Michoacán:** Parque Nal. Barrancas de Cupatitzio, estación forestal SARH Uruapan, 1750 m, 12 Dec 1991, *Estrada et al. 2354* (BRIT).

Habitat.—This taxon is found climbing over shrubs in pine-oak forests, on moist north slopes with *Calliandra* and *Baccharis* and in red lutite soils.

Comments.—Known only from few specimens which resemble *P. marechalii* in most characteristics except that the leaves are not as densely pubescent and with the upper surface covered by minute pits which impart a silvery sheen.

D.II.3.—Phaseolus nodosus Freytag & Debouck, sp. nov. (**Figs. 41, 46**). TYPE: MÉXICO. JALISCO: in barrancas below Canoa de Leoncito, NE slopes of Nevada de Colima, (19°30'N, 103°35'W), 2250–2550 m, 11 Oct 1952, *McVaugh 13433* (HOLOTYPE: MICH).

Herba magna lignosa scandens, foliolis parvulis albidis hispidis ovatis, paniculae parvulae ramulis secundariis ad processibus noduliformibus brevibus in axillem bractearum reductis. Crecit modo in clivis vulcanicis Colimae Novae Galiciae rarus.

Aerial shoot a woody, indeterminate vine, climbing 5–6 m high on trees. ***Root*** unknown. ***Stems*** terete, 1.5 mm in diam., striate; internodes 5–8 cm long, densely covered by closely appressed, reflexed- and short, white hispid hairs, purplish. ***Stipules*** lanceolate to ligulate, acute, 4 mm long, 1 mm wide, 3- to 4-nerved, glabrous adaxially and covered by short hispid hairs abaxially. ***Leaves*** 5–10 cm long; petioles longer than leaflets, delicate, covered by short white hispid hairs; petiolule 5–11 mm long; pulvini 2–2.5 mm long, covered by densely white hispid hairs, dark; stipels ovate, 0.75–1.25 mm long, acute, minute white hispid hairs on abaxial surface; terminal leaflet ovate, 2–4 cm long, 1–3 cm wide at 1/4 from base, acuminate, not apiculate, white veined, adaxial surface puberulent of scattered minute white hispid hairs, surface pitted by minute white spots (scrobiculate), abaxial surface densely covered with white hispid hairs. ***Inflorescence*** a small panicle, the lateral branches reduced to short nodular growths in the axils of the bracts; peduncle 3–6 cm long, densely covered by minute reflexed-hispid and uncinate hairs; rachis 3–6 cm long, of many nodes mostly 2–3 mm apart, the basal ones 1 cm apart; primary bracts ovate-acuminate, 2–3 mm long, obscurely 2- to 3-nerved, covered by short white hispid hairs, persistent; pedicel delicate, 4–5 mm long becoming 7 mm long at mature pod, densely covered with short white hispid hairs, purple. ***Bracteoles*** ovate-acute to ligulate, 1 mm long, 0.5–0.5 mm wide, covered with minute white hispid hairs, obscurely 1-nerved. ***Flower*** striped purple; calyx campanulate, purplish, puberulent of minute, white hispid hairs, the upper lobes rounded, 0.5 mm long, 2.5 mm wide, lower 3 lobes subequal, dentate, acute, 0.75 mm long, 1.5 mm wide, a few longer hairs on middle lobe; standard distinctly striped purple, a few short white hispid hairs at apex in bud, 4 mm to flexure and 5 mm more to emarginate apex, rounded, 7 mm wide; wings, the claw 4.5 mm long, 0.25 mm wide, the blade purple and showy, cupped and spreading, rounded, 9 mm long, 5 mm wide, the auricle pronounced; keel 5 mm long to bend and 4–4.5 mm more to the terminal 1 3/4 coils of 2.5 mm diam., the pointed knobs pronounced at 2 mm from base; ovary straight, 5 mm long, glabrous, with thickened sutures, 4 ovules; style 6 mm long to the terminal thickened coil of 2 mm diam.; stigma narrow linear, introrse, 1 mm long. ***Pod*** nearly straight, broadest near tip, 3 cm (immature) long, 8 mm wide; valves glabrous, covered with minute, white speckles (lenticels?); sutures strong, tip rounded; beak 1.5 mm long, reflexed downward. ***Seed*** unknown. ***Seedling*** unknown.

Habitat.—This species is reported to be abundant on steep mountainsides in pine-oak forest (zone of broad-leaved trees in barrancas).

Comments.—It is known from only a single collection (not reported in 'Flora Novo-Galiciana', McVaugh 1987) which closely resembles *P. jaliscanus* but differs in having the secondary branches of the inflorescence reduced to short knobby extensions and in many minor details of the flower structure.

D.II.4.—Phaseolus albinervus Freytag & Debouck, sp. nov. (**Figs. 42, 46**). TYPE: MÉXICO. CHIHUAHUA: E of Maguarichi on road to the main route between Basaseachi and San Juanito, headwaters of the Río Oteros, (27°55'N, 107°55'W?), 17 May 1985, *Lavin et al. 5426* (HOLOTYPE: TEX; ISOTYPES: MO, UCR).

Herba scandens suffrutescens, foliolis parvulis viridibus ovatis infra manifeste albinervis, inflorescentiis brevibus ut videtur eramosis, leguminibus parvulis fibrosis falcatisque. Habitat solum in desertis prope Maguarichum Chihuahuae rarus.

Aerial shoot a scandent, shrubby and wooody, indeterminate vine, (to 1 m?). ***Root*** unknown. ***Stem*** woody, straight, twining at tips, terete, striate, sparsely covered by white, strigose and minute unci-

Fig. 41. Illustrations of *Phaseolus nodosus* Freytag & Debouck.–c. Vine tip with leaves and inflorescence with flowers and immature pods; note lateral branches of inflorescence reduced to stubby nodes.–e. Exploded view of the flower showing all parts, including–s. Style tip and stigma as seen under the microscope; note the scale-like bracteoles and well-developed spur on the stamen tube. All drawings from the type specimen *McVaugh 13433* from NE slopes of Nevado de Colima, Jalisco, México.

nate hairs, the internodes 10–15 cm long. ***Stipules*** broad triangular, 2 mm long, 1.5 mm wide, densely covered with strigose hairs on abaxial surface, 5- 7-nerved. ***Leaves*** 12.6–8.3 cm long; petiole 2.5–3 cm long, delicate, striate, densely covered with white uncinate and strigose hairs; petiolule 6–7 mm long; pulvini 2–3 mm long, green, densely covered by uncinate hairs; stipels ovate 1.25 mm long, 0.5 mm wide, indistinctly 2- to 3-nerved, densely covered by strigose hairs; terminal leaflet ovate, 3.5–4 cm long, 1.75–2 cm wide, acute, minutely apiculate, green, distinct white veined abaxially with long, white strigose hairs mostly on veins, short hispid hairs on adaxial surface, somewhat variegated; lateral leaflets similar, inequilateral. ***Inflorescence*** a short stout raceme; peduncle 6–7.5 cm long, striate, sparsely covered with white strigose and uncinate hairs; rachis 1.5–5 cm long, 3–8 nodes, sparsely covered with white strigose and uncinate hairs; primary bract broad lanceolate, 3 m long, 1.4 mm wide, indistinctly 3- to 5-nerved, covered by white strigose hairs; pedicel 5–6 mm long, stout, covered by long, white strigose hairs; pedicellar bract linear 2 mm long strongly 1-nerved sparsely covered

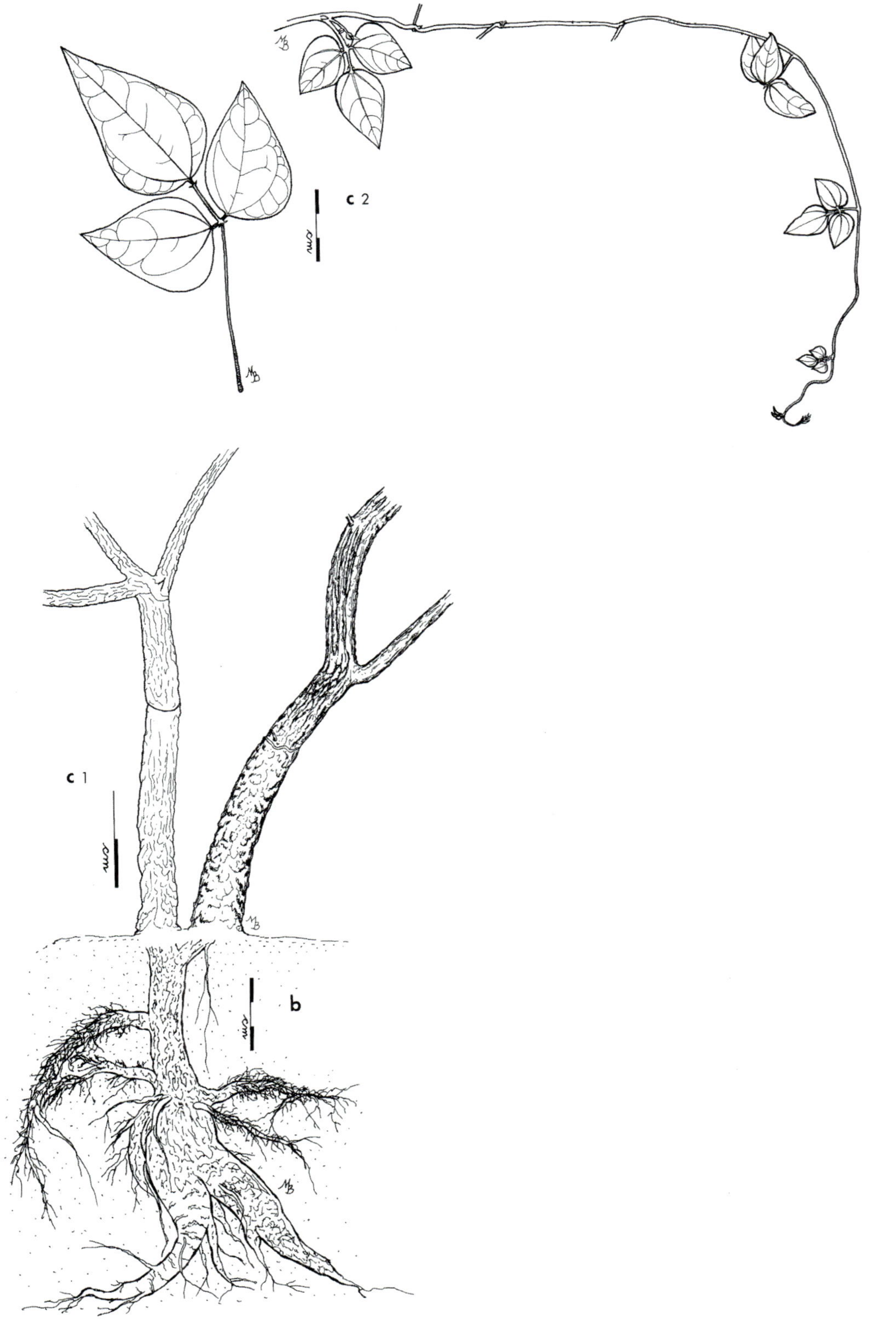

Fig. 42. Illustrations of *Phaseolus albinervus* Freytag & Debouck.—**b.** Mature root.—**c.-1.** Lower portion of stem and, separately a mature leaf.—**c.-2.** Vine tip and leaves. All drawings from living material grown in greenhouse in Mayagüez from seed of TARS #460 (*Lavin 5426*) collected near Maguarichi, Chihuahua, México.

FIG. 43. Illustrations of *Phaseolus marechalii* Delgado,–**a.** Seedling a few weeks after germination; note very small eophylls.–**b.** Root of mature plant.–**c.** Lower stem and sections of vine from base to tip with mature leaves and inflorescences; note well-developed lateral secondary branches of inflorescences and terminal inflorescences at tip of vine.–**d.** Flowers, side view and front view.–**e.** Exploded view of the flower showing all parts, including–**s.** Style tip and stigma as seen under the microscope; note the scale-like bracteoles and striped standard.–**f.** Pods, side view and dehisced carpels.–**g.** Seeds, side view and view from the hilum. All drawings from living material grown in greenhouse at Mayagüez of TARS #124 (NI 402) collected by INIA, from Morelos, México.

with white strigose hairs. ***Bracteoles*** caducous(?). ***Flower*** (known only in bud); calyx broad flaring, scabrous, covered by scattered long strigose hairs at base and lower lobes, ciliate, tube 2–2.5 mm long, upper lobes scarcely elongate, 2 mm wide, lower lateral lobes dentate, 1 mm long, 2 mm wide, middle lobe 2 mm long, 2 mm wide. ***Pod*** medium, falcate, 4–5 cm long, 7–8 mm wide; sutures pronounced; valves, fibrous, brittle, glabrous; beak stout, 4–5 mm long, straight. ***Seed*** oval to nearly spherical, 4.5–6 mm long, 3.9–4.5 mm wide, 2–2.5 mm thick, solid tan to reddish brown, nerved, black ring around hilum; hilum 1.2–1.8 mm long, 0.7–0.8 mm wide. ***Seedling*** unknown.

Habitat.—It was found growing as a scandent shrub hanging from cliff faces on steep mountain slopes in a narrow canyon.

Comments.—This species is known from only a single collection and from the most northern location for any species of this section. It resembles *P. marechalii* in having small leaves densely covered with whitish pubescence but differs with strikingly prominent veins most visible from the lower surface. Its location and fruiting period seem not correspond with those of *P. marechalii* (Chihuahua and May, and Trans-Mexican Volcanic Belt and December, respectively). Given the woody aspect of stems it could be a perennial with second growth after rainy periods.

D.II.5.—Phaseolus marechalii Delgado, Syst. Bot. 25:427–429. 2000. (**Figs. 43, 46**). TYPE: MÉXICO. MORELOS. Mpio. Tlanepantla, a 3 km al S de la desviación a Tlanepantla, sobre la carr. Zochimilco-Oaxtepec, (18°15'N, 99°W), 2,400 m, 17 Nov 1975, *Delgado et al. 115* (HOLOTYPE: MEXU n.v.; others to be distributed).

Aerial shoot a perennial, woody, shruby, climbing, indeterminate vine, 1–3 m long. ***Root*** a perennial, much branched thick, fleshy and woody, 8–30 cm long, 5 or more cm wide, the 8–10 branches 1–2 cm thick and tapered. ***Stems*** terete, at base of plant 8–10 mm thick, 3–5 mm thick in upper portions of plants, a few secondary branches often terminating in determinate inflorescences, densely covered with white strigose pubescence. ***Stipules*** triangular, acute, 3–4 mm long. ***Leaves*** small; petioles 5–7 cm long, fairly stout, densely covered with white strigose hairs; petiolule 2–9 cm long; basal pulvinus 6–8 mm long and swollen, the terminal and lateral pulvini 2–3 mm long, densely covered with white strigose hairs; terminal leaflet broadly ovate, 4 cm long, 3 cm wide, adaxial surface green, the abaxial surface nearly white from dense strigose pubescence; lateral leaflets similar but slightly inequilateral. ***Inflorescence*** a stout panicle with most nodes producing secondary branches, some to 4 cm long, the peduncle 2–3 cm long, the rachis 8–15 cm long; primary bract ovate, 2–3 mm long; pedicel 4–5 mm long; pedicellar bracts 1 mm long narrowly triangular 0- 1-nerved, all parts of inflorescence densely covered with white strigose hairs. ***Bracteoles*** scale-like, ovate, 1–2 mm long, acute, 1-nerved. ***Flower*** purple; calyx campanulate, the united tube 1.5–2 mm long, 2 upper lobes rounded, united, emarginate, lower 3 lobes equal, broadly acute, 2 mm long, 1.75 mm wide; standard distinctly striped purple, broadly rounded-squarose, reflexed, 5 mm from base to bend and 8–10 mm more to emarginate tip, 10–12 mm broad, lateral edges enrolled, the auricles well-developed 0.5–1 mm long; wings purple, spatulate, unequal, clasping, the claw 4 mm long, the blade 12–15 mm long, 5–6 mm broad, the spur pronounced 2–3 mm long; keel, the claws 3 mm long, 3 mm more to bend and 8–9 mm more to base of the 1 1/2 terminal coils of 2 mm diam.; vexillary stamen, the claws 1.5 mm long, the thickened geniculate portion 3–4 mm long to filaments; united stamen tube 1 mm long to the pronounced geniculate ridges and 10 mm more to separated filaments; basal collar 1 mm long, minutely denticulate; ovary 3–4 mm long, 1.5 mm wide, glabrous, 3–5 ovules; style 10 mm long to terminal coil of 2 mm in diam.; stigma lateral introrse oblique, oblong, 1.2 mm long. ***Pod*** small, 3 cm long, 7 mm wide, 4–5 mm thick, falcate, ligulate restriction at pedicel end, somewhat flattened; valves tough fibrous, obscurely reticulate, somewhat glaucous, slightly purple striate marked, slightly restricted between seeds, twisted 1/2–1 turn at dehiscence; sutures thickened, covered with minute glandular knobs, whitish; beak slightly curved downwards, 1 mm long; 2–3 seed. ***Seed*** elongate squarish, 5–7.5 mm long, 4–4.5 mm wide, 2.8–4 mm thick, somewhat angular ends, often flattened, brown and black speckled on tan or brown with a black ring around hilum; hilum orbicular, 1.25 mm long, 0.6 mm wide; lens large and raised. ***Seedling*** from hypogeal germination; epicotyl delicate, 6–8 cm long, the next internode 10–18 cm long; primary leaves small and simple, petiole 3 cm long, stipels 2 mm long, the blade broadly ovate, 3 cm long, 2 cm wide, base cordate, tip acute, distinctly nerved.

Specimens examined: **MÉXICO. México:** 12 km SE de Villa del Carbón, 19°41'N, 99°25'W, 2520 m, 6 Nov 1987, *Debouck et al. 2377* (CHAPA, MICH, US). **Morelos:** Cuernavaca, *Schubert 623* (K) (grown at the Faculté des Sciences Agronomiques de l'Etat, Gembloux, Belgium, under NI 402, herbier no. 2386); Sierra Morelos, Cuernavaca, 2000 m, 7 Nov 1969, *J.Hinton 17462* (K, MEXU); barranca just below Km 62 marker, about 14 km N of Cuernavaca, (19°N, 99°12'W), 27 Oct 1947, *Norvell HM215* (UC). **Puebla:** cerro 4 km NE de Pueblo Acajete, 19°08'N, 97°54'W, 2400 m, 11 Nov 1987, *Debouck et al. 2389* (CHAPA, MICH, US).

Habitat.—This species is found growing rather sparsely and localized in forests of pine, pine-oak, oak or *Alnus*, in the central part of the "Eje Volcánico," on both northern and southern slopes, with frequent fog and sometimes with evidence of grazing or fire and, rarely, frost damage. It is generally climbing over thickets of Compositae, *Agave, Juniper, Desmodium* and grasses and on loose dark brown organic soils derived from volcanic ash, riolites or shists with good drainage.

Diseases and pests.—Damage from *Apion* has been reported.

Comments.—Norvell, who first collected this species in 1947, recognized the uniqueness of his collection by naming it "*glaucocarpus*," but did not publish it. Delgado (1985) omitted the description and type for this taxon from his thesis. Maréchal et al. (1978b) mentions a seed collection (NI-402 collected by *Schubert 623* from Morelos, México) of this taxon as having a great similarity (94%) with *P. polystachyus* and that he had made crosses to it with both *P. lunatus* and *P. ritensis*, producing sterile F_1s. This would indicate that all of these species are closely related thus reinforcing their placement in this section with *P. lunatus*. Other results (seed storage proteins and allozyme: Maquet et al. 1999; PCR-RFLPs: Fofana et al. 1999; ITS DNA sequencing: Delgado et al. 1999; Gaitán et al. 2000) support this further. This species has a rather stout and pubescent inflorescence but the flower is very similar to others of the section (see Color Plate I, photo 12).

D.II.6.—Phaseolus rotundatus Freytag & Debouck, sp. nov. (**Figs. 44, 46**). TYPE: MÉXICO. JALISCO: cerca del crucero de la carretera Tepatitlán-Yahualica, a 20 km NW de la ciudad de Tepatitlán, en ambos lados del Río Verde en laderas con pendiente de 45°a 60°, (20°50'N, 102°55'W, 2000 m), 23 Dec 1997, *Lépiz, R. #410511.1* (HOLOTYPE: US; ISOTYPES: ARIZ, CSU, F, IBUG, TEX).

Similis *Phaseolo marechalii* in herba valida lignosa scandens partibus dense strigosis vel tomentosis trichomatibus viridi- vel flavi-albis, sed ob eo foliis majoribus rotundatis ovatis et paniculis longioribus ramulis secundariis deminutis vel ad nodos gongylodibus reductis differt. Habitat solum in vallibus prope Tepatitlanum Novae Galiciae rarus.

Aerial shoot a stout, woody and shrubby, climbing, indeterminate vine, somewhat perennial on lower portions, 3–4 m long, to 0.8 cm thick, some 4–6 arising from the crown of the root. ***Root*** a perennial, very large, thick, fleshy and woody, 20–30 cm long, 5–6 cm wide, with 5–6 branches 10–30 cm long, 1–4 cm thick, prominent narrow lenticels and a corky and scaly cortex. ***Stems*** terete, stout, woody, 5–8 mm in diam., glabrous and corky cortex at base, densely covered with white hispid and strigose pubescence above, the internodes 11–19 cm long. ***Stipules*** triangular, acute, 2–3 mm long, covered by white hispid hairs. ***Leaves*** 12–26 cm long, greenish to yellowish-white from dense pubescence; petioles 6–12 cm long; petiolules 1.5–4 cm long; pulvini, the lower 4–8 mm long, the upper 2–5 mm long; stipules ligulate, acute, 3–4 mm long, striate, pubescent; stipels ligulate, 2–3 mm long, early caducous; terminal leaflet broadly ovate to orbicular, 4–11 cm long, 4–10 cm wide, broadest at 1/3 from base, obtuse, apiculate, distinctly 3-nerved, veinlets whitish, dark green above, lighter green below, densely covered by whitish strigose to tomentose hairs, especially abaxially; lateral leaflets similar but very slightly inequilateral. ***Inflorescence*** a panicle usually with 1–2 short branches at base, to 21 cm long, the peduncle 5–11 cm long, stout, tomentose, slightly curved, the rachis 5–10 cm long with some 5–10–15 knobby flowering nodes irregularly spaced, tomentose, subtended by primary bracts aciculate 1.5–2 mm long, strongly 3-nerved, densely covered with whitish pubescence; pedicels slightly recurved 5 mm long densely covered with whitish strigose pubescence and short uncinate hairs; pedicellar bracts linear 1 mm long narrow 1-nerved densely covered with whitish pubescence. ***Bracteole*** linear 1mm long strongly 1-nerved densely covered with whitish strigose pubescence. ***Flower*** deep purple fading blue violet; calyx green campanulate densely covered with short whitish pubescence, the united tube 1.5–2 mm long, 2 upper lobes rounded, united, emarginate, lower 3 lobes equal, broadly acute, 1.5 mm long, 1.75 mm wide; standard slightly striped purple, broadly rounded-squarose, reflexed, covered with a few short whitish hairs on pale lilac adaxial surface, 5 mm from base to bend and 7–8 mm more to emarginate tip, 9–11 mm broad, lateral edges enrolled, the auricles poorly

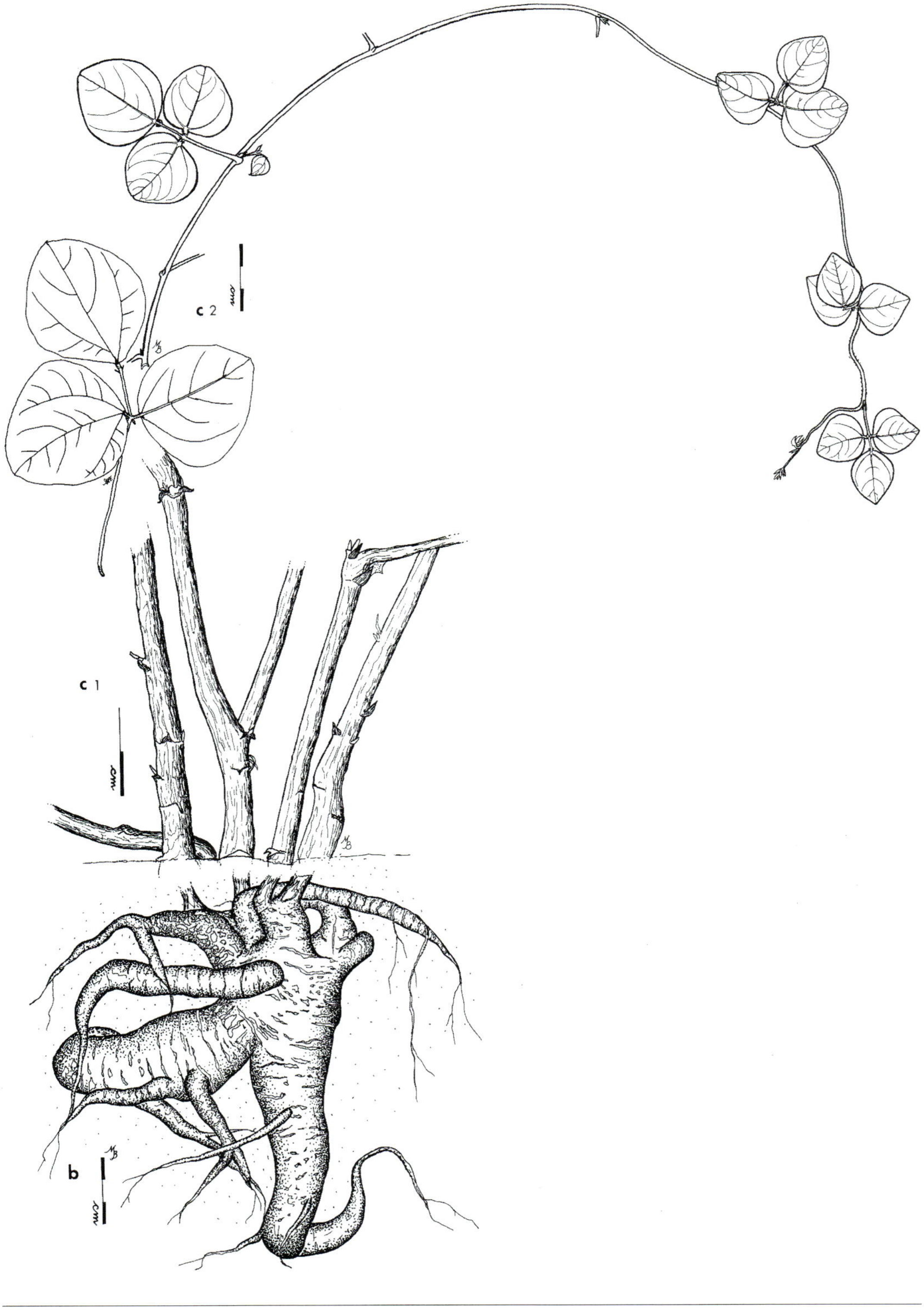

FIG. 44. Illustrations of *Phaseolus rotundatus* Freytag & Debouck.—**b.** Root of mature plant.—**c.-1.** Basal stems.—**c.-2.** Vine tip with separate mature leaf; note the rounded shape of leaflets. Drawings **b, c-1** & **c-2** from living material grown in greenhouse at Mayagüez of TARS #334 (NI 1046); Drawing **c-3** from field collection by Dr. R. Lépiz from near Tepatitlán, Jalisco, México.

developed 0.5–1 mm long; wings purple, spatulate, unequal, spreading, the claw 4 mm long, the blade 8–10 mm long, 5–6 mm broad, firmly adhering to the keel, the spur pronounced 2–3 mm long; keel pink lilac tip yellow, the claws 3 mm long, 3 mm more to bend and 8–9 mm more to base of the 1 1/2 terminal coils of 2 mm diam.; vexillary stamen, a thin sheath 0.75 mm high at its insertion, the claw 1.5 mm long, the thickened cupulate knob, 4–5 mm long to filament; united stamen tube 1 mm long to the geniculate ridges and 10 mm more to separated filaments; basal collar 1 mm long, minutely denticulate; ovary 4–5 mm long, 1.5 mm wide, glabrous, 3–5 ovules; style 10 mm long to terminal coil of 2 mm in diam.; stigma lateral introrse oblique, oblong, thickened, 1 mm long. ***Pod*** small, 35–42 mm long, 7–9 mm wide, 4–5 mm thick, falcate, somewhat flattened; valves tough fibrous, glaucous, slightly white streaked, slightly restricted between seeds, twisted 1–1 1/2 turn at dehiscence; sutures thickened, whitish; beak curved downwards, 1 mm long; 3–4 seed. ***Seed*** round oblong 6–7 mm long 5–6 mm wide 2–3 mm thick, finely black speckled on cream background with a black ring around hilum; hilum irregularly ellitpic, 1.2 mm long, 0.6 mm wide; lens small and slightly raised.

PARATYPES. **MÉXICO. Jalisco:** near Tepatitlán, Río Verde, young leaves only, grown in screenhouse in Mayagüez, PR, 15 Aug. 1989, from seed of NI 1046 (= TARS #334 collected by *Lépiz s.n.*, 22 Oct 1978), as Study Collection, *Freytag GF #SC-334* (CSU, WIS, US, MO, ARIZ, F). **Oaxaca:** Telixtlahuaca, (17°50'N, 97°W), 1737 m, 4 Dec 1967, *Gentry 22509* (NA).

Habitat.—This species was found twice by Dr R. Lépiz growing in the valley of the Río Verde, some 25 km NW from Tepatitlán, Jalisco. Soils are rocky, with much organic matter, and red clays. It grows on very steep slopes (more than 30%) with spiny Mimosoideae and Caña Brava grasses and with other species of the genus.

An apparently similar plant was collected by H.S. Gentry in a thicket of *Arundo donax* along an arroyo in Oaxaca. Because of the great distance between these two locations it is possible that the collection from Oaxaca may be another species, but is placed here because of the lack of adequate material to make a definitive identification.

Comments.—This species seems to be rare with only two collections known from quite disjunct areas. It closely resembles *P. marechalii* but differs in having somewhat larger, rounder leaflets and is a considerably larger plant (to 10 m). Physiologically it is also quite different as, under greenhouse conditions in Mayagüez, PR, it is much more difficult to get to bloom and although flowering has never set seed. In its native habitat it blooms in Oct.-Nov. and fruits in Nov.-Dec. In PCR-RFLPs of cpDNA studies (Fofana et al. 1999), it falls close to *P. polystachyus*, *P. ritensis*, and *P. reticulatus*. In ITS DNA sequencing studies (Gaitán et al. 2000), it has been found close to *P. jaliscanus* and *P. acinaciformis*.

D.II.7.—Phaseolus acinaciformis Freytag & Debouck, sp. nov. (**Fig. 46**). TYPE: MÉXICO. OAXACA: Mpio. Nochixtlán, Km 127 carr. Mex-190, 1 km E de Llano Verde, Asunción Nochixtlán, 17°19'N, 97°05'W, 2140 m, 24 Oct 1987, *Debouck & Muruaga 2331* (HOLOTYPE: US; ISOTYPES: CHAPA, COL, M, MICH).

Herba decumbens vel scandens, stipulis lanceolatis mediocribus foliolis obscuris olivaceis petiolorum tenuum, paniculae ramulis generaliter 3 vel 4 secundariis ad 10–15 cm longis vel ad nodos gongyloides reductos, bracteolae deminutae squamosae, vexillum ad apicem pubescentem, legumen curvatim usque ad 3 cm longem glaucum fibrosum 2 vel 3 seminibus. Crecit ad ripas rivulorum in nemoribus pinetis Oaxacae rarus.

Aerial shoot a decumbent or climbing, indeterminate vine, 2–4 m long. ***Root*** conical, thick, fleshy, 5 cm wide, 30 cm long. ***Stems*** terete, 1.5–2 mm diam., slightly striate, sparsely covered with short, reflexed-hispid and minute uncinate hairs; internodes 5–40 cm long, purple above, green below. ***Stipules*** lanceolate, 2.5–3 mm long, obscurely 1- to 3-nerved, covered with short, white hispid hairs. ***Leaves*** 6.8–17 cm long, dark olive-green above; petioles 2–7 cm long, often delicate, puberulent mostly of minute uncinate hairs; petiolules 0.5–2 cm long; pulvini, at base of petiole 4–5 mm long, those of the lateral leaflets 2–3.5 mm long, dark, white hispid on upper surface; stipels ligulate, 2 mm long, obtuse, sparsely white hispid; terminal leaflets broadly rounded ovate to ovate, 3.7–7.5 cm long, 2.2–6.7 cm wide, acute, apiculate, covered by yellowish strigose and uncinate hairs especially on veins; lateral leaflets similar but very slightly inequilateral. ***Inflorescence*** a panicle, the lateral secondary branches generally greatly reduced to knobs or to 3–14 mm long, but some terminal inflorescences greatly branched with 3–4 secondary branches to 10–15 cm long; peduncle 2.5–7–21 cm long, puberulent to glabrous, curved; rachis 2.5–10 cm long, covered by minute uncinate hairs; primary bract 1.25–2 mm long, ovate, acute, 1- to 3-nerved, scabrous, minute white hispid; pedicel 3 mm long, becoming 6–7 mm long at mature pod, densely covered with white hispid and minute uncinate hairs, purple. ***Bracteoles*** scale-like, ovate, 0.75 mm long, acute, covered with minute, white hispid hairs, purple. ***Flower*** purple; calyx campanulate, tube 2.25 mm long, upper lobes rounded, joined into 1 emarginate, 0.5 mm long, the lower 3 lobes subequal, dentate, acute, 0.75 mm long, 1.25 mm wide, the whole calyx covered by minute white hispid hairs, purple; standard purple, rounded, many minute white hispid hairs at apex in bud, 3 mm long to flexure, 4 mm more to emarginate apex, 8 mm wide; ovary of 3 ovules. ***Pod*** scimitar shaped, short and broad, 2.5–3 cm long, 7–8 mm wide; valves fibrous, glaucous, covered by many speckled white (lenticels?), dark green, drying tan; sutures well defined; beak minute, straight, less than 1 mm long. ***Seed*** unknown. ***Seedling*** unknown.

Habitat.—It was reported as locally abundant, rare elsewhere, in moist low places in partial shade in disturbed pine-oak forests with frequent mists with *Bromelia*, grasses, *Desmodium* and composites and growing in soils of loose organic and rocky structure derived from gneiss and metamorphic schists.

Diseases and pests.—Damage was reported as caused by powdery mildew, beetles and weevils.

Comments.—This species is named for the broad, scimitar shaped pods, and comes from an area (central Oaxaca) between *P. xolocotzii* in the south and *P. marechalii* in Morelos, western México and southern Puebla to the north. In ITS DNA sequencing studies (Gaitán et al. 2000), it has been found close to *P. jaliscanus* and *P. rotundatus*.

D.II.8.—Phaseolus xolocotzii Delgado. Syst. Bot. 25:429–431. 2000. (**Figs. 45, 46**). TYPE: MÉXICO. GUERRERO: Mpio. Chilpancingo de los Bravos, 5.3 km al N de la desviación a Filo de Caballo para Chichihualco, (17°35'N, 99°30'W), 1270 m, 4 Nov 1985, *R. Torres 7700* (HOLOTYPE: MEXU n.v.; others to be distributed).

Aerial shoot a large climbing, indeterminate vine, to 3–5 m long. ***Root*** a thick, fleshy, taproot about 80 cm long. ***Stems*** terete, 2–3 mm thick; internodes 7–10 cm long, covered with stout, appressed, reflexed-strigose, whitish or yellowish hairs. ***Stipules*** spatulate-triangular, 3 mm long, indistinctly 3-nerved, sparsely covered with white strigose hairs. ***Leaves*** 10.5–14.7 cm, parallel veined, dark green; petiole 3–4 cm long, somewhat striate, densely covered by appressed, white strigose hairs; petiolule 1–2 cm long, covered with white strigose pubescence; pulvini dark, 5–6 mm long, densely covered with white strigose hairs; stipels narrowly lanceolate, 2 mm long, 0.5–0.75 mm wide, acute, covered with white strigose hairs; terminal leaflet ovate to broadly ovate, 5.5–8 cm long, 3.5–6 cm wide, broadest at about midpoint, acuminate, apiculate, distinctly 3-nerved at base and all principal secondary veins from midrib and tertiary veins at leaf base nearly straight and parallel, ultimate veinlets dark brown and reticulate, distinctly yellowish pubescent and sparsely covered with appressed white strigose hairs especially on veins, somewhat farinaceous in older leaves, adaxial surface dark green to dark olive green, abaxial surface densely covered with appressed, yellowish (young leaves) or whitish (older leaves) strigose hairs; lateral leaflets similar but somewhat inequilateral. ***Inflorescence*** a panicle usually with 2 branches from base nearly as long as main raceme, with many knobby nodes; peduncle 3–6 cm long; rachis 3–8 cm long of many nodes, densely covered with white strigose and uncinate hairs; primary bract aciculate, curved adaxially and cupped, 2–3 mm long, covered with white strigose hairs; nodes, knobby from short reduced branches; pedicels 5 mm long to 17 mm long in pod, covered with white strigose hairs; pedicellar bracts linear narrowly lanceolate 1 mm long covered with white strigose hairs. ***Bracteoles*** minute, triangular, 0.5–1 mm long, 0.25 mm wide, acute, white strigose. ***Flower*** light pink to purple; calyx flaring campanulate, 3 mm long, upper 2 lobes united into 1 emarginate, 3 mm wide, 0.75 mm long, the lower 3 subequal, dentate, 1 mm wide, 0.75 mm long, all the calyx covered with white strigose hairs, ciliate on margins; standard light pink, the claw 1.5 mm long, 0.5 mm at base to 2.5 mm wide above, the blade 2.5 mm long from base to flexure and 5 mm more to emarginate apex, 10 mm wide, the auricles 1–1.5 mm long, somewhat pointed; wings light pink (purple), the claws 3.5 mm long, the spur 1.5 mm diam, the blade obovate, 9 mm long, 5–6 mm wide, cupped; keel, the claws 4 mm long, 1 mm more to upward bend and 4 mm more to base of the terminal 1 3/4 coils of 2.5 mm diam.; vexillary stamen, the claw 0.75 mm long, the rounded geniculate

C

Fig. 45. Illustrations of *Phaseolus xolocotzii* Delgado.–c. Short piece of vine with mature leaf and inflorescence with bud, flower and immature pods. Note parallel veination of leaflets. All drawings from the type specimen *Debouck et al. 2346*, collected near Pochutla, 38 km S of San José Pacífico, Oaxaca, México.

knob 0.75 mm long, 1 mm broad, and 3.5 mm more to end of the thickened portion; stamen tube 1.5-2 mm wide at base, 5 mm long to upward bend and 4 mm more to end of united filaments, the distinct knobs at about 2 mm from base are about 0.5 mm high; basal collar about 1 mm long, obliquely tapered, not denticulate; ovary smooth and straight, about 5 mm long, style 6 mm long to base of the terminal thickened coil 2 mm in diam.; stigma introrse, lateral, pointed, 0.75 mm long. ***Pod*** curved, 6 cm long, 1 cm wide, flattened (immature) and narrow at tip; valves fibrous, glabrous (glaucous?), wrinkled; sutures thickened; beak, short, stout 2–3 mm long, recurved downward; 4 ovules. ***Seed*** ovoid, (immature) 6–7 mm long, 4.5–5 mm wide, thick, solid brown or black. ***Seedling*** unknown.

Specimens examined: **MÉXICO. México:** Mpio. Tejupilco, Volcán de Tejupilco, (18°53'N, 100°10'W), 1510 m, 3 Nov 1932, *Hinton 2484* (K); Mpio. Ixtapan del Oro, 3 km W of Ixtapan del Oro on road to Zitácuaro, 19°15'N, 100°16'W, 3 Dec 1983, *Solheim et al. 1074* (NY). **Oaxaca:** 38 km S de San José Pacífico, Pochutla, Candelaria Loxicha, 15°59'N, 96°31'W, 1780 m, 27 Oct 1987, *Debouck & Muruaga 2346* (CHAPA, COL, M, MICH, SI, US).

Habitat.—This species is reportedly scarce and found mostly in barrancos and sides of canyons, but growing locally abundant in partial shade in pine-oak forests with long (3–5 m) stems climbing over bushes. It grows in loose, yellow organic soils derived from metamorphic schists. On the basis of the few collections reported by Delgado (2000), it may be the variant of *P. marechalii* on the northern slope of Sierra Madre del Sur.

Comments.—The specimen of *Debouck & Muruaga 2346* at US was identified by Dr Delgado as *P. xolocotzii*, and by the senior author as possibly another species (unpublished, *P. ulmifolius*), thus this collection may be of more than one species. The specimen *Hinton 2484* at K was annotated "*Phaseolus sousae* A. Delgado sp. nov." The species resembles others of this group by having densely pubescent foliage and a shrubby growth habit; however, the leaves are unusual in distinctly resembling the leaf of the american elm tree with nearly parallel lateral veins instead of the more typical ovate acuminate leaves with the variously curved and branched veination of most legumes. The examination of cpDNA restriction site variation (Delgado et al. 1993) shows *P. xolocotzii* forming a group with *P. polystachyus* and *P. sonorensis*, itself linked to *P. lunatus*.

D.II.9.—Phaseolus sonorensis Standl., Publ. Field Mus. Nat. Hist., Bot. Ser. 22:31. 1940. (**Fig. 46**). Type: MÉXICO. Sonora: on moist canyon slope, Sierra Chirivo, Tepopa, Río Mayo, 7 Mar 1935, *Gentry 1396* (holotype: F; isotypes: BM, DES, GH, K, MEXU n.v., MICH, MO, UC, US).

Aerial part a long, woody, climbing, indeterminate vine. ***Root*** possibly a perennial (but otherwise unknown). ***Stem*** to 2.5 mm thick, terete, somewhat striate, all parts of plant covered by dense white uncinate and strigose pubescence; internodes to 15 cm long. ***Stipules*** inconspicuous, lanceolate-oblong, 3–4 mm long, densely covered by pilose hairs. ***Leaves*** 7.7–13.8 cm long; petioles shorter than the leaflets, 4–4.5 cm long; petiolules 7–8 mm long; pulvini 2.5 mm long, very densely covered with white pilose and uncinate hairs; stipels minute, rounded ovate, 1.25 mm long, 1–1.25 mm wide, obscurely 1- to 2-nerved, acute, ciliate; terminal leaflets ovate-acuminate to very broad rhombic-ovate, 2.5–7–8 cm long, 1.5–3.5 cm wide at 1/3 from the base, the base rounded, at times hastate, heavily veined, acute to acuminate, minutely apiculate, densely covered by white pilose and minute uncinate hairs, slightly light green or white variegated on midvein adaxially, paler abaxial surface; lateral leaflets somewhat inequilateral and broader. ***Inflorescence*** a raceme to 35 cm long, mostly shorter, sometimes with a few small leaves in lower portions, the peduncle 15–21 cm long, the rachis 3–5–(15) cm long, 12–30 flowered, nodulose, densely covered by long uncinate hairs; primary bract, ovate in the lower part of the stems, 2–3 mm long, 3- to 5-nerved, those in the upper parts of the plant nearly linear, densely pubescent; pedicel slender, about 5 mm long in bud to 9–10 mm long at flowering, very densely covered by long white uncinate and hispid hairs; pedicellar bract linear 1.5–2 mm long 1-nerved covered with white hispid hairs. ***Bracteoles*** oblong-lanceolate to linear, 1–2 mm long, 0.7 mm wide, hyaline, 1-nerved, densely pubescent, early caducous. ***Flower*** purple, large; calyx broadly campanulate, somewhat oblique, 3–4.25–5 mm long, the upper two teeth united into a short, rounded, emarginate lobe, the lower three, rounded-triangular, acute, about 1.75 mm long, all sparsely covered with pilose hairs and ciliate on margins; standard purple, the claw 1.5 mm long, rounded and emarginate, reflexed at 3.5 mm from base and 11 mm more to apex, 16 mm wide, recurved; wings large and broad, spreading, the claw 5.5 mm long, the spur 1.75 mm diam., the blade rounded, 15 mm long. 11 mm wide; keel large, the claws 5 mm long, 5 mm more to bend and 10 mm more to base of the terminal 2 full coils of 5–6 mm in diam.; vexillary stamen geniculate, the claw 1.5 mm long, 6 mm more to filament, the knob raised about 0.75 mm, 0.5 mm wide, slightly bent; stamen tube narrow, the ridges raised about 0.,5 mm at 1.5 mm from base; basal collar 1 mm long; ovary slightly curved, 8 mm long, 1.75 mm wide, glabrous, heavy sutures, 6 ovules; style 5 mm long to base of the terminal 2 coils, the

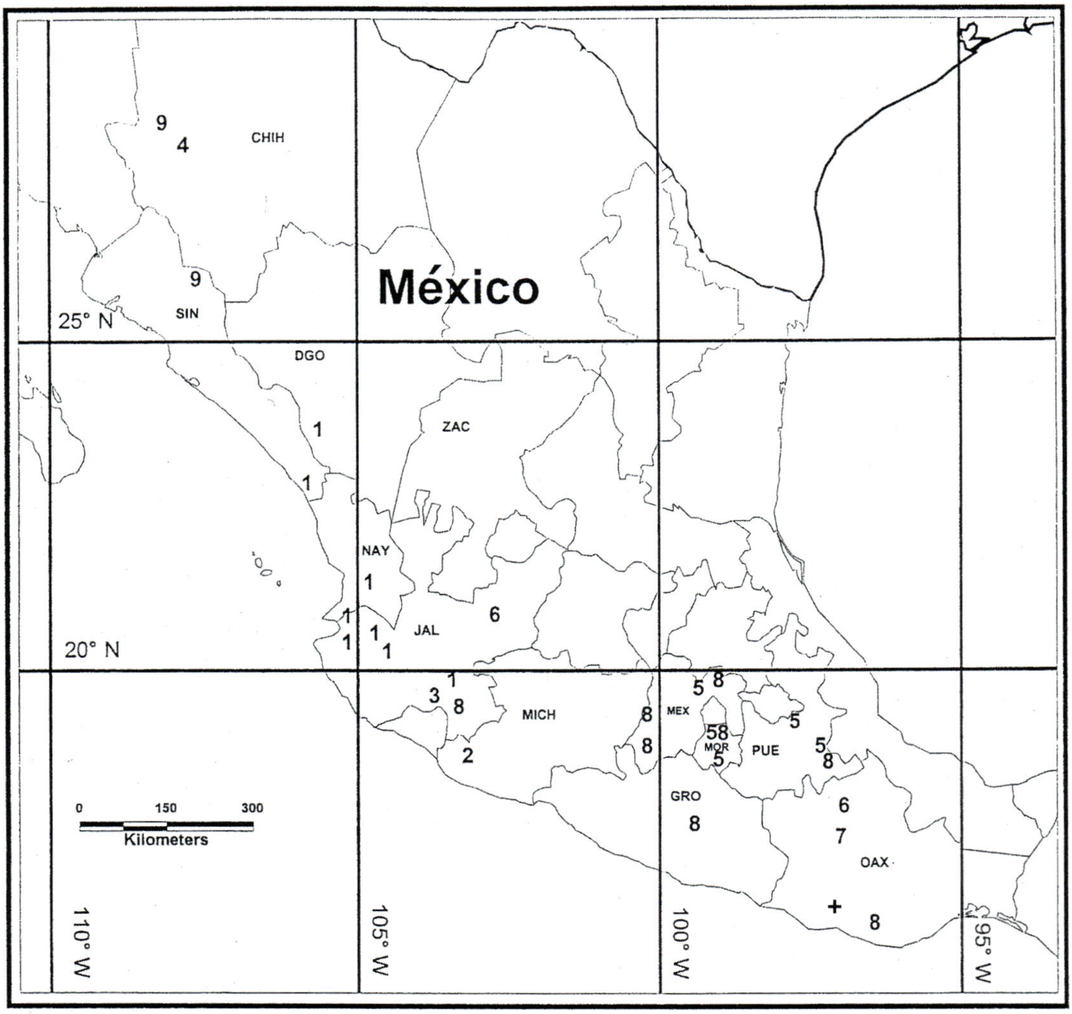

Fig. 46. Distribution of species of section D. *Paniculati*, subsection II. *Lignosi*, as follows: 1 = *P. jaliscanus*; 2 = *P. scrobiculatifolius*; 3 = *P. nodosus*; 4 = *P. albinervus*; 5 = *P. marechalii*; 6 = *P. rotundatus*; 7 = *P. acinaciformis*; 8 = *P. xolocotzii*; 9 = *P. sonorensis*; + = *P. juquilensis*.

1st coil 4.75 mm in diam., and the 2nd coil 5.75 mm in diam.; stigma terminal, capitate oblique introrse, 0.75 mm long, 0.3 mm wide. ***Pod*** slightly curved to nearly straight, 6–7 cm long, 6–7 mm wide, constricted over seeds, heavy sutures, glabrous and almost shiny; beak stout, recurved, 3–5 mm long. ***Seed*** none on specimens examined; given by Delgado as: oblong, 5.7–6.1 mm long, 4.5–5.5 mm wide; hilum, ovate 1–1.2 mm long; lens prominent, divided; mottled black or light brown and black punctate with brown halo around the hilum. ***Seedling*** unknown.

Specimens examined: **MÉXICO. Chihuahua:** Arroyo Hondo, Sierra Charuco, Sierra Madre Occidental, (28°10W, 108°20'N) 1524 m, 16–30 Apr 1948, *Gentry 7989* (CAS-DS, MEXU, MICH, UC, US). **Sinaloa:** Sierra Surotato, Los Ornos, along the road to Surotato 53 mi E of Mocorito, 1910 m, 27 Mar 1967, *Breedlove 15591* (MEXU); Mpio. Sinaloa y Vela, side road above La Joya in the upper reaches of the Cañon de Tarahumares, Sierra Surotato, (25°50'N, 107°40'W)?, 1829 m, 9 Mar 1971, *Breedlove 19269* (MICH); Puerto a Tamiapa, 1500 m, 6–8 Mar 1940, *Gentry 5827* (DES, GH, MEXU, MICH, MO, NA); above la Jolla, Sierra Surotato, 1676 m, 17 Mar 1945, *Gentry 7290* (DES, F, GH, MICH, UC, US). **Sonora:** Curohuí, Río Mayo, 4 Apr 1938, *Gentry 3636* (DES, F, GH, K, MO, NA, UC).

Habitat.—This species was found as an abundant, large, common vine sprawling and climbing over shrubs and herbs and on shrubbery around milpas. It grows on steep igneous or brushy clay or loose black loam, and in rocky canyon slopes and along rocky streams in Upper Sonoran pine-oak forest with *Alnus, Ilex, Molinadendron, Pinus, Quercus, Rapanea* and palms. It has been reported in the "Río Mayo plants" (Gentry 1942).

Common names.—Guirote, wirote (as indicated on the type), uirote (this name by Martínez 1979, for southern Sonora).

Comments.—In the type publication the common name was given by Standley as "Huirote" and collection date as March 8, however the collection label on the type herbarium sheet at US gives the collection date as March 7, the vernacular name as "Guirote," and the location as Tepopa, Río Mayo. This species is distinctive in having a large woody and densely tomentose vine, and large showy, purple, reddish purple, lavender, or pink flowers blooming in March and April, with a keel with very large (5–6 mm diam.) coils, much larger than any other species. Consequently, it does not fit very well in this section, but perhaps better here than anywhere else. On the basis of cpDNA analysis, it has been confirmed as a true *Phaseolus* sensu stricto species (Delgado Salinas et al. 1993), and close to the group of *P. ritensis* and *P. maculatus* (Delgado Salinas et al. 1999). Maréchal et al. (1978b) considered it with several affinities with *P. ritensis*. Nabhan (1990) considered it as a synonym of *P. salicifolius*.

D.II.10.—Phaseolus juquilensis Delgado, Syst. Bot. 25:421–427. 2000. (**Fig. 46**). TYPE: MÉXICO. OAXACA: Dist. Juquila, a 25 km al S de Juchatengo, 1,850 m, 20 Oct 1976, *Sousa et al. 6353* (HOLOTYPE: MEXU n.v.; ISOTYPE: MO n.v.).

Comments.—Unfortunately in our revision of herbaria we did not come across any collections of this species and are thus dependent on Dr. Delgado for the validity of this species. This species might be valid, and is apparently closely related to *P. marechalii* and *P. xolocotzii* and other species of the *P. polystachyus* group (Delgado et al. 1999).

Section E.—Bracteati Freytag, sect. nov. TYPE SPECIES: *Phaseolus macrolepis* Piper, Contr. U.S. Natl. Herb. 22:698. 1926.

Herba mediocris vel grandis volubilis scandens, radix carnosa cylindracea longa et perennis, bractea grandissima et foliacea, legumina linearia glabra differt.

Plant a medium to large climbing vine; root fleshy, long cylindrical and perennial, or fibrous and annual; primary bracts very large and foliaceous; pods linear and glabrous.

Comments.—This section is composed of only two species, both of which are distinctive in having very large, nearly foliaceous bracts on the pedoncule of inflorescence. ITS DNA sequencing data (Gaitán et al. 2000) indicate that these two species may indeed form a natural entity together with Section L. *Brevilegumeni*, while the junior author (Debouck 2000a) has considered them as individual taxa.

KEY TO SPECIES

1. First bract on inflorescence large, longer than broad (15 mm long 3 8–9 mm wide), leaves small to medium; scarce, in dense pine forests of Guatemala; 1500–3000 m ____ E.1. **P. macrolepis**
1. First bract on inflorescence medium, broader than long (8–9 mm long 3 9–11 mm wide), leaves large; rare, in moist dense forests of Costa Rica; 1800–1900 m ____ E.2. **P. talamancensis**

E.1.—Phaseolus macrolepis Piper, Contr. U.S. Natl. Herb. 22:698. 1926. (**Figs. 47, 48**). TYPE: GUATEMALA. SACATEPÉQUEZ: on a ridge above Calderas, Volcán de Fuego, (14°30'N, 90°45'W), 2490 m, 20 Oct 1873, *Salvin s.n.* (HOLOTYPE: K; photographs GH(2), US).

Aerial shoot a perennial, slender climbing, indeterminate vine, 2–5 m long. ***Root*** long cylindrical, fleshy, about 2 cm thick, tapering slowly to 1 m long, the crown near soil surface to 10 cm below, with 1–3 smaller branches. ***Stems*** terete, near base not dying back yearly, cortex rough from many papillae or tuberosities (warts) of 1–2 mm diam. on both stem and root crown; internode length at middle of vine 10–12 cm long, the younger stems heavily covered by appressed, yellowish, stiff, reflexed-strigose hairs, the largest hairs to 1 mm long, especially at nodes, becoming sparse on old stems. ***Stipules*** foliaceous, basifix, 4 mm long, 1.5–2 mm wide, ovate to oblong, 7-nerved, somewhat extended, moderately covered by hirsute to strigose hairs on adaxial surface. ***Leaves*** 17 cm long; petiole 4.2–7.2 cm long, petiolule 1.3–2.6 cm long, nearly glabrous; basal pulvinus 2–2.5 mm long, lateral and terminal pulvini 3 mm long, heavily pubescent; stipels 5-nerved, lanceolate, 2–2.5 mm long, heavily pubescent; terminal leaflet ovate acuminate, 5.2–7.5 cm long, 2.5–4 cm wide at 1/4 from base, nearly glabrous

Fig. 47. Illustrations of *Phaseolus macrolepis* Piper.–**b.** Mature root.–**c.** Portion of lower stem with mature leaf and inflorescence in fruit, and, separately mid-portion of stem with leaf and inflorescence, and vine tip.–**f.** Mature pod. All drawings from field collection of *Freytag et al. 78-Guat-60* from near Santa Lucía Utatlán, Guatemala.

on abaxial surface, appressed-strigose hairs on veins and leaf edges, scattered strigose hairs on adaxial surfaces with more on veins and leaf margins, some pubescence with scaly and glandular bases; lateral leaflet ovate, inequilateral, 4.2–6.5 cm long, 2.3–3.8 cm wide at base, acuminate, lightly nerved, medium thickness, pubescence similar to terminal leaflet. ***Inflorescence*** a raceme to 17–18 cm long; peduncle 11–12 cm long; rachis 5.3 cm long, moderately pubescent below, somewhat less above, with mostly simple hairs but also hooked hairs 0.1–0.2 mm long; primary bract very large and foliaceous, 10–30 mm long, 8–10 mm wide, broadly ovate, strongly 11-nerved at base to 19–21 nerves at upper mid-portion, adaxial surface moderately covered by strigose hairs, abaxial surface sparsely covered by strigose hairs, purplish at tip; the secondary bracts sessile, ovate to lanceolate, 1.5–2.5 mm long, sparsely pilose abaxially, nearly glabrous adaxially; pedicel 7–9 mm long at anthesis and 8–11 mm long at young pod stage, moderately covered by pubescence near base of flower. ***Bracteoles*** sessile, narrowly ovate, 1–1.5 mm long, 0.5 mm wide, not nerved, moderately pubescent, early caducous at anthesis or before. ***Flower*** purple to lavender; calyx 5 mm long, the 2 upper lobes 0.6–1 mm long, 3 mm broad, very obtusely dentate, the 2 lower lateral teeth 0.8–1.25 mm long, 2–2.25 mm wide, the central 1.5 mm long, 1.75–2.25 mm wide, acute, moderately covered by strigose and pilose hairs adaxially dorsal surface, hairs to 0.75 mm long; standard slightly purple abaxially and greenish adaxially, nearly round, 10–11 mm long, hooded forward over the keel, depressed on top and slightly notched, short campanulate base, 2 glands 1.25 and 2.25 mm long at sides, glabrous; wings purple to lavender with dark veins, nearly equal, the blade nearly round, 9 mm long, 8.5 mm wide, rolled somewhat lengthwise, slightly spreading, glabrous, the basal claw 5–6 mm long, 0.75 mm wide, the lateral spur 4–5 mm long, 2–2.5 mm wide, moderately adhering to keel, glabrous; keel of 1 1/2 coils, tip greenish, the basal claws 4 mm long, the lateral knobs equal, raised 0.5 mm. ***Pod*** straight to somewhat falcate, rounded on lower suture, restricted at upper stem end, 5 cm long, 12–13 mm wide, flattened; valves reticulate by raised veins, fibrous, rugose; beak straight, weak, 2–3 mm long. ***Seed*** rounded to oblong, 6.1–6.5 mm long, 5.2–5.8 mm wide, 2.5 mm thick, black speckled and streaked on light brown, no ring around hilum; hilum oblong, 1.1–1.3 mm long, 0.7–0.8 mm wide, no placental tissue over hilum scar; lens prominent and raised. ***Seedling*** from hypogeal germination, epicotyle terete 40–52 mm long green-reddish densely covered by uncinate hairs; eophylls opposite deltoid lengthily acuminate 35 mm long 19 mm wide, base slightly cordate auricles poorly developed, stipules entire to bifid 3–3.5 mm long; petiole 12 mm long no stipels; first true leaf trifoliolate.

Specimens examined: **GUATEMALA. Chimaltenango:** 6 km NW de Tecpan, en camino a Aserradero Santa Helena, 14°47'N, 91°02'W, 2620 m, 11 Dec 1987, *Debouck et al. 2447* (BR, COL, MICH, UC, US, USCG). **Jalapa:** 6 mi S of Miramundo, between Miramundo and summit of Montaña Miramundo, between Jalapa and Mataquescuintla, (14°30'N, 90°10'W), 2000–2500 m, 5 Dec 1939, *Steyermark 32707* (F). **Quetzaltenango:** Volcán Santa María, between Santa María de Jesús, Los Mojadas, and summit of volcano, (14°40'N, 91°35'W)?, 1500–3000 m, 12 Jan 1940, *Steyermark 34037* (F). **Sacatepéquez:** Volcán de Agua, 3000 m, 13 Nov 1967, *Molina 21034* (EAP, F, US). **Sololá:** 6 km E of Nahuala, 14°47'N, 91°19'W, 2440 m, 10 Dec 1985, *Debouck et al. 1634* (BR, K, SI, US, USCG, WIS); Km 153, on secondary road to Sta. Lucía Utatlán and Sta. Clara La Laguna, at highest point on road, 14°45'N, 91°17'W, 2330 m, 7 Oct 1978, *Freytag et al. 78-Guat-60* (ARIZ, BR, EAP, F, GH, K, MEXU, MICH, MO, UC, US(2: Sheet 1 = plant, fl, pod; Sheet 2 = root only).

Habitat.—This species is found growing on brush and low trees in old-growth pine or pine-oak forests of very large trees with Asteraceae, Compositae, Gramineae, Labiateae, Lamiaceae, Melastomataceae, and in deep humus soil from volcanic ash and rock.

Diseases and insects.—Thrips, *Gargaphia.*

Common names.—Atzij and frijolillo.

Comments.—*Phaseolus macrolepis* seems to be endemic to the volcanic mountainous ranges of central and western Guatemala (Debouck 1991; Delgado Salinas 1985; Standley & Steyermark 1946), and to date has not been reported from Chiapas (Breedlove 1986). This species is distributed rather irregularly, and is found only in old pine-oak forests which are rather rare in present-day Guatemala, but forms pretty good populations where left undisturbed. Searches for it at the type locality have been unsuccessful so far, due to the extending of pastures and vegetable growing higher up (personal observations by the junior author in 1985, 1987, 1995). Piper (1926) reported it as "perhaps nearest related to *P. vulgaris* L.," a conclusion with which we cannot agree. DNA sequencing (Gaitán et al. 2000) shows it related to *P. tuerckheimii*, *P. talamancensis* and *P. oligospermus*, far away from the Phaseoli.

E.2.—Phaseolus talamancensis Debouck & Torres, Novon 11:280–286. 2001. (**Fig. 48**). TYPE: COSTA RICA. SAN JOSÉ: San Isidro El General, orillas del Río Blanco, al pie de la Fila Ojo de Agua, 5 km N de Herradura, 9°31'N, 83°37'W, 1890 m, 13 Jan 1987, *Debouck et al. 2130* (HOLOTYPE: CR; ISOTYPES: BR, COL, K, MICH, MO, US).

Aerial shoot a small, climbing, indeterminate vine, to 1.5–2 m long. ***Root*** fibrous; adventitious roots often formed at lower nodes of stems. ***Stems*** slender, terete, striate, glabrous to puberulent; internodes 3.3–10.8 cm long, often dark reddish, covered by yellowish uncinate hairs 0.5–0.8 mm long, on all vegetative parts. ***Stipules*** very large, deltoid to rounded ovate, 5–7 mm long, 2–5 mm wide, 6–12 nerved, foliaceous, glabrous. ***Leaves*** 6.8–25 cm long, dark green; petiole 1.8–11.5 cm long, terete, striate, glabrous to puberulent; petiolule 0.4–2.3 cm long, terete, striate, glabrous to puberulent; pulvini, the lower large, 7 mm long, sparsely covered with strigose hairs, the upper 4–5 mm long, densely covered with strigose hairs on adaxial surface; stipels, the lower lanceolate, 4 mm long, 1–2 mm wide, 2–3-nerved, glabrous, the upper broad ovate, 2 mm long, 1.5 mm wide, 1-nerved, glabrous; terminal leaflet broadly ovate, 3.5–10 cm long, 2–7.5 cm wide, deltoid at base, acuminate, dark green adaxially, paler abaxially, puberulent, veins prominent abaxially purplish; lateral leaflets similar but inequilateral. ***Inflorescence*** long raceme, extending much above foliage; peduncle 5–25 cm long, rachis 18–20 cm long, 4–8–16 nodes of 2 flowers each, covered by minute uncinate hairs, the first node subtended by a large foliaceous, broadly ovate bract, more or less persistent, 8–9 mm long, 9–11 mm wide, glabrous, multi-nerved, the upper bracts early deciduous, broadly rounded, purplish, cupped and enclosing each pair of flowers, first flowers fertile at base of inflorescence, later flowers not producing pods; pedicel 5–12 mm long at anthesis to 15 mm long in young pod; pedicellar bracts triangular 1 mm long 0-nerved caducous. ***Bracteoles*** ovate to lanceolate, acuminate, scale-like, 0.75 mm long, 0.5 mm wide, acute, hyaline, ciliate, early caducous. ***Flower*** violet quickly fading; calyx broadly campanulate, tube 3 mm long, scabrous and slightly farinaceous, purple, the upper 2 lobes united into one scarcely elongate, rounded, emarginate, the lower 3 subequal, broadly rounded, 1–2 mm long, 1.5–2 mm wide, acute, glabrous, ciliate; standard violet, sharply reflexed at 5 mm from base and 10–11 mm more to emarginate apex, broadly rounded, 13–15 mm wide, the auricles prominent, 5 mm long, 2 mm wide; wings violet, long and broadly spatulate, 20 mm long, 10 mm wide, cupped; keel 8 mm to bend and 5 mm more to base of terminal 2 3/4 coils of 2.5–3 mm diam.; stigma lateral, introrse, linear 1 mm long; ovary straight, 5 mm long, 6–7 ovules. ***Pod*** nearly straight to slightly falcate, 3.6–5 cm long, 6–7 mm wide, the valves thin and chartaceous, sparsely pubescent, strong sutures; beak stout, recurved, 2–3 mm long. ***Seed*** suborbicular convex, 3–5 × 3–4 mm, black speckles on brown, hilum elliptic ringed by black, lens slightly raised. ***Seedling*** from hypogeal germination; epicotyl 28–32 mm long, terete, greenish to deep purple; eophylls ovate-lanceolate, prominently 3-veined beneath, acuminate, 17–30 mm long 10–23 mm wide, membranaceous, slightly reddish; stipules at eophyll node, 1–2 mm long, united, bifid.

Specimens examined: **COSTA RICA. Limón:** Cordillera de Talamanca, Atlantic slope, canyon of the Río Siní, (9°13'N, 82°59'W), 1800–1900 m, 15 Sep 1984, *Davidse et al. 29127* (CR, MO). **Puntarenas:** Cantón de Buenos Aires Ujarrás, El Carmen, Sabanas de Murur Bisuk, slope of Cerro Amú, 9°20'N, 83°17'W, 1900 m, 29 Sep 1989, *Herrera et al. 3590* (CR, MO).

Habitat.—Scarce, and found growing on shrubs and composites at elevations of 1800–1900 m, in humid montane forest with epiphytes in friable rocky soils with humus, derived from volcanic rock.

Diseases and insects.—Damage from spider mites, aphids and *Apion* was found by collectors.

Comments.—This species seems to be rare since only three collections are known so far. Perhaps it has a limited distribution in the Sierra de Talamanca of Costa Rica extending south to Chiriquí Province in Panama. The spiciform aspect of the young inflorescences and the large bracts on the inflorescence falling early in the course of floral ontogenesis, an anomaly under the present definition of the genus (Maréchal et al. 1978b), separate it from all other *Phaseolus* species. From DNA sequencing data (Gaitán et al. 2000), it may be related to *P. macrolepis*, and for this reason these are placed together in the section *Bracteati*.

Section F.—Minkelersia (Mart. & Gal.) Maréchal, Mascherpa & Stainier, Taxon 27:199. 1978a. (sensu Lackey, Iselya 2:21–64. 1983). *Minkelersia* Mart. & Gal., Bull Acad. Roy. Sci. Brux. 10:200. 1843. TYPE SPECIES: *Phaseolus galactioides* (Mart. & Gal.) Maréchal, Mascherpa & Stainier (now *P. pauciflorus* Sessé & Mociño ex G. Don).

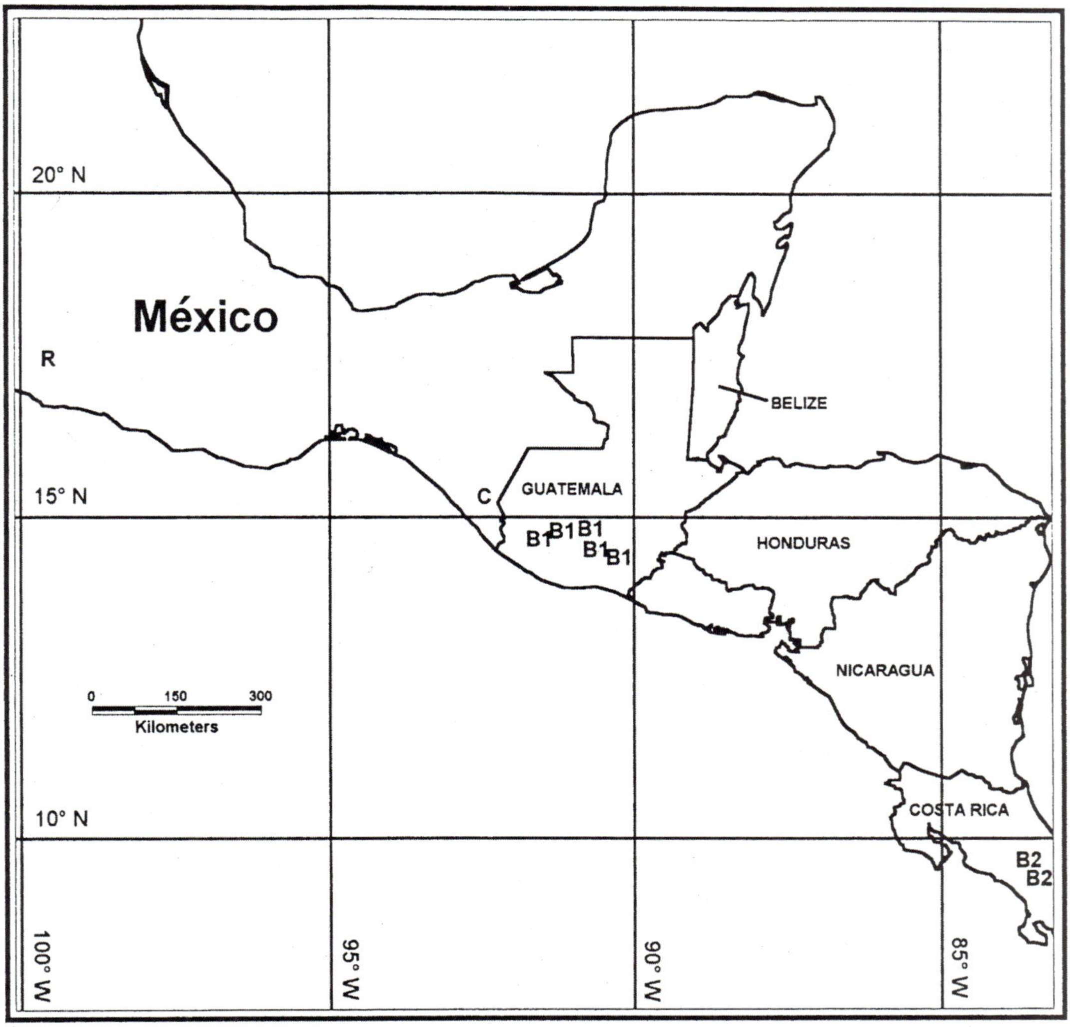

Fig. 48. **Distribution of species of Section E.** ***Bracteati*****, as follows: B.1 =** ***P. macrolepis*****; B.2 =** ***P. talamancensis*****; Section H.** ***Revoluti*****, R =** ***P. leptophyllus*****; and Section N.** ***Chiapasana*****, C =** ***P. chiapasanus*****.**

Alepidocalyx Piper, Contr. U.S. Natl. Herb. 22:672. 1926. *Phaseolus* sect. *Alepidocalyx* (Piper) Maréchal, Mascherpa & Stainier, Taxon 27:199. 1978a. Type Species: *Phaseolus parvulus* Greene.

Aerial shoot a small, mostly prostrate vine. Root perennial, small globose, fleshy. Pedicels shorter than calyx. Bracteoles minute or lacking (early caducous). Flower somewhat tubular; calyx lobes as long as or longer than tube. Pods mostly linear. Seed mostly very small, spherical, numerous, black.

Comments.—A small group of species mostly growing in grasslands at higher altitudes, frequently bearing variously lobed leaflets and elongate flowers possibly pollinated by butterflies. Maréchal et al. (1978b) reduced the genus *Alepidocalyx* of Piper to a section of *Phaseolus*. Since the species in this section are very similar to those of *Minkelersia* (i.e. sharing shorter pedicels, longer calyx lobes, and vestigial or early caducous bracteoles), Lackey (1983) placed the genus *Alepidocalyx* within the section *Minkelersia* of *Phaseolus*, a treatment followed by Delgado (1985). Thus all are placed here under a single section showing a clear common group of similar characters, from *P. anisophyllus* closely resembling the typical *Phaseolus* species, to the most unique species, *P. pauciflorus*.

KEY TO SPECIES

1. Flower bud distinctly pointed at upper portion of apex; inflorescence with many nodes; pod falcate and reflexed at maturity.
    2. Calyx lobes considerably longer than calyx and subequal, acute at apex; leaflets ovate-lanceolate, uniformly tapered to apex, mostly not lobed at base but may sometimes be deeply lobed; widely distributed in NW and central México; 1500–2900 m ________ F.1. **P. pluriflorus**
    2. Upper two calyx lobes broader and larger than others, slightly obtuse at apex; leaflets ovate to broadly lanceolate, usually constricted at mid-length and obscurely lobed at base; scarce, in S México: Oaxaca and Chiapas; 1440–3000 m ________ F.2. **P. nelsonii**
1. Flower bud rounded at apex; inflorescence usually with few (1–4) nodes; pod linear, sometimes slightly falcate, and extended at maturity.
    3. Inflorescence of several nodes; calyx lobes mostly obtuse; standard variously shaped, not tubular in bud; pod linear or slightly falcate, few to 10 seed.
        4. Upper two lobes of calyx united, emarginate.
            5. Calyx lobes obtuse or rounded, upper two short, 0.5 mm long; leaflets small, variously ovate to diamond shaped, sometimes slightly lobed at base; scarce, in central México; 2100–2200 m ________ F.6. **P. tenellus**
            5. Calyx lobes at least some acute, upper two at least 2–3 mm long.
                6. Calyx lobes subequal, to 3–4 mm long.
                    7. Plants small to medium to 1 m long, erect or prostrate but only slightly vining, the leaflets small to medium, lanceolate, sometimes lobed at base.
                        8. Stipules small, to 5 mm long, plants small of less than 10 nodes, leaflets sometimes slightly lobed at base; common, in SW US and NW México; 830–2800 m) ________ F.7. **P. parvulus**
                        8. Stipules large, to 10 mm long, plants medium of many nodes, lateral leaflets usually distinctly lobed at base; rare, in SW corner of Chihuahua and neighboring Durango; 2600–2800 m ________ F.8. **P. anisophyllus**
                    7. Plants fairly large and climbing, to 2–3 m long, the leaflets medium sized, upper leaflets distinctly lobed at base; flowers large, about 24 mm long; rare, from W Chihuahua; 1660 m ________ F.9. **P. amabilis**
                6. Lower central calyx lobe much longer than others, 4–5 mm long; leaflets variously ovate, deeply 3-lobed; rare, in mountains W of Monterrey, Nuevo León; 900 m ________ F.4. **P. plagiocylix**
        4. Upper two lobes of calyx separate or distinctly bifid.
            9. Calyx lobes short, about 2 mm long; leaflets broadly ovate or rounded, usually less than 2 cm long; rare, in Durango; 2300–2800 m ________ F.5. **P. amblyosepalus**
            9. Calyx lobes long, 3.5–6 mm long; leaflets variously lanceolate, usually about 5 cm long; scarce, in central México; 1450–2200 m ________ F.3. **P. perplexus**
    3. Inflorescence with a single node; calyx lobes long, subequal, acute; standard long and nearly straight tubular in bud; pod long linear of 10–20 seeds; leaflets variously ovate, usually 2–2.5 cm long; common, in western mountains of N and central México; 600–3040 m ________ F.10. **P. pauciflorus**

F.1.—Phaseolus pluriflorus Maréchal, Mascherpa & Stainier, Taxon 27:199. 1978a. (**Figs. 49, 50**).

Minkelersia multiflora Rose, Contr. U.S. Natl. Herb. 5:142. 1897. TYPE: MÉXICO. DISTRITO FEDERAL: Pedregal (lava beds), Valley of México, (19°20'N, 99°10'W), 2470 m, 1 Sep 1896, *Pringle 6471* (LECTOTYPE: US 304095, designated by Maréchal et al. (1978b); ISOTYPES: BM, BR n.v., G, GH, K, MEXU n.v., MO, P n.v., UC).

Phaseolus vulcanicus (Piper) Maréchal, Mascherpa & Stainier, Taxon 27:199. 1978a. *Minkelersia vulcanica* Piper, Contr. U.S. Natl. Herb. 22:671–672. 1926. TYPE: MÉXICO. GUANAJUATO: Volcán Batea, (21°N, 101°W)?, 2000–2200 m, 1872, *Guillemin-Tarayre s.n.* (HOLOTYPE: P n.v.; ISOTYPE: G n.v.).

Phaseolus anisotrichos subsp. *incisus* Piper, Contr. U.S. Natl. Herb. 22:700. 1926. TYPE: MÉXICO: ZACATECAS: near Plateado, 4 Sep 1897, *Rose 2801* (HOLOTYPE: US 301734).

Aerial shoot an annual, prostrate and slightly climbing, indeterminate vine, to 1–2 m long. ***Root*** a perennial, globose, thick, fleshy, somewhat elongate, about 2 cm long and 1.75 cm wide at 1 year of age. ***Stems*** terete, somewhat striate, sparsely covered with pubescent hairs and slightly scabrous; internodes 9–13 cm long. ***Stipules*** broadly ovate, 10 mm long, 4.5 mm wide, multi-nerved, acute, membranous, pubescent abaxially, glabrous adaxially. ***Leaves*** 8.6–19 cm long; petioles 3.7–5–8 cm long; petiolule 0.5–1.5 cm long, hispid adaxially and nearly glabrous abaxially; pulvini 2 mm long, densely covered by hispid and hooked pubescence; stipels, the lower lanceolate, 3 mm long, 0.75 mm wide, 1–2 nerved, ciliate, the upper stipels smaller 1 mm long aciculate or lacking; terminal leaflet ovate-acuminate to lanceolate, 4–6–9 cm long, 1.2–3 cm wide, the base broad, usually slightly lobed, cuneate, rather strongly veined, acute, apiculate, slightly pubescent; lateral leaflet similar but strongly inequilateral with a broad truncate base. ***Inflorescence*** a many-flowered (15 or more) raceme, 20–38

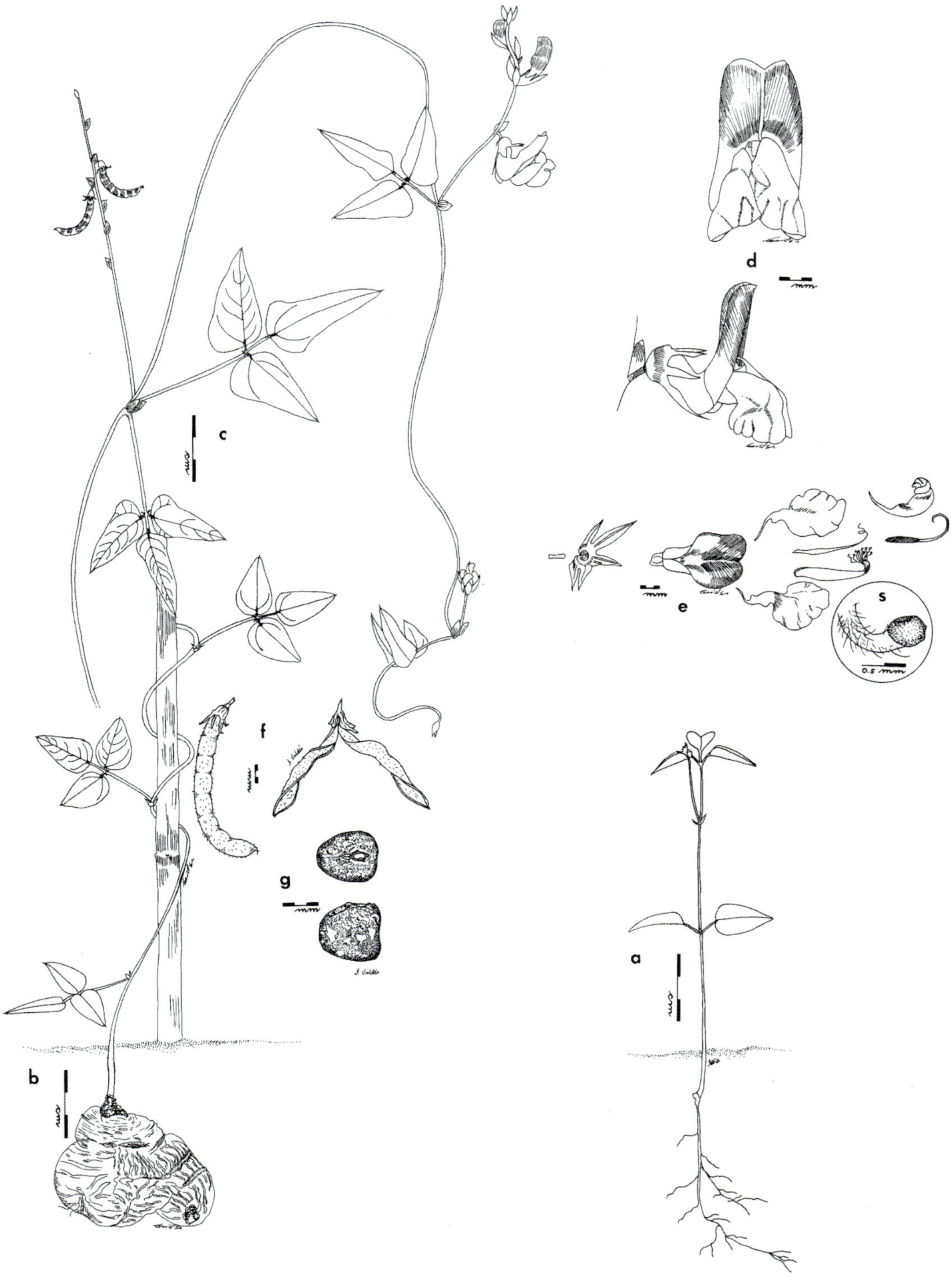

Fig. 49. Illustrations of *Phaseolus pluriflorus* Maréchal, Mascherpa & Stainier.—**a.** Seedling plant a few weeks after germination.—**b.** Root after 1 year.—**c.** Lower portion of stem with mature leaves, and, separately, vine tip with mature leaves, inflorescence with pods, and inflorescences with flowers.—**d.** Flowers, side view and front view.—**e.** Exploded view of flower showing all parts, including—**s.** Style tip showing stigma as seen under the microscope; note long sepals, narrow, erect nature of standard and large, capitate stigma.—**f.** Mature pods showing side view and dehisced carpels.—**g.** Seeds, view from hilum and side view; note somewhat angular, nearly round seed. All drawings of living material grown in greenhouse at Mayagüez of seed of *Freytag & Sullivan 81-21* (TARS #131) collected at La Calaverna, Jalisco, México, except stigma from plants of seed of *Debouck et al. 320* (TARS #91) collected near Mariano Matamoros, Durango, México.

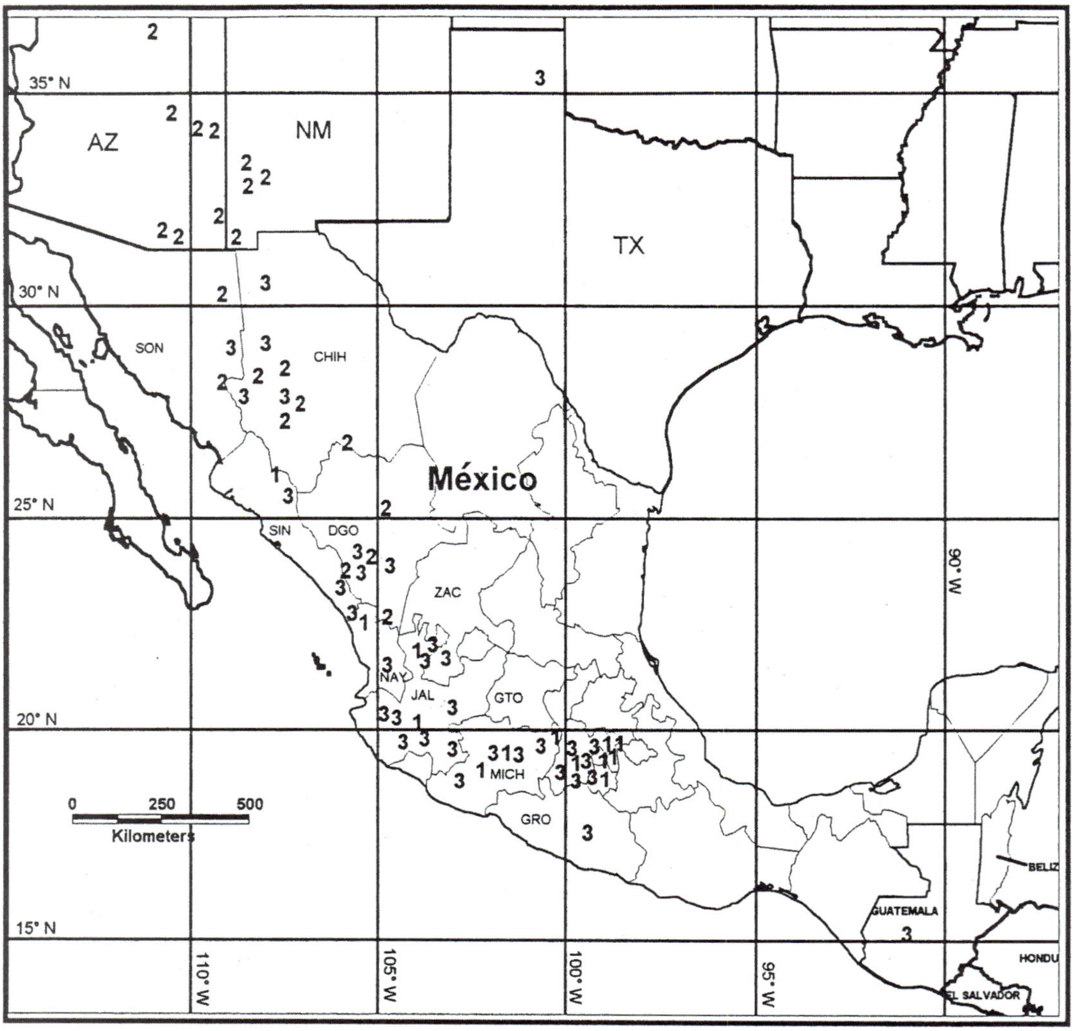

Fig. 50. Distribution of major species of Section F. *Minkelersia*, as follows: 1 = *P. pluriflorus*; 2 = *P. parvulus*; 3 = *P. pauciflorus*.

cm long, the peduncle 7.5–12 cm long, striate, puberulent of minute hooked hairs, the rachis 20 cm long of 10 or more evenly spaced flowering nodes, about 1 cm apart, densely covered with minute uncinate hairs; primary bract broadly ovate, 10 mm long, 7 mm wide, many-nerved (10–15 and over), acute, nearly glabrous, purplish, persistent; pedicel 2–2.5 mm long, stout, covered with minute hooked pubescence adaxially; pedicellar bracts minute less than 1 mm long linear scale-like hyaline. ***Bracteoles*** absent. ***Flower*** violet, the flower buds erect, somewhat elongate and acute, flowers spreading, reflexed sharply downward after anthesis, 12–15 mm long; calyx broadly campanulate, the tube about 3 mm long, glabrous to slightly pubescent at base of minute hooked hairs, purplish, the lower lateral lobes broadly lanceolate, 6 mm long, acute, the center lobe broader and 8 mm long, all lobes fimbriate pubescent; standard dark violet, 20 mm long, 12–15 mm wide, erect, the lateral edges recurved towards base, the claw somewhat tubular, 6 mm long; wings dark violet, somewhat clasping, 18–20 mm long, very broad to 15 mm wide; keel straight, 7–8 mm long to the base of the terminal 1 1/2 coils of 3 mm in diam.; stigma terminal, transverse, elongate, 0.75 mm long. ***Pod*** immature pendant, densely covered by hirsute hairs becoming yellowish hirsute at maturity, mature pod slightly curved, inflated, 25 mm long, 4 mm wide, reflexed and pendant, the beak sharply recurved, weak, 1 mm long; 8 ovules. ***Seed*** oblong to reniform, 2–6 mm long, 2–6 mm wide, hilum oblong, 0.5 mm long with no epihilum; lens reduced; light brown mottled black. ***Seedling*** from hypogeal germination; hypocotyl not elongated; epicotyl 3.2–4.2 cm long, the next internode shorter; eophylls opposite, simple, the petiole with basal and apical pulvini, stipels absent, the blade ovate to widely ovate, base truncate to subcordate, acute.

Specimens examined: **MÉXICO. Distrito Federal:** Pedregal, San Geronimo, Vallée de México, 2 Aug 1866, *Bourgeau 576* (G, GH, K, US); al O de la Ciudad Universitaria, Pedregal de San Angel, 1 Sep 1954, *Gold 598* (US); San Andrés, Aug 1930, *Lyonnet 694* (GH, K, MO, US); near Guadalupe, Valley of México, 23 Sep 1903, *Rose 7299* (US); vertiente NW del Cerro Chiquihuite, cerca de Cuautepec, 2450 m, 10 Aug 1970, *Rzedowski 27425* (MICH); Cerro Ahumada, cerca del Rancho Nuevo, 4 km al N de Huehuetoca, 2350 m, 11 Aug 1971, *Rzedowski 28387* (MICH). **Durango:** Fco. I. Madero, camino a José Librado Rivera, cerro El Frasco, 24°30'N, 104°19'W, 2550 m, 6 Oct 1978, *Debouck et al. 313* (CHAPA, COL). **Jalisco:** Mazamitla, 4 km NE Mazamitla, 19°57'N, 103°02'W, 1950 m, 9 Nov 1978, *Debouck et al. 436* (CHAPA, COL); Km 31, 2 km E of Tapalpa, Cerro de Talcozagua on Sierra de Tapalpa, 19°57'N, 103°45'W, 2000 m, 5–7 Aug 1960, *Iltis et al. 765* (MICH, US, WIS(2)); Mpio. de San Martín de Bolaños, Las Treinta Vueltas, 10 km al NW de El Platanar, (21°45'N, 103°50'W), 2200 m, 2 Sep 1968, *Rzedowski 26220* (MICH). **México:** Mpio. de Texcoco, Cerro Tetzcutzingo, 7 km E of Texcoco, (19°35'N, 98°45'W), 2550 m, 7 Oct 1983, *Barrie 446* (TEX); W of Encinillas, 70 mi NW of México City along Hwy 57, 2500 m, 26 Aug 1977, *Croat 44173* (MO); Texcoco, San Nicolas Tlamincas, slope S cerro Baños de Netzahualcoyotl, 19°31'N, 98°49'W, 2430 m, 16 Nov 1986, *Debouck et al. 2083* (CHAPA, COL); lote experimental "La Siberia," 1.5 km al Oriente de Huexotla, (19°30'N, 98°55'W), 2240 m, 6–13 Aug 1972, *García 108* (CAS); Km 41, Puebla Hwy., (19°20'N, 98°45'W), 4 Aug 1940, *Langman 2544* (NA); hills above El Oro, 3100 m, 12 Sep 1901, *Pringle 955* (F, GH, K, MO, US); Toluca, (19°5'N, 99°45'W), 4 Sep 1903, *Rose et al. 6771* (MICH, US); Mpio. Texcoco, San Pablo Ixtayoc, 2500 m, 3 Sep 1983, *Ventura 1315* (CAS); Mpio. Texcoco, San Nicolás Tlamincas, 2400 m, 9 Sep 1983, *Ventura 1369* (ARIZ, ASU). **Michoacán:** Morelia, Punguato, 2100 m, 9 Aug 1909, *Arsène 2875* (MO, US); Morelia, Senguio, 7 km NW de Los Reyes, 19°47'N, 100°17'W, 2340 m, 17 Nov 1987, *Debouck et al. 2403* (CHAPA, MICH); Mpio. Tancitaro, road from Tancitaro to Apatzingan, (19°N, 102°15W), 1650 m, 17 Aug 1940, *Leavenworth 614* (MO); 3.2 mi S of Pátzcuaro, (19°25'N, 101°35'W), 2400 m, 16 Sep 1962, *Ugent s.n.* (WIS). **Morelos:** El Parque, (18°55'N, 99°15'W)?, 31 Aug 1910, *Orcutt 3835* (F, GH, K, TEX, US(2)). **Nayarit:** Mpio. Acaponeta, Cangrejo, 3 mi W of Mesa del Nayar, (22°30'N, 105°30'W), 31 Jul 1970, *Norris et al. 14705* (MICH); Mpio Nayar, 28 Km entronque a San Juan Peyotán sobre carr. Jesús María-Huejuquilla, 2290 m, 3 Sep 1991, *Ramírez et al. 944* (MEXU, MO). **Sinaloa:** Ocurahuí, Sierra Surotato, (26°N, 107°45'W), 1980 m, 27–30 Aug 1941, *Gentry 6172* (ARIZ).

Habitat.—This species is found growing sparsely and localized in open oak or pine-oak forests with understory of grasses (*Aristida, Bouteloua, Muhlenbergia*), shrubs (*Acacia, Crataegus, Lonicera*), and composites. Soils are red, organic, friable and derived from basalt or igneous rocks.

Comments.—Delgado (1985) considers *P. vulcanicus* as synonymous with this species, with which we would agree since the type and few specimens which are available show little difference, and cites a specimen at Geneva as well as at Paris, neither of which have we seen.

The roots of this species are the most spherical of the section and quite distinctive from the others (see Color Plate III, photo 30).

F.2.—Phaseolus nelsonii Maréchal, Mascherpa & Stainier, Taxon 27:199. 1978a. (**Fig. 53**). *Minkelersia pauciflora* Rose, Contr. U.S. Natl. Herb. 5:142. 1897. (not *P. pauciflorus* Sessé & Moc. ex G. Don). TYPE: MÉXICO. OAXACA: 18 mi SW of the City of Oaxaca, (17°N, 96°55'W), 2140–2950 m, 10–20 Sep 1894, *Nelson 1362* (HOLOTYPE: US 764858).

Aerial shoot a short, climbing, indeterminate vine, 1–2.5 m long. ***Root*** a perennial, thick, fleshy, fusiform. ***Stems*** terete, 1–2 mm thick; internodes to 15 cm long, covered with minute uncinate and sparsely covered by hirsute hairs. ***Stipules*** broadly ovate, 5–7 mm long, 4 mm wide, acute, membranous, multinerved (8–10), puberulent abaxially, glabrous adaxially, ciliate margins. ***Leaves*** 5.5–9.4 cm long; petioles 2–4 cm long, canaliculate, covered with minute uncinate hairs; petiolule 6–11 mm long canaliculate; pulvini 2 mm long, densely covered with uncinate hairs; stipels the lower narrow lanceolate, 2 mm long, faintly 2-nerved, sparsely ciliate, the upper ones minute less than 1 mm long scale-like or lacking; terminal leaflet lanceolate to oblong-ovate, 2.5–4.5 cm long, 0.7–2 cm wide, the base hastate to cuneate, indistinctly lobed, obtuse, apiculate, sparsely covered with glandular ciliate hairs adaxially, somewhat more puberulent and lighter green abaxially; lateral leaflet similar but strongly inequilateral and more or less oblique, 1.5–2.2 cm wide, truncate with 1 squarish lobe at base, dark green above, glabrate. ***Inflorescence*** a short, erect raceme; peduncle 6.5–12 cm long, somewhat striate and sparsely covered with minute uncinate hairs; rachis 2–6 cm long, of 2–5 flowering nodes, densely covered with minute uncinate hairs; primary bract orbicular, 5–6 mm long, 2 mm wide, acute, puberulent below, glabrous above; pedicellar bracts linear, 1.5–3 mm long, 0.5 mm wide, faintly 3-nerved, ciliate margins; pedicel 2–3 mm long, densely covered with uncinate hairs. ***Bracteoles*** sessile to 1 mm below calyx, minute, spatulate to linear, 0.5–2 mm long, 0.5 mm wide, hyaline, ciliate, early caducous. ***Flower*** violet; calyx open campanulate, the tube 2–3 mm long, densely covered with strigose, hirsute and uncinate hairs on both adaxial and abaxial surfaces, the lobes broad and subequal, the upper 2 lobes spatulate, 4 mm long, 2 mm wide, united 1 mm at bases, rounded at apex, puberulent, purple speckled, the lower 3 lobes dentate, acute, 5–6 mm long, puberulent to pubescent; standard violet, horizontal through anthesis, 12–15 mm long, the claw 1.5 mm long, 1 mm wide, the blade squarish, reflexed at 10 mm from base and the terminal erect portion 5–7 mm long, 8–10 mm wide, deeply emarginate, lateral edges recurved; wing violet, the blade rounded, 10 mm long, 7–8 mm wide, the narrow basal portion 3 mm long, 2 mm wide, the claw narrow, 7 mm long, the spur not developed; keel nearly straight, the divided claws 6 mm long, 8 mm more to bend and 4 mm more to base of the terminal 1 3/4 coils of 3 mm in diam.; stamen tube straight, 4 mm long from base to the minute, rounded ridges of 0.4 mm in diam. and about 7 mm more to end of the united portion; ovary straight, 13 mm long, 1.5 mm wide, densely covered with appressed, white strigose hairs, 8–10 ovules; style 6 mm long to the terminal thickened coil of 2.25 mm diam.; stigma linear, 1 mm long, lateral, introrse. ***Pod***, the immature ones strongly reflexed parallel to ground, densely tomentose of long yellowish-brown strigose hairs, at maturity slightly curved, inflated, 3 cm long, 4 mm wide, beak short, recurved, 1 mm long, covered by bristly strigose hairs. ***Seed*** nearly spherical, 1 mm in diameter, black; hilum oblong, 0.3–0.4 mm long. ***Seedling*** unknown.

Specimens examined: **MÉXICO. Chiapas:** Mpio. Motozintla de Mendoza, Cerro Mozotal below the microwave tower along the road from Huixtla-El Porvenir and Siltepec, (15°25'N, 92°10'W), 3000 m, 19 Sep 1976, *Breedlove 40272* (CAS); Mpio. Motozintla, 24 km NW of Motozintla, camino a Siltepec, 2270 m, 22 Nov 1986, *Martínez et al. 19271* (MO). **JALISCO:** Cerro Santa María, 8–10 km SW of Jiquilpán and 5 km NE of Quitupán, (19°50'N, 103°W), 2000 m, 8–9 Aug 1959, *Feddema 235* (MICH); 3 mi S of Mazamitla, Sierra del Tigre, (19°45'N, 103°15'W), 2100–2200 m, 21 Sep 1952, *McVaugh 13119* (G, K, MICH, US); Sierra Madre, W of Bolaños, (21°45'N, 103°50'W), 15–17 Sep 1897, *Rose 2967* (US). **México:** Dist. Temascaltepec, Mpio. Tejupilco, Acatitlán, (19°02'N, 100°02'W), 24 Sep 1933, *Hinton 4799* (G, GH, K, US). **Michoacán:** Dist. Coalcomán, Mpio. Coalcomán, Pto. Las Cruces, (18°45'N, 103°10'W), 1440 m, 13 Sep 1938, *Hinton et al. 12191* (F, GH, K, MICH(2), US). **Oaxaca:** cerro de microondas de San Juan Teposcolula, aproxim. 26 km SE de Tamazulapan, 17°36'N, 97°24'W, 2530 m, 24 Oct 1987, *Debouck et al. 2329* (CHAPA, MICH, MO); en El Marquez, San Andrés Ixtlahuaca, Nochixtlán, 17°13'N, 97°02'W, 2180 m, 24 Oct 1987, *Debouck et al. 2332* (CHAPA, MICH); 10 km N de Villa Ordaz hacia Cuajimoloyas, Tlacolula, 17°03'N, 96°24'W, 2270 m, 25 Oct 1987, *Debouck et al. 2337* (CHAPA, MICH); Sta. Lucía Mixtepec, 19 Km S de Miahuatlán, 16°15'N, 96°32'W, 2160 m, 27 Oct 1987, *Debouck et al. 2344* (CHAPA, US); Cumbre de Istepec, Sep 1842, *Liebmann 5317* (F, US); Dist. Ixtlán, Rancho La Teja, 5 km al E de Ixtlán, sobre el camino a Yavesía, (17°20'N, 96°25'W), 2180 m, 26 Sep 1982, *Sousa et al. 12640* (CAS, MO, WIS); Dist. Zaachilá, 2 km W de Santiago Clavellina, hacia San Antonio Huitepec, 18 Sep 1982, *Torres et al. 1317* (CAS, MO). **Zacatecas:** near Plateado, Sierra de los Morones, (21°55'N, 103°05'W), 2621 m, Sep 1897, *Rose 2722* (US).

Habitat.—This species is found climbing over shrubs on steep mountain slopes of scattered oak, pine, or pine-oak with *Asplenium*, *Begonia*, *Clethera*, Compositae, *Cornus*, *Desmodium*, *Drimys*, Labiatae, *Photinia*, *Valeriana densiflora*, and grasses. Soils are yellow or red clays and rocky with abundant organic matter.

Disease and pests.—This taxon is reported as sometimes damaged by frosts.

Comments.—This species has been difficult to classify correctly because of the paucity of specimens and the close resemblance to other species of the section. Nevertheless, we believe Delgado (1985) has correctly identified the distinguishing caracteristics of the species, the most useful of which are the deeply lobed calyx, the distinctly pointed standard (in bud), very erect keel shape, and mature terminal leaflet which is quite broad and slightly narrowed at the middle. The type of the species, *M. pauciflora* Rose, required a new epithet because of the prior use of this name by G. Don (1832) for another taxon; therefore Maréchal et al. (1978b) proposed *P. nelsonii* to replace it. Maréchal, Piper before him, and even Delgado have confused specimens of other species with this taxon, perhaps trying to simplify the distribution patterns of each, which in our estimation considerably overlap. Although *P. nelsonii* seems to be often confused with *P. perplexus*, Delgado (1985) believes it is most closely related to *P. pauciflorus* on the basis of pollen structure. We find *P. nelsonii* most closely resembles *P. pluriflorus* in plant and flower structure.

F.3.—Phaseolus perplexus Delgado, Syst. Bot. 25:418–421. 2000. (**Figs. 51, 53**). TYPE: MÉXICO. JALISCO. Mpio. Talpa de Allende, a 24 km al S de Talpa, camino a La Cuesta (Cumbre), (20°13'N, 104°45'W), 8 Sep. 1979, *J.A.S. Magallanes 1965* (HOLOTYPE: MEXU n.v.; others to be distributed).

Similar to *P. nelsonii* with the following exceptions: ***Leaves*** terminal leaflet narrowly elongate, ob-

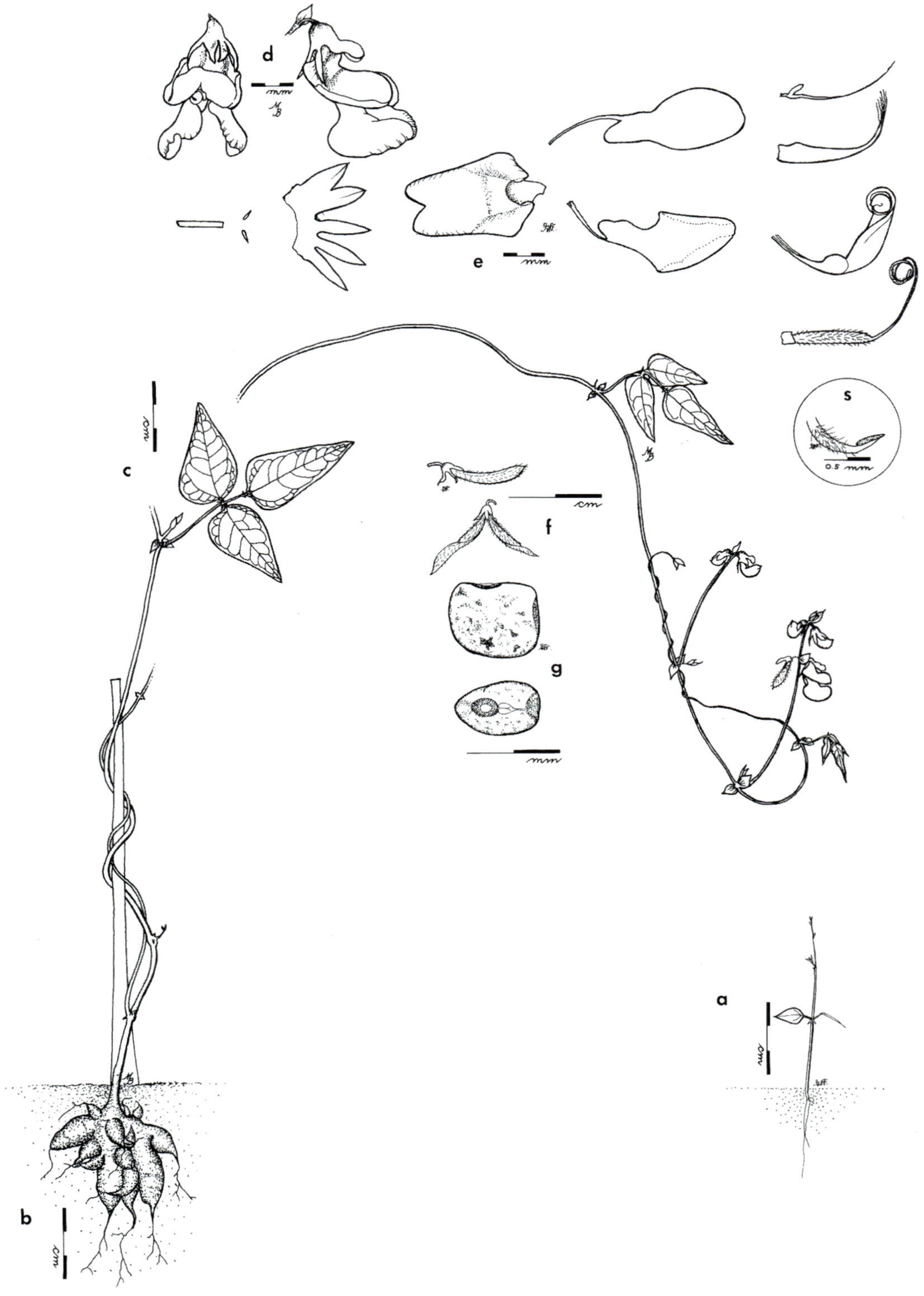

Fig. 51. Illustrations of *Phaseolus perplexus* Delgado.–**a**: Seedling plant several days after germination; note very small size.–**b.** Root after 1 year.–**c.** Lower part of stem with mature leaf, and, separately, vine tip with inflorescences.–**d.** Flowers, front view and side view.–**e.** Exploded view of flower showing all parts, including–**s.** Style tip showing stigma as seen under the microscope; mote the very long and narrow lower three sepals, the long finger-like knob at base of vexillary stamen.–**f.** Mature pods, side view and dehisced.–**g.** Seeds, lateral view and view from hilum. All drawings from living material grown in greenhouse at Mayagüez from seed of *Debouck & Muruaga 2366* (TARS #468) collected near Coatepec, México, México.

long to lanceolate; lateral leaflet ovate-oblong to oblong, often somewhat 1-lobed at base. ***Inflorescence*** 8–22 cm long, with 5–11 flowering nodes; primary bract orbicular, 5–7 mm long, 2.5–4 mm wide, acute, 5–10-nerved, glabrous; pedicellar bracts linear, 4 mm long, 1.5 mm wide, faintly 1-nerved; pedicel 2.5–4 mm long. ***Bracteoles*** linear 1–2 mm long not nerved hyaline caducous. ***Flower*** buds spreading, short and rounded, 1.5–1.7 cm long; calyx 7–8 mm long, deeply lobed more than half the length of calyx, the lobes narrow, oblong to lanceolate, the upper two 2 mm long, the lower three about 4 mm long.

Specimens examined: **MÉXICO. Jalisco:** 4 mi N-NE of Talpa de Allende, 1450–1500 m, 12–13 Oct 1960, *McVaugh 20143* (CAS, MICH, MO); 1–2 mi E of Tapalpa, (19°55'N, 103°44'W), 2100–2200 m, 1 Nov 1960, *McVaugh 20586* (CAS, MICH, MO). **México:** Mpio. Valle de Bravo, 6 km N of Temascaltepec on the road to Valle de Bravo, (19°05'N, 100°02'W), 1940 m, 12 Oct 1985, *Bartholomew et al. 2951* (CAS); Chiltepec, 15 km NW de Coatepec, 18°55'N, 99°47'W, 2120 m, 1 Nov 1987, *Debouck et al. 2366* (CHAPA, MICH, MO, US); Dist. Temascaltepec, Nanchititla, 7 Sep 1933, *Hinton 4714* (K); Temascaltepec, Acatitlán, 24 Sep 1933, *Hinton 4799* (G, K). **Michoacán:** Coalcomán, Puerto Las Cruces, 13 Sep 1938, *Hinton 12191* (K); Mpio. Coalcomán, 0.3 mi NE of Km 50 on road from Coalcomán-Tepalcatepec, 14.1 mi NE of Coalcomán, (18°55'N, 103°W), 13 Sep 1985, *Luckow et al. 2924* (MO, TEX).

Habitat.—It is reported as scarce, growing on steep slopes of pine-oak forests and climbing on well-developed understory growth of *Desmodium*, Lamiaceae and grasses. Soils are rocky, friable, organic, derived from volcanic ash.

Disease and pests.—Damage from pod weevils has been reported.

Comments.—According to Delgado (1985), this is the taxon that was included by Piper (1926) with *M. pauciflora* (now *P. nelsonii*). However, this latter species has pointed buds, while *P. perplexus* has smaller rounded buds.

F.4.—Phaseolus plagiocylix Harms, Notizbl. Bot. Gart. Berlin 7:508. 1921. (**Fig. 53**). TYPE: MÉXICO. NUEVO LEÓN: Monterrey, Cerro del Obispado, (25°45'N, 100°30'W), 12 Oct 1895, *Seler 1042* (HOLOTYPE: B (lost); neotype: GH (photo) here designated by Freytag; ISOTYPE: GH).

Aerial shoot is an annual, climbing, indeterminate vine, to 0.5–1 m long. ***Root*** unknown. ***Stems*** terete, delicate, striate, puberulent; internodes 1.5–8 cm long. ***Stipules*** ovate to spatulate, 4–5 mm long, 5-nerved, pubescent. ***Leaves*** 4.7–7.7 cm long, deeply lobed; petioles delicate, 2–2.5 cm long, striate, densely covered by minute uncinate hairs; petiolules striate, 7–12 mm long, covered by minute uncinate hairs; stipels lower lanceolate, 2.5 mm long, covered with minute hispid hairs, upper missing (or early caducous); very short pulvini; terminal leaflet 2–4 cm long, 1.5–3 cm wide, deeply 3-lobed (rarely subentire), the terminal lobe nearly spatulate, 1.7 cm long, acute and apiculate, the lateral rounded, about half as long as the central lobe, obtuse, scabrous and hispidulose adaxially, puberulent of minute uncinate hairs abaxially, ciliate margins, dark green adaxially, lighter abaxially; lateral leaflets obliquely inequilaterally tri-lobate to obliquely bi-lobate, otherwise similar. ***Inflorescence*** a long, curved pseudoraceme, extending considerably beyond foliage, 8–20 cm long; peduncle curved, 5–8 cm long, strongly striate, densely puberulent of minute uncinate hairs; rachis 2–6 cm long of 3–4 nodes loosely many-flowered; primary bract lanceolate to ovate-lanceolate, 4–4.5 mm long, 1.75 mm wide, 5-nerved, acute, pubescent; pedicel stout, 3–5 mm long, covered with uncinate hairs. ***Bracteoles*** lacking (or early caducous). ***Flower*** violet; calyx campanulate, restricted at base, distinctly bilobed, the tube 2.5 mm long, scabrous, puberulent, purple, the upper lobes rounded, distinctly separate, 1.5 mm long, 2 mm wide, the 2 lower lateral lobes rounded to broadly spatulate, 2.5 mm long, 1.5 mm wide, glabrous, densely ciliate at rounded "crotch" between upper lobes, the lower central lobe very large, nearly foliaceous, broadly ovate to obovate, 5–6 mm long, 2–3 mm wide, acute, scattered dentate and ciliate on margin, scabrous, purplish; standard violet to purple, sharply reflexed at 4 mm from base, 6 mm more to apex, erect, somewhat longer than broad, 9–10 mm wide, the claw short and distinctly narrow, with the short auricles above the base, glabrous; wings violet, the blade broadly spatulate, 10 mm long, the claw 3 mm long; keel narrow, 8 mm to bend and 3 more to base of the terminal 2 coils of 2 mm diam.; vexillary stamen, the claw 0.75 mm long, the knob very small and sheath 0.5 mm long, the thickened portion to 1.5 mm from base; stamen tube long linear, 9 mm to bend and 3 more to the divided filaments; ovary linear, 5 mm long, compressed, densely covered with white canescent hairs, 8 ovules; style with terminal thickened coil of 1.75 mm diam.; lower part of stigma puberulent with a few long hairs, stigma terminal, lateral, somewhat extrorse, 1 mm long. ***Pod*** falcate, immature

3–4 cm long, 4–5 mm wide, strong sutures, covered with white strigose and minute uncinate hairs; beak thick, recurved, 1.5 mm long. ***Seed*** unknown. ***Seedling*** unknown.

Specimens examined: **MÉXICO. Nuevo León:** 2.6 mi N of Grutas García (W of Monterrey), (25°52'N, 100°31W), 914 m, 5 Sep 1967, *Weedon 4074* (MICH).

Habitat.—It is apparently very rare on rocky, limestone soil found in the mountains west of Monterrey, Nuevo León, México. Only a few collections are known.

Comments.—This is perhaps the most striking species of the section, with fairly small but mostly lobed leaves and flowers in which the lower calyx lobe is large and well-developed and somewhat foliaceous.

F.5.—Phaseolus amblyosepalus (Piper) Morton, Contr. U.S. Natl. Herb. 29:85. 1944. (**Fig. 53**). *Alepidocalyx amblyosepalus* Piper, Contr. U.S. Natl. Herb. 22:672. 1926. TYPE: MÉXICO. DURANGO: (no location or date), *Ibarra 456* (in part, #2) (HOLOTYPE: US 1013387).

Aerial shoot a very small, short (less than 0.5 m long), annual, erect, (often looking determinate) vine. ***Root*** a perennial, globose to fusiform, thick, fleshy, about 1 cm in diameter and 2.5 cm long. ***Stems*** terete, striate, 1 mm thick; internodes 2–5 cm long, puberulent of reflexed hairs. ***Stipules*** broad, round-ovate to round, 5–10 mm long, acute to obtuse, 15- to 20-striate-nerved, puberulent to glabrous. ***Leaves*** 2.3–3.9 cm long, few, none produced at the 2–3 terminal stem nodes; petioles 1–2 cm long, a little longer than the leaflets, minutely puberulent; petiolules, the terminal one 3.3–4.2 mm long, the lateral ones 1.7–3.3 mm long; leaflets broadly ovate, 1–1.5 cm long, truncate at base, 1- to 3-nerved, acute to obtuse, densely covered by minute, basally glandular uncinate hairs and sparsely strigillose on adaxial surface, nearly glabrous abaxially, margins ciliate. ***Inflorescence*** a pseudoraceme, 1- to 4-flowered; peduncles densely covered with minute uncinate hairs, 2 to 3 times as long as the leaves, 7–10–(28) cm long; rachis 2–5 cm long; primary bract broadly oval, 5–6 mm long, acute, glabrous, greenish; pedicel 2 mm long, densely covered with minute uncinate hairs. ***Bracteoles*** not present. ***Flower*** violet; calyx wide campanulate, tube 2 mm long, the two broad upper lobes nearly round, 1 mm long, 2 mm wide, the 3 lower lobes subequal, 1.5 mm long, 1.5 mm wide, triangular, obtuse, covered by minute uncinate hairs, some strigose hairs at base; standard violet, the basal claw 5 mm long to base of blade, the blade reflexed, round, erect, 7 mm long, 12 mm wide, evenly flaired, not enrolled; wings violet, spreading, the oval blade 9 mm long, 10 mm wide, the claw 7 mm long; keel 8 mm from base to reflexed curve and 5 mm to base of the terminal 2 1/2 coils about 2 mm in diam., tip whitish; ovary linear, 6–7 mm long, densely covered by strigillose hairs, ovules 12, the stigma lateral, oblique, 1 mm long, with 2 flat, lateral, laminar extensions with fimbriate edges, each 0.25 mm wide. ***Pod*** immature, slightly falcate, densely covered with short silvery hairs. ***Seed*** unknown. ***Seedling*** unknown.

Specimens examined: **MÉXICO. Durango:** 123 km by road W of Durango on Hwy 40 to Mazatlán, (23°45'N, 105°45'W), 2600 m, 10 Sep 1983, *Anderson 12579* (CAS, MICH); Mpio. Durango, 55–60 km SW of Durango City on road to La Flor, (23°55'N, 105°5'W), 2400 m, 16 Sep 1979, *Breedlove 44126* (CAS); El Salto, S along lumber road toward Pueblo Nuevo (about 60 air mi SW of Cd. Durango), (23°45'N, 105°30'W), 2500–2650 m, 26–29 Aug 1952, *Maysilles 7801* (MICH); parque El Tecuan, 57 km SW de Durango, Mpio. Durango, 2520 m, 12 Sep 1985, *Tenorio et al. 9601* (MO). **Michoacán:** 15 km W of Ciudad Hidalgo along road to Morelia (3.5 mi E of "La Corbu" Industrial Plant), 19°40'N, 100°40'W, 2300 m, 11 Aug 1960, *Iltis et al. 868* (WIS). **Sinaloa:** Cerro de la Sandía, NE of Panuco, (23°25'N, 105°50). 1800–2000 m, 29–30 Aug 1935, *Pennell 20061* (US).

Habitat.—This species is found growing in open pine or pine-oak forests with tussock grasses (*Stipa, Andropogon*), *Arctostaphylos* and *Garrya* and in soils that are rocky and derived from basalts.

Comments.—This distinctive species is a very small plant quite similar to *P. parvulus* but differing by the large stipules, small ovate to rhombic leaflets, rounded buds and small flowers with nearly 3 coils on the keel, and falcate pods. The upper part of the stem often does not develop further once 1–2 axillary racemes are developed, resulting in a determinate looking. It is known from only a few collections from the mountains in western Durango and neighboring Sinaloa. Both Piper (1926) and Morton (1944) reported the type as Ibaña García.

F.6.—Phaseolus tenellus Piper, Contr. U.S. Natl. Herb. 22:696. 1926. (**Fig. 53**). TYPE: MÉXICO. MICHOACÁN: Campanario, near Morelia, (19°45'N, 101°10'W), 2200 m, 12 Sep 1912, *Arsène 8310* (HOLOTYPE: US 1003657; ISOTYPES: BM, F, GH, MO).

Aerial shoot a small, prostrate, indeterminate vine, 15–60 cm long, nearly glabrous throughout except for minute uncinate hairs. ***Root*** a perennial, thick, fleshy, about 3 cm long. ***Stems*** terete, much branched, very slender; internodes straight, 10–15–22 cm long. ***Stipules*** broadly oblong-ovate, 5–6 mm long, 3 mm wide, 3- to 7-nerved, acute. ***Leaves*** 4.8–7.3 cm long; petioles 2.5–3.5 cm long; petiolules 4–7 mm long; stipels lanceolate to linear, 2–3 mm long, 0.75 mm wide; pulvini 2 mm long, puberulent of minute hispid and uncinate hairs; all 3 leaflets similar, the terminal somewhat larger, rhombic-ovate, 1.5–2.7 cm long, broadly cuneate at base, sometimes obscurely to distinctly rounded lobed at base, 3-nerved, obtuse-acutish and short-apiculate, distinctly fine reticulate below, minutely scabrous on the nerves and margins, paler beneath, coriaceous. ***Inflorescence*** a few-flowered raceme, the peduncle mostly 8–12 cm long, the rachis 25–40 mm long, of 1–3 nodes with 2–6 flowers; primary bract oblong, 3 mm long, acute, faintly 3-nerved only the midnerve prominent, the secondary bracts linear, minute; pedicel 5 mm long, puberulent of minute white uncinate hairs. ***Bracteoles*** minute hyaline scales, 0.25–0.5 mm long, broadly ovate, oblong, obtuse, caducous. ***Flower*** violet; calyx campanulate, oblique, the tube 3 mm long, the upper lobes rounded, 1 mm long, the lower lobes with 3 broad acute teeth, the center 2 mm long, the lateral 1.5 mm long, glabrous to scarcely puberulent, ciliate margins; standard violet, broadly oblong, 12–18 mm long, notched at apex, somewhat reflexed at 5 mm from base, the claw very broad, 3 mm long, the 2 triangular auricles on the inside near the base; wings violet, the blade spatulate, 12–18 mm long, the spur pronounced, 1 mm wide, the claw 4 mm long; keel straight from base, 10 mm long to the 1 1/2 coils of 2 mm in diam.; vexillary stamen much thickened at base but not geniculate; stamen tube united for 2/3 their length; anthers oblong; basal collar very elongate; ovary straight, linear, 5 mm long, flattened, scabrous, with distinct sutures, about 10 ovules; style, glabrous to the coils, then long-hairy nearly to the stigma; stigma terminal, oblong, slightly inflexed. ***Pod*** immature linear, puberulent. ***Seed*** unknown. ***Seedling*** unknown.

Specimens examined: **MÉXICO. México:** Dist. Temascaltepec, Nanchititla, (19°02'N, 100°02'W), 8 Sep 1933, *Hinton 4719* (GH, K, US). **Michoacán:** Cerro Azul, Morelia, 2200 m, 1900, *Arsène 6579* (US); Campanario, vicinity of Morelia, 2200 m, 14 Sep 1911, *Arsène 6782* (MO, US). **Zacatecas:** Brecha Jalpa-Tlaltenango, (21°45'N, 103°05'W), 2100 m, 2 Aug 1971, *Díaz 2352* (MICH).

Habitat.—Reported to be found growing beneath cliffs and in forests of pine-oak.

Comments.—Delgado (1985) correctly recognized this species as belonging to Section F. *Minkelersia* even though the calyx lobes are quite short and rounded. Nevertheless, the elongate flower structure and many other minor characteristics of flower and plant show its natural affinity with the other species of this section.

F.7.—Phaseolus parvulus Greene, Bot. Gaz. 6:217. 1881. (**Figs. 50, 52**). *Alepidocalyx parvulus* (Greene) Piper Contr. U.S. Natl. Herb. 22:672. 1926. TYPE: UNITED STATES. NEW MÉXICO: abundant in deep wood of *Pinus ponderosa* in the Pinos Altos Mts., (33°N, 108°W), 14 Aug 1880, *Greene s.n.* (HOLOTYPE: NDG n.v.; ISOTYPES: F, G, GH, K, MO(2; # 12792), NA n.v., UC).

Aerial shoot an annual, very short, indeterminate vine. ***Root*** perennial, small and fleshy, spherical, 1–2 cm in diam., about 5 cm under the soil surface. ***Stems*** solitary, rarely branched, slender, erect but sometimes prostrate, usually 10–20 cm long, rarely to 0.5 m; the first internode at base of plant 2.5–5 cm long, most others 0.8–1.3 cm long, only 3–5 internodes per plant. ***Stipules*** ovate to ovate-lanceolate, sometimes nearly linear, 5 mm long, 2.5 mm wide, 5- to 6-nerved. ***Leaves*** 3.8–6.9 cm long; petiole 1–3 cm long; petiolule 3–4 mm long; stipels 1–2.5 mm long; terminal leaflet long ovate to narrow-lanceolate, mucronate, 2.5–3.5 cm long, 4–11 mm wide, entire or with more or less distinct lobes at the base, obtuse, covered by minute glandular, ciliate hairs adaxially especially at margins, glabrous abaxially; lateral leaflets similar but lobes larger. ***Inflorescence*** 1–2-flowered; peduncle 2–8 cm long; rachis 1–1.2 cm long, mostly 1- to 2-8-flowered; primary bracts ovate apiculate 4–5 mm long 2.5–3 mm wide 3–7-nerved; pedicel 1–3 mm long, puberulent of minute hooked hairs; pedicellar bracts 1 mm long narrowly lanceolate hyaline. ***Bracteoles*** minute, scale-like, broadly ovate to oblong, 0.75 mm long, 0.5 mm wide, 1-nerved, obtuse, hyaline, glabrous, early caducous. ***Flower*** violet, somewhat extended or elongate, especially in bud; calyx tubular, 5–7 mm long, the lobes sub-equal, slightly longer than basal tube, acute to obtuse, sparsely covered with minutely ciliate hairs especially on the inner surface of tube; standard deep violet, the blade erect, narrow and flared, 12–20 mm long,

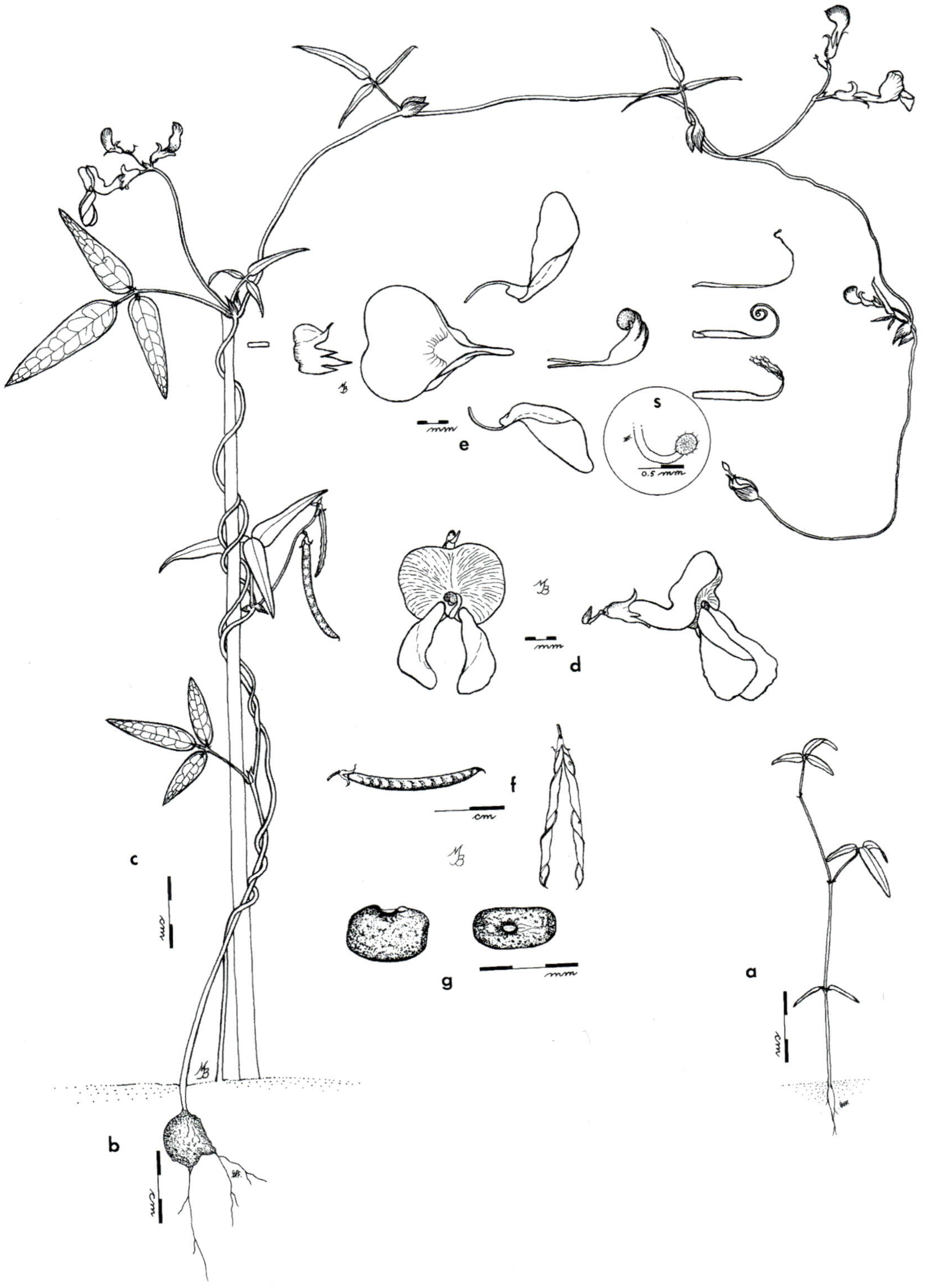

FIG. 52. Illustrations of *Phaseolus parvulus* Greene.—**a.** Seedling plant several weeks after germination.—**b.** Root after 1 year; note small spherical shape.—**c.** Plant with mature leaves, inflorescences, pods, flowers, and vine tip.—**d.** Flowers, side view and front view.—**e.** Exploded view of flower showing all parts, including—**s.** Style tip showing stigma as seen under the microscope; note the fairly short sepals, long claw on standard, lack of knob at base of vexillary stamen and the capitate stigma.—**f.** Mature pods, side view and dehisced.—**g.** Seeds, side view and view from hilum; note the oblong shape of seed. All drawings from living material grown in greenhouse at Mayagüez from seed of L-128 from University of California at Riverside (TARS #246) collected from Pima Co., Arizona.

10–12 mm wide, reflexed at about the middle, the claw about 1.5 mm long, 7 mm to bend and 6 mm more to deeply emarginate tip, the auricles not developed; wings deep violet, the blade broadly obovate, 12 mm long, 7 mm wide, the basal claw very delicate 7.5 mm long, less than 0.2 mm wide, the spur pronounced, 1 mm long; keel 10–12 mm to the 1 3/4 coils of about 2 mm in diam., the 3 claws separated for 6 mm long, purple and tip whitish; vexillary stamen only slightly thickened, the claw 1 mm long, about 0.5 mm wide; stamen tube straight, tubular, 11 mm long, barely inflated, no ridges; basal collar 0.5 mm long; ovary straight, 5–6 m long, 0.7 mm wide, minutely puberulent, 8–10 ovules; style, the terminal thickened coil of 2 mm in diam.,; stigma lateral, introrse, very narrow, 0.5 mm long. ***Pod*** linear, straight, 2.5–3.5 cm long, 3–4 mm wide, sutures somewhat thickened, the beak recurved, 1–2 mm long, young pod sparsely covered with minute ciliate hairs and at maturity nearly glabrous. ***Seed*** short-reniform, 1–3.7 mm long, 1.7–2.6 mm wide, smooth, speckled with purple or black on light brown; hilum oblong, 0.4–0.5 mm long; lens pronounced. ***Seedling*** from hypogeal germination; epicotyl 3–4 cm long, glabrous; stipule bifid; eophylls opposite, simple, the petioles with only a lower pulvinus, the stipels absent, the blade ovate, rounded-truncate at base, acute at tip.

Specimens examined: **MÉXICO. Chihuahua:** Mpio. Guerrero, Arroyo Ancho, S of Mesa Colorado, (28°30'N, 107°30'W), 2100 m, 28 Aug 1978, *Bye 8871a* (UCR); Hidalgo del Parral, Los Altares, 23 km W of Parral, 26°52'N, 105°55'W, 2000 m, 1 Oct 1978, *Debouck et al. 280* (CHAPA, COL, K); Memelichi, Río Mayo, 19 Sep 1936, *Gentry 2796* (ARIZ, F); Mojarachic, SW of San Juanito, 27°52'N, 108°W, 2140 m, 13 Aug 1954, *Knobloch 1251* (MICH, MSC); Chuhuichupa, Aug–Sep 1936, *Le Sueur 696* (ARIZ, BRIT, F, GH, MO, TEX); 15 mi S of Creel, (27°40'N, 107°30'W), 2195 m, 20 Aug 1967, *Marin CF18* (UCR); 2 km SE of Cocheno, on road to Mina Tarahumara, 28°18'30"N, 108°12'30"W, 2200 m, 28 Aug 1986, *Martin et al. s.n.* (ARIZ); Mpio. de Madera, Cerro La Concha, Laguna de Babicora, 2400 m, 18 Aug 1994, *Quintana et al. 3212a* (BRIT); Mpio. Batopilas, 9 mi toward Quirire S of Jct. of La Bufa Rd. and Roroachic Rd. between Creel and La Bufa, (27°15'N, 107°30'W), 17 Aug 1984, *Randolph 425* (UCR); Mpio. de Ocampo, Parque Nac. Cascada de Basaseachic, 1950 m, 5 Aug 1984, *Yen et al. 2856* (BRIT). **Durango:** 123 km by road W of Durango on Hwy 40 to Mazatlán, 2600 m, 10 Sep 1983, *Anderson 12578* (CAS, MICH); 11 km SW of La Ciudad near Buenos Aires, 2900 m, 13 Aug 1974, *Breedlove 36472* (CAS); Mpio. of Durango, 65–75 km SW of Durango City on road to La Flor, (23°50'N, 105°15'W), 2620 m, 17 Sep 1979, *Breedlove 44280* (CAS); Mpio. Pueblo Nuevo, 20 km NE of La Ciudad on Hwy 40, 2770 m, 21 Aug 1980, *Breedlove et al. 45857* (CAS); El Salto, 30 km W of Durango, 2623 m, 19 Aug 1061, *Dettling 8562* (US); La Providencia, Sierra Madre and Sierra Santa Barbara, (26°45'N, 105°50'W), 1981 m, 11–12 Sep 1898, *Nelson 4991* (US); Las Cruces, 52.7 km W of Tarahumar, camino Tepehuanes-Tabahueto, 2200 m, 28 Aug 1983, *Torres et al. 3501* (MO); Mpio. Santiago Papasquiaro, 22 km WNW of Santiago Papasquiaro, 25°4'N, 105°51'W, 2667 m, 25 Aug 1983, *Worthington et al. 11447* (UCR). **Nayarit:** Terr. de Tepic, Sierra Madre, Santa Teresa, (22°30'N, (104°45'W), 7 Aug 1897, *Rose 3425* (US). **Sinaloa:** Cerro de la Sandía, NE of Panuco, 1800–2000 m, 29–30 Aug 1935, *Pennell 20062* (GH, US); Rancho Liebre Barranca, 1 mi S of El Palmito on Hwy 40 and 1 mi W of the Hwy., Sierra Madre Occidental, 23°36'N, 105°51'W , 2150 m, 19 Aug 1988, *Sanders et al. 8077* (UCR). **Sonora:** 11 mi E of Santa Ana, on old Rd. to Yecora (3/4 mi W Rancho El Puerto, or 21 kms WSW of Yecora on old Rd. to Nuri, (28°10'N, 109°10'W), 7 Sep 1964, *Lowe et al. s.n.* (ARIZ); 3 km S of La Lobera, Ejido Zahuarivo, Mpio. Alamos, 27°09'N, 108°58'W, 1550 m, 25 Aug 1986, *Tenorio et al. 11922* (MO); El Tigre Canyon and Mt. above El Tigre Mine, E of Esqueda and Lago Angostura, (30°15'N, 109°10'W), 1676 m, 9 Sep 1961, *Turner et al. 2124* (ARIZ). **Zacatecas:** Plateado, Sierra de Los Morones, 1 Sep 1897, *Rose 2723* (US).

UNITED STATES. ARIZONA: Apache Co.: White Mts., about 10 mi S of McKays Peak, (34°5'N, 109°30'W)?, 2286 m, 31 Aug 1948, *Gould et al. 5016* (ARIZ, GH, UC); head of Lone Pine Creek, 24 Aug 1968, *Kobetich 68-134* (ASU); Fort Apache Indian Reservation, 5 mi NE of McNary, (34°5'N, 109°45'W), 2347 m, 8 Aug 1976, *McLaughlin 1312* (ARIZ). **Cochise Co.:** Barfoot Park, Chiricahua Mts., (32°5'N, 109°15'W), 2438 m, Sep 1907, *Blumer 114* (US); Chiricahua Mts., Onion Saddle, 2316 m, 29 Aug 1966, *Cazier 629* (ASU); upper Hunt Canyon, Chiricahua Natl. Monument, 14 Aug 1939, *Clark 8586* (ARIZ); Chiricahua Natl. Monument, 2 mi E of Rustlers Park on dirt road, 19 Sep 1958, *Fearing 2063* (TEX); Fort Huachuca, Huachuca Mts., (31°45'N, 110°20'W), Aug 1882, *Lemmon 2682* (F(2), G, GH, US); Chiricahua Mts., Pinery Canyon, 9 Aug 1974, *Reeves 966* (ASU); Huachuca Mts., 1/4 mi S of Reef Road, Reef Mine, 2225 m, 31 Aug 1982, *Reichenbacher 1071* (ARIZ); 7 mi W of Portal, Rustlers Camp, 17 Aug 1981, *Sundberg 1267* (TEX). **Coconino Co.:** 100 yards W of Woods Canyon Lake Jct. on Ariz 160, (36°25'N, 111°W)?, 2316 m, 13 Aug 1966, *Lehto 7212* (ASU); Woods Canyon Lake, 23 Sep 1967, *Pinkava et al. 4438* (ASU). **Navajo Co.:** 0.3 to 1.9 mi E of the Gila Co. line, about 27 mi NE (airline) of Payson, Mogollon Rim, along Hwy 277, (34°30'N, 110°30'W), 2278 m, *Sanders et al. 3202* (UCR). **Pima Co.:** near the top of Mt. Lemmon along the highway from Tucson, 32°25'N, 110°45'W, 30 Aug 1966, *Adams s.n.* (US, AZ, UCR); Spud Ranch, Rincon Mts., 2260 m, 23 Aug 1909, *Blumer 3306* (ARIZ, CAS-DS, F, GH, MO, UC); Santa Rita Mts., Madera Canyon, 12 Aug 1914, *Rapson s.n.* (ASU). **Santa Cruz Co.:** Santa Rita Mts., Madera Canyon, (31°45'N, 110°45'W), 1981 m, 8 Aug 1943, *Darrow s.n.* (ARIZ). **NEW MÉXICO: Catron Co.:** 18 mi NE of Mogollon on St. Rd. 78, trail from Willow Creek Campground to Whitewater Baldy Mt., (33°20'N, 108°30'W), 2970 m, 19 Aug 1966, *Hess 914* (BRIT, MICH); Socorro Co. (Catron Co.), Mogollon Mts., West Fork of the Gila, 6 Aug 1900, *Wooton s.n.* (US). **Grant Co.:** Gila National Forest, Cherry Creek Canyon, 12 mi NE of Silver City on Hwy 25, (32°50'N, 108°15'W), 2370 m, 30 Jul 1967, *Hess 1313* (ARIZ, BRIT). **Hidalgo Co.:** Coronado Natl. Forest, Peloncillo Mts., Clanton Draw, 2.3 mi W of the E side of the Natl. Forest by Forest Rd. 63, (31°30'N, 108°50'W), 1676 m, 29 Aug 1986, *Worthington 14957* (ARIZ).

Habitat.—*P. parvulus* is found growing at high altitudes in open pine-oak forests with other trees and shrubs (spruce, fir, juniper, and *Arbutus*) and grasses (see also Buhrow 1983), and in soils that are deep and rich with lots of humus and derived from sandstone or volcanic ash, and often rocky and andesitic.

Comments.—Piper (1926) placed this species in a new genus *Alepidocalyx* based mostly on the basis of the lack of bracteoles and small globose roots which Lackey (1983) found insufficient to maintain separate from Section F. *Minkelersia*. All specimens are from cool climates, usually found at high elevations in grasslands between or among sparse forests of pine and oak.

Under this species, Delgado (1985) recognizes two varieties: a) *parvulus* and b) *anisophyllus*, which are based on vegetative and inflorescence differences. These differences seem to be significant to us and correlate with the site locations in SW Chihuahua and neighboring Durango and we believe they could be maintained as separate species pending on new data.

F.8.—Phaseolus anisophyllus (Piper) Freytag & Debouck, comb. nov. (**Fig. 53**). *Alepidocalyx anisophyllus* Piper, Contr. U.S. Natl. Herb. 22:672. 1926. TYPE: MÉXICO. CHIHUAHUA: Mt. Morinora (Mohinora), SW Chih., (25°56'N, 107°04'W), 1 Sep 1898, *Nelson 4911* (HOLOTYPE: US 332953).

Aerial shoot a prostrate, indeterminate vine, to 1 m long. ***Root*** a perennial, probably small, thick, fleshy, globose, 1 dm below soil surface. ***Stems*** terete, glabrous; internodes 6.7–15 cm long, 8–9 per plant. ***Stipules*** broadly ovate-oblong, acute, glabrous, 10- to 11-nerved, 8–10 mm long, 4–6 mm wide. ***Leaves*** with petioles to 45 mm long, glabrous; petiolules puberulent; stipels lanceolate, acute, 2–3 mm long; terminal leaflets lance-oblong, 2–5 cm long, the upper ones slightly lobed at base, obtuse, apiculate, glabrous; lateral leaflets similar but distinctly 1- to 2-lobed at base, the lobes 3.3–6.7 mm long. ***Inflorescence*** a pseudoraceme, the peduncles 8–13 cm long, the rachis 5–8 cm long of 2 to 3 flowering nodes, puberulent of minute retrorse hairs; bract ovate, acute, 3–4 mm long, nerved, persistent; pedicel fairly stout, 2–4 mm long. ***Bracteoles*** lacking or unknown. ***Flower***, calyx campanulate, puberulent, the lobes 2 mm long, acute; wings 3 mm long. ***Pods*** linear, 4 cm long, 4 mm wide; carpels twisting tightly at maturity, glabrous; beak short; 6 ovules. ***Seed*** ellipsoid-cylindric, 4 mm long, reddish brown, shiny; hilum minute, circular. ***Seedling*** unknown.

Specimens examined: **MÉXICO. Durango:** Mpio. Santiago Papasquiaro, Sierra Madre Occidental, about 22 air km WNW Santiago Papasquiaro, 3.3 mi by Topia Rd. W from Jct. Rd. to antenna at E edge crest of sierra, 25°04'N, 104°51'W, 2667 m, 25 Aug 1983, *Worthington et al. 11447* (ARIZ, UCR).

Comments.—Our reasons for maintaining species rank are given above under *P. parvulus*. More collecting in this very interesting part of México is needed to confirm the distribution and habitat preferences of this species.

F.9.—Phaseolus amabilis Standl., Publ. Field Mus. Nat. Hist., Bot. Ser. 22:28. 1940. (**Fig. 53**). TYPE: MÉXICO. CHIHUAHUA: Río Mayo, Guasaremos, (27°40'N, 108°42'W), 1660 m, 26 Aug 1936, *Gentry 2471* (HOLOTYPE: F; ISOTYPES: DES, GH, K, MO, UC).

Aerial shoot a large, climbing, indeterminate vine. ***Root*** unknown. ***Stems*** thin, terete, very sparsely pubescent. ***Stipules*** broadly ovate, large, 5–8 mm long, 3 mm wide, lightly 10-nerved, acute, glabrous, spreading at 90°to stem. ***Leaves*** 6–8.5 cm long; petiole 3–4 cm long, petiolules 7–9 mm long, the leaves on lower portions of plant with entire, ovate-acuminate leaflets, the upper leaflets 3-lobed; terminal leaflet 3–4 cm long, 3-lobed, the lower 2 lobes broadly rounded, the terminal lobe long, nearly spatulate, acute, apiculate, glabrous; lateral leaflet similar but unequally 3-lobed. ***Inflorescence*** a few-flowered, delicate raceme; peduncle 4–8 cm long, striate, glabrous; rachis with 2–3 flowering nodes separated by 1–2.5 cm, glabrous; primary bract ovate-lanceolate, 3–4 mm long, 1.75 mm wide, acute, hyaline, glabrous; pedicel 4–5 mm long, glabrate except for a few minute cilia; pedicellar bracts linear 1 mm long or less hyaline glabrous. ***Bracteoles*** lacking. ***Flower*** violet turning purple, relatively large, 24 mm long; calyx tube somewhat narrow campanulate, 3.5 mm long, with a small basal protuberance about 0.25 mm long, the calyx lobes subequal, lanceolate, 4–5 mm long, the upper two connate, nearly glabrous; standard purple, 10 mm long, with a long narrow base, the banner reflexed, squarish, 7 mm long, slightly notched at center; wings purple, the blade rounded, 14 mm long, 10 mm wide, the claw 10 mm long; keel purple, 13–14 mm long, with nearly 2 coils of 2 mm diam.; stigma terminal, capitate. ***Pod*** unknown. ***Seed*** unknown. ***Seedling*** unknown.

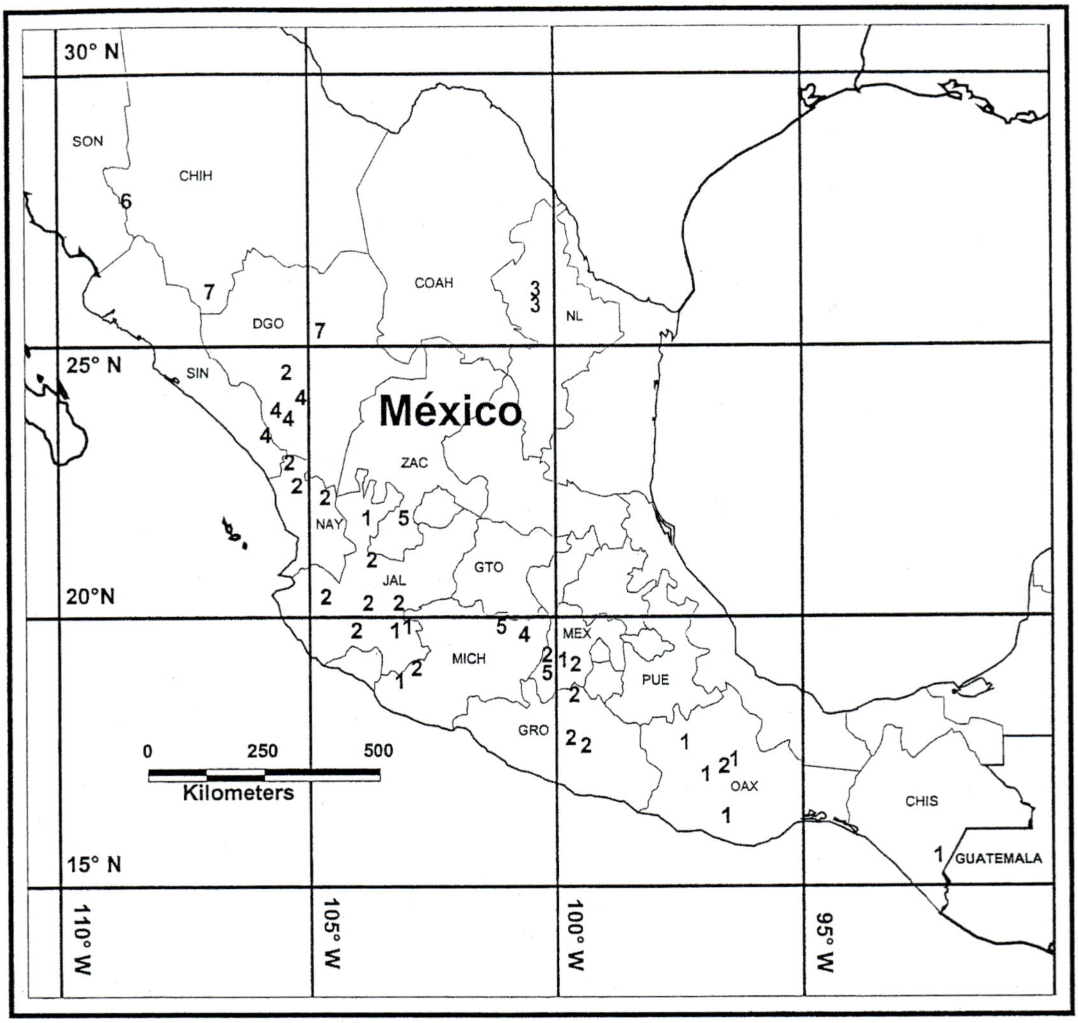

FIG. 53. Distribution of minor species of Section F. *Minkelersia*, as follows: 1 = *P. nelsonii*; 2 = *P. perplexus*; 3 = *P. plagiocylix*; 4 = *P. amblyosepalus*; 5 = *P. tenellus*; 6 = *P. anisophyllus*; 7 = *P. amabilis*.

Habitat.—This species was found growing on a moist mossy bank in pine-oak forest.

Comments.—Delgado (1985) places this taxon as a synonym of variety *P. anisophyllus* apparently in part because of their similar distribution in Chihuahua. However, the senior author sees this species as having very distinctive and much larger flowers than all the rest of the Section F. *Minkelersia* and with deeply lobed leaflets on the upper portions of the long climbing vine. More collections are badly needed since only a single collection is known.

F.10.—Phaseolus pauciflorus Sessé & Mociño ex G. Don, Gen. Hist. 2:356. 1832. (**Figs. 50, 54**). TYPE: MÉXICO. GUERRERO: in Tixtla montibus, (17°30'N, 99°25'W), Sep 1789, *Sessé & Mociño s.n.* (LECTOTYPE: OXF designated by Sousa & Delgado (1981) n.v.; ISOTYPES: F (M 3714), MA n.v.).

Minkelersia galactioides Mart. & Galeot., Bull. Acad. Roy. Brux. 10:200. 1843. *Phaseolus galactioides* (Mart. & Gal.) Maréchal, Mascherpa & Stainier, Taxon 27:199. 1978a. TYPE: MÉXICO. OAXACA: Bois près l'O Pacif. de Fevo. Rare nov Apres Cordillera, (16°30'N, 96°30'W)?, Nov–Apr 1840, *Galeotti 3175* (HOLOTYPE: K; ISOTYPE: G ; photos: GH, MICH, MO).

Phaseolus lambertianus Dietr., Syn. Pl. 4:1196. 1847.

Minkelersia biflora Hemsl., Diag. Pl. Nov. Mex. 3:48. 1880. TYPE: MÉXICO. DISTRITO FEDERAL: Vallée de México, *Schaffner 126* (HOLOTYPE: K; photo GH).

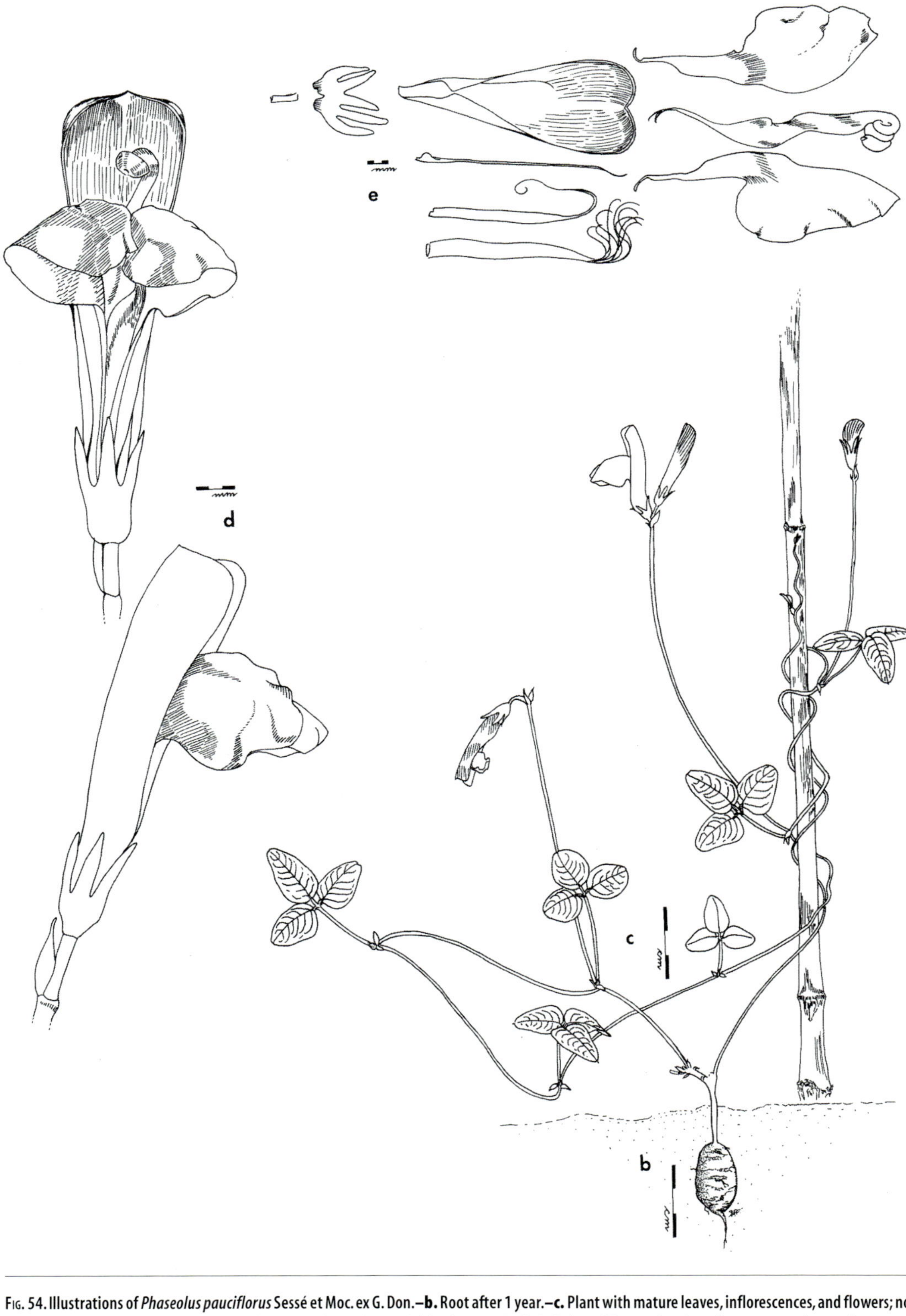

Fig. 54. Illustrations of *Phaseolus pauciflorus* Sessé et Moc. ex G. Don.—**b.** Root after 1 year.—**c.** Plant with mature leaves, inflorescences, and flowers; note the erect position of buds and flowers.—**d.** Flowers, front view and side view.—**e.** Exploded view of flower showing all parts; note the very long sepals, elongate standard and nearly straight keel. All drawings from living material grown in greenhouse at Mayagüez from seed collected near UNAM, Pedregal, D.F., México).

Aerial shoot an annual, prostrate, seldom climbing, indeterminate vine, to 1 m long, sometimes forming mats to 1/4 m diam. ***Root*** a perennial, thick, fleshy, globose to pyriform, to 2-3 cm in diam. ***Stems*** terete, obscurely striate, 1.5-2 mm thick, sparsely covered by strigose hairs. ***Stipules*** conspicuous, broad ovate, acute, 6-10 mm long, 6-8 mm wide, many-nerved, glabrous. ***Leaves*** 4-9 cm long; petioles 1.5-5 cm long; petiolule 3-10 mm long; stipels aciculate, 1.5 mm long, glabrate; pulvini dark, 1.5-2 mm long, covered by strigose and hooked pubescence; terminal and lateral leaflets similar, broadly ovate, to orbicular to somewhat rounded to elongate, 2-2.8 cm long, 1.6-1.8 cm wide at 1/4 from base, distinctly nerved, acute to obtuse, minutely apiculate, light green adaxially, densely covered with minute glandular hairs and sparsely hispidulous adaxially and on margins, abaxial surface a lighter color, not glandular, but sparsely covered with strigose hairs. ***Inflorescence*** erect, tall, extending much above foliage, 10-15 cm long with only 2 terminal, vertically erect flowers; primary bract ovate, 2-4 mm long, acute, many-branch-nerved, glabrous except for fimbriate margins; pedicel 2-5 mm long, puberulent of minute hooked hairs, erect through maturity; pedicellar bracts narrowly ovate 4-5 mm long many-branch-nerved greenish glabrous sparsely ciliate margins. ***Bracteoles*** caducous in bud, hyaline, oblong, 2 mm long, 0.6 mm wide, scarcely 1-nerved, glabrous, ciliate margins. ***Flower*** vertically erect, nearly tubular, pink to purple; calyx subtubular, 10-13 mm long, covered with strigose hairs, the lobes longer than the basal tube, spatulate, obtuse, glabrous, ciliate on margins; standard pink at tip, greenish adaxially and at base, elongate, erect, straight, widest near the apex, obtuse to retuse, slightly hooded, about 30 mm long, 10-15 mm wide, slightly nerved; wings purple to pink, the blade, broad spatulate, about 30 mm long, 10-18 mm wide, slightly spreading near apex; stamen tube straight, about 25 mm long, 2.5 mm diam.; vexillary stamen, thickened portion about 30 mm long, somewhat flattened with small genicular flat at base, terminal filament about 5 mm long; keel nearly straight, 25 mm long with nearly 3 close coils, the first 2 coils smaller than the last one of 3.75 mm diam.; ovary linear, 1 cm long, densely covered with villous silvery hairs, 20-25 ovules; style, the terminal thickened coil of 3.25 mm diam.; stigma lateral introrse, somewhat twisted to the side, linear, 1 mm long. ***Pod*** straight, nearly linear, erect through maturity, 5 cm long, 3-4 mm wide, the beak 3-6 mm long, recurved, slightly thickened sutures, densely covered with hispidulose hairs and sparsely covered by strigose hairs, tightly twisted at dehiscence. ***Seed*** spherical, 1.5-2 mm in diameter, flattened ends, solid black; hilum round, 0.4 mm in diameter. ***Seedling*** from hypogeal germination; epicotyl 16-17 mm long; stipules bifid; eophylls opposite, simple, the petioles with minute basal and apical pulvini, the stipels absent, the blade ovate-oblong, rounded at base, acute at tip.

Specimens examined: **GUATEMALA. Quiché:** (no location), (15°N, 91°W)?, 1942, *Aguilar 1472* (F).

MÉXICO. Chihuahua: Creel, (27°50'N, 107°30'W), 31 Aug 1973, *Bye 4848* (UCR); Tecolote, 10 mi W of Chinatu, 1920 m, 10 Oct 1959, *Correll et al. 23001* (TEX-LL); Río Mayo, Sierra Canela, (27°50'N, 108°35'W)?, 27 Aug 1936, *Gentry 2483* (F, GH, US); Loreto, Río Mayo, 2 Sep 1936, *Gentry 2563* (F, K, GH, MO, UC); Marsh lake, Sierra Madre Mts., 2134 m, 19 Sep 1903, *Jones s.n.* (US); Guayanopa Canyon, Sierra Madre Mts., 2310 m, 24 Sep 1903, *Jones s.n.* (MO); Basigochi, on Creel-La Bufa Rd., 15 Sep 1957, *Knobloch 483* (MSC, TEX); Mojarachic, 21 Aug 1938, *Knobloch 5532* (F); Mpio. Temosachi, Nabogame, 28°30'N, 108°30'W, 1800 m, 12 Sep 1987, *Laferriere 1133* (UCR); Chuhuichupa, Aug-Sep 1936, *Le Sueur 702* (F, GH, MO, TEX, UC, US); Mpio. de Madera, Mts. about Garabato, (29°5'N, 108°W)?, 28 Sep 1939, *Muller 3544* (GH, MICH(2), TEX-LL, UC); Mpio. Casas Grandes, El Pinal, 1 km al SW del Ejido Hernández, (30°30'N, 108°W)?, 2100 m, 26 Sep 1982, *Tenorio et al. 1745* (CAS, MO); Colonia García, Sierra Madre, 2475 m, 11 Sep 1899, *Townsend et al. 319* (GH, K, MO, US(2)). **Distrito Federal:** Pedregal, San Angel, (19°20'N, 99°12'W), Sep 1927, *Lyonnet 117* (in part) (K, US); 3 km al SW de Tlalmanalco, sobre el camino a Santo Tomás Atzingo, 2500 m, 25 Aug 1968, *Pineda 492* (MICH); Pedregal de San Angel, cerca de Cuicuilco, 7 Sep 1952, *Rzedowski 1741* (MSC); Delegación de Contreras, Cerro del Judío, 2600 m, 11 Sep 1977, *Ventura 3046* (MICH). **Durango:** 32 km W of Durango on Hwy 40 to Mazatlán, (23°50'N, 105°W), 2000-2500 m, 9 Sep 1966, *Anderson et al. 3573* (GH, MICH, US); Estación Coyotes, 2600 m, 13 Aug 1974, *Breedlove 36500* (CAS); along Hwy 40, between Ciudad and Palmito, 2700 m, 1 Sep 1969, *Clarke et al. 690901-56* (UCR); 7 km E of El Salto, 2455 m, 18 Aug 1961, *Detling 8588* (US); 20 mi W of Durango just W of the Río Pino Valley, 22 Aug 1958, *Fearing 1141* (TEX); 6 mi SW of Palmito, 1890 m, 22 Oct 1966, *Gentry et al. 22036* (US); 5.3 km W of (Ejido) Los Mimbres turnoff, at km 54.2 WSW of Durango and ca halfway to El Salto, 23°45'N, 105°30'W, 2725 m, 16 Sep. 1978, *Iltis et al. 219* (MICH, US); Sierra Madre Occidental, plateau 8 mi N of Coyotes, about 80 km W of Cd. Durango, (24°N, 105°30'W), 2430 m, 28 Sep 1962, *McVaugh 21689* (MICH, TEX-LL); 25.5 mi SW of El Salto on road to Santa Lucía, (22°40'N, 105°45'W), 22 Sep 1953, *Ownbey et al. 1879* (MICH); empalme Purísima, 2480-2500 m, 27 Aug 1934, *Pennell 18252* (US); Las Cruces, 52.7 km W of Tarahumar, camino Tepehuanes-Tabahueto, 2200 m, 28 Aug 1983, *Torres et al. 3489* (MO). **Guerrero:** (location unknown), 1787-1795-1804, *Sessé et al. s.n.* (F (M 3633)); (location unknown), 1787-1795-1804, *Sessé et al. s.n.* (F (M 3716); Malinaltepec, Mpio. Malinaltepec, 17°14'N, 98°40'W, 1700 m, 31 Oct 1989, *Wagenbreth 282* (MO). **Jalisco:** 48 km W of Ayutla on road to Talpa, 1850 m, 21-22 Sep 1983, *Anderson 12749* (CAS, MICH); Mexticacán, 18 km NE of Mexticacán,

21°24'N, 102°45'W, 1900 m, 23 Nov 1978, *Debouck et al. 460* (CHAPA); Las Vidrieras, El Platanar, San Martín de Bolaños, 2450 m, 1 Sep 1968, *Díaz 892* (MICH); Mpio. Volcanes, La Campana, 1800 m, 29 Aug 1969, *Díaz 1380* (MICH); Mpio. de Talpa, 9 km al N de Ayutla, camino a Talpa de Allende, (20°15'N, 104°30'W), 7 Sep 1979, *Magallanes 1914* (ASU, CAS); Mpio. de Talpa, 17 km al S de Talpa camino a la Cuesta, (20°20'N, 104°50'W), 8 Sep 1979, *Magallanes 1951* (ASU, CAS); 5 km al SO de Tapalpa, camino a Venustiano Carranza, (19°45'N, 103°45'W), 13 Sep 1980, *Magallanes 2582* (MICH, TEX); Sierra del Tigre, 3 mi S of Mazamitla, 2100–2200 m, 18 Sep 1952, *McVaugh 13042* (MICH, US); Km 58 road from Zapotlanejo, 7 mi WNW of Tototlán, (20°30'N, 103°W), 1800 m, 24 Aug 1958, *McVaugh 17267* (G, MICH, US); Mts. above Ahuacapán, road to Corralitos, 10–12 mi SSE of Autlán, (19°40'N, 104°20'W), 1500–1800 m, 29 Sep 1960, *McVaugh 19623* (G, K, MICH, TEX, US); 32 km E of Arandas, 2100 m, 15 Nov 1970, *McVaugh 24405* (MICH); 8 km SE of Hwy 110 on a lumber road leaving the Hwy. 12 km SSW of Tecalitlán and extending to San Isidro, 19°20'N, 103°10'W, 2100 m, 24–25 Sep 1965, *Roe et al. 2089* (MICH, US, WIS); Mpio. San Martín de Bolaños, Los Yerbanis, 8 km al NW de El Platanar, (21°40'N, 103°45'W), 2000 m, 1 Sep 1968, *Rzedowski 26125* (MICH); Mpio. Tamazula, Cerca de Agua Hedionda, aproxim. 45 km al E de El Aserradero, sobre el camino a Manuel M. Dieguez, (19°30'N, 103°W)?, 1900 m. 26–27 Oct 1973, *Rzedowski et al. 1127* (MICH). **México:** 35 mi NW of Toluca, (19°25'N, 100°W), 14 Aug 1947, *Barkley et al. 2810* (F, TEX); Dist. Temascaltepec, Comunidad, 2480 m, 7 Sep 1932, *Hinton 1555* (G, GH, K, MO, US); Dist. Temascaltepec, Cumbre de Tejupilco, 2000 m, 12 Sep 1932, *Hinton 1588* (G, GH, K, US); Dist. Temascaltepec, Rincón, 2000 m, 19 Sep 1933, *Hinton 4653* (K, TEX-LL, US); Dist. Temascaltepec, Salitre, Cañitas, 1060 m, 6 Sep 1933, *Hinton 4708* (G, GH, K, US); Mesa de Nanchititla, 1800 m, 15–16 Sep 1958, *Matuda et al. 32794* (CAS); 8 km S of Temascaltepec on Hwy 130, 18°57'N, 100°5'W, 1950 m, 4 Sep 1965, *Roe et al. 1766* (MICH, US, WIS); en Villa Victoria, carr. Toluca–Zitacuaro, 21 Aug 1978, *Soto et al. 855* (BM, CAS); Mpio. de Nicolás Romero, Caja de Agua Progreso, 2400 m, 6 Sep 1980, *Ventura 3652* (MO). **Michoacán:** Caracol, 10 Oct 1902, *Altamirano 859* (US); Cerro de las Nalgas, Morelia, 1900 m, 9 Sep 1909, *Arsène 2579* (MO, US); Cerro Azul, Morelia, 2200 m, 15 Sep 1909, *Arsène 3372* (GH, MO, US); vers la Huerta, Morelia, 1950 m, 1 Sep 1910, *Arsène 5130* (GH, MO, US); Loma Santa María, Morelia, 1950 m, 28 Aug 1910, *Arsène 5841* (K, MO, US); desviación a Tanaco, por la carr. a Uruapan, 2100 m, 23 Sep 1976, *Delgado et al. 262* (MICH); 12 mi E of Uruapan, (19°25'N, 101°55'W), 27 Aug 1957, *Fearing 1103* (MSC, TEX); Mpio. Villa Escalante, Ejido Cuitzian, Potrero Huaniqueo, 8 km al SE de Villa Escalante, carr. hacia Ario de Rosales, (19°22'N, 101°38'W), 2300 m, 4 Sep 1981, *García 1494* (TEX); Sierra Naranjillo, Coalcomán, 23 Aug 1939, *Hinton et al. 15137* (GH, TEX, TEX-LL, US); "Cerritos de Agua," 3 mi below the lumber camp at Dos Aguas, W of Aguililla, 18°45'N, 102°56'W, 2000–2100 m, 15 Sep 1958, *McVaugh 17848* (G, MICH(2), US(2)); 7 km al NW de Cd. Hidalgo, carr. a Morelia, 2050 m, 20 Aug 1980, *Soto et al. 2349* (CAS). **Morelos:** along the México City–Cuernavaca road, Cuernavaca, 1676 m, 28 Aug 1935, *MacDaniels 269* (F); W of Tres Marías, 54 km from México City, (19°10'N, 99°15'W), 3048 m, 29 Oct 1930, *Reddick 203* (US). **Nayarit:** partie occidentale, Sierra del Nayarit, (21°30'W 104°45'W)?, *Diguet s.n.* (MICH); 9 mi N of Compostela, 1000 m, 29 Aug 1957, *McVaugh 16546* (MICH, US); Mpio. Nayar, 12 km NW of Linavista, camino a Santa Teresa, 22°27'W 104°54'W, 2200 m, 26 Sep 1989, *Tenorio et al. 16471* (MO). **Sinaloa:** Mpio. Concordia, 1–2 km N of the Mazatlán–Durango road at Liberas, (23°20'N, 106°W), 1850 m, 2 Oct 1985, *Bartholomew et al. 2546* (CAS); Mpio. Badiraguato, Sierra Surotato, 3 mi N of Los Ornos along road to Ocurahuí, (25°30'N, 107°25'W)?, 1981 m, 2 Oct 1970, *Breedlove et al. 18383* (MICH); Mpio. La Concordia, 6–8 km SE of El Palmito on Hwy 40, 2000 m, 11 Sep 1979, *Breedlove 43853* (CAS, MO, TEX); Mpio. Concordia, Loberas Microwave Station, 2000 m, 18 Oct 1983, *Breedlove et al. 58854* (CAS, MO); Tres Hermanos, Concordia, 600 m, Sep 1919, *Dehesa 1548* (US); 9.9 mi SW of El Paraíso, on road between Villa Unión and El Salto, 1981 m, 27 Sep 1953, *Ownbey et al. 1905* (F, GH, MICH, UC, US); Cerro de la Sandía, NE of Panuco, 1800–2000 m, 29–30 Aug 1935, *Pennell 20057* (GH); Cerro Quemado, NE of Panuco, 1400–1500 m, 30 Aug 1935, *Pennell 20145* (GH, US); El Batel, 70 km NE of Mazatlán, 1554 m, 21 Oct 1946, *Pitelka 268* (UC). **Sonora:** 8 km NW of Trigo on road from Yecora–Sahuaripa, (28°30'N, 109°W), 1280 m, 25 Aug 1984, *Breedlove 61087* (CAS); 3 km S of La Lobera, Ejido Zahuarivo, Mpio. Alamos, 27°09'N, 108°58'W, 1550 m, 25 Aug 1986, *Tenorio et al. 11926* (MO). **Zacatecas:** Plateado, road from Colotlán, (21°50'N, 103°25'W)?, 31 Aug 1897, *Rose 2696* (US); aproxim. 38 km al O de Jalpa, sobre la carr. a Tlaltenango, 2550 m, 21–23 Oct 1973, *Rzedowski et al. 1022* (MICH, MO).

Habitat.—This species grows in open grassy areas of pine, oak, or pine-oak forests, along streams or in moist open flat areas. It is reported to be associated with *Acacia, Arctostaphylos, Alnus, Arbutus, Buchnera, Cacalia, Conyza confusa, Crusea longifolia, Eryngium lemmonii, Juglans, Juniperus, Mahonia, Molinadendron, Penstemon barbatus, Piptochaetium, Polyanthes, Psacalium pachyphyllum, Spiranthes aurantiaca, Stipa* and *Styrax argenteus*. The soils are black, yellow, red or brown and are usually rocky and gravelly, often shallow and clayey, derived from calcareous, volcanic or sedimentary rock.

Common names.—Frijolillo or Jícama Cimarrona or Jopita.

Comments.—The appearance of the plant, partly because of the short petiolules, is a bit like that of a clover with entire leaflets. The species blooms in Aug.-Sep.-Oct. with upright buds and very showy and quite large flowers (see Color Plate II, photo 13). As aptly explained by Maréchal et al. (1978b) and Lackey (1983), this species probably represents the evolutionary tendency of the section towards an end product of reduction of the inflorescence to a single terminal node producing only two flowers, borne erect for pollination by an unknown pollinator, perhaps a moth or butterfly. The terminal part of stems often does not develop further, with a well-developed raceme at one of the last axils, making the stems looking determinate. The roots are spherical to somewhat elongate and quite typical for the section (see Color Plate III, photo 31). A few collectors (e.g. Fuentes, Tenorio) have reported white flowered plants mixed with the normally colored ones.

Distribution seems to be mainly NW and W central México with only a single collection from Oaxaca, none from Chiapas and only one from W Guatemala.

Section G.—Xanthotricha Delgado stat. & sect. nov. PhD thesis 1985 (unpublished). TYPE SPECIES: *Phaseolus xanthotrichus* Piper, Contr. U.S. Natl. Herb. 22:698–699. 1926.

Herbae parvae scandentes volubiles indeterminatae, radix plerumque parva perennis crassa carnosa, foliolae parvae ad basim aliquantum lobatam, flos albus caesius vel lilacinius, bracteolae parvae ad minutae caducae, carina bispiralis primitus ad basim et magnam segundis parvis et ad apicem 2-spiralem, stigmate laterali introrso, interdum cleistogamo differt.

Small, indeterminate climbing vines; root usually perennial, small, thick, fleshy; leaflets small, usually somewhat lobed at base; flower white, light blue or violet, the keel of two coil types (1 large basal and 2 small terminal coils), the stigma lateral, introrse; at times cleistogamous.

Comments.—A special keel, unique to this section, consists of one large basal coil at the tip of which are two smaller coils (see Color Plate II, photo 15). According to Delgado (1985) these two series of coils spiral in opposite directions, however we find that all the coils spiral in the same direction (counter-clockwise when viewed towards the apex). These are twisted so that the stigma emerges near the base of the first coil, and provides an exceptionally sensitive mechanism for the exsertion of the stigma by a very light downward pressure on the wings. The species also produce copious quantities of powdery pollen under favorable cool and dry weather conditions with the result that most flowers produce pods if tripped by insects. If not tripped, some species will also produce selfed pods, and if temperatures are high many flowers will also become cleistogamous, but under these conditions many flowers may not produce pods. Lackey (1983) only recognizes *P. xanthotrichus* and includes it within section *Minkelersia*.

Allozyme data have shown a similarity among species of the section (Jaaska 1996). ITS DNA sequencing data (Delgado et al. 1999; Gaitán et al. 2000) confirm that similarity, also a possible relationship with *P. chiapasanus*.

KEY TO SPECIES

1. Terminal leaflets mostly small, 2–5 cm long.
   2. Terminal leaflets narrow lanceolate to nearly linear, 2.5–5 cm long; flower dark rose-violet; rare, in central México: E San Luís Potosí and bordering Hidalgo; 1550–2100 m ______ G.4. **P. gladiolatus**
   2. Terminal leaflets broader.
      3. Terminal leaflets mostly entire to slightly lobed.
         4. Terminal leaflets ovate to lanceolate, 4.5–5 cm long; flower purplish; seed oblongoid; scarce, in W central México; 1000–2130 m ______ G.2. **P. hintonii**
         4. Terminal leaflets ovate-rhomboid, entire to very slightly lobed, 2–3 cm long; flower violet; seed slightly angular; scarce, in central and S México, S to Guatemala and Costa Rica; 1400–2430 m) ______ G.1. **P. xanthotrichus**
      3. Terminal leaflets often deeply lobed though some may be entire, ovate to narrowly lanceolate, 2.5–5 cm long; flower purplish (lilac); scarce, in central México; 890–2460 m ______ G.3. **P. zimapanensis**
1. Terminal leaflets usually larger, 5–9.2 cm long.
   5. Terminal leaflets ovate-rhomboid with large broad lobes at base; seed large and usually solid black, round, lenticular, 3.7–4.9 mm long; scarce, in W central México; 2130 m ______ G.5. **P. magnilobatus**
   5. Terminal leaflets with pronounced narrow basal lobes, broad lanceolate; flower white; rare, known only from Chiapas; 1080 m ______ G.6. **P. esquincensis**

G.1.—Phaseolus xanthotrichus Piper, Contr. U.S. Natl. Herb. 22:698–699. 1926. (**Figs. 55, 61**). TYPE: GUATEMALA. JALAPA: Laguna de Ayarza, (14°25'N, 90°05'W), 2438 m, Oct 1892, *Heyde et Lux 4171* (in part) (HOLOTYPE: US 247544).

Aerial shoot an herbaceous, usually small, slender, climbing, indeterminate vine, 2–3 m long. ***Root*** a perennial, thick, fleshy, nearly spherical, 2.5–3 cm in diam. ***Stems*** terete; internodes covered with yellow-brown, reflexed-pilose hairs. ***Stipules*** broadly triangular-ovate, 3 mm long, 1.5–2 mm wide, prominently 7-nerved, acutish, sparsely covered by strigose hairs adaxially, glabrous abaxially. ***Leaves*** 5–11 cm long; petioles 2–2.5–6 cm long, covered with pilose hairs; petiolule 7–12 mm long; lower pair of stipels ovate, 1-nerved, the upper ones much narrower; terminal and lateral leaflets similar, ovate-rhomboid to triangular, 2–3 cm long, broadly rounded at base to truncate, 3 main veins and veinlets

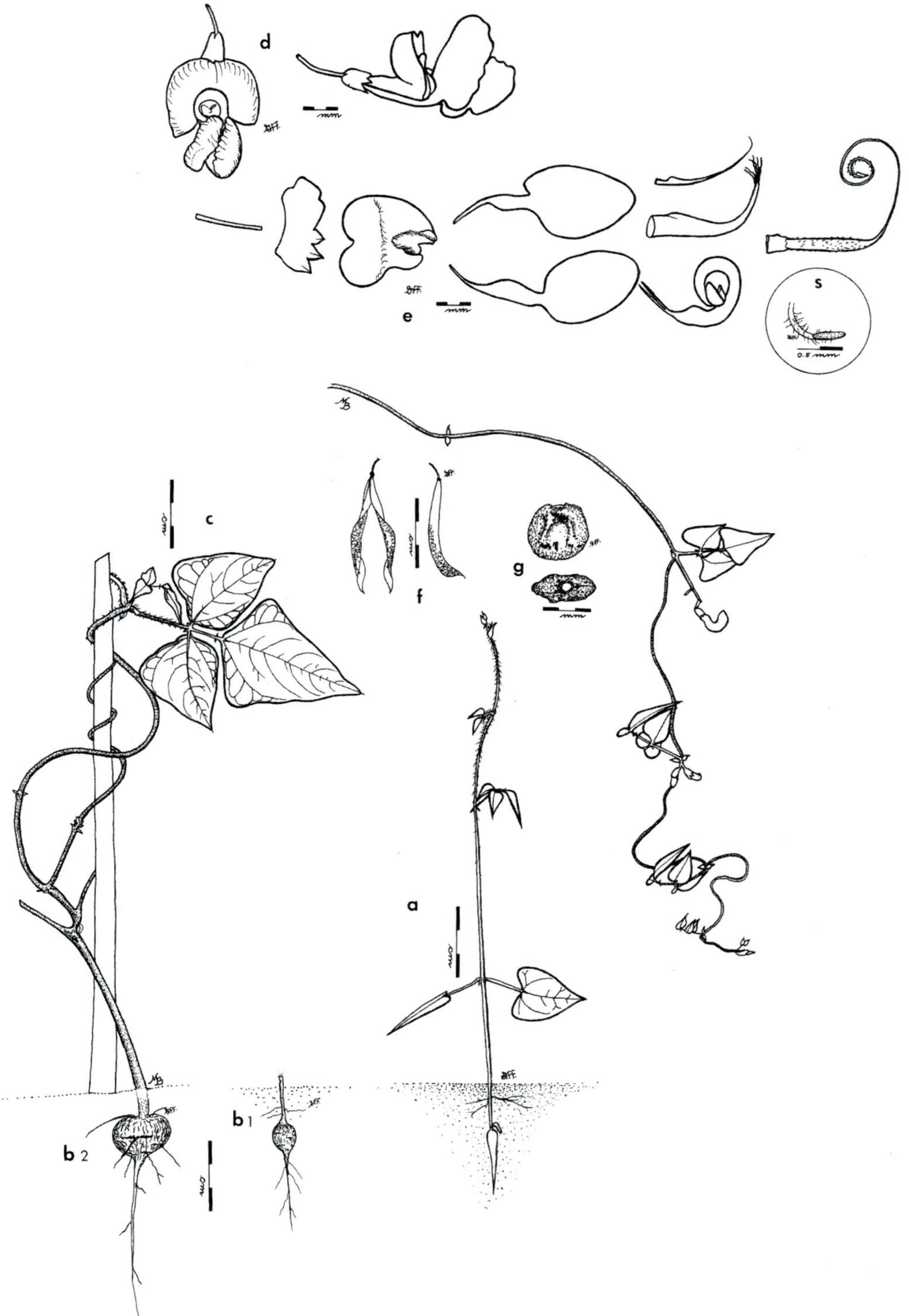

FIG. 55. Illustrations of *Phaseolus xanthotrichus* Piper.—**a.** Seedling a few weeks after germination.—**b.1.** Root after 1 year.—**b.2.** Root after several years.—**c.** Plant, basal portion with mature leaf, and, separately vine tip with small inflorescences; note triangular shape of leaflets.—**d.** Flowers, side view and lateral view.—**e.** Exploded view of flower showing all parts, including—**s.** Style tip and stigma as seen under the microscope; note two types of coils in keel tip.—**f.** Pods, side view and dehisced.—**g.** Seeds, side view and view from hilum. All drawings made from living material grown in greenhouse at Mayaguez from seed of *Debouck et al. 1604* (TARS #375) collected near Sta. María de Jesús, Guatemala, except a. Seedling, from seed of *Debouck et al. 1620* (TARS #392) collected near San Miguel Dueñas, Guatemala, and **b.2.** Old root, from seed of *Kretchmer 738* (TARS #388) collected from Costa Rica.

distinct, acuminate to obtuse, apiculate, membranous, covered with strigose-pilose hairs on both surfaces but more densely near the margins adaxially; lateral leaflets similar but inequilateral and more rounded. ***Inflorescence*** a short pseudoraceme loosely 6- to 10-flowered; peduncle slender, 2-9 cm long, covered with pilose hairs; rachis 3-6 cm long of 3-6 nodes; primary bract elliptic, 2 mm long, 1- to 3-nerved, acute, ciliate, covered with a few strigose hairs; pedicel slender, 5 mm long. ***Bracteoles*** minute, early deciduous, spatulate, 0.5-0.75 mm long, hyaline, 1-nerved, glabrous, early caducous. ***Flower*** light to dark violet; calyx flaring campanulate, 3 mm long, the upper lip very short, emarginate, the lower 3 lobes subequal, rounded dentate, 0.75-1 mm long, 0.75-1 mm wide, the median acute, sparsely bearded, with pilose hairs especially on upper portions and lower central lobe; standard violet, elongate-orbicular, claw 0.5 mm long, the blade 13 mm long, reflexed at 5 mm from base and 4 mm more to broadly emarginate apex, 7-9 mm wide and broadly rounded, without auricles; wings violet, the blade oblong, enrolled, 7-9 mm long, 3-6 mm wide, the claw narrow, 4 mm long, the spur poorly developed, 0.5 mm diam.; keel narrow tubular, the claws 4 mm long and 2 mm more to bend and first large nearly complete coil to the right of about 3-4 mm diam. at the end of which are the second smaller 1 1/2 coils of 1.5-2 mm diam.; stamen tube straight, the ridges scarcely developed, 5 mm to bend and 4-5 mm more to the divided stamens; vexillary stamen with distinct knob 0.5 mm high at about 1 mm from base, tapering about 3 mm more to filament; ovary straight linear, 4-6 mm long, 0.5-1 mm wide, glabrous to puberulent; stigma terminal, lateral oblique. ***Pod*** linear, falcate, 4.5-5 cm long, 5 mm wide, compressed; valves reticulate veined, glabrous; beak short, few to 8-10 seed. ***Seed*** slightly angular, flat-squarish to lenticular, 3-4.2 mm long, 2.8-3.6-4.3 mm wide, 1.3-1.9 mm thick, speckled yellowish-brown and/or black on brown, shiny, black ring around hilum; hilum round, 0.8 mm diam. ***Seedling*** from hypogeal germination; hypocotyl not elongated; epicotyl 4-5 cm long; the next internode, 3.5-5-8 cm long, about half as long as that of the following of 8-9-11.5 cm long; eophyll ovate, the petiole 1.25-2.5 cm long, the blade 2-3.5 cm long, 1.5-2 cm wide, the base truncate, acuminate.

Specimens examined: **COSTA RICA. Alajuela:** camino a San Juan Norte, 2 km N desvío ruta al Volcán Poás, 10°08'N, 84°14'W, 1550 m, 9 Jan 1987, *Debouck et al. 2105* (CR, US). **Cartago:** environs de Cartago, Versant atlantique, (9°35'N, 83°55'W), 1417 m, Oct 1894, *Biolley 8999* (US); barranco Río Reventado, 0.3 km W de Turbina, 9°53'N, 83°55'W, 1550 m, 7 Jan 1987, *Debouck et al. 2090* (CR). **Heredia:** San José de la Montaña, 4 km NNE de San José de la Montaña hacia Sacramento, 10°05'N, 84°07'W, 1720 m, 3 Feb 1998, *Debouck et al. 3104* (CR). **San José:** Alajuelita, pie del Cerro San Miguel, 9°52'N, 84°07'W, 1620 m, 8 Jan 1987, *Debouck et al. 2101* (CR); al pie de la Piedra de Aserrí, 1.5 km E de Aserrí, 9°52'N, 84°07'W, 1560 m, 11 Jan 1987, *Debouck et al. 2112* (CR, UC, US); Frailes, 2 km E de Frailes hacia Sn Cristobal Sur, 9°45'N, 84°02'W, 1700 m, 4 Feb 1998, *Debouck et al. 3109* (CR, MO).

GUATEMALA. Chimaltenango: 2.5 Km E of Acatenango, 14°33'N, 90°56'W, 1790 m, 7 Dec 1985, *Debouck et al. 1623* (K, US, USCG); 6 km N de San José Poaquil, 14°52'N, 90°54'W, 1970 m, 12 Dec 1987, *Debouck et al. 2449* (MICH, US, USCG); Finca El Rosario, approxim. 8 km de Chimaltenango, 14°42'N, 90°47'W, 1660 m, 13 Dec 1987, *Debouck et al. 2454* (COL, G, SI, UC, US, USCG). **Guatemala:** 2 km NE de Concepción Las Lomas (al oriente de Cd. Guatemala), 14°37'N, 90°30'W, 1470 m, 17 Dec 1987, *Debouck et al. 2470* (MICH, US, USCG); 5 km E de San José Pinula. 14°32'N, 90°23'W, 1690 m, 17 Dec 1987, *Debouck et al. 2476* (BR, MICH, US, USCG). **Sacatepéquez:** 10 km S of Santa María de Jesús, 14°27'N, 90°42'W, 1560 m, 4 Dec 1985, *Debouck et al. 1609* (US, USCG); 5 km W of San Miguel Dueñas, 1820 m, 6 Dec 1985, *Debouck et al. 1620* (BR, COL, UC, US, USCG). **Santa Rosa:** orillas lado N de la Laguna de Ayarza, San Rafael Las Flores, 14°26'N, 90°08'W, 1430 m, 7 Dec 1987, *Debouck et al. 2438* (MICH, US, USCG). **Sololá:** 1.8 km N de Panajachel, 14°46'N, 91°10'W, 1640 m, 10 Dec 1985, *Debouck et al. 1630* (US, USCG).

MÉXICO. Chiapas: Mpio. Las Margaritas, 8 km E of Las Margaritas along road to La Soledad, (16°20'N, 91°50'W), 1700 m, 15 Sep 1974, *Breedlove 37930* (CAS-DS); Mpio. de Teopisca, S edge of Teopisca, (16°30'N, 92°30'W), 1798 m, 13 Oct 1965, *Breedlove et al. 13078* (BM, CAS-DS, F, MEXU, MICH).

Habitat.—This species is found growing in clearings in forests of pine, oak, pine-oak, or juniper with undergrowth of composites, Bambusoideae, *Dahlia*, *Desmodium*, *Ipomea*, Labiatae, *Rubus*, Solanaceae, *Stizolobium*, *Vigna* and grasses. Soils are friable with much humus and are dark brown, rocky and sandy, often derived from volcanic ash or basalt (see also Araya et al. 2001).

Diseases and insects.—The leaflets are damaged by anthracnose and leafminers, *Diabrotica* beetles, and caterpillars, and birds attack the pods.

Comments.—The type at US is apparently the only sheet of this species with this collection number since duplicates at F and GH and two others at US are of *P. lunatus* under which name the collection was distributed. Piper (1926) did not note the unusual characteristics of the double spiral of the keel of the flower, possibly because the specimens he examined were poorly pressed. He indicated the standard as being sharply reflexed, which it is not, and the keel with 2 1/2 close coils (true of the apex), but does not mention the 1st large basal coil, first reported by Delgado (1985), at the end of which are the 1 1/2 to 2 smaller coils. Piper also gives no flower color, which might have been a useful diagnostic characteristic since it is more violet than purple in fresh material.

G.2.—Phaseolus hintonii Delgado, Syst. Bot. 25:431–432. 2000. (**Figs. 56, 61**). TYPE: MÉXICO. MÉXICO: Mpio. Tejupilco, Dist. Temascaltepec, Cañitas-Salitre, (18°55'N, 100°05'W), 1000–1350 m, 8 Oct 1934, (locality and date from Hinton & Rzedowski 1972), *Hinton 6728* (HOLOTYPE: K; ISOTYPE: GH).

Similar to *P. xanthotrichus* except as follows: ***Aerial shoot*** a long, scandent and climbing, annual, indeterminate vine. ***Root*** a perennial, thick, fleshy, somewhat oblongoid, 5–8 cm long, 5.5 cm wide, 1 cm wide at crown. ***Stems*** slender, branched in upper portions; internodes 10–18 cm long, densely covered by long, yellow, reflexed-strigose hairs. ***Stipules*** broadly ovate-triangular, 3.5 mm long, 1.75 mm wide, strongly 6- to 7-nerved, acute, a few hispid and strigose hairs especially on margin. ***Leaves*** 4–6.7 cm long; petiole 2–2.5 mm long, covered by yellowish strigose hairs; petiolule 1 mm long, sparsely covered by yellowish strigose hairs; stipels ovate, 1.5 mm long. 0.75 mm wide, acute, strongly 2-nerved, hispidulose to nearly glabrous adaxially; terminal leaflet near base of plant ovate to triangular-ovate, upper leaflets narrowly triangular to somewhat rhomboid, 2–5 cm long, 2–2.7 cm wide at base, often slightly lobed, especially in the upper leaves, acute, apiculate, variegated or not, hispidulose adaxially, glabrous abaxially. ***Inflorescence*** a long pseudoraceme exceeding foliage; peduncle slender, 2–6 cm long, densely covered by minute yellowish uncinate hairs; rachis 6–9 cm long of 6–8 flowering nodes, covered with minute yellowish uncinate hairs; primary bract oblongate, 2.5 mm long, 0.9 mm wide, strongly 3-nerved, acute, puberulent hispidulose and ciliate; pedicel very delicate, 5–8 mm long, very sparsely pubescent of very minute hooked hairs 0.1 mm long. ***Bracteoles*** minute and scale-like, early dehiscent, ovate, 0.5–0.75 mm long, 0.3 mm wide, 1 inserted at base of calyx and the other 0.25–1 mm below, indistinctly 1-nerved, acute, hyaline, ciliate. ***Flower*** lavender-purplish with a somewhat elongate shape at anthesis fading yellowish green; calyx tubular-campanulate, 3.5–4.5 mm long, the upper 2 lobes united into 1 rounded emarginate, 0.5 mm long, 2.5–3 mm wide, the lower 3 subequal, dentate, acute, 0.8–1.25 mm long, 1 mm wide, sparsely covered with dark, curved hispidulose hairs; standard cream to yellowish-green, round, emarginate, 9–10 mm long, cupped forward over the keel, lateral edges not enrolled, the claw 0.75 mm long, the auricles vestigial, not thickened at midpoint, waxy, lustrous; wings lavender-purplish, the blade oblong-spatulate, enrolled lengthwise, 6–7 mm long, 4 mm wide, the claw 3–4 mm long, the spur well-developed 0.5 mm in diam.; keel tube fairly (2 mm) broad, about 3 mm long to first large coil to the right (counter clockwise) and about 3 mm diam., followed by 2 1/2 smaller coils to the right, about 1.75–2 mm in diam., the claws separated 1.75 mm long, cream to greenish, green tip; stamen tube 4 mm long to bend and 3 mm more to divided filaments, minutely evident ridges; vexillary stamen, the claw 1.5 mm long to base of the slightly thickened central portion about 1.5 mm long and tapering 3 mm more to the filament; ovary slightly curved, 4–5 mm long, 1 mm wide, covered with fine white pubescence with a few dark hispid hairs at base, 11–12 ovules; style 6 mm long to the terminal coil; stigma lateral, lanceolate, 0.8 mm long. ***Pod*** linear, nearly straight to slightly falcate, subterete 3.5–4.5 cm long, 3–3.5 mm wide, pendent at maturity; valves chartaceous, glabrous, twisting several times at dehiscence. ***Seed*** oblongoid, flattened to lenticular, 2.4–3–4.5 mm long, 1.8–4.3 mm wide; 1.2–2.5 mm thick, speckled black on brown, shiny, a black ring around hilum; hilum oblong, 0.4 mm long. ***Seedling*** from hypogeal germination; epicotyl 4–5.5 cm long; the next internodes subequal (4–8.3 cm long); eophyll petiole 1 cm long, the blade simple, entire, nearly triangular, 2 cm long, 1.2–1.7 cm wide, truncate base, acute.

Specimens examined: **MÉXICO. Durango:** microondas San Pedro, Fco. I. Madero, 24°22'N, 104°19'W, 2130 m, 7 Oct 1978, (grown in screenhouse at Mayagüez, PR from seed of TARS #120), *Debouck et al. 314* (in part) (CHAPA, US); El Troncón, camino a San Juan, Mezquital, 23°30'N, 104°26'W, 1940 m, 28 Oct 1978, *Debouck et al. 391* (in part) (BR, CHAPA, K, MICH, US) (grown in screenhouse at Mayagüez, PR from seed of TARS #119B). **Jalisco:** near Veludero, 19°45'N, 103°10'W, 1650 m, 2 Dec 1981, (grown in screenhouse at Mayagüez, PR from seed of TARS #132B), *Freytag et al. 81-22 B* (in part) (US); near Puerto de los Mazos on road Autlán-Barra Navidad, 19°30'N, 104°30'W, 1110 m, 4 Dec 1981, (grown in screenhouse at Mayagüez, PR from seed of TARS #133-1 & 2), *Freytag et al. 81-26* (US). **Nayarit:** Mts. 10 mi SE of Ahuacatlán, on road to Barranca del Oro, 1100–1300 m, (21°N, 104°30'W), 11–12 Aug 1959, *Feddema 333* (MICH).

FIG. 56. Illustrations of *Phaseolus hintonii* Delgado,–**a.** Seedling a couple of weeks after germination.–**b.1.** Root after 1 year.–**b.2.** Mature root after several years.–**c.** Plant portions showing separately lower stem with mature leaf, mid-plant stem with mature leaf and inflorescence, and upper vine tip with separate mature leaf.–**d.** Flowers, side view and front view; note clasping nature of standard.–**e.** Exploded flower showing all parts, including–**s.** Style tip and stigma as seen under the microscope; note enrolled nature of standard and wings and thickened keel with two types of coils at tip.–**f.** Pods, side view and dehisced.–**g.** Seeds, side view and view from hilum. All drawings of living material grown in greenhouse at Mayagüez from seed of *Freytag et al. 81-22* (TARS #132B) collected near Veludero, Jalisco, México.

Habitat.—This species is found growing in loam soils and in partial shade on precipitous wooded slopes of pine or oak.

Comments.—It seems to grow in mixed associations with other species of the section, but is fairly scarce or at least has been relatively poorly collected throughout its range. It is quite distinct from the others by having small leaves, fairly broad throughout the length and with minor lobing at the base. The flower and seed are very similar to the other species of the section (see Color Plate II, photo 14 and Color Plate IV, photo 44).

G.3.—**Phaseolus zimapanensis** Delgado. Syst. Bot. 25(3):432–435. 2000. (**Figs. 57, 61**). TYPE: MÉXICO. HIDALGO: Mpio. Zimapán, a 6 km NE de Zimapán, La Placita, 10 km al SO de Jacala, nopalera con *Prosopis*, (20°45'N, 99°30'W), 1820 m, 7 Sep 1979, *Delgado 1113* (HOLOTYPE: MEXU n.v.; ISOTYPES: CHAPA n.v., TEX-LL n.v.; others to be distributed).

Similar to *P. xanthotrichus* except as follows: ***Aerial shoot*** a small, prostrate and climbing, delicate, indeterminate vine, to 1–2 m long. ***Root*** thick, fleshy, fusiform to spherical, 6 cm long, 3 cm thick. ***Stems*** slender, terete, glabrous; internodes 2–12 cm long. ***Stipules*** broadly triangular, 2.5–3 mm long, 1–1.5 mm wide, 5- to 6-nerved, obtuse, hyaline margin, glabrous, stiffly extended and often revolute. ***Leaves*** 5.3–9 cm long; petiole 2–2.5 cm long, glabrous; petiolule 0.5–1 cm long; stipels narrow spatulate, 1–1.5 mm long, strongly 1-nerved; pulvini 1.5–2 mm long, covered by uncinate hairs; terminal leaflets near base of plant ovate to triangular-ovate, upper leaflets narrowly lanceolate to nearly linear, 2.5–5 cm long, 0.5–1–1.3 cm wide, often deeply basally lobed to 0.5–1 cm long especially in the upper leaves, distinctly variegated several shades of lighter green, dark green adaxially, lighter abaxially, glabrous, ciliate along margins; lateral leaflets similar but broader and inequilateral. ***Inflorescence*** 5–30 cm long. ***Flower*** purplish (lilac); keel with a double spiral, the first coil somewhat elongate and open. ***Seed*** lenticular to squarish or rounded, 2.5–3.2–3.4 mm long, 2.5–3.4 mm wide, 1.1–2.3 mm thick, speckled black (gray) on brown, black ring around hilum. ***Seedling*** from hypogeous germination; epicotyl 2–5.5 cm long; eophyll petiole 0.5–1.5 cm long, the blade simple, entire, triangular-ovate, 1.8–3.2 cm long, 1.5–2.5 cm wide, slightly cordate, acuminate, slightly variegated.

Specimens examined: **MÉXICO. Durango:** 25 mi S of Cd. Durango on road to Mezquital, Sierra Registro on road to Mezquital, (23°45'N, 104°20'W), 2027 m, 3 Nov 1966, *Gentry 22086* (US). **Hidalgo:** Mpio. Jacala, El Piñon, 15 km SO de Jacala en la carr. Mex 85 a Tamazunchale, 20°57'N, 99°13'W, 1870 m, 1 Nov 1986, *Debouck et al. 2038* (BR, CHAPA, MICH, MO, US, WIS); Km 185 de Mex 85 a Tamazunchale, 0.5 km antes de Puerto de Horcones, 5 km NE de Jacala, 21°N, 99°12'W, 1550 m, 1 Sep 1986, *Debouck et al. 2040* (in part) (BR, CHAPA, MO, US); Mpio. de Zimapán, 300–400 m al E del Hotel Posada del Rey, 6 km al E de Zimapán, 1900 m, 5 Sep 1979, *Hernández 3686* (CAS); La Placita, Mpio. Pacula, 33 km al NE de Zimapán, 1800 m, 7 Sep 1979, *Hernández 3718* (MEXU, MO); between P. Colorada and P. Zorra, at Km 285 from D.F., 19 Oct 1948, *Norvell et al. s.n.* (CAS); 4 km from Puerto de Piedra on road to Jacala, 1700 m, 30 Oct 1966, *Rudd 1047* (US). **Nuevo León:** 6 km W de Milpillas, camino de San José del Jilguero a Milpillas, Aramberri, 24°22'N, 99°49'W, 1780 m, 7 Sep 1985, *Debouck et al. 1504* (CHAPA, COL, MICH, UC, US); 2.5 km N de Santa Lucía en la terraceria hacia La Siberia, Doctor Arroyo, 23°47'N, 99°55'W, 1910 m, 9 Sep 1985, *Debouck et al. 1511* (CHAPA, MICH, SI, UC, US(2)); Arroyo de La Tijera, 3.0 km N de Las Jarillas, Doctor Arroyo, 23°46'N, 100°15'W, 2000 m, 13 Sep 1985, *Debouck et al. 1515* (CHAPA, US). **Queretaro:** 0.5 km SE de El Madroño, Landa de Matamoros, 21°16'N, 99°08'W, 1720 m, 2 Nov 1986, *Debouck et al. 2047* (CHAPA, MICH, US); barranco cerca de El Carrizal, camino hacia El Guayabo, 1 km desde entronque con Mex 120, Landa de Matamoros, 21°19'N, 99°27'W, 890 m, 3 Nov 1986, *Debouck et al. 2049* (BR, CHAPA, MO, UC, US); 10 mi S of Zimapán, 20 Aug 1957, *Waterfall et al. 14114* (BRIT). **San Luís Potosí:** Xilitla, carr. Mex 120 a Jalpán, 7 km SW de Ahuacatlán, 21°17'N, 99°06'W, 1430 m, 2 Nov 1986, *Debouck et al. 2045* (BR, CHAPA, MO, US); 2 km SW de Puerto de Lobos, Km 10 de Mex 80 a Ciudad Mante, Ciudad del Maíz, 22°27'N, 99°34'W, 1170 m, 13 Nov 1986, *Debouck et al. 2073* (BR, CHAPA, MO, US); Minas de San Rafael, Nov 1910, *Purpus 4855* (UC). **Tamaulipas:** 29 km SW de Ciudad Victoria, Km 150 de Mex 101 a San Luís Potosí, 23°36'N, 99°14'W, 1470 m, 12 Nov 1986, *Debouck et al. 2065* (BR, CHAPA, US); Miquihuana, 23°42'N, 99°45'W, 2460 m, 8 Aug 1941, *Stanford et al. 774* (ARIZ, GH, MO).

Habitat.—This species grows as an understory vine in semi-shady places in open oak or pine forest with *Acacia, Arctostaphylos*, cactus, composites, *Cupressus*, juniper, maguey, mints, and grasses. Soils are deep organic, often rocky and usually brown or red clays derived from igneous or limestone rocks.

Diseases and pests.—Angular leaf spot is reported as well as leaf damages from leafminers. *Apion* pod weevil damages the pods.

Comments.—This species differs from others of the section in that the terminal leaflets on the upper portions of the vine are deeply lobed at the base, inflorescences are quite long (to 30 cm) and flowers are pale violet. It is also limited in distribution to central and NE México (see also Rodríguez-Cabrera et al. 1987), extending into southern Durango (Delgado 1985; Nabhan 1990; and this study). The sample *Hernández 3718* (MO) is reported as of *Phaseolus aderens*. In his thesis, Delgado (1985)

Fig. 57. Illustrations of *Phaseolus zimapanensis* Delgado,–**a.** Seedling several weeks after germination.–**b.1.** Root after 1 year.–**b.2.** Mature root after several years.–**c.** Separate portions of plant including basal cluster of stems, mature leaves from base and middle of plant and upper leaf with vine tip; note lobing of all leaflets in upper portion of plant.–**d.** Flowers side and front views; note clasping standard.–**e.** Exploded view of flower showing all parts, including–**s.** Style tip and stigma as seen under the microscope; note very narrow and sharply bent keel.–**f.** Pods, side view and dehisced.–**g.** Seeds, side view and view from hilum. All drawings made from living material grown in greenhouse at Mayagüez from seed of *Debouck & Muruaga 2047* (TARS #399) collected near El Madroño, Queretaro, México.

recognizes this taxon as a new variety of *P. xanthotrichus*, without providing however a full description of it but the major differences with the type species. Jaaska (1996) after an isozyme study found a constant three-loci allozymic divergence between the two taxa, and thus raised var. *zimapanensis* to the species rank. Under the rule of anteriority, the correct name for this taxon should be *Phaseolus zimapanensis* (A. Delgado) Jaaska (see Debouck 1999).

G.4.—Phaseolus gladiolatus Freytag & Debouck, sp. nov. (**Figs. 58, 61**). TYPE: MÉXICO. SAN LUÍS POTOSÍ: Cerro 2 km SE de Realejo, cerca de "Minas de Plata," Guadalcazar, 22°39'N, 100°24'W, 2100 m, 7 Nov 1986, *Debouck & Muruaga 2057* (in part) (HOLOTYPE: US; ISOTYPES: CHAPA, UC).

Similis *Phaseolo xanthotricho*, sed foliolis terminalibus anguste lanceolatis vel fere linearibus, 2.5–5 cm longis, flore obscurio roseo-violaceo differt. Habitat in locis Nova Hispania centralis in provincis San Luis Potosium et Hidalgum rarus.

Similar to *P. xanthotrichus* except as follows: ***Plant*** a small prostrate and climbing, indeterminate vine, to 1–2 m long. ***Root*** a perennial, thick, fleshy, fusiform, 6 cm long, 4 cm thick, somewhat branched near crown. ***Stems*** slender, terete, glabrous; internodes 2–8 cm long; stipules broadly triangular, 2.5–3 mm long, 1–1.5 mm wide, 5- to 6-nerved, obtuse, hyaline margin, glabrous, stiffly extended and often revolute. ***Leaves*** 5.3–9 cm long, somewhat lobed especially in upper portions of plant; petiole 2–2.5 cm long, glabrous; petiolule 0.5–1 cm long; stipels narrow spatulate, 1–1.5 mm long, strongly 1-nerved; pulvini 1.5–2 mm long, covered with uncinate hairs; terminal leaflet narrow, lanceolate to nearly linear, acute, 2.5–5 cm long, 0.5–(1)–1.3 cm wide, slightly lobed at base to 0.5 cm long, dark green adaxially, lighter green abaxially, glabrous, ciliate along margins; lateral leaflets similar but broader and inequilateral. ***Inflorescence*** quite long to 6 cm. ***Flowers*** dark rose-violet; keel, the claws 4 mm long, the basal coil very elongate to open and twisted, the upper 2 coils 2–2.5 mm diam.; vexillary stamen with no appreciable knob; ovary slightly curved, 6 mm long, 1 mm wide, pubescent. ***Pod*** somewhat falcate, 4 cm long, 0.7 cm wide, purplish, twisting 2–3 times at dehiscence. ***Seed*** mostly lenticular to round, 3.8–5.4 mm long, 2.9–3.8 mm wide, 1.6–2.1 mm thick, black pinto on brown, black ring around hilum. ***Seedling*** from hypogeal germination; epicotyl 3–4 cm long, the next internode (4.5–5.5 cm long) about half the length of the subsequent internode (9.5 cm long); eophyll, the petiole 1 cm long, the blade ovate 2 cm long, 2 cm wide, base slightly cordate, acuminate.

PARATYPE. **MÉXICO. Hidalgo:** Km 185 de Mex 85 a Tamazunchale, 0.5 km antes de Puerto de Horcones, 5 km NE de Jacala, 21°N, 99°12'W, 1550 m, 1 Sep 1986, (grown in screenhouse at Mayagüez, PR from seed of TARS #396), *Debouck et al. 2040* (in part) (CHAPA, US).

Habitat.—This species is an understory vine found growing on deep organic, igneous rock derived soil, in semi-shady places in pine forests with oak and *Arctostaphylos*.

Comments.—It is very rare in N Hidalgo and E San Luís Potosí and growing mixed with other species of the section. Nevertheless, the small, narrow, sword-shaped terminal leaflets characteristic of this species are very different from other species of the section and do seem to breed true to type in the greenhouse.

G.5.—Phaseolus magnilobatus Freytag & Debouck, sp. nov. (**Figs. 59, 61**). TYPE: MÉXICO. JALISCO: Km 69 antes del Veludero, Mazamitla a Masula, 19°45'N, 103°10'W, 1650 m, 2 Dec. 1981, *Freytag GF et al. 81-22 A* (in part), grown in screenhouse at Mayagüez, PR from seed of TARS #132A Sep.-Dec., 1985 as Study Collection, *Freytag GF #SC-132-A* (HOLOTYPE: US; ISOTYPES: ARIZ, CSU, F, GH, MO, TEX, UC, WIS).

Similis *Phaseolo xanthotricho*, sed foliolis terminalibus ovato-rhomboideis 5–9.2 cm longis cum lobis magnis latisque, seminibus lenticularibus 3.7–4.9 mm longis plerumque atris differt. Crecit in locis Nova Hispania centralis occidentalibus rarus.

Similar to *P. xanthotrichus* except as follows: ***Root*** oblongoid, thick, fleshy, 9–10 cm long, 5–6 cm wide. ***Leaves*** with large rounded lobes; terminal leaflets at base of plant broadly ovoid to pyramidal, upper leaflets ovate-rhomboid and distinctly lobed at base, 5–7 cm long, the lobes to about 1 cm long and rounded. ***Inflorescence*** to 16 cm long. ***Pod*** 4.5 cm long, 0.8 cm wide, slightly curved, flattened. ***Seed*** large, squarish, flattened, 3.7–4.9 mm long, 3.5–4.5 mm wide, 1.7–2.3 mm thick, solid black or pinto speckled black on brown, black eye around hilum. ***Seedling*** from hypogeal germination; a very long epicotyl, 8–10 cm long, next internodes subequal 4–6 cm long; eophyll petiole 1.5–2 cm long, the blade ovate-triangular, 3–3.5 cm long, 2–2.5 cm wide, slightly cordate at base, acuminate.

FIG. 58. Illustrations of *Phaseolus gladiolatus* Freytag & Debouck.–**a.** Seedling plant a few weeks after germination.–**b.1.** Root after 1 year.–**b.2** Mature root after several years.–**c.** Plant showing separately lower stem and mature leaf, mature leaves and an inflorescence with pods from the middle of the plant, and vine tip with inflorescences; note the long narrow terminal leaflets.–**d.** Flowers, side view and view from front.–**e.** Exploded view of flower showing all parts, including–**s.** Style tip and stigma as seen under microscope; note very large and broad standard and wings and the thick, erect keel with long claws.–**f.** Mature pods, side view and dehisced.–**g.** Seeds, side view and view from hilum. All drawings made from living material grown in greenhouse at Mayagüez from seed of *Debouck & Muruaga 2040* (TARS #396) collected near Jacala, Hidalgo, México.

Fig. 59. Illustrations of *Phaseolus magnilobatus* Freytag & Debouck.—**a.** Seedling plant a few weeks after germination.—**b.1.** Root after one year.—**b.2.** Mature root after several years growth.—**c.** Lower portion of plant with mature leaves and inflorescence with pods; note pyramidal and lobed shapes of terminal leaflets and separately a vine tip with leaves and inflorescences.—**d.** Flowers, front and side views.—**e.** Exploded view of flower showing all parts including—**s.** Style tip and stigma; note the very narrow and large first coil of keel.—**f.** Pods, side views and dehisced; note the reticulate veination of carpels.—**g.** Seeds, side view and view from hilum. All drawings made from living material grown in greenhouse at Mayagüez from seed of *Freytag et al. 81-22* (TARS #132A) collected near Veludero, Jalisco, México, except **b.1** Root, 1 year from seed of *Debouck & Acosta 403* (TARS #121) collected near Fco.I.Madero, Durango, México, and **b.2.** Mature root from seed of *Debouck 391* (TARS #119A) collected near Mezquital, Durango, México.

PARATYPES. **MÉXICO. Durango:** Cerro Alto, 1.5 km lado camino Castillo Najera, 3.5 km N of Carlos Real, 2020 m, 24°19'N, 104°27'W, 28 Sep 1978, *Debouck et al. 264* (CHAPA, COL, K, US); San Pedro, Fco. I. Madero, near microondas, 24°22'N, 104°19'W, 2130 m, 7 Oct. 1978, *Debouck et al. 314* (in part), grown in screenhouse at Mayagüez, PR from seed of NI 716 (= TARS #120-B), Sep.-Dec. 1985 as Study Collection, *Freytag GF #SC-120-B* (US); El Troncón, camino a San Juan, Mezquital, 23°30'N, 104°26'W, 1940 m, 28 Oct. 1978, *Debouck et al. 391* (in part; K, MO)), grown in screenhouse at Mayagüez, PR from seed of NI 707 (= TARS #119A, Sep.-Dec., 1985 as Study Collection, *Freytag GF #SC-119A* (CSU, US, WIS); near Cerro La Cantera, Las Huertas, Fco. I. Madero, 1950 m, 28 Oct. 1978, *Debouck et al. 403* (in part) (CHAPA, COL, K, US), grown in screenhouse at Mayagüez, PR from seed of NI 806 (= TARS #121) Sep.-Dec. 1985 as Study Collection, *Freytag GF #SC-121* (CSU, WIS, US).

Habitat.—This species is very scarce, growing in dry, open oak forest with understory of Agaves, cactus, composites and grasses. Soils are shallow, rocky and sloping.

Diseases and insects.—Foliage has been reported to be damaged by caterpillars, spider mites, and chrysomelid beetles.

Comments.—This species is very rare and there are no early collections. Apparently it grows only a short time in the xerophytic habitat and is a very small, inconspicuous vine among the brush. This species is easily recognized by the relatively large leaflets with large rounded lobes at the base. It seems to be found frequently with *P. hintonii* in the same areas, from which it is easily distinguished by not only leaflet shapes but also the seed which is a very small and somewhat oblongoid and pinto in *P. hintonii* and a larger, squarish and usually solid black seed in *P. magnilobatus*. The root is also a large and globose (see Color Plate III, photo 32).

G.6.—Phaseolus esquincensis Freytag, sp. nov. (**Figs. 60, 61**). TYPE: MÉXICO. CHIAPAS: 2-3 km on branch road to Gabriel Esquinca, from Km 41 on Hwy. Tuxtla Gutiérrez-Chicoasén, 16°55'N, 93°10'W, 1080 m, 25 Sep 1978, *Freytag & Vakili 78-Mex-75* (HOLOTYPE: US; ISOTYPES: ARIZ, BR, EAP, F, GH, IBUG, K, MEXU, MICH, MO, NA, UC, WIS).

Similis *Phaseolo xanthotricho*, sed foliolis terminalibus ovato-lanceolatis 5-9.2 cm longis lobis ad basem distinctum angustis, flore candido differt. Habitat in locis Chiapasanis centralis occidentalibus rarus.

Similar to *P. xanthotrichus* except as follows: ***Aerial shoot*** a strongly twining and climbing, apparently annual, indeterminate vine, to 3-4 m long. ***Root*** a perennial, somewhat fibrous to fleshy, narrow and elongate, to 2 cm thick. ***Stems*** at base of plant, terete, somewhat nerved, 1.75 cm thick at near ground level, fairly heavily covered with yellowish strigose hairs which are somewhat wavy, to 1.5 mm long, with more at nodes and nearly perpendicular to stem; maximum internode length 13 cm long on older portions. ***Stipules*** broadly foliaceous, nearly round, 4.5 mm long, 4.5 mm wide, strongly 9-veined, acute at tip, somewhat appressed forward, pubescent. ***Leaves*** 10.7-15 cm long; petioles 4.5 cm long; petiolules 2.5 mm long; basal pulvinus 6 mm long, lateral and terminal pulvini 3.5 mm long; terminal leaflet triangular to lanceolate, 5-9.2 cm long, 3-6.5 cm wide near base, the base distinctly hastately lobed, the lobe to about 13 mm long, the apex acute, apiculate about 1 mm long, older leaves fairly thickened, moderately pubescent on both surfaces of strigose and some hooked hairs; lateral leaflet similar but 7.4 cm long, 4.8 cm wide near base, slightly 1- to 2-lobed. ***Inflorescence*** a horizontal to upright multi-flowered raceme, to 20-40 flowers; peduncle 3-4.2 cm long; rachis to 13-15 cm long or more, heavily pubescent with mostly hooked hairs extending from axis in all directions with many perpendicular and somewhat yellowish; primary bract oblong-ovate, 3.5 mm long, 1 mm wide, strongly 3-nerved, slightly strigose, the secondary bracts to 1 mm long; pedicel 5-6 mm long, slightly short hooked pubescent. ***Bracteoles***, sessile, a few subsessile, hyaline, 1 mm long, 0.3 mm wide, 1- to 2-nerved, glabrous, caducous after anthesis. ***Flower*** white; calyx 4.5 mm long, the 2 upper teeth rounded 0.75 mm long, the central 1.5 mm long, 1 mm wide, the lateral 1 mm long, 1.25 mm wide, moderately pubescent; standard white to cream, 9.5 mm long, 8.5 mm wide, nearly round and very broad, slightly indented at tip and somewhat hooded over keel, somewhat green adaxially at tip, glabrous, the basal claw about 1.75 mm long, inconspicuous lateral glands; wings white, the blade nearly round, 7 mm long, 6.5 mm wide, somewhat rolled and extended to spreading, the basal claw 5 mm long, 0.25 mm wide, the spur 2.5 mm long, 1 mm wide, weakly adhering to stamen tube; keel with 1 1/2 coils, tip white, the basal claws split 2 mm long, the lateral knobs raised less than 0.5 mm; vexillary stamen 13 mm long, no basal knob, 0.25 mm wide; stamen tube united 9.5 mm; anthers 0.6 mm long, 0.3 mm wide, yellow; basal collar 0.75 mm long, 0.75 mm wide; ovary 5.75 mm long, 0.75 m wide, glabrous, 9 ovules; style, the thin basal portion 11 mm long, the

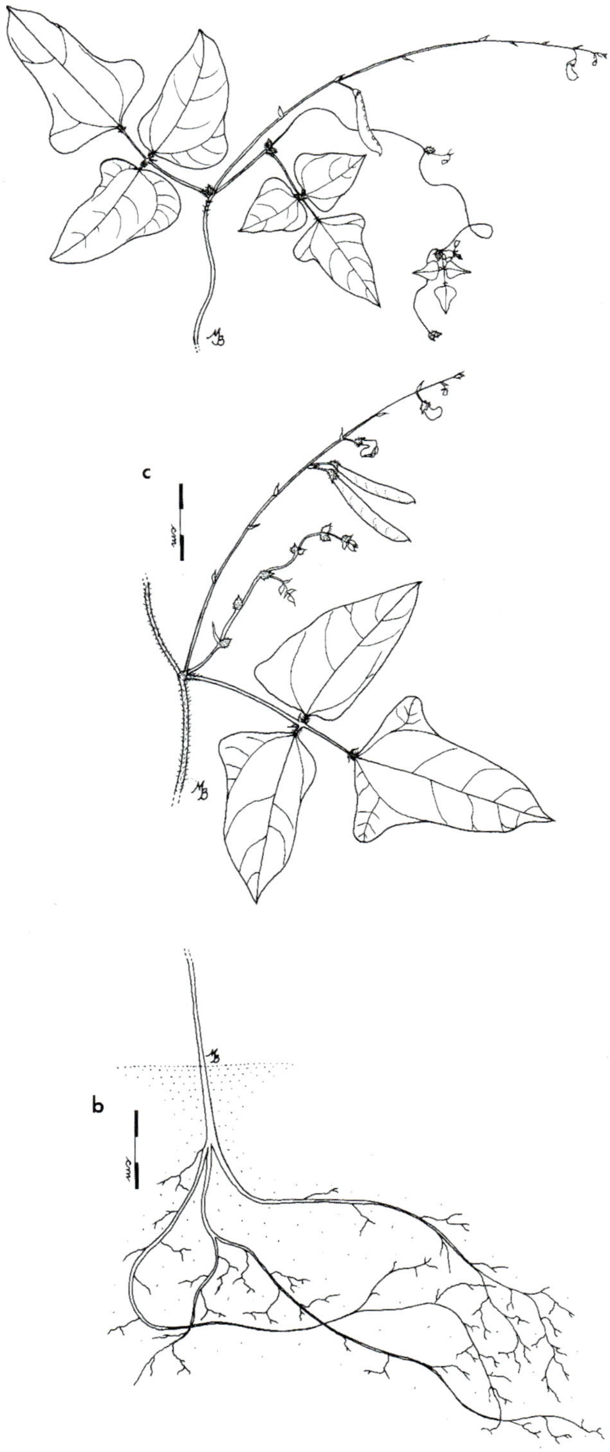

Fig. 60. Illustrations of *Phaseolus esquincensis* Freytag.—**b.** Root from young stem or plant.—**c.** Portions of vine from middle of plant with mature leaf and inflorescence and separately a vine tip with leaves and inflorescence; note the nearly hastate base of leaf with lateral lobes. All drawings made from type collection *Freytag et al. 78-Mex-75* collected near Gabriel Esquinca, Chiapas, México.

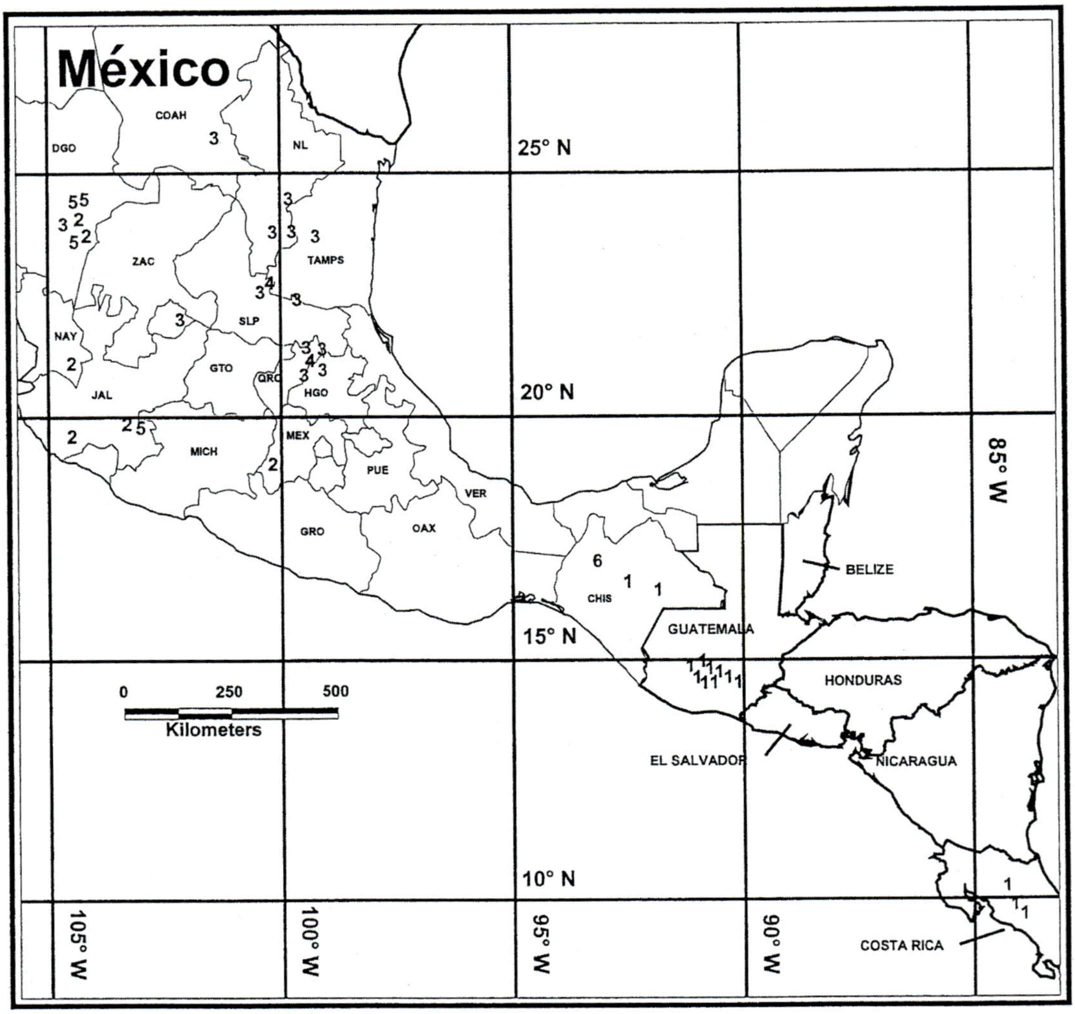

FIG. 61. Distribution of species of Section G. *Xanthotricha* as follows: **1** = *P. xanthotrichus*; **2** = *P. hintonii*; **3** = *P. zimapanensis*; **4** = *P. gladiolatus*; **5** = *P. magnilobatus*; **6** = *P. esquincensis*.

thickened terminal portion of 1 full coil, 1.75 mm in diameter; stigma lateral, introrse, straight, 0.75 mm long, narrow. ***Pod***, immature slightly hirsute becoming sparsely ciliate and thickly covered with minute glandular hairs 0.04 mm long and small hooked hairs 2 mm long; sutures becoming more pronounced in green mature pods and with fine reticular veining 0.5–1 mm size over pod surface, 4 cm long, 0.6 cm wide, 1 mm thick, nearly straight to slightly curved, the beak curved, 1.75 mm long. ***Seed*** unknown. ***Seedling*** unknown.

Habitat.—This species is found growing over shrubs and weeds on a nearly level stretch of road with cliffs and a steep slope nearby, at roadside, dense oak and shrub hills; soil is limestone with sand and humus between rocks.

Comments.—This taxon is named for the small town closest to the collection locality. Very rare as only one collection has been made, and an effort was made to locate the frequency and local distribution but no other specimens were found close by. This species has quite large leaves with pronounced basal lobes. It differs from all others of this section in apparently not having a fleshy, tuberous root and not having a keel with two types of coils.

Section H.—Revoluti Freytag, sect. nov. TYPE SPECIES: *Phaseolus leptophyllus* G. Don, Gen. Hist. 2:350. 1832.

Herba volubilis scandens parvissima, foliola parva 5–7 cm longae linearia marginibus revolutis, 3-venis venae parallelae costis laterale alatis, legumen angustum, 10-seminibus.

Plant a very small, climbing vine; leaflets small, 5–7 cm long, linear, with revolute margins, 3-nerved of equally long, parallel veins, the mid-rib laterally winged; pod narrow with 10 seeds.

Comments.—this section is named after Piper's (1926) epithet for the only species in this section which describes most accurately the outstanding characteristic now known for this species, that is, the leaflet margins which fold downward at the two lateral veins until they cover the under surface of the leaflet (see Figure 63). Presumably this is an adaptation for reducing water loss in the very xerophytic habitat of the species. The single species seems now to be extinct as no recent collections have been made since its first collection in 1789 in spite of determined efforts by botanists.

H.1.—Phaseolus leptophyllus G. Don, Gen. Hist. 2:350. 1832. (**Figs. 48, 62, 63**). *Phaseolus linearis* Sessé & Mociño, Pl. Nov. Hispaniae, ed 1:116. 1887. (not *P. linearis* HBK). *Phaseolus revolutus* Piper, Contr. U.S. Natl. Herb. 22:697–698. 1926. TYPE: MÉXICO. GUERRERO: in Chilpantzingi montibus, (17°24'N, 99°28'W)?, Jul 1789, *Sessé et Mociño s.n.* (HOLOTYPE: OXF (in Herb. Lambert) designated by Delgado (1985) n.v.; ISOTYPES: F (fragment), MA n.v., P n.v.).

Aerial shoot apparently a prostrate, indeterminate vine, to 1 m long. ***Root*** unknown. ***Stems*** slender, twining, terete, 1 mm diam., puberulent with spreading minute hooked hairs. ***Stipules*** triangular, 2–3 mm long, 1.5 mm wide, broadly triangular, acute, nearly glabrous. ***Leaves*** 7–9 cm long; petiole 1.5 cm long, nearly terete, glabrous; petiolule 3 mm long; pulvini 1.5 mm long, sparsely covered with hooked pubescence; stipels linear, 1 mm long; terminal and lateral leaflets similar, linear, 5–7 cm long, 2.5–4 cm wide, obtuse and minutely short-apiculate, 3-nerved, the 2 lateral veins running parallel to the mid-rib but at 0.5–1 mm from the leaf margin, the narrow margin revolute and closely appressed to the abaxial surface of the blade when wilted or dried, the blade dark green adaxially and very sparsely covered with short hispid pubescence, light green and puberulent abaxially, the midrib with a 2 lateral narrow wings 0.25 mm wide its whole length, veinlets very sparse. ***Inflorescence*** a pseudoraceme to 16 cm long; peduncle 8 cm long, terete, puberulent; rachis 7 cm long of 2–8 flowers; primary bract ovate, 1–2 mm long, acute, puberulent of minute hooked hairs; pedicel 4 mm long in bud becoming 10 mm long in fruit, puberulent of minute hooked hairs. ***Bracteoles*** scale-like, one inserted at base of calyx and the other 0.75 mm below, ovate, acute, 0.7 mm long, 0.4 mm wide, glabrous, ciliate, early caducous. ***Flower*** 12 mm long; calyx campanulate, 5 mm long, the upper 2 teeth united into one, rounded, emarginate, 1.25 mm long, the lower 3 subequal, ovate rounded-dentate, acute, 1.25 mm long, 1.25 mm wide, the tube, puberulent of minute hooked hairs, ciliate at margins; standard orbicular, 8 mm wide, in mature bud sharply reflexed, 3 mm to bend and 4 mm more to tip, stipitate, thickened, a pair of intramarginal auricles near the base; wings oblong, stipitate, a short lobe on the upper basal angle; keel narrow, stipitate, 4 mm to bend, the tip with two complete coils of 2.5 mm in diam.; (additional flower structures are unknown since the fragment seen was not dissected). ***Pod*** (immature) linear, 2 cm long, 1.5 mm wide, densely covered by hispid and short, white strigose hairs, 10 ovules. ***Seed*** unknown. ***Seedling*** unknown.

Habitat.—This species was apparently collected somewhere near Chilpanzingo, Guerrero, but has not been collected since the original specimen was obtained, in spite of field collecting by botanists. Judging from the leaflet morphology, this taxon must have originated in extremely dry, desert locations such as still exist in the area from Iguala to Chilpanzingo, but may be very scarce, if not extinct, due to the heavy grazing of goats and other animals in this area.

Comments.—Apparently Piper (1926) only saw the specimen at Paris with a label indicating it came from the Herbarium of Pavon, 1898, No. 68 and which led him to believe the specimen came from Peru. Standley borrowed the complete Sessé & Mociño collection from Madrid and apparently was permitted to retain fragments for the Field Herbarium, one of which perhaps was of the original collection of *P. leptophyllus* and which shows on the label "Herb. Horti Botanici Matritensis #3715." According to the itinerary of the 2nd Expedition of Sessé and Mociño in 1789, as given by McVaugh (1977) from a study of the voucher payment records in Mexican archives, the expedition was at the following localities on the indicated days:

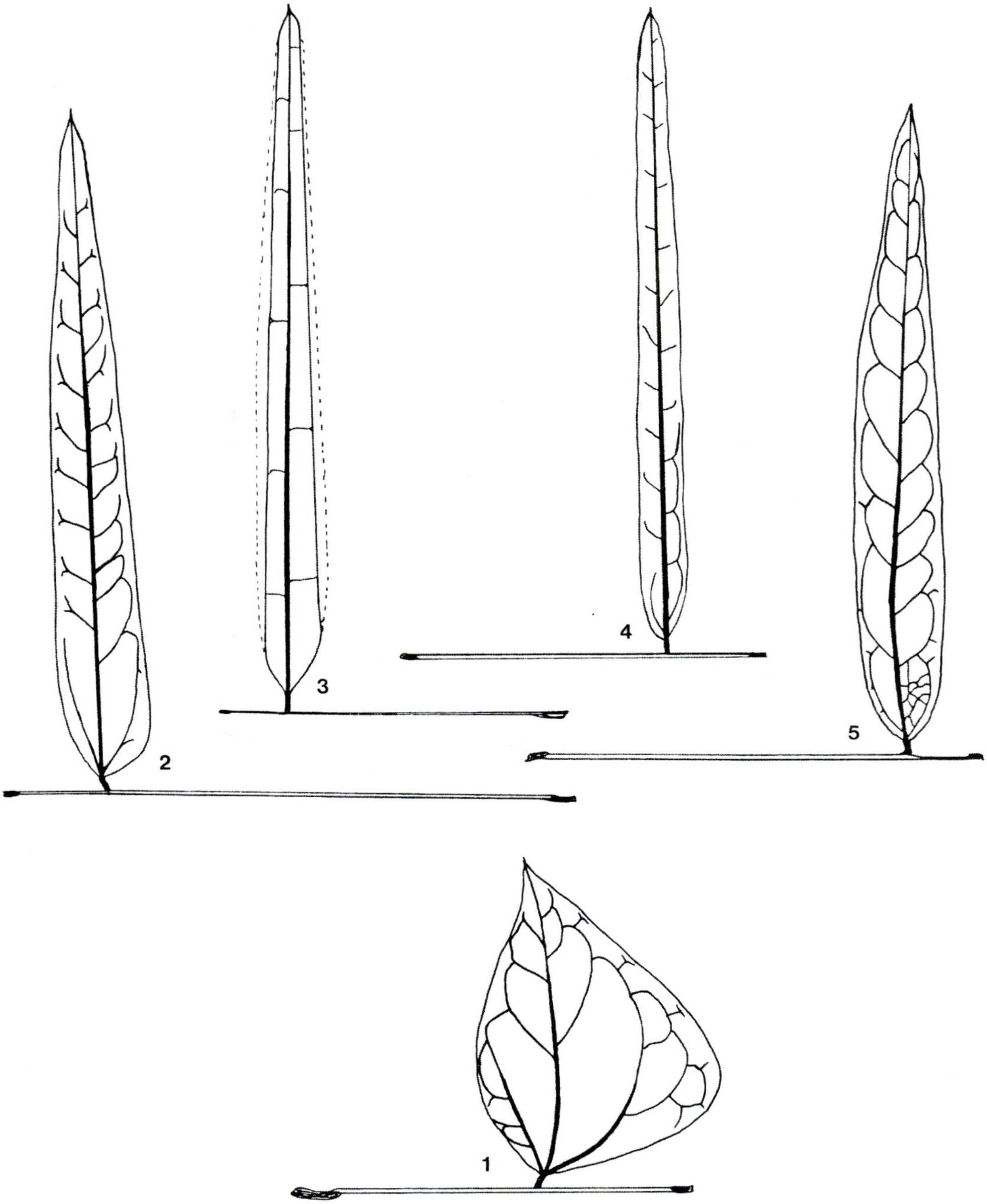

Fig. 62. Illustrations of leaflets showing comparison of shape of *Phaseolus leptophyllus* G. Don. and other similar wild *Phaseolus* species, as follows: narrow leaflet shapes with attached petioles and petiolules for species:–**1.** *P. vulgaris*;–**2.** *P. acutifolius* var. *tenuifolius*;–**3.** *P. leptophyllus*;–**4.** *P. angustissimus*;–**5.** *P. parvifolius*. Note the folded leaf margins and presence of uncinate hairs only on the lower surface of species *P. leptophyllus*.

Cuernavaca, Mar. 19–May 16, 1789
Hacienda Mazatlán (a large farm near Chilpancingo), Jun. 2–Jun. 24
Acahuizotla, Jul. 2–Jul. 20
Mazatlán, no dates
Chilpancingo, Aug. 5
Chilapa (Aihualtempa), Aug. 30

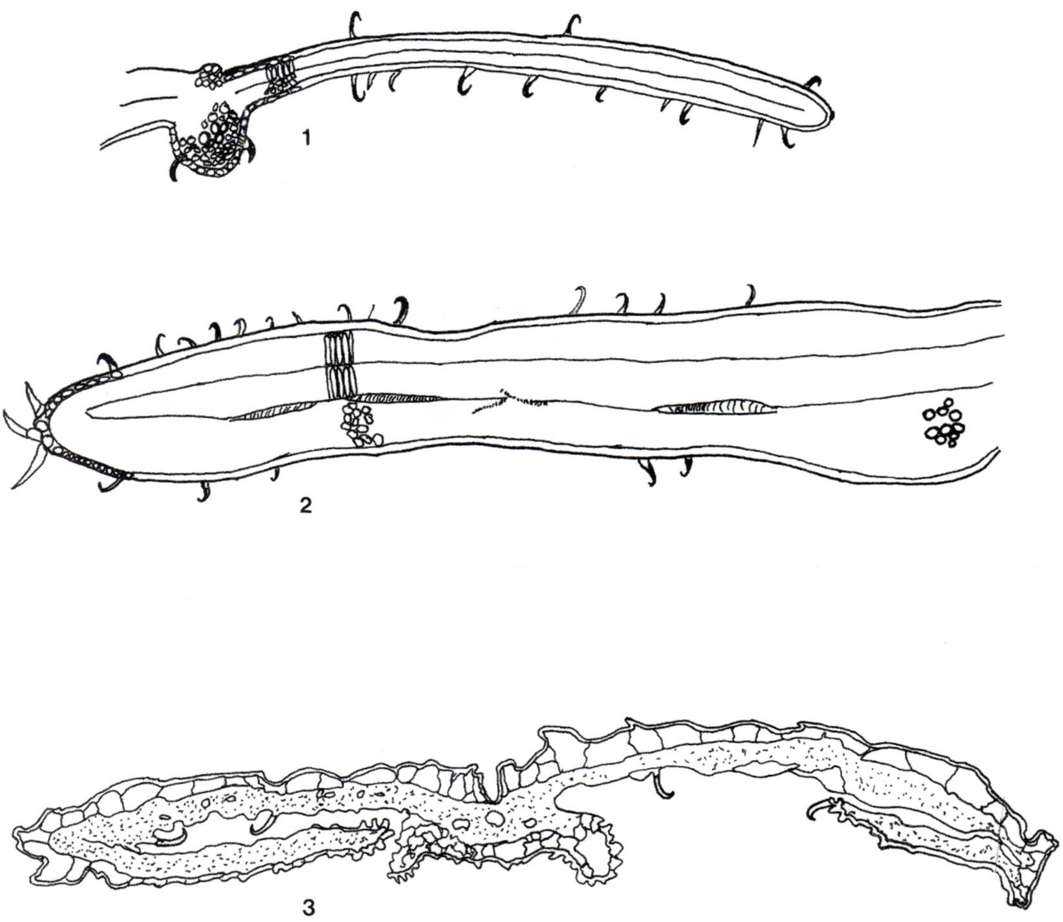

Fig. 63. Comparison of internal structure of *Phaseolus leptophyllus* G. Don. and other similar wild *Phaseolus* species, as follows: cross sections of leaflets for species:–1. *P. vulgaris*,–2. *P. angustissimus*,–3. *P. leptophyllus*.

Therefore, it would seem that the specimen was probably either collected in Acahuizotla in July or in Chilpancingo in June (indicated as the date of flowering on the specimen described as "*leptophyllus*," Plate Fl. Mex. Ic. 297), or August. Delgado (1985) gives the collection date as 1786 according to the label on the sheet in OXF, but this must be in error since the expedition took place in 1789 according to McVaugh (1977).

This species most closely resembles those of Section A. *Acutifolii* and *P. angustissimus* of Section J. *Rugosi*.

Section I.—Digitati Freytag, sect. nov. Type Species: *Phaseolus neglectus* Hermann, J. Wash. Acad. Sci. 38:238. 1948.

Herba scandens vel prostrata volubilis, radix perennis crassa carnosa, foliola terminales modice triangularia rhomboidea et trilobata, flos violascens vel albus in sicco flavescentem, gemma longissima in formam digitatum ad apicem, alae grandes rotundatae et patentae, carina longa 2-spiralibus grandibusque, stigma terminale oblongum et laterale tortam, stylus pubescentis prominens.

Plant a climbing or trailing herbaceous vine, root perennial, thick, fleshy, terminal leaflets medium sized, rhomboidal or somewhat triangular and 3-lobed, flower pale violet to white, drying yellow, the bud very long and shaped like the last joint of a finger just before opening; wings large, rounded and spreading; keel long with a rather large, double coil; stigma oblong, terminal and lateral, twisted to the external side, the style brush prominent.

Comments.—This is a group of closely related species all from the mountains of northeast México, poorly collected prior to the 1980's, which is difficult to realize since they are very showy with large flowers with fan-like wings and strikingly different from other wild *Phaseolus* beans. Additional collecting is required to adequately determine the extent of species distributions.

KEY TO SPECIES

1. Primary bracts entire.
   2. Primary bracts large, 5–7 mm long, linear–lanceolate, glabrescent; bracteole linear–lanceolate, 1.5–2 mm long; flowers violet; rare, in forests W of Montemorelos, Nuevo León ____ I.1. **P. neglectus**
   2. Primary bracts small, 1–3 mm long.
      3. Bracteole less than 1 mm long, aciculate, glabrous; bracteole ovate to lanceolate; flower white; scarce, mountains of S Nuevo León; 1380–2060 m ____ I.2. **P. albiflorus**
      3. Bracteole 2–3 mm long.
         4. Flower standard twisted or crumpled, white to pale purple; plant a prostrate vine (never climbing); bract lanceolate, 2–3.25 mm long; bracteole lanceolate, puberulent; rare, mountains SW of Monterrey, Nuevo León; 1850 m ____ I.5. **P. altimontanus**
         4. Flower standard not twisted, pale purple; plant a prostrate or climbing vine; bract ovate-lanceolate, 2–2.5 mm long; bracteole ovate lanceolate; scarce, mountains of S Nuevo León and Tamaulipas; 900–1430 m ____ I.3. **P. albiviolaceus**
1. Primary bracts trifid, 7–9 mm long, white hispid; plant a climbing vine; bracteoles oblong, l.75–2.25 mm long; flowers pale lavender; rare, mountains of S Nuevo León; 1900 m ____ I.4. **P. trifidus**

I.1.—Phaseolus neglectus Hermann, J. Wash. Acad. Sci. 38:238. 1948. (**Fig. 68**). TYPE: MÉXICO. NUEVO LEÓN: Mpio. Montemorelos, in oak woods along trail up Sierra de la Cebolla from La Trinidad, (25°N, 100°W)?, 20 Aug 1939, *Muller 2881* (HOLOTYPE: US 2216110).

Aerial shoot a prostrate, somewhat climbing, indeterminate vine. ***Root*** a perennial, thick, fleshy, conical. ***Stems*** slender, sparsely puberulous to glabrate. ***Stipules*** linear-oblong to lanceolate, 5–6 mm long, rigid, nearly perpendicular to stem, 3- to 5-nerved, glabrous. ***Leaves*** 7–13 cm long; petioles 3–6 cm long, nearly glabrous, petiolules 6–10 mm long; stipels linear-oblong, rigid, lower 3 mm long, upper 2–2.5 mm long, 1- to 2-nerved, ciliate; terminal leaflets triangular to lanceolate, 3.3–6 cm long, broadly 3-lobed, the two lateral lobes rounded to 1 cm long, cuneate, acuminate, apiculate, nearly glabrous, short appressed hispid adaxially, puberulent on veins abaxially, ciliate, dark green adaxially, slightly paler abaxially, membranous; lateral leaflets similar but inequilateral, slightly 3-lobe. ***Inflorescence*** a slender panicle to 15–20 cm long of up to 20 flowers with 1–2 branches at base; peduncle 4–8–12 cm long; rachis 3–5 cm long of 11–25 flowers; primary bract, linear to lanceolate, 4–7 mm long, 1–1.5 mm wide, mostly 3-nerved, glabrous to sparsely pubescent, green, the margins ciliate and purplish, persistent; pedicel, slender, short, 5 mm long, glabrous; pedicellar bracts narrowly lanceolate 1 mm long 1-nerved purplish margin. ***Bracteoles*** linear-lanceolate, 1.5–2 mm long, 0.5 mm wide, faintly 1-nerved, glabrous, hyaline, caducous. ***Flower*** to 20 mm long, light violet fading to pale salmon; calyx short campanulate, 3.5 mm long, upper 2 lobes united into 1 scarcely elongate, 6 mm wide, 3 lower lobes subequal, 1 mm long, 1.5 mm wide, acute, sparsely pubescent, ciliate on margins; standard light violet, obovate, broadly rounded, 16 mm long, 12 mm wide, reflexed, 10 mm to flexure, 6 mm more to deeply emarginate tip, the claw 0.5 mm long, the auricles flap-like, 1 mm long, 0.75 mm wide; wings light violet, the blade orbicular, 15–20 mm long, spreading; keel 13 mm from base to upward bend and 9 mm more to base of 1 1/2–2 coils of 3 mm diam.; vexillary stamen with a reniform enlargement; ovary 2 mm long, pubescent; style, the terminal thickened coil of 2 mm in diam.; stigma linear, 1 mm long, lateral, twisted to nearly extrorse, 1 mm long. ***Pod*** unknown. ***Seed*** unknown. ***Seedling*** from hypogeal germination; epicotyl 6.5–7.5 cm long; stipules bifid; eophyll, the petiole 2.5–3 cm long, with basal and apical pulvini, stipels present, the blade simple, ovate to broadly ovate-triangular, obtuse-truncate at base, acute.

Specimens examined: **MÉXICO. Nuevo León:** Hacienda Pablillo, Galeana, (24°50'N, 100°5'W), 18 Aug 1936, *Taylor 194* (in part) (TEX); Mpio. Linares, Los Pinos, 1750–1800 m, Sep 1980, *Villanueva 9* (MEXU). **Tamaulipas:** Mpio. Victoria, upper valley of Río San Marcos, El Picacho, 1500 m, 13 Oct 1986, *Martínez 1346* (MEXU; MO); Gómez Farias region, (23°N, 99°15'W), Sep 1965, *Webster et al. 223* (TEX).

Habitat.—This species is found growing in oak and pine-oak forests.

Comments.—Delgado (1985) recognized two races in *P. neglectus*: race A with lavender flowers including the type *Muller 2881* and *Meyers & Rogers 2899*, of which the latter has very distinctly trifid bracts and stouter racemes and is recognized here as a distinct species *P. trifidus*, and race B with white flowers which is probably also a distinct species (perhaps *P. albiflorus*) but uncertain since no specimens are cited.

I.2.—Phaseolus albiflorus Freytag & Debouck, sp. nov. (**Figs. 64, 68**). TYPE: MÉXICO. NUEVO LEÓN: General Ignacio Zaragoza, Palo Bola, 11 km S de Zaragoza, terraceria a la Encantada, 23°56'N, 99°47'W, 2060 m, 8 Sep 1985, *Debouck et al. 1510* (HOLOTYPE: US(3); ISOTYPES: CHAPA, COL, G, K, M, MICH, SI, UC, WIS).

Similis *Phaseolo neglecto*, sed bractea primaria 2.5–3 mm longa glabra, bracteola ovata vel lanceolata minus quam 1 mm longis, floribus albidis differt. Habitat in montibus meridionalibus Novoleonensis rarus.

Aerial shoot a long trailing and climbing, indeterminate vine, to 3 m long. ***Root*** a perennial, thick, fleshy, long tapering, 1–2 cm thick above, 20 cm long, the crown at soil surface, with a few branch roots. ***Stems*** terete, striate, 2–3 mm thick, covered with minute, white uncinate hairs and sparsely covered with appressed reflexed-pilose hairs, purple with minute green streaks; internodes 8–10 cm long. ***Stipules*** triangular, 3 mm long, 1.5–2 mm wide, obtuse, ciliate. ***Leaves*** 8–12.5 cm long; petiole delicate to normal, 3–4 cm long, glabrous; petiolule 1.5–2 cm long, glabrous; pulvini, the lower 2.5–4 mm long, glabrous, purple, upper 2.5 mm long, glabrous below, densely covered by white hirsute hairs adaxially; stipels lanceolate, 2 mm long, 1-nerved, glabrous; terminal leaflet broadly triangular, 3–6 cm long, entire to deeply 3-lobed, the central lobe linear to triangular, 2–3 cm long, 1–2 cm wide, acute, minutely apiculate, white raised veins, sparsely covered with hirsute and glandular hairs adaxially, covered by minute uncinate hairs abaxially, margins ciliate, the 2 lateral lobes rounded to sub-acute, 1–1.5 cm long, glabrous; lateral leaflets similar but the upper of the two lateral lobes much reduced and more acuminate. ***Inflorescence*** a slender pseudoraceme with up to 16 nodes, the lateral branches only 1–2 mm long; peduncle 5.5–9 cm long, distinctly striate and minutely uncinate; rachis 6–17 cm long of 2–11–25 flowers; primary bract, aciculate to lanceolate, 1–3 mm long, 1- to 3-nerved, mostly 3-nerved, glabrate to glabrous; pedicel 4–6 mm long, covered with minute uncinate hairs. ***Bracteoles*** scale-like, ovate to lanceolate, 0.75–1 mm long, 0.5 mm wide, 1-nerved, acuminate, puberulent and ciliate. ***Flower*** white, fading yellowish salmon, large in bud, elongate and bent in the middle; calyx campanulate, the tube 4 mm long, glabrous, the upper lobes rounded, 1.5 mm long, 4 mm wide, ciliate, the lower 3 lobes, subequal, rounded dentate, acute to obtuse, 2 mm long, 2.5 mm wide, puberulent, ciliate; standard white, the claw curved into about a 90°angle, 2.5 mm long, the blade broadly squarish, slightly reflexed at 6–10 mm from base and 12 mm more to deeply (2 mm) emarginate apex, 11–15 mm wide, scarcely enrolled at lateral margins, the auricles small flaps, at base 2 mm long and 1.5 mm wide; wings white, both similar, one more enrolled lengthwise than the other, the blade obovate, rounded, 17 mm long, 7–9 mm wide, the claw 6–9 mm long, 1 mm wide, the spur 2 mm long; keel, the divided claws 3.5–5 mm long, 9–13 mm more to sharp upward bend and 9 more to base of the terminal 1 3/4 coils of 4 mm in diam., the spur not raised and tightly adhering to wings; vexillary stamen, the claw 0.75 mm long, the genicular flap cupped, squarish, 1.5 mm long, 1.25 mm wide, the thickened portion straight and slowly tapering for 15 mm more; stamen tube straight and narrow, 14 mm to bend and 7 more to end of united portion, the ridges at base rounded, 0.5 mm long; stamens oblong 1.25 mm long, 0.5 mm wide, basally afixed; basal collar of 12 large teeth, 1 mm long, smooth; ovary straight, 8–10 mm long, 1–2 mm wide, densely covered with long white cannose hairs, 8 ovules; style 16–20 mm long to base of terminal thickened 1 1/4–1 1/2 coils of 3 mm diam., stylar brush densely covered by long strigose hairs throughout last 3/4 of coil; stigma terminal, oblique, extrorse, laterally twisted, narrow oblong to 1 mm long. ***Pod*** falcate, 4.5–5 cm long, 6.8–7.2 mm wide, 5–5.3 mm thick; valves puberulent of minute white hirsute hairs, green drying greenish tan, fading light tan, somewhat restricted between seed, weak fibrous, twisting once at dehiscence; sutures slightly thickened; beak straight 5–6 mm long; 6–8 seed. ***Seed*** rhomboid to oblongoid, angular, much variation in size and shape, considerably flattened opposite the hilum, 4.6–6.7 long, 4.5–5.3 mm wide, 2.8–3.5 mm thick, dark brown from heavily speckled black on light tan, black eye around hilum; hilum oblong,

FIG. 64. Illustrations of *Phaseolus albiflorus* Freytag & Debouck.–**a.** Seedling a few weeks after germination.–**c.** Plant showing separately a lower portion of stem with mature basal leaf and vine tip with a mature leaf and inflorescences with pods and flowers; note ovate basal leaflets and lobed upper leaflets.–**d.** Flowers, side view and front view.–**e.** Exploded view of flower showing all parts including–**s.** Style tip and stigma as seen under the microscope.–**f.** Mature pods, side view and dehisced.–**g.** Seeds, side view and view from hilum. All drawings made from living material grown in greenhouse at Mayagüez from seed of *Debouck et al. 1522* (TARS #426) collected near Santiago, Nuevo León, México.

1.25 mm long, 0.75 mm wide, covered by placental tissue; lens large, raised, rounded, scarcely divided. ***Seedling*** from hypogeal germination; epicotyl about 3 cm long; eophyll, the petiole 1.3 cm long, the blade rhomboidal, 2.75 cm long, 2.5 cm wide, acute.

PARATYPES. **MÉXICO. Coahuila:** 1 mi N of "La Laguna," Cañon del Mulato, 65 mi NW of Sabinas, Serranias del Burro, (26°45'N, 100°55'W), 1067 m, 8 Sep 1963, *Gould 10592* (MICH). **Nuevo León:** near top of Chipinque, 1100 m, 7 Sep 1978, *D'Arcy 11801* (MO); 1 km E de La Colorada, límite entre Iturbide y Galeana, 24°44'N, 99°56'W, 1650 m, 5 Sep 1985, *Debouck et al. 1503* (CHAPA, COL, US); al lado del Parque recreativo El Salto, 1.5 km S de General Ignacio Zaragoza, 23°57'N, 99°46'W, 1410 m, 8 Sep 1985, *Debouck et al. 1506* (CHAPA, US); Galeana, Cañon de las Placetas, 0.6 km N of San Lucas, 24°56'N, 100°11'W), 1860 m, 16 Sep 1985, *Debouck et al. 1520* (CHAPA, COL, G, K, M, WIS); Puerto Genovevo, 14 km N de Laguna de Sánchez, Santiago, 25°22'N, 100°12'W, 1380 m, 18 Sep 1985, (grown in screenhouse at Mayagüez, PR from seed of TARS #426), *Debouck et al. 1522* (CHAPA, COL, K, M); Cerro de Las Mitras, cañon de la Mina de la Voladora, abajo del Périco, 8 km W de Monterrey, 25°50'N, 100°36'W, 1040 m, 22 Sep 1985, *Debouck et al. 1527* (CHAPA, COL, K, M, MICH, WIS); Mpio. Iturbide, El Charco, 2000 m, 12 Oct 1989, *Estrada 1869* (BRIT, MEXU); Mpio. Iturbide, on the road from La Purísima (32 km S of Iturbide) up the E face of Sierra Cieneguillas, 42 km S of Iturbide, 2000 m, 27 Oct 1982, *Grimes et al. 2355* (TEX); Mpio. Zaragoza, Cerro del Viejo, 15 mi W of Dulces Nombres, (24°N, 100°W)?, 2000 m, 20 Aug 1948, *Meyer et al. 3046* (BM, BR, G, MO); 1 mi from Horse Tail Falls, 25°21'N, 100°09'W, 22 Oct 1981, *Poole 2401* (MEXU); Hacienda Pablillo, Galeana, 17 Aug 1936, *Taylor 194* (in part) (F). **Tamaulipas:** Rancho del Cielo, Nov 1964, *Webster et al. 87* (TEX).

Habitat.—This species is found climbing over low vegetation in small isolated groups on steep wooded slopes of oak or pine-oak, in association with *Arctostaphylos*, cactus, *Comelina*, *Cupressus*, *Desmodium*, *Ipomea*, Maguey, Nogal, palms with epiphytes (*Usnea*?), *Populus*, *Solanum*, *Tilia*, and *Ugnadia*. Soils are of rocky clay derived from schists or limestone, well drained, and with humus.

Diseases and pests.—The green pods are attacked by weevils. Also damaged by thrips, beetles, flea beetles and with some galls possibly caused by insects (*Cynipoidea*?).

Comments.—This species seems to be the most widely distributed of the section and has only white flowers. Though it resembles *P. neglectus* it can be distinguished by its much smaller bracts and bracteoles.

I.3.—Phaseolus albiviolaceus Freytag & Debouck, sp. nov. (**Figs. 65, 68**). TYPE: MÉXICO. TAMAULIPAS: 1 km antes de Altas Cumbres, 17 km SW de Ciudad Victoria, Km 162 Mex 101 a San Luís Potosí, 23°36'N, 99°08'W, 900 m, 11 Nov 1986, *Debouck et al. 2063* (HOLOTYPE: US(2); ISOTYPES: BR, CHAPA, G, M, MICH, WIS).

Similis *Phaseolo neglecto*, sed bractea ovati-lanceolata 2–2.5 mm longi, bracteola lanceolata 2–3 mm longi, floribus purpurascentibus differt. Crecit in montibus meridionalibus provinciae Novoleonensis et Tamaulipasensis rarus.

Aerial shoot a prostrate and climbing vine. ***Root*** a perennial, thick, fleshy branched, 10–20 cm long, 4–5 cm wide, with a fissured and corky cortex, and large nodules on lateral roots. ***Stems*** slender, with internodes 12–14 cm long in median part of stem, epidermis slightly striate dark purple and green, sparsely and shortly ciliate. ***Stipules*** very small, triangular. ***Leaves*** 3.7–9.8 cm long; petioles 1.3–3.4 cm long, petiolules 8–14 mm long; stipels 0.8–1.3 mm long; terminal leaflets triangular (on lower parts of stems) to lanceolate and 3-lobed (upper parts of stems), 1.6–5.4 cm long, 1.3–5 cm wide, cuneate and with deep rounded lateral lobes, apiculate, nearly glabrous, dark green adaxially (sometimes variegated) and somewhat paler abaxially, membranous; lateral leaflets similar but inequilateral. ***Inflorescence*** an erect, short, stout, slightly curved pseudoraceme, extending slightly above foliage; peduncle 3–5–10 cm long, 1.25–1.5 mm thick at base, purple striate, glabrous to puberulent; rachis 3–4 cm long of 2–6 flowering nodes, purple, ciliate; primary bract ovate-lanceolate, 2–2.5 mm long, 0.75 mm wide, strongly 3-nerved, covered by minute uncinate hairs; pedicel 0.5–1 cm long to 1 cm long, in mature pod rather stout, puberulent to glabrous; pedicellar bracts linear 1 mm long 1-nerved margins hyaline. ***Bracteoles*** 1–3 mm long, lanceolate, persistent, attached anywhere from midpoint on pedicel to sessile on calyx, ciliate, 1-nerved. ***Flower*** light purplish; upper calyx lobes united, emarginate, the lower 3 subequal, 1.7–2 mm long, acute; standard purplish, broadly squarish, 9–10 mm long, 9–11 wide, deeply emarginate apex, the well-developed auricles at base, 3 mm long; wings light purple, the blade rounded, flaring and twisted, the claw 4 mm long, the spur 1.5 mm long; keel, the claws 3mm long, 5 mm more to bend and 7 mm more to the terminal 1 1/2 coils of 2 mm diam.; stamen tube straight; vexillary stamen, the claw 1.3 mm long to the well-developed knob of 0.75 mm long, 4 mm more to the filament; stigma lateral, extrorse, narrow elongate, 1 mm long. ***Pod*** nearly straight to very slightly falcate, 5–6 cm long, 7–8.5 mm wide, 5–6 mm thick; valves indistinctly reticulate veined, green drying olive-green and bleaching tan and slightly wrinkled at maturity, strongly

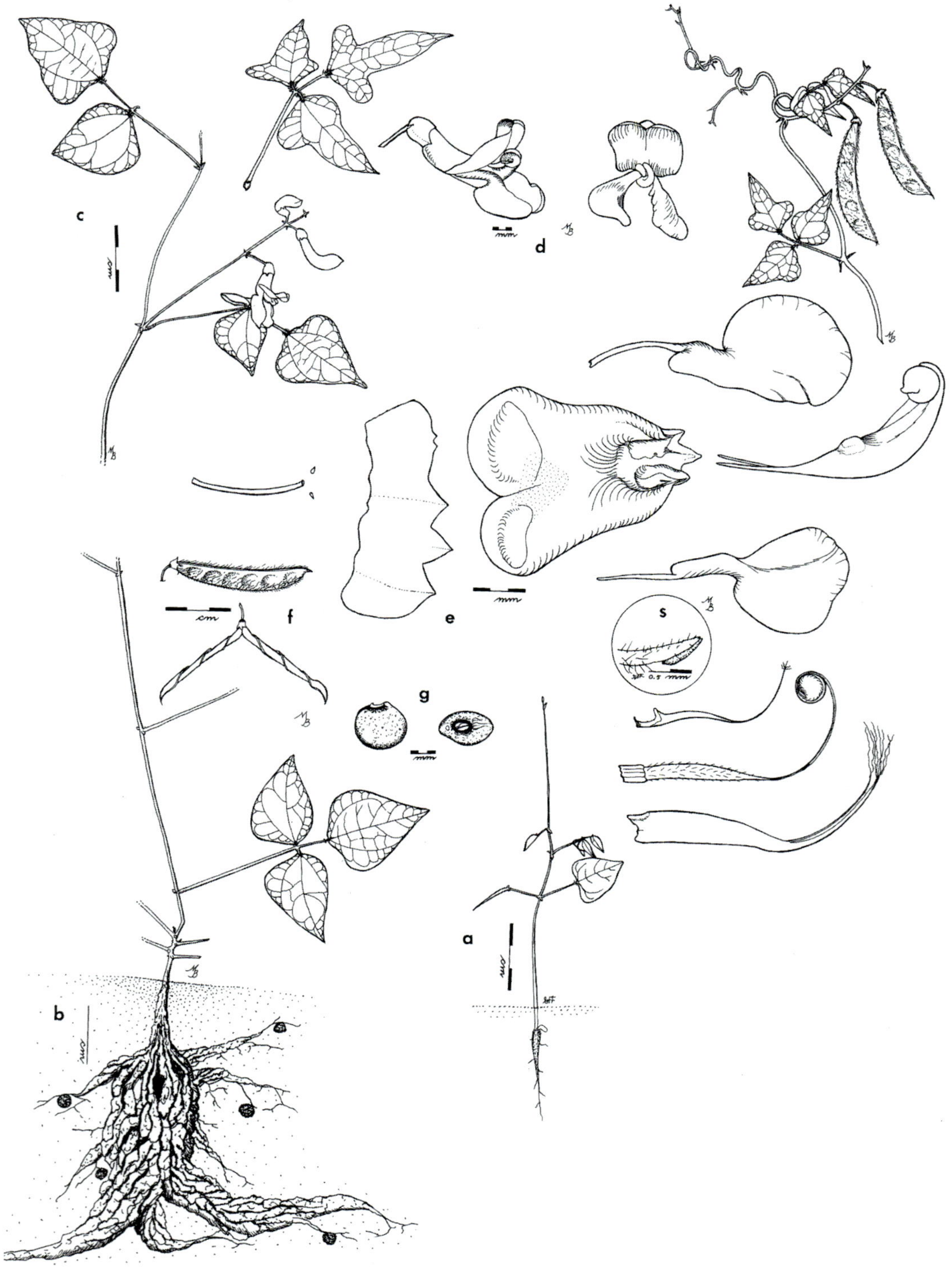

FIG. 65. Illustrations of *Phaseolus albiviolaceus* Freytag & Debouck.–**a.** Seedling a few weeks after germination.–**b.** Mature root after several years growth; notice corky, wrinkled aspect.–**c.** Plant showing separately a lower portion of stem with mature basal leaf, mid-portion of stem with mature leaves and inflorescence and a vine tip with leaves and inflorescence with pods; notice lobed upper leaflets.–**d.** Flowers, side view and front view.–**e.** Exploded view of flower showing all parts including–**s.** Style tip and stigma as seen under the microscope.–**f.** Pods, side view and dehisced.–**g.** Seeds, side view and view of hilum. All drawings made from living material grown in greenhouse at Mayagüez from seed of *Debouck et al. 2063* (TARS #400) collected near Cd. Victoria, Tamaulipas, México.

twisted 1–2 times at dehiscence, the immature pods densely covered with long yellowish strigose hairs, at maturity covered with minute white uncinate and scattered long white strigose hairs; sutures fairly well defined; beak short and slightly recurved downward, about 3 mm long; with 5–7 seed. ***Seed*** globose to oblongoid and slightly flattened opposite the hilum, 4.4–6.5 mm long, 3.6–5.3 mm wide, 3–3.4 mm thick, densely speckled black (gray) on tan, black ring around hilum, shiny; hilum oval, 1.5 mm long, 0.8 mm wide, nearly devoid of placental covering; lens large, rounded, raised, scarcely divided. ***Seedling*** from hypogeal germination; epicotyl 42–45 mm long, the 2nd of the next internodes is about half (7–8 mm long) the length of the 1st and 3rd (17–25 mm long).

PARATYPES. **MÉXICO. Nuevo León:** 1 km S de Zaragoza, en el camino hacia La Encantada, 23°57'N, 99°47'W, 1430 m, 8 Nov 1986, *Debouck et al. 2059* (BR, CHAPA, COL, US(2)); Sierra between Jaumave and Victoria, (22°35'N, 99°25'W), 14 Oct 1931, *von Rozynski 139* (MICH).

Habitat.—This species is found along streams and growing over shrubs (composites and Lamiaceae) in dry oak forests and growing in rocky clay soils derived from limestone and with organic material.

Diseases and pests.—Damage caused by weevils has been reported.

Comments.—This species with pale purple (pink) flowers (see Color Plate II, photo 16) is similar to *P. neglectus* which has larger bracts and purple flowers; and is similar to *P. albiflorus* which has smaller bracteoles and white flowers; and all of which have the same finger or thumb-shaped floral buds. The location of bracteoles well below the calyx insertion is noteworthy.

I.4.—Phaseolus trifidus Freytag, sp. nov. (**Figs. 66, 68**). TYPE: MÉXICO. NUEVO LEÓN: Dulces Nombres, E side of Cerro Linadero, 24°N, 99.5°–100.5°W, 1900 m, 9 Aug 1948, *Meyer & Rogers 2899* (HOLOTYPE: MO 1597351; ISOTYPES: BM, BR, G).

Similis *Phaseolo neglecto*, sed bractea primaria trifida 7–9 mm longa albida hispida, bracteola oblonga 1.75–2.25 mm longi, floribus lavandulescentibus differt. Crecit in montibus meridionalibus Novoleonensis rarus.

Aerial part an annual, climbing and trailing, indeterminate vine. ***Root*** unknown. ***Stems*** terete, striate, 2–3 mm thick; internodes 8–9 cm long, covered with minute uncinate and long, yellowish strigose hairs. ***Stipules*** long ligulate to lanceolate, 7 mm long, 1.75 mm wide, 4- to 5-nerved, acute, densely pubescent of long, white strigose hairs to 1 mm long. ***Leaves*** 11–12.2 cm long; petiole 4–5 cm long, covered with long white strigose hairs; petiolule 1.5 cm long; pulvini lower 5 mm long, densely covered with long, white strigose hairs, upper 2 mm long, very densely covered with white strigose hairs; stipels narrowly spatulate, 2.5 mm long, 2-nerved, acute, covered with white hirsute hairs; terminal leaflet somewhat triangular, 3-lobed, the center lobe large, 5 cm long, 2.5 cm wide, acute to slightly acuminate, apiculate, the 2 lateral lobes narrow and rounded, 1–1.5 cm long, 1 cm wide, sparsely covered with white hispid to strigulose hairs on both surfaces, mostly on veins, membranous, dark green; lateral leaflets similar but inequilateral. ***Inflorescence*** a very long, many-flowered pseudoraceme; peduncle 14–20 cm long, fairly stout, puberulent; rachis 4–10–18 cm long, densely covered with white uncinate and strigose hairs, of 12 or more flowering nodes; primary bracts deeply trifid, 7–9 mm long, 2–3 mm wide, lateral lobes slightly above middle, dentate, acute, 0.5–0.75 mm long, distinctly veined, covered with white hispid hairs; pedicel delicate, 7–8 mm long, sparsely white hispid; pedicellar bracts narrowly lanceolate 4 mm long faintly 1-nerved covered with white hispid hairs. ***Bracteoles*** oblong, 1.25–2.25 mm long, 0.7–1 mm wide, 1-nerved, covered with white hispid and strigose hairs. ***Flower*** pale lavender, drying yellowish; calyx campanulate, upper 2 lobes united into 1 emarginate, 4 mm wide, scarcely elongate to 0.5 mm long, 3 lower lobes subequal, dentate, acute, 1 mm long, 1.5 mm wide, the central covered with white strigose hairs, calyx tube 4 mm long, sparsely covered with white strigose hairs; standard pale lavender, 6 mm long to flexure, 5 mm more to emarginate tip, 12 mm wide; wing pale lavender, the blade broadly obovate-rounded, 10 mm long, 9 mm wide, the claw 3.5 mm long, the spur 1 mm diam.; keel 6 mm to flexure up and 6 mm more to base of terminal 1 3/4 coil of 3 mm diam.; style, the terminal thickened coil of 2.5 mm diam.; stigma oblong, terminal and lateral, 1 mm long, with a heavy beard. ***Pod*** straight 5 cm long (immature), 0.5 cm wide, sparsely covered with appressed-hispid and strigose hairs; beak short, recurved, 2 mm long; 6 ovules. ***Seed*** unknown. ***Seedling*** unknown.

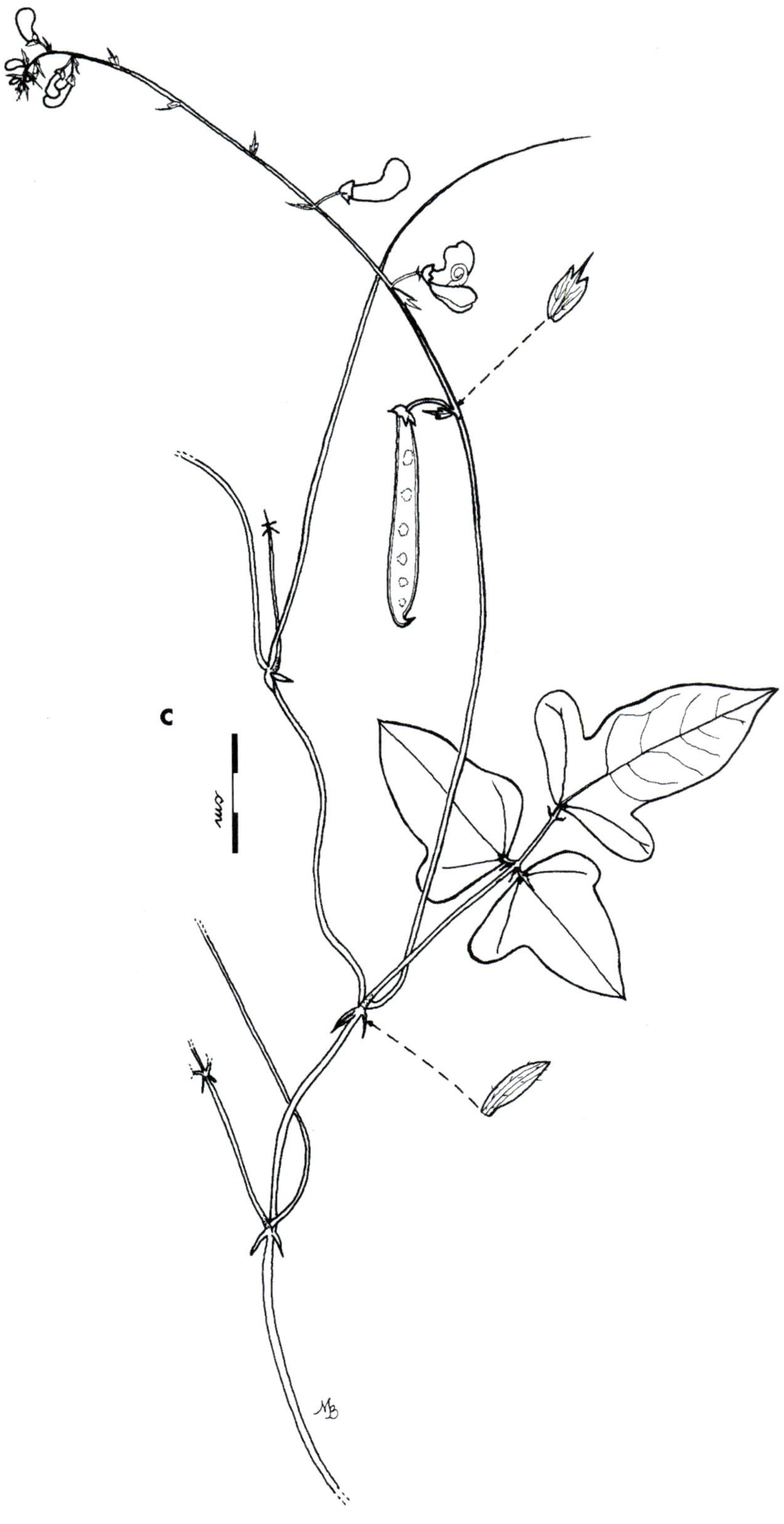

Fig. 66. Illustrations of *Phaseolus trifidus* Freytag.–**c.** Portion of plant with mature leaf and inflorescences with flowers and immature pod; note the large stipule and the trifid primary bract. Drawing from type specimen *Meyer & Rogers 2899* collected near Dulces Nombres, Nuevo León, México.

Habitat.—This species is reported to be a climbing vine scrambling over low growing vegetation in woods in a canyon.

Comments.—It is most striking in having large, trifid primary bracts and white or pale lavender flowers.

I.5.—Phaseolus altimontanus Freytag & Debouck, sp. nov. (**Figs. 67, 68**). TYPE: MÉXICO. NUEVO LEÓN: 2.5 km S de Laguna de Sánchez, camino hacia Puerto El Tejocote, Santiago, 25°20'N, 100°15'W, 1850 m, 18 Sep 1985, *Debouck et al. 1525* (HOLOTYPE: US(2 sheets); ISOTYPES: CHAPA, G, K, M, SI, UC).

Similis *Phaseolo neglecto*, sed volubilis prostrati (nec scandenti), bracteola lanceolati puberuli, vexillo torsivo vel corrugato albido vel violacentia differt. Crescit in montibus austro-occidentalibus citatis Monterregis Novoleonensis rarus.

Aerial part a trailing, prostrate, much branched, indeterminate vine, to 2 m long. ***Root*** a perennial, thick, fleshy, tapering, multi-branched, 15 cm long, 1 cm diam., the crown often 3–4 cm under soil surface. ***Stems*** angled and striate, delicate, nearly straight, puberulent; internodes 2.5–6 cm long. ***Stipules*** triangular, 2–2.5 mm long, 1 mm wide at base, 3-nerved, puberulent. ***Leaves*** 4.3–8.3 cm long, dark green; petiole 2–4 cm long, delicate, puberulent; petiolule 0.5–1.3 cm long, puberulent; pulvini dark green, 2 mm long below, 1.5 mm long above, glabrous to glaucous; stipels, the lower lanceolate, 2 mm long, glabrous, the upper lanceolate to linear, 1 mm long, glabrous, 1-nerved; terminal leaflet rhomboidal to ovate, 1.5–3 cm long, 1.5–2 cm wide at near midpoint, upper leaves becoming 3-lobed, the 2 lateral lobes rounded 0.5–1 cm diam., the central lobe triangular 1.5–2 cm long, 0.5–1 cm wide, obscurely nerved, acuminate, minutely apiculate, dark green and covered with scattered apressed-hirsute hairs adaxially, lighter to silvery and glabrous abaxially. ***Inflorescence*** a curving, erect, pseudoraceme 3–12–(25) cm long; peduncle 3.5–18 cm long, striate, glabrous; rachis 3–7 cm long, white pubescent; nodes knobby; primary bract lanceolate, 2–3.25 mm long, 0.5 mm wide, 3-nerved, puberulent; pedicel delicate, 5 mm long, glabrous, becoming 8 mm long and rather stout at mature pod, pedicellar bract 0.5–0.7 mm long scale-like hyaline 1-nerved. ***Bracteole*** lanceolate, 1–3 mm long, scale-like, hyaline, puberulent. ***Flower*** white to light purple; calyx short to 3 mm long; standard light purple, broadly rounded, 15 mm long, 19 mm wide, emarginate; wings white, the blade rounded, 15 mm long; keel stout, 4–5 mm wide; vexillary stamen with a minute knob nearly at base, the claw about 1 mm long; stigma introrse, linear, 1 mm long. ***Pod*** falcate, broader near tip, 5 cm long, 7–8 mm wide, 5–6 mm thick; valves fibrous, obscurely reticulate, puberulent of minute white uncinate hairs, dark green (immature) drying brown to tan, somewhat constricted between seeds, loosely twisting 1–2 times at dehiscence; sutures slightly thickened; beak straight, pointed, 5 mm long; 6–7 seed. ***Seed*** oblongoid, slightly flattened at side opposite hilum, 5.7–6.3 mm long, 4.8–5 mm wide, 2.8–3 mm thick, streaked and speckled on tan, black eye around hilum, shiny; hilum oval, 1.3 mm long, 0.7 mm wide, hilum ring distinctly raised; lens pronounced but indistinctly divided. ***Seedling*** from hypogeal germination; epicotyl 5.5 cm long; eophyll with petiole 3 cm long and blade broadly triangular, 3–3.2 cm long, 3–3.2 cm wide, truncate base, acute.

PARATYPE. **MÉXICO. Nuevo León:** El Manzano, 2.5 km N de La Cienaga, camino a Laguna de Sánchez, Santiago, 25°22'N, 100°12'W, 1460 m, 18 Sep 1985, *Debouck et al. 1523* (CHAPA, G, K, M, MICH, UC, US(2), WIS); 15 mi up Cola de Caballo, S of Monterrey, 27 Aug 1984, *Lavin 4889* (MEXU).

Habitat.—This species is found in inaccessible areas of fairly dense forests of pine or oak with *Arbutus, Cornus, Crataegus*, and cactus, composites, and Maguey and growing in rocky, organic and red clay soils derived from schists or limestone.

Diseases and pests.—It is damaged by spider mites and powdery mildew.

Comments.—This species has a decidedly prostrate growth habit and does not climb even when tied to stakes. It also has small leaflets (compared to the other species of the section), many of which are ovate and not lobed. The flower standard is twisted to one side rather than facing forward as is normal with species of this section, though this is difficult to recognize in pressed specimens (see Color Plate II, photo 17).

Section J.—Rugosi Freytag, sect. nov. TYPE SPECIES: *Phaseolus filiformis* Benth., Bot. Voy. Sulph. 13. 1844.

Radix annua vel perennis, saepe fibrosa vel carnosa elongata angusta, bracteolae minimae, flore mediocri vel minutissimo, stigma laterale introrsum, apiculatum, legumen parvum chartaceum, seminum rugosum undulatum papillosum vel porcatum radiatum.

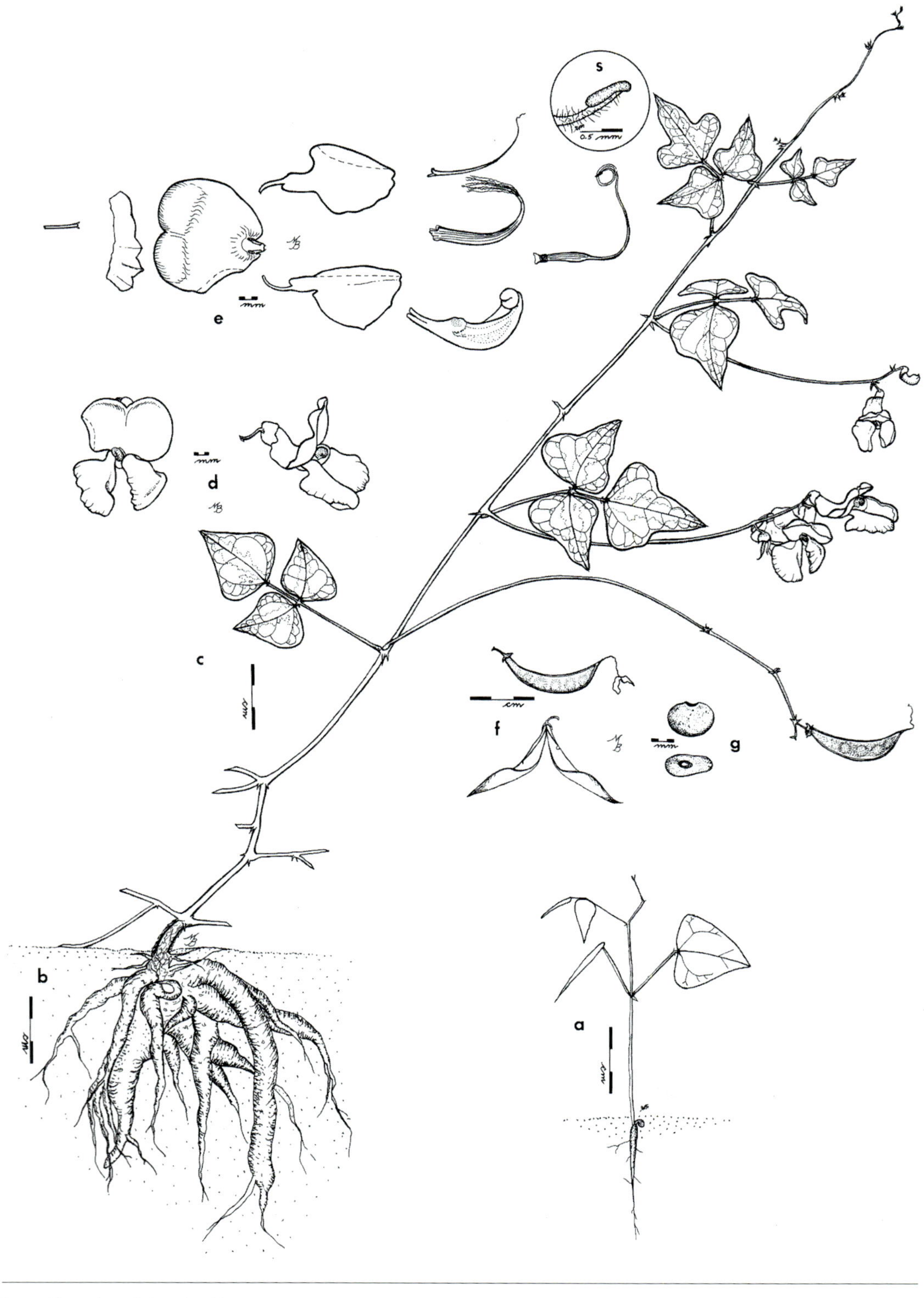

FIG. 67. Illustrations of *Phaseolus altimontanus* Freytag & Debouck.–**a.** Seedling a few weeks after germination.–**b.** Root after several years of growth; note the many branches.–**c.** Plant with mature leaves and inflorescences with immature pod and flowers; note lobed leaves in upper portions of plant.–**d.** Flowers, side view and front view.–**e.** Exploded view of flower showing all parts including–**s.** Style tip and stigma as seen under the microscope.–**f.** Pods, side view and dehisced.–**g.** Seeds, side view and view of hilum. All drawings made from living material grown in greenhouse at Mayagüez from seed of *Debouck et al. 1523* (TARS #427) collected near Santiago, Nuevo León, México.

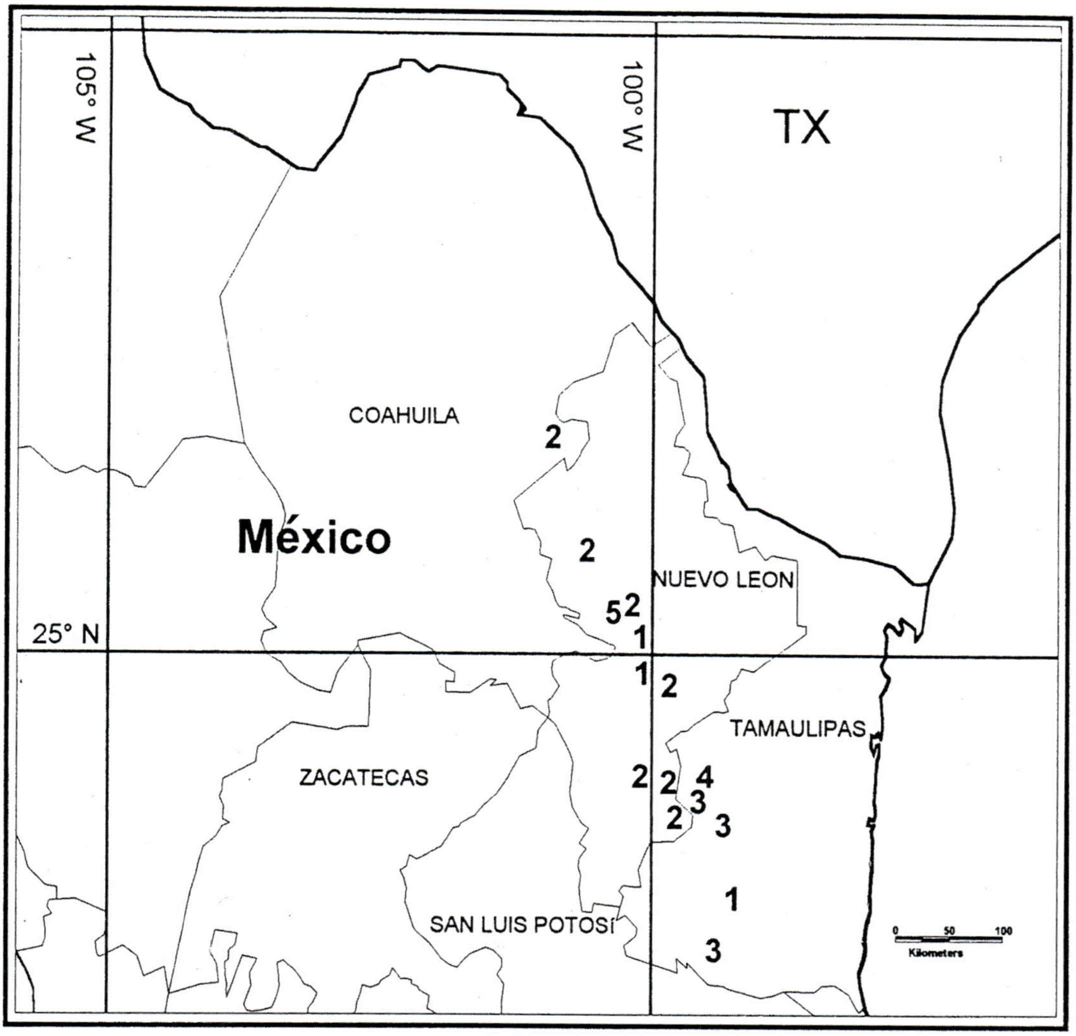

Fɪɢ. 68. Distribution of species of Section I. *Digitati*, as follows: 1 = *neglectus*; 2 = *albiflorus*; 3 = *albiviolaceus*; 4 = *trifidus*; 5 = *altimontanus*.

Root annual or perennial, usually fibrous but may be elongate and narrow fleshy; bracteoles minute; flowers medium to very small; stigma lateral, introrse, pointed; pod small, chartaceous; seed rugose, undulate knobby, sometimes with radiating ridges.

Comments.—This is a natural grouping of small-seeded, desert habitat species from N and NW México and SW United States. Piper (1926) describes this group as having linear pods which they certainly do not; with few ovules the pods are rather short, somewhat falcate and chartaceous. Maréchal et al. (1978b) mention the close similarity of the rugose-seeded species *P. filiformis* and *P. angustissimus* (a distinctive trait also recognized by Lackey 1983: "seed coat ... tuberculate"), and the close apparent relationship between *P. filiformis* and *P. acutifolius* (87%). They keep *P. wrightii* separate, as did Lackey (1983), on the basis of scarcity of material for study. The type of *P. wrightii* has more of a perennial root but otherwise seems to be *P. filiformis*; on cultivation in Mayagüez all species in this section seem to have perennial tendencies and with continued growth form larger and larger roots, though none have been found with storage (thick or fleshy) growth tendencies.

The group based around *P. filiformis* (excluding *P. microcarpus*) has been characterized by allozyme data (Jaaska 1996) and amplified products of the leghemoglobin gene (Skroch et al. 1993) and by work with PCR-RFLPs on cpDNA (Fofana et al. 1999), showing the affinity between *P. filiformis* and *P. angustissimus*, although these two species do not cross easily (Katanga & Baudoin 1987b). It also separates from the other taxa on the basis of ITS DNA sequencing data (Delgado et al. 1999; Gaitán et al. 2000). In the latter, the authors were able to include *P. carteri* that goes together with *P. filiformis*.

KEY TO SPECIES

1. Flowers very small, the standard about 3–4 mm wide; pods rhomboid, 1–2 seeded, early dehiscent, flower 8 9 mm long, purple; in dry habitats from Durango to Nicaragua; from sea level to 2060 m ____________ **P. microcarpus** (see at the end of section)

1. Flowers medium sized, the standard about 6–10 mm wide; pods variously falcate, 3–6 seeded
   2. Plants short erect becoming prostrate pendant nearly straight vines; leaflets ovate-elongate to linear, often slightly lobed at base; bracteole ovate; pods of 3–4 seed; common, SW US and N México; 790–2400 m ____________ J.2. **P. angustissimus**
   2. Plants short to long climbing vines.
      3. Seedling eophylls broad ovate, deeply cordate; leaflets at base of plant seldom slightly lobed; bracts lanceolate, 1.5 mm long; pedicel densely pubescent; flower purple, the standard 10 mm wide; pods of 4–6 seed; common, in desert habitats of SW US, Baja California and NW México; from sea level to 2000 m ____________ J.1. **P. filiformis**
      3. Seedling eophylls ovate-elongate, slightly cordate; leaflets at base of plant deeply lobed, upper leaflets slightly lobed or entire; bracts ovate, 1 mm long; pedicel puberulent; flower white, the standard 6 mm wide; pods of 4–7 seed; rare, only found near La Paz, Baja California; 200 m ____________ J.3. **P. carteri**

J.1.—Phaseolus filiformis Benth., Bot. Voy. Sulph. 13. 1844. (**Figs. 69, 73**). TYPE: MÉXICO. BAJA CALIFORNIA: Bay of Magdalena, (24°45'N, 111°55'W), 1841, *Barclay & Hinds s.n.* (HOLOTYPE: K (= *Hinds s.n.*, 1845) designated by Delgado (1985), GH (photograph).

Phaseolus wrightii A. Gray, Pl. Wright. 1:43. 1850. TYPE: UNITED STATES. TEXAS: declivity of a mountain, near El Paso, (31°45'N, 108°10'W)?, 12 Sep (Oct?) 1849, *Wright 108* (Gray's 111) (HOLOTYPE: GH designated by Delgado (1985); ISOTYPES: GH, K, US).
Phaseolus sanctorum Jones, Contr. West. Bot. 15:140–141. 1929. TYPE: MÉXICO. BAJA CALIFORNIA: Todos Santos, (22°30'N, 110°10'W), 19 Feb 1928, *Jones 24141* (HOLOTYPE: POM n.v.; ISOTYPES: CAS, F, GH, MICH, MO).

Aerial shoot an herbaceous, short-lived annual or biannual, prostrate to climbing, indeterminate vine. ***Root*** an annual to pluriannual, often fibrous with a long narrow taproot, 0.5–1 m long, 0.2–2.5 cm thick, quickly forming a thick, multi-branched crown at ground level. ***Stems*** terete, delicate; internodes 10–12 cm long, densely covered with minute, white uncinate hairs. ***Stipules*** lanceolate, 2.5 mm long, 3- to 4-nerved, somewhat hispid. ***Leaves*** 5.2–8.4 cm long, membranous, green, sometimes variegated, extremely variable in size and lobation, the lobes shape from nearly entire and ovate on lower portions of stems to deeply lobed and sublinear above; petiole 1.5–3 cm long, slightly pubescent; petiolule 0.5–1 cm long, slightly pubescent; pulvini 1–2 mm long, densely covered by uncinate hairs; stipels acicular to lanceolate 0.25–1 mm long, covered by minute uncinate hairs; terminal leaflet broadly triangular to nearly linear, 3–4 cm long, 2.5–3 cm wide at about 1/4 from base, entire to variously shaped from rounded to narrow above, 5–7 mm deep, obtuse, minutely apiculate, sometimes variegated, sparsely pubescent of minute hooked hairs mostly on margins and veins; lateral leaflets similar but somewhat inequilateral. ***Inflorescence*** a short, erect pseudoraceme, the peduncle 0.7–4 cm long, the rachis 0–4 cm long, usually of 1–3 (up to 7) flowering nodes, sparsely pubescent of minute hooked hairs; primary bract lanceolate, 1.5 mm long, 1-nerved, margins ciliate; pedicel 5 mm long, densely covered by minute hooked hairs. ***Bracteoles*** minute, linear to spatulate, 1 mm long, 0.25–0.3 mm wide, pubescent and ciliate fringed, faintly 1-nerved. ***Flower*** light purple or pink; calyx upper lip truncate, entire to slightly emarginate, purplish, the lower teeth subequal, triangular, acute, 0.75 mm long, 1.5 mm wide, puberulent; standard pink, greenish adaxially, squarish-rounded, deeply emarginate, 7 mm long, 10 mm wide, very slightly enrolled at sides, the claw 0.75 mm long, the auricles 0.5 mm long, reflexed at 2 mm from base and thickened at flexure; wings purple, the blade broadly oval, 12 mm long, 7 mm wide, cupped and rolled lengthwise, the claw 2.5 mm long, the spurs well-developed, 1 mm diam., strongly adhering to keel; keel slightly pink, tip green, 3 mm to sharp bend and 3 more to the terminal 1 3/4 coils of 2 mm diam., the claws 2 mm long, separated for 1 mm from base; vexillary stamen, the claw 0.75 mm long, the geniculate knob 0.75 mm wide, 0.5 mm high; stamen tube 3.5 mm to bend and 2.5 mm to the divided filaments, the ridges hardly developed; basal

FIG. 69. Illustrations of *Phaseolus filiformis* Benth.—**a.** Seedling several days after germination; note broad ovate eophylls.—**b.** Root and basal stem from mature plant.—**c.** Vine tip with flowers and inflorescences and detached mature leaf from middle of plant; note the variable lobing of leaflets.—**d.** Flowers, side view and front view.—**e.** Exploded view of flower showing all parts including—**s.** Style tip and stigma as seen under the microscope.—**f.** Pods, side view and dehisced; note reticulation and pubescence of carpels.—**g.** Seeds, side view and view of hilum. All drawings made from living material grown in greenhouse at Mayagüez from seed of *UC L-318* (TARS #256) collected from Pima Co., Arizona, except the detached leaf from a plant similarly grown from seed of *Buhrow ATWR-4* (TARS #117) collected from near Canyon Lake, Arizona, USA.

collar 0.5 mm long; ovary straight, 3.75 mm long, 1 mm wide, covered with minute, white pubescence, 5–6 ovules; style 5 mm to the terminal thickened coil of 1.75 mm diam.; stigma lateral introrse, linear, 0.75 mm long. ***Pod*** pendant, compressed falcate, 2.5–3 cm long, 5–6 mm wide, tip slender cuspidate to straight, 2.5 mm long, the valves membranaceous, reticulate nerved, puberulent of minute uncinate hairs, early dehiscent by 1 loose twist. ***Seed*** somewhat rounded to squarish, flattened, 2–4.3 mm long, 2.3–4 mm wide, rugose, brownish, a darker brown ring around the hilum; hilum ovate, 0.3 mm long; lens prominent, divided. ***Seedling*** from epigeal germination; hypocotyl 0.9–1 cm long; epicotyl 3–3.5 cm long; stipules united, entire; petioles with basal and apical pulvini; stipels absent; eophyll petiole 1.8 cm long, the blade simple, entire, broadly ovate, 1.8–2 cm long, 2 cm wide, cordate at base, acute.

Specimens examined: **MÉXICO. Baja California:** Miraflores, 14 Oct 1890, *Brandegee 157* (CA); Santa Margarita Is., (24°20'N, 111°55'W), Mar 1889, *Brandegee s.n.* (UC); Bosca of Santo Domingo, (25°35'N, 112°W)?, 29 Mar 1889, *Brandegee s.n.* (UC); San Benito Is., 28 Mar 1897, *Brandegee s.n.* (UC); E Base of Sierra de Placeros, 40 km SE of San José de Castro, 366 m, 24 Mar 1984, *Breedlove 60863* (CAS); Puerto Chale on the Magdalena plain along the Río Salado, 15 m, 20 Oct 1977, *Breedlove 43111* (CAS); 5.1 mi NE of San Carlos water tower via Hwy 22, then 0.5 mi NW via dirt road, Magdalena Plain, 24.9°N, 112.1°W, 15 m, 28 Oct 1981, *Burgess et al. 6167* (ARIZ); NW of Portezuelo de la Cuesta de los Dolores (westerly from the N end of Isla de San José), Sierra de la Giganta, 25°7'N, 110°57'W, 700 m, 19 Oct 1964, *Carter 4792* (BM, CAS, MICH); summit of Cerro Teombo, N of Portezuelo Gavilán, 25°51'N, 111°25'W, 1080 m, 1 Oct 1965, *Carter 5072* (ARIZ, CAS, MICH, US); S edge of Llano de Buenos Aires, 11 km SE of San Agustín on road to Cataviña, 7 Nov 1947, *Carter et al. 1875* (BM, CAS, MICH, GH, US); W of Bahía de la Concepción, 40.6 km S of Mulegé, (26°45'N, 111°55'W), 17 Nov 1947, *Carter et al. 1989* (MO,TEX); Llano de Magdalena, 4.5 km N of Pozo Grande, 6 Dec 1947, *Carter et al. 2130* (MEXU, WIS); E side of San Juanico Bay, 48 km NW of La Purísima, (26°15'N, 112°30'W), 6 Jan 1948, *Carter et al. 2488* (CAS, LIL, MEXU, MSC, TEX-LL, US); Cañon del Diablo, to the N and W of Picacho del Diablo (Cerro La Encantada), E flank of Sierra San Pedro Mártir, 31°5'N, 115°23'W, 13 Jun 1954, *Chambers 529* (CAS-DS, UC); San Francisco Mts, Viscaino Desert, 10 mi W of Rancho San Francisco, E of Guerrero Negro and Scammon's Lagoon, 1524 m, 29 Mar 1986, *Charlton 94* (UCR); 25.1 mi S of San Felipe, (30°45'N, 114°45'W), 30 m, 29 Aug, *Clarke et al. 19915* (UCR); Cañon Carrizo, E side of Sierra de Juárez, the next major canyon N of Cd. Guadalupe, (31°55'N, 115°W)?, 305 m, 30 Mar 1985, *Dallman et al. 22* (UCR); Rancho El Coyote on W side of Bahía Concepción, 17.1 mi S of Mulegé, 61 m, 7 Jan. 1983, *Daniel 2388* (ASU, MICH); Puerto Escondido, 11 Feb 1940, *Dawson 1085* (F); 16 mi SE of Guerrero Negro, 23 Mar 1981, *Dillon et al. 1939* (F, UCR); Mpio. de Comondú, cerca de La Purísima, 900 m, 21 Feb 1989, *Domínguez 902* (US); S of Santa Catarina Landing, 16 Apr 1949, *Dressler 600* (BRIT, GH(2), MO); 7 km N of Guerrero Negro, along Rt 1 5 mi N of La Pinta Hotel, (28°N, 114°5'W), 27 Jun 1988, *Elias et al. 10855* (UCR); San Quintín, 7 Apr 1936, *Elmore s.n.* (CAS-DS, F, MICH, UC, US); 2 mi E of Pozo Aleman near Calmalli, 5 Feb 1935, *Epling et al. s.n.* (CAS-DS, UC); San Agustín, 12 Nov 1938, *Gentry 4009* (ARIZ, CAS-DS, K, MO, UC); Viscaino Depression, S and W of Scammon's Lagoon, 9 and 16 Mar 1947, *Gentry 7365* (ARIZ, CAS-DS, UC); Sierra Giganta above Pto. Escondido, 610 m, 21 Apr 1938, *Gentry 37454* (ARIZ); Vizcaino Desert, 5 mi S of Laguna del Gallo, 24 Apr 1952, *Gentry et al. 11750* (MICH, NA, TEX); Cedros Island, 1 mi W of the village, (28°10'N, 115°20'W), 90 m, 11 Feb 1939, *Haines et al. s.n.* (ARIZ, CAS, CAS-DS, MEXU, MICH, UC, US); 29 mi N of Mesquital, 27 Sep 1941, *Hammerly 78* (CAS, CAS-DS, US); La Paz Bay, about 15 mi W of La Paz, 7 Oct 1941, *Hammerly 211* (CAS, CAS-DS, GH, US); Barril, 28 Mar 1947, *Harbison 41610* (CAS-DS); Purísima, (26°15'N, 112°5'W), 19 Sep 1955, *Harbison s.n.* (UC); 2.3 mi NE of Guerrero Negro, 30 m, 26 Oct 1967, *Hastings et al. 67-95* (ARIZ, CAS-DS); 2 km N of Pichilingüe, 16 km N of La Paz, (24°15'N, 110°15'W), 46 m, 12 Mar 1966, *Henrickson 2216* (MICH); Las Animas ranch, 5 km from Las Teristis ranch, road to Todos Santos, El Picacho Mt., 500–1000 m, 22 Nov 1976, *Hoover 321* (GH); 3 mi N of Puertocitos, 13 Dec 1967, *Humphrey 3* (ARIZ); Punta Estrella, 20 mi S of San Felipe, 4 Mar 1967, *Humphrey s.n.* (ARIZ, ENCB); Las Animas Bay, 8 May 1921, *Johnston 3513* (CAS); San Francisquito Bay, 10 May 1921, *Johnston 3554* (CAS, GH, K, US); Angel de la Guarda Island, opposite Pond Island, 30 Jun 1921, *Johnston 4220* (CAS, GH, K, SI, US); Arroyo Undo Ranch, Loreto, 26 Oct 1930, *Jones 27180* (MO); Isla Magdalena, across Magdalena Bay from Mantacitas, 3 m, 14 Jan 1975, *Kipping 344* (CAS); approxim.1 mi NW of San Felipe, (31°5'N, 113°W), 305 m, 5 Dec 1972, *Kniffen s.n.* (UCR); between Villa Insurgentes and Santo Domingo (just N of Colonia Purísima), 30 m, 19 May 1978, *Lane et al. 2345* (TEX); lado SW de Isla San Lorenzo Sur, (28°35'N, 112°50'W), 6 May 1985, *Lott et al. 2465* (CAS, MO); 7 mi NW of El Rosarito on Road to San Borja Mission, 2 Jul 1969, *McGill 46* (ASU); Cerralvo Is., Ruffo Ranch Canyon, 24°13'N, 109°54'W, 4 Apr 1952, *Moran 3623* (CAS-DS); Portezuelo de Jamau, Sierra Juárez, 31°38'N, 115°38'W, 1300 m, 2 Oct 1966, *Moran 13655* (CAS-DS); 4 mi NE of Abreojos, 26°45'N, 113°34'W, 5 m, 4 Feb 1973, *Moran et al. 19727* (US); El Paraiso to El Triunfo, 508 m, 30 Jan 1906, *Nelson et al. 7473* (US); El Cien, Apr 1979, *Nisbe 47* (DES); San Quintín, 22 Apr 1886, *Orcutt 1332* (BM, F, GH, K, MICH, MO(2), UC(2), US(3)); Carmen Island, *Palmer 5(?)* (GH, US); Lagoon Head, 6–15 Mar 1889, *Palmer 819* (GH, K, MEXU, MICH, UC, US); 11.8 mi NE of Santa Rosalia on Hwy 1, 29 Oct 1983, *Perrill et al. 5509* (ARIZ); 2.1 mi N of San Ignacio off Rt 1, (27°30'N, 113°W), 8 Jan 1984, *Pinkava et al. 3073* (ASU, DES); 4.8 mi E of Rt 1 and of Rosario, (30°5'N, 115°40'W), 2 Jun 1973, *Pinkava et al. 11156* (ASU); 16 mi W of Vizcaino junction on the road to Bahía Tortugas, 7 Jan 1991, *Rebman et al. 1027* (ASU); 1 mi W of Km 31 on dirt road N of Bahía Asunción to Rancho San Miguel, 7 Jan 1991, *Rebman et al. 1048* (ASU); Cañon, 10 mi N of Santa Rosalía on Gulf Coast, 183 m, 28 Jan 1929, *Reed s.n.* (CAS-DS); 6 mi W of Hwy #1 on the road to San Xavier Mission, 75 m, 24 Mar 1979, *Reeder et al. 7073* (ARIZ, UC); 8 mi N of Puertocitos, 30 m, 23 Dec 1961, *Roos s.n.* (UCR); San Bartolome Bay, 3 Mar 1911, *Rose 16194* (US); NW of La Paz about Km 60, 13 Feb 1972, *Rudd 3394* (MEXU, US); Santa María Bay, near Cabo San Lucas, 14 Feb 1972, *Rudd 3396* (US); 8 mi NW of San Ignacio, 24 Mar 1980, *Sanders 1179* (UCR); Guerrero Negro airport, 24 Mar 1980, *Sanders 1213* (UCR); 40 mi (65 km) W of Hwy 1 at Viscaino Jct., Viscaino Peninsula,

27°27'N, 113°57'W, 24 Mar 1980, *Sanders 1225* (ASU); El Caimancito, 5 km N of La Paz along Hwy 11 toward Pichilingüe, 3 m, 12 Dec 1982, *Sanders et al. 3281* (UCR); Ammonite Canyon at Jct. with Arroyo Santa Catarina, 5.4 mi E of Sta. Catarina Landing on the road to Rancho Sta. Catarina, 61 m, 29 Mar 1985, *Sanders et al. 5560* (UCR); 20 mi W of Bahía Los Angeles, (29°N, 113°55'W), 21 Feb 1935, *Shreve 6908* (ARIZ, MICH, US); W end of San Bartolo, on Mex. Hwy 1, 200 m, 4 Nov 1983, *Starr 696*, (ARIZ); Punta Refugio, N end of Isla Angel de la Guarda, (29°30'N, 113°40'W), 21 Feb 1977, *Sylber s.n.* (ARIZ); Mpio. La Paz, 18 km al N de Todos Santos, 180 m, 13 Oct 1985, *Tenorio et al. 10416* (CAS, MO); 3 mi SE of Todos Santos along road to La Burrera, 304 m, 15 May 1959, *Thomas 7867A* (CAS-DS, US); 6 mi N from Cuarenta on road between San Ignacio and La Purísima, 30 m, 26 Oct 1959, *Thomas 8361* (CAS-DS, ENCB, MICH, US); Km 40 on Hwy between La Paz and Santo Domingo, 366 m, 30 Oct 1959, *Thomas 8466* (CAS-DS); Angel de la Guarda Island, 29°10'N, 113°10'W, 10 m, 21 Apr 1983, *Turner 83-36* (ARIZ); 4 mi S of Guadalupe, 21 Mar 1935, *Whitehead 836* (CAS-DS); El Marmol, 6 Mar 1930, *Wiggins 4372* (CAS-DS, US); 6 mi above the Mission, Río Santo Domingo, 12 Sep 1930, *Wiggins et al. 4801* (ARIZ, CAS, CAS-DS, ENCB, F, GH, MICH, UC(2), US); between Medano and Venancio, 29 Apr 1931, *Wiggins 5533* (CAS-DS(2), GH, MICH, UC, US); 24 mi S of Punta Prieta, (28°45'N, 114°15'W), 23 Feb 1935, *Wiggins 7734* (CAS-DS(2), F, GH, MICH, UC, US); Las Posas, 3 mi SW of Portesuela, between Los Emes and San Antonio del Duarte, SW Sierra San Pedro Mártir, 14 May 1941, *Wiggins 9961* (CAS-DS, UC, US); 7.3 mi W of Los Planes, 250 m, 21 Dec 1958, *Wiggins 14463* (ARIZ, CAS, CAS-DS, TEX); 1.5 mi NW of San Bartolo, 9 Jan 1959, *Wiggins 14761* (CAS, CAS-DS); Isla Espíritu Santo, old pearl culture tanks, Bahía de San Gabriel, 24°25'N, 110°25'W, 1 Nov 1959, *Wiggins 15229* (CAS, CAS-DS, GH, TEX, UC); 8 mi S of Puertocitos, 22 Mar 1960, *Wiggins et al. 15876* (CAS-DS, G, MEXU, MICH, TEX(2), US); 7 mi SE of Rancho La Cantina, 80 m, 1 Apr 1961, *Wiggins 16210C* (CAS-DS); Arroyo Estatón, W side of Is. la Guarda, 30 m, 15 Mar 1962, *Wiggins 17009* (CAS-DS); Isla de la Vivera, Pond Island, 29°2'N, 103°6'W, 17 Mar 1962, *Wiggins 17072* (CAS-DS, ENCB, MEXU, MICH); S end of Isla San Esteban, 22 Mar 1962, *Wiggins 17240* (CAS-DS); Isla Partida, N of Isla Raza, 28°53'N, 113°3'W, 26 Mar 1962, *Wiggins 17301* (CAS-DS, MEXU, US); Isla Ildefonso, 26°37'N, 111°26'W, 2 Apr 1962, *Wiggins 17418* (CAS-DS, MEXU, MICH); S end of Isla Coronados, 26°7'N, 111°17'W, 3 Apr 1962, *Wiggins 17482* (CAS-DS, US); SW side of N end of Isla Danzante, 24°45'N, 111°15'W, 7 Apr 1962, *Wiggins 17559* (CAS-DS); Sulphur mine, about 22 mi S of San Felipe, 12 Nov 1967, *Wiggins 20854* (CAS-DS, MICH); Agua del Mezquitito waterhole, 31 mi S of Puertocitos, 13 Nov 1967, *Wiggins 20874* (CAS-DS, ENCB); Cape St. Lucas, Aug. 1859–Jan. 1860, *Xantus 23* (GH, US); Cabo San Lucas, 30 m, 5 Mar 1981, *Zwinger 428* (ASU). **Chihuahua:** Km 84, NWRR, Sierra Madre, 2 Oct 1911, *Barlow s.n.* (F); Majalca, 24 Jun 1936, *Le Sueur 682* (F). **Coahuila:** Picachos Colorados, 11 Aug 1940, *Johnston et al. 113* (GH); San Antonio de los Alamos, Sierra de San Antonio, 1290 m, 2–3 Sep 1940, *Johnston et al. 901* (GH, TEX-LL); Ejido la Trinidad, Sierra de Jimulco, 25°8'N, 103°22'W, 1900–2000 m, 25 Aug 1988, *Rodríguez et al. 4418* (ARIZ, ASU); Rancho de Las Uvas, Mts. along the E margin of the Valle de Acatita, 1060 m, 21 Sep 1942, *Stewart et al. 2692* (GH). **Sonora:** Cholla Bay, 7 mi NW of Puerto Peñasco, (31°25'N, 113°40'W), 7–8 Apr 1963, *Bell 17552* (MICH, TEX-LL(2)); 58 mi Mex. Hwy #2 E of San Luís, Río Colorado, Sierra de las Tenajas Altas, 274 m, 29 Feb 1968, *Breedlove 15995* (CAS, MICH); San Agustín Bay, 28°27'N, 111°40'W, 22 Mar 1978, *Burke 79-31* (ARIZ, UC); SW part of Isla Tiburón, Ensenada Blanca landing field, 7 Apr 1963, *Felger 7126* (ARIZ(2)); 2.8 mi by road S of Papago Tanks, Pinacate region, 150 m, 7 Mar 1970, *Felger 18734* (ARIZ); Isla Tiburón, Sierra Kunkook, about 3 to 5 mi W of Punta Narragansett, 23 Mar 1963, *Felger et al. 6990* (ARIZ); E playa of Molina Crater, Pinacate region, 29 Sep 1964, *Felger et al. 10636* (ARIZ); 1.5 mi E of estero at Santa Rosa, estrecho Sanjiemillo, 15 Feb 1965, *Felger et al. 12578* (ARIZ); N side of Island San Esteban, 10 Apr 1968, *Felger et al. 17563*, (ARIZ); 1 mi S of Pinacate Peak, Sierra Pinacate, 875 m, 23 Mar 1970, *Felger et al. 19302* (ARIZ); Elegante Crater, NE part of Pinacate region, 290–300 m, 20 Apr 1970, *Felger et al. 19679* (ARIZ); Sykes Crater, NW part of Pinacate region, 155 m, 8 Dec 1970, *Felger et al. 20018* (ARIZ, ENCB); SE corner of Tiburón Island, El Monumento, 22 Dec 1966, *Felger et al. 15523B* (ARIZ, UCR); Cañon de Nacapules, about 6 km NE of Bahía San Carlos, 19 Oct 1984, *Felger et al. 84-126* (ARIZ); Estero Soldado, 9 km NW of Guaymas, 27°57'42"N, 110°58'42"W, 21 Nov 1985, *Felger et al. 85-1444* (ARIZ); Cerro El Vigía, Guaymas, 420 m, 5 Sep 1980, *Felger et al. 880-19B* (ARIZ); Cholla Bay, near and N of Punta Peñasco, 0 m, 27 Jan 1947, *Gould 4137* (ARIZ, CAS, UC); island in bay, Guaymas, 14 Apr 1921, *Johnston 3089* (CAS); Tiburón Is., Freshwater Bay, (29°N, 112°20'W), 23 Apr 1921, *Johnston 3265* (CAS); Sierra de San Emeterio, 6 mi N of Quitovac on road to Sonoyta, Paso San Emeterio, 595 m, 13 Mar 1936, *Keck 4120* (CAS, CAS-DS(2), F, GH, MO, UC, US); Bahía Kino, 15 Jul 1973, *Knobloch 2513* (ARIZ, ENCB); 45 mi W of Sonoita on San Luis road, 13 Mar 1959, *Lowe 2954* (ARIZ); Punto Cirio, Sierra Bacha, S of Pto. Libertad, 24 Feb 1978, *Martin s.n.* (ARIZ); Suvuk, SE of Pinacate Peak, 28 Aug 1982, *Nabhan et al. 367* (ARIZ); Cerro Prieto, E of Bahía de la Cholla, E coast of Sea of Cortes, N of Puerto Peñasco, 61 m, 13 Mar 1984, *Parker s.n.* (ARIZ); Bahía Kino, 29 Dec 1966, *Patton et al. s.n.* (ARIZ); Caborca, 2 Apr 1884, *Pringle 14* (GH); Cholla Bay, 15 Jan 1966, *Redden s.n.* (ASU); Mina Santa Cruz, 25 mi SW of Caborca, Sierra El Viejo, 30°22'N, 112°22'W, 549 m, 8 Mar 1983, *Sanders et al. 3542* (TEX, UCR); 4.1 mi NW of Kino Bay on road to Punta Chueca, (29°N, 112°5'W), 27 Dec 1983, *Van Devender et al. 83-93* (ARIZ, UCR); 0.2 mi above Gulf of California in Coloradito Canyon, Sierra Bacha, 29 Dec 1988, *Van Devender et al. 88-875* (ARIZ).

UNITED STATES. Arizona: Maricopa Co.: Hidden Water, Boulder Creek, 594 m, 15 Oct 1951, *Blakley 766* (DES); Apache Trail near Canyon Lake between Mileposts 205 and 206, 518 m, 19 Aug 1982, *Buhrow 179* (ARIZ); Usury Mts. County Park, N of Eramadas, 10 Mar 1975, *Earle s.n.* (DES); Roosevelt Dam, (33°30'N, 111°W), 19 Apr 1917, *Eastwood 6254* (CAS); 2 mi below Bartlett Dam, Verde River, 4 May 1975, *Engard 459* (DES); Harquahala Mts. Blue Tanks Canyon, (near Yuma Co. line, SE side near base), (33°30'N, 113°W), 915 m, 23 Mar 1936, *Keck 4287* (CAS-DS); 3 mi N of Sunflower, along Sycamore Creek, 762 m, 4 May 1968, *Lehto et al. 12818* (ASU); below Roosevelt Dam, 25 Sep 1948, *Norvell s.n.* (CAS). **Pima Co.:** Saguaro National Monument, Safford Peak, Tucson Mts., (32°20'N, 110°30'W), 914 m, 15 Apr 1990, *Bertelsen 90-135* (UCR); Safford Canyon, 792 m, 11 Aug 1990, *Bertelsen 90-173* (UCR); 2 mi S of Coffeepot Mt., N end of Sikort Chuapo Mts., 823 m, 22 Apr 1981, *Butterwick et al. 7405* (CAS); Mt. Arroyas, S of Ajo, (32°15'N, 112°40'W)?, 31 Oct 1943, *Clark 11151* (GH); Puerto Blanco Mts., Dripping Springs, 610 m, 18 Mar 1945, *Darrow et al. 2438* (ARIZ); Saguaro Lake, 30 Apr 1942, *Dearing et al. 4822* (DES); Quijotoa Mts., 19 Oct 1945, *Goodding 213-45* (ARIZ); Bates Well, 18 Nov 1939, *Harbison 26158* (ARIZ); Ajo Mts., Organ Pipe National Monument, Alamo Wash, (32°N, 112°35'W), 11 Sep 1946,

Haskell et al. 1092 (UCR); N end of Sikort Chuapo Mts., 2 mi S of Coffeepot Mt., 823 m, 22 Apr 1981, *Hillyard et al. 7405* (ASU); Ajo Mts., Ajo Mt. Drive, 1/8 mi N of Estes Canyon, 610 m, 16 Apr 1983, *Hodgson et al. H2039* (DES); Organ Pipe Cactus National Monument, along E Loop, 3.3 mi from Rte 87, *Nash et al. 9973* (US); Quitovaquito, (32°N, 113°W), 330 m, 5 Mar 1940, *Peebles 14554a* (ARIZ, GH); San Juan Mts., 27 Dec 1925, *Peebles et al. 648* (ARIZ, US); Papago Indian Reservation, Ventana Cave, 7 mi SSE of Ventana, (32°30'N, 112°W), 17 Nov 1974, *Van Devender s.n.* (ARIZ(2)); Organ Pipe Cactus National Monument, Alamo Canyon, 7 Mar 1987, *Van Devender et al. 87-6* (ARIZ, UCR); Saguaro National Monument, Tucson Mts., Safford Peak Canyon, 24 Feb 1989, *Van Devender et al. 89-26* (ARIZ); Raggged Top Peak, about 4 mi N of Silver Bell, 914 m, 9 Sep 1988, *Wiens s.n.* (ARIZ). **Pinal Co.:** Superstition Mts., Superstition Wilderness Area, 23 Jun 1965, *Crutchfield 282* (TEX-LL); Plateau between Cañon Lake and Roosevelt, 853 m, 11 Oct 1931, *Gillespie 8559* (CAS-DS, US); Le Barge Canyon in Superstition Mts., 713 m, 28 Apr 1963, *Lehto 1847* (ASU); Picacho, (32°30'N, 111°15'W), 26 Mar 1926, *Peebles et al. 1468* (ARIZ); Superstition Mt., 22 Mar 1927, *Peebles et al. 3721* (ARIZ); N of Kearny, 3 km SE of Granite Mt., 792 m, 6 Sep 1990, *Shaari et al. 7989* (ASU); Picacho Peak, 28 Apr 1983, *Van Devender et al. s.n.* (ARIZ, UCR). **Yavapai Co.:** Peeple's Canyon in the vicinity of S Peeple's Spring, (34°15'N, 112°30'W), 823 m, 5 May 1980, *Butterwick et al. 6339* (ARIZ). **Yuma Co.:** Kofa Mts., Kofa Game Range, High Tank, Upper Burro Canyon, 975 m, 31 Mar 1970, *Furlow 20* (ARIZ); Kofa Mts., Palm Canyon, 22.5 airline mi SSE of Quartzite, (33°30'N, 113°45'W), 640 m, 30 Mar 1973, *Holmgren et al. 6584* (ASU); Cabeza Prieta Game Range, Cabeza Prieta Tank, 6 Apr 1979, *Lehto et al. L23538* (ASU); on Pinacate lava flow near edge of Pinta Sands, S side of Camino del Diablo, 2 mi W of Jct 81 jeep trail (benchmark 681), 8 mi W of O'Neill Grave, 5 Mar 1977, *Mason et al. 3238* (ARIZ); Castle Dome Mts., (33°15'N, 114°W), 9 Mar 1955, *Monson 14* (CAS); Kofa National Wildlife Refuge, wet tank at Little White Tanks, 475 m, 14 Feb 1986, *Russo 503* (ASU); Castle Dome Mts., Kofa National Wildlife Refuge, 1/4 mi above Figueroa Tank, 475 m, 15 Feb 1986, *Russo 547* (ASU); Kofa Mts., Burro Canyon, 549 m, 1 Dec 1972, *Van Devender s.n.* (ARIZ). **County Unknown:** Coachella Valley at the foot of the Santa Rosa Mts., Martínez Canyon, W end of Ave. 74, 2.6 mi W of Polk St., 122 m, 19 Dec 1982, *Sanders 3451* (UCR). **New México: Dona Ana Co.:** Aden Crater, 15–17 mi SW of Las Cruces, (32°5'N, 107°W), 1 Sep 1974, *Van Devender s.n.* (ARIZ). **Texas: Brewster Co.:** Big Bend National Park, Burro Mesa, 23 Sep 1966, *Correll 33850* (GH, TEX-LL); Chisos Mts., Boot Spring, 26 Aug 1937, *Warnock 441* (F). **El Paso Co.:** Franklin Mts., Altura Park, near El Paso, 10 Jul 1911, *Barlow s.n.* (in part) (F); Franklin Mts., El Paso, McKelligon Canyon Park, 28 Sep 1948, *Norvell H340* (CAS(2)). **Presidio Co.:** Ojo Escondido, 1 mi W of Big Bend Ranch headquarters, 26 Sep 1975, *Butterwick et al. 1675* (TEX); Panther Canyon, N of Hwy 170 and Río Grande River, in vicinity of Colorado Canyon, 25 Jun 1975, *Butterwick et al. B-1007* (TEX); 13 mi from Redford on road to Lajitas, (29°15'N, 103°55'W), 26 Oct 1959, *Correll 23336* (TEX-LL); 19 mi SE of Redford on Camino del Río #170, 24 Sep 1966, *Correll 33892* (TEX-LL); Tapia Canyon, 12.4 mi E of Redford, 8 Nov 1964, *Correll et al. 30555* (TEX-LL); Capote Creek near Rancheria, S of Marfa, (30°5'N, 103°55'W), Oct 1883, *Havard 23* (GH, US); Sierra Vieja Mts., Musgrave Canyon, (30°25'N, 104°30'W), 1200 m, 6 Jul 1941, *Hinckley 1943* (BRIT, GH, TEX); S to old Madrid (Madril) Ranch, about 80 mi S of Marfa, Arroyo Primero, a tributary of Fresno Canyon, 914 m, 13 Dec 1941, *Hinckley 2300* (ARIZ, GH); W of Fresno creek, 5.5 mi above the mouth ("Bluebonnet Hill"), 914 m, 1 Mar 1959, *Johnston et al. 3702* (TEX); Charro Canyon, 19 Jun 1974, *Powell 2764* (MO, TEX-LL); 2.8 mi by Tx 170 NW of Lajitas, 2 Apr 1980, *Worthington 5645* (ARIZ).

Habitat.—This species is found mostly at lower elevations, from near sea level to 1200 m or more, along sand dunes, in rocky outcrops and talus slopes, throughout the desert shrubs and low tree forests, especially in moist areas along dry wash stream beds and canyons, and associated with *Acacia, Agave, Bursera, Cercidium, Encelia, Euphorbia, Fouquieria, Jatropha, Larrea, Mimosa, Platanus, Prosopis, Yucca*, and many cactus species and grasses. Soils may be clays, or silty to sandy, rocky, and frequently alluvial loams from basalt, volcanic ash or limestone rocks.

Diseases and pests.—Only spider mite damage is reported.

Common names.—The Papago name is given as "frijolillo"; Wright's bean, slender-stemmed bean (Isely 1993; Niehaus et al. 1984).

Ethnobotany.—Eaten both as green pods and as dry seed, namely in the Pinacate region of NW Sonora (Buhrow 1983).

Comments.—The flower of *P. filiformis* is somewhat smaller and of slightly different shape than other species with which it might be confused (see Color Plate II, photo 23) and the seed is also somewhat smaller and both ridged and rugose-papillate (see Color Plate IV, photo 47). It is also highly variable in growth habit and vegetative parts from small nearly bush semi-vines to large climbing vines, and in leaflet lobing from nearly entire to deeply incised. It is commonly found at low elevations in the lower SW US, Baja California and neighboring Sonora, México, and on all the islands in the Gulf of California, but does not extend very far either north or south from this area. According to Isely (1993), being a floristic element of the Sonoran desert, it extends to SE California, although Lackey (1983) reports the genus absent from California. Given its distribution, not surprisingly, good sources of tolerance to salinity have been found in this species at different stages of development (Bayuelo et al. 2002a). *P. filiformis* seems also interesting for adaption to extreme temperatures (Balasubramanian et al. 2000; Buhrow 1980, 1983). The reader should consult the work by Buhrow (1983) about the synonymy of *P. wrightii*.

J.2.—Phaseolus angustissimus A. Gray, Pl. Wright. 2:33. 1853. (**Figs. 63, 70, 73**). TYPE: UNITED STATES. NEW MÉXICO: stony hillsides, at the crossing of the Río Grande above Doña Ana, Doña Ana Co., (19) Jul 1851; and between San Pedro and Santa Cruz, Sonora, (32°30'N, 106°55'W), Sep, *Wright 68* (in part = *951* of *Gray*, see Shaw 1987) (HOLOTYPE: GH (designated by Delgado (1985); ISOTYPES: GH, K, MO, US).

Phaseolus angustissimus var. *latus* Jones, Contr. West. Bot. 12:14. 1908. TYPE: UNITED STATES. ARIZONA. Navajo Co.: on the Little Colorado River, northern Arizona, below Winslow, (35°N, 110°30'W), 9 Jun 1890, *Jones s.n.* (HOLOTYPE: POM designated by Delgado (1985) n.v.; F photograph n.v., UC photograph).

Phaseolus dilatatus Woot. & Standl., Contr. U.S. Nat. Herb. 16:139. 1913. TYPE: UNITED STATES. NEW MÉXICO. Catron Co.: in the Mogollon Mts., (33°30'N, 108°45'W), 30 Aug 1881, *Rusby 2024* (HOLOTYPE: US 138616).

Aerial shoot an annual, early, erect and then becoming prostrate or pendant, trailing, indeterminate vine. ***Root*** a perennial, somewhat woody and thickened, narrowly elongate, deep to 1–3 m long, 1–2 cm thick, somewhat branched, the elongate crown 5–10 cm long. ***Stems*** terete, 2–3 mm wide at base, 1–2 m long, with multiple branches from the first aerial node, each branch a short trailing, delicate stem, usually pendant, only slightly twining at tips; internodes 7–8 cm long, covered by minute hooked pubescence. ***Stipules*** triangular to oblong, 1.5–2 mm long, 0.5–0.75 mm wide, obscurely 3-nerved, covered by minute ciliate hairs, purplish. ***Leaves*** 6.8–10 cm long; petiole shorter to as long as leaflets, 2–3 cm long, sparsely covered with minute, white uncinate hairs; petiolule 0.5–1 cm long; pulvini 1.5–2.5 mm long, densely covered by minute uncinate hairs; stipels aciculate to linear, 1.5–2 mm long, covered by minute uncinate hairs; terminal leaflet broad and entire at base of plant to narrowly linear and sometimes lobed above, 4–5.5 cm long, 2–3–17 mm wide, variously lobed with mostly broad leaflets slightly lobed (northern range) to narrow leaflets with larger lobes to 9 mm long (Texas) to narrow leaflets 6 cm long by 4–8 mm wide (S. Arizona), tip obtuse, sparsely hispidulose of short hooked hairs and lustrous adaxially, glabrous abaxially, somewhat thickened or fleshy; lateral leaflet similar but more linear, 2.5–3 cm long, 2–3 mm wide. ***Inflorescence*** an erect, delicate pseudoraceme to 15 cm long; peduncle 6.5–12 cm long; rachis 3–8 cm long of 2–8 flowers; primary bracts minute, ovate-acuminate to oblong-linear 1.25 mm long, 0.25–0.5 mm wide, covered by minute puberulent and ciliate hairs, fairly persistent, pedicel 5–7 mm long, covered with minute hooked pubescence. ***Bracteoles*** small, ovate to ovate-elongate, 0.8–1.25 mm long, 0.5 mm wide, obscurely 1- to 3-nerved, acute, puberulent and ciliate, persistent through young pod stage. ***Flower*** purple, numerous; calyx campanulate, 2.5–3 mm long, the upper lobes united, barely emarginate, the lower lobes subequal, acute, 1 mm long, 1.5 mm wide, slightly reflexed, covered by minute pubescence and ciliate; standard purplish, broadly squarish-rounded, reflexed at 2.5 mm from base, erect tip, emarginate, 6–11 mm long, 10 mm wide, slightly enrolled at sides, the claw 0.5 mm long, the auricles small, 0.5 mm long; wings purple, the blade obovate, rolled lengthwise, 8–10 mm long, 6 mm wide, the claw 2.5 mm long, the spur pronounced, 1 mm long; keel short, the claws 1 mm long, 3 mm from base to bend and 4 mm more to the terminal 1 3/4 coils of 2.5 mm diam., tip greenish; vexillary stamen, the claw 0.5 mm long, the geniculate knob 0.75 mm wide, 0.4 mm high; stamen tube 4 mm to bend and 1.5 mm more to the filaments, the ridges barely developed; anthers with copious pollen but tripping is usually required to get pollination; basal collar 0.5 mm long; ovary 3.5–4 mm long, 1 mm wide, covered with minute white pubescence, 3–4 ovules; style 6 mm long to the terminal thickened coil of 2 mm diam.; stigma lateral introrse, slightly twisted laterally with a terminal pointed narrow tip, about 1 mm long. ***Pod*** falcate, 2–3 cm long, 0.5 cm wide, thin walled, reticulate, minutely puberulent, early dehiscent, with a relatively long, straight beak, 1–2 mm long. ***Seed*** oblong, 3.5–5 mm long, 3–4 mm wide, 1.3 mm thick, flattened, rugose to reticulate, brownish at times mottled black, black halo around hilum; hilum ovate, 0.5 mm long; lens globose, partially divided. ***Seedling*** from hypogeal germination; epicotyl 7–8 cm long; primary leaf petiole 0.75 cm long, the blade ovate-oblong, 2.8–3 cm long, 1.4 cm wide, slightly cordate base, acute.

Specimens examined: **MÉXICO. Sonora:** N of Horconcitos, Río Bavispe, (30°N, 109°W)?, 6 Sep 1940, *Phillips 838* (ARIZ, GH, MICH).

UNITED STATES. Arizona: Apache Co.: 10 mi N of Conche, (34°35'N, 109°35'W), 1981 m, 29 Jul 1963, *Lehto 2090* (ASU, MSC); 15 mi N of Ganado, (35°45'N, 109°20'W), 1890 m, 10 Jun 1937, *Peebles et al. 13482* (ARIZ, CAS). **Cochise Co.:** Walnut Wash, 1 mi N of Tombstone, (31°45'N, 110°5'W), 1295 m, 22 Jul 1953, *Gardner et al. 109* (US). **Coconino Co.:** Flagstaff, Sunset Crater Road, 1 mi E of visitor center, (35°25'N, 111°30'W), 2134 m, 8 Aug 1978, *Buhrow 12* (ARIZ); 1 mi SE of Arizona State Teachers College Campus, (35°15'N, 111°45'W)?, 2103 m, 21 Sep 1945, *Deaver s.n.* (ARIZ); Cemetery, S rim Grand Canyon National Park, (36°N, 112°15'W)?, 2134

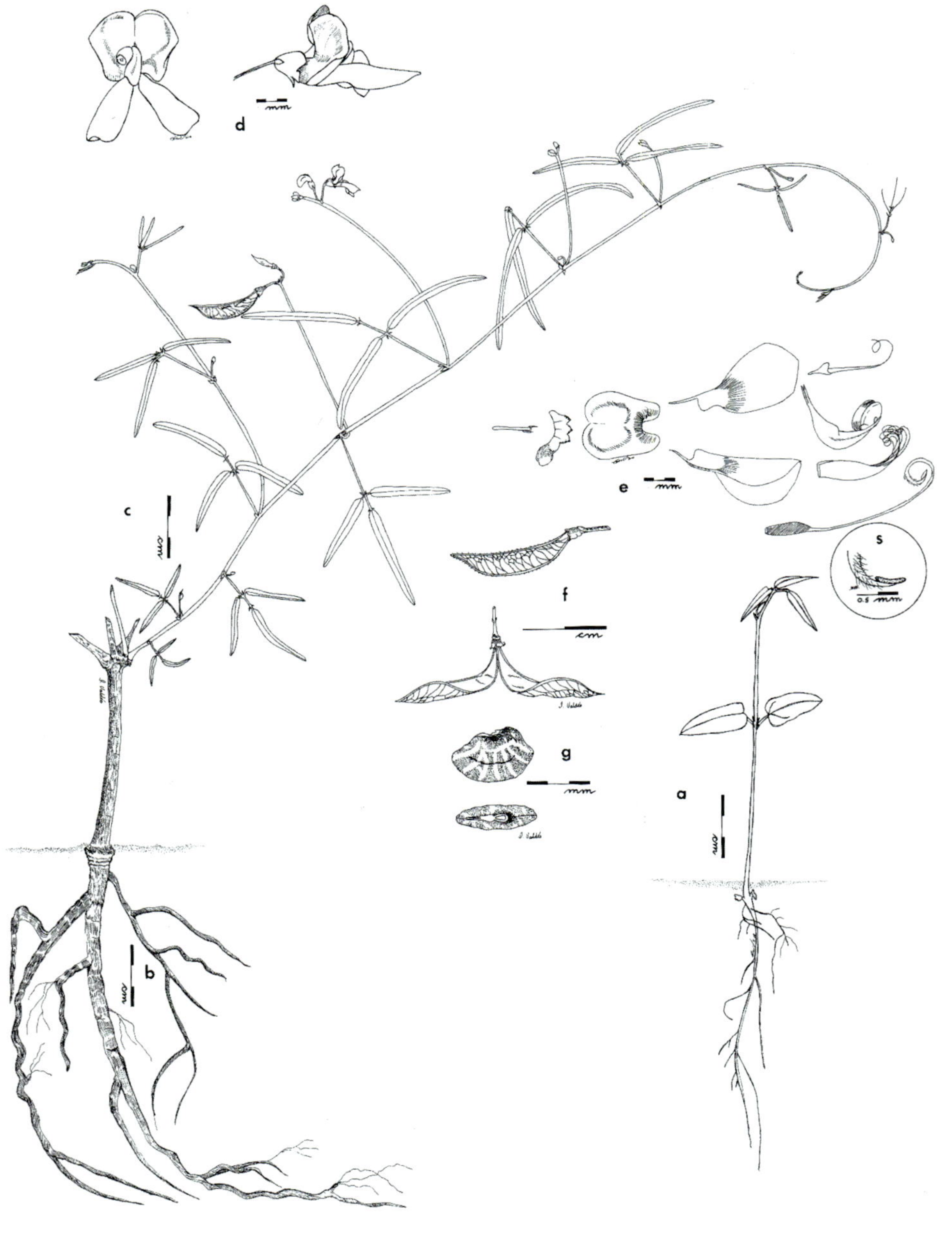

Fig. 70. Illustrations of *Phaseolus angustissimus* Gray.–a. Seedling a few weeks after germination; note the hypogeal germination; note short petiole and oblong-ovate eophyll.–b. Root of mature plant.–c. Portion of plant with leaves and inflorescences with pods and flowers.–d. Flowers, side view and front view.–e. Exploded view of flower showing all parts including–s. Style tip and stigma as seen under the microscope.–f. Pods, side view and dehisced; note reticulate venation.–g. Seeds, side view and view of hilum; note partially rounded and flattened shape and ridges. All drawings made from living material grown in greenhouse at Mayagüez from seed of *Buhrow SAAN*, (TARS #166) collected in the Organ Mts., New México, USA.

m, 25 Jun 1942, *Collom 1256* (US); Martin Budggelin's Ranch, Grand Canyon to Grand View, (36°N, 112°W)?, 2080–2150 m, 25 Aug 1919, *Eggleston 15695* (US); S side San Francisco Mts. N of Flagstaff, Schultz Pass road ca 2 1/2 mi E of US 180, (35°25'N, 111°50'W), 2134 m, 25 Jul 1983, *Ertter et al. 4970* (ASU, CAS, TEX); Williams, (35°20'N, 112°10'W), 8–25 Jul 1903, *Griffiths 4943* (MO, US); Tolchaco, 1524 m, 11 Jun 1922, *Hanson 80* (F, MO, MSC, TEX); Superior, (33°25'N, 111°W), 16 May 1926, *Harrison 1875* (ARIZ, US); Sedona, (34°50'N, 112°W), 26 May 1922, *Jones et al. 363* (GH, K, MO); Lake Mary, 30 Jun 1973, *Lehto et al. 17770* (ASU); 50 mi N of Williams, (35°50'N, 113°15'W), 2073 m, 7 Jun 1937, *Peebles et al. 13324* (ARIZ, US); Flagstaff, (31°10'N, 111°20'W), 1 Sept 1883, *Rusby 583* (in part) (MICH). **Gila Co.:** Strawberry Hill, Mogollon Rim N of Pine, (34°30'N, 111°30'W), 1798 m, 10 Sep 1935, *Collom 415* (ASU, DES); Apache Mts., 23 mi E of Rice, 6 Jul 1918, *Ferris 1288* (CAS-DS); Pinal Mts., S of the town of Globe, FS Road 580 to Madera Peak, (33°25'N, 110°35'W), 1920 m, 21 Sep 1980, *Forbes 1495* (ASU); Tontozona, Robert's burn area, 28 Jul 1965, *Pinkava*

2355 (ASU); Tonto National Forest, Rose Creek drainage, Sawmill Flats in the Sierra Ancha along Hwy 288, 2.4 mi S of Workman Creek, 1768 m, 4 Jul 1983, *Sanders et al. 3902* (UCR). **Graham Co.:** ? Pinaleño Mts., (32°42'N, 109°45'W)?, 1524 m, 30 Oct 1985, *Buhrow 85-1* (ARIZ); Goat Canyon, tributary to Bonita Creek, Upper Gila River drainage, 20 Apr 1978, *McGill 2365* (ASU, TEX); Galiuro Mts., (32°45'N, 110°20'W)?, 28 Jul 1894, *Toumey s.n.* (ARIZ, GH). **Greenlee Co.:** ? Coronado, Clifton, (33°10'N, 109°20'W), 1887, *Davidson 274* (UC); 2 mi N of Blue, (33°40'N, 109°5'W), 1900 m, 4 Sep 1948, *Gould et al. 5168* (ARIZ, BRIT, GH, UC); Blue Range Primitive Area, FSR 567, just above Blue Crossing, 1935 m, 11 Aug 1984, *Hodgson 3181* (DES). **Maricopa Co.:** Seven Springs Canyon, 914 m, 25 Aug 1951, *Blakley 663* (DES); 1/2 mi W of springs in Seven Springs wash, 27 Apr 1974, *Engard 196* (DES); Ariz. 87 at Bushnell Tanks (N of Sunflower), 30 Apr 1966, *Lehto et al. 6084* (ASU); Tonto National Forest, Cave Creek, 1066 m, 6 Oct 1958, *Pase 1020* (US). **Mohave Co.:** Upper Francis Creek, 1189 m, 14 Jun 1979, *Butterwick et al. 5123* (ASU, DES); Grand Canyon National Monument, Toroweap, Mt. Trumbull, Junction of Bob Sullivan's Ranch Road, (36°20'N, 113°10'W), 1981 m, 13 Jun 1954, *Cottam 13958* (US); Bull Canyon, 1341 m, 17 May 1979, *Hillyard et al. 4910* (ASU); Willows Ranch Rd at I-40, about 30 mi E of Kingman at Cottonwood Cliffs, (35°10'N, 113°30'W), 1524 m, 10 Oct 1982, *Sanders et al. 3241* (ARIZ, CAS, UCR). **Navajo Co.:** 7.3 mi from Taylor on road to Pinedale, (34°30'N, 110°15'W), 1859 m, 9 Oct 1966, *Bohrer 1158* (ARIZ). **Pinal Co.:** Tonto National Forest, Superstition Wilderness, 3/4 mi W of J.F. Ranch along Fraser Canyon, 914 m, 22 Mar 1984, *Hodgson et al. 2774* (DES); Tonto Forest, White Ridges, 1067 m, 16 Oct 1934, *Johnson 40* (ASU). **Yavapai Co.:** Phoenix Dist., Santa María Canyon, 792 m, 30 Aug 1978, *Butterwick 4042* (ASU, DES); Prescott National Forest, Milk Creek, 7 mi N of Wagoner, 1219 m, 28 Apr 1968, *Carpenter s.n.* (ASU); S of Prescott, (34°25'N, 112°35'W)?, 1 Jun 1944, *Clark 11737* (GH); Bloody Basin, Willow Springs Trail, 7 Sep 1974, *Engard 375* (DES); 2–5 mi N of road to Bartlett Dam, along Seven Springs Road, (33°50'N, 111°25'W), 26 Apr 1984, *Hodgson 2881* (DES); Lynx Creek, 23 Sep 1967, *Keil 2706* (ASU); Thumb Butte Campground, 8 May 1967, *Lehto et al. 8282* (ASU); Granite Basin Lake, 13 May 1967, *Lehto et al. 8321* (ASU); Prescott National Forest, Milk Creek, 7 mi N of Wagoner, 1219 m, 28 Apr 1968, *Lehto et al. s.n.* (DES); Kendrick Mts, 2500 m, 7 Jul 1901, *Leiberg 5660* (CAS-DS, US); 1.2 mi W of Hillside Mine, W of Bagdad (W of Prescott), (34°30'N, 113°5'W), 914 m, 11 Apr 1976, *McManus et al. 19* (ARIZ); Sunset Crater Dist., NE of Flagstaff, 2134 m, 29 Jun 1935, *Whiting 756* (ARIZ, GH). **County Unknown:** Hwy 87, 3 mi NW of Payson, just SE of the confluence of Sycamore Creek and East Verde River, (34°25'N, 111°5'W), 1378 m, 9 Jun 1991, *Baker 8527* (ASU); Salt River Crossing Hwy 60, (33°50'N, 110°25'W), 20 Sep 1938, *Goodding A9466* (ARIZ); 10 mi S of Black River, White Mts., 1890 m, 22 Jun 1930, *Goodman et al. 1293* (CAS, CAS-DS, F, GH, MICH, MO, UC); Elder Mts., May–Oct 1902, *Purpus 34* (MO, UC, US); Red Mt., 2438 m, May–Oct 1901, *Purpus 8009* (MO, US). **New México: Catron Co.:** Hwy 12, Continental Divide just E of Aragon, (33°55'N, 108°30'W), 17 Jul 1973, *Higgins 7775* (ASU); 31 mi N of Cottonwood Campground, Rte 180, 14 Aug 1974, *Pinkava et al. P12478* (ASU, US); Fullerton Ranch, Bat Cave, S(W) edge Plains of San Agustin (14 mi SW of Horse Springs), (33°45'N, 108°20'W), 15 Aug 1948, *Smith 156* (ARIZ, GH, US); 11 mi WSW of Quemado on U.S. Rt 60 near Red Cone, (34°15'N, 108°40'W), 1890 m, 7 Aug 1984, *Van Devender et al. 840315* (ARIZ). **Dona Ana Co.:** San Augustin Pass, E of Organ, (32°30'N, 106°30'W), 1753 m, 17 Aug 1977, *Nabhan GN677* (ARIZ); Van Patten's, Organ Mts., 11 Jun 1906, *Standley s.n.* (MO, US); Soledad Canyon, Organ Mts., 11 Jul 1904, *Wooton s.n.* (US); Organ Mts., Aguirre Springs Recreational Area, Sotol Creek at Pine Tree Trail, 1768 m, 18 May 1985, *Worthington 13140* (UCR). **Grant Co.:** 4 mi E of Gila, (32°50'N, 108°30'W), 23 May 1935, *Maguire et al. 11664* (ARIZ); Glenwood, 13 Aug 1935, *Moeller 244* (ARIZ). **Luna Co.:** Florida Mts., 8 mi SE of Deming, 1610 m, 3 May 1969, *Hess 2485* (BRIT); E side of Florida Mts. at W end of the Three Little Hills, (31°50'N, 107°35'W), 1433 m, 5 May 1985, *Worthington 13094* (UCR); Florida Mts., 0.5 mi NW from top of Baldy Peak, at spring, 1676 m, 25 Aug 1990, *Worthington 18632* (UCR). **Santa Fé Co.:** Los Frijoles, 5 Sep 1929, *Whitehouse 8623* (CAS, F, MICH, TEX). **Valencia Co.:** 20 mi SW of Grants, (35°N, 108°W), 12 Aug 1967, *Turner 5692* (BRIT, TEX); Craters, (34°45'N, 107°W)?, 28 Jul 1906, *Wooton s.n.* (US). **County Unknown:** Lake Valley, (36°5'N, 108°25'W), 1914, *Beals s.n.* (US); Cottonwood Canyon, Stevens Mill, 17 Aug 1938, *Goodding et al. A9438* (ARIZ); Florita Mts., 1615 m, 7 Sep 1903, *Jones s.n.* (CAS-DS(3), US); Mesa #3, México Springs, (35°45'N, 108°45'W), 19 Jun 1934, *Klinger A208(?)* (ARIZ); Burro Mts., (32°33'N, 108°30'W)?, Jun 1881, *Rusby 2479* (F, MO, US); Castle Rock, Datil Forest, 2347 m, 10 Sep 1915, *Talbot 112* (US). **Texas: Brewster Co.:** El Solitario, 10 May 1928, *Cory 1813* (GH); 20 mi S of Alpine, (30°N, 103°30'W), 12 Sep 1959, *Turner et al. 4594* (TEX); Chisos Mts., Bobbs Hut on Pinnacle Mt., (29°15'N, 103°15'W), 22 Jul 1937, *Warnock 652* (TEX, US); Calamity Creek, 6 Aug 1941, *Warnock 21018* (TEX). **Culberson Co.:** 7 mi N of Van Horn, E side of Beach Mt., (31°15'N, 104°40'W), 13 Jul 1943, *Waterfall 5118* (GH, MO). **El Paso Co.:** Franklin Mts., El Paso, E end of Kenyon Joyce Rd., 1 block E of Stanton St., 31°48'N, 106°30'W, 1219 m, 4 Oct 1982, *Sanders et al. 3106* (ARIZ, UCR). **Mojave Co.:** Willows Ranch Road at I-40, 30 mi E of Kingman at Cottonwood Cliffs, 1524 m, 10 Oct 1982, *Sanders et al. 3241* (UCR). **Presidio Co.:** 10 mi from Ruidosa on road to Marfa, Pinto Canyon below Scheele's Goat Ranch, (29°50'N, 104°25'W), 11 Sep 1961, *Correll et al. 24409* (BRIT, GH, TEX-LL); Chinati Mts., 10 Aug 1936, *Hinckley 796* (ARIZ, BRIT, F, GH, TEX). **County Unknown:** Tarlinga Cr. and Grapevine Cañon, Sep (18)83, *Havard s.n.* (US); Doney Park, 2012 m, 25 Jun 1916, *Wyman s.n.* (ARIZ).

Habitat.—This species seems to require eroded areas and gullies on steep, rocky hillsides where it becomes trailing and pendant. It is reported as commonly found in grassy areas of piñon-juniper, open (Ponderosa) pine or oak forest, or chaparral associations with *Acacia, Amelanchier, Artemisia, Berberis, Celtis, Ephedra, Euphorbia, Fraxinus, Penstemon, Rhus,* and *Yucca.* Soils are rocky and sandy derived from limestone, sandstone, conglomerate or volcanic rock. This species ranges farther north in Arizona than any other species of the genus in this part of the United States (see also Buhrow 1983).

Common names.—It has been reported as the "Slimleaf Lima bean" by Niehaus et al. (1984) for the Southwestern US and Texas.

Comments.—Piper (1926) recognized the perennial nature of the species—in reality a trailing vine from a very long, linear, somewhat thickened (to a few cms), very deep tap root with a narrow crown (see Color Plate III, photo 33). Usually the plant branches from the first internode some centimeters (to 1 dm) above ground level and is somewhat erect at first, becoming a pendant vine in eroding gullies and is seldom found in non-eroded areas nearby. The flower is flattened and facing upwards which may be a response to its prostrate habitat (see Color Plate II, 18). It can easily be distinguished from *P. filiformis* and other species with which it might be confused by the pod which is somewhat chartaceous and scimitar shaped, the heavier stems and somewhat waxy leaves, and the seed which are less rugose and somewhat larger (see Color Plate IV, photo 48).

Cluster analysis of electrophoretic patterns of seed proteins has indicated a possible close relationship between this species and *P. leptostachyus* (= *P. anisotrichus*) (Sullivan & Freytag 1986); however larger samples should be analysed and interspecific hybridization should be attempted to clarify this relationship. *P. angustissimus* has been found recently with some tolerance to freezing (Balasubramanian et al. 2002). Although Delgado (1985) listed the original publication as per 1852, it seems to be of 1853.

J.3.—Phaseolus carteri Freytag & Debouck, sp. nov. (**Figs. 71, 73**). TYPE: MÉXICO. BAJA CALIFORNIA: 3 mi SW of San Pedro on La Paz–Todos Santos road, (23°50'N, 110°15'W), 213 m, 14 Oct 1966, *Gould 12181* (HOLOTYPE: MICH).

Persimilis *Phaseolo filiformi*, sed plantulis foliolis elongatis, floribus modice grandibus, leguminibus plerumque 6–7 seminibus differt. Occurrit nunc solum prope San Pedrensis in locis australibus La Pax Bassi Californici multo rarus.

Aerial shoot a trailing and climbing, indeterminate vine, to 1–2 m long, branched once or twice at base. ***Root*** annual, fibrous with a somewhat thickened taproot, small, poorly developed. ***Stem*** terete, 2–3 mm in diam.; internodes mostly 6–10–(15) cm long, puberulent of minute uncinate hairs, slightly ribbed. ***Stipules*** triangular, 1–1.5 mm long, 0.5–0.75 mm wide, acute, lightly 1- to 3-nerved, slightly pubescent. ***Leaves*** 9–12.5 cm long; petiole 4–5 cm long, ribbed, heavily covered by uncinate hairs; petiolule 1.5–2 cm long, covered by uncinate hairs; pulvini dark, 2–2.5 mm long, densely covered by uncinate hairs; stipels acicular, 1.25 mm long, 1-nerved, nearly glabrous; terminal leaflet triangular to broadly ovate, very slightly lobed at base, the basal leaves much more lobed than the upper ones, to 5 cm long, 5 cm broad at about 1/4–1/3 from base, obscurely nerved, acute to obtuse, thin membranaceous drying nearly hyaline, nearly glabrous on both surfaces, somewhat puberulent, ciliate at margins, mostly light green. ***Inflorescence*** a delicate, erect to spreading, slightly curved pseudoraceme, the basal racemes to 23 cm long; peduncle 2–4.5–13 cm long, heavily covered by minute uncinate hairs; rachis 5.5–8–12 cm long of 3–7 flowering nodes, heavily covered by minute uncinate hairs; primary bracts ovate, 1–1.5 mm long, acute, 1(3)-nerved, covered with minute uncinate hairs; pedicel 3.5 mm long, covered by minute uncinate hairs; pedicellar bracts scale-like narrowly lanceolate 1 mm long or less 1-nerved margins hyaline ciliate. ***Bracteoles*** lanceolate to spatulate, 0.8–1 mm long, 0.25–0.5 mm wide, 1-nerved, nearly hyaline, ciliate, persistent. ***Flower*** very small to 15 mm long, white to pale purple; calyx campanulate, tube 1.5 mm long, the upper two teeth rounded, scarcely elongate, 1.25 mm wide, joined into 1 emarginate, 0.3 mm long, 2.5 mm wide, the lower 3 subequal, dentate and acute to elongate, 0.5 mm long, 1.23 mm wide, covered with minute uncinate hairs and ciliate; standard white (pale purple?), the claw 1.25 mm long, 0.5 mm wide, sharply reflexed at 2.5–2.75 mm from base and 2.5 mm more to emarginate apex, 6–8.5 mm wide, the auricle flaps 0.4 mm wide, 0.4 mm long; wings white (to purple), subequal, the blade ovate, rolled lengthwise and cupped, spreading, 8–12 mm long, 5 mm wide, the claw 3 mm long, 0.3 mm wide, the spur well-developed, angular, 1 mm high; keel, the claws 2 mm long, the ridges well-developed, 1 mm diam., 4 mm to bend and 2.5 mm more to base of the terminal 1 3/4 coils of 1.5–2 mm diam.; vexillary stamen, the claw 1 mm long, the geniculate flap 0.3 mm long, 0.5 mm wide, 2 mm more to end of thick portion; united stamen tube 3.5 mm to bend, 2.5 mm more to end of united portion, the small ridges at 1.5 mm from base, 0.25 mm high; basal collar 0.5 mm long; ovary straight 3.5 mm long, 0.75 mm wide, densely appressed hirsute, with 5–7 ovules; style 5 mm long to the terminal thickened coil of 1.5 mm diam.; stigma terminal, introrse, linear, 0.7 mm long. ***Pod*** straight to slightly curved, 2–3 cm long, 5 mm wide, somewhat squarish at tip; sutures somewhat thickened, covered with uncinate hairs; valves reticulate, puberulent, light yellowish green, drying tan; beak straight to recurved downward, delicate, 2–2.5 mm long;

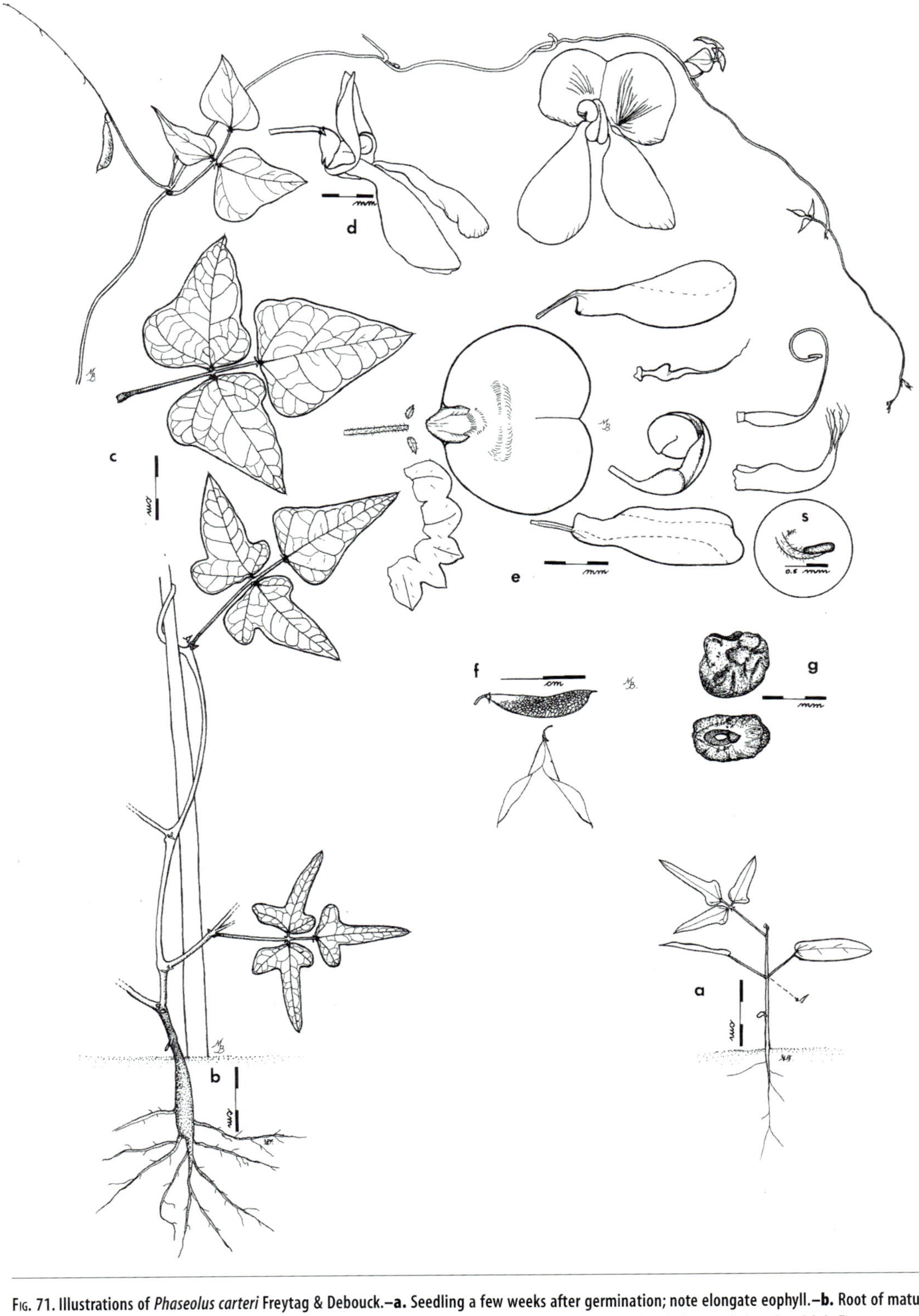

Fig. 71. Illustrations of *Phaseolus carteri* Freytag & Debouck.—**a.** Seedling a few weeks after germination; note elongate eophyll.—**b.** Root of mature plant.—**c.** Lower portion of stem with deeply lobed leaflets and separately a mature leaf from middle of plant and vine tip with inflorescence.—**d.** Flowers, side view and front view; note very long wings.—**e.** Exploded view of flower showing all parts including—**s.** Style tip and stigma as seen under the microscope; note very broad standard.—**f.** Pods, side view and dehisced.—**g.** Seeds, side view and view of hilum; note ridged seed surface. All drawings made from living material grown in greenhouse at Mayagüez from seed of *León de la Luz 3751* (TARS #535) collected near San Pedro, S of La Paz, Baja California, México.

4–7 seeds. ***Seed*** rhomboidal to squarish, slightly flattened, 3.4 mm long, 3 mm wide, 2.3 mm thick, somewhat undulate-rugose, speckled black on light tan, black ring around hilum; hilum oval, 0.5 mm long. ***Seedling*** from epigeal germination; hypocotyl 16.4 mm long; epicotyl 16.4 mm long; eophylls with petiole 1.7–2 cm long, with basal and upper pulvini, the blade simple, entire, narrowly elongate, 3.2 cm long, 1 cm wide, cordate at base, obtuse, apiculate.

PARATYPES. **MÉXICO. Baja California:** Miraflores, (23°13'N, 109°46'W), 14 Oct 1890, *Brandegee 159* (UC); cerca de San Pedro, 12 km S de La Paz, *León de la Luz 3751* (US); Mpio. La Paz, inmediaciones de la Presa Sta. Inés, 23°26'N, 110°14'W, 200 m, 5 Oct 1989, *León de la Luz 4137* (US); Km 10 carr. San José del Cabo–La Rivera, Los Cabos, 23°5'N, 109°36'W, 14 m, 6 Oct 1989, *León de la Luz 4188* (US).

Habitat.—This species is found growing with and under cactus and shrubs in the sandy soils of dry washes and gulleys of the low tree and saguaro type cactus desert. Apparently it is scarce and may not develop into flowering plants every year.

Comments.—The species is named to honor Dr. Anneta Carter (deceased) who was active for many years in collecting the flora of Baja California and without whose assistance it would not have been possible to validate the presence of the species through recent collections by the local botanists. Thanks to the recent collections of plants and viable seed made by Biol. José Luís León de la Luz stationed at La Paz, it was possible to grow it out in the USDA screenhouse in Mayagüez, Puerto Rico and in CIAT. The distinguishing elongate seedling leaves (see Color Plate V, photo 59), puberulent pedicels, relatively large flowers, and the pods with usually 6–7 seeds, separate this taxon from *P. filiformis*. *P. carteri* seems to be rare and now known only from near San Pedro just S of La Paz, Baja California; however, it seems likely that the range for this species might involve Sierra El Novillo, Sierra de La Laguna and perhaps Sierra Santa Clara.

The case of *Phaseolus microcarpus* Mart.

Comments.—Here follows another case of divergence between morphological and molecular data. While morphologically *P. microcarpus* would fall within section *Rugosi* (although its seed testa are not exactly rugose as they are in the three species belonging to the section), molecular evidence using different kinds of markers (Delgado et al. 1999; Fofana et al. 1999; Gaitán et al. 2000; Jaaska 1996; Jacob et al. 1995) has shown it quite distant from it, even completely isolated, thus filling in a strong justification to create another section. The junior author (Debouck 2000a) has proposed a 'phylum of *P. microcarpus*.'

Phaseolus microcarpus Mart., Auswahl. Merkw. Pfl. 18–19, pl. 12. 1831. (**Figs. 72, 73**). TYPE: MÉXICO. [MICHOACÁN]: Tlalpujahua, (19°50'N, 100°10'W), 1827 or 1828, *Karwinski s.n.* (?, see below) (HOLOTYPE: BR designated by Piper (1926) n.v.).

Phaseolus monospermus Robins. & Greenm., Proc. Amer. Acad. Arts & Sci. 29:385–386. 1894. TYPE: MÉXICO. JALISCO: barranca of Tequila, (20°55'N, 103°45'W), 13 Oct 1893, *Pringle 5446* (HOLOTYPE: GH designated by Delgado (1985); ISOTYPES: F n.v., K, MEXU n.v.).

Aerial shoot a climbing and trailing, indeterminate vine, to 2–3 m long. ***Root*** an annual, fibrous, extensively branched. ***Stems*** branched at most lower internodes; internodes fairly short to 13 cm long; lower stem terete, to 2–3 mm in diameter, sparsely covered with strigose hairs and densely covered with very minute hooked hairs, slightly ribbed, the younger stems somewhat more pubescent. ***Stipules*** broadly ovate, 3–3.5 mm long, 2 mm wide, heavily 5- to 6-nerved, acute, glabrous, stiff. ***Leaves*** 3.5–6 cm long, mostly light green leaves, very resistant to drying; petiole 11–13 mm long, glabrescent to densely covered with minute hooked and scattered strigose hairs; pulvini 2 mm long, dark, with a few to many minute hooked hairs; terminal leaflet broadly ovate at base of plant and triangular and lobed above, to 4 cm long, 4 cm broad at 1/3 from base, obscurely to clearly squarish lobed at base, membranaceous; the lateral leaflet to 3.6 cm long, 3.2 cm wide at 1/3 from base, obscurely 3 nerved with smaller lateral veins frequently perpendicular to the main veins, the tip acute, foliaceous, abaxially nearly glabrous to covered with minute, glandular hairs 0.04–0.06 mm long and minute hooked hairs, adaxially nearly glabrous to covered with scattered, basally spreading, glandular-hispid hairs, ciliate on leaf margins. ***Inflorescence*** an erect, slightly curved panicle to 20 cm long, the peduncle to 11 cm long, the rachis with 5–11 or more flowering nodes, the nodes early 2-flowered, the small secondary branches developing a few millimeters from base and sometimes flowering, cov-

FIG. 72. Illustrations of *Phaseolus microcarpus* Mart.—**a.1** & **a.2** Seedling plants several weeks after germination; note broad ovate eophylls.—**b.** Root of mature plant.—**c.1** Lower portion of plant with leaves and inflorescences with pods.—**c.2** Upper portion of vine with mature leaf and separately a mature leaf showing lobing.—**d.** Flower, side view and front view.—**e.** Exploded view of flower showing all parts including—**s.** Style tip and stigma as seen under the microscope.—**f.1** Mature pods, side view and dehisced; note reticulate and pubescence of carpels.—**f.2** Mature pod of unusual black-seeded plant shown side view.—**g.1** Light colored seed, side view and view of hilum.—**g.2** Black seed shown side view. All drawings of living material grown in greenhouse at Mayagüez from seed of *Freytag et al. 78-Mex-65* (TARS #85B) collected at Km 82, Rt. 190, Oaxaca, México, except: black seed, pod, and seedling **a.2** grown from seed of *Debouck 389* collected near Mariano Matamoros, Durango, México; lobed leaf from the field collection of *Freytag et al. 78-Mex-67* collected at Km 140, Rt. 190, Oaxaca, México.

ered with minute hooked hairs to 0.3 mm long, more densely adaxially; primary bracts ovate, 3 mm long, 2–2.5 mm wide, strongly 3- to 5-nerved, glabrous, ciliate, the secondary bracts minute linear scale-like hyaline 1-nerved; pedicel delicate, 5 mm long, purple, heavily pubescent with a few strigose and many minute hooked hairs. ***Bracteoles*** scale-like, triangular-lanceolate, 1.5 mm long, 1–1.5 mm wide, 1- to 3-nerved, sparsely pubescent, often persistent. ***Flower*** very small, 8–9 mm long, light lavender to purple, mostly fertile; calyx 1.5–1.75 mm long, the upper 2 teeth broadly obtuse, 0.25–0.75 mm long, the lower teeth acute, equal, 0.75 mm long, 1 mm wide, sparsely pubescent of minute hairs, a few longer strigose hairs, ciliate; standard light purple, broad, erect and reflexed, 1.75 mm long, 3.5 mm wide, the claw 1.5 mm long; wings purple, subequal, the blade 4–7 mm long, 2.5 mm wide, cupped, twisted to nearly vertical, the basal claw 2 mm long, 0.25 mm wide, the spur 0.5 mm long, 1.5 mm wide; keel extending out for about 4 mm with 1 1/2 terminal coils, white to yellowish, the basal claw separated for about 1 mm, the lateral knobs about 1 mm in size; vexillary stamen minute, 5.5 mm long, the basal knob about 0.5 mm from the base, 0.5 mm in width, somewhat longer than wide and inflated, no sheath projecting; united stamen tube about 4 mm long, the anthers 0.35 mm long, 0.2 mm wide, yellow; basal collar about 0.25 mm long; ovary 2.5 mm long, 1 mm wide, densely covered by short, silky pubescence, beginning development while still in flower, 1–2 ovules; style 4.25 mm to the thickened 1 1/3 coils of 1.25 mm in diam.; stigma lateral introrse, about 0.65 m long, pubescent at base. ***Pod*** rounded-rhomboid or diamond shaped, 10 mm long, 7 mm wide, 2.5 mm deep, heavily pubescent when immature with a few long canescent hairs on the central area of each valve, carpels covered by minute hooked hairs, when mature the sutures thickened, the tip usually pointed, rarely rounded, easily dehiscent with the pedicel or falling from the plant before full maturity, usually dark brown, sometimes speckled with black. ***Seed*** rounded rhomboid or pyramidal, 6.25 mm long, 4.4 mm wide, 2 mm deep, the surface somewhat angular and undulate, colored by black speckles on a tan background, shiny, a black ring around the hilum; hilum at the angle opposite the rounded side, oval, 0.6 mm long, 0.3 mm wide. ***Seedling*** from epigeal germination; hypocotyl 15–35–44 mm long; epicoytl 15–22–35 mm long; eophyll, the petiole 1.5–2 cm long, the blade simple, entire, broadly ovate, 2.75–3.5 cm long, 2.75–3.5 cm wide, cordate base, acute.

Specimens examined: **EL SALVADOR. Ahuachapán:** San Benito, al E de la cima del cerro La Olla, 13°49'N, 89°56'W, 13 Oct 1992, *Sandoval et al. 763* (MO).

GUATEMALA. Zacapa: vicinity of Zacapa, (14°45'N, 89°30'W), 200 m, 7–16 Oct 1940, *Standley 74200* (F).

MÉXICO. Chiapas: Mpio. Berriozabal, 5 km E of Berriozabal along Hwy 190, (16°47'N, 93°15'W), 800 m, 11 Oct 1971, *Breedlove 20412* (CAS-DS, MO, TEX); 11 mi SE of Tapanatepec on Hwy 190, 701 m, 20 Oct 1965, *Breedlove et al. 13708* (CAS-DS, F, MEXU, MICH, TEX-LL, US); Mpio. Tuxtla Gutiérrez, Mirador Tepehuaje en Cañón del Sumidero, 21 km N de Tuxtla Gutiérrez, 1270 m, 20 Oct 1989, *Reyes et al. 1285* (MO); Mpio. Trinitaria, Campana, a 16 km al SE de la Trinitaria, (16°5'N, 92°W), 820 m, 1 Nov 1976, *Sousa et al. 6820* (MO). **Durango:** Rodeo, El Resbalón, Km 146 carr. 45, 25°08'N, 104°35'W, 1390 m, 4 Oct 1978, *Debouck et al. 294* (CHAPA, COL, K); Mezquital, approxim. 12 km SSE Mariano Matamoros, 23°44'N, 104°26'W, 2060 m, 27 Oct 1978, *Debouck 389* (CHAPA, COL, G, K, M, MICH); 1/2 camino El Mezquital a El Salitre, 1420 m, 23°28'N, 104°24'W, 28 Oct 1978, *Debouck 392* (CHAPA, COL, G, K, M, MICH, UC); Mpio. Durango, Puente Nombre de Dios, Nombre de Dios, 47 km de la frontera en camino a La Ciudad Zacatecas, 1970 m, 6 Nov 1978, *Delgado et al. 1025* (CAS). **Guanajuato:** Valle de Santiago, cañada 0.5 km W de Volcán Batea, 20°19'N, 101°12'W, 1850 m, 16 Nov 1987, *Debouck et al. 2399* (CHAPA, MICH, US). **Guerrero:** 9 km NE de Xaltianguis, 17°08'N, 99°39'W, 320 m, 30 Oct 1987, *Debouck et al. 2353* (CHAPA, MICH, MO); 7 km N de Zumpango del Río, 17°42'N, 99°33'W, 860 m, 31 Oct 1987, *Debouck et al. 2358* (CHAPA, MICH); entre Acapulco y Puerto Marqués, (16°50'N, 99°50'W), *Miranda 8418* (GH); 9 km al N de La Unión, (18°N, 101°50'W), 200 m, 24 Oct 1983, *Soto et al. 6014* (CAS, MO, TEX). **Jalisco:** La Huerta, 0.8 km W de Chamela, 19°32'N, 108°06'W, 8 m, 8 Nov 1978, *Debouck et al. 424* (CHAPA, COL, G, K, M, MICH); Tecolotán, 3 Km NE of Linda Vista, 20°15'N, 103°56'W, 1650 m, 26 Nov 1978, *Debouck et al. 470* (CHAPA, COL); 2 km SE de San Agustín, Unión de Tula, 1350 m, 26 Nov 1979, *Debouck et al. 472* (CHAPA, COL); Playa 'El Paraiso', Chamela, (19°30'N, 105°W), 0 m, 19 Sep 1976, *Delgado et al. 129* (in part) (ARIZ, ENCB, MEXU); 5 km seaward from the main Hwy., along the side-road to El Tecuan, Barra de Navidad, (19°10'N, 104°40'W), 100 m, 10–12 Nov 1971, *Dieterle 4158* (MICH); Tonalá, near Río Santiago, 900 m, grown in screenhouse in Mayagüez, PR from seed of NI 709 (= TARS #80 collected 23 Nov 1978 by *Debouck et al. 449*), Sep.-Dec, 1985 as Study Collection *Freytag GF #SC-89* (CSU, WIS); Mpio. La Huerta, 11 km al E de la carr. Puerto Vallarta–Barra de Navidad, antiguo camino a Nacastillo, 9 Oct 1982, *Lott et al. 1408* (CAS, MICH, MO); Mpio. La Huerta, Rancho Cuixmala, SW of Laguna de Corte on road to Casa Jeanette, 2 km N of mouth of Río Cuixmala, 5–10 m, 24 Oct 1990, *Lott et al. 2881* (UCR); Mpio. La Huerta, Arroyo El Colorado, a 100 m dentro del lindero Este de la Estación Exp. de Chamela, 1 Apr 1977, *Magallanes 592* (DES, MEXU); 8 km E of Chamela, 30–50 m, 8 Dec 1970, *McVaugh et al. 25088* (CAS, MICH, US); Chapala, (20°20'N, 103°10'W), 5 Oct 1903, *Rose et al. 7650* (GH, US). **México:** Dist. Temascaltepec, Acatitlán, (19°N, 100°W)?, 21 Oct 1935, *Hinton et al. 8584* (BM, F, G, GH, K, MEXU, US). **Michoacán:** Mpio. Aquila, La Ticla, (18°35'N, 103°30'W), 200 m, 12 Oct 1979, *Guerrero 594* (TEX); Mpio. de Tzitzio, Iranche, 29 Sep 1977, *Soto 369* (DES). **Nayarit:** Mpio. Nayar, Km 2–5 del Cangrejo a la

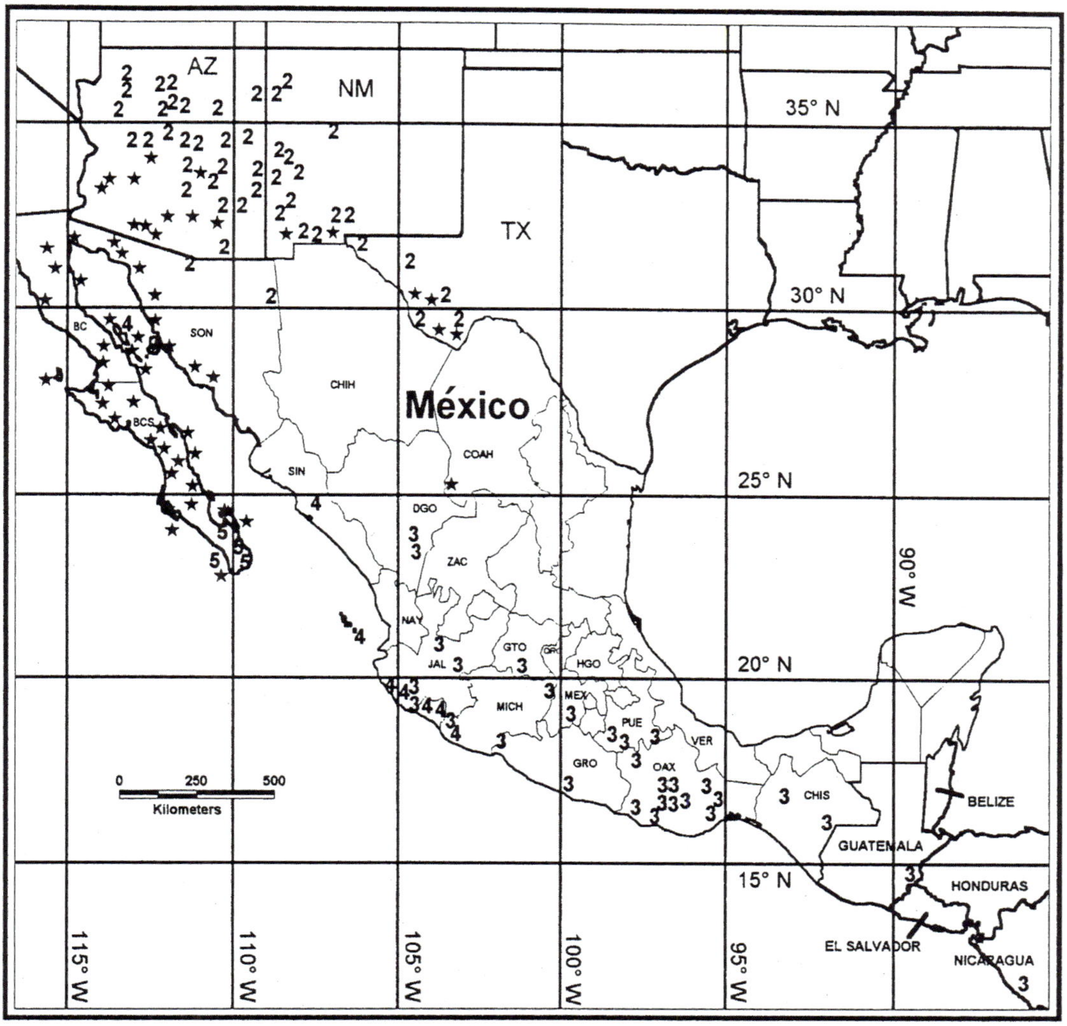

Fig. 73. Distribution of species of Section J. *Rugosi* as follows: ★ = *P. filiformis*; 2 = *P. angustissimus*; 5 = *P. carteri*. Other small-seeded species: 3 = *P. microcarpus*; 4 = *P. macvaughii*.

Mesa del Nayar, 22°14'N, 104°39'W, 1100 m, 17 Sep 1989, *Téllez 12215* (MEXU, MO). **Oaxaca:** cerro al lado del desvío a Macuilxochitl, Teotitlán del Valle, Tlacolula, 17°01'N, 96°33'W, 1570 m, 25 Oct 1987, *Debouck et al. 2339* (CHAPA, MICH, US); Cerro El Labrador, 8 km N de El Portillo, 16°38'N, 96°35'W, 1570 m, Coatecas, Ejutla, 26 Oct 1987, *Debouck et al. 2341* (CHAPA, US); 6 km NW de Puerto Escondido, San Pedro Mixtepec, Juquila, 15°58'N, 97°5'W, 90 m, 28 Oct 1987, *Debouck et al. 2350* (BR, CHAPA, MICH, SI, UC, US); 3 km SE de Jamiltepec, 16°16'N, 97°48'W, 190 m, 29 Oct 1987, *Debouck et al. 2352* (CHAPA, MICH); Km 4 Hwy 175, Oaxaca-Tuxtepec, 17°03'N, 95°36'W, 1580 m, 22 Sep 1978, *Freytag et al. 78-Mex-58* (MEXU, MO, US); Km 82 Hwy 190, Oaxaca-Tehuantepec, 16°39'N, 96°14'W, 890 m, 22 Sep 1978, *Freytag et al. 78-Mex-65* (MEXU, MO, US); Km 109 Hwy 190 Oaxaca-Tehuantepec, 16°40'N, 95°05'W, 780 m, 22 Sep 1978, *Freytag et al. 78-Mex-66* (EAP, MEXU, MO, US); Km 140 Hwy 190, Oaxaca-Tehuantepec, 16°34'N, 96°56'W, 940 m, 22 Sep 1978, *Freytag et al. 78-Mex-67* (BR, EAP, MEXU, MO, US); 18 mi S of Mitla Junction along main Hwy, 1219 m, 21 Oct 1967, *Gentry 22358* (GH, MEXU, NA); 4 mi NE of Hwy Jct along road to Ixtlán de Juárez, 1676 m, 2 Nov 1967, *Gentry 22397* (MEXU, MICH, NA); Monte Albán, (17°5'N, 96°55'W), 2040 m, 11 Oct 1895, *Smith 931* (GH, K); Mpio. Teotitlán del Valle, Dist. Tlacolula, Km 5 _ camino Teotitlán-Benito Juárez, adelante de la presa Piedra Azul, 1820 m, 27 Oct 1978, *Solano et al. 208* (F, MO); Desierto a 18 km al SE de Huajuapan de León, (17°45'N, 97°45'W), 1800 m, 18 Oct 1976, *Sousa et al. 6205* (BM, MEXU, MO(2)); Dist. Tlacolula, ruinas arqueológicas de Yagul, 2750 m, 17 Sep 1978, *Sousa et al. 9395* (BM, CAS(2), CR, MO); Dist. Tehuantepec, 9 km al W de Marilú, (16°20'N, 95°20'W), 490 m, 18 Sep 1980, *Sousa et al. 9492* (CAS); Dist. Tlacolula, 7 km al N-NE de Díaz Ordaz (Sto. Domingo del Valle), 1800 m, 23 Sep 1978, *Sousa et al. 9685* (ASU, CAS); 7 km al NW de Yautepec, 875 m, 19 Sep 1976, *Téllez et al. 133* (BM, CAS, MEXU, MO). **Puebla:** Puerto Las Palmas, 17 km SE de Izucar de Matamoros, 18°28'N, 98°23'W, 1430 m, 21 Oct 1987, *Debouck et al. 2326* (CHAPA, MICH, MO); Puente Marquéz, 6 km NW de Tehuitzingo, 1000 m, 23 Oct 1987, *Debouck et al. 2327* (CHAPA,

MICH); Acatlán, 1 km SE de El Papayo, 14 km NW de Amatitlán, 18°15'N, 98°10'W, 1300 m, 23 Oct 1987, *Debouck et al. 2328* (CHAPA, COL, MICH, MO); Km 8 Hwy 135 Tehuacán-Huautla, 18°08'N, 96°59'W, 1480 m, 20 Sep 1978, *Freytag et al. 78-Mex-38* (BR, EAP, F, GH, MEXU, MO, UC, US); Coxcatlán, 1650-1980 m, 9 Sep 1909, *Purpus 4203* (F, GH, MO, UC, US); Tehuacán, Sep 1911, *Purpus 5780* (BM, F, GH, MO(2), UC, US).

NICARAGUA. Chinandega: Volcán Chonco, 12°41'N, 87°02'W, 700-1100 m, 30 Oct 1984, *Moreno 25030* (MO). **Estelí:** Mesas Moropotente, 11.7-13.7 km NE of Hwy 1 at Estelí on road to Yalí, 13°10'N, 86°17'W, 1200-1265 m, 15 Nov 1979, *Stevens et al. 15702* (MO). **León:** Volcán Momotombo, 200 m, 29 Jan 1980, *Araquistain et al. 1090* (MO). **Managua:** S slope and rim of Apoyeque, 12°13'N, 86°20'W, 220-350 m, 18 Nov 1980, *Stevens et al. 18447* (MO). **Masaya:** Parque Nacional Volcán Masaya, W shore of Laguna de Masaya, (11°45'N, 86°5'W), 140 m, 29 Nov 1977, *Neill 3058* (MO, US).

Habitat.—This species is mostly found growing along streams or on steep talus slopes in deserts or dry, deciduous tropical, or open pine-oak forests, or even tropical moist forests, in a wide range of altitudes, climbing over shrubs of and under *Acacia, Agave, Bursera, Cardiospermum, Ceiba, Guapira, Heliocarpus, Jatropha, Lysiloma, Mimosa, Pithecellobium*, and various cacti and grasses. Soils are sandy, rocky, clays or silts derived from schist, basalts, granite, limestone or volcanic rock.

Diseases and pests.—Reported to be attacked by virus and powdery mildew. Damage is reported from thrips, caterpillars, chrysomelids and grazing.

Comments.—In the type description, Martens (1831) indicates to have received seeds from Mr. Keerl, not W. Karwinski. Maréchal et al. (1978b) report the type as *H. Galeotti* s.n. at BR. The species is distinctive by the very small, chartaceous, densely pubescent pods with 1 to 3 triangular or rhomboidal seed, and small flowers (see Color Plate II, photo 23). The pods seem to be slightly bigger in some Nicaraguan materials, although the specimens match fully with the type description. Before his work at MO, the junior author (Debouck 2000a) considers the southern limit of distribution as eastern Guatemala, obviously it goes further southeast; this species is reported in 'Flora de Nicaragua' (Delgado 2001).

Section K.—Falcati Freytag, sect. nov. TYPE SPECIES: *Phaseolus leptostachyus* Benth., Comm. Legum. Gen. 72. 1837.

Herba perennis volubilis parva prostrata pendulata vel scandens multiramularis et indeterminata vel determinata, radix plerumque crassa carnosa, stipulae elongatae, foliola pusilla mediocria ovata saepe integra, flos parvus purpureus vel violaceus, legumen parva falcata turgida pilosis longis vel strigosis.

Plant a prostrate, pendant or scandent, multi-branched, small, herbaceous, indeterminate or determinate vine; root usually perennial, branched, thick, fleshy, branched; with prominent, extended stipules and small to medium, ovate, mostly entire leaflets; flower small, purple or violet; pod small and falcate, somewhat inflated and covered with long pilose or strigose hairs.

Comments.—A section composed of species with falcate pods and scandent vines and which have been reported as having a chromosome count of 2n=2x=20 (Lackey 1980; Maréchal 1970; Mercado-Ruaro & Delgado Salinas 1996). ITS DNA sequencing data (Delgado et al. 1999; Gaitán et al. 2000) have shown *P. leptostachyus* grouping together with *P. macvaughii* and *P. micranthus*, with a remote link to the tertiary gene pool of Lima bean (see also Debouck 2000a,b). The most common species *P. leptostachyus* has an exceptionally wide distribution in México and Central America.

KEY TO SPECIES AND VARIETIES

1. Plant a determinate, prostrate or pendant vine, the leaves often reduced or absent near the apex of the stems, pedicels very short about 2 mm long, bracteoles ovate, 1 mm long; common, throughout highland México and S to Costa Rica; 250–2700 m ______ K.2.1. **P. leptostachyus** var. **leptostachyus**
    2. At least some leaves 5-pinnate, the terminal leaflets deeply divided or sessile pinnate.
        3. Flowers purple; rare, Jalisco and Michoacán; 200 m ______ K.2.3.1. **P. leptostachyus** Benth. var. **pinnatifolius** forma **purpureus**
        3. Flowers white; rare, Oaxaca; 2080 m ______ K.2.3.2. **P. leptostachyus** var. **pinnatifolius** forma **albus**
    2. None of the terminal leaflets deeply divided or pinnate.
        4. Lateral leaflets with a large basal lobe, terminal leaflets elongate nearly broad linear; rare, Jalisco; 1400 m ______ K.2.5. **P. leptostachyus** var. **lobatifolius**
        4. Lateral leaflets not lobed.
            5. Some of the terminal leaflets slightly lobed at base, more noticeable near apex of stems; scarce, throughout central México and S to Honduras; 800–2190 m ______ K.2.2. **P. leptostachyus** var. **intonsus**
            5. Terminal leaflets not lobed, usually small ovate, acute; rare, Durango and Jalisco; 2000–2700 m ______ K.2.4. **P. leptostachyus** var. **nanus**

1. Plant an indeterminate, prostrate or climbing vine, the leaves only slightly smaller near the apex.
   6. All flower parts persisting on tip of pod until late pod state, all leaflets broad ovate, dark olive green, pedicels 5–7 mm long, bracteoles broad ovate, 5–6 mm long; rare, Volcán Acatenango, Guatemala; 1820 m ____ K.4. **P. persistentus**
   6. Flower parts early deciduous after anthesis.
      7. Leaflets dark olive green, ovate-elongate, medium to 8 cm long, plant upright but not twining; rare, Veracruz and Tamaulipas; 1000 m ____ K.3. **P. opacus**
      7. Leaflets not dark olive green, ovate to oblong, plants mostly prostrate vines with erect, climbing tips.
         8. Plants mostly prostate and rooting from nodes touching soil, wings nearly flattened and touching one another, lower calyx lobes acute, 1 mm long; scarce, along streams in oak or pine-oak forests, Sonora to Michoacán; 60–2100 m ____ K.1. **P. micranthus**
         8. Plants prostrate at base and then climbing, wings enrolled and spreading, lower calyx lobes broadly acute, 0.5 mm long; scarce, in coastal plains of Sinaloa to Michoacán; from sea level to 550 m ____ K.5. **P. macvaughii**

K.1.—Phaseolus micranthus Hook. & Arn., Bot. Beech. Voy. 287. 1838. (**Figs. 74, 80**). TYPE: MÉXICO. JALISCO: Talisco, (20°30'N, 105°20'W)?, Dec–Feb 1827–1828, *Lay & Collie s.n.* (= *Beechey s.n.*) (HOLOTYPE: K designated by Delgado (1985), US 1326443 (fragment of K), GH (photograph of K), US (photograph of K).

Phaseolus brevicalyx Micheli, Mém. Soc. Phys. Hist. Nat. Genève 34:261–262 (+ pl. 12). 1903. TYPE: MÉXICO. [MICHOACÁN or GUERRERO]: Sierra Madre, (18°45'N, 103°10'W)?, 1750 m, Jan 1899, *Langlassé 791* (HOLOTYPE: G; ISOTYPES: GH, K).

Phaseolus falciformis Piper, Contr. U.S. Natl. Herb. 22:693–694. 1926. TYPE: MÉXICO. NAYARIT: Tepic, near Compostela, (21°15'N, 105°W), (1500–1800 m), 7–8 Apr 1897, *Nelson 4173* (HOLOTYPE: US 763711; ISOTYPE: GH).

Aerial shoot a slender, scandent, indeterminate vine. ***Root*** fibrous, annual, somewhat woody, with adventitious rooting from horizonal stems. ***Stems*** near the base terete, striate, 1–3 mm thick, somewhat woody, mostly prostrate or climbing by the stipules, younger stems densely covered by long, yellowish, somewhat reflexed-strigose and hispid hairs to 1.5 mm long and by minute hooked hairs. ***Stipules*** oblongate, 4–6 mm long, 1.75 mm wide, 5- to multi-nerved, glabrous, ciliate at margin. ***Leaves*** 9.4–16 cm long; petioles delicate, 5–6.5 cm long, similarly pubescent as the stem, the petiolules 1.5 cm long; pulvini 2–2.5 mm long, densely covered with long, yellow strigose hairs; stipels linear, 1.25–1.5 mm long, 1-nerved, ciliate; terminal leaflet ovate to rhombic-ovate to lanceolate, 2.5–7.5 cm long, 2–5 cm wide at 1/3 from base, indistinctly 3-nerved, mostly acuminate but rarely obtuse, apiculate, thin-membranous, dark green to olive-green adaxially, very sparsely covered with long hispid to hirsute hairs, somewhat silvery reflective below, ultimate veinlets well defined; lateral leaflets similar and only slightly inequilateral. ***Inflorescence*** a pseudoraceme, most 5–10–20 cm long; peduncle 4–10 cm long; rachis 3–4–10 cm long of 10–12 flowers, densely covered with long yellowish strigose and uncinate hairs; primary bract narrowly ovate, 2.3–3.5 mm long, 3-nerved, densely pubescent on abaxial surface, nearly glabrous adaxially; pedicel 2–3.5 mm long, densely covered by pilose hairs; pedicellar bracts 1.5–2 mm long narrowly lanceolate purplish 1- or 2-nerved sparsely ciliate margin. ***Bracteoles*** lanceolate, 1–1.5 mm long, 0.5 mm wide, strongly 1-nerved, covered with minute ciliate hairs, purplish, often inserted 1 mm below the calyx. ***Flower*** lilac to purple, often drying yellowish or brownish; calyx flaring campanulate, 2–3 mm long, upper lobes united into 1 rounded, emarginate, scarcely elongated, puberulent, the lower lobes triangular, acute, 0.5 mm long, the center one 0.75–1 mm long, sparsely bearded by long pilose hairs, sometimes black-punctate; standard purple, strongly reflexed at 2.5 mm from base and 4.5 mm more to apex, thickened at flexure, enrolled ventrally at sides, deeply emarginate at apex, 6–7 mm long, 7–8 mm wide, the claw 0.75 mm long, the auricles 0.5 mm long; wings light violet, the blade oblong, 10–12 mm long, 5 mm wide, rolled somewhat lengthwise, the claw 2.5 mm long, the spur 0.5 mm long, tightly adhering to the keel; keel, the claws 2.5 mm long, 5 mm more to bend and 1 mm more to tip terminal 1 3/4 coils of about 2 mm diam., greenish tip; vexillary stamen, the claw 0.5 mm long, the geniculate knob globose, 0.7 mm wide; stamen tube 3.5 mm to bend and 3 mm more to divided filaments, 1.5 mm wide, the ridges raised 0.25 mm; basal collar not developed; ovary straight, 3 mm long, 1 mm wide, covered by long hispid pubescence, 3–5 ovules; style 4 mm long to the terminal thickened coil of 1.5 mm diam.; stigma lateral introrse, linear, 0.8 mm long, pointed. ***Pod*** immature densely covered with minute ciliate hairs, valves with sparse, long pilose pubescence, with fine purplish streaks by mid development, older pods nearly glabrous, somewhat stipitate and falcate, 2–3 cm long, 0.6 cm wide, valves chartaceous, compressed, reticu-

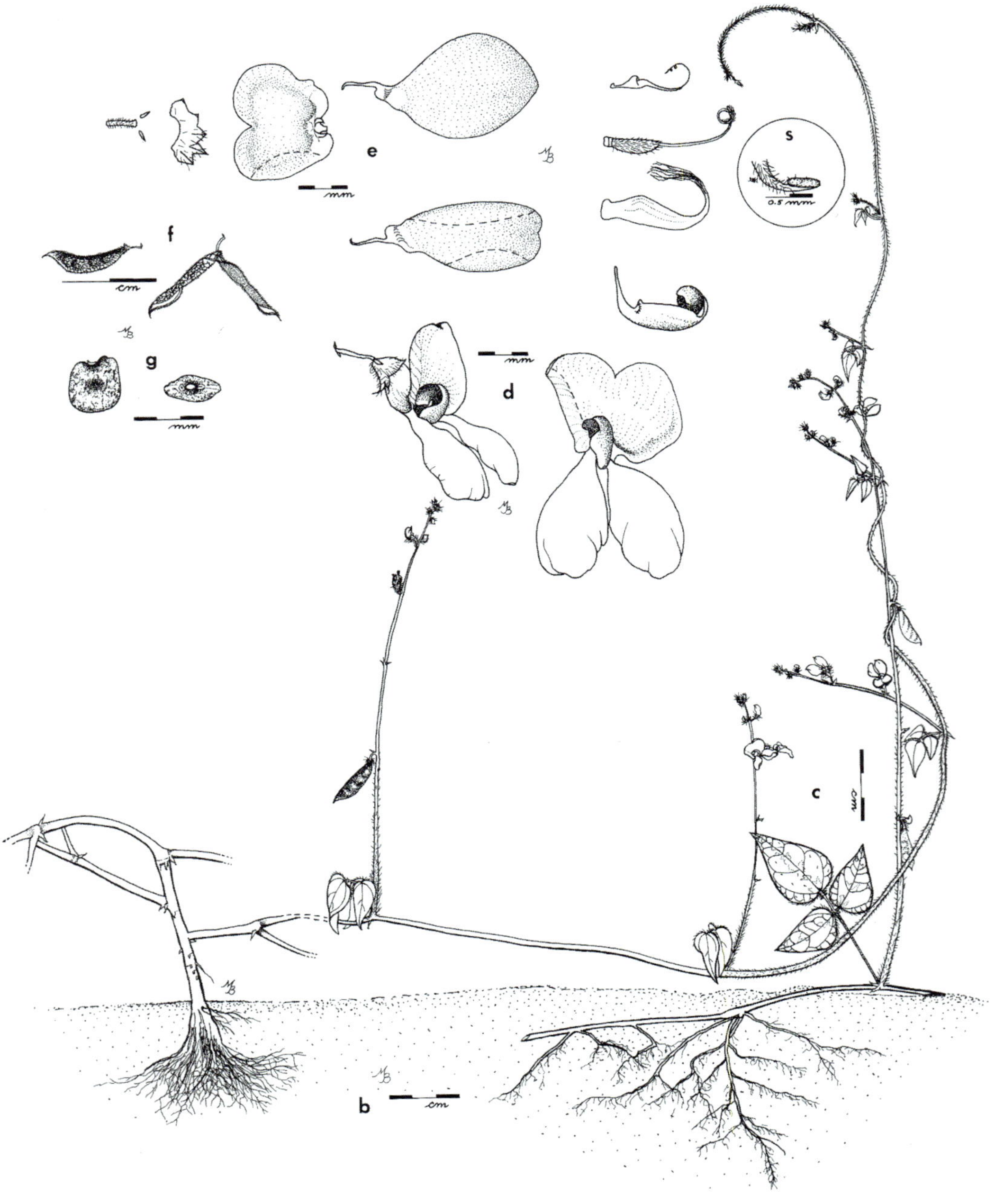

Fig. 74. Illustrations of *Phaseolus micranthus* Hook. & Arn.–**b.** Two different types of roots, one from germination and one from stem rooting.–**c.** Prostrate and climbing stems showing leaves and inflorescences; note pubescence of stems.–**d.** Flowers, front view and side view.–**e.** Exploded view of flower showing all parts including–**s.** style tip and stigma as seen under the microscope.–**f.** Pods, side view and dehisced; note pubescence on carpels.–**g.** Seeds, side view and view of hilum. All drawings made from living material grown in greenhouse at Mayagüez from seed of *Buhrow M28* (TARS #424) collected near El Tuito, Jalisco, México.

late, dehiscent with 1 loose twist, sutures finely pronounced, the beak short, bluntly curved, 1–2 mm long, weak. ***Seed*** rhomboid to squarish, flattened, 2 mm long, 2–2.5 mm wide, speckled black on light tan, black ring around hilum; hilum oval, 0.3 mm long. ***Seedling*** from epigeal germination; hypocotyl 12–14 mm long; epicotyl 5–7 mm long; eophylls opposite broadly ovate triangular, 15 mm long 12–14 mm wide,slightly cordate at base, acute to slightly acuminate; petiole 1–2 cm long, glabrous, canaliculate with basal and distal pulvini, no stipels; first true leaf trifoliolate.

Specimens examined: **MÉXICO. Jalisco:** 4–8 mi E of El Tuito, (20°20'N, 105°20'W), 1000 m, 13 Aug 1983, (Grown in Campbell Ave. Farm greenhouse, Pima Co, AZ) *Buhrow M28* (ARIZ); Mpio. Talpa, entre Cumbre del Tejamanil y Cuale, 1600 m, 3 Mar 1971, *González 83* (MICH); Mpio. Cabo Corrientes, 3–10 km E on the road to Mina Del Cuale, from the Jct. 5 km NW of El Tuito, (20°20'N, 105°35'W), 850–1150 m, 16–19 Feb 1975, *McVaugh 26359* (MICH); Arroyo del Chorillo, Quimixto, 60 m, 1 Dec 1926, *Mexía 1207* (F, G, GH, US); San Sebastian, Hacienda del Cura, Sierra Madre Occidental, (20°45'N, 104°51'W), 1425 m, *Mexía 1361a* (UC, US); 8 km SE of Hwy 110 on a lumber road leaving the Hwy. 12 km SSW of Tecalitlán and extending to San Isidro, 19°20'N, 103°10'W, 2100 m, 24–25 Sep 1965, *Roe et al. 2081* (MICH); Casimiro Castillo, 1–2 km al SE base del Cerro La Petaca, 19°35'16"N, 104°24'48"W, 700 m, 11 Nov 1992, *Santana et al.5967* (WIS); Mpio. Talpa, Talpa a Desmoronado, 700 m, 6 Dec 1981, *Vázquez 1134* (WIS). **Michoacán:** Dist. Coalcomán, Sierra Naranjillo, (18°55'N, 103°W), 13 Mar 1941, *Hinton et al. 15781* (TEX-LL, UC). **Nayarit:** Mpio. Compostela, 2 km NW of Sayolita, 20°50'N, 105°20'W, 20 m, 22 Mar 1989, *Flores et al. 909* (MO); Tiger Mine, Acaponeta, (22°30'N, 105°30'W), 1 Mar 1927, *Jones 23096* (MO); 9 mi N of Compostela, (21°15'N, 105°W), 1000–1200 m, 12 Nov 1959, *McVaugh et al. 484* (MICH); Tepic, 5–6 Jan 1892, *Palmer 17* (US); Mpio. Nayar, N of Mesa del Cangrejo, 22°16'N, 104°38'W, 1350–1500 m, 14 Nov 1990, *Ramírez et al. 796* (MO); Mpio. Tepic, Las Tierritas, 2 km NE of Izote, cerro San Juan, 21°31'N, 104°59'W, 1200 m, 23 Mar 1989, *Tenorio et al. 15586* (MO). **Sinaloa:** Sianori, 1924, *Ortega 5297* (US); 7.5 mi SW of El Palmito by Hwy 40, 8.2 mi SW of Durango-Sinaloa state line across Hwy 40, (23°30'N, 105°50'W), 610 m, 6 Jan 1983, *Worthington et al. 9300* (UCR). **Sonora:** Arroyo Tejas, Sierra Charuco, 1372 m, 14 Oct 1966, *Gentry 22002* (GH, NA, US); Las Piedras Canyon, Sierra Alamos, (27°N, 109°W), 750 m, 12 Mar 1984, *Starr 746* (UCR).

Habitat.—This species is found growing in rocky stream valleys in oak or pine-oak forests on decomposed granitic or volcanic soils, with *Echinocereus scheeri*, *Podocarpus*, *Russelia furfuracea* and other deciduous trees.

Comments.—*Phaseolus micranthus* though scandent is distinct from others of the section in having annual fibrous roots, small, boat-shaped, reticulate, chartaceous pods covered with long yellowish pilose and uncinate pubescence which makes them prickly to the touch, broad ovate to lanceolate leaflets, and much long, yellow strigose pubescence on stems and inflorescences (see Color Plate V, photo 56). The black color of the quite long, bristly hairs on the calyx, as mentioned by Piper (1926) as a distinctive character, is more often found in bud stage which quickly fades as the flower matures; in addition it is not displayed by all populations. Most of the pubescence, which is quite long, yellowish strigose or pilose, is found on young stems and peduncles. A useful diagnostic character is the insertion of the bracteoles on the pedicel often well below the calyx (as nicely reported by Micheli, 1903). The chromosome number is reported as n=x=10 by Mercado-Ruaro and Delgado Salinas (1996). *P. micranthus* is relatively rare and found only in W Mexico.

K.2.1.—Phaseolus leptostachyus Benth. var. **leptostachyus**, Commen. Legumin. Gener. 72. 1837. (**Figs. 75, 80**). TYPE: MÉXICO. (1827 or 1828), *Karwinski s.n.* (HOLOTYPE: M n.v.; GH, MICH, MO (photographs of M)).

Phaseolus anisotrichos Schlecht., Linnaea 12:326–327. 1838. TYPE: MÉXICO. VERACRUZ: Xalapa, in dumetis prope Jalapam, (19°30'N, 96°45'W)?, Aug 1828, *Schiede s.n.* (HOLOTYPE: HAL 55314 designated by Delgado (1985) n.v.).

Phaseolus fulvus Brandeg., Univ. Calif. Publ. Bot. 4:87. 1910. TYPE: MÉXICO. PUEBLA: Cerro del Gavilán, (2300–2500 m), Aug 1909, *Purpus 3889* (HOLOTYPE: UC 137610; ISOTYPES: ARIZ, CAS-DS, F, GH, MO, UC, US).

Aerial part a perennial, determinate vine, prostrate or rarely climbing over grass and low shrubs by means of the large, stiff, extended stipules. ***Root*** a perennial, thick, fleshy, tapering, much branched, orange-brown, to 25 cm or more long, seldom over 2 cm thick. ***Stems*** often forming a dense mat of much branched stems from the root crown, terete at base, mostly 2 mm or less in diameter, sometimes only 15 cm but usually to 2–3 m long; internodes 5–10 cm, near the vine tip to 2 cm long, mostly shorter with the ultimate leaves reduced or missing, often densely covered with yellowish or brownish strigose hairs 1–2.25 mm long with many minute hooked hairs beneath. ***Stipules*** broadly ovate to rhomboedric, enrolled lengthwise, stiff, extended or somewhat reflexed, 6–7 mm long, 5–6 mm wide, strongly 10- to 12-nerved, slightly covered with strigose hairs especially on margins. ***Leaves*** 6–11.5 cm long; petiole 2–4 cm long; petiolule 1–1.9 cm long, nearly terete, deeply canaliculate, nearly glabrous to quite pubescent, mostly moderately covered with long strigose and many minute hooked hairs; pulvini 2–4 mm long, densely covered with long strigose hairs; terminal and lateral leaflets similar, although lateral ones slightly inequilateral, broadly ovate to oblong-ovate, sometimes slightly lobed at base, to 4.8 cm long, 4 cm wide at about 1/3 from base, acuminate, acute or somewhat obtuse, not apiculate, moderately nerved, somewhat coriaceous, nearly glabrous to densely strigose on both surfaces with minute hooked hairs and many minute granular hairs on the adaxial surface, considerably lighter in color abaxially, infrequently mottled adaxially. ***Inflorescence*** an erect

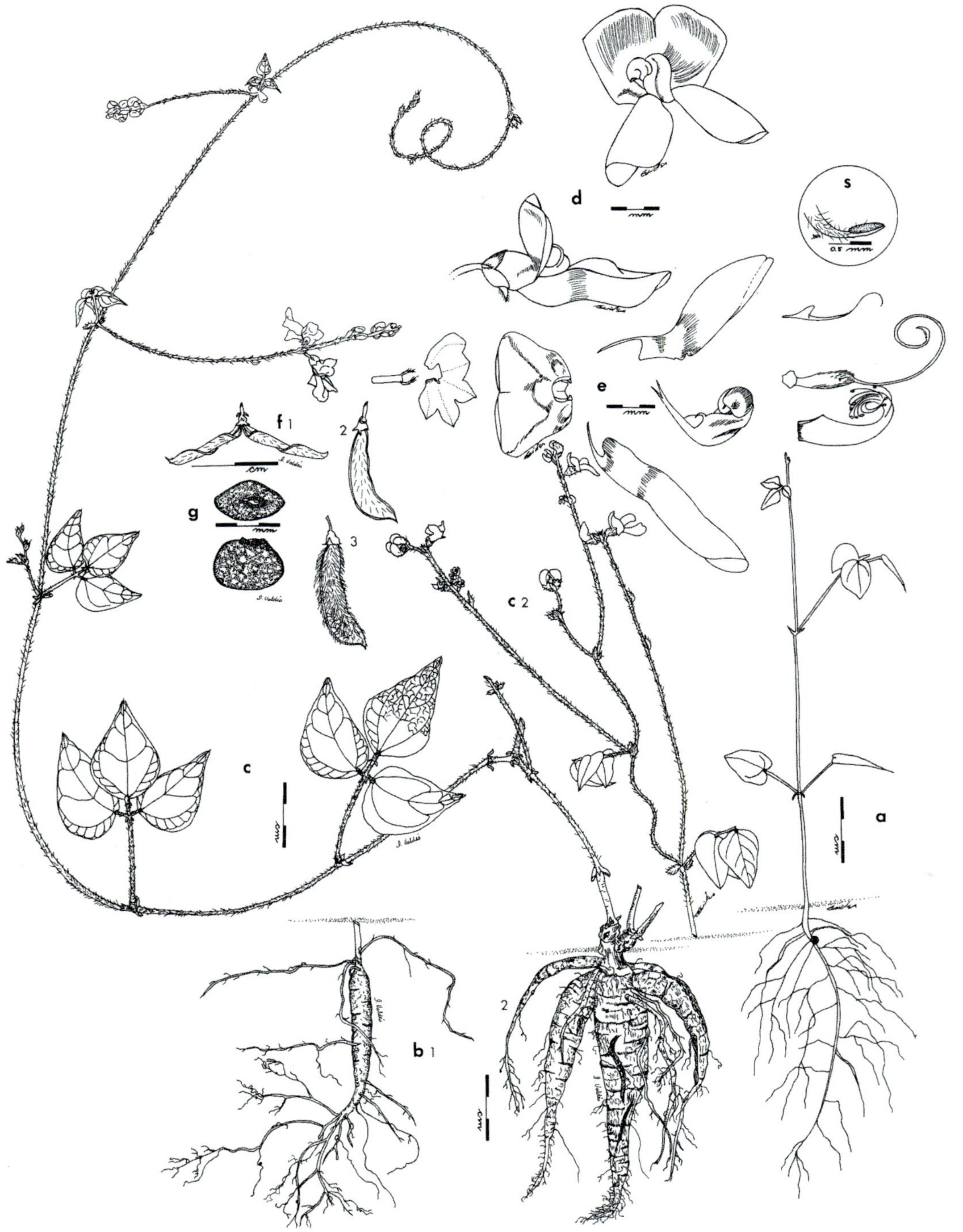

FIG. 75. Illustrations of *Phaseolus leptostachyus* Benth. var. *leptostachyus*.–**a.** Seedling a few weeks after hypogeal germination.–**b.1.** Root after one year.–**b.2.** Root after several years growth.–**c.1.** Young lateral stem with mature leaves and inflorescences; note the heavy pubescence of stems and inflorescences.–**c.2.** Vine tip of old stem showing determinate growth habit of several nodes of inflorescences.–**d.** Flowers, front view and side view; note sharply reflexed standard and flaring wings.–**e.** Exploded view of flower showing all parts including–**s.** Style tip and stigma as seen under microscope.–**f.1.** Dehisced pod.–**f.2.** Lateral view of mature pod with sparce pubescence.–**f.3.** Lateral view of mature pod with heavy pubescence.–**g.** Seeds, side view and view of hilum; note the heavily speckled black surface. All drawings made from living material grown in greenhouse at Mayagüez as follows: **a, b.2, f.1, f.2,** & **g.** from seed of *Freytag et al. 81-33* (TARS #20) collected near El Chante, Jalisco, México; **b.1** from seed of *Hernández 224* (TARS #21) near UNAM, Cd. México, D.F., México; **c.2** from seed of *Debouck & Muruaga 410* (TARS #23) collected near Km 30 carr. Durango–Mazatlán, Durango, México; **f.3** from seed of *Debouck 480* (TARS #163) collected from near Tequesquitlán, Jalisco, México; **c.1, d.** & **e.** from seed of *Cojulún 78-Guat-3B* (TARS #182) collected at Km 34, Jalapa–San Pedro Pinula, Guatemala.

pseudoraceme usually slightly curved, to 30 cm long; peduncle 12–20 cm long; rachis 10 cm or more long, often producing 15 or more flowering nodes, moderately to heavily pubescent of minute hooked hairs; primary bracts broadly ovate to oblong-lanceolate, 4–9 mm long, 2.5–4.5 mm wide, strongly many-nerved, strigose hairs mostly on borders, persistent, often dark purple; the pedicellar bracts only 1–2 mm long scale-like 1-nerved hyaline puberulent; 2–3 nearly sessile flowers produced at each node, most setting pods with often many pods per raceme; pedicel 2 mm long, very slightly pubescent of minute hooked hairs. ***Bracteoles*** minute scalelike, ovate, sessile to nearly 0.5 mm below calyx, 0.75–1 mm long, 0.25–0.5 mm wide, cupped inward, 1-nerved, moderately covered with minute strigose and hooked pubescence and ciliate-fimbriate on margins, weakly persistent. ***Flower*** light to dark purple (ageing to light brown); calyx campanulate, 3–3.5 mm long, the 2 upper teeth united into a single rounded emarginate lobe 0.75–1 mm long, 4–4.5 mm wide, the lower central lobe acute, 1.5 mm long, 1.5 mm wide, the lateral lobes 1–1.25 mm long, 1.5–2 mm wide, slightly covered with strigose hairs and fimbriate on the lower teeth, purplish; standard pale pink to deep purple with green base, 6 mm long, 7.5 mm wide, reflexed and thickened at 3 mm from base, the end portion recurved, erect and hooding the keel, emarginate, the claw 1 mm long, the auricles well-developed, 0.75 mm long; wings purple, the blade obovate, 12–13 mm long, 5–6 mm wide, enrolled lengthwise, spreading, the basal claw 3–3.5 mm long, 0.4 mm wide, the lateral spur pronounced, 1.25 mm long, 1.25 mm wide, strongly adhering to keel; keel white, the basal claw 2.5 mm, the long, lateral knobs poorly developed, with the terminal 1 3/4 coils of 2 mm diam., yellowish or greenish tip; vexillary stamen to 10 mm long, the claw 1 mm long, the knob pronounced geniculate, rounded, raised 1 mm, about 0.5 mm wide; stamen tube, united portion 6.75 mm long, reflexed at 4 mm, the auricles not developed; anthers 0.5 mm long, 0.25 mm wide; basal collar 0.4 mm long; ovary straight, oblongoid, 3–4 mm long, 0.5–0.9 mm wide, covered laterally with dense fine cannose hairs about 1–2 mm long, 6 ovules; style 6 mm long to the thickened terminal coil of 1.75 mm in diameter; stigma lateral, introrse, terminally pointed, 0.75 mm long, 0.01 mm wide. ***Pod*** falcate, 1.7–3.3 cm long, 6–7 mm wide, inflated, sometimes reticulate, moderately appressed-strigose with hairs 1–2 mm long only on valves, covered with many minute hooked hairs abaxially, tip acute, the valves dehiscent and moderately twisted at maturity with 1–2 turns, brown when dry, often very many pods per pseudoraceme. ***Seed*** squarish to rounded 2.75–3 mm long, 2.75–3 mm wide, 1.3 mm deep, smooth, shiny, somewhat reddish when immature, speckled black or dark brown, or rarely solid black, a black ring around the hilum; hilum round, 0.4 mm in diameter. ***Seedling*** from hypogeal germination; epicotyl 4–4.7 cm long; petiole 2.3 cm long with basal and upper pulvini; eophyll ovate, 3 cm long, 1.3 cm wide, cordate at base, acute.

Specimens examined: **COSTA RICA. Alajuela:** 0.3 km S de Alto Villegas, San Antonio de Barranca de San Ramón, Naranjo, 10°9'N, 84°26'W, 1350 m, 9 Jan 1987, *Debouck et al. 2107* (BR, CR, US). **Guanacaste:** Finca Taboga, Cañas, (10°18'N, 85°10'W), 1 Dec 1969, *Daubenmire 302* (CR, F); Comelco G, W of Bagaces, (10°25'N, 85°25'W), 17 Nov 1971, *Heithaus 438* (MO). **San José:** 2 km S de Monterrey, 9°45'N, 84°7'W, 1210 m, 10 Jan 1987, *Debouck et al. 2109* (BR, CR, US); San Francisco de Guadalupe, 1150 m, Oct 1892, *Pittier 7159* (G).

EL SALVADOR. San Salvador: San Salvador, (13°30'N, 89°10'W)?, 1922, *Calderón 1255* (GH, US).

GUATEMALA. Baja Verapaz: Km 115 Hwy 5 Cobán–Cd. Guatemala between Rabinal and El Chol, 15°N, 90°14'W, 1360 m, 12 Oct 1978, *Freytag et al. 78-Guat-132-1* (MEXU, US). **Chimaltenango:** Aldea Buena Vista, 2 km W de Chimaltenango, 14°40'N, 90°51'W, 1940 m, 9 Dec 1985, *Debouck et al. 1626* (COL, K, UC, US, USCG); 1 km N de aldea Hacienda Vieja, 12 km N de San José Poaquil, 14°52'N, 90°57'W, 1630 m, 12 Dec 1987, *Debouck et al. 2450* (BR, MICH, SI, US, USCG). **Chiquimula:** Camotan, 13 Km NE of Camotan, 14°53'N, 89°18'W, 800 m, 4 Dec, *Debouck et al. 2420* (G, USCG). **El Progreso:** San Agustín Acasaguastlan, 2 Km N of Chanrayo, 15°02'N, 89°56'W, 750 m, 3 Dec 1987, *Debouck et al. 2415* (COL, G, USCG). **Guatemala:** Palencia, Río de los Ocotes, 14°38'N, 90°26'W, 1160 m, 1 Dec 1987, *Debouck et al. 2410* (COL, G, M, USCG); Fraijanes, 2 km SE de Lo de Diegues, 14°30'N, 90°25'W, 1690 m, 16 Dec 1987, *Debouck et al. 2464* (COL, M, UC, US, USCG); 2 km NE de Concepción Las Lomas, Cd. Guatemala, 14°37'N, 90°30'W, 1470 m, 17 Dec 1987, *Debouck et al. 2471* (COL, MICH, US, USCG); Finca El Naranjo, San José El Naranjo, Mixco, 14°38'N, 90°35'W, 1530 m, 18 Dec 1987, *Debouck et al. 2477* (BR, MICH, US, USCG); 1 km N de Ciudad Quetzal en antiguo camino de Guatemala a San Raimundo, Mixco, 14°39'N, 90°34'W, 1480 m, 18 Dec 1987, *Debouck et al. 2478* (BR, MICH, SI, US, USCG); 3 km N de San Juan Sacatepéquez, 14°44'N, 90°38'W, 1680 m, 18 Dec 1987, *Debouck et al. 2481* (MICH, US, USCG); between Campanales and Villalobos, 1300 m, 29 Sep 1972, *Molina et al. 27693* (EAP, ENCB, F); Chillaní, 1500 m, 20 Jun 1921, Rojas 56 (US); Finca Bretaña, between Guatemala and Fiscal, 1200 m, 12 Dec 1938, *Standley 59703* (F). **Huehuetenango:** La Cruz de la Lacha, 5.5 km S de Malacatancito, 15°14'N, 91°30'W, 1880 m, 15 Dec 1985, *Debouck et al. 1673* (COL, UC, US, USCG); 0.5 km de San Sebastian Huehuetenango, 15°23'N, 91°36'W, 16 Dec 1985, *Debouck et al. 1677* (US, USCG); Km 242 Hwy CA-1 Quetzaltenango–Huehuetenango, 15°11'N, 91°32'W, 1720 m, 9 Oct 1978, *Freytag et al. 78-Guat-99* (BR, EAP, MEXU, MO, UC, US); Km 7 Hwy CA-1 to San Pedro Necta, 15°30'N, 91°46'W, 1520 m, 9 Oct 1978, *Freytag et al. 78-Guat-101* (MEXU, MO, US); Km 339 Hwy 7W Huehuetenango-Cobán, 15°21'N, 91°25'W, 2000 m, 11 Oct 1978, *Freytag et al. 78-Guat-114* (MEXU, MO, US); Río San Juan Ixtan, E of San Rafael Petzal, 1730 m, 9 Jan 1941, *Standley 83050* (F); San Sebastian Huehuetenango, Río Selegua, 2000–2100 m, 13 Aug 1942, *Steyermark 50466* (F); barranco "Palo Negro," 10 km W of Aguacatán, 2200 m, 28 Nov 1962, *Williams et al. 21834* (EAP, F). **Jalapa:** Km 124 Hwy 18 Jalapa-San Pedro Pinula, 14°40'N, 89°56'W, 1180 m, 5 Oct 1978, Freytag et al. 78-Guat-29 (BR, EAP, MEXU, MO, UC, US). **Jutiapa:** Volcán Flores, 825 m, Oct 1892, *Shannon 3678* (G, US). **Quiché:** 4 Km S de Chichicastenango, 1 Nov 1970, *Harnon 4610* (ENCB). **Sacatepéquez:** Ciudad Vieja, Nov 1914, Tejada 270 (US). **Santa Rosa:** Santa Cruz Naranjo, 1 km SE de Salitre, 14°26'N, 90°22'W, 1190 m, 16 Dec 1987, *Debouck et al. 2468* (MICH, US, USCG); Cenaguilla, 1219 m, Sep 1892, *Heyde et al. 3754* (G, GH, K, US). **Sololá:** San Andrés Semetabaj, 1 km N de Godínez, 14°44'N, 91°07'W, 2170 m, 9 Dec 1985, *Debouck et al. 1627* (BR, COL, UC, US, USCG); 1 km N de Panajachel, 14°46'N, 91°10'W, 1610 m, 10 Dec 1985, *Debouck et al. 1629* (COL, K, SI, US, USCG); Km 99 Hwy 11 Patzicia-Panajachel, 1750 m, 6 Oct 1978, *Freytag et al. 78-Guat-46* (MEXU, MO, US); Lake Atitlán, 3–5 km W of Panajachel, 2100 m, 6–7 Dec 1963, *Williams et al. 25311* (EAP, F, US). **Zacapa:** Río Hondo, Sierra de las Minas, (15°5'N, 89°35'W)?, 250–900 m, 11 Oct 1939, *Steyermark 29541* (F); between Río Hondo and Finca Alejandría, Sierra de las Minas, 1000–1500 m, 1 Oct 1939, *Steyermark 29698* (F).

HONDURAS. Comayagua: 5km N de La Libertad, Montaña, 835 m, 6 Feb 1980, *Amador 126* (MO); Siguatepeque, (14°30'N, 87°45'W)?, 1200 m, 11 Jul 1936, *Yuncker et al. 5789* (F, G, GH, K, MO). **Distrito Central:** N of Tegucigalpa, (14°N, 87°15'W)?, 17 Jun 1970, *Barkley et al. 40690* (GH); Tamara, 19 Jun 1970, *Barkley et al. 40753* (GH). **El Paraíso:** Güinope to Manzaragua, (13°40'N, 86°55'W), 1440 m, 5 Jan 1947, *Standley et al. 2113a* (F); between La Vistada de San José and Fatima, 10 km N of Yuscarán, 800 m, 9 Dec 1973, *Williams et al. 42808* (EAP, F). **Fco. Morazán:** 12 km E of Tegucigalpa along road to Zamorano, (13°50'N, 87°10'W), 1143 m, 26 Oct 1965, *Gentry 21466* (ASU, US); entre Agua Amarilla y Piedras Gordas, Río Yeguare, 14°N, 87°W, 1300 m, 2 Nov 1948, *Molina 1380* (EAP); entre Cuesta de Los Muertos y Monte Oscuro, Cordillera Azacualpa, 1500 m, 14 Jul 1964, *Molina 14535* (EAP, F); Paul C. Standley tomb in San Antonio de Oriente cemetery, 1400 m, 25 Jul 1979, *Molina 31749* (EAP, MO); Río Agua Amarilla, El Zamorano, 1000–1200 m, Oct–Nov 1948, *Standley 12862* (EAP, F); between Hoya Grande and Valle Encantado, Cerro de Uyuca, 1400–1500 m, 2 Dec 1948, *Standley 15248* (F); entre Peña Blanca y Lo de Ponce, 1600 m, 5 Feb 1950, *Williams et al. 17135* (EAP).

MÉXICO. Chiapas: Mpio. Tuxtla Gutiérrez, 2 mi S of Tuxtla Gutiérrez along road to Villa Flores, (16°45'N, 93°15'W), 853 m, 16 Oct 1965, *Breedlove et al. 13320* (CAS-DS, F, MEXU, MICH, US); Mpio. Chiapa de Corzo, El Chorreadero, (16°45'N, 93°W), 800 m, 15 Oct 1971, *Breedlove et al. 20483* (CAS-DS, MO); Mpio. Motozintla de Mendoza, 25–27 km NE of Huixtla along road to Motozintla, SW of Tolimán, (15°20'N, 92°30'W), 700 m, 7 Oct 1972, *Breedlove 28662* (CAS-DS); Mpio. Angel Albino Corzo (Jaltenango), 3–5 km above Jaltenango along road to Finca Prusia, (15°53'N, 92°46'W), 900 m, 11 Oct 1974, *Breedlove 38620* (CAS-DS); Mpio. Frontera Comalapa, 6–8 km E of Frontera Comalapa along road to Ciudad Cuauhtémoc, (15°40'N, 92°7'W), 1000 m, 23 Oct 1974, *Breedlove 39016* (CAS-DS); Mpio. Teopisca, Teopisca, (16°30'N, 92°30'W), 1800 m, 17 Oct 1980, *Breedlove et al. 46372* (CAS); Mpio. of Villa Corzo, 65 km S of Mexican Hwy 190 on road from Tuxtla Gutiérrez-Nueva Concordia, (16°N, 93°30'W), 850 m, 19 Oct 1980, *Breedlove et al. 46514* (CAS); Mpio. of Villa Corzo, Colonia Vincente Guerrero on road to Finca Cuxtepec, 1100 m, 20 Oct 1980, *Breedlove et al. 46531* (CAS); Mpio. Pueblo Nuevo Solistahuacán, 3 mi N of Pueblo Nuevo Solistahuacán, (17°10'N, 92°54'W), 1615 m, 3 Aug 1967, *Clarke 487* (UCR); Mpio. Amatenango de la Frontera, El Regadío, a 1 km de Tapizula y a 2 km de Guerrero, 670 m, 28 Nov 1977, *Delgado et al. 814* (MO); Km 6 Hwy 211 Comalapa-Huixtla, 15°30'N, 92°10'W, 800 m, 29 Sep 1978, *Freytag et al. 78-Mex-100-1* (MEXU, MO, US); Sierra de Tonalá, (16°10'N, 93°45'W), Oct 1913, *Purpus 6650* (MO; mixed); Chinkultic, 18 Aug 1896, *Seler 2682* (GH); Mpio. Pueblo Nuevo Solistahuacán, 3 km NW of Pueblo Nuevo Solistahuacán, 17°30'N, 92°40'W, 1646 m, 8 Oct 1971, *Thorne et al. 41232* (CAS-DS). **Chihuahua:** Arroyo Hondo, Sierra Charuco, 11 Sep 1935, *Gentry 1772* (ARIZ, F, GH, K, MO, UC, US); La Cieneguita, Río Mayo, 10 Sep 1936, *Gentry 2637* (ARIZ, F, GH, MO, UC); 15.9 mi W of General Trías on Chihuahua 16, 42.9 mi W of Chihuahua, (28°15'N, 106°30'W), 27 Aug 1971, *Keil et al. 8302* (ASU); 5 km W of Rancho La Lobera, 3 km E of Chinacas, Sierra Guaribo, 28 Oct 1984, *Mittleman et al. 524* (ASU, DES); Sacramento River, 15 Sep 1886, *Pringle 727* (BM, F, G, GH, K, MEXU, MICH, MO, MSC, NA, UC, US(2)); Nuevas Delicias along Río El Alamo, Sierra Del Nido, 29°5'45"N, 106°29'W, 1775 m, 7 Sep 1981, *Worthington 7721* (ARIZ, BRIT). **Colima:** Rancho El Jabali, 22 km (airline) N of Colima, 19°27'N, 103°42'W, 1300 m, 4 Oct 1991, *Vazquez 1335* (MO). **Distrito Federal:** Pedregal de San Angel, (19°19'N, 99°12'W), 2200 m, Sep 1927, *Lyonnet 117* (in part) (MO); Tlalpan, Valley of México, (19°17'N, 99°9"W), 20 Aug 1903, *Rose et al. 6454* (GH, US); Santa Marta Ixtapalapa, (19°22'N, 99°1'W), 2245 m, 13 Oct 1975, *Sousa et al. 5106* (CAS, MEXU, US(2), WIS); El Pedregal, near ruinas Cuilco, 19 Aug 1962, *Webster et al. 13099* (MEXU, MO). **Durango:** Medina, Canatlán, 24°34'N, 104°36'W, 2250 m, 30 Sep 1978, *Debouck et al. 278* (CHAPA, COL, K, MICH); Km 156 carr. a Mazatlán, Revolcaderos, El Salto, 23°37'N, 105°50'W, 1960 m, 2 Nov 1978, *Debouck et al. 412* (CHAPA, COL, K, MICH, UC); city of Durango, (24°N, 104°55'W), Apr–Nov 1896, *Palmer 483* (in part) (BM, F, K, MO, US); Carpintero, Sierra Madre Occidental, 2000–2100 m, 25 Aug 1934, *Pennell 18209* (GH). **Guanajuato:** Valle de Santiago, 0.5 Km W Volcan Batea, 20°19'N, 101°12'W, 1850 m, 16 Nov 1987, *Debouck et al. 2400* (CHAPA, COL, G, M); abajo de la Presa Allende, 10 Sep 1977, *Kishler 47* (MEXU). **Guerrero:** Taxco, Tehuilotepec, 28 Sep 1937, *Abbott 460* (ENCB, GH); 9 km E de Chilpancingo de los Bravos, Tixtla, 17°34'N, 99°29'W, 1620 m, 31 Oct 1987, *Debouck et al. 2357* (CHAPA, MICH, MO); Cerro Tuxpan, 13 km NE de Iguala, 18°23'N, 99°28'W, 1630 m, 31 Oct 1987, *Debouck et al. 2360* (CHAPA, MICH); 4 km W de Huajojutla, 8 km NE de Taxco, 18°34'N, 99°34'W, 1630 m, 1 Nov 1987, *Debouck et al. 2363* (CHAPA, MICH); Pungarabato, Coyuca, 22 Sep 1934, *Hinton 6642* (G, GH, K, US); Dist. Mina, Guayameo (South), 900 m, 11 Sep 1936, *Hinton et al. 9382* (BM, F, GH, K, US); Vallecitos, Montes de Oca, 14 Jul 1937, *Hinton et al. 10601* (BM, GH, K, TEX-LL, US); Dist. Galeana, Atoyac, (17°15'N, 100°30'W)?, 660 m, 19 Nov 1937, *Hinton et al. 10947* (BM, K, MEXU, TEX-LL, US); Mpio. de Teloloapan, 13 km al NE de Almoloya, carr. Altamirano-Iguala, (18°30'N, 100°15'W), 1250 m, 13 Oct 1981, *Soto et al. 3305* (CAS); S de Xochipala, 1600 m, 2 Aug 1967, *Sousa 3130* (GH, MEXU). **Hidalgo:** 4 km NW desde entronque Mex 85 (Hda. La Estancia) en camino a San Francisco del Oro, Zimapán, 20°48'N, 99°20'W, 2100 m, 31 Oct 1986, *Debouck et al. 2032* (CHAPA, US); Jacala, (19°N, 99°15'W), 27 Aug 1938, *Kenoyer A668* (F); valley of Tula, (20°5'N, 99°30'W), 2240 m, 6 Sep 1902, *Pringle 9768* (BRIT, CAS, F,

GH, K, MICH, MO, US); Dublan, 2 Jul 1901, *Rose et al. 5295* (US). **Jalisco:** Purificación, Puerto Los Mazos, 19°42'N, 104°25'W, 1150 m, 26 Nov 1978, *Debouck et al. 467* (CHAPA, COL, G, K, M); 1.5 km E of El Deposito, 10 km WSW of Cd. Guzmán, 19°39'N, 103°32'W, 1750–1800 m, 24 Sep 1978, *Iltis et al. 583* (WIS): 4 mi W of Chapala, (20°20'N, 103°25'W), 4 Jan 1973, *Johnson 49-73* (MO); Mpio. Atenguillo, 12 km al NE de Los Volcanes, camino a Talpa, (20°10'N, 105°W)?, 7 Sep 1979, *Magallanes 1932* (CAS); Puente San Pedro, 5 mi SW of Tecalitlán, (19°30'N, 103°20'W), 1200 m, 22 Sep 1958, *McVaugh 18115* (CAS, MICH(2)); Guadalajara, (20°40'N, 101°40'W), Sep 1893, *Pringle 4544* (in part) (ENCB, MEXU, SI, US); Chapala, 5 Oct 1903, *Rose et al. 7630* (GH, US); Mpio. Tamazula, Agua Hedionda, 45 km al E de El Aserradero, camino a Manuel M. Dieguez, (19°30'N, 103°5'W, 1900 m, 26–27 Oct 1973, *Rzedowski et al. 1131* (MICH). **México:** 9 km NW de Coatepec, 18°55'N, 99°46'W, 1980 m, 1 Nov 1987, *Debouck et al. 2364* (CHAPA, MICH); 10 km N de Malinalco, 2370 m, 4 Nov 1987, *Debouck et al. 2373* (CHAPA, MICH); Mpio. Temascaltepec, Nanchititla, (19°N, 100°W)?, 1900 m, 28 Oct 1978, *Delgado et al. 1008* (F, MO); Km 35 Hwy 136 Texcoco-Apitzaco, 19°32'N, 98°43'W, 16 Sep 1978, *Freytag et al. 78-Mex-6* (BR, EAP, MEXU, MO, UC, US); Dist. of Temascaltepec, Tejupilco, 1340 m, 24 Oct 1932, *Hinton et al. 2287* (BM, G, K, US); Cerro de Ixtapan de la Sal, (18°45'N, 99°45'W), 1920 m, 26 Sep 1954, *Matuda et al. 31701* (CAS, ENCB, MEXU, US(2)); Rancho Renundadero, Apr 1931, *Purpus 145006* (UC); 2 km N of Ixtapan at Km 145 on Hwy 55, 18°50'N, 99°25'W, 1930 m, 9–10 Sep 1965, *Roe et al. 1914* (MICH); Victoria, 4 Apr 1926, *Runyon 5707* (TEX). **Michoacán:** Chemin des Parcs, Morelia, 1850 m, 1 Aug 1909, *Arsène 2415* (MEXU, MO, US); Cerro Azul, Morelia, (19°45'N, 101°15'W)?, 2200 m, 1910, *Arsène 5768* (GH, MC, US); Punguato, Morelia, 1950 m, 25 Aug 1910, *Arsène 10043* (MO, US); 5 mi SW of Quiroga, 13 Aug 1947, *Barkley et al. 2727* (TEX); 1.7 mi NW of Tuxpan, Km 129 on Hwy 15 to Morelia, (19°40'N, 100°30'W), 1780 m, 19 Aug 1975, *Davidse et al. 9810* (MO); Zacapu, 2 Km W of Los Llanos, 19°48'N, 101°50'W, 1940 m, 10 Nov 1978, *Debouck et al. 437* (CHAPA, COL, G, K); 20 km S of Uruapan on Hwy 37, 1300 m, 11 Sep 1972, *Dieterle 4388* (ENCB, MICH); 4 mi S of Carapan along road to Uruapan, (19°45'N, 102°W), 2134 m, 6 Nov 1966, *Gentry 22141* (US); 7 mi E of Uruapan along road to Cirimicuaro, (19°30'N, 101°55'W), 1524 m, 5 Nov 1966, *Gentry et al. 22151* (US); Dist. Zitacuaro, Zitacuaro-Coyota, (19°25'N, 100°25'W)?, 1650 m, 25 Aug 1938, *Hinton et al. 13163* (ENCB, G, K, MICH, TEX-LL, UC, US); Tancitaro, Uruapan, (19°20'N, 102°05'W), 1750 m, 4 Nov 1940, *Hinton et al. 15631* (in part) (ARIZ, ENCB, TEX-LL, US); Mpio. Tuzantla, 6 km al S de Tuzantla por la carr. a Huetamo, 60 km al S de Zitacuaro, (19°10'N, 100°35'W), 670 m, 9 Nov 1983, *Koch et al. 8358* (TEX); Mpio. Coalcomán, 0.3 mi NE of Km 50 on road from Coalcomán-Tepacaltepec, 14.1 mi NE of Coalcomán, (18°50'N, 103°W), 13 Sep 1985, *Luckow et al. 2926* (TEX); 12 mi W of Quiroga, (19°39'N, 101°31'W), 2134–2743 m, 3 Oct 1953, *Sohns 783* (US); Guanoro, 19 km al SW de Zitácuaro, (19°20'N, 100°30'W), 27 Oct 1978, *Soto 1207* (MO); Mpio. Tuxpan, Puente Turundeo, 11 km al SE de Cd. Hidalgo, 22 Aug 1978, *Soto et al. 877* (MO); Mpio. San Lucas, en Malpaso, 4 Oct 1978, *Soto et al. 1143* (CAS, MO); Mpio. Caracuaro, 4 km de Eréndira, 8 Oct 1978, *Soto et al. 1171* (MO); Arteaga, carr. a Nueva Italia, 900 m, 21 Sep 1979, *Soto et al. 1704* (TEX); Las Trincheras, 12 km al N de Huetamo, carr. a Zitacuaro, (18°35'N, 100°55'W), 500 m, 12 Oct 1979, *Soto et al. 1802* (TEX, WIS); Mpio. Gabriel Zamora, Barranca Honda, carr. Uruapan-Nueva Italia, (19°5'N, 102°5'W)?, 12 Oct 1979, *Soto et al. 1832* (CAS, MO). **Morelos:** San José de los Laureles, 2 mi NW of Tlayacapan, 1800 m, 6 Oct 1988, *Anderson 13508* (MO); Cuernavaca, 9 Sep 1903, *Rose et al. 6902* (GH, US); cerro Tetillas, 7 Sep 1968, *Vazquez 2016* (MEXU). **Nayarit:** 9.5 mi W of Tepic, 1000–1100 m, 25 Sep 1960, *McVaugh 19386* (CAS, MICH); between Aguacala and Dolores, Sierra Madre, Tepic, 6 Aug 1897, *Rose s.n.* (US); between Tepic and the turnoff to San Blas, (21°35'N, 105°5'W), 930 m, 4 Oct 1963, *Schubert et al. 2015* (GH); Mpio. Nayar, 6 km NW de La Mesa del Nayar, brecha a Santa Teresa, 22°11'N, 104°42'W, 1900 m, 27 Sep 1989, *Tenorio et al. 16522* (MO). **Nuevo León:** 2.5 km S de La Caballada hacia Milpillas, Aramberri, 24°20'N, 99°47'W, 1360 m, 7 Sep 1985, *Debouck et al. 1505* (CHAPA, COL, G, K, M, MICH, UC); Cañon Jazmines, 4 km S de Cuevas, Iturbide, 24°34'N, 99°46'W, 1010 m, 15 Sep 1985, *Debouck et al. 1519* (CHAPA, MO, US); Puerto Genoveo, 16 km N de Laguna de Sánchez, Santiago, 25°21'N, 100°11'W, 1120 m, 18 Sep 1985, *Debouck et al. 1521* (CHAPA, COL, K, G, MICH); Cañon de La Boca, 7 km N de La Cienaga, camino hacia Laguna de Sánchez, Santiago, 25°23'N, 100°19'W, 1510 m, 18 Sep 1985, *Debouck et al. 1524* (CHAPA, COL, SI, US); El Ebanito, 2 km E de Los Altares, Iturbide, 24°43'N, 99°50'W, 1020 m, 10 Nov 1986, *Debouck et al. 2062* (BR, CHAPA, COL, G, US(2)); IRF Iturbide, Linares, Río Iturbide, 740 m, 30 Sep 1979, *Hinton et al. 17697* (TEX); Río Ramos, 30 Dec 1947, *Lacas 544* (F); Mpio. de Montemorelos, La Trinidad, (25°15'N, 99°55'W), 19 Aug 1939, *Muller 2829* (F, GH, MICH, MO, NA, TEX-LL, UC); Cerro La Silla, Monterrey, 1 Sep 1937, *White et al. 130* (MICH). **Oaxaca:** Valley of Etla, (17°10'N, 96°55'W), Sep 1895, *Alvarez 734* (GH); 6 mi SE of Matatla, 32 mi SE of Oaxaca along Hwy 190 to Tehuantepec, 13 Sep 1971, *Clarke 20486-21* (UCR); Dist. del Centro, Cerro San Antonio, 1800 m, 8 Sep 1921, *Conzatti 4232* (US); Asunción Nochixtlan, Km 120 Mex 190 to San José Sosola, 17°20'N, 97°08'W, 2030 m, 24 Oct 1987, *Debouck et al. 2330* (CHAPA, G, M, MICH); Cerro El Labrador, 8 km N de El Portillo, Coatecas, Ejutla, 16°38'N, 96°35'W, 1820 m, 26 Oct 1987, *Debouck et al. 2340* (CHAPA, MICH, US); Candelaria Loxicha, 38 km S de San José Pacífico, Pochutla, 15°59'N, 96°31'W, 1780 m, 27 Oct 1987, *Debouck et al. 2347* (BR, CHAPA, MICH, US); 1 km NW de Luz de Luna, 12 km N de San Gabriel Mixtepec, Juquila, 16°07'N, 97°04'W, 860 m, 28 Oct 1987, *Debouck et al. 2351* (CHAPA, MICH, US); Mpio. Sola de Vega, 16 km al O de Sola de Vega, (16°30'N, 97°10'W), 2080 m, 22 Nov 1977, *Delgado et al. 625* (CAS); Mpio. de Juquila, 9 km al SO de San Pedro Juchatenango, 1210 m, 22 Nov 1977, *Delgado et al. 650* (CAS); Km 6 Hwy 131, Oaxaca-Tehuacán, 17°20'N, 96°53'W, 1750 m, 21 Sep 1978, *Freytag et al. 78-Mex-39* (ARIZ, BR, CSU, EAP, F, GH, IBUG, K, MEXU, MICH, MO, NA, TEX, UC, US, WIS); Km 149.5 Hwy 190 Huajuapan-Oaxaca, 17°16'N, 97°55'W, 1950 m, 21 Sep 1978, *Freytag et al. 78-Mex-41* (BR, EAP, MEXU, MO, US): Km 143 Hwy 190 Huajuapan-Oaxaca, 2050 m, 21 Sep 1978, *Freytag et al. 78-Mex-42* (MEXU, US); Km 102 Hwy 190 Huajuapan-Oaxaca, 17°18'N, 97°06'W, 2080 m, 21 Sep 1978, *Freytag et al. 78-Mex-45* (BR, EAP, F, GH, K, MEXU, MICH, MO, UC, US); Km 4 Hwy 175 Oaxaca-Tuxtepec, 17°03'N, 95°36'W, 1580 m, 22 Sep 1978, *Freytag et al. 78-Mex-59* (MEXU, US); 7 mi E of Mitla along road to Ayutla, (17°N, 96°5'W), 1981 m, 18 Oct 1967, *Gentry 22342* (GH, MEXU, MICH, NA); 5 mi NE of Santa María Albarradas along road to Ayutla, 2286 m, 19 Oct 1967, *Gentry 22348* (GH, MEXU, MICH, NA); 25–30 km WSW of summit along trail from Santo Domingo Albarradas-Mitla, Cerro Zempoaltépetl, (17°5'N, 96°W), 1900 m, 18 Aug 1950, *Hallberg 1037* (CAS-DS, ENCB, MICH, US); Cerro San Felipe, N de la Ciudad de Oaxaca, (17°15'N, 96°45'W), 1800 m, 19 Aug 1976, *Hernández 2648* (GH, MO); Hierve el Agua, travertine spring near San Lorenzo, 1524–1676 m, 7 Jul 1968, *Kitchen 96* (ASU); Mpio. Jamiltepec, 5 km al NO de Jamiltepec, 27 km al SE de Pinotepa Nacional por carr. a Puerto Escondido, (16°15'N, 97°55'W), 310 m, 9 Nov 1979, *Koch et al. 79449*

(CAS, MO, TEX); 12.3 mi N of Jct. 190 and 175 on 175 from Oaxaca to Tuxtepec, 2591 m, 15 Aug 1975, *Ledoux et al. 2238* (ASU); Cumbre de Ixtepec, Sierra de Oaxaca, 6 Jan 1842, *Liebmann 5316* (F); Mpio. Albarradas, Dist. Tlacolula, 17 km E of Mitla on road to Ayutla, 2030 m, 17 Sep 1980, *Martin 177* (MICH, US); 18 mi SW of the City of Oaxaca, (16°55'N, 97°W), 2286–2896 m, 10–20 Sep 1894, *Nelson 1350* (US); Tierra Azul, 8 km NE de Tlaxiaco, (17°25'N, 97°40'W), 2000 m, 22 Aug 1976, *Sousa et al. 5939* (K, MEXU, MICH, MO(2)); Dist. Etla, 2 km al N de Las Sedas, carr. Oaxaca-Cuicatlán, 1950 m, 25 Aug 1976, *Sousa et al. 6113* (TEX-LL); Dist. Nochixtlán, 4 km al SE de Llano Verde, (17°30'N, 97°25'W), 2200 m, 18 Oct 1976, *Sousa et al. 6243* (ARIZ, BM, CAS, DES, MEXU, MO); Dist. Cuicatlán, 15 km al ENE de San Juan Bautista Cuicatlán, carr. a Concepción Papalo, (17°55'N, 96°55'W), 1550 m, 5 Aug 1977, *Sousa et al. 7758* (CAS, MEXU, MO, UC); Dist. Mixe, 5 km al NW de San Pedro y San Pablo Ayutla, 2000 m, 8 Aug 1977, *Sousa et al. 7867* (CAS, DES, MEXU, MO, UC); en el NE del Dist. Centro, El Estudiante, 1950 m, 10 Aug 1977, *Sousa et al. 7889* (CAS, MEXU, UC); Dist. Tlacolula, 17 km al N de Totolapan, 1750 m, 17 Sep 1978, *Sousa et al. 9418* (CAS); Dist. Zimatlán, Santa Cruz Mixtepec, (16°55'N, 96°55'W), 1400 m, 23 Jun 1979, *Sousa et al. 10470* (ARIZ, CAS, TEX); El Zacatal, 7 km al NO de Sta. María Albarradas, 1900 m, 17 Sep 1976, *Téllez et al. 83* (BM, MEXU, MO); 8 km al NO de Yautepec, 1100 m, 19 Sep 1976, *Téllez et al. 142* (CAS, K, MEXU); Dist. Miahuatlán, 25.4 km al O de San Pablo Coatlán, camino a Piedra Larga, (16°25'N, 96°30'W)?, 1670 m, 25 Oct 1982, *Torres et al. 1735* (CAS); Dist. Tehuantepec, 13.6 km al SW de la Reforma hacia Sta. María Ecatepec, (16°20'N, 95°5'W), 1265 m, 26 Jul 1984, *Torres et al. 5647* (CAS); Tlapacoyan-Sola de Vega, nach Puerto Escondido, Sierra Madre del Sur, 23 Sep 1969, *Vogel 126a* (US). **Puebla:** Puebla, Mayorazgo sur l'Atoyac, 2120 m, 10 Sep 1907, *Arsène 1271* (MO, US); Puebla, barrancas de l'Alseseca, Hacienda Batan, près Totimehuacan, (18°58'N, 98°15'W), 2120 m, 8 Aug 1907, *Arsène 7107* (US); Puebla, Tepoxuchilt, 16 Jul 1910, *Arsène et al. 5283* (US); Puebla, Manzanilla, 2250 m, 18 Sep 1910, *Arsène et al. 5493* (GH, MO, US); 4 km al NE de Huauchinango, Sierra Norte, 20°7'N, 97°33'W, 5 Nov 1978, *Basurto et al. 86* (CAS, MO); Barranco Puente Los Molinos, 4 km NE de Atlixco, Santa Isabel Cholula, 18°58'N, 98°23'W, 1880 m, 22 Oct 1987, *Debouck et al. 2324* (CHAPA, MICH, MO). **Queretaro:** Queretaro, Escuela Libre, 1900 m, 29 Jun 1975, *Arguelles 101* (MEXU). **San Luís Potosí:** 13.2 mi W of Cerritos on road to Hwy 57, (22°30'N, 100°20'W), 7 Aug 1983, *Buhrow et al. M16* (ARIZ); 25 mi NW of El Salto, 96.2 mi SE of El Huizache along Hwy 80, (22°45'N, 99°35'W), 1134 m, 25 Jul 1967, *Clarke et al. s.n.* (UCR); 7 km SW de Ahuacatlán en carr. Mex 120 a Jalpan, Xilitla, 21°17'N, 99°6'W, 1430 m, 2 Nov 1986, *Debouck et al. 2044* (CHAPA, COL, G, US); Km 10 de Mex 80 a Cd. Mante, 2 km SW de Puerto de Lobos, Cd. del Maíz, 22°27'N, 99°34'W, 1170 m, 13 Nov 1986, *Debouck et al. 2072* (BR, CHAPA, US); Km 139 Mex 80 a Cd. Mante, 4 km W de Platanito, Cd. del Maíz, 22°29'N, 99°29'W, 1150 m, 13 Nov 1986, *Debouck et al. 2074* (BR, CHAPA, COL, G, MO, US); Rascón, Aug 1911, *Purpus 5509* (US). **Sinaloa:** 49.8 mi E of Villa Unión on Hwy 40, 2 Aug 1983, *Buhrow et al. M1* (ARIZ); Quebrado de Mansana, Sierra Surotato, (26°N, 107°45'W), 1219 m, 10–14 Sep 1941, *Gentry 6493* (MO, US); 8 mi W of Palmito along road to Durango, 1828 m, 21 Oct 1966, *Gentry et al. 22029* (NA, US); 23 km al SSO de Revolcaderos, carr. Durango-Mazatlán, 1850 m, 17 Sep 1978, *Grether et al. 1089* (MICH); Mpio. Sinaloa de Leyva, Ejido de Cuitanaca, 35 km al N de Agua Caliente de Zevada, (26°N, 108°5'W), 14 Oct 1976, *Pérez 73* (BM, CAS, MEXU, MO). **Sonora:** 7 mi N of Guirocoba, 13 Apr 1989, *Salmon s.n.* (ARIZ); 18.3 mi E of the Río Yaqui bridge near Tonichi, on the road to Carrizal and Santa Rosa, Sierra Madre Occidental, 28°31'N, 109°21'W, 975 m, 27 Mar 1983, *Sanders 3710* (ARIZ, UCR); Las Piedras Canyon on the E side of Sierra Alamos, 26°59'N, 108°57'30"W, 750 m, 12 Mar 1984, *Starr 746* (ARIZ); 21 mi E of Río Yaqui crossing at Tonichi along road to Yecora, 914 m, 7 Apr 1979, *Toolin et al. 298* (ARIZ); Río de los Alisos, 31 mi S of Nogales on Hwy to Magdalena, (30°55'N, 111°W), 8 Sep 1934, *Wiggins 6995* (ARIZ, F, UC, US). **Tamaulipas:** La Vegonia, San José, Sierra de San José, 24°35'N, 99°W, 1036 m, 6 Jul 1930, *Bartlett 10137* (F, MICH, TEX-LL, US); Km 150 de Mex 101 a San Luís Potosí, 29 km SW de Cd. Victoria, 23°36'N, 99°14'W, 1470 m, 12 Nov 1986, *Debouck et al. 2066* (BR, CHAPA, COL, G, US); 11 mi by road W of Victoria toward Jaumave, 914 m, 29 Sep 1959, *Johnston et al. 4115* (MICH, TEX). **Veracruz:** Cerro de Macuiltépetl, Cd. Xalapa, (19°30'N, 96°55'W)?, 9 Jun 1976, *Calzada 2410* (F); Zacuapan, Nov 1908, *Purpus 3678* (UC); Río de los Pescados near Apasapan, (19°19'N, 96°43'W), Oct 1912, *Purpus 6169* (BM, F, UC); Orizaba, (19°N, 97°W)?, 1219 m, 31 Jul 1891, *Seaton 137* (F, GH); Mpio. Dos Ríos, El Aguaje, 300 m, 9 Oct 1973, *Ventura 9101* (CAS, MEXU). **Zacatecas:** 2 km NW of Corrales, Sombrerete, 23°20'N, 103°36'W, 2150 m, 17 Oct 1978, *Debouck 365* (CHAPA, COL, G, K, M, MICH).

NICARAGUA. Chontales: Hda. Veracruz including Cerro La Batea, 12°11'N, 86°22'W, 120–375 m, 30 Oct–2Nov 1984, *Stevens 23227* (MO). **Estelí:** 3 km N de San Nicolás, 12°56'N, 86°90'W, 1100–1200 m, 23 Sep 1981, *Moreno 11417* (MO); Estanzuela creek, 8 kms W of Estelí, (12°55'N, 86°30'W), 1000 m, 4 Nov 1968, *Molina 23094* (EAP, F); San José, 10 km SE de Estelí, 5 Nov 1981, *Téllez-Valdés et al. 4792* (MO).

UNITED STATES. Arizona: Cochise Co.: Fairbanks, (31°45'N, 110°10'W), 24 Sep 1939, *Morris et al. A11584* (ASU).

Habitat.—This species is found growing in a multitude of habitats from dry to very moist in mostly forests of pine-oak, oak-juniper, tropical deciduous, low thorny shrub to nearly desert and usually on rocky and steep forest slopes of 30–45° (see Color Plate V, photo 54) and in open grassy, rolling foothills. Major associated species are: *Acacia, Bursera, Carpinus, Cnidoscolus, Cordia, Croton, Diospyros, Guazuma, Heliocarpus, Jatropha, Juglans, Juniperus, Liquidambar, Luehea, Magnolia, Mimosa, Pinus, Platanus, Plumeria, Podocarpus, Quercus, Rhus, Tabebuia, Taxodium,* and *Tilia*. Agaves, yucca, cacti and grasses are also commonly reported. Many soil types are reported and are commonly sandy and rocky with clay and considerable organic matter, derived from limestone, shists, metamorphic, volcanic, granitic or basalt rocks.

Diseases and pests.—Diseases reported are: black spot(?), powdery mildew, rust, golden mosaic virus and rugose virus. Insects include: *Apion* weevil, beetles, aphids, leafminers, Mexican bean beetle, and spider mites.

Common names.—Jícamo, Pega-pega, Chorequito, Shut chenek mut (Tzeltal).

Comments.—Unfortunately, Piper (1926) did not trace down the original publication by Bentham of species *leptostachyus* as did Delgado (1985) to whom we are indebted for pointing out Bentham's prior publication of 1837. However, Piper did recognize that it was conspecific with *P. anisotrichos* Schlecht. At all events the type collection as given by Piper as *Karwinski s.n.*, from somewhere in México in 1827 or 1828, is correct. A fragment at US sent by Dr. Ross from Münich to Dr. Piper confirms this (with the exception of the number on the tag affixed to the specimen). The specimen *Purpus 7099* (MO) is given as "*Phaseolus humistratus*" Piper ined. 9/15/24.

The senior author finds the very small seed size, as given by Maréchal et al. (1978b) who believed *Galeotti 3169* to be the type (indicated as such at K), not to be the distinctive characteristic of this species, but rather the other characters cited by him, especially the large, extended stipules and the long strigose pubescence of the stems and also the flower characteristics of shape and the short pedicel (see Color Plate II, photo 20). It also has a large and branched fleshy root (see Color Plate III, photo 34). Additionally, on a collecting trip to México in 1978, Dr. Nader Vakili found the species to be always of determinate plant habit (anthotelic), with terminal nodes producing reduced leaves or no leaves, and the apex of the plant terminating in an inflorescence. This determinate growth habit is only found in this species. The senior author later found the internode production controlled by daylength (see Color Plate V, photo 55), and some collections, from Oaxaca at least, will continue to flower indefinitely under tropical environments (in the screenhouse in Puerto Rico) producing pseudoracemes of a half meter or more in length until daylength is long enough to cause the renewed production of vegetative internodes by lateral buds.

This species was found by Lackey (1979) to have a chromosome number of 20 (n=x=10), confirmed by Mercado-Ruaro & Delgado Salinas (1996).

Interestingly, *P. leptostachyus* has not been reported previously from the S USA (Correll & Johnston 1970; Kearney & Peebles 1960; Turner 1959; Wooton & Standley 1915), and its extension to the north is worth documenting. Not surprisingly given its huge range of distribution (from Arizona down to Costa Rica, and from 170 to 2250 masl), this species is extremely variable, especially in size and shape of foliage, inflorescence, bract, pod, and pubescence, and the senior author has drawn attention to the more extreme of these by assigning them varietal status. Delgado (1985) places *P. opacus* and *P. intonsus* as synonyms of *P. leptostachyus*, but the senior author finds that species *opacus* differs in a number of characteristics in foliage and pubescence and has many characteristics similar to *P. micranthus* found on the west coast of México. *P. opacus*, with its distinctive leaf, stem, and flower characteristics, is only found on the eastern coast and thus deserves to be maintained as a different species. The variant described as species *intonsus* by Piper (1926) is difficult to distinguish from the type but does often have a somewhat indistinct basal lobing of the terminal leaflet and a greater tendency to a climbing habit on the terminal portions of the stems. The senior author has found both characters to be inherited in the majority of the offspring of these varieties over several years in plantings in Puerto Rico.

Recent molecular genetic work has shown the group to be distinctive on the basis of a-amylase inhibitor data (Pueyo & Delgado Salinas 1997), allozyme data (Jaaska 1996), amplified products of the leghemoglobin gene (Skroch et al. 1993), cpDNA restriction site variation (Delgado Salinas et al. 1993), and PCR-RFLPs on cpDNA (Fofana et al. 1999).

K.2.2.—Phaseolus leptostachyus Benth. var. **intonsus** (Piper) Freytag, comb. & stat. nov. (**Figs. 76, 80**). *Phaseolus intonsus* Piper, Contr. U.S. Natl. Herb. 22:696. 1926. TYPE: MÉXICO. GUANAJUATO: Sierra de Guanajuato, (21°10'N, 101°15'W), 2000 m, 1872, *Guillemin-Tarayre s.n.* (HOLOTYPE: P n.v., US (photo & fragment of P).

Descriptive characteristics are the same as for var. *leptostachyus* except: ***Aerial shoot*** a vine, the apex with more of a climbing tendency. ***Leaves*** the terminal leaflet ovate-lanceolate, obscurely to distinctly hastately lobed, usually more noticeable near the apex of the vine. ***Pod*** more narrow and curved than var. *leptostachyus*. ***Seed*** shorter and deeper and more flattened than for var. *leptostachyus*.

Specimens examined: **HONDURAS. El Paraíso:** between La Vistada de San José and Fátima, 10 km N of Yuscarán, (13°45'N, 86°50'W), 800 m, 9 Dec 1973, *Williams et al. 42808* (MICH).

Fig. 76. Illustrations of *Phaseolus leptostachyus* Benth. var. *intonsus* (Piper) Freytag.–a. Seedling a week after germination.–c. A portion of lower stem with mature leaf and separately an upper vine tip with mature inflorescence and leaves; notice the slightly hastately lobed upper leaves.–d. Flowers, side view and front view.–e. Exploded view of flower showing all parts.–f. Pods, side view and dehisced.–g. Seeds, side view and view of hilum; rhomboidal and flattened shape. All drawings made from living material grown in greenhouse at Mayagüez from seed of *Buhrow M44* (TARS #324A) collected east of Navajoa on the state border of Chihuahua and Sonora, México.

MÉXICO. Colima: Mpio. de Comala, 1 km al S de San Antonio, carr. a Comala, (19°20'N, 103°45'W)?, 4 Sep 1979, *Magallanes 1865* (ASU, CAS). **Durango:** Mpio. of Durango, Río Chico, 34 km W of Durango City, (23°50'N, 103°W), 2190 m, 21 Aug 1980, *Breedlove et al. 45777* (CAS); Durango, (24°N, 104°55'W), Apr–Nov 1896, *Palmer 483* (in part) (GH, UC); 9 km E of Tabahueto on road to Tepehuanes, 31 Aug 1983, *Torres et al. 3590* (MO). **Guanajuato:** 3 km NE de Ojo de Agua de Calvillo, 24 km W de Dolores Hidalgo, 21°7'N, 101°6'W, 2040 m, 15 Nov 1987, *Debouck et al. 2395* (BR, CHAPA, MICH, UC, US(2)). **Jalisco:** Huejotitlán, Oct 1912, *Diguet s.n.* (MICH); Guadalajara, (20°40'N, 103°30'W)?, 12 Oct 1903, *Holway 5147* (US); Mpio. Atenguillo, 12 km al NE de Los Volcanes, Camino a Talpa, (20°10'N, 105°W)?, 7 Sep 1979, *Magallanes 1932* (in part) (ASU); Huejuquilla, 25 Aug 1897, *Rose 2546* (US); Etzatlán,

2 Oct 1903, *Rose et al. 7529* (in part) (US); Mpio. Talpa, Los Paredones, 12 km S de Talpa camino a la Cuesta, (20°15'N, 103°W), 8 Sep 1979, *Solís et al. 1948* (MICH, TEX). **Michoacán:** Morelia, (19°45'N, 101°W), 1900 m, 16 Aug 1910, *Arsène 6556* (US); Tancitaro, Uruapan, (19°20'N, 102°5'W)?, 1750 m, 4 Nov 1940, *Hinton et al. 15631* (in part) (GH, TEX, US). **San Luís Potosí:** 15 mi E of Ciudad del Maíz along Hwy to Antiguo Morelos, (22°30'N, 99°30'W), 1372 m, 4 Sep 1958, *Fearing 2041* (TEX). **Sinaloa:** Mpio. Concordia, 4 km SW of Liberas and Km 220 from Durango on the Mazatlán-Durango road, (23°20'N, 106°W), 1540 m, 2 Oct 1985, *Bartholomew et al. 2585* (CAS). **Sonora:** Río de los Alisos, 31 mi S of Nogales, (30°55'N, 111°W), 8 Sep 1934, *Shreve 6608* (ARIZ, US).

Habitat.—The same habitat is usually given for this variety as for the type variety.

Comments.—The scarcely lobate terminal leaflets are not always easy to distinguish and this variety, though showing little segregation, usually is found in mixed collections growing as a small portion of the total population and often have some variation in seed shape and color (see Color Plate IV, photo 46). For these reasons the senior author has reduced Piper's placement of the taxon as a species to varietal rank. This variety has a strong tendency to climb whereas the type variety is almost always prostrate or scandent. Neither the climbing tendency nor the determinate plant habit has been mentioned by plant collectors nor by previous scholars of the taxonomy of the genus.

K.2.3.1.—Phaseolus leptostachyus Benth. var. **pinnatifolius** Freytag forma **purpureus** Freytag, var. et forma nov. (**Figs. 77, 80**). TYPE: MÉXICO. JALISCO: near Piedra Ancha, Km 20 on Hwy from Cd. Guzmán to El Grullo, 19°38'N, 103°35'W, 2000 m, 3 Dec 1981, *Freytag & Sullivan 81-25* (in part) (HOLOTYPE: US).

Persimilis *Phaseolo leptostachyo*, sed foliis saepe pinatis foliolis 5 infimis usque ad compluriens longis differt. Habitat in locis modo meridionalibus Novo Galiciae rarus.

Very similar to type *leptostachyus* except: ***Plant*** a very large prostrate and climbing vine. ***Stem*** terete, 1–1.5 mm thick, covered with long yellow strigose hairs mostly 2–2.5 mm long; internodes 12–14–17 cm long. ***Stipules*** broadly rounded-ovate, 4–5 mm long, 3.5–4 mm wide, obtuse, strongly 7- to 8-nerved, glabrous, long ciliate margins. ***Leaves*** larger than for type, 3- to 5-pinnate, but only the 3 upper pulvini are present; petioles 2–4–6 cm long; petiolules 1 cm long; pulvini 2 mm long; stipels obovate to lanceolate to somewhat spatulate, 2.5 mm long, acute, glabrous, ciliate; terminal leaflet mostly oblong, 2.5–4 cm long, often deeply divided at base, forming 2 orbiculate or obovate lobes, 1–2 cm long, 0.8–1 cm wide; lateral leaflets similar but broader and somewhat lobate at base. ***Inflorescence*** densely covered with long yellowish strigose hairs. ***Flowers*** purple.

PARATYPE. **MÉXICO. Michoacán:** Cerro Santa María, 8–10 km SW of Jiquilpan and 5 km NE of Quitupan, (20°N, 102°50'W), 2000 m, 5-7 Aug 1959, *Feddema 60* (MICH).

Habitat.—This variety and form of the species was found growing in patches on an old road following a ravine through grassland with occasional shrubs and tall weeds, in pine-oak forest, but is not found growing on surrounding hillsides.

Comments.—The type collection for this taxon contained a lot of variability in plant growth habit with not only many plants much like the type for the species, but also plant habits much smaller and others which are very rare with much larger foliage and pecularly pinnate leaves.

K.2.3.2.—Phaseolus leptostachyus Benth. var. **pinnatifolius** Freytag forma **albus** Freytag, forma nov. (**Fig. 80**). TYPE: MÉXICO. OAXACA: Grown in screenhouse at Mayagüez, PR from seed of TARS #176, (collected at Km 102 on Hwy 190 N of Cd. Oaxaca, 17°18'N, 97°6'W, 2080 m, 21 Sep. 1978 by *Freytag et al. 78-Mex-45* (in part)), Sep.-Dec., 1985 as Study Collection *Freytag GF #SC-176* (HOLOTYPE: US; ISOTYPES: ARIZ, CSU, F, MO, TEX, UC, WIS).

Similis *Phaseolo leptostachyo* var. *pinnatifolio*, sed floribus albis differt. Habitat in locis modo centralibus regionis Oaxacae rarus.

Very much like *P. leptostachyus* var. *pinnatifolius* forma *purpureus* except: ***Flower*** white.

Habitat.—This form of the variety was found growing in open spaces in oak forest on gentle slopes with schist soil.

Comments.—The population collected had variable flower colors from purple to white. White flower color is extremely rare in the species, being found only in this central Oaxaca collection.

K.2.4.—Phaseolus leptostachyus Benth. var. **nanus** Freytag, var. nov. (**Figs. 78, 80**). TYPE: MÉXICO. JALISCO: near Piedra Ancha, grown in greenhouse from seed of TARS #18 (collected at Km 20 on Hwy from Cd. Guzmán to El Grullo, 19°38'N, 103°35'W, 2000 m, 3 Dec 1981,by *Freytag & Sullivan 81-25* (in part)), Sep.-Dec. 1985 as Study Collection *Freytag GF #SC-18* (HOLOTYPE: US; ISOTYPES: MO, CSU, ARIZ, F, TEX, UC, GH).

FIG. 77. **Illustrations of *Phaseolus leptostachyus* Benth. var. *pinnatifolius* Freytag. forma *purpureus* Freytag.—b. Root after 1 year.—c. Vine tip, leaves and inflorescences with flowers and pods and separately a mature leaf from basal portion of plant; note the pinnate nature of leaf due to two small leaflets at base of terminal leaflet.—g. Seeds, side view and view of hilum. All drawings made from living material grown in greenhouse at Mayagüez from seed of *Freytag et al. 81-25* (TARS #18, in part) collected from near Piedra Ancha, Jalisco, México.**

Similis *Phaseolo leptostachyo*, sed foliis deminutis, internodiis brevibus usque ad 10 cm longis differt. Habitat in locis modo meridionalibus regionis centrali Novo Galiciae et montibus Serra Mater occidentali Durangensi rarus.

Very similar to type species *leptostachyus* except: ***Plant*** a short, nearly prostrate vine, to 0.5 m long; internodes short, to 10 cm long. ***Leaves*** small, with very small leaflets, variegated dark green along main veins. ***Inflorescence*** short and erect, to 10 cm long. ***Pod*** small, to 1.5–3 cm long, 2–3 seed.

PARATYPES. **MÉXICO. Durango:** Metates, N of Cueva, Sierra Madre Occidental, (23°45'N, 105°20'W)?, 2650–2700 m, 29–30 Aug 1934, *Pennell 18392* (US). **Sinaloa:** S of San Blas, S of Mazatlán, 10 m, 18 Mar 1976, *Shaw 20* (MO).

FIG. 78. Illustrations of *Phaseolus leptostachyus* Benth. var. *nanus* Freytag.—**a.** Seedling a few weeks after germination.—**b.** Root after one year.—**c.** Portion of lower vine near ground level and separately, vine tip showing mature leaf and inflorescence.—**d.** Flowers, front view and side view.—**e.** Exploded view of flower showing all parts including—**s.** Style tip and stigma as seen under microscope.—**f.** Lateral view of pod on mature inflorescence and separately, a dehisced pod.—**g.** Seeds, side view and view of hilum. All drawings made from living material grown in greenhouse at Mayagüez of seed (TARS 18) from type collection *Freytag et al. 81-25* from near Cd. Guzmán, Jalisco, México.

Habitat.—This taxon is found in grassy pine forests.

Comments.—Apparently a rare dwarf form of the type species and growing mixed with other varieties of the species, but was found not to segregate in greenhouse plantings.

K.2.5.—Phaseolus leptostachyus Benth. var. **lobatifolius** Freytag, var. nov. (**Figs. 79, 80**). TYPE: MÉXICO. JALISCO: 13 km by road S of El Chante on road to Manantlán, (19°35'N, 104°15'W), 1400 m, 20 Sep 1983, *Anderson 12731* (HOLOTYPE: MICH; ISOTYPE: CAS).

Similis *Phaseolo leptostachyo*, sed lobatis ad foliolarem basem foliolis lateralibus fere triangularibus differt. Habitat in locis regionis meridionali et centrali Novo Galiciae rarus.

Very similar to type species *leptostachyus* except: ***Plant*** a short, prostrate vine. ***Root*** unknown. ***Stems*** rooting from nodes; internodes 5–11 cm long. ***Stipules*** broadly rounded, 6–7 mm long, 4 mm wide, multi-nerved, slightly acute apex, glabrous, ciliate margin, enrolled. ***Leaves*** 10.6–12.7 cm long; petiole delicate, 3–3.5 cm long, heavily covered with uncinate hairs with a few scattered, long strigose or hirsute hairs; petiolule 1–1.5 cm long; pulvini 3–3.5 mm long, sparsely covered with long strigose hairs and densely covered with minute uncinate hairs; stipels spatulate to oblanceolate, 3.25 mm long, 1 mm wide, acute, glabrous, ciliate margin; terminal leaflet broadly linear to ligulate, 6–7 cm long, 1.5–2 cm wide, rounded at base, acute, minutely apiculate, indistinctly farinaceous-glandular adaxially, nearly glabrous abaxially, obscurely veined, sparsely covered by strigose hairs mostly on veinlets, ciliate of short, stout, hispid hairs on margins, faintly variegated adaxially and lighter green abaxially; lateral leaflets similar and inequilateral, somewhat broader and to nearly triangular, 5–6 cm long, 3–3.5 cm wide, one rounded lobe on lower margin at base. ***Inflorescence*** erect, exceeding foliage, 21–24 cm long; peduncle 15–18 cm long; rachis 6–9 cm long; primary bracts obovate, 7 mm long, 4 mm wide, strongly white 9- to 11-nerved, glabrous, heavily covered with long strigose hairs and ciliate at margin, persistent; pedicels 2–2.5 mm long at anthesis, becoming 3 mm long at full pod, stout, heavily covered with minute white uncinate hairs. ***Bracteoles***, persistent, minute, scale-like, sessile, white, 1-nerved, 0.7–0.75 mm long. ***Flower*** dark purple, very fertile, at times producing 2 pods at each flowering node. ***Pod*** slightly falcate mostly near apex, 1.75 cm long, 3 mm broad; reticulate valves, densely covered with long, white strigose hairs; beak stout, 2 mm long. ***Seed*** unknown. ***Seedling*** very small; epicotyl 4.5 cm long; eophyll blade ovate-triangular, 2 cm long, 1.5 cm wide.

PARATYPE. **MÉXICO. Jalisco:** Guadalajara, (20°40'N, 101°30'W), Sep 1893, *Pringle 4544* (in part) (CAS, GH, TEX-LL, US).

Habitat.—This variety is found growing in open grassy oak forest.

Comments.—This taxon is named for the very large basal lobes on the lateral leaflets, very unusual for species in this section.

K.3.—Phaseolus opacus Piper, Contr. U.S. Natl. Herb. 22:695–696. 1926. (**Fig. 80**). TYPE: MÉXICO. VERACRUZ: Barranca de Tenampa, Zacuapán, Mar 1914, *Purpus 7081* (HOLOTYPE: US 567441; ISOTYPES: BM, F, GH, MO).

Aerial shoot an herbaceous prostrate or stiff upright, determinate vine. ***Root*** unknown. ***Stems*** slender, covered with long, straggly yellowish pilose hairs; internodes 10–15 cm long. ***Stipules*** triangular-lanceolate, 3–4 mm long, 7- to 9-nerved, acute, sparsely covered by strigose hairs abaxially. ***Leaves*** 4.5–15.5 cm long; petioles as long as terminal leaflet, 1.5–6 cm long, covered with long strigose hairs; petiolules 1–1.5 cm long, covered with strigose hairs; stipels linear, 2–2.5 mm long, ciliate; terminal leaflets ovate to long ovate, 2–8 cm long, rounded at base, obtuse to acute, apiculate, sparsely covered with long appressed-pilose hairs adaxially, glabrous abaxially, dark olive green on adaxial surface, paler (grayish) on abaxial surface; lateral leaflets similar and only slightly inequilateral. ***Inflorescence*** a pseudoraceme of 10–12 flowers; peduncle 5–6.5 cm long, exceeding the leaves, covered with pilose hairs; rachis 3–10 cm long of 6–8 flowering nodes, densely covered with pilose or strigose hairs; primary bract lanceolate, 3–4 mm long, 1.75 mm wide, strongly 3-nerved, acute, sparsely covered with long white pilose hairs, persistent, projecting beyond the buds at the tip; pedicel short, 2.5–3 mm long, densely covered with minute yellow uncinate hairs; pedicellar bracts 1–1.5 mm long narrowly lanceolate 1-nerved covered with a few yellow strigose hairs. ***Bracteoles*** linear-lanceolate, 1–1.75 mm long, 0.2 mm wide, acute, strongly 1-nerved, sparsely covered with hispid hairs, ciliate on margins.

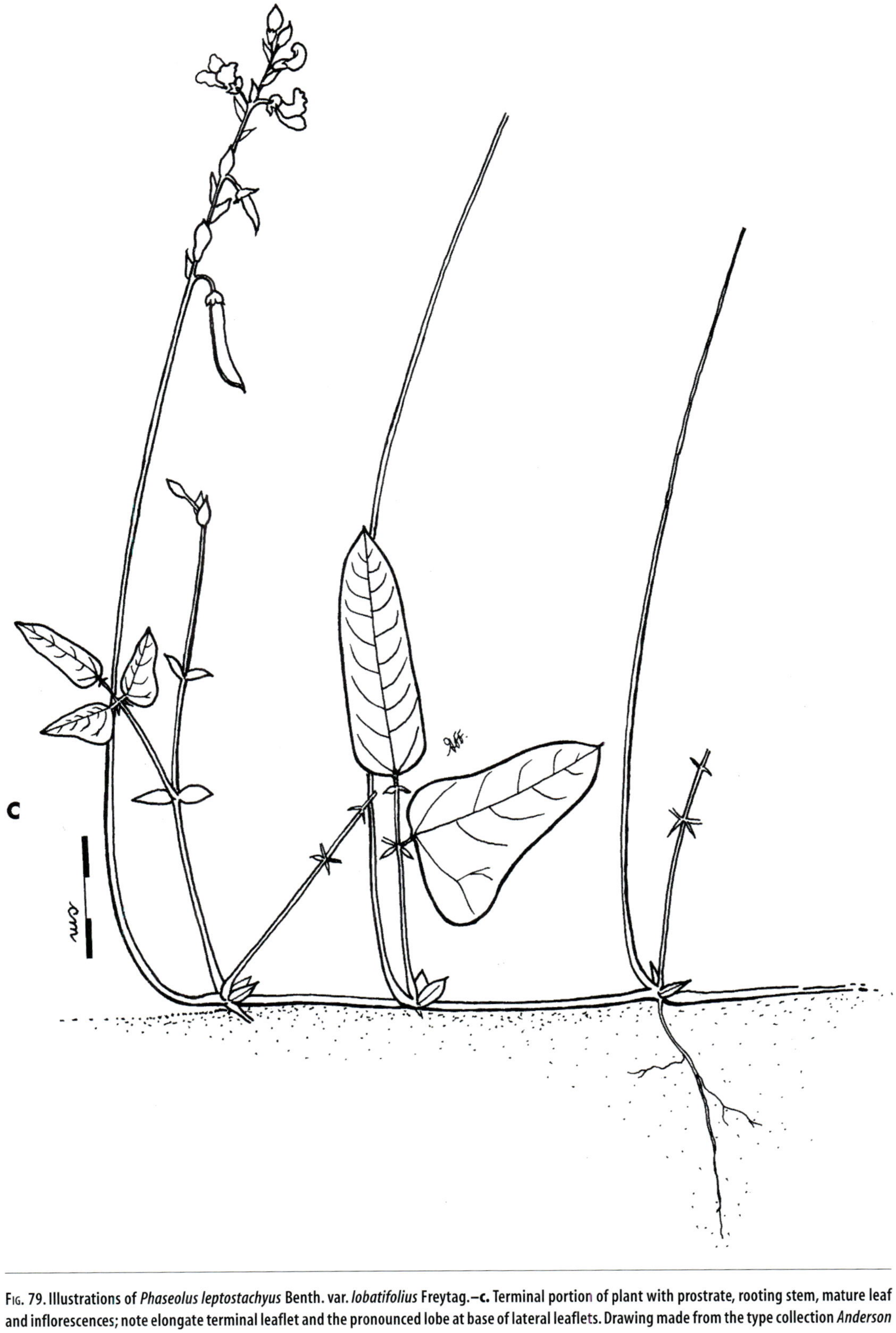

Fig. 79. Illustrations of *Phaseolus leptostachyus* Benth. var. *lobatifolius* Freytag.—**c.** Terminal portion of plant with prostrate, rooting stem, mature leaf and inflorescences; note elongate terminal leaflet and the pronounced lobe at base of lateral leaflets. Drawing made from the type collection *Anderson 12731* collected near El Chante, Jalisco, México.

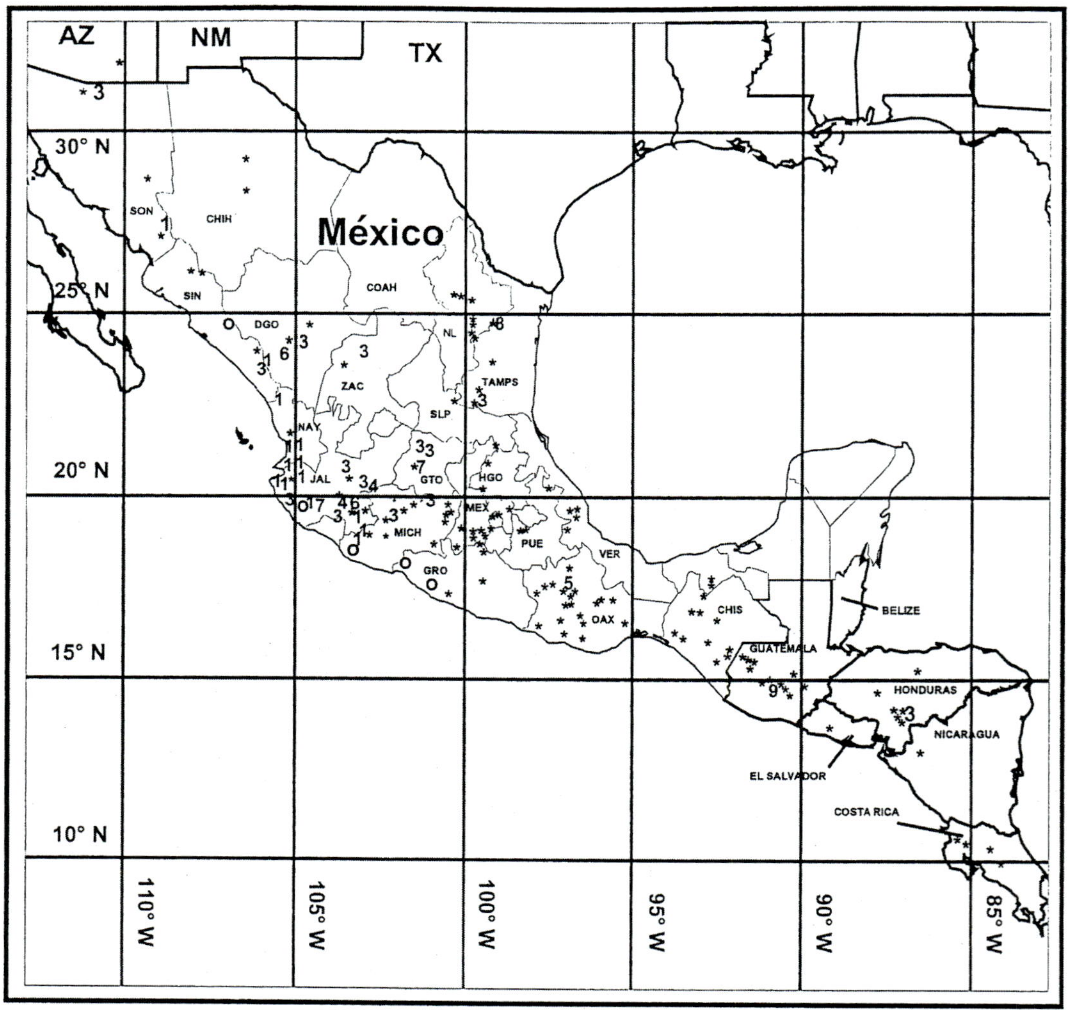

Fig. 80. Distribution of species and varieties of Section K. *Falcati* as follows: 1 = *P. micranthus*; ★ = *P. leptostachyus* var. *leptostachyus*; **3** = *P. leptostachyus* var. *intonsus*; **4** = *P. leptostachyus* var. *pinnatifolius* f. *purpureus*; **5** = *P. leptostachyus* var. *pinnatifolius* f. *albus*; **6** = *P. leptostachyus* var. *nanus*; **7** = *P. leptostachyus* var. *lobatifolius*; **8** = *P. opacus*; **9** = *P. persistentus*; **o** = *P. mcvaughii*.

Flower small, purple; calyx flaring campanulate, upper 2 lobes scarcely developed, sparsely covered with pilose hairs, ciliate margins, the lower lobes dentate, 0.75 mm long, sparsely covered with pilose hairs, bearded on longer lower central lobe with strigose hairs 1–3 mm long; standard purple with the upper lip emarginate, reflexed at 3 mm from base and 3 mm more to tip, thickened at the point of flexure, stipitate, not auriculate, blade 4–5 mm long, 4–5 mm wide; wings purple, the blade oblong, 10–12 mm long, the claw 3 mm long, the spur obscure, about 1 mm long; keel divided, the claws 2.5 mm long, 2.5 mm more to bend and 1.5 mm more to base of the terminal 2 coils of 1.75 mm in diam., tip yellow; stamen tube 4 mm to bend and 1.25 mm more to divided filaments; ovary 3 mm long, 0.5 mm wide, covered with long white canescent hairs, 6 ovules; style beard short, extending halfway around the terminal thickened coil of 1.25 mm in diam.; stigma linear, 0.75 mm long, introrse, lateral. ***Pod*** somewhat stipitate, scimitar shaped, falcate, 2 cm long, compressed, sutures finely marked; beak broad, recurved, 2 mm long, nearly glabrous to covered by a few minute hispid hairs and a few long yellow hairs. ***Seed*** oblongoid, nearly spherical, 3 mm long, 2.25 mm wide, 1.5 mm thick, speckled black on tan, a black ring around hilum; hilum oblong, 0.75 mm long, 0.4 mm wide. ***Seedling*** unknown.

Specimens examined: **MÉXICO. Tamaulipas:** La Vegonia, San José, Sierra de San Carlos, (24°35'N, 99°W?), 1036 m, 6 Jul 1930, *Bartlett 10137* (in part) (CAS-DS, F, GH, US).

Habitat.—Unknown.

Comments.—We found no black bristly hairs on the calyx as indicated by Piper (1926), rather there are numerous very long, strigose to pilose yellowish or brownish hairs on the lower lobes of the calyx and throughout the plant, especially inflorescences and stems. The black color referred to by Piper on the type specimen could be due to fungi or algae scattered on these hairs. The drab olive-green upper surface of the leaves and the somewhat silvery appearance of the underside of the leaves could be distinctive. Delgado (1985) places this species as synonymous to *P. leptostachyus.* Additional collections in that part of Veracruz are obviously needed, since *Purpus 8009* collected in Zacuapan in 1917 was identified as var. *leptostachyus* by the senior author and Delgado Salinas.

K.4.—Phaseolus persistentus Freytag & Debouck, sp. nov. (**Fig. 80**). TYPE: GUATEMALA. SACATEPÉQUEZ: Volcán Acatenango, 5 km W of San Miguel Dueñas, 14°33'N, 90°50'W, 1820 m, 6 Dec 1985, *Debouck & Soto 1621* (HOLOTYPE: US; ISOTYPES: BR, K, USCG).

Similis *Phaseolo leptostachyo*, sed bracteolis late ovatis, petalis marcescentibus ad legumen maturam apicem differt. Habitat modo in montibus vulcanicis Acatenangum Guatemalae rarus.

Aerial shoot a small, climbing vine. ***Root*** unknown. ***Leaves*** 8–12 cm long; petiole 3–4 cm long; petiolules 8–16 mm long; stipels lanceolate, 2 mm long; pulvini 1.5–2 mm long; terminal leaflet ovate, 4–6 cm long, 3–4.5 cm wide at 1/3 from base, acute, apiculate, dark olive green, densely pubescent. ***Inflorescence*** stout; peduncle 2–4 cm long; rachis 0–1 cm long, of 1–2 floral nodes and 2–4 flowers; bract broadly ovate-acuminate, 2 mm long; pedicel 5–7 mm long. ***Bracteoles*** broadly ovate, 5–6 mm long, 3–4 mm wide, heavily 9-nerved, acute to obtuse, glabrous, persistent. ***Flower*** color unknown; calyx nearly tubular, tube 2.5 mm long, 2.5 mm in diam., upper 2 teeth united into 1 entire, less than 0.25 mm long, the lower 3 acute, 0.5 mm long, the center one somewhat longer 1 mm wide, slightly pubescent; standard round, entire, 6 mm long, 7 mm wide, slightly reflexed at 2 mm from base, the claw 0.5 mm long, somewhat lobed at base, no auricles, heavily reddish nerved at base, usually remaining partially closed at suture and clasping the pod nearly to maturity; wings, the blade obovate, 7–8 mm long, 4–5 mm wide, heavily nerved at base, the claw 1.752 mm long, the spur 1 mm long; keel short, 4 mm from base to terminal single coil of 3 mm diam., the claws 1.5 mm long, separate for 0.75 mm, the ridges not developed; vexillary stamen minute, the claw 0.2 mm long, the spur 1 mm long, the knobs scarcely developed, 0.5 mm wide, thickened portion 2 mm long; stamen tube broad at base, 6 mm long, 3 mm wide, heavily reddish-nerved; ovary straight, glabrous, the terminal thickened coil of 1.75 mm diam.; stigma terminal, capitate and introrse linear, 0.75 mm long, persistent nearly to mature pod stage with all flower parts adhering, 5–8 ovules; cleistogamous. ***Pod*** (immature) falcate, 3–4 cm long, 6–7 mm wide, narrow at base and wider at apex; beak nearly straight, 2 mm long. ***Seed*** unknown. ***Seedling*** unknown.

Habitat.—This species was found growing in a cut-over forest of oak with an understory of Mimosoideae, Lamiaceae, Compositae, and Graminae. The soil was sandy and rocky from volcanic ash. Other *Phaseolus* species were found near by.

Comments.—The name of this taxon refers to the persistence of flower parts on the apex of the pod even when nearing maturity. Only a single plant in green pod stage was found on the side of the volcano, so very little is known about this taxon and no other similar specimens were found in the herbaria reviewed.

K.5.—Phaseolus macvaughii Delgado, Syst. Bot. 25:414–418. 2000. (**Figs. 73, 80, 81**). TYPE: MÉXICO. JALISCO: Mpio de La Huerta, Playa El Paraiso, (19°30'N, 105°W), 1 m, 19 Sep 1976, *Delgado et al. 129* (in part) (HOLOTYPE: MEXU n.v.; ISOTYPES: CAS n.v., CHAPA n.v., IBUG n.v., MICH, MO).

Aerial shoot an annual slender, scandent, trailing or climbing indeterminate vine, up to 1.5–3 m long. ***Root*** annual, thin, fibrous. ***Stems*** near the base terete, striate, 1–3 mm thick, sparingly setulose with fine, straight-appressed, retrorse hairs and minute whitish hooked hairs. ***Stipules*** narrowly triangular to triangular, 2–3(4.7) mm long, 3–5-nerved, ciliate margins. ***Leaves*** medium 6–10 cm long thin membranous; petioles delicate, 2–5 cm long, sparsely pubescent of white long hairs and minute

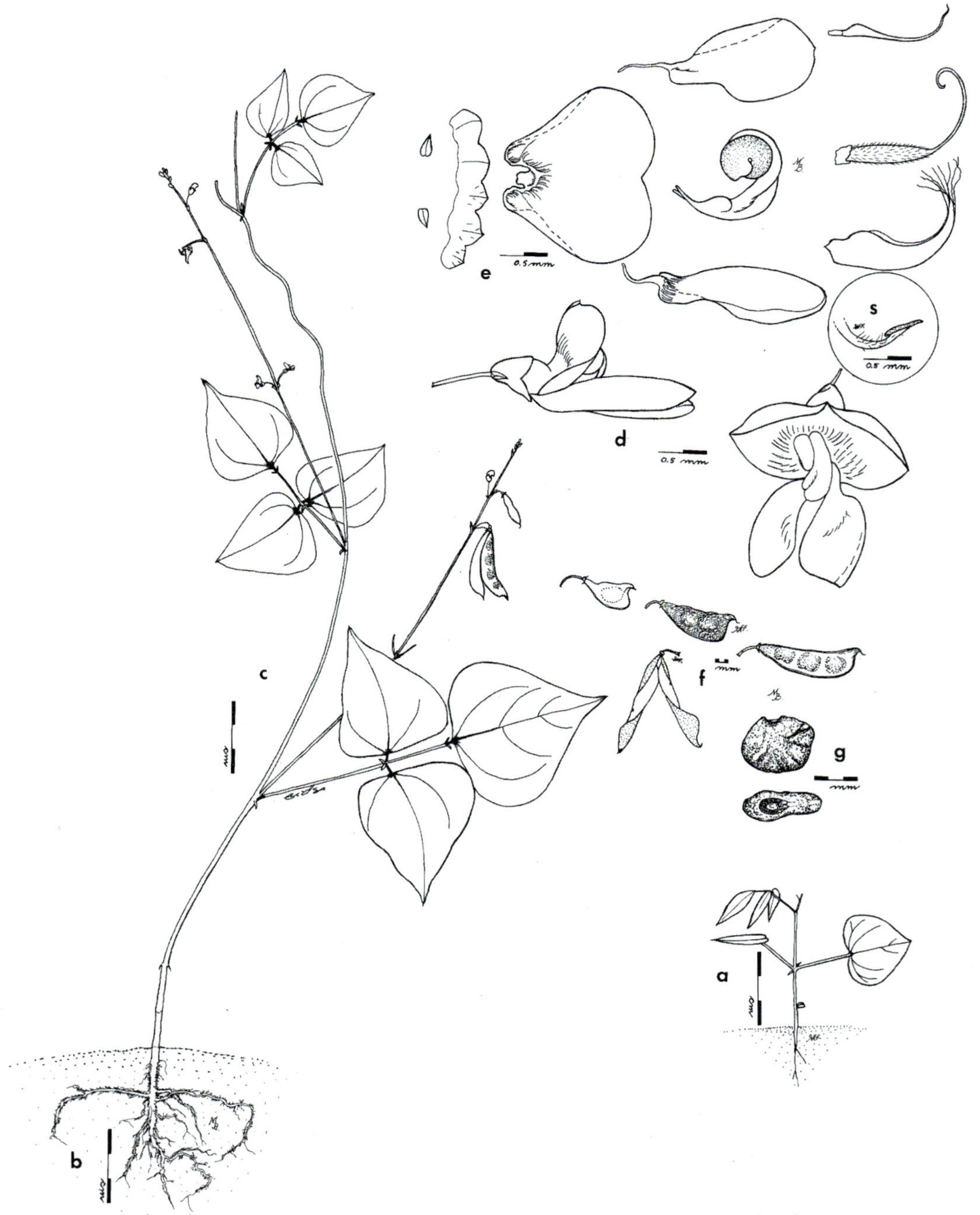

FIG. 81. Illustrations of *Phaseolus macvaughii* Delgado.–**a.** Seedling a few weeks after germination; note broad ovate eophylls.–**b.** Root of mature plant.–**c.** Lower portion of plant showing mature leaves and inflorescences; note very small buds and flowers.–**d.** Flowers, side view and front view.–**e.** Exploded flower showing all parts including–**s.** Style tip and stigma as seen under the microscope.–**f.** Pods, side views of three pods and dehisced full-sized pod.–**g.** Seeds, side view and view of hilum; note flattened shape. All drawings made from living material grown in greenhouse at Mayagüez from seed of *Norvell M-6999* (TARS #532) collected in México, except c. plant with leaves and inflorescences from field collection *Brandegee s.n.* from near Culiacán, Sinaloa, México.

hooked hairs; petiolules (0.5)1–1.5 cm long; stipels linear, 1–1.5 mm long, 0.25–0.5 mm wide, 1-nerved, sparsely ciliate tip; pulvini 2 mm long heavily covered with uncinate hairs adaxially, nearly glabrous abaxially; terminal leaflet ovate to broadly ovate, 2.5–5.5(7) cm long, 3–4.2(5.5) cm wide, mostly acuminate, very sparsely covered with straight and uncinate hairs; lateral leaflets similar and slightly inequilateral. ***Inflorescence*** a straight pseudoraceme with 3–6 or more secondary racemes, most 3–10(14) cm long; peduncle 3–7 cm long; rachis 3–4–10 cm long of 3–6(8) nodes, densely covered with uncinate hairs; primary bract triangular-lanceolate, 2–2.5(2.8) mm long, strongly 3-nerved; pedicel 2–5 mm long, covered by pilose hairs; pedicellar bracts minute 0.5 mm long almost linear scale-like hyaline ciliate margins. ***Bracteoles*** linear to lanceolate, 0.5–0.6(1) mm long, strongly 1-nerved, sparsely covered with minute ciliate hairs. ***Flower*** minute 5–6 mm long, white or pale lilac, often drying yellowish or brownish; calyx flaring campanulate, 1–2 mm long, upper lobes united into 1 rounded, emarginate, scarcely elongated, puberulent, the lower lobes triangular, acute, 0.5 mm long; standard, strongly reflexed at 1 mm from base and 2 mm more to apex, thickened at flexure, enrolled ventrally at sides, deeply emarginate at apex, 4–5 mm long, 3.5–5.5 mm wide, the claw 0.75 mm long, the auricles 0.6–0.7 mm long; wings white or light violet, the blade oblong, 5 mm long, 3 mm wide, rolled somewhat lengthwise, the claw 2 mm long, the spur 0.5 mm long; keel, the claws 2 mm long, 2 mm more to bend and 1 mm more to the terminal 1 3/4 coils of about 1.5 mm diam.; vexillary stamen, the claw 0.5 mm long, the geniculate knob globose, 0.7 mm wide; basal collar not developed; ovary straight, 2.5 mm long, 0.7 mm wide, covered by long hispid pubescence, 3–5 ovules; style 4 mm long to the terminal thickened coil of 0.8 mm diam.; stigma lateral introrse, narrowly ovate to lanceolate, 0.5 mm long. ***Pod*** often dimorphic somewhat stipitate and falcate, when 3–4 seeded 2 cm long, 4.5–6 mm wide, or when 1–2 seeded boat prow shaped 1–1.5 cm long 5–6 mm wide, valves chartaceous, compressed, reticulate, puberulent, dehiscent with 3 loose twists, the beak short, straight, less than 1 mm long, weak. ***Seed*** rhomboid to squarish, flattened, 3–4.5 mm long, 3–4 mm wide, surface rugose, speckled black on light tan, black ring around hilum; hilum oval, 0.3 mm long. ***Seedling*** from epigeal germination, hypocotyl 3 cm long, epicotyl 7–8 mm long, stipules entire, stipels absent, eophylls with 3-part petioles 2 cm long, the blade broadly ovate cordate at base 2.5 cm long 2.5 cm wide obtuse.

Specimens examined: **MÉXICO. Baja California:** Punta Norte de la Isla Angel de La Guarda, Mpio. Ensenada, Puerto Refugio, 29°33'N, 113°34'W, 50 m, 7 Feb 1986, *Tenorio et al. 10857* (ASU). **Colima:** La Cumbre, 2 km NW del pueblo, Río El Salado, 19°07'N, 103°42'W, 320 m, 9 Nov 1978, *Debouck et al. 427* (CHAPA, US); Cuyutlán Lagoon, Manzanillo, (19°N, 104°20'W), 28 Nov 1925, *Ferris 6091* (CAS-DS). **Guerrero:** Mpio. Zihuatanejo, Ixtapa, 17°40'N, 101°37'W, 10 m, 20 Dic 2001, *Acosta 211* (COL, G); Acapulco, presqu'île Griffon, Oct 1866, *Le Jolis s.n.* (G). **Jalisco:** Mpio. La Huerta, vereda a Rincón de Ixtán, viejo camino a Nacastillo, 9 Oct 1982, *Lott et al. 1420* (MO); Mpio. La Huerta, Estación de Biología Chamela (UNAM), Cuenca 2, 19°30'N, 105°3'W, 550 m, 11 Oct 1982, *Lott 1449* (MO, TEX); Mpio. La Huerta, carretera a Melaque, 12 km al SE de la Estación, 13 Oct 1982, *Lott et al. 1490* (CAS, MICH, MO); Estación de Biologia-Chamela, UNAM, *Magallanes 830* (CAS); Chamela, Playa de Cuastecomate, 8 km by road NW of Barra de Navidad, Playa de Cuastecomate, (19°15'N, 104°50'W), 50 m, 6 Nov 1960, *McVaugh 20747* (MICH). **Michoacán:** Mpio. de Aquila, 6 km al O de Río Cachán, carr. Tecomán–Playa Azul, (18°30'N, 103°30'W)?, 26 Sep 1983, *Lott et al. 1975* (TEX-LL). **Sinaloa:** Mazatlán, (21°10'N, 106°15'W), 8 Oct 1893, *Brandegee s.n.* (UC); Culiacán, (24°45'N, 107°30'W), 17 Sep 1904, *Brandegee s.n.* (UC); Labradas, 20 Sep 1925, *Ferris et al. 5199* (CAS-DS); Mazatlán, (grown in experimental gardens of Stanford Univ. from seed), 8 Sep 1948, *Norvell M-249* (CAS-DS).

Habitat.—Apparently rare and in isolated populations, growing on rocky beaches, in sandy clayey soils derived from metamorphic and conglomerate rocks, often along small streams in openings in mid-altitude subdeciduous oak or oak-pine forests with by *Acacia, Acalypha, Astronium, Bursera, Croton, Psychotria, Tabebuia*, and cacti.

Comments.—An uncommon species growing at or near sea level on the east shore of the Gulf of California and south to Michoacán, Colima and Guerrero. Though in plant type it resembles *P. filiformis* (see Color Plate V, photo 59), it has very small, white flowers, the smallest of all the genus (see Color Plate II, photo 19), pods with a boat-shaped tip, and many other minor characteristics which set it apart from *P. filiformis*. It was first recognized as a distinct species by Oliver Norvell (annotated as "*minimiflorus*" on UC 82386), and therefore we have during this work named specimens under that name (see Debouck 1991, 2000a). It is reported to have 2n=2x=20 chromosomes (Mercado-Ruaro & Delgado Salinas 1998), and evidence from ITS DNA sequencing suggests that it is closely allied to *P. leptostachyus* and *P. micranthus* (see Delgado et al. 1999; Gaitán et al. 2000). Recent studies (Bayuelo et al. 2002b) have shown its potential for salinity tolerance.

Section L.—Brevilegumeni Freytag, sect. nov. TYPE SPECIES: *Phaseolus oligospermus* Piper, Contr. U.S. Natl. Herb. 22:698. 1926.

Herba volubilis grandissima scandens, radix perennis globosa grandis crassa carnosa, foliola grandia ovata vel lanceolata (in *Phaseolum oligospermum* basim truncata et lobata) saepe pubescentissima, flos purpureus vel azuri-violaceus, legumen parvum latum pleurumque strigosum aureo-fuscum longum et 4–6 seminibus.

Plant a very large climbing vine; root perennial, globose, large, thick, fleshy; leaflets large, ovate to lanceolate (in *P. oligospermus* often basally truncate and somewhat lobed), usually densely pubescent; flower purple or bluish-violet; pod short, broad, usually covered with long yellowish-brown, strigose pubescence and of 4–6 seed.

Comments.—The three species of this section are easily distinguished from all other species since they are large climbing vines all parts of which are usually covered with long yellowish or whitish pubescence and growing in the humid cloud forests on the tops of mountains in S México, Central America and Panama. They also have very large inflorescences producing short, broad pods with very few (3–6) seeds. On the basis of ITS DNA sequence data (Delgado et al. 1999), they form a group, itself linked to the Xanthotricha group; our results (Gaitán et al. 2000) show in addition a link to *P. macrolepis* and *P. talamancensis*.

KEY TO SPECIES

1. Leaflets ovate to lanceolate, frequently truncate and basally lobed, membranaceous, both surfaces equally pubescent, often variegated; inflorescences large but sparsely flowered.
   2. Bracteoles lanceolate, to 2 mm long; flowers purplish; pods 4–6 cm long, usually 3–4–5 seeds; seeds black speckled; scarce, S México and Central America S to Panama; 730–2100 m ________ L.1. **P. oligospermus**
   2. Bracteoles minute, 1.25–2 mm long; flower purple; pod unknown, but presumably 4–5–6 seeds; rare, in pine forests of W central México: mountains of Nayarit; 1300 m ________ L.2. **P. campanulatus**
1. Leaflets all ovate, nearly coriaceous, densely yellowish or brownish pubescent giving a silvery sheen below; inflorescences large and many flowered; bracteoles minute scales; flowers bluish-violet; pods 3–4 cm long, usually 3–4 seeds; seeds brown speckled; scarce, S México and S to Panama; 1300–2600 m ________ L.3. **P. tuerckheimii**

L.1.—Phaseolus oligospermus Piper, Contr. U.S. Natl. Herb. 22:698. 1926. (**Figs. 82, 83**). TYPE: COSTA RICA. Cartago, (9°35'N, 83°55'W), 1857, *Oersted 30* (HOLOTYPE: K, photographs GH(2), US(2)).

Aerial shoot a pluriannual, climbing, indeterminate vine, to 10 m long. ***Root*** a perennial, thick, fleshy, long turbinate, to 10 cm or more in diam., much branched with some lateral roots producing smaller enlargements. ***Stems*** terete, 3–5 mm thick, covered with short, spreading, rusty hairs when young, when mature nearly glabrous. ***Stipules*** narrowly triangular, 3 mm long, acute, pubescent. ***Leaves*** 6.7–28 cm long; petioles 2.5–8 cm long, nearly as long as the leaflets, covered with rusty-pilose hairs; petiolules 1.2–4.5 cm long; stipels similar to stipules, 2 mm long; terminal leaflet ovate-lanceolate to broad linear, 3–5–15 cm long, truncate to broadly cuneate at base, acuminate, the apex acute, coriaceous, the secondary nervation distinct, pubescent with sparse short hairs on both surfaces, more or less variegated along midvein adaxially, paler abaxially; lateral leaflets similar, inequilateral, often with a pronounced rounded basal lobe. ***Inflorescence*** a few- to many-flowered pseudoraceme; peduncles 5–11 cm long, about as long as the leaves; rachis 5–29 cm long, rusty puberulent; primary bracts triangular-lanceolate, 5–6 mm long, 5-nerved, acute, rusty pubescent, persisting; pedicel 4–9 mm long, as long as the calyx or in fruit to 1 cm long; pedicellar bracts narrowly lanceolate 3–4 mm long 3-nerved sparsely covered with strigillose rusty hairs. ***Bracteoles*** ovate to lanceolate, strongly 1-nerved, 1.5–2 mm long, two thirds as long as the calyx, sparsely covered with strigillose hairs, persistent. ***Flower*** purple and greenish, fading lilac and cream; calyx campanulate, pubescent with reddish hairs, the lobes subequal, the upper broad and emarginate, the lower with 3 triangular acute teeth, 1.5–4 mm long, the middle one longest; standard mostly greenish, striate with darker veins, orbicular, 12 mm long, sharply reflexed and cupped, short stipitate, bearing a transverse callosity near the angle of reflexion and a flaplike auricle on each side of the base back from the margin; wings lilac-purple, the blade obovate, 15 mm long, truncate at base, the stipe 4 mm long, not auricled; keel tubular, the claws 4 mm long, the tube not broadened in the middle, the ridges well-developed, 5 mm to bend and 4 more to the terminal 2 coils of 3 mm in diam.; vexillary stamen with a globose enlargement on the filament near the base, 1.5 mm long, but not geniculate; stamen tube 2 cm long; basal

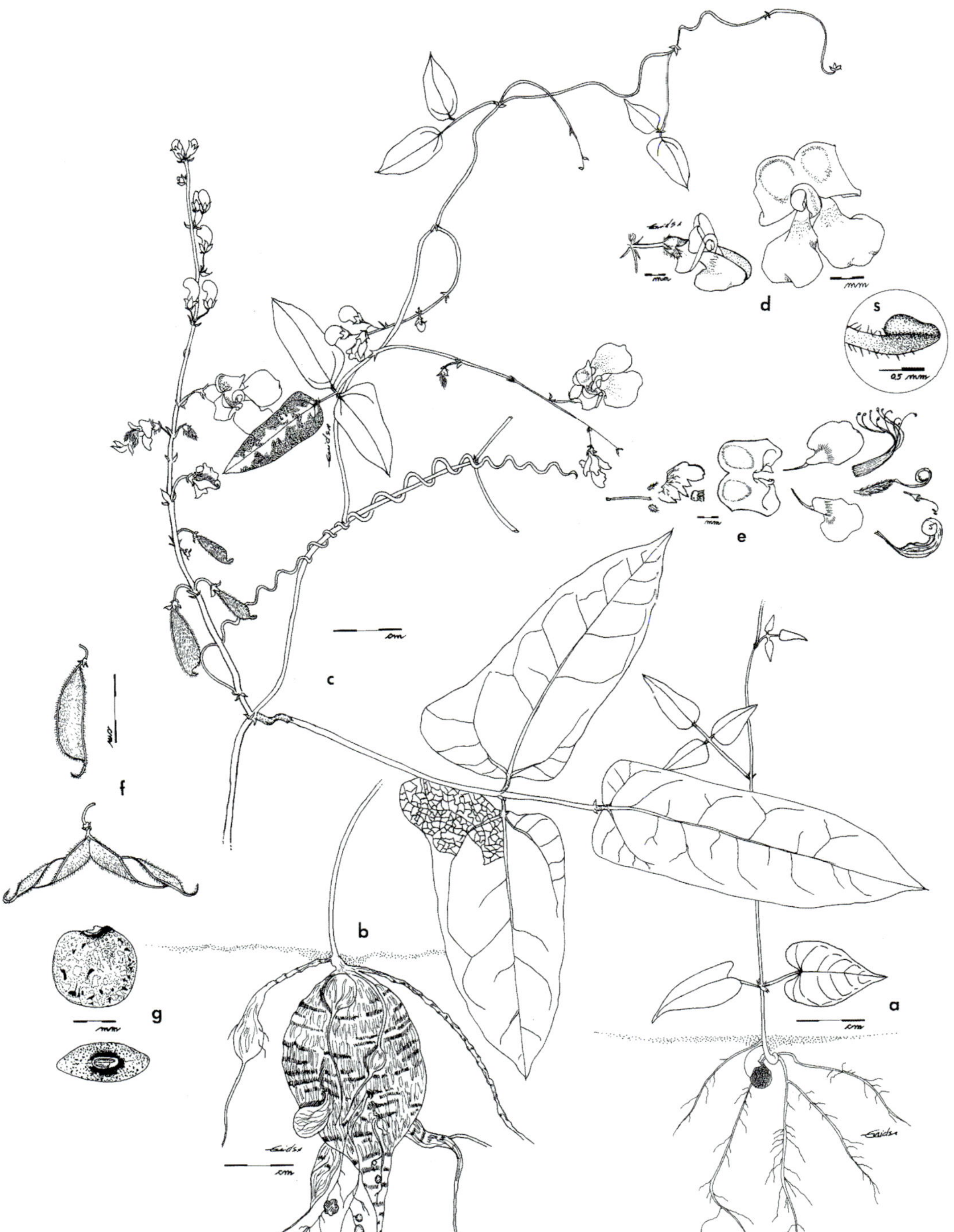

Fig. 82. Illustrations of *Phaseolus oligospermus* Piper.—**a.** Plantlet a few weeks after germination of seed; note short epicotyl.—**b.** Root after first year of growth.—**c.** Portion of stem with mature leaf, full-sized inflorescence, and small lateral branch with flowering nodes.—**d.** Flowers, showing lateral and front views at slightly different scales.—**e.** Exploded view of flower showing all parts including—**s.** Style tip and stigma as seen under the microscope.—**f.** Pods, side view and dehisced.—**g.** Seed, side view and view from hilum. All drawings made from living plants grown at Mayagüez from seed of *Cojulún 78-Guat-1* (TARS #101), collected at microwave tower, San Juan Acatempa, Guatemala, except **c.** which was drawn from the field collection *Freytag et al. 78-Guat-18* collected at the same location.

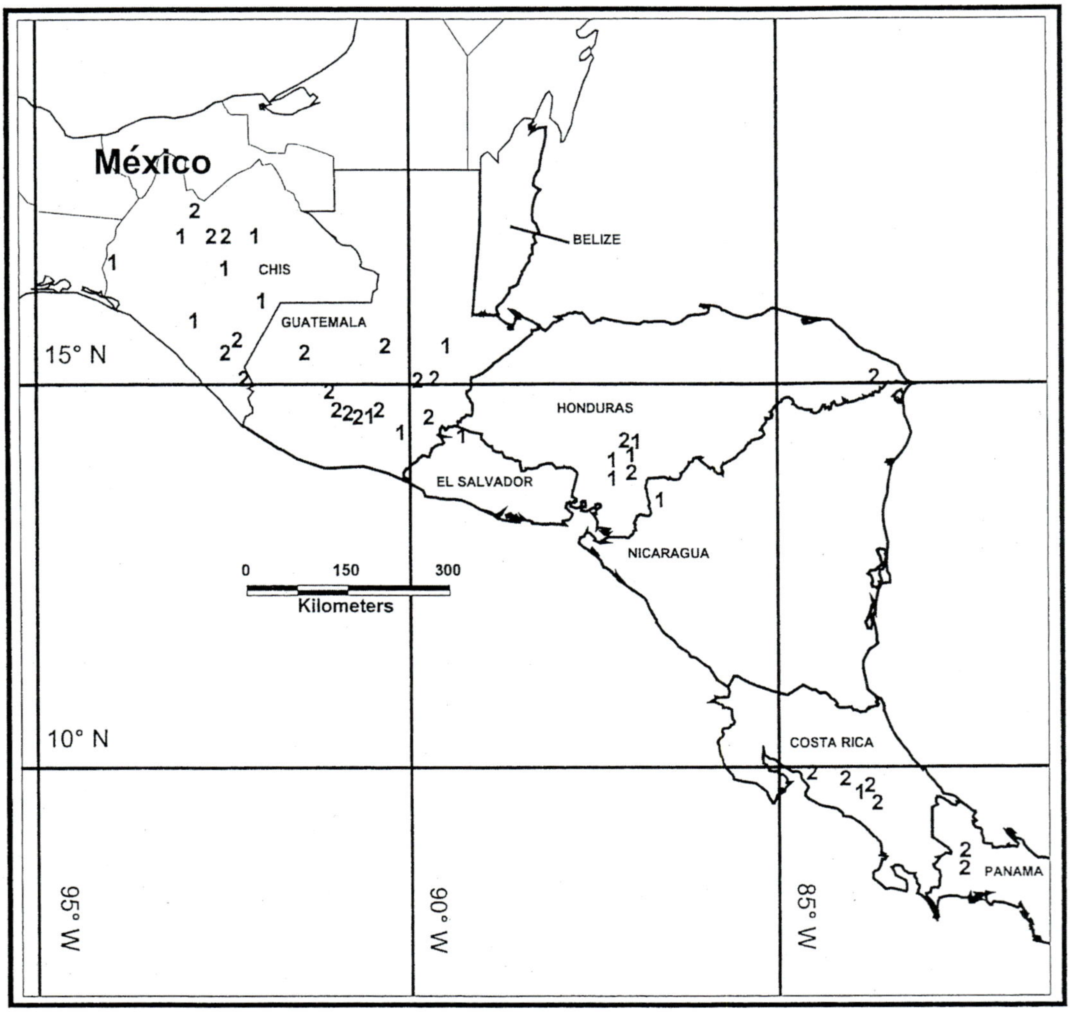

Fig. 83. Distribution of species of Section L. *Brevilegumeni* as follows: **1** = *P. oligospermus*; **2** = *P. tuerckheimii.*

collar 1.5-2 mm long; ovary linear-oblong, 7 mm long, densely covered with white pubescence, 3-5 ovules; style bearded on the inner side a short distance below the stigma; stigma lateral behind the sterile conical tip of the style, 1.2 mm long. ***Pod*** immature straight, broad, green, densely covered with ferruginous long strigose hairs, the mature pod 4-6 cm long, 11-12 mm wide, inflated, distinctly covered with long strigose hairs, obscurely reticulate, brittle, carpels dehiscent by 1-2 loose twists, the sutures heavy, the beak 5 mm long, stout, recurved. ***Seed*** roundish, compressed to somewhat lenticular, 6.5 mm long, 6.2 mm wide, 3 mm thick, colored by black flecks on a tan or brown background, with a black hilum ring, shiny; hilum round, 1.7 mm in diam., placental tissue mostly absent. ***Seedling*** from hypogeal germination; epicotyl 2.5-3.2-4.7 cm long; eophyll, the petiole 1.7-2 cm long, the blade ovate-acuminate, 2.3-3.7 cm long, 1.6-3 cm wide at near base, base somewhat auriculate to cordate, tip acuminate, minutely apiculate, membranous, glabrous except sparsely covered with hispid hairs on main veins and margins, sparsely glandulose.

Specimens examined: **COSTA RICA. Cartago:** barranco del Río Reventado, 0.3 km W de Turbina, 9°53'N, 83°55'W, 1550 m, 7 Jan 1987, *Debouck et al. 2091* (CR). **San José:** San Cristobal Norte, 4 Km W de San Cristobal Norte hacia Llano Los Angeles, 9°46'N, 84°01'W, 1400 m, 4 Feb 1998, *Debouck et al. 3107* (CR).

GUATEMALA. Guatemala: 4 km SE de Concepción Las Lomas al oriente de la Cd. de Guatemala, 14°34'N, 90°30'W, 1500 m,

17 Dec 1987, *Debouck et al. 2472* (BR, COL, G, M, MICH, SI, US, USCG). **Jutiapa:** microwave tower above San José Acatempa, Km 82 Hwy CA-1 Guatemala-Jutiapa, 14°18'N, 90°8'W, 1500 m, 4 Oct 1978, *Freytag et al. 78-Guat-18* (BR, EAP, F, GH, K, MEXU, MO, UC, US). **Santa Rosa:** San Rafael Las Flores, Laguna de Ayarza, orillas lado N de la Laguna, 14∞26'N, 90∞08'W, 1430 m, 7 Dec 1987, *Debouck et al. 2439* (USCG). **Zacapa:** Rillito del Volcán de Monos, (15°25'N, 89°30'W)?, 1150–2100 m, 10 Jan 1942, *Steyermark 42412* (F, UC).

HONDURAS. El Paraíso: Cerro Monserrat, cerca de Yuscarán, 2000 m, 6–9 Oct 1977, *Nelson et al. 4290* (MO). **Fco. Morazán:** Cerro Uyuca, 1300 m, 20 Oct 1978, *Bustillo 189* (MO); Quebrada Suyapa near Suyapa, (13°55'N, 87°15'W), 1100 m, 20 Dec 1947, *Molina 707* (EAP); Río Rancho Quemado, SE of Tegucigalpa, Kms 20–25 road to Sabana Grande, (13°45'N, 87°15'W), 1300 m, 9 Nov 1966, *Molina 18680* (EAP, F); Los Angeles creek, 2 kms SE of Valle de Angeles, (14°N, 87°W), 1500 m, 3 Jan 1969, *Molina 23271* (EAP); Rancho Quemado above San Juancito, (14°10'N, 87°W), 2100 m, 27 Oct 1951, *Williams 18533* (EAP(2), GH). **Lempira:** Montaña Puca entre Guatan y Cuabanos, 1600 m, 25 Sep 1963, *Molina 12908* (EAP, F(2), US). **Ocotepeque:** Mt. Cocal of Cordillera Merendón, 20 kms NW of Ocotepeque, (14°20'N, 89°15'W), 1800 m, 25 Aug 1968, *Molina 22147* (EAP, F, MO, NY).

MÉXICO. Chiapas: Mpio. La Trinitaria, 11 mi S of La Trinitaria, along Hwy 190, (16°N, 92°W), 1066 m, 14 Oct 1965, *Breedlove et al. 13246* (CAS-DS, F, MICH, US); Mpio. Cintalapa, Oaxaca-Chiapas border near La Cienega de León, (16°30'N, 94°W)?, 1080–1230 m, 1 Dec 1980, *Breedlove et al. 48132* (CAS); Mpio. Las Rosas, 3 km S of Aguacatenango along road to Las Rosas, (16°25'N, 92°30'W), 1768 m, 4 Oct 1981, *Breedlove 53246* (CAS, MO, NY); Mpio. La Trinitaria, near Cinco Lagos, Montebello National Park, 1372 m, 5 Oct 1981, *Breedlove 53275* (MEXU); Mirador El Tepehuaje, Km 21.5 Hwy Tuxtla Gutiérrez-El Sumidero, 16°50'N, 93°5'W, 1150 m, 25 Sep 1978, *Freytag et al. 78-Mex-74* (BR, EAP, F, GH, K, MEXU, MO, UC, US); Mpio. Ocosingo, 2 km de Cushul-ja, 17 km O de Ocosingo, (16°50'N, 92°5'W), 1200 m, 6 Dec 1980, *Sousa et al. 11359* (BM, CAS, MEXU, MO); Mpio. Angel Albino Corzo, Rancho Viejo of the Finca Prusia, (15°45'N, 92°55'W), 732 m, 23 Jan 1968, *Ton 3533* (CAS-DS, MICH).

NICARAGUA. Estelí: El Delirio, entrada a El Zacatón, camino a Miraflor, 13°13'N, 86°14'W, 1400 m, 26 Dec 1982, *Moreno 19370* (MO); Llano Almaciguera, 8.4 Km S of Hwy 1 on road through Estanzuela, 13°00'N, 86°21'W, 1260 m, 13 Nov 1978, *Stevens 10723* (MEXU, MO). **Jinotega:** Hwy between Santa Lastenia and El Ocotillero, 1500 m, 31 Oct 1968, *Molina 22924* (BM, CAS-DS, EAP, F, G, K). **Madríz:** Volcán Somoto, about 10 kms S of Somoto, (13°20'N, 86°40'W), 1400 m, 12 Jan 1967, *Williams et al. 20259* (EAP, F, US).

Habitat.—This species is found growing on the edge of or in tropical deciduous cloud forests ("evergreen seasonal forest" according to Gómez Pignataro, 1986 for Costa Rica and "temperate subtropical moist forest" according to de la Cruz S. & Sagastume L., 1983 for Guatemala) or pine-oak forests at 720–2100 m altitude, and is often at the tops of hills or bluffs, usually in rocky soils derived from limestone or volcanic ash or basalts and with copious leafmold. The large vine (2–4 m) is often scrambling over shrubs and small trees or in open, sunny clearings and is often associated with *Acacia, Alnus, Ficus, Gravilea, Juniperus, Liquidambar,* and *Ostrya.* It is also found growing with *P. xanthotrichus* Piper in both Costa Rica and Guatemala.

Common names.—Frijol de culebra, or frijolillo de culebra.

Diseases and pests.—It is a long lived perennial but susceptible to diseases (root rots, BCMV and other virus, Xanthomonas blight, Ascochyta leaf spot) and pests (scales, red spiders, leaf webber). Sources of resistance to *Phoma exigua* causing Ascochyta blight have been reported in this species (Lepoivre & Baudoin 1994).

Comments.—This species is relatively uncommon and found only in small populations. It prefers well drained, fertile soils with moisture on the dry side, especially when nearing flowering and at maturity. The plant produces a large multi-branched root (see Color Plate III, photo 35), and dies back to the main stems during the dry season after which new growth initiates at the beginning of the next rainy season. Flowering is in response to distinctly short days and cool weather. Pollination is mostly by tripping but will set a few seed without.

This species was not reported by Standley and Steyermark (1946) in their "Flora of Guatemala," nor by Maréchal et al. (1978b), nor by Lackey (1983). It has been however reported in the "Flora of Costa Rica"(Standley 1937) and in the "Flora of Nicaragua" (Delgado Salinas 2001). It is a true *Phaseolus* sensu stricto species, perhaps going long unnoticed because of its relative rarity in herbaria. Analyses of allozyme diversity (Jaaska 1996) and RFLP polymorphisms of rRNA genes (Jacob et al. 1995) have shown it to be a distinct branch in *Phaseolus* somewhat intermediate between *P. vulgaris* and *P. lunatus* based on analyses of a collection from Jutiapa, Guatemala (not from Durango, México as reported) and apparently with an affinity with species of the Section G. *Xanthotricha* on the basis of allozyme diversity (Jaaska 1996), PCR-RFLPs on cpDNA (Fofana et al. 1999), and ITS DNA sequencing (Delgado et al. 1999; Gaitán et al. 2000).

Delgado Salinas (1985) reported a collection supposedly of this species from southern Nayarit and another from western Jalisco (with apparently one more ovule in comparison to collections in the southeastern range of the species as the only distinctive character). Curiously enough McVaugh (1987) does not report either of these in his "Flora of Nueva Galicia." We have seen one specimen (*McVaugh 16341*, at MICH) that we suspect to be of this material; however we find it distinctive enough to propose as a new species, *P. campanulatus*.

L.2.—Phaseolus campanulatus Freytag & Debouck, sp. nov. (**Figs. 9, 84**). TYPE: MÉXICO. NAYARIT: Arroyo de la Fundición, 5 mi SE of Ahuacatlán on road to Barranca del Oro, (20°55'N, 104°35'W), 1300 m, 25 Aug 1957, *McVaugh 16341* (HOLOTYPE: MICH).

Affinis *Phaseolo oligospermo*, sed bracteolis minutissimis 1–2 mm longis et floribus majusculis calycibus maxime insigniter campanulatis (non bilabiatis) differt. Habitat in montibus Nayaritae rarus.

Aerial shoot apparently a perennial, indeterminate, climbing vine to 3–4 m long. ***Root*** unknown. ***Stems*** terete, 1–1.5 mm thick, densely covered by yellow, long strigose hairs, the internodes 10–17 cm long. ***Stipules*** triangular, 4–5 mm long, 2 mm wide at base, 4- to 5-nerved, dark, acute, sparsely covered by hirsute hairs. ***Leaves*** 18–21 cm long; petiole delicate, 6.5.-7.5 cm long, with scattered yellowish long strigose and short uncinate hairs; petiolules 2 cm long; pulvini dark, on stem 7 mm long, on leaflets 3 mm long, covered by densely long yellow-brown strigose hairs; stipels, the lower ones narrowly lanceolate, 2.5 mm long, 0.5 mm wide, dark, 2-nerved, acuminate, with hirsute hairs, the upper linear, 1–2 mm long, 1-nerved, covered with hirsute hairs; terminal leaflet ovate, 9–10 cm long, 6 cm wide at 1/4 from base, acuminate, apiculate, lightly veined, membranous, green and smooth to puberulent adaxially, minute white uncinate hairs abaxially, especially on veins; lateral leaflets similar and inequilateral. ***Inflorescence*** a fairly long, slender pseudoraceme; peduncle 11–13 cm long, with small, white uncinate hairs; rachis 10–16 cm long of 8–10 flowering nodes, densely covered by minute, white uncinate hairs; primary bract ovate, 3–4 mm long, acuminate, 3-nerved, with scattered hirsute and ciliate hairs; secondary bracts spreading, lanceolate, 1-nerved; pedicel 10 mm long, densely covered with white uncinate hairs. ***Bracteoles*** minute, ovate-lanceolate to lanceolate, 1.25–2 mm long, 0.4–0.5 mm wide, obscurely 1-nerved, acute, hyaline with white hirsute hairs. ***Flower*** light purple (pink); calyx campanulate, 5 mm long, tube inflated, 3–3.5 mm long, purplish, the lobes all dentate and subequal, 2–2.5 mm long, 2–2.5 mm wide, acute, densely covered by minute white uncinate hairs and with a few scattered white strigose hairs, ciliate margins; standard dull greenish purple, claw stout and broad, 2 mm long, 1.5–2 mm wide, blade rounded, reflexed at 5 mm from base and 4–6 mm more to deeply emarginate apex, 10–14 mm wide, enrolled laterally, strongly dark multi-nerved, the auricles well-developed flaps, 1 mm long, 2 mm wide; wings purple, obovate, spreading, unequal, lengthwise rolled or slightly cupped, claw 5 mm long, 0.75 mm wide, the blade obovate, 11–12 mm long, 6–7 mm wide, distinctly dark multi-veined, the spur pronounced, 0.75–2 mm diam., firmly affixed to keel; keel with claws 4 mm long, 5.5 mm more to upward bend and 4 mm more to base of terminal 1 3/4 coils of 3.5 mm diam., the ridges well-developed, about 1.5 mm diam.; vexillary stamen, claw 2 mm long, geniculate flap small, 0.5 mm long, 2 mm wide, 3–4 mm more to end of thickened portion; stamen tube, narrow tubular, 8–9 mm to bend and 4 mm more to end of united filaments, about 2 mm wide at base, the ridges at 2.5 mm from base 0.3 mm high, the basal collar irregular denticulate, 2 mm long, darkened cream; ovary straight, 5.5–8 mm long, 1.5 mm wide, densely covered with long white canescent hairs, 4 (rarely more?) ovules, the style 10 mm long to the terminal thickened coils of 2.75–3 mm diam., the stigma terminal introrse, linear-oblong, 1 mm long, 0.25 mm wide. ***Pod*** unknown. ***Seed*** unknown. ***Seedling*** unknown.

Habitat.—This taxon was found at only one location among pine trees on steep wooded ravines with semi-permanent streams at 1300 m elevation.

Comments.—This species is named for the nodding, bell-like flowers (calyx with 5 equal, dentate, acute sepals) which is quite unique in the genus since most species have bilabiate flowers which face horizontally or upwards. Only a single specimen has been seen so it is assumed that this species is rare. Additional collections and living germplasm are definitively needed to confirm placement in this section (the senior author thinks it belongs to the *Phaseoli*, but germination habit not seen), and to evaluate relationships with *P. oligospermus*.

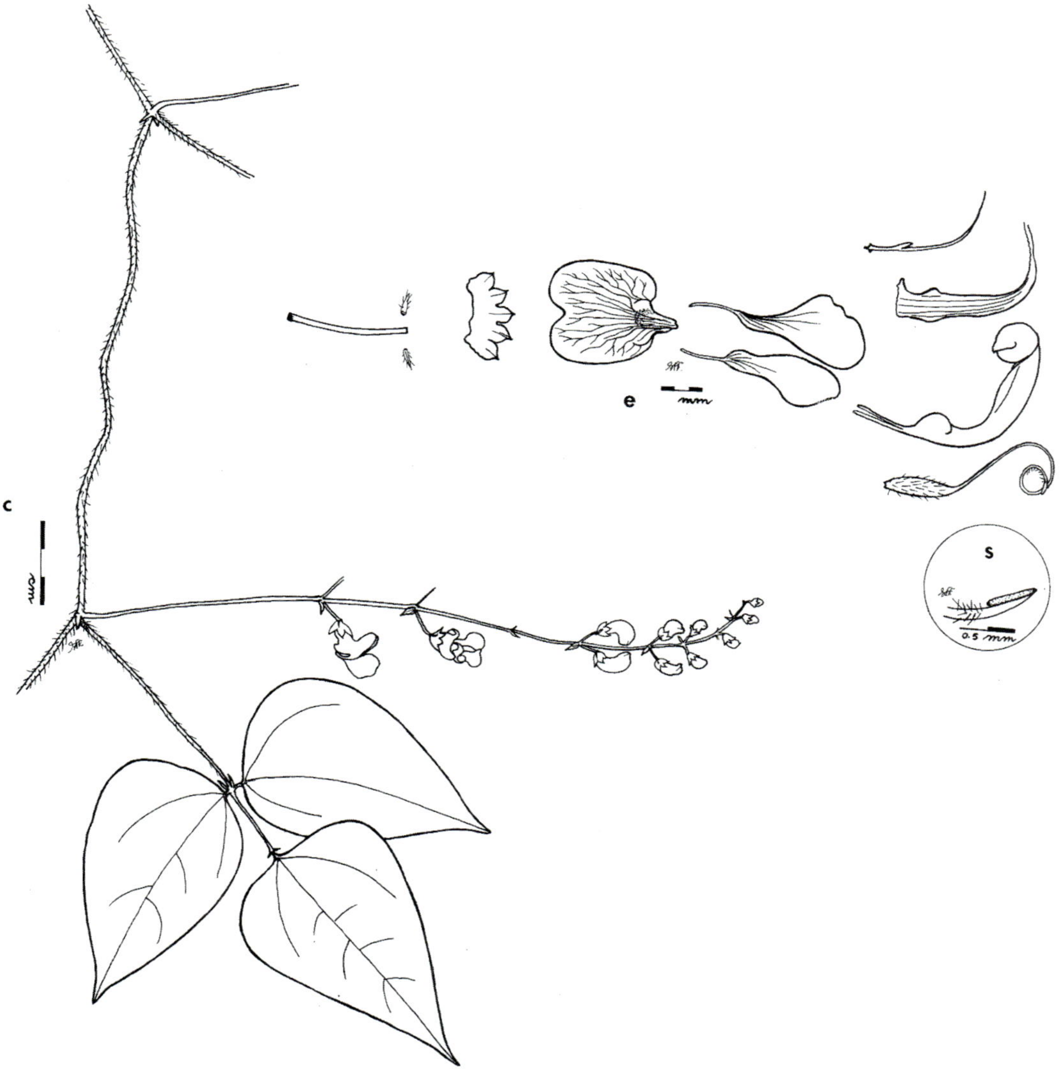

Fig. 84. Illustrations of *Phaseolus campanulatus* Freytag & Debouck.—c. Portion of stem with mature leaf and flowering inflorescence.—e. Exploded view of the flower showing all parts, including—s. Style tip and stigma as seen under the microscope. All drawings from type specimen *McVaugh 16341* from Nayarit, México.

L.3.—Phaseolus tuerckheimii Donnell-Smith, Bot. Gaz. 56:54. 1913. (**Figs. 83, 85**). Type: GUATEMALA. Alta Verapaz: (Auf den Bergen zwischen) ad montem inter Tactic et Cobán, (15°25'N, 90°20'W), 1800 m, Dec 1907, *H. von Tuerckheim II 1536* (holotype: US 942247; isotypes: BR n.v., US(2)). SYNTYPE: Costa Rica. Comarca de Puntarenas: Cuesta de las Borucas, (9°45'N, 84°45'W), Jan 1897, *Pittier 10539* (US 942246).

Phaseolus chiriquinus Standl., Contr. U.S. Natl. Herb. 18:109. 1916. Type: PANAMÁ. Chiriquí: Cuesta de Cerro Quemado, E slope of Chiriquí Volcano, (8°35'N, 82°30'W), 1800–2160 m, 10–13 Mar 1911, *Pittier 3111* (holotype: US 677501; isotype: NY n.v.).

Aerial shoot a climbing, indeterminate vine, to 10 m long. ***Root*** unknown. ***Stipules*** oblong, 4–5 mm long, 1.5–2 mm wide, 4- to 7-nerved, acute, both surfaces densely hirsute. ***Leaves*** 11–22 cm long; petiole 4.5–9 cm long, densely covered with strigose hairs; petiolule 1.8–2.3 cm long, densely covered with strigose hairs; pulvinus of petiole 5 mm long, of leaflets 4 mm long, densely covered with strigose hairs; stipels oblong to lanceolate somewhat rubro-maculate, 2.5–3 mm long, 0.75 mm wide, 4- to 5-

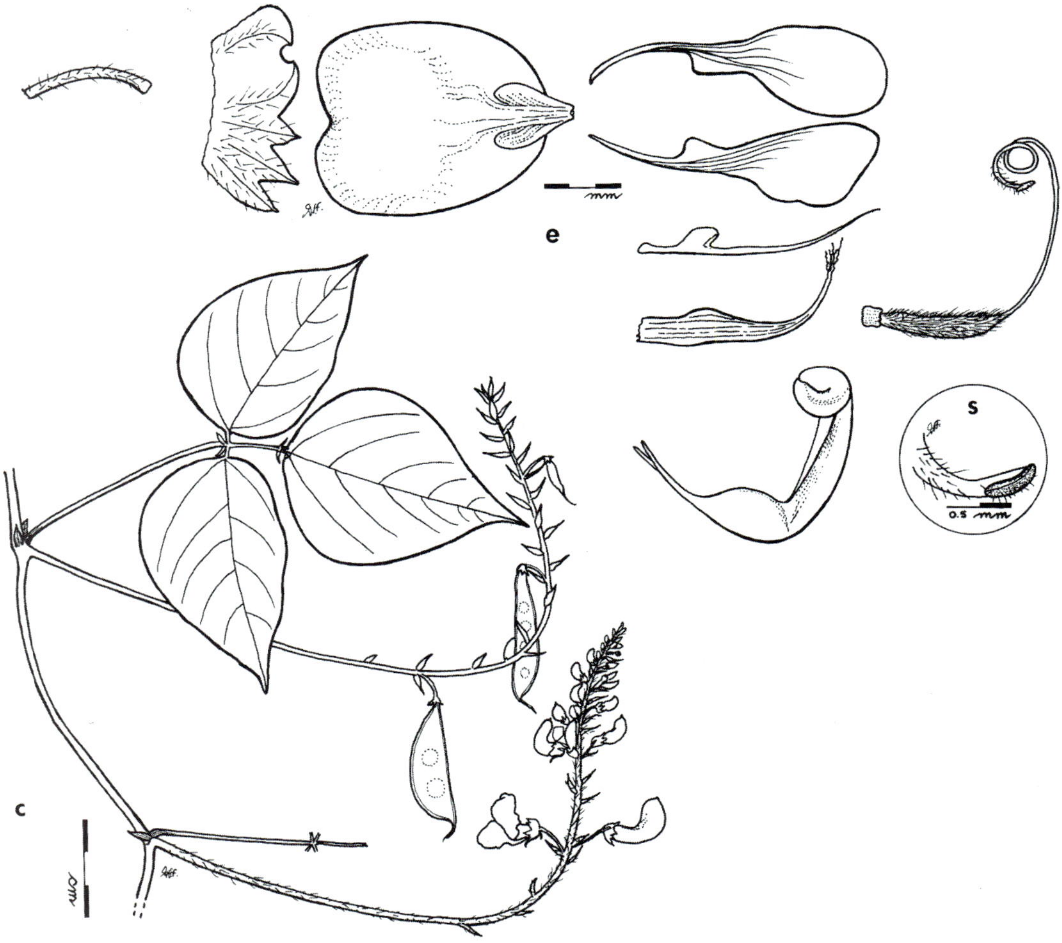

Fig. 85. Illustrations of *Phaseolus tuerckheimii* Donn. Smith.–**c.** Portion of stem with mature leaf and inflorescences with flowers and immature pods; note the many flowering nodes with large bracts.–**e.** Exploded view of flower showing all parts including–**s.** Style tip and stigma as seen under the microscope. Drawings made from dried herbarium specimens of field collections: **c.** from *Debouck et al. 2474* collected near Santa Catarina Pinula, Guatemala, Guatemala; e. from *Debouck et al. 1617* from near San Miguel Dueñas, Sacatepéquez, Guatemala.

nerved, sparsely covered with hirsute hairs, ciliate on margins; terminal leaflet rounded ovate, 4–10 cm long, 4–8 cm wide at 1/3 from base, moderately heavily nerved, short-acuminate, not obviously apiculate, densely covered with short white strigose hairs abaxially giving a silvery sheen, sparsely covered with strigose hairs adaxially except on veins and margins; lateral leaflet very similar but slightly inequilateral. ***Inflorescence*** a large panicle, 1.5–3 dm long, the few lower branches only a few mm long, all parts densely covered with minute uncinate and yellowish brown strigose hairs; peduncle 4–8 cm long; rachis 7–15–20 cm long of many (20–30) nodes, internodes 5–10 mm long striate; primary bract linear-lanceolate, 4–8 mm long, faintly 3-nerved, covered by scattered hispid hairs; pedicel 7–9 mm long, puberulent of minute hooked hairs; pedicellar bracts aciculate 1 mm long faintly 1-nerved. ***Bracteoles*** minute aciculate scales, 0.75 mm long, 0.25 mm wide, covered with minute hispid hairs. ***Flower*** violet or blue; calyx tube 2.5–3 mm long, puberulent, upper lobes rounded, 0.5 mm long, 1.75 mm wide, glabrescent, lower 3 lobes subequal, dentate, the lower central elongate 2 mm long, 1 mm wide, covered with scattered strigose and hirsute hairs, ciliate on margins; standard violet, the claw 1 mm long, the blade broadly rounded, 10 mm long, 11 mm wide, slightly reflexed at 6 mm from base, not thickened, adaxial side greenish, the auricles flap-like 1.5 mm long, 1.25 mm wide;

wings violet-blue, the blade spatulate-obovate 13–18 mm long, 5–8 mm wide, the claw 4 mm long, the spur pronounced at 5 mm from the base, 0.5–1 mm long; keel 15 mm long, linear-oblong, 2 mm wide, the claws 2.5 mm long, the lateral ridges inconspicuous, 5 mm from base to bend and 6 mm more to base of the terminal 2 coils of 2 mm diam.; vexillary stamen very delicate, the claw 1–1.5 mm long, the knob 0.5 mm wide, somewhat less than 1 mm raised, the thickened portion 2–3 mm long to the filament; stamen tube narrow tubular, curved, 5 mm to bend and 5 mm more to filaments, 1.5 mm high, 1 mm wide, the ridges inconspicuous; basal collar 0.6 mm long; ovary straight aciculate, 4–5 mm long, 0.7 mm wide, heavy sutures, covered with long white or light yellow cannose-sericeus hairs mostly on sutures, 3–6 ovules; style 7 mm long to the terminal thickened coil of 2 mm in diam.; stigma terminal and extrorse oblique, 1 mm long. ***Pod*** straight, 3–4 cm long, 1 cm wide, inflated, densely covered with short yellowish strigose hairs; sutures, enlarged; beak stout, slightly curved, 4–5 mm long. ***Seed*** oblong, 5–6 mm long, 4–5 mm wide; hilum oblong, 1.5 mm long; lens prominent; mottled and speckled dark brown on reddish-brown; placental tissue absent. ***Seedling*** from epigeal germination, hypocotyle terete 17–20 mm long, epicotyle 5–7 mm long, both deep purple streaked and covered with minute whitish uncinate hairs; eophylls opposite triangular lengthily acuminate base slightly auriculate 27 mm long 15 mm wide, stipule slightly bifid 2–3 mm long; petiole 5–6 mm long no stipels; first true leaf trifoliolate.

Specimens examined: **COSTA RICA. San José:** Fila La Legua abajo del cerro Caraigres, en camino a San Francisco, 1.2 km S de La Legua, 9°46'N, 84°7'W, 1800 m, 10 Jan 1987, *Debouck et al. 2110* (CR, US); San Isidro El General, 0.5 km N de Piedra de Rivas, 2 km N de Buenavista, 9°31'N, 83°40'W, 1550 m, 13 Jan 1987, *Debouck et al. 2125* (BR, CR, MICH, US); San Isidro El General, 3.5–4 km N de Herradura, al pie de la Fila Ojo de Agua, Río Blanco, 9°31'N, 83°37'W, 1820 m, 13 Jan 1987, *Debouck et al. 2129* (BR, CR, US); San Isidro El General, Km 113 carr. Interamericana, 9°28'N, 83°42'W, 2020 m, 14 Jan 1987, *Debouck et al. 2131* (BR, CR, MICH, SI, UC, US, WIS); San Isidro El General, 1 km NNE de Providencia a orillas del Río Brujo, 9°34'N, 83°51'W, 1920 m, 14 Jan 1987, *Debouck et al. 2133* (CR, MICH, US); San Isidro El General, 3 km NW de Providencia a Copey, 9°34'N, 83°52'W, 2350–2550 m, 14 Jan 1987, *Debouck et al. 2134* (CR, UC, US, WIS); División, en Jardín de Paramo, 9°29'N, 83°43'W, 2180 m, 4 Feb 1998, *Debouck et al. 3110* (CR, MO).

GUATEMALA. Chimaltenango: Acatenango, 4 km N San Antonio Nejapa, 14°35'N, 90°56'W, 1830 m, 7 Dec 1986, *Debouck et al. 1624* (G, K, MICH, SI, UC, US, USCG, WIS). **El Progreso:** San Agustín Acasaguastlán, 9 Km NE of Albores to Las Torres, 15°05'N, 89°56'W, 1930 m, 3 Dec 1987, *Debouck et al. 2418* (G, M, USCG); between Finca Piamonte and slopes SE of Finca Piamonte, Sierra Las Minas, (15°N, 90°W)?, 2400–2500 m, 4 Feb 1942, *Steyermark 43396* (EAP, F). **Guatemala:** 4 km SE de Concepción Las Lomas (al oriente de Cd. Guatemala), 14°34'N, 90°30'W, 1500 m, 17 Dec 1987, *Debouck et al. 2473* (BR, MICH, SI, US, USCG); Santa Catarina Pinula, 1 km SE de Piedra Parada, 14°33'N, 90°30'W, 1800 m, 17 Dec 1987, *Debouck et al. 2474* (MICH, US, USCG). **Huehuetenango:** tributary of Río Blanco, 5 km W of Aguacatán, (15°20'N, 91°25'W), 2000 m, 4 Dec 1962, *Williams et al. 22323* (EAP, F, GH, US, WIS). **Jalapa:** Cerro Alcoba, E of Jalapa, (14°30'N, 89°45'W), 1300–1700 m, 2 Dec 1939, *Steyermark 32551* (F). **Quetzaltenango:** Zunil, 2.5 Km NE de la Estancia de la Cruz, Balneario de Aguas Amargas, 14°45'N, 91°31'W, 24 Jan 1995, *Debouck et al. 3085* (USCG). **Quiché:** El Molino, 4 km from Chichicastenango, (14°50'N, 91°5'W), 2100 m, 14 Jan 1974, *Molina et al. 30282* (EAP, F, MO); above Cotzal, 1890 m, 9 Feb 1945, *Sharp 45191* (F); barranco S of Chichicastenango, 1850 m, 11 Jan 1939, *Standley 62387* (F). **Sacatepéquez:** 5 km de San Miguel Dueñas, 14°33'N, 90°50'W, 1820 m, 6 Dec 1985, *Debouck et al. 1617* (BR, K, MICH, SI, US, USCG); above Las Calderas, 1800 m, 15 Dec 1938, *Standley 59991* (F); Finca El Hato, NE of Antigua, (14°30'N, 90°42'W), 1950–2040 m, 28 Dec 1938, *Standley 61137* (F). **Sololá:** 6 Km S de Pamezabal, Santa Clara La Laguna, 14°43'N, 91°18'W, 2230 m, 26 Jan 1995, *Debouck et al. 3097* (USCG). **Zacapa:** along Río El Repollal to summit of Mt., Sierra las Minas, (15°N, 89°40'W)?, 2100–2400 m, 12–13 Jan 1942, *Steyermark 42471* (F).

HONDURAS. Fco. Morazán: La Tigra, Mt. San Juancito, (14°10'N, 87°W), 2100 m, 4 Feb 1968, *Molina et al. 21483* (BM, CAS-DS, EAP, ENCB, F(2), GH, K, TEX-LL, UC, US); Mt. Uyuca, drainage of the Río Yeguare, (13°50'N, 87°W), 1800 m, 21 Feb 1949, *Williams et al. 15597* (EAP, F). **Gracias a Dios:** (no location), (15°N, 83°40'W)?, 4 Oct 1961, *Semple 77* (US).

MÉXICO. Chiapas: Mpio. Chamula, Toh Tik, along the road to Chenalho, (16°50'N, 92°41'W), 1463 m, 21 Jan 1965, *Breedlove et al. 8188* (CAS-DS, F, US, WIS); La Trinitaria, wooded slope along Mexican hwy 190 11 mi south of La Trinitaria, 1155 m, 14 Oct 1965, *Breedlove 13246* (MEXU); Mpio. Pueblo Nuevo Solistahuacán, Clínica Yerbabuena, 3 km NW of Pueblo Nuevo Solistahuacán, (17°10'N, 92°54'W), 1700 m, 14 Dec 1971, *Breedlove 23205* (CAS, F, MICH, MO); Mpio. Motozintla de Mendoza, Cerro Mozotal, 11 km NW of the Jct of road to Motozintla along road to El Porvenir and Siltepec, (15°30'N, 92°20'W), 2100 m, 21 Nov 1976, *Breedlove 41672* (CAS-DS, MO, TEX); Mpio. Unión Juárez, Talquián, SW slope of Volcán Tacaná, (15°N, 92°10'W), 2200 m, 13 Dec 1976, *Breedlove 42582* (CAS-DS); Mpio. of Motozintla de Mendoza, Niquivil, Cerro Boquerón, 2600 m, 16 Dec 1976, *Breedlove 42717* (CAS-DS, TEX); Mpio. Tenejapa, Paraje Krus Chen, (16°50'N, 92°29'W), 2220 m, 21 Dec 1980, *Breedlove 48849* (CAS, MO, TEX); Mpio. Tenejapa, Paraje Yash'anal, 2460 m, 28 Jan 1981, *Breedlove 49658* (CAS); Mpio. Siltepec, Siltepec on the road to Huixtla, (15°20'N, 92°30'W), 2000–2400 m, 18 Jan 1973, *Breedlove et al. 31916* (CAS-DS, TEX); Carelas, Motozintla, 2176 m, Apr 1945, *Matuda 15522* (MEXU, US); Mt. Ovando, Escuintla, 4 Nov 1945, *Matuda 6098* (MEXU, US); Mpio. Tenejapa, Paraje of Yash'anal, 1828 m, 21 Feb 1967, *Ton 2093* (CAS-DS, F, MEXU, MO, TEX-LL, US); Angel Albino Corzo, slope near the Rancho Viejo of the Finca Prussia, 790 m, 23 Jan 1968, *Ton 3533* (MEXU). Mpio. Tenejapa, Cerro Cruz Chen, 2100 m, 16 Nov 1982, *Ton 5078* (CAS); Mpio. Tenejapa, along Río Yash zanal, 1500 m, 2 Jan 1983, *Ton 5307* (G, MEXU, MO).

PANAMÁ. Chiriquí: Boquete District, Volcán Chiriquí, 19 Nov 1975, *D'Arcy 9777* (MO); El Baru above Boquete, 1200–1800 m, 21 Nov 1975, *D'Arcy 9881* (MO); between Cerro Punta and Las Nubes, E side Cerro Punta, (8°40'N, 82°30'W), 15 Jan 1971, *Wilbur et al. 13248* (CAS).

Habitat.—The species is found growing mostly in montane rain forests, often at edge of clearings or in the pine-oak on the edges of the cloud forest over shrubs and herbaceous weedy plants such as: *Clethra, Dahlia, Matudaea, Oecopetalum, Rubus, Saurauia, Stizolobium, Styrax, Symplocos,* and Melastomataceae and Bambu. Soils are high in organic matter, light to dark brown and derived from volcanic ash and other igneus rocks.

Ethnobotany.—In places of Chiapas, it is reported as "yashal chenek'mut" or as "shut chenek."

Comments.—In the original description published in the Botanical Gazette in 1913, John Donnell Smith did not give the herbarium and number for the type. At the time of publication he apparently distributed a sheet to the U.S. National Herbarium with a label printed with "Flora von Guatemala" followed by genus and species and with "n.sp" and the annotatation "Berge Zweischen inter Coban und Tactic." In his personal herbarium he apparently kept three sheets, later donated to the U.S. National Herbarium, several of which bear labels similar to those distributed, while the third bears a label "Herbarium of John Donnell Smith" and which is the only sheet with a packet containing a dissected flower and other plant parts. This latter sheet also contains the "n.sp." after the genus and species name, and so the senior author concludes it is most certainly the one used for providing the very detailed description of the flower parts in the published description. For these reasons the senior author believes this sheet (US 942247) must be considered the holotype sheet.

The taxon described by Standley (1916) as species *P. chiriquinus* has been considered synonymous with *P. tuerckheimii* since there are only slight differences in morphology of plant, flower, and pod. Additionally, *chiriquinus* extends the range of *tuerckheimii* to the area of Volcán Chiriquí in Panama and neighboring Costa Rica, thus this is a wide ranging species in the high mountains of S México and throughout most of Central America. Delgado (2001) mentioned it as possible for the flora of Nicaragua. The junior author has collected a few plants with pure white flowers (*Debouck et al. 2131*, deposited at US and WIS).

Section M.—Pedicellati (Benth.) Freytag, sect. nov. TYPE SPECIES: *Phaseolus pedicellatus* Benth., Commen. Legumin. Gener. 73. 1837.

Herba volubilis scandens generaliter 1–2 m longa vel usque ad 5 m longa, radix perennis crassa carnosa, folia pleurumque parva foliola lobata profunda vel non profunda, panicula longa vel pseudoracema, flores parvi purpurei, stigma introrsum laterale, legumen parvum propre strictum glabrum aliquantum compressum, 4–6 seminibus.

Plants are climbing vines generally 1–2 m long but some are larger to 5 m long; roots perennial, thick, fleshy; leaves mostly small, the leaflets slightly to deeply lobed; inflorescence a long panicle or pseudoraceme; flowers small, purple, with lateral introrse stigma; pod small, nearly straight, glabrous, somewhat compressed, 4–6 seed.

Comments.—This is a large group of very closely related species all with variously lobed leaflets, from central and northeastern México. At first glance it would seem that they are possibly a collection of varieties rather than species, especially since some are not well known from only a few collections. However, for those which we have seen in the field or managed to obtain viable seed (*P. pedicellatus, P. esperanzae, P. oaxacanus* and *P. grayanus*) and grown out under greenhouse conditions, we have found them to differ in a number of characteristics in root, vegetative parts, flower, pod and seed (see Debouck 1991).

This section contains several commonly occurring and wide ranging species and a number of rather rare, apparently locally isolated species, most located on the limits of distribution of the more common species. Although Piper (1926) recognized this group as a natural collection of species with small to medium vines from thick, fleshy perennial roots, lobed leaflets, small, mostly dark purple flowers and small pods and seed, he included species of *Phaseolus* which definitely do not fit here, viz. *salicifolius, viridis, acutifolius, falciformis* and *lunatus* (and also *mollis, rosei* and *bolivianus*, from South America, and thus not treated here).

The group seems to be distributed mostly in the northern Sierra Madre Oriental of Mexico and the neighboring mountain ranges in the drier central plateau, though a couple of species such as *P. grayanus* and *P. scabrellus* are found in northwest México. Speciation in this group seems to be accentuated by isolation since each separate mountain range or climatic area appears to have its own distinct taxon. This is a section which requires a good deal more field collecting and study.

KEY TO THE SPECIES AND VARIETIES

This key is intended to be used for determining species using principally vegetative characteristics since that is about all that is generally found on most herbarium specimens. Even then one must have a nearly complete sampling of leaves from the base to the tip of the plant at the time of late flowering or after, to see the range of leaf types found in each species since usually the more entire leaves are found at the base of the plant and the more divided or lobed leaves are found near the tips of the vines. Additionally, characters are used which are most distinctive for each species, such as: inflorescence, bracteoles, flower, pod, and seed. Unfortunately, all of these are seldom found on a single herbarium sheet. Also note that distribution can be a helpful key character where ranges do not overlap.

1. Terminal leaflets not lobed or only slightly lobed.
    2. Lateral leaflets not lobed or only slightly lobed.
        3. Leaflets ovate, bright green, bracteoles indistinctly 1- to 3-nerved; common, throughout E central México and N into the Huasteca; 1540–3100 m ____ M.1. **P. pedicellatus**
        3. Leaflets narrowly triangular to triangular-ovate, very thin membranaceous, bracteoles ovate, not nerved; very rare, in moist forests of E central México; 2120–2550 m ____ M.11. **P. laxiflorus**
    2. Lateral leaflets slightly or at least some deeply lobed.
        4. Lateral leaflets very deeply lobed, narrowly lanceolate or nearly linear, terminal leaflet linear; rare, in forests near (Charcas)?, SLP; (2100)? m ____ M.6. **P. purpusii**
        4. Lateral leaflets slightly lobed.
            5. Terminal leaflets broadly squarish-lobed; inflorescence very large, much exceeding leaves; bracteoles nerveless; pod straight and flat, covered by long pubescence; 3–4 large seeds, nearly spherical and solid black; rare, in forests W of Volcán Orizaba and in forests of S Hidalgo; 2190–2670 m ____ M.3. **P. esperanzae**
            5. Terminal leaflets slightly lobed, inflorescence not much exceeding leaves, young pods scarcely pubescent.
                6. Primary bract never 3-dentate; bracteoles minute and ovate-acuminate; leaflets seldom or only slightly variegated, petioles short; common, in forests throughout central and NE México; 1540 3100 m ____ M.1. **P. pedicellatus**
                6. Primary bract often 3-dentate; bracteoles minute and broadly ovate; leaflets distinctly variegated, petioles as long as or longer than terminal leaflet; rare, only in mountains NE of Cd. Oaxaca; 2100 2800 m ____ M.2. **P. oaxacanus**
1. Terminal leaflets distinctly lobed.
    7. Vine prostrate, seldom twining and climbing, leaflets extremely variable on the same plant from lower leaflets entire to upper leaflets deeply squarish-acute and 2- to 3- or more-lobed, at least some of the major lateral lobes pointed.
        8. Terminal portion of pseudoraceme a loose, elongate series of buds; common, in pine forests at high altitudes in SW US and NW México; 1280–1900 m ____ M.7. **P. grayanus**
        8. Terminal portion of pseudoraceme a conical or pyramidal mass of buds; very rare, in W Chihuahua; 2100 m ____ M.10. **P. pyramidalis**
    7. Vine twining and climbing, leaflets somewhat variable on the same plant from more entire at the base to more extremely lobed above.
        9. Terminal and lateral leaflets deeply 3-lobed, the lateral lobes pointed; few and large flowers; broad pods; very rare, only found near Concepción del Oro in N Zacatecas; (2500)? m ____ M.5. **P. palmeri**
        9. Terminal leaflets 3-lobed and lateral leaflets slightly or not lobed.
            10. Center lobe of terminal leaflet quite narrow and often much longer than lateral lobes, lateral lobes very large compared to central lobe and rounded or squarish, leaves small to medium, primary bracts small, flowers small, pods narrow.
                11. Flowers purple; common, type found in mountains just W of Cd. San Luís Potosí but ranges through high plateau of central Coahuila, E to west slopes of Sierra Madre Oriental, and S to Hidalgo; 800 3050 m ____ M.4.1. **P. polymorphus** var. **polymorphus**
                11. Flowers white; very rare, found only in mountains just S of Big Bend National Park; 2250–2450 m ____ M.4.2. **P. polymorphus** var. **albus**
            10. Central lobe of terminal leaflet broadly lanceolate or triangular.

12. Seed lenticular; primary bracts small; terminal leaflet lateral lobes rounded; rare, in forests of W Zacatecas; 2200–2400 m ________ M.9. **P. teulensis**
12. Seed nearly spherical; primary bracts very small; terminal leaflet lateral lobes rounded to squarish; bracteole 1-nerved; rare, in forests of Sonora; 1800–2400 m ________ M.8. **P. scabrellus**

M.1.—Phaseolus pedicellatus Benth., Commen. Legumin. Gener. 73. 1837. (**Figs. 86, 87**). TYPE: MÉXICO (no location): *Karwinski 13061* (HOLOTYPE: M (designated by Ross, date ?) n.v., US (fragment & photo of M)).

Phaseolus floribundus Piper, Contr. U.S. Natl. Herb. 22:690. 1926. TYPE: MÉXICO. SAN LUÍS POTOSÍ: Alvárez, (22°10'N, 100°45'W), 5–10 Sep 1902, *Palmer 121* (HOLOTYPE: US 397669; ISOTYPES: BM, GH, MO).

Aerial shoot a perennial, many branched, strongly climbing, slender, indeterminate vine, to 3 m long. ***Root*** a perennial, branched, thick, fleshy, fusiform. ***Stem*** terete, 1.25 mm thick; internodes mostly 4–6 to 10 cm long; very sparsely covered by reflexed-strigose hairs, to about 0.5 mm long, often dark purple. ***Stipules*** broad triangular to nearly spatulate to oblongate, 2.5–3.5 mm long, 1.5 mm wide, strongly greenish 3- to 6-nerved, acute to obtuse, purple, moderately covered by minute strigose hairs to glabrous, extended. ***Leaves*** 7–14 cm long; petiole 1.5–2.5–5 cm long, slender, glabrous; petiolule 0.8–1.4 cm long, delicate, slightly ribbed, nearly glabrous; pulvini 2.5–3 mm long, darkened, ridged, heavily covered by strigose and uncinate hairs on abaxial surface; stipels lanceolate, 1-nerved, glabrous; terminal leaflets broadly ovate to ovate-acuminate, frequently slightly 3-lobed to deeply rounded lobed in northern range, 4.4–7 cm long, 2–5–6 cm wide at about midway from base, often smaller, base hastate with rounded lobes, slightly acuminate, obtuse, slightly apiculate, somewhat coriaceous, puberulent to covered with strigose pubescence, ciliate on margin, moderately nerved, the ultimate veinlets darkened brownish, moderately thickened, bright green with a few sparse hispid hairs adaxially, somewhat mottled along midvein, lighter green and glabrous adaxially; lateral leaflet similar, also often variegated and slightly lobed on each side at base, 3.6 cm long, 2.7 cm wide at about midpoint. ***Inflorescence*** a short to long, slender, erect pseudoraceme mostly less than 10 (rarely 20) cm long and exserted above foliage; peduncle 2–14 cm long, nearly glabrous; rachis 3.5–10 cm long, mostly 2 but may have up to 8 flowering nodes, almost completely covered by minute hooked pubescence; primary bract ovate-lanceolate, 2.5–4 mm long, 1.25 mm wide, strongly 1- to 5-nerved, purplish, minutely pubescent; pedicellar bract thin, reaching 1 mm long or less, greenish along the central vein, translucid on both sides; pedicel delicate, 6–10 mm long, glabrous. ***Bracteoles*** ovate-acuminate, minute, often inserted a little below the calyx, 1.25–2 mm long, 0.3–0.75 mm wide, indistinctly 1- to 3-nerved, acute, covered with minute pubescence, sparsely ciliate on margins, purplish, persisting through young pod stage. ***Flower*** purple fading light brown; calyx campanulate, 5 mm long, purplish, the upper two lobes short, rounded, 1 mm long, 3.5 mm wide, the lower lobes acute, center lobe 1.75 mm long, 3 mm wide, lateral 1.25 mm long, 2.75 mm wide, puberulent, border ciliate; standard dark purple, large, rounded, erect, 11 mm long, 12–13 mm wide, the lower lobes well-developed; wings very large, the blade cupped, 10–17 mm long, 12 mm wide, unequal with one larger than the other, the basal claw 5 mm long, 0.6 mm wide, the well-developed spur 1.5 mm long, 1.5 mm wide and strongly adhering to the keel; keel 10 mm long to the bend and 5 mm more to the base of the 2 terminal coils of 3 mm diam., tip whitish, the basal claws separated for 5 mm, the lateral knobs well-developed; vexillary stamen delicate, with a well-developed knob 0.75 mm wide about 0.5 mm from base and extending as a hooded sheath forward about 1 mm; stamen tube united about 11 mm, with 2 well-developed knobs near base, 0.75 mm in size; anthers 1 mm long, 0.4 mm wide, yellow; basal collar 0.75 mm long, 1 mm wide; ovary 5 mm long, 1.25 mm wide, densely covered with long canescent hairs, 4 ovules; style about 9 mm long to the terminal 1 1/2 coils of about 2 mm in diameter; stigma lateral, introrse, about 1 mm long, the stylar hairs extending to the base of the stigma. ***Pod*** nearly straight to falcate, slightly compressed, 2.5–5 cm long, 0.8–1 cm wide, 4.7 mm thick, nearly stipitate at base, immature solid green or speckled and striped black or purple; valves reticulate, dark, heavily covered with whitish or brownish strigose pubescence becoming nearly glabrous at maturity, especially at ends and sutures, some hairs glandular at base, slightly restricted between seeds; fairly pronounced suture; beak straight, about 1.5–4 mm long; maturing 1–2 pods per inflorescence and mostly 3 seed/pod. ***Seed*** round and flattened, slightly angular, 4.7–5.7 mm long, 4–5.3 mm wide, 1.7–2.4 mm thick, speckled black on brown, shiny, black ring around hilum. ***Seedling*** from hypogeal

FIG. 86. Illustrations of *Phaseolus pedicellatus* Benth.—**a.** Seedling a few weeks after germination.—**b.** Mature root; note elongate crown.—**c.** Basal portion of plant with mature leaves and separately vine tip with mature leaf and inflorescence with flowers.—**d.** Flowers, side view and front view.—**e.** Exploded view of flower showing all parts including—**s.** Style tip and stigma as seen under the microscope.—**f.** Pods, side view (attached to inflorescence) and dehisced.—**g.** Seeds, side view and view of hilum; notice nearly spherical shape. All drawings made from living material grown in greenhouse at Mayagüez from seed of *Freytag et al. 81-25* (TARS #18, in part) collected from near Piedra Ancha, Jalisco, México.

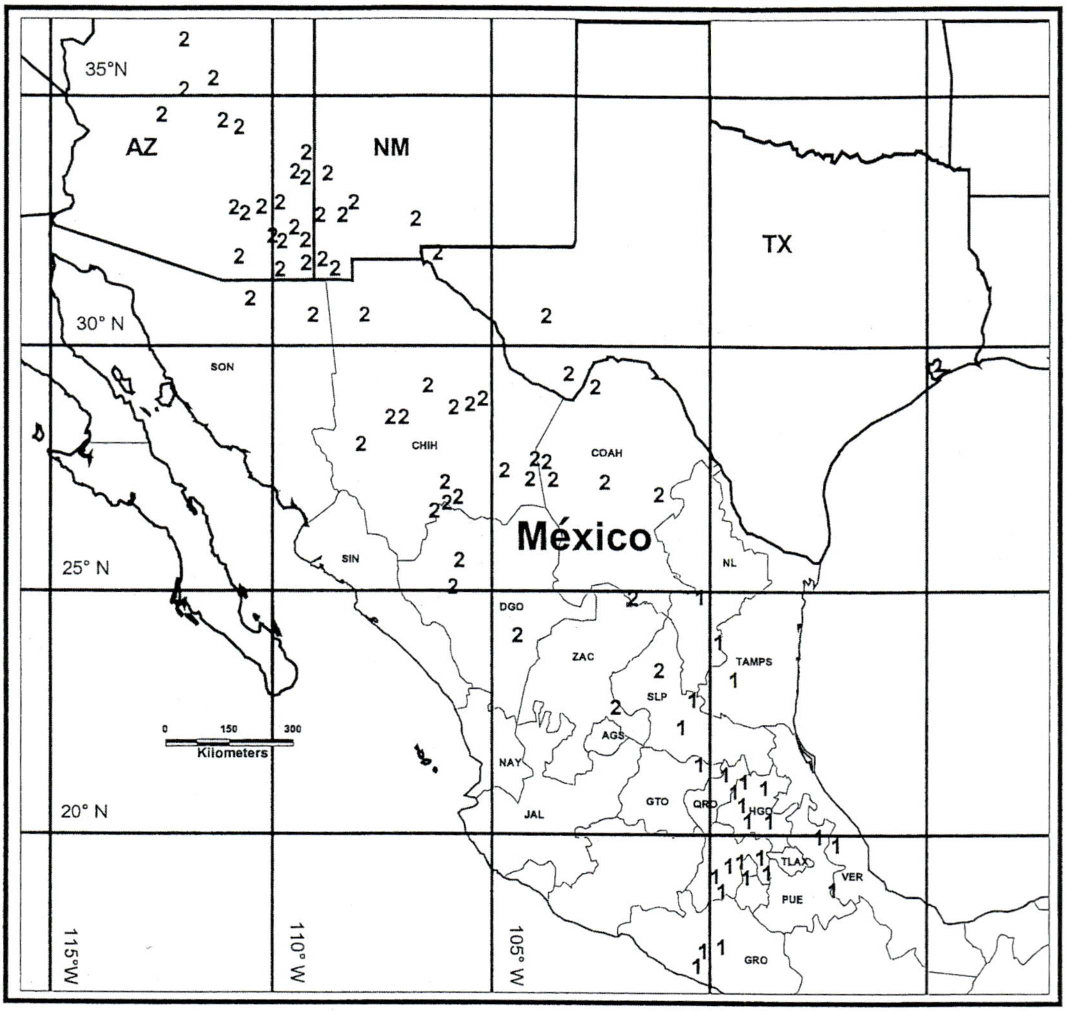

Fig. 87. Distribution of major species of Section M. *Pedicellati*, as follows: 1 = *P. pedicellatus*; 2 = *P. grayanus*.

germination; hypocotyl not elongated; epicotyl 28–59 mm long reddish; the 2nd of the next 3 internodes shorter (10–18 mm long) than the 1st or 3rd (16–27 mm long); eophylls opposite broadly triangular bluntly acuminate base subcordate 34 mm long 33 mm wide often variegated along midvein, petiole 15–19 mm long; first true leaf trifoliolate.

Specimens examined: **MÉXICO. Distrito Federal:** Desierto, San Nicolas, Vallée de México, (19°20'N, 99°18'W), 27 Sep 1865, *Bourgeau 940* (F, G, K(2), SI, US(3)); Dínamo de Contreras, Valle de México, (19°15'N, 99°35'W), 3000 m, 29 Oct 1950, *Matuda 18641* (CAS, ENCB, MEXU, US); base of Sierra de Ajusco, 2640 m, 23 Oct 1896, *Pringle 6616* (BM, CAS, F, G, GH, K, MEXU, MICH, MO, UC, US(2)); Dción. Xochimilco, Rancho El Conejo, 2350 m, 17 Oct 1976, *Ventura 2319* (ENCB, MICH). **Guanajuato:** Xichu Road, (21°20'N, 100°15'W)?, 19 Aug 1947, *Kenoyer 2309* (GH). **Guerrero:** Omiltemi, 32–40 km W of Chilpancingo, (17°35'N, 99°45'W), 2100–2325 m, 4 Dec 1966, *Anderson et al. 4377* (CAS, MICH); Mpio. Tlacotepec, Agua Fría, 40 km N of Coyuca de Benítez, Cerro Tlacotepec, (17°N, 100°15'W), 2650 m, 4 Dec 1963, *Dieterle 3236* (CAS, MICH, MO); Dist. Galeana, Teotepec, (17°30'N, 100°10'W), 2540 m, 26 Dec 1937, *Hinton 11142* (CAS-DS, F, GH, K, MICH, US). **Hidalgo:** 4.9 mi S of Durango, 13.2 mi N of northern turnoff to Zimapán along Hwy 85, (20°45'N, 99°25'W), 2350 m, 31 Jul 1978, *Daniel 389* (MICH); Km 8 de la carr. Mex 105 a Tampico, 1 km antes Entronque a Mineral del Monte, Pachuca, 20°07'N, 98°43'W, 2670 m, 29 Oct 1986, *Debouck et al. 2028* (BR, COL, CHAPA, MICH, MO, US, WIS); Mpio. Mineral del Monte, 0.5 km NE de Guerrero antes de Omitlán, Km 17 carr. Mex 105 a Tampico, 20°9'N, 98°38'W, 2380 m, 29 Oct 1986, *Debouck et al. 2029* (CHAPA, COL, MICH, MO, US(2)); Mpio. Nicolás Flores, 3 km E desde entronque con Mex 85 camino a Nicolás Flores, 20°49'N, 99°14'W, 2180 m, 30 Oct 1986, *Debouck et al. 2030* (BR, CHAPA, COL, G, M, MICH, SI, US(2)); Mpio. Zimapán, 4 km NW del entronque Mex 95 y camino a San Fco. del Oro, 20°48'N, 99°20'W, 2100 m, 31 Oct 1986,

Debouck et al. 2033 (CHAPA, M, MO, US); Mpio. Zimapán, Barranca de Los Mármoles, 4 km NE de Trancas a Cd. Victoria, 20°50'N, 99°14'W, 2320 m, 1 Nov 1986, *Debouck et al. 2035* (BR, CHAPA, COL, M, SI, UC, US(2)); Mpio. Jacala, La Placita, 9 km SW de Agua Fría en Mex 85 a Tamazunchale, 20°58'N, 99°13'W, 1780 m, 1 Nov 1986, *Debouck et al. 2039* (CHAPA, COL, MICH, MO, US); Acaxotchitlán, cerro 4 km E de la Presa El Tejocotal, 20°09'N, 98°08'W, 2180 m, 7 Nov 1987, *Debouck et al. 2380* (CHAPA, COL, M, US); Mpio. Molango, 5 km al N de Molango, rumbo a Huejutla de Reyes, (20°50'N, 98°45'W), 1700 m, 13 Nov 1975, *Delgado et al. 71* (CAS, CHAPA); Mpio. Tlanchinol, San Bernardino, 14 Nov 1975, *Delgado et al. 104* (CAS, CHAPA, MEXU); San Vicente, 16 Aug 1937, *Edwards 872* (TEX); Mpio. Acaxochitlán, 8 km al occidente de Acaxochitlán, 2300 m, 7 Sep 1980, *Hernández 4940* (CAS, MEXU); Mpio. Zimapán, Los Nogales, 30 km al NE de Zimapán, 1900 m, 13 Sep 1981, *Hernández et al. 6540* (CAS, MEXU, MO); Mpio. Real del Monte, 2 km al WSW de Real del Monte, 15 Aug 1976, *Medina 1534a* (ASU); Km 231-2 from DF along Pan American Hwy. (No 1), N of Zimapán, 17 Oct 1948, *Norvell et al. HM 898* (CAS); Ixmiquilpan, (20°30'N, 99°20'W), Aug 1905, *Purpus 1365* (F, GH, MO, UC); 3 mi N of Puerto Ignacio Isidro Díaz along Mex 85, N of Zimapán, 20 Aug 1971, *Vaughan et al. 1062* (MO); Barranca de Mármoles, Km 145.5 Rt 85, 15 mi NE Villa Carranza, 2300 m, 12 Oct 1981, *Warnock 2469* (TEX). **México:** 4 km N de Chiltepec, 18°55'N, 99°52'W, 2470 m, 1 Nov 1987, *Debouck et al. 2368* (CHAPA, COL, MICH); Temascaltepec, Tequesquipan, (19°N, 100°W)?, 22 Apr 1932, *Hinton 542* (GH, K); Temascaltepec, Cajones, 2480 m, 8 Nov 1932, *Hinton 2375* (GH, K, MEXU); Cerro de Pinal, Otzoloapan, 2300–2500 m, 18–22 Oct 1954, *Matuda et al. 31844* (US); 5 km NE de Amecameca, Cerro La Coronilla, 2650 m, 22 Sep 1968, *Pineda 570* (F, MICH); Mpio. Tlalmanalco, 1 km al E de Santo Tomás Atzingo, (19°15'N, 98°50'W)?, 2450 m, 28 Oct 1981, *Ramos 260* (GH, MO); Mpio. Amecameca, 1 km al E de San Antonio, (19°8'N, 98°45'W), 2550 m, 31 Oct 1968, *Rzedowski 26451* (CAS-DS, ENCB, MICH); Mpio. Tepozotlán, Sierra Alcaparrosa, 2800 m, 21 Sep 1973, *Rzedowski 31272* (CAS, ENCB). **Morelos:** Km 52, 3–5 mi S of Cuajomulco along toll road to Cuernavaca, (19°N, 99°10'W), 2286 m, 30 Nov 1967, *Gentry 22489* (GH, MICH); Vallée del Tepeite, Dec 1932, *Lyonnet 876* (MEXU, US). **Nuevo León:** 9 km S de Zaragoza, 1 km antes de llegar a Palo Bola, 23°57'N, 99°47'W, 1960 m, 8 Sep 1985, *Debouck et al. 1508* (CHAPA, COL, M, MICH, SI, UC, US); General Ignacio Zaragoza, 2 km de desviación a Joya de San Diego, 6 km N de La Siberia, 23°50'N, 99°49'W, 2550 m, 9 Sep 1985, *Debouck et al. 1512* (CHAPA, COL, M, MICH, SI, US(2)); Mpio. Galeana, Las Placetas, 7 km N de 18 de Marzo camino a San Rafael, 24°54'N, 100°11'W, 2140 m, 14 Sep 1985, *Debouck et al. 1516* (CHAPA, COL, G, K, M, MICH, MO, SI, US); Mpio. Galeana, Cerro El Potosí, 3100 m, 21 Jun 1869, *Hinton et al. 17107* (ENCB, US); Mpio. Zaragoza, Cerro del Viejo, 15 mi W Dulces Nombres, 2400 m, 19 Aug 1948, *Meyer et al. 3029* (G, MO); Picacho Onofre, Cerro Peña Nevada, 12 km NE of San Antonio Peña Lavada, 30 km E of Doctor Arroyo, Jul 1977, *Wells et al. 313* (TEX-LL). **Puebla:** Zacatlan, 2 km S of Zacatlan, 19°55'N, 97°59'W, 1950 m, 12 Nov 1987, *Debouck et al. 2390* (BR, CHAPA, G, M, MICH); Contadero, 30 Aug 1901, *Pringle 9623* (US); Mpio. Zacapoaxtla, Los Manzanos, Zacapoaxtla, (19°50'N, 97°30'W), 1800 m, 21 Aug 1975, *Ramos s.n.* (CAS, ENCB, MEXU). **Queretaro:** Cerro, 2 km SE de Realejo, cerca de 'Minas de Plata', Guadalcazar, 22°39'N, 100°24'W, 2100 m, 7 Nov 1986, *Debouck et al. 2057* (CHAPA, COL, MICH, US, WIS); 16 km SW de Escanelilla, carr. Mex 120 a Cadereyta, Pinal de Amoles, 21°7'N, 99°40'W, 1950 m, 14 Nov 1986, *Debouck et al. 2080* (CHAPA, MICH, US); 0.5 km S de San Joaquín, 20°54'N, 99°34'W, 2360 m, 15 Nov 1986, *Debouck et al. 2081* (BR, CHAPA, COL. MICH, US). **San Luís Potosí:** 21.1 mi E of Ciudad San Luís Potosí on Hwy 70, Sierra Alvárez, 2380 m, 5 Aug 1983, *Buhrow et al. M12* (ARIZ); 0.2 Km N de Alvárez, Zaragoza, 22°2'N, 100°38'W, 2400 m, 3 Nov 1986, *Debouck et al. 2051* (BR, CHAPA, COL, MICH, US, WIS); Zaragoza, 1 km E de El Cincuenta y Ocho, 22°01'N, 100°38'W, 2440 m, 4 Nov 1986, *Debouck et al. 2052* (BR, CHAPA, COL, UC, US); 2 km SE de Realejo, cerca de "Minas de Plata," Guadalcazar, 22°39'N, 100°24'W, 2100 m, 7 Nov 1986, *Debouck et al. 2057* (CHAPA, MICH, US). **Tamaulipas:** Mpio. Tula, 19 km E de Tula en el camino hacia Ocampo, 22°54'N, 99°37'W, 1540 m, 12 Nov 1986, *Debouck et al. 2070* (BR, CHAPA, US); Rancho del Cielo area, road from Agua Linda to La Mina, 29 Aug 1968, *Richardson 890* (TEX). **Veracruz:** Mpio. Huayacocotla, Viborillas tramo San José, 2200 m, 9 Sep 1976, *Calzada et al 2637* (F); Paso del Viento on old road, Km 30 Hwy 150, Orizaba-Tehuacán, 18°45'N, 97°10'W, 2200 m, 26 Nov 1981, *Freytag et al. 81-6* (BR, EAP, GH, MEXU, MO, UC, US); Mpio. Huayacocotla, Rancho Quemado, Huayacocotla, 1950 m, Oct 1970, *Hernández et al. 886* (F, MEXU); Mpio. Tatatila, La Mancuerna, 2250 m, 28 Nov 1977, *Ventura 14780* (CAS); Mpio. Perote, Cruz Verde, (19°40'N, 97°5'W), 2350 m, 23 Aug 1980, *Ventura 17760* (ASU, MO).

Habitat.—This species is found in dense, humid, scrub oak-hackberry, oak-alder and oak-pine forests intermixed with alder and associated with *Berberis, Juniperus, Rubus* and *Solanum*, frequently on clayey and organic soils derived from limestone, schists or igneous rocks. It is flowering and fruiting with cool night temperatures in the range of 10°C and extremely sensitive to high temperatures (25–30°C) which cause the plant to go into heat shock with immediate wilting and subsequent loss of foliage, flowers and buds.

Diseases and pests.—Only bean rust is reported as a disease. Insects include chrysomelid beetles, leafminers, thrips and weevils (probably *Apion*).

Common names.—Frijol de ratón and frijolillo.

Comments.—Delgado (1985) found an earlier (1837) publication of Bentham's species than those cited by both Piper (1926) and Maréchal et al. (1978b) as 1840 and 1838, respectively. Piper also received a photograph and fragment of the type at Munich from Dr. Hermann Ross, who clearly indicated that the fragment was taken from the type. It is possible to distinguish from the photograph (apparently not seen by Delgado) that there is a tag affixed to the type specimen with the number 13061 which the senior author presumes would have been placed there by the collector at the time of collection.

Delgado (1985) has taken what we consider an overly conservative stance, placing as synonyms of the type species the quite different entities of *P. scabrellus*, *P. esperanzae*, *P. floribundus* and *P. laxiflorus*, each of which are quite distinct entities with several unique characteristics and with separate habitat and distribution ranges.

According to Stainier (Maréchal et al. 1978b) *P. pedicellatus* has the same pollen type as *P. polystachyus*.

M.2.—Phaseolus oaxacanus Rose, Contr. U.S. Natl. Herb. 8:48, plate IX. 1903. (**Figs. 88, 92**). TYPE: MÉXICO. OAXACA: Sierra de San Felipe (Sierra Juárez?), (17°15'N, 96°45'W)?, 2550 m, 24 Aug 1894, *Pringle 5814a* (HOLOTYPE: US 372011 designated by Delgado (1985); ISOTYPES: ENCB, MEXU, MO).

Aerial shoot a strongly climbing, long, indeterminate vine, to 2–3 m. ***Root*** perennial, somewhat thick, fleshy, to 30 cm or more long, 1 cm thick, variously branched, surface rough and light to dark brownish, internal flesh whitish, the crown to some 20 cm below the soil surface with multiple stems. ***Stems*** somewhat perennial and branched at base, terete, 1.25 mm thick, slightly striate; internodes on lower portion of plant to 10 cm long, nearly glabrous, younger stems sparsely covered with appressed reflexed-strigose hairs. ***Stipules*** oblong, foliaceous, 4.5 mm long, 2 mm wide, somewhat spreading, 5- to 7-nerved, nearly glabrous. ***Leaves*** 16.5–14 cm long; petiole 4–6.2 cm long, petiolule 1.5–1.8 cm long; basal pulvinus 4 mm long, lateral and terminal pulvini 2.5 mm long, nearly glabrous; terminal leaflet mostly hastate to ovate, faintly lobed at base, 4.5–5.5 cm long, 2.8–3.2 cm wide at near midpoint, acute to slightly acuminate, apiculate to about 0.5 mm, sparsely covered by strigose hairs on adaxial surface with hairs to 1 mm long, almost glabrous abaxially, a few minute glandular hairs scattered on veins, membranaceous, lightly veined, very delicately mottled on upper surface; lateral leaflet similar but slightly inequilateral, 3.5–4.3 m long, 2.7–3 cm wide at 1/4 from base. ***Inflorescence*** an erect pseudoraceme; peduncle 11–12 cm long; rachis to 30 cm long of 1–3 flowering nodes bearing 2–6 flowers; primary bract 4 mm long, 2–2.5 mm wide, more or less 3-dentate and often deeply divided to 1.5 mm, slightly 5- to 6-nerved, covered by glandular ciliate hairs on adaxial surface; nodes 2-flowered; pedicel 7–8 mm long, moderately pubescent of small hooked hairs; pedicellar bracts scale-like less than 1 mm long hyaline. ***Bracteoles*** minute scales, broadly ovate, 0.25–0.75 mm long, slightly ciliate on margins, early caducous. ***Flower*** light purple; calyx large, flaring campanulate, 5 mm long, the 2 upper teeth 1 mm long, 2.75 mm wide, inequilaterally rounded dentate, the 3 ventral teeth with the central 1.25 mm long, 2 mm wide, the lateral 1 mm long, 1.5 mm wide, moderately minutely glandular and slightly ciliate, persistent; standard light purple, nearly round, reflexed at about midpoint and hooded over keel, 9 mm long, 9.5 mm wide, slightly notched at tip, slightly green on abaxial surface, the auricles pronounced, about 2 mm long, 1–1.5 mm wide; wings light purple, the blade slightly oblong, 8 mm long, 5 mm wide, spreading, the basal claw 4.5 mm long, 0.5–0.7 mm wide, the spur pronounced, 2.5 mm long, 1.5–2 mm wide; keel 13.5 mm long with 8 mm to bend, the basal claw about 5 mm long, with 1 to 1 1/4 coils of 2 mm diam.; vexillary stamen 10 mm long, the basal knob pronounced, 0.75 mm from base and 0.75 mm long, 1.25 mm high; stamen tube 10.5 mm long, about 6 mm to bend, the basal ridges pronounced, about 1 mm from base, 0.75–1 mm high, filaments 6–7 mm long; anthers 0.6 mm long, 0.25 mm wide, yellow; basal collar 1.25 mm long, 1 mm wide; ovary 5 mm long, 1 mm wide, moderately pubescent, 3–4 ovules; style 8 mm long, the thickened terminal portion of 1 1/4 coils of 1.5 mm in diameter; stigma lateral, straight, 0.75 mm long. ***Pod***, immature densely covered with strigose hairs to 0.75 mm long mixed with many hooked hairs, yellowish, at maturity becoming sparse to nearly glabrous, 4 cm long, 1.1 cm wide, 5 mm thick, with large persistent campanulate calyx; beak straight, acute, 2 mm long; sutures somewhat pronounced; valves slightly twisted at dehiscence, reticulate, light tan. ***Seed*** round to oblong and somewhat flattened, 6.4 mm long, 6.3 mm wide, 2.5 mm thick, (100 seed weight = 6.464 g), mottled with fine black specks on brown background, the hilum ring black, smooth, shiny; hilum round, 1 mm long; micropyle level; lens ridges raised 0.25 mm. ***Seedling*** from hypogeal germination; epicotyl 8 cm long; eophyll, the petiole 3 cm long with upper and lower pulvini, the blade broadly triangular, 2.8 cm long, 2.4 cm wide, cordate, acute, mottled light and dark green; stipules at eophyll node, united, stipels present.

Specimens examined: **MÉXICO. Oaxaca:** Tlacolula, Villa Díaz Ordaz, 14 km N of V. Díaz Ordaz to Cuajimoloyas, 17°04'N, 96°24'W,

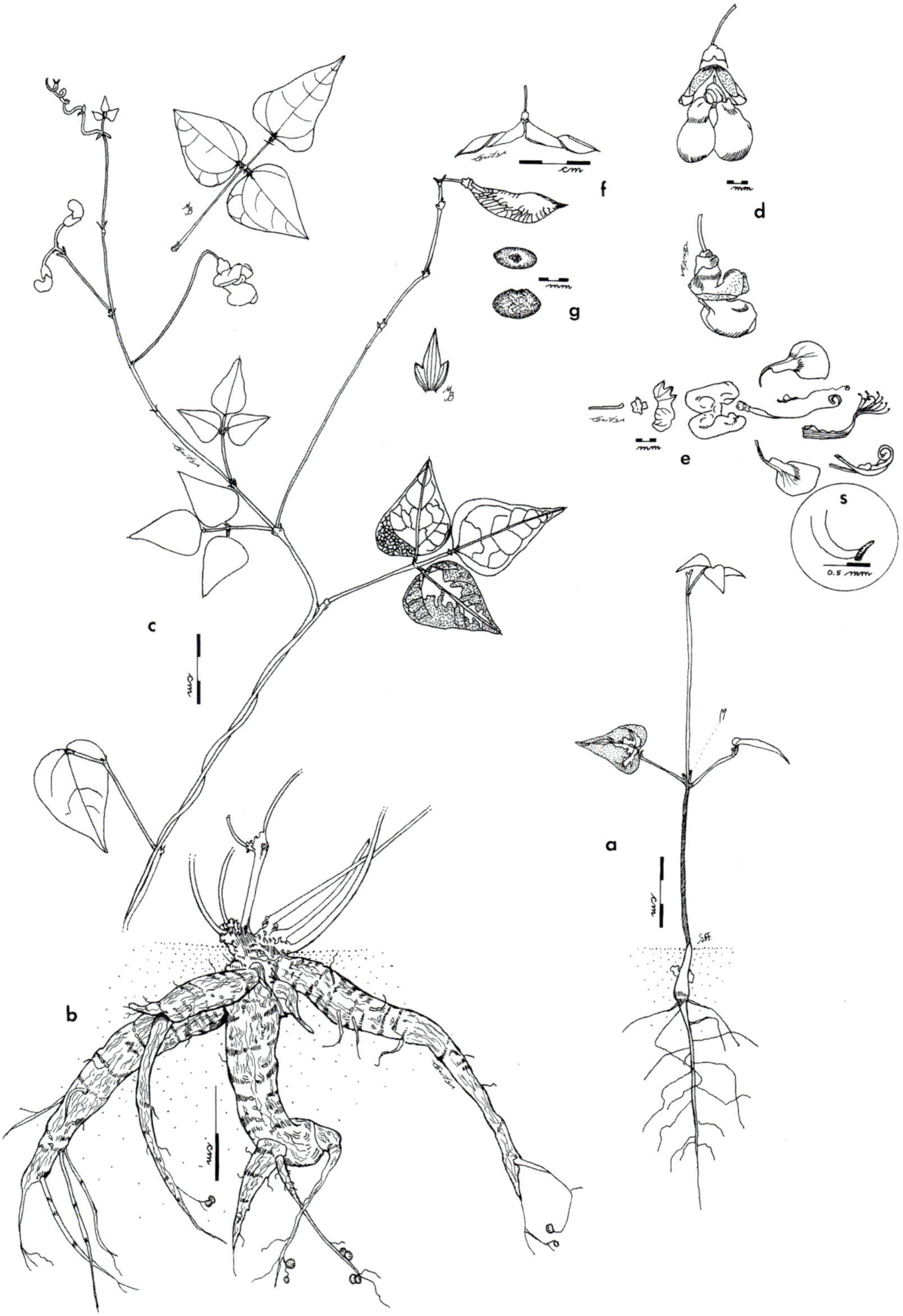

Fig. 88. Illustrations of *Phaseolus oaxacanus* Rose.—**a.** Seedling a few weeks after germination; note nearly triangular eophylls.—**b.** Root after several years.—**c.** Portion of plant with mature leaf, inflorescences with flowers and pod, and vine tip.—**d.** Flowers, side view and front view.—**e.** Exploded view of flower showing all parts including—**s.** Style tip and stigma as seen under the microscope.—**f.** Pods, side view and dehisced.—**g.** Seeds, side view and view of hilum. All drawings made of living material grown in greenhouse at Mayagüez from seed of *Freytag et al. 81-11* (TARS #261) collected at Km 15 of road from Oaxaca to Tuxtepec, Oaxaca, México.

2550 m, 24 Oct 1987, *Debouck et al. 2335* (CHAPA, COL, G, M, MICH); cerro El Labrador, 8 km N de El Portillo, Coatecas, 16°38'N, 96°35'W, 1900 m, 26 Oct 1987, *Debouck et al. 2342* (CHAPA); Km 15 Hwy 175, Oaxaca-Tuxtepec, 17°8'N, 96°35'W, 2250 m, 22 Sep 1978, *Freytag et al. 78-Mex-61* (BR, EAP, F, GH, K, MEXU, MO, UC, US) (also collected at the same locality on 27 Nov 1981 as *Freytag et al. 81-11*(MO)); Yavila-Gueletao, 2600 m, 21 Aug 1976, *Hernández 2670* (MEXU); 2 mi NE of Ixtlán de Juárez, (17°30'N, 96°30'W), 2200 m, 4 Dec 1940, *Krueger et al. 22* (CAS, GH, MO); La Parada, 2286 m, 19 Aug 1894, *Nelson 996* (GH, US); Cerro San Felipe, 2134–2896 m, 1 Sep 1894, *Nelson 1181* (US); Sierra de Clavellinas, 2743 m, 19 Oct 1894, *Smith 566* (EAP, F, MEXU).

Habitat.—This species is found growing in dense shade under pine or pine-oak forest, usually on very steep slopes, along mountain streams and growing over small bushes and trees. Soils are very moist with rocks and lots of humus.

Diseases.—Bean rust and virus have been reported.

Comments.—Maréchal et al. (1978b) recognized this taxon as close to *P. pedicellatus* and felt that the trifid bract character, quite constant for this species, was unique for the genus. In reality the trifid primary bracts, though unusual and relatively constant where found, are characteristically present in a few taxa of several different sections, viz. in Section C. *Coccinei*, *P. coccineus* var. *tridentatus* and var. *splendens*; in Section I. *Digitati*, *P. trifidus*; and in Section M. *Pedicellati*, *P. oaxacanus* and *P. grayanus*. Delgado (1985) treats *P. oaxacanus* as simply a variety of *P. pedicellatus*, nevertheless there are many differences between these species in plant, leaf, raceme, flower, seed and pod (see Color Plate IV, photo 45), and root, which with the limited distribution of *P. oaxacanus* in the mountains around Cd. Oaxaca, we believe enough to justify species rank, as did Lackey (1983). Using ITS DNA sequencing, Delgado et al. (1999) have found that *P. oaxacanus* (there recognized as a separate species!) forms a clade together with *P. glabellus*, *P. grayanus*, *P. pedicellatus*, and *P. neglectus*.

M.3.—Phaseolus esperanzae Seaton, Proc. Amer. Acad. Arts Sci. 28:118. 1893. (**Figs. 89, 92**). TYPE: MÉXICO. PUEBLA: wooded hills near Esperanza, (SW of) Mt. Orizaba, (19°N, 96°20'W), 2440 m, 15 Aug 1891, *Seaton 371* (HOLOTYPE: GH; ISOTYPES: F, US).

Aerial shoot an annual, scandent or climbing, indeterminate vine, to 2–5 m. ***Root*** perennial, thick, fleshy, fusiform, 7–8 cm long, 4 cm thick, cortex fissured and corky, with many thickened lateral roots. ***Stems*** terete, 1–4 mm thick, striate, sparsely covered with minute retrorse-hispid hairs, mostly dark colored (purple), the internodes 4–8–10 cm long. ***Stipules*** triangular, stiffly extended, 3–5 mm long, 1–2 mm wide, 3- to 5-nerved, obtuse, sparsely covered with strigose hairs, to glabrous. ***Leaves*** 6–11.3 cm long; petiole 1.5–2.5–4.5 cm long; petiolule 0.6–1.2 cm long; pulvini 2–3 mm long, sparsely covered with white hispid hairs; terminal leaflet nearly triangular to broadly ovate, 3.5–5 cm long, 2.5–4 cm wide, with two large, obtusely rounded basal lobes, 1–2 cm in size, the central lobe acute, apiculate, sparsely covered with white hispid and strigose hairs, slightly variegated and dark green adaxially, lighter green abaxially; lateral leaflet similar but inequilateral, ovate, 3.5–4.5 cm long, 2–3.5 cm wide, scarcely to distinctly large, round-lobed, 1–2 cm in size. ***Inflorescence*** a stout pseudoraceme, mostly 10–15–(38 on type) cm long, often shorter; peduncle 2.5–16–28 cm long, striate, sparsely covered with uncinate hairs; rachis 1.5–12 cm long, bearing 6–20 flowers or more; primary bract oblong-acuminate sometimes divided at tip with 1–3 teeth, 2–5 mm long, 1–1.5 mm wide, obscurely 3-nerved, strigose to glabrous; pedicel 5–8 mm long, becoming 6–9 mm long at young pod stage, puberulent of uncinate hairs; pedicellar bracts narrowly triangular 1 mm long hyaline 0-nerved. ***Bracteoles*** minute, lanceolate, 1–1.5 mm long, 0.4 mm wide, 0- to 1-nerved (the type specimen is not nerved), hyaline, ciliate, early deciduous. ***Flower*** purple; calyx broadly campanulate, 3–3.5 mm long, the 2 upper lobes rounded, 0.5 mm long, fimbriate-ciliate, the 3 lower lobes triangular acute, 0.75 mm long, sparsely covered with uncinate hairs especially on margins and lower lobes, ciliate; standard purple, the blade broad, rounded, 10 mm long, reflexed 5 mm from base, the erect portion 4 mm long, 9 mm wide, the claw not well defined, 1.5 mm long; wings purple, the blade 15 mm long, broad, the basal claw 5–6 mm long, the spur pronounced, 1.5 mm long, loosely adhering to keel; keel 10 mm long, the claws 4.5 mm long and separated for 1.5 mm, 6 mm from base to bend and 5 mm more to base of the terminal 1 3/4 coils of 2.5 mm diam.; vexillary stamen, the claw 1 mm long, the knob 0.5 mm high, the thickened portion 3 mm more to filament; stamen tube straight, 7 mm to bend and 2 mm more to filaments, 2 mm wide, the ridges 0.3 mm high at 1.5 mm from base; basal collar 0.75 mm long; ovary straight, 5 mm long, 1 mm wide, densely covered by white strigose hairs, 3–6 ovules; style

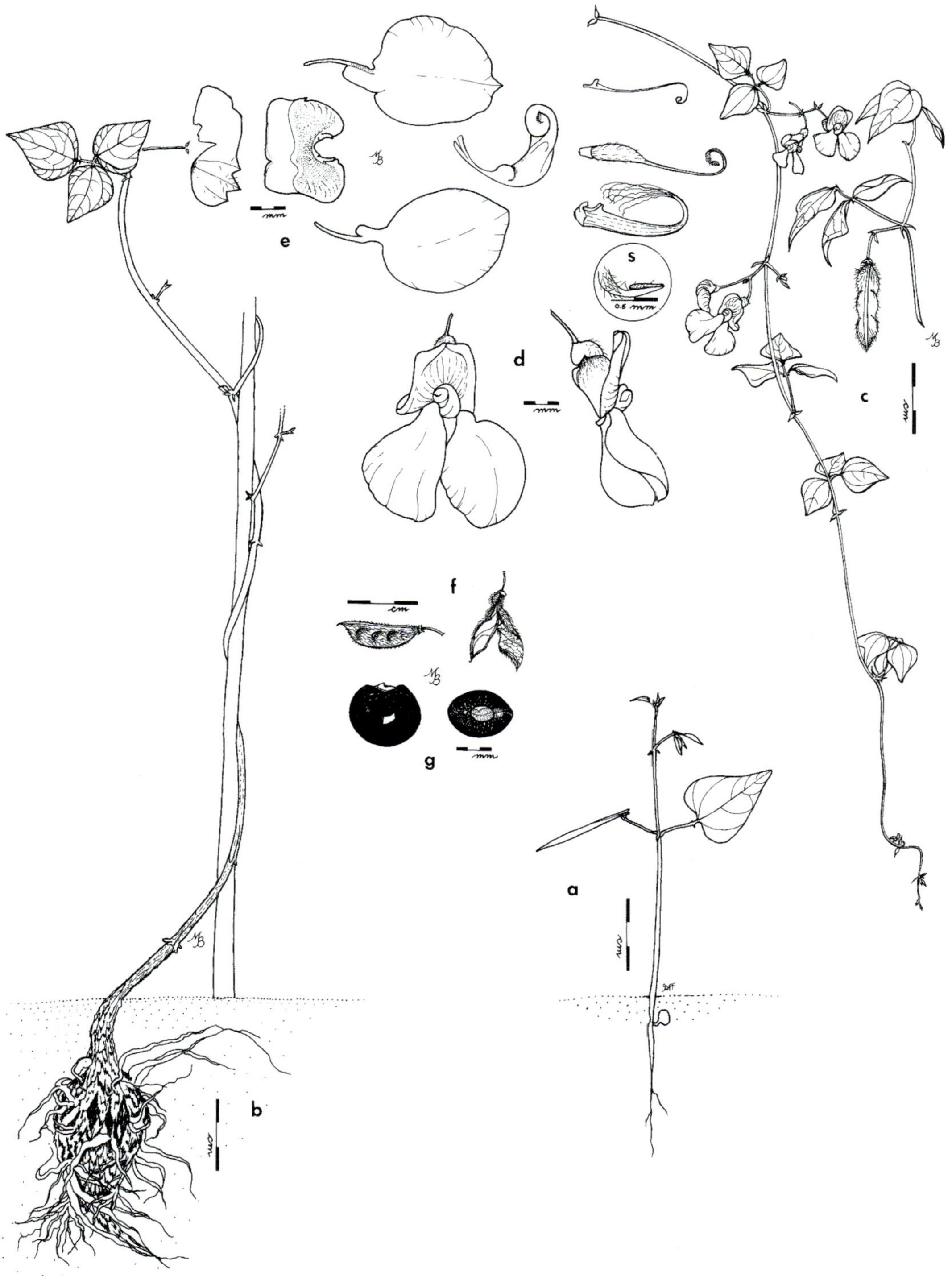

FIG. 89. Illustrations of *Phaseolus esperanzae* Seaton.–**a.** Seedling a few weeks after germination; note ovate eophylls.–**b.** Root after 1 year and lower stem with mature leaf; note thickened lateral roots.–**c.** Vine tip with leaves and inflorescences with flowers and separately a detached portion of stem with mature pod.–**d.** Flowers, side view and front view; note sharply reflexed standard and nearly vertical position of flower.–**e.** Exploded view of flower showing all parts including–**s.** Style tip and stigma as seen under the microscope.–**f.** Pods, side view and dehisced.–**g.** Seeds, side view and view of hilum; note spherical shape and solid black color. All drawings made from living material grown in greenhouse at Mayagüez from seed of *Debouck et al. 2027* (TARS #406) collected near Tepeapulco, Hidalgo, México.

with the terminal thickened coil of 2 mm diam.; stigma lateral, pointed, 0.75 mm long. ***Pod*** straight to slightly falcate with blunt ends, when young densely covered with long yellowish strigose hairs, the mature pod 2–4 cm long, 0.5–1.2 cm wide, 5–6 mm thick; sutures thickened; valves wrinkled and restricted between seeds, covered with long, white yellowish strigose pubescence; beak straight to recurved, very short, 1 mm long. ***Seed*** nearly spherical to slightly flattened, 6.2–8 mm long, 6.1–7 mm wide, 3.7–4.2 mm thick, solid black, shiny; hilum round to slightly oval, 2 mm long; lens pronounced and divided. ***Seedling*** from hypogeal germination; epicotyl 6–7 cm long, the 2nd of the next 3 internodes somewhat shorter (28 mm long) than the 1st or 3rd internodes (36–49 mm long); eophyll petiole 1.5–2 cm long, the blade ovate, 2.9–4 cm long, 2.8–3 cm wide, nearly truncate at base, acute, upper and lower pulvini present, stipels present.

Specimens examined: **MÉXICO. Hidalgo:** cerro 6 km NE de Tepeapulco, 19°48'N, 98°31'W, 2650 m, 29 Oct 1986, *Debouck et al. 2027* (BR, CHAPA, US); 3 km NE de El Abra, Tulancingo, 20°07'N, 98°19'W, 2200 m, 7 Nov 1987, *Debouck et al. 2379* (CHAPA, MICH); Acaxotchitlán, 3 km E de San Pedro, 20°08'N, 98°12'W, 2190 m, 7 Nov 1987, *Debouck et al. 2381* (CHAPA, MICH, US). **México:** 9 km E de Espíritu Santo, Ciudad Adolfo López Mateos, 19°32'N, 99°21'W, 2520 m, 6 Nov 1987, *Debouck et al. 2376* (CHAPA, MICH); 12 km SE de Villa del Carbón, 19°41'N, 99°25'W, 2520 m, 6 Nov 1987, *Debouck et al. 2378* (CHAPA, COL, MICH); Mpio. de Texcoco, 2 km al SE de San Pablo Ixayoc hacia Ojo de Agua, 12.5 km al SE de Texcoco, 2670 m, 21 Oct 1976, *García 168* (MO). **Michoacán:** cerro 0.5 km S Tlapujahua, 19°48'N, 100°10'W, 2500 m, 17 Nov 1987, *Debouck et al. 2406* (CHAPA, MICH). **Puebla:** Tlalanpaleca, 3 Km SE de Santa Rita, 19°18'N, 98°38'W, 2450 m, 22 Oct 1987, *Debouck et al. 2322* (CHAPA, COL, MICH); En el sitio El Acocote, 1 km E de San José, Esperanza, 18°51'N, 97°17'W, 2400 m, 10 Nov 1987, *Debouck et al. 2382* (CHAPA, COL, MICH, US); 6 km N de Esperanza en camino a Xochiloma, 18°56'N, 97°19'W, 2590 m, 10 Nov 1987, *Debouck et al. 2386* (CHAPA); cerro 4 km NE de Pueblo Acajete, 19°08'N, 97°54'W, 2400 m, 11 Nov 1987, *Debouck et al. 2387* (CHAPA); Puente del Mirador 2, Km 74 autopista México–Puebla, (19°22'N, 98°35'W)?, 3 Aug 1977, *Martínez et al. MA1713* (F, MO); Mpio Cholula, Cerro Teoton, lado E San Pedro Yancuictlapan, 19°04'N, 98°28'W, 2450–2600 m, 29 Sep 1987, *Tlapa et al. 641* (MEXU).

Habitat.—This species is reported as growing in oak and pine-oak forests with yucca, *Opuntia* cactus and juniper and in areas sometimes disturbed by fire, mainly along the Trans-Mexican Volcanic Belt. Soils are dark with much organic matter and clay to sandy with rocks, derived from volcanic ash, basalt, or metamorphic schists.

Diseases and pests.—The seed are often attacked and damaged by *Apion* weevils.

Common name.—Frijol Cimarron morado.

Comments.—The leaves of *P. esperanzae* are small and similar (although a bit more coriaceous) to those of *P. pedicellatus*, the species it most closely resembles, but the seed are somewhat larger, more rounded and solid black while those of the latter are smaller, somewhat angular and speckled black on brown. In addition, pod sutures are thickened and valves are restricted between the seeds in *P. esperanzae*. It has not been mentioned by Maréchal et al. (1978b) nor by Lackey (1983). In the study of Gaitán and co-workers (2000) using ITS DNA sequencing, it came in the group of *P. pedicellatus*, although different from it, confirming early observations (Debouck 1991). We cannot agree with Morton (1944) listing *P. floribundus* as synonym of *P. esperanzae*.

M.4.1.—Phaseolus polymorphus S. Wats. var. **polymorphus**, Proc. Amer. Acad. Arts Sci. 17:346. 1882. (**Fig. 92**). TYPE: MÉXICO. SAN LUÍS POTOSÍ: in arenosis circa urbem (SLP), 1876, *Schaffner 811*, (LECTOHOLOTYPE: GH (designated by Delgado 1985); ISOTYPE: K). SYNTYPE: MÉXICO. SAN LUÍS POTOSÍ: 22°N, (101°W), 1829–2438 m, (1878), *Parry & Palmer 186* (in part) (SYNTYPE: GH; ISOTYPE: K). SYNTYPE: MÉXICO. SAN LUÍS POTOSÍ: Chiefly in the region of San Luís Potosí, 22°N, 1829–2438 m, 1878, *Parry & Palmer 188* (SYNTYPE: US 49596; ISOTYPES: F, GH, K, MO, NA).

Phaseolus schaffneri Piper, Contr. U.S. Natl. Herb. 22:690–691. 1926. TYPE: MÉXICO. SAN LUÍS POTOSÍ: in arenosis circa urbem (SLP), 1879, *Schaffner 606* (HOLOTYPE: US 939707; ISOTYPES: BM, G, K, NY n.v.).

Aerial shoot a large, much branched, scandent, annual, indeterminate vine. ***Root*** thick, fleshy, perennial, 2–3 cm long. ***Stems*** woody at base, slender, terete, striate, nearly glabrous; internodes 2.5–10 cm long. ***Stipules*** ovate to lance-ovate, 2–4 mm long, acute, 3- to 5-nerved, glabrous, ciliate on margin. ***Leaves*** 5–13.8 cm long; petiole 1.3–5 cm long, glabrous except for a few hispid hairs at base, the petiolule 0.5–2 cm long with a few scattered small hispid hairs; stipels ovate-lanceolate, 1-nerved, 1–1.5 mm long, sparsely hispid; pulvini 1.5 mm long, dark, small hispid on upper surface, glabrous below; terminal leaflet nearly triangular and usually deeply squarish-rounded lobed at base, the leaflets of leaves on the basal portion of plants are more entire, the center lobe oblong-lanceolate to nearly linear, 2.5–6.5 cm long, 3–5–13 mm wide, obtuse and apiculate, longer than the broader rounded or squarish lateral lobes of 0.6–2 cm long and 12 mm wide, membranous, glabrous to puberulent except for the scabrous margin, green adaxially and scarcely paler abaxially; lateral leaflets similar and inequilateral. ***Inflorescence*** a pseudoraceme; peduncle 5–7–(40) cm long, glabrous; rachis 1.5–4–25 cm long, few to many-flowered, covered with minute hispid and uncinate hairs; primary bract broadly lanceolate, 1–4 mm long, 1 mm wide, 3- to 4-nerved, glabrous, sometimes ciliate; pedicel delicate 4–7 mm long, puberulent appressed-uncinate; pedicellar bracts narrowly lanceolate 1 mm long 1-nerved. ***Bracteoles*** oblong-obtuse to broadly obovate, 2 mm long, 1- to 3-nerved, glabrous, sometimes inserted to 1 mm below calyx. ***Flower*** purple medium sized; calyx campanulate, 2.5 mm long, puberulent to glabrous except the ciliate margin, the broad upper lip emarginate, 0.5 mm long, 3 mm wide, the lower lip with 3 broad triangular acute teeth, 1 mm long; standard 8–10 mm long, reflexed at 3 mm from base and 3 mm more to apex, 8 mm wide, the claw 0.5 mm long, the auricles small, 0.5 mm long; wings, the blade broad, cupped and spreading, 14 mm long, the claw 3 mm long, the spur pronounced 0.75–1 mm long; keel with 1 1/2–2 coils of 2 mm in diam.; vexillary stamen, the claw 0.75 mm long, the genicuclate sheath 0.5 mm long, the thickened portion extending 1.5 mm more; stamen tube narrow, 5 mm to bend and 3 mm more to the filaments, the slightly developed knobs at 2 mm from base; basal collar 0.5 mm long, slightly denticulate; ovary 4 mm long, 0.5 mm wide, covered with minute pubescence, 4 ovules; style 7.5 mm long to the terminal thickened coil of 2 mm diam.; stigma terminal, lateral, linear 0.75 mm long. ***Pod*** straight, compressed, 4 cm long, 1 cm broad; valves puberulent and sparsely covered with short white strigose hairs, tightly twisted at maturity; beak straight, 1 mm long or less, acute. ***Seed*** rounded oblongoid, somewhat flattened, 6.3–8 mm long, 4.4–6.6 mm wide, 2.4–3.5 mm thick, solid black or speckled black on tan, shiny, black ring around hilum; hilum ovate to orbicular, 0.8–1 mm long, 0.5–0.75 mm wide; lens pronounced. ***Seedling*** unknown.

Specimens examined: **MÉXICO. Aguascalientes:** near the summit of Sierra del Laurel, 10 mi SE of Calvillo, (21°45'N, 102°40'W), 2500 m, 4 Nov 1959, *McVaugh et al. 210* (MICH); Mpio. de San José de García, 12 km al SW de La Congoja, 2700 m, 16–17 Oct 1973, *Rzedowski et al. 740* (CAS, MICH). **Coahuila:** E of Rancho El Caballo, Sierra del Jardín, (29°3'N, 102°37'W), 2250–2450 m, 16 Sep 1972, *Chiang et al. 9250* (CAS); Mpio. Ocampo, Sierra Maderas del Carmen, 2360 m, 17 Sep 1989, *Estrada 1783* (BRIT); Canyon de Centinela, S side of Pico de Centinela, Sierra del Carmen, 29°03'N, 102°35'W, 1798 m, 31 Jul 1973, *Henrickson 11668* (TEX-LL); Potrero de la Mula, about 20 km NW of Ocampo, (27°40'N, 102°30'W), 18 Sep 1941, *Johnston 9235* (GH); Del Carmen Mts., 26 Aug 1936, *Marsh 593* (F, GH, TEX(2)); Muzquíz Palm Canyon, (27°55'N, 101°30'W), 19 Sep 1936, *Marsh 987* (TEX); S of Muzquíz, Santa Rosa Mts., 13 Jul 1938, *Marsh 1331* (F, GH, TEX); SE of Monclova, Sierra de la Gloria, (26°50'N, 101°10'W), 4 Mar 1939, *Marsh 1935* (in part) (GH, TEX); Mpio. Cuatro Ciénagas, Cañon del Agua, W of Monclova, Sierra de la Madera, (27°N, 101°45'W)?, 8 Sep 1939, *Muller 3223* (GH, NA, TEX-LL, UC); 21 mi SE of Monclova, Caracol Mts., Aug 1880, *Palmer 2122* (GH, K, N, n.v., US); Mpio. Cuatro Ciénegas and Ocampo, Cañon Desiderio, 9.8 mi W from Rancho Cerro de la Madera, 27°7'25"N, 102°31'45"W, 2280 m, 29 Sep 1976, *Wendt et al. 1836* (ASU, TEX). **Durango:** Canyon Cantero, Sierra Gamon, 2000 m, 21 Sep 1948, *Gentry 8376* (GH, MEXU, MICH, UC, US). **Guanajuato:** 8 km E de Carbajal, San José Iturbide, 21°04'N, 100°11'W, 1970 m, 4 Nov 1987, *Debouck et al. 2393* (BR, CHAPA, MICH, US); 8 km W de Mesa de San José, Dolores Hidalgo, 21°04'N, 101°11'W, 2520 m, 15 Nov 1987, *Debouck et al. 2398* (BR, CHAPA, US); 14.5 mi from Guanajuato on the road to Dolores Hidalgo, (21°10'N, 101°W), 2499 m, 5 Jul 1955, *Johnston 2631D* (TEX). **Jalisco:** Paso de la Troje, near Km 36 SW of Ojuelos on road to Aguascalientes, 2100–2300 m, 9–12 Aug 1958, *McVaugh 16760* (MO, NY). **Nuevo León:** Mpio. Galeana, Cerro El Potosí, (24°55'N, 100°15'W), 3050 m, 21 Aug 1969, *Hinton 17245* (ENCB, K, US); 1.5 mi below the microwave station on Cerro Potosí, N of Galeana, Sep 1970, *Norris 17688* (CAS). **Queretaro:** entre Cadereyta y Vizarrón, (20°50'N, 99°45'W), 22 Aug 1905, *Altamirano 1660* (US); Cadereyta, 21 Aug 1905, *Rose et al. 9723* (MEXU, US). **San Luís Potosí:** 7 km S de la Presa del Peaje, Cerro de Los Fieros, Sierra de El Potosí, 22°02'N, 101°06'W, 2300 m, 6 Nov 1986, *Debouck et al. 2056* (CHAPA, MICH, UC, US, WIS); 1 km W of Real de Catorce on road to Estación Catorce, Socavón La Purísima, 23°41'40"N, 100°53'50"W, 2400–2450 m, 17 May 1973, *Johnston et al. 11068D* (TEX-LL); 2 mi NNW of mine, Charcas, (23°08'N, 101°11'W), 18 Jul 1934, *Lundell 5317* (MICH, US); Potrero, Mt. E of Santa Ana, Sierra de Catorce, 2800–3000 m, 24 Jul 1934, *Pennell 17553* (US); Mts. W of San Luís Potosí, Sierra Madre Oriental, 1900–2100 m, 27–28 Jul 1934, *Pennell 17619* (GH); in arenosis circa urbem (SLP), 1879, *Schaffner 605* (NY). **Tamaulipas:** San José, Cerro Tres Vetas, Sierra de San Carlos, (24°45'N, 99°W)?, 823 m, 15 Jul 1930, *Bartlett 10384* (CAS-DS, F, GH, MICH, NY, US). **Zacatecas:** 15 mi NE of Estación Camacho on NW slope of Pico de Teyra, 24°34'N, 102°11'W, 2012 m, 23 Sep 1973, *Henrickson 13405a* (TEX-LL); on Mt. 18 km W of Concepción del Oro on Coahuila-Zacatecas border, 24°54'N, 101°45'W, 2850 m, 22 Jul 1941, *Stanford et al. 563* (CAS-DS, GH, MO, UC).

Habitat.—This species is found growing as an understory vine in the semi-shady or sunny places in the open parks of pine or pine-oak forests and associated with *Acacia, Agave, Arbutus, Arctostaphylos, Ceanothus, Crataegus, Cupressus, Dalea, Fouquieria, Jatropha, Mimosa, Opuntia, Prosopis, Prunus,* and *Yucca*. Soils are gravely clay loam with abundant organic matter and derived from igneous rock, volcanic ash, or diorite.

Diseases and pests.—Damage is reported from *Apion* weevils and frosts.

Comments.—Apparently there was a manuscript mixup of some kind in Piper's (1926) work since he indicates there are only two specimens of *P. polymorphus*: the type *Parry & Palmer 188*, and *Purpus 5196* the type of *P. purpusii*. Since these two sheets are completely different, it is inconceivable to the senior author that Piper would have knowingly made this great an error. On the other hand, the type of *P. schaffneri* (*Schaffner 606*) is almost identical to the type of *P. polymorphus*. Thus it seems that Piper may have intended to recognize the species *polymorphus* and *purpusii* as distinct, but by "accidentally" placing *purpusii* as a synonym he then created the superfluous species *schaffneri*.

The date given by Piper (1926) for publication of *P. polymorphus* as 1822 is also in error since both *P. polymorphus* and *P. scabrellus* were published at the same time in Proc. Amer. Acad. Arts & Sci. 9 of 1882 and were cited on the same page in Piper's manuscript. Apparently this error was followed by Maréchal et al. (1978b) who also confused *P. purpusii* as a synonym for *P. polymorphus*.

The original publication (Watson 1882) of *P. polymorphus* gives the type as "*Schaffner 811* mainly; also *186* and *188* of *Parry & Palmer*." Piper has changed this and mentions only *Parry & Palmer 188* as the type and gives *Purpus 5196* as the type of *P. purpusii* equal to *P. polymorphus*. On examining these specimens the senior author finds the first type cited in the original publication for each species correct. However, the type of *P. schaffneri* (*Schaffner 606*) looks the same as the type of *P. polymorphus* (*Schaffner 811*). Species *purpusii* is quite different and only represented by *Purpus 5196*. *Parry & Palmer 188* and *186* are also correct as syntypes of *P. polymorphus*. Species *schaffneri* is not mentioned by Maréchal et al. (1978b).

M.4.2.—Phaseolus polymorphus S. Wats. var. **albus** Freytag, var. nov. (**Fig. 92**). TYPE: MÉXICO. COAHUILA: E of Rancho El Caballo, High Western ridge of Sierra del Jardín, 29°03'N, 102°37'W, 2250–2450 m, 16 Sep 1972, *Chiang et al. 9342* (HOLOTYPE: TEX-LL; ISOTYPE: CAS).

Similis *Phaseolo polymorpho*, sed folias lobatis minus profunde, inflorescentiis grandioribus et floribus plus numerosis, floribus albidibus maturascentibus flavis differt. Habitat modo in locis regionis boreo-occidentali Coahuilae rarus.

This variety is similar to type species *polymorphus* except: ***Flower*** white, fading yellow.

Habitat.—It is found growing on grassy areas in pine-oak forest on steep slopes of igneous rock, and in a sandy, gravely soil and associated with juniper and grasses.

Comments.—This variety is recognized because the white flower is extremely rare for the species in this section, and because the taxon seems to be found at the extreme north edge of the range for the type variety with no intermediates having been encountered. Additional collecting in this region is needed to obtain a better representation of the morphological variation of the taxon, and to understand the complete range of distribution and whether genetic intermixing takes place or not.

M.5.—Phaseolus palmeri Piper, Contr. U.S. Natl. Herb. 22:691. 1926. (**Fig. 92**). TYPE: MÉXICO. ZACATECAS: Concepción del Oro, (24°40'N, 101°30'W)?, (2500 m), 11–14 Aug 1904, *Palmer 294* (HOLOTYPE: US 471163; ISOTYPES: GH, NY n.v.).

Aerial shoot an herbaceous, annual, climbing, indeterminate vine. ***Root*** unknown. ***Stems*** slender, terete, purple, with minute green speckles and streaks, sparsely puberulent; internodes 2–9 cm long. ***Stipules*** ovate, 2 mm long, strongly 5-nerved, obtuse, glabrous. ***Leaves*** 4.8–5.8 cm long, with deeply lobed leaflets; petioles slender, 2–2.5 mm long, pubescent; petiolules 8 mm long, glabrous, purple and green speckled; stipels oblong, 1 mm long, 1-nerved, acute; all leaflets similar, squarely truncate at base, 2–2.5 cm long, 2–2.5 cm wide, very deeply 3-lobed, the median lobe oblong-ovate, 2.5–3 cm wide, the lateral lobes 1–1.5 cm deep, triangular-ovate, each sometimes again obscurely lobed, all apices obtuse and apiculate, membranous, minutely puberulent, nearly glabrous adaxially and slightly green abaxially. ***Inflorescence*** a pseudoraceme exceeding the leaves; peduncle 5–9–11 cm long, puberulent; rachis 1–4 cm long 4- to 12-flowered with 2 buds at a node and up to 8 nodes; primary bract broadly ovate, 1 mm long, 1-nerved, obtuse; pedicel 5–9 mm long, puberulent. ***Bracteoles*** broadly ovate, 1 mm long, 1-nerved, acute, ciliate. ***Flower*** large, violet; calyx broadly campanulate, the tube 3 mm long, the 2 upper lobes rounded 0.5 mm long, 2.5 mm wide and united into one shallowly emarginate, ciliate, the lower lip with 3 broad acute teeth 1.25 mm long, 1.5 mm wide, minutely puberulent and ciliate; standard orbicular, 1 cm long, 1 cm broad, shallowly emarginate at apex, short-stipitate, cuneate at base, without auricles, reflexed from below the middle, thickened along the zone of flexure, at each end of which is a double fold; wings, the blade 15 mm long, broadly obovate, truncate at base, inrolled at margin, the spur poorly developed, the slender stipe 4 mm long; keel cylindrical, with 1 to 1 1/2 terminal coils of 3 mm in diam., 2-lobed at orifice; ovary linear, appressed puberulent, 3-4 ovules; stigma lateral, introrse, linear 1 mm long. ***Pod*** slightly falcate, 4.5 cm long on type, 1 cm broad, much compressed, short-beaked, nearly glabrous, ***Seed*** unknown. ***Seedling*** unknown.

Habitat.—Unknown, but probably this species is found growing in pine-oak forest similar to that for other species of this section found in the same general area.

Comments.—This species is quite similar in most aspects to *P. polymorphus* and *P. grayanus* but easily distinguished from them by its much larger flowers and broad pods. It also has small leaves most all of which are deeply divided, producing quite long and broad lobes on both the terminal and lateral leaflets. Curiously Piper (1926) keys it out as having bracteoles linear to oblong while his species description gives them as ovate (which they are), thus the key must have typographical errors. This species is known only from the type collection. Additional collecting must be done to confirm the validity of the species, the habitat and range.

M.6.—Phaseolus purpusii Brandegee, Univ. Calif. Publ. Bot. 4:271. 1912. (**Fig. 92**). TYPE: MÉXICO. SAN LUÍS POTOSÍ: Minas de San Rafael, (23°5'N, 101°13'W), Jul 1911, *Purpus 5196* (HOLOTYPE: UC 149341 n.v.; ISOTYPES: BM, G, GH, MO, US).

Aerial shoot an annual, long scandent or climbing, indeterminate vine, 1 m long or more. ***Root*** thickened, perennial. ***Stems*** terete, nearly glabrous, sparsely covered with short reflexed-hispid and hooked hairs, purple. ***Stipules*** broadly lanceolate, 2.5 mm long, 0.75 mm wide, reflexed, heavily 3-nerved, acute, nearly glabrous, ciliate especially on margins, purple. ***Leaves*** 3.3–7 cm long; petioles 1–2 cm long, ciliate, purplish, petiolules 3–6 mm long, minutely puberulent; terminal leaflet long lanceolate to nearly linear, 2–4.5 cm long, 3–5 mm wide, not lobed (in the lower parts of shoots but lobed in the upper parts, with lobes forming a right angle), obtuse, apiculate, scaberulous, glabrous, ciliate margin; lateral leaflets deeply lobed, with a long basal lobe at right angles to main central lobe, the lobes lanceolate to linear, 1/2 to 3/4 the length of the leaflet, the longer 2–3 cm long, the shorter 1–2 cm long, linear or broad lanceolate, the younger leaves narrower, leaflet tips often obtuse, apiculate, the adaxial surfaces distinctly reticulate veined, minutely ciliate on margins, dark green, the abaxial surface glabrous and lighter green. ***Inflorescence*** a short pseudoraceme, 2–3 cm long with 1–4 flowering nodes; peduncle 2–4 cm long, striate, puberulent; rachis 1–1.5 cm long, puberulent; primary bract ovate-acuminate to lanceolate, 2 mm long, heavily 3-nerved, glabrous, purple; pedicel slender, 6–11 mm long, minutely ciliate; pedicellar bracts narrowly lanceolate 1 mm long strongly 1-nerved hyaline. ***Bracteoles*** ovate to oblongate, 0.8 mm long, 0.5 mm wide, strongly 1-nerved, obtuse, ciliate margins, purplish. ***Flower*** purple; calyx broadly campanulate, 3.5 mm long, the lower lobes broadly acute, 1.25 mm long, ciliate margins, glandular, purplish; standard purple, very broad, rounded, sharply reflexed at 4.5 mm from base, 5 mm more to apex, about 10 mm wide, the claw 1 mm long, the auricles well-developed, 1 mm long; wings purple, the blade broad, cupped, spreading, 13–15 mm long, the claw about 5 mm long, the spur 1 mm long; keel 5 mm to bend and 5 mm more to base of the terminal 1 1/2 coil of 3 mm diam.; basal collar 0.75 mm long, minutely denticulate; vexillary stamen, the claw 1 mm long with a sheath 1 mm long on the geniculate knob, the thickened portion 3 mm long; stamen tube 6 mm long to bend and 4 mm more to filaments, a small knob 0.5 mm long at 1.5 mm from base; ovary 4–5 mm long, densely covered with fine white pubescence, 2–4 ovules; style 9 mm long to the terminal thickened coil of 2 mm in diam.; stigma lateral, introrse, elongate, 1 mm long. ***Pod*** falcate, 4 cm long, 9-10 mm wide, fibrous and brittle, glabrescent; beak acute to obtuse. ***Seed*** small, oblongoid, 4.5 mm long, 3.7 mm wide, 2.4 mm thick, smooth, solid black, shiny; hilum oblong, 1 mm long, 0.5 mm wide, open—no placental tissue. ***Seedling*** unknown.

Specimens examined: **MÉXICO. San Luís Potosí:** Mojarras Río Verde, Cd. Valles, *Grether 782* (MEXU); Minas de San Rafael, Nov 1910, *Purpus 4854* (UC).

Habitat.—Unknown, but it probably grows in pine-oak forests similar to that for other species of the section found growing in the same general area.

Comments.—The exact location of the type is in doubt since Purpus apparently collected at two major locations (stations along the railroad, the principal mode of transportation in those days), one east of San Luís Potosí at Rascon and the other north of SLP near Charcas. At the latter location there is a small village named San Rafael which is surrounded by a number of mines, possibly the location where species *purpusii* was collected. Delgado (1985) believes the location is SE of San Luís Potosí.

It is difficult to understand how Piper (1926) (who wrote '*purpusi*') could consider *Purpus 5196* as the same taxon as the type of *P. polymorphus* (*Schaffner 811*) (see discussion under *Comments* for species *polymorphus* above), since the latter has slightly lobed leaflets while the other has very deeply lobed leaflets with all lobes narrowly lanceolate to linear. Delgado (1985) places this taxon as a variety of *P. pedicellatus*, again an impossible treatment since *P. pedicellatus* has the least lobed leaves of species of the section.

M.7.—Phaseolus grayanus Woot. & Standl., Contr. U.S. Natl. Herb 16:139–140. 1913. (**Figs. 87, 90**).
Phaseolus wrightii Gray var. *grayanus* (Woot. & Standl.) Kearney & Peebles, J. Wash. Acad. Sci. 29:485. 1939. TYPE: UNITED STATES. NEW MÉXICO: base of San Luís Mountains, (31°30'N, 108°50'W)?, 1829 m, 5 Sep 1893, *Mearns 2124* (HOLOTYPE: US 232982).

Phaseolus foliaceus Piper, Contr. U.S. Natl. Herb. 22:689–690. 1926. TYPE: MÉXICO. DURANGO: Sierra Madre and Sierra Santa Barbara, near La Providencia, (26°45'N, 106°W)?, 1981–1829 m, 11–12 Sep 1898, *Nelson 4990* (HOLOTYPE: US 333018; ISOTYPES: K, NY).

Aerial shoot an annual, prostrate, trailing and weakly climbing, indeterminate vine. ***Root*** a perennial, woody, narrow, 1–2 branched, to 50 cm long, 2 cm thick, covered by a black corky bark. ***Stipules*** broad triangular to 3–4 mm long, 5-nerved, glabrous, ciliate margin. ***Leaves*** 3.5–13.6 cm long; petioles 1–5 cm long, about as long as leaflets; petiolules 1–2.8 cm long; terminal leaflets rhomboidal to nearly triangular, to 1.5–5.8 cm long, 3–5 cm wide, entire to variously 3-lobed, sometimes squarish but always pointed lobes, one shorter than the other, sometimes the larger again slightly lobed, nearly glabrous or ciliate on veins and margins; variegation of leaflets common. ***Inflorescence*** a very long pseudoraceme to 30 cm, 10- to 14-flowered; primary bract ovate-oblong, sometimes trifid, 2–5 mm long, 2–7-nerved, nearly glabrous, ciliate margins; pedicel fairly stout, 5–6 mm long, heavily pubescent; pedicellar bracts 0.5 mm long lanceolate hyaline puberulent scale-like. ***Bracteoles*** ovate-lanceolate, 1 mm long, inserted any place on pedicel from 3 mm below (about midpoint) to base of calyx, 1-nerved or none apparent, ciliate, sometimes deciduous; pedicellar bracts triangular less than 1 mm long scale like. ***Flower*** purple; calyx small, campanulate, short acute lobes, pubescent to ciliate, upper lobes often dark purple; standard light purple and greenish, broadly rounded, 9 mm long, 15 mm wide, emarginate, sharply and heavily recurved, edges enrolled, nearly glabrous; wings purple, the blade obovate, 9–11 mm long, 12 mm wide, slightly enrolled, spreading, the claw 3.5 mm long, the spur pronounced, 2.8 mm diam.; keel, the claws 2.5 mm long, the tube 3 mm thick, 4 mm more to bend and 6 mm more to base of the terminal 1 1/2 coils of 2 mm diam.; stamen tube thickened at base, 3 mm wide, slightly curved, 11 mm to separated filaments; vexillary stamen, the claw 2 mm long, the geniculate knob well-developed, 1.5 mm high; ovary straight and heavily covered by yellow strigose hairs, 6 mm long, 1.5 mm wide, pubescent, 2–4 ovules; stigma lateral, introrse, linear, 1.2 mm long, fairly broad. ***Pod*** slightly falcate, inflated, 2.5–3 cm long, 0.5–0.8 cm wide, somewhat restricted between seeds, pubescent. ***Seed*** squarish and flattened, 3–7 mm in diam., often somewhat ridged, speckled black on tan, black ring around hilum; hilum oblong, 1 mm long; lens pronounced. ***Seedling*** from hypogeal germination; epicotyl 3–11 cm long; eophyll, the petiole 1.5 cm long with upper and lower pulvini, the blade nearly triangular, 2.2 cm long, 2 cm wide, acute, slightly mottled green.

Specimens examined: **MÉXICO. Chihuahua:** Sierra de Las Pampas, W of Hda. El Berrendo, 27°20'N, 104°43'W, 1600–1850 m, 25 Aug 1972, *Chiang et al. 8834* (CAS, MO, TEX-LL); 4 km SSW of Cerro del Gringo in the Sierra del Diablo, 27°9'–27°10'N, 104°8'–104°9'W, 1800–2050 m, 30 Aug 1972, *Chiang et al. 9014A* (TEX-LL); Mpio. Fco. del Oro, Las Altares, 23 km después de Parral, 26°52'N, 105°55'W, 1990 m, 1 Oct 1978, *Debouck 279* (BR, CHAPA, COL, K, US); 31 (air) mi NW of Julimes above Rancho El Recuerdo, Sierra de Carrasco (S of Sierra de Chorreras), 28°47'N, 105°9'W, 1829 m, 15 Sep 1973, *Henrickson 12970a* (TEX-LL); W side of Velardena arroyo above Santo Domingo mine, Santa Eulalia, 28°37'N, 105°53'W, 2012 m, 9 Sep 1948, *Hewitt 355* (GH); N and NW of Rancho El Sauz, Sierra del Roque, N of Julimes, 28°39'–28°41'N, 105°18'–105°20'30" W, 1450–2150 m, 24 Aug 1973, *Johnston et al. 12332* (CAS, TEX-LL); Mojarachic, 7 Aug 1938, *Knobloch 5204* (F, MSC); Mojarachic, SW of San Juanito, 27°52'N, 108°W, 2140 m, 11 Aug 1954, *Knobloch 1197* (MICH); El Cima, (28°25'N, 107°5'W), 29 Jun 1936, *Le Sueur 713* (F, UC); Sta. Clara Canyon, 17 Aug 1936, *Le Sueur 731* (ARIZ, BRIT, CAS, F, GH, US); N of Chuhuichupa, 1 Sep 1937, *Le Sueur 1339* (F, GH, MO); Río Gavilán, 7 mi SW of Pacheco, 1829 m, 12 Aug 1948, *Leopold 183* (UC); 10.4 km NNE of El Vergel, 112 km SSW of Hidalgo del Parral on Hwy 24, (26°30'N, 106°20'W), 2700 m, 23 Aug 1983, *Nesom 4885* (TEX); Barranca Colorada, 35 km SW of Minaca, Sierra Gazachic, Sierra

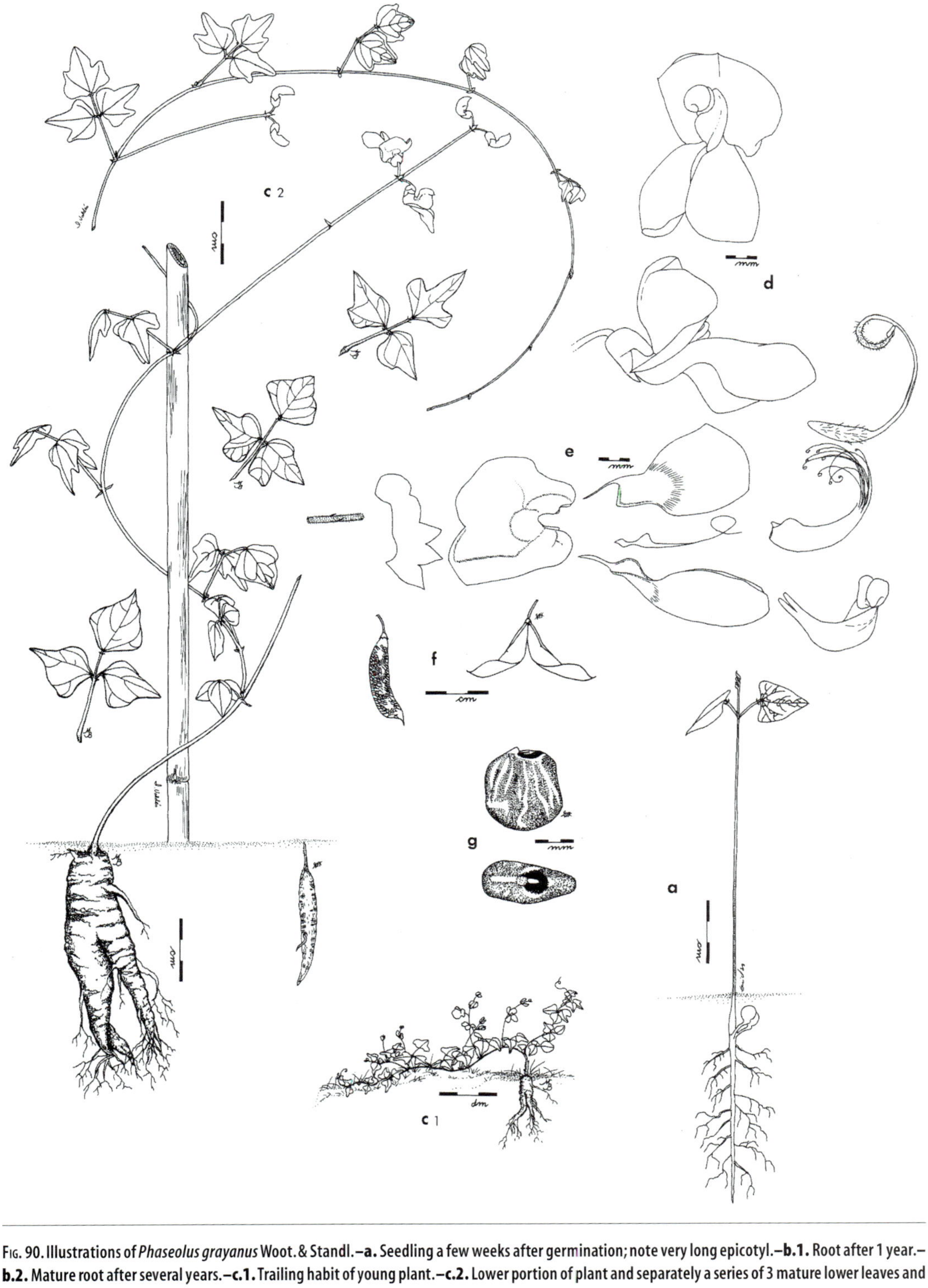

FIG. 90. Illustrations of *Phaseolus grayanus* Woot. & Standl.—**a.** Seedling a few weeks after germination; note very long epicotyl.—**b.1.** Root after 1 year.—**b.2.** Mature root after several years.—**c.1.** Trailing habit of young plant.—**c.2.** Lower portion of plant and separately a series of 3 mature lower leaves and a vine tip; note increased lobing towards apex of plant.—**d.** Flowers, side view and front view.—**e.** Exploded view of flower showing all parts.—**f.** Pods, side view and dehisced.—**g.** Seeds, side view and view of hilum. All drawings made from living material grown in greenhouse at Mayagüez from seed of *Buhrow DRGR* (TARS #332) collected in the Dragoon Mts., Arizona, USA.

Madre Occidental, 2200–2400 m, 16–17 Sep 1934, *Pennell 18932* (MEXU, US); Mpio de Madera, camino Nicolas Bravo, Las Varas, 2150 m, 18 Aug 1994, *Quintana et al. 3064* (BRIT); 16 km W of Cuauhtémoc, (28°25'N, 107°20'W), 2074 m, 30 Aug 1965, *Rzedowski 9727* (in part=A) (ASU); Mpio. Casas Grandes, El Pinal, 1 km al SW de El Ejido Hernández, (30°30'N, 107°55'W), 2100 m, 26 Sep 1982, *Tenorio et al. 1748* (TEX-LL); Colonia García, Sierra Madre, 2500 m, 3 Jul 1859, *Townsend et al. 80* (BM, F, G, GH, K, MICH, MO, UC, US); Cerro Del Nido complex, 7.5 mi W of Bella Vista, Mesa La Boquilla, 20–25 mi NW of Chihuahua, 29°4'30"N, 106°29'30"W, 2020 m, 15 Jul 1981, *Worthington 7321* (ARIZ)); Mpio. Guerrero, Km 246 carr. Tomochí-Basaseachic, 2100 m, 5 Aug 1994, *Yen et al. 2806* (BRIT); Mpio. Cuauhtémoc, Km 123 carr. Cuauhtémoc-La Junta, 2050 m, 12 Aug 1994, *Yen et al. 2952* (BRIT); Km 8–9 carr. San Juanito-Bocoyna, 2250 m, 12 Aug 1994, *Yen et al. 2966* (BRIT). **Coahuila:** E of Rancho El Caballo, Sierra del Jardín, 29°3'N, 102°37'W, 2250–2450 m, 16 Sep 1972, *Chiang et al. 9350* (TEX-LL); Sociedad Cooperativa Minera, S of Esmeralda, Sierra Mojada, 27°16'N, 103°41'W, 1600–2463 m, 1 Sep 1972, *Chiang et al. 9078a* (TEX-LL); Hacienda of Sierra de la Madera, 35 mi W of Cuatro Cienegas, 27°5'N, 102°25'W, 1829 m, 6 Aug 1973, *Henrickson et al. 12004* (TEX(2)); Rancho El Almagre, Sierra Almagre, Coahuila-Chihuahuan Boundary, (27°30'N, 103°50'W), 11 Sep 1940, *Johnston et al. 1153* (GH, TEX-LL); Campo Cinco on Madera del Carmen Road, Sierra del Carmen, 2400 m, 11 Aug 1981, *Lowry et al. 3142* (MO, TEX); Monclova, Gloria Mts., (26°50'N, 101°10'W), 4 Aug 1939, *Marsh 1935* (in part) (F); Tinaja Blanca, 6 km W of Santa Elena Mines, Sierra de las Cruces, 31 Oct 1940, *Stewart 339* (GH); Cañon del Gringo, 6 km due N of the town of Sierra Mojada, Sierra Planchada, (27°30'N, 103°45'W), 4 Aug 1941, *Stewart 1026* (GH); San Salvador Mine, near Esmeralda, Cañon de Hidalgo, Sierra Mojada, 4 Aug 1941, *Stewart 1056* (GH); Picacho de San José, Santa Elena Mines, Sierra de las Cruces, 9 Aug 1941, *Stewart 1114* (GH); 2 km up Cañon de San Enrique, 5 km W of Rancho Buena Vista, Sierra de la Encantada, 4 Sep 1941, *Stewart 1389* (GH, TEX-LL); Cañon del Milagro, 12 km W of Hacienda de la Encantada, Sierra de los Guajes, 20 Sep 1941, *Stewart 1535* (GH); Cañon de Ibarra, Sierra del Pino, 22 Sep 1941, *Stewart 1820* (GH); Campo Tres, Sierra Madre del Carmen, 2600 m, 6 Aug 1974, *Wendt et al. 518B* (TEX). **Durango:** Mpio. Santiago Papasquiaro, 3.3 mi by Topia Road N of crest of sierra, 20 air km WNW Santiago Papasquiaro, Sierra Madre Occidental, 25°5'N, 105°48'W, 2400 m, 25 Aug 1983, *Corral et al. 643* (TEX(2), UCR); Mpio. El Ojito, 17 km después de El Ojito camino a Guadalupe y Calvo, 26°37'N, 106°03'W, 2580 m, 1 Oct 1987, *Debouck et al. 282* (CHAPA, COL, K, US); Mpio. Durango, ESE de Minerva, Sierra El Registro, 23°58'N, 104°25'W, 2340 m, 10 Oct 1978, *Debouck 325* (CHAPA, K, US); Canatlán, Venustiano Carranza, cerro El Sombreretillo, 24°28'N, 104°36'W, 2230 m, 26 Oct 1978, *Debouck et al. 396* (CHAPA, COL, K); Tepehuanes, (25°30'N, 105°45'W), 2640 m, 30 Jul 1944, *Fisher 44272* (MO). **San Luís Potosí:** Charcas, (23°15'N, 101°10'W), Jul–Aug 1934, *Lundell 5521* (in part) (ARIZ). **Sonora:** El Tigre Canyon and Mts. above El Tigre Mine, E of Esqueda and Lago Angostura, (30°30'N, 109°5'W), 1676–2286 m, 9 Sep 1961, *Mason et al. 2125* (ARIZ); San José Mts., 2134 m, 3 Aug 1893, *Mearns 1626* (US); Cananea, (30°50'N, 110°30'W), 20 Aug–1 Sep 1914, *Murdock s.n.* (F); Puerto de los Aserraderos, Río Bavispe, 4–9 Aug 1940, *White 3148* (GH, MICH); Las Tierritas de El Temblor, Río Bavispe, Sierra de El Tigre, 18–24 Aug 1940, *White 3419* (ARIZ, GH, MICH). **Zacatecas:** La Bufa, 12 Aug 1948, *Dressler 204* (MO); Paso de la Troje, Km 36 SW of Ojuelos on road to Aguascalientes, Cerro La Campaña, (22°30'N, 102°10'W), 9–12 Aug 1958, *McVaugh 16760* (CAS, MICH); 18 km W of Concepción del Oro on Coahuila-Zacatecas border, 24°54'N, 101°45'W, 2850 m, 22 Jul 1941, *Stanford et al. 563* (ARIZ, MO).

UNITED STATES. Arizona: Cochise Co.: Chiricahua Mts., Onion Saddle Road near E Turkey Creek, (32°N, 109°15'W), 1981 m, 14 Aug 1964, *Barr 64-411* (ARIZ); Cooper Peak, Dos Cabezas Mts., (32°15'N, 109°30'W), 2134 m, 21 Aug 1976, *Bingham 2535* (ASU); Dragoon Mts., (32°10'N, 110°W)?, Aug 1979, *Buhrow 141* (ARIZ); Chiracahua Mts., S Fork Cave Creek near Southwestern Research Station (Amer. Museum), 16 Sep 1966, *Clarke et al. 660916-56* (UCR); Dragoon Mts., Stronghold Canyon E along USFS Trail 279, (32°N, 109°50'W)?, 1524 m, 7 Sep 1983, *Daniel 3031* (ASU); Huachuca Mts., Reef Mine, Carr Canyon, (32°40'N, 110°15'W), 1981 m, 9 Sep 1944, *Darrow et al. 1420* (ARIZ, UC); Chiricahua National Monument area, 1 mi SW of Paradise, 20 Sep 1958, *Fearing 2069* (TEX); Huachuca Mts., The Reef, 26 Sep 1949, *Goodding 753-49* (CAS); Johnston's Ranch (9 mi S Bisbee), near Monument No 88, Mexican boundary line, (31°25'N, 109°50'W), 12 Aug 1893, *Mearns 1806* (US); Chiricahua Mts., Pinery Canyon, 831 m, 9 Aug 1974, *Reeves 963* (ASU); Chiricahua Mts., W of Rodeo, 15 Aug 1967, *Turner 5723* (TEX); Huachuca Mts., Upper Carr Canyon, 22 Aug 1975, *Van Devender et al. s.n.* (ARIZ). **Coconino Co.:** Rim Road at Fulton's Point, (36°N, 112°W), 2316 m, 14 Sep 1964, *Lehto 4201* (ASU); Oak Creek, S of Flagstaff, (35°N, 112°W), 30 Jul 1891, *McDougal 475* (US); Mogollon Rim, 16.1 mi WNW of Hwy 260, 34°23'N, 111°7'W, 2286 m, 24 Aug 1986, *McLaughlin et al. 3816* (ARIZ); Walnut Canyon, Flagstaff, (35°10'N, 111°20'W), Oct 1912, *Thornber s.n.* (ARIZ). **Gila Co.:** Rim above Young, 18 Aug 1938, *Goodding et al. 343* (ARIZ); Cherry Creek, 20 mi N of Young, (34°15'N, 110°45'W), 1737 m, Aug 1950, *Weber s.n.* (DES). **Graham Co.:** Santa Teresa Mts., 21 Aug 1981, *Buhrow 163* (ARIZ); Pinaleno Mts., Shake Spring trail, 2210 m, 18 Aug 1982, *Buhrow 172* (ARIZ); Mt. Graham, (32°45'N, 109°50'W), 2591 m, 12 Aug 1934, *Kearney et al. 9865* (ARIZ, CAS, US). (**Greenlee Co.**), Milepost 179 N of Morenci on US 666, 2103 m, 7 Aug 1978, *Buhrow 13* (ARIZ). **Greenlee Co.:** Apache National Forest, 8 mi N of S entrance on Hwy 666, (33°15'N, 109°15'W)?, 11 Aug 1986, *Landrum et al. 5213* (ASU); grown in a UCR greenhouse from seed (collected at Greenlee Co., 19 mi from Luna on the road to Blue, (33°45'N, 109°5'W)?, 11 Sep 1976, *Waines L284*), 20 May 1982, *Sanders et al. 2450* (UCR); 40 mi N of Clifton along Hwy 666, (33°25'N, 109°20'W), 13 Aug 1967, *Turner 5710* (BRIT, TEX). **Mohave Co.:** Pine Valley, 1892, *Toumey 565* (US). **Pima Co.:** Summit of Mt. Bigelow, Sta. Catalina Mts., (32°30'N, 110°40'W), 2591 m, 23 Aug 1939, *Benson 9706* (ARIZ, CAS-DS); Santa Catalina Mts., Oracle Ridge Trailhead, (32°40'N, 110°45'W)?, 2408 m, 3 Aug 1978 *Buhrow s.n.* (ARIZ(2)); W side San Luís Mts., (31°30'N, 109°5'W)?, 1753 m, 2 Oct 1893, *Mearns 2534* (US). **Pinal Co.:** 1/2 mi W of Chuichu on Hwy 93, then SSE 7.9 mi on PMC Mine-White Horse Pass Road to Game & Fish Water-Catchment Tank #632, 560 m, 18 Dec 1971, *Oxford 357* (ASU, ENCB). **Santa Cruz Co.:** Santa Rita Mts., Trail from Madera Canyon to Mt. Wrightson (Baldy), (31°40'N, 110°45'W), 1890 m, 12 Aug 1945, *Parker 5828* (ARIZ). **Yavapai Co.:** Prescott Pines Baptist Camp, (34°30'N, 112°30'W)?, 1829 m, 8 Oct 1966, *Keil 742* (ASU). **Unknown County:** White Mts., 6–15 Aug 1903, *Griffiths 5364b* (MO, US(2)); Elder Mts., May–Oct 1902, *Purpus 33* (MO, UC,US); White Mts., summer 1938, *Schroeder s.n.* (ARIZ). **New México:** Socorro (**Catron**)? **Co.:** Mogollon Mt., Mogollon Creek, (33°20'N, 108°45'W), 2438 m, 18 Jul 1903, *Metcalfe 259* (ARIZ, CAS, CAS-DS, G, GH, K, MO, UC, US). **Doña Ana Co.:** Valley of Río Grande, Dona Ana, (32°25'N, 106°45'W), *Bigelow et al. 207* (US). **Grant Co.:** Forest Nursery, Fort Bayard Watershed, 1981 m, 20 Aug 1905, *Blumer 9* (GH); Pinos Altos Mt., Aug–Sep 1880, *Greene s.n.* (F, K, MO); Gila

Nat. Forest, 15 mi N of Silver City, (32°45'N, 108°5'W), 2134 m, 30 Jul 1938, *Hitchcock et al. 4380* (CAS, CAS-DS, UC); Hanover Mts., 10 mi of GOS Ranch house, 29 Jul 1911, *Holzinger s.n.* (US); Signal Peak, 2103 m, 10 Sep 1972, *Klue 21* (ARIZ); Mangas Springs, Aug 1901, *Metcalfe s.n.* (ARIZ, US); Mogollon Mts., 11 Aug 1881, *Rusby 1071/2* (F, MICH, US); Pinos Altos Mts., 0.2 mi S of the Gila National Forest boundary, 8 mi N of Silver City and 1.6 mi N of Pinos Altos on Rt 15, 2073 m, 8 Oct 1982, *Sanders et al. 3177* (UCR). **Hidalgo Co.:** San Luís Mts., (31°26'N, 108°35'W)?, 1280 m, 31 Jul 1940, *Hershey 2133* (ARIZ); Coronado National Forest, Peloncillo Mts., Upper Clanton Draw, (32°30'N, 108°55'W)?, 1646 m, 3 Oct 1982, *Sanders et al. 2987* (UCR); Peloncillo Mts., Road to Douglas from Animas Valley between the headwaters of Cottonwood Creek and Clanton Draw, 1829 m, 3 Oct 1982, *Sanders et al. 3047* (UCR). **Socorro Co.:** Gila Forest, White Creek Ranger Station, 2540 m, 1 Aug 1920, *Eggleston 16891* (US). **Unknown County:** Gila Forest, Burro Mts., W. C. Selby's ranch, (32°30'N, 108°25'W), 1880–2180 m, 22 Oct 1919, *Eggleston 16450* (US); Animas Peak, 21 Sep 1939, *Gooddinget al. A9771* (ARIZ, ASU); Piney Gohe, Aug 1851, *Thurber 1124* (F, GH(2)); (sides of stony hills, at the copper mines, Aug) 1851, *Wright 952* (GH(2), K, US). **Texas: Brewster Co.:** Big Bend National Park, Lost Mine Peak, 2225 m, 12 Sep 1961, *Correll et al. 24550* (TEX-LL); Juniper Canyon, 16 mi NNE of Castolon, 9 May 1937, *Cutler 992*, (MO); Chisos Mts., Big Bend National Park, between Boot Springs and pass overlooking Inner Basin on Juniper Flat Trail, 2134 m, 1 Aug 1959, *Kruckeberg 4780* (MICH); Chicos area, Aug 1935, *Marsh 2093* (US); Pine Canyon, 24 Sep 1977, *Powell et al. 3235* (TEX-LL); Chisos Mts., Pine Canyon, 29°16'N, 103°14'W, 1646 m, 6 Oct 1982, *Sanders et al. 3148* (TEX); BBNP, Gulliam Peak and top of Green Gulch, 1615–1768 m, 24 Aug 1970, *Semple 414* (MO); Chisos Mts., Green Gulch and Blue Creek canyon, 7 Sep 1936, *Sperry 441* (US); Upper Green Gulch of Chisos Mts., 31 Aug 1938, *Warnock C642* (ARIZ, GH, TEX, US); Chisos Mts., Canyon, 6 Sep 1915, *Young 124* (MO). **El Paso Co.:** Franklin Mts., El Paso, Altura Park, (31°45'N, 106°15'W), 10 Jul 1911, *Barlow s.n.* (in part) (F). **Jeff Davis Co.:** Davis Mts., Sawtooth Mt., (30°30'N, 103°45'W), 11 Sep 1967, *Correll 34980* (GH, TEX); Davis Mts., Tricky Gap, Buffalo Trail Scout ranch, 1676 m, 8 Aug 1948, *Warnock et al. 8086* (BRIT, TEX-LL).

Habitat.—This species is found growing under or in open rocky or grassy areas in oak or pine-oak forests at high altitudes, mostly on the dry side, and is associated with *Acacia, Arbutus, Cercocarpus, Cupressus, Dasylirion, Fouquieria, Fraxinus, Garrya, Juniperus, Platanus, Rhus, Ungnadia, Yucca,* and *Zaluzania*. Generally growing prostrate on the forest floor and seldom climbing up the brush, grass or small trees, it often forms dense stands, which in the more open places completely cover the ground with vines (see also Buhrow 1983). Soils are usually sandy and gravelly from limestone, granite, or igneous rocks.

Common names.—It has been reported as the "Silverspot Lima bean" by Niehaus et al. (1984) for the Southwestern US and Texas.

Comments.—Delgado (1985) considers *P. grayanus* a variety of *P. pedicellatus*. As did Lackey (1983), we see great differences between these two taxa. All *P. pedicellatus* seem to be vining and climbing plants with only slightly lobed leaves, growing at mid-elevations and with a more southern distribution in central México, while *P. grayanus* is definitely a trailing or prostrate vine with only slight tendencies for climbing, growing at high altitudes and with a more northern distribution in northern México and the southwestern USA (see also Buhrow 1983). Moreover, the leaflets of *P. grayanus* are more coriaceous than in *P. pedicellatus* and, if lobed, the lobes are more or less pointed. Additionally, Delgado (1985) places *P. palmeri* and *P. foliaceus* as synonyms of *P. grayanus*. We agree with the latter, but the former is definitely distinct. On the other hand, *P. grayanus* has often been confused with *P. filiformis* (better say with *P. wrightii*, now synonym of *P. filiformis*, see above), due to some similarity in leaflet shape of poor specimens. The leaves of *P. grayanus* are perhaps the most variably lobed of all the species of this section and usually are more entire at the base of the plant and more deeply lobed at mid- and upper-portions of the vines, and the lobes are always pointed. Variegation is often seen. But pods and seeds are very different when compared with those of *P. filiformis*. In addition, *P. grayanus* is often a much larger plant, with stronger sprawling guides.

Recent work using PCR-RFLPs on cpDNA (Fofana et al. 1999) has also shown that *P. pedicellatus* and *P. grayanus* although distinct are related; this has been confirmed by allozyme data (Jaaska 1996), and ITS DNA sequencing (Delgado et al. 1999; Gaitán et al. 2000).

M.8.—Phaseolus scabrellus Benth. ex S. Wats., Proc. Amer. Acad. Arts. 17:346–347. 1882. (**Fig. 92**). TYPE: MÉXICO. SONORA: Sonora Alta, (1854), *Coulter s.n.* (HOLOTYPE: K, GH (photo K); ISOTYPE: GH).

Aerial shoot an annual, medium sized, delicate, indeterminate vine. ***Root*** unknown. ***Stem*** terete, 1 mm thick, somewhat striate, covered with reflexed-hispidulose hairs to glabrous; internodes 2.5 to 16 cm long. ***Stipules*** oblong, 3 mm long, 1.5 mm wide, acute, heavily 3- to 8-nerved, puberulent. ***Leaves*** 7.3–9.3 cm long; petiole 3–3.5 cm long; petiolule 8–12 mm long, slightly ribbed, hispidulose; pulvini 2.5 mm long, darkened, ridged, covered by hispid and uncinate hairs; stipels, the lower linear, 2 mm long, heavily 1-nerved, the upper broader; terminal leaflets nearly lanceolate, distinctly 3-lobed, 3–4 cm long, 2–3 cm wide at base, the lobes squarish to rounded, 7–10 mm in size, slightly apiculate, somewhat coriaceous, sparsely covered with hispid and somewhat scabrous-glandular hairs and darker green with slight variegation adaxially, lighter green abaxially, ciliate on margin, indistinctly veined, the ultimate veinlets darkened; lateral leaflet similar and slightly lobed on each side at base. ***Inflorescence*** a curving pseudoraceme; peduncle 5–9 cm long, somewhat striate and glabrous; rachis mostly 4–5–15 cm long, to 9 flowering nodes, nearly glabrous; primary bract lanceolate, 3 mm long, 1 mm wide, heavily 3-nerved, puberulent; pedicel very delicate, 9–12 long, glabrous; pedicellar bracts linear aciculate 1 mm long hyaline glabrous. ***Bracteoles*** lanceolate, 1.25 mm long, 0.3 mm wide, heavily 1-nerved, acute, glabrous, early caducous. ***Flower*** purple; calyx campanulate, 4–5 mm long, the upper 2 lobes joined into one emarginate, 4 mm wide, ciliate, the lower lobes acute, subequal, 1 mm long, 2 mm wide, ciliate, the center sparsely covered with strigose hairs; standard purple, large, rounded, sharply reflexed at 3 mm from the base, 4–5 mm more to apex, 12–15 mm wide, the lower lobes well-developed, the claw 1/2 mm long; wings purple, the blade, cupped, 12 mm long, 5–6 mm wide, unequal with one larger than the other, the claw 4 mm long, 0.5 mm wide, the well-developed spur 0.75 mm wide and adhering to keel; keel 5 mm long to bend and 3 more to base of the terminal 1 1/2 coils of 2.25 mm diam., the tip whitish, the claws separated for 2 mm, the lateral knobs 0.75 mm wide; vexillary stamen short, with a well-developed knob 0.4 mm from base and extending forward as a hooded sheath about 1 mm, the thickened portion extending forward about 2 mm from base; stamen tube united 8 mm long, with the two small knobs about 1 mm from base; anthers 0.7 mm long, 0.4 mm wide, yellow; ovary 3.5 mm long, 0.75 mm wide, densely covered with long white canescent hairs, 2–3 ovules; style 3.5 mm long to base of the terminal thickened coil of about 1.5 mm diam.; stigma lateral, introrse, 0.75 mm long, linear and wide. ***Pod*** falcate, 3.5 cm long, 0.8 mm wide. ***Seed*** unknown. ***Seedling*** unknown.

Specimens examined: **MÉXICO. Durango:** 68.2 mi E of Villa Unión, 10.1 mi E of Sinaloa-Durango border on Hwy 40, (23°45'N, 105°50'W), 671 m, 3 Aug 1983, *Buhrow et al. M8* (ARIZ); 40 mi SW of El Salto at Arroyo De Agua, Sierra Madre Occidental, 23°45'N, 105°24'W, 2438 m, 20 Aug 1982, *Worthington 8926* (TEX, UCR). **Sinaloa:** Ocurahuí, Sierra Surotato, (26°N, 107°40'W), 1980–2310 m, 1–10 Sep 1941, *Gentry 6322* (CAS-DS, F, GH, MICH, MO).

Habitat.—This species grows in moist, shady canyon slopes of pine forest at high altitude and in rocky soil.

Comments.—In the original description, Watson (1882) cites two collections for this species: *Palmer 2122* and *Coulter s.n.* (in Gray herbarium). However, according to Piper (1926) the former is *P. pedicellatus* due to the 3-nerved bracteoles while the latter is *P. scabrellus* with 1-nerved bracteoles. Piper had carefully dissected a flower of *Coulter s.n.* (GH), a specimen the senior author finds carries a label, apparently handwritten by Bentham, which reads "*P. scabrellus* n.sp." A similar label is also on the specimen at K. The label on the sheet of *Palmer 2122* is in a different hand, probably that of Palmer; at all events the senior author finds this sheet to be of *P. polymorphus* (not *P. pedicellatus* as indicated by Piper), a species also described by Watson in the same publication and from SLP and which ranges in the central highlands of México from Monclova in the N, to San Luís Potosí in the S, E to Saltillo and W to Torreón. Other specimens cited by Piper (1926) as being of *P. scabrellus*, the senior author finds to be of other species.

M.9.—Phaseolus teulensis Freytag, sp. nov. (**Fig. 92**). TYPE: MÉXICO. ZACATECAS: Cienega San Pedro, 8 km SSW of Jiménez de Teul, 23°20'N, 103°55'W, 2400 m, 19 Oct 1978, *Debouck 380* (HOLOTYPE: US; ISOTYPES: CHAPA, COL, G, K).

Smilis *Phaseolo pedicellato*, sed volubiliibus longis gracilibus scandens, foliolis lanceolatis ad basim lobatam profunde, seminibus lenticularibus fuscis vittatis atris differt. Crescit modo in locis regionis occidentali Zacatecum rarus.

Similar to *P. pedicellatus* except: ***Plant*** a slender, climbing, indeterminate vine. ***Root*** thick, fleshy, (immature) clavate. ***Stems*** terete, striate, puberulent; internodes short, 2.5–5 cm long. ***Stipules*** oblong-lanceolate, 3.5 mm long, 0.75 mm wide, 3-nerved, hyaline margin, ciliate. ***Leaves*** 5–6.4 cm long, with deeply lobed leaflets; petioles 1.5–2.5 cm long, slender; petiolule 1 cm long; stipels minute, linear, upper 1 mm long, 1-nerved, hyaline, the lower 0.5 mm long; pulvini 2 mm long, covered by minute white hooked hairs; terminal leaflet deeply 3-lobed, the center one lanceolate, 2–2.5 cm long, 1–2 cm wide, lateral lobes rounded, 0.8–1 cm long, slightly obtuse, apiculate, adaxial surface dark green, slightly variegated along midvein, paler abaxially, puberulent, ciliate on margins. ***Inflorescence*** a long, slender pseudoraceme; peduncle, 9–12 cm long, glabrous; rachis 4–5 cm long of 2–6 flowers; primary bract ovate, 2.5 mm long, 1 mm wide, 1-nerved, glabrous; pedicel slender, 1 cm long at pod, covered with minute uncinate hairs. ***Flower*** unknown. ***Pod*** short and broad, nearly straight, 3–3.5 cm long, 0.8 cm wide, valves reticulate, sparsely long white strigose, the sutures somewhat thickened, blunt ends; beak short, straight, 1 mm long, 3–4 seed. ***Seed*** lenticular, 5.6–6.5 mm long, 5–5.5 mm wide, 1.8–1.9 mm thick, brown and black on black, black ring around hilum. ***Seedling*** from hypogeal germination; epicotyl 7–8 cm long, slender; eophyll, the stipule at eophyll node lanceolate, 3 mm long, the petiole 13 mm long, the stipels minute, the blade cordate at base, ovate, acuminate, 3.5–4 cm long, 3.3–3.5 cm wide at about 1/4 from base, obtuse.

PARATYPE. **MÉXICO. Durango:** Cerro SE de Medina, Canatlán, 2275 m, 24°34'N, 104°36'W, 30 Sep 1978, *Debouck et al.* 277 (CHAPA, K, US).

Habitat.—This species is an understory vine climbing on *Arctostaphylos* in open parks in oak forest mixed with *Acacia* and *Agave*. It is also in semi-shady places on soil derived from igneous rocks, with good drainage and abundant organic matter.

Diseases and pests.—*Alternaria*, anthracnose and rust are reported, as well as pod weevil.

Comments.—This species has peculiar lens-shaped, brown and black pinto seed, and comes from an area apparently not collected previously. In a study using ITS DNA sequencing, Gaitán et al. (2000) have found this taxon to be close to *P. pedicellatus*, *P. glabellus*, *P. grayanus*, *P. scabrellus*, *P. polymorphus*, and separating from them at about the same level.

M.10.—Phaseolus pyramidalis Freytag, sp. nov. (**Figs. 91, 92**). TYPE: MÉXICO. CHIHUAHUA: El Cima, (28°25'N, 107°5'W)?, 29 Jun 1936, *Le Sueur 713* (in part) (HOLOTYPE: GH).

Similis *Phaseolo pedicellato*, sed caulibus strictis erectis, foliolis omnibus 3–5 lobatis, inflorescentiis compactis et pyramidalibus in alibastrum. Crecit modo in locis regionis occidentali centrali Chihuahuae rarus.

Similar to *P. pedicellatus* except: ***Aerial shoot*** a scrambling, prostrate, indeterminate vine. ***Root*** unknown. ***Stems*** straight, stiff and erect to trailing; internodes to 6 cm long. ***Stipules*** very large and spreading, triangular, 7–9 mm long, acute. ***Leaves*** 4.6–6 cm long, variously 3–5 lobed as in *P. grayanus*, but with at least some lateral lobes rounded, olive green to grayish; petiole 1.5–2 cm long; petiolule 6–8 mm long; terminal leaflet 2.5–3.2 cm long, 2–3.5 cm wide, dark reticulate veined, farinose to scabrous above, covered with scattered strigose hairs especially on veins adaxially, more so abaxially, ciliate margins. ***Inflorescence*** an erect pseudoraceme; peduncles 14 cm long, striate, covered with uncinate hairs; rachis 5–15 cm long, of many nodes, densely covered with uncinate hairs, the terminal portion with dense buds forming a conical or pyramidal top; bract broadly ovate-acuminate, 3–3.5 mm long, 1.6 mm wide, lightly 5- to 6-nerved, covered with white, strigose hairs, pedicel 5 mm long, densely covered with uncinate hairs. ***Bracteoles*** lanceolate, 1 mm long, 0.4 mm wide, lightly 1-nerved, ciliate, hyaline. ***Flower*** white or light pink(?); calyx tube 2 mm long, upper lobes rounded, 0.5 mm long, 1.25 mm wide, lower lobes, subequal, dentate, acute, 1 mm long, 1.25 mm wide, densely covered by strigose hairs adaxially, a few hairs on abaxial surface; standard 2.5 mm to flexure, 5 mm more to apex, 8 mm wide, emarginate. ***Pod*** unknown. ***Seed*** unknown. ***Seedling*** unknown.

PARATYPE. **MÉXICO. Chihuahua:** 30 mi W of the La Junta Jct. on Hwy 16, (28°55'N, 107°45'W)?, 2118 m, 19 Jul 1975, *Ellis et al. 1040* (TEX-LL).

Habitat.—This taxon is found growing in oak, pine and madrone forests.

Comments.—It is named for the pyramidal shape of the bud clusters on the young inflorescences, an unusual characteristic for the species of this section. It is very rare and found only near the summits of the mountains west of Cd. Chihuahua.

M.11.—Phaseolus laxiflorus Piper, Contr. U.S. Natl. Herb. 22:692. 1926. (**Fig. 92**). TYPE: MÉXICO. HIDALGO: near Honey Station (20°14'N, 98°14'W), Trinidad Iron Works, 21 Aug 1905, *Pringle 13690* (HOLOTYPE: US 462384; ISOTYPES: GH, MICH).

Aerial shoot an herbaceous, annual, delicate, climbing, indeterminate vine, to 3 m long. ***Root*** apparently perennial, thick, fleshy. ***Stems*** slender, terete, 1.25–1.5 mm thick, covered by retrorse-strigillose

FIG. 91. Illustrations of *Phaseolus pyramidalis* Freytag.–**c.** Portion of basal stem with mature leaves and inflorescence and separately the vine tip; note the deeply lobed leaflets and the pyramidal shape of tip of inflorescence. Drawing made from the type collection *Le Sueur 713* from near El Cima, Chihuahua, México.

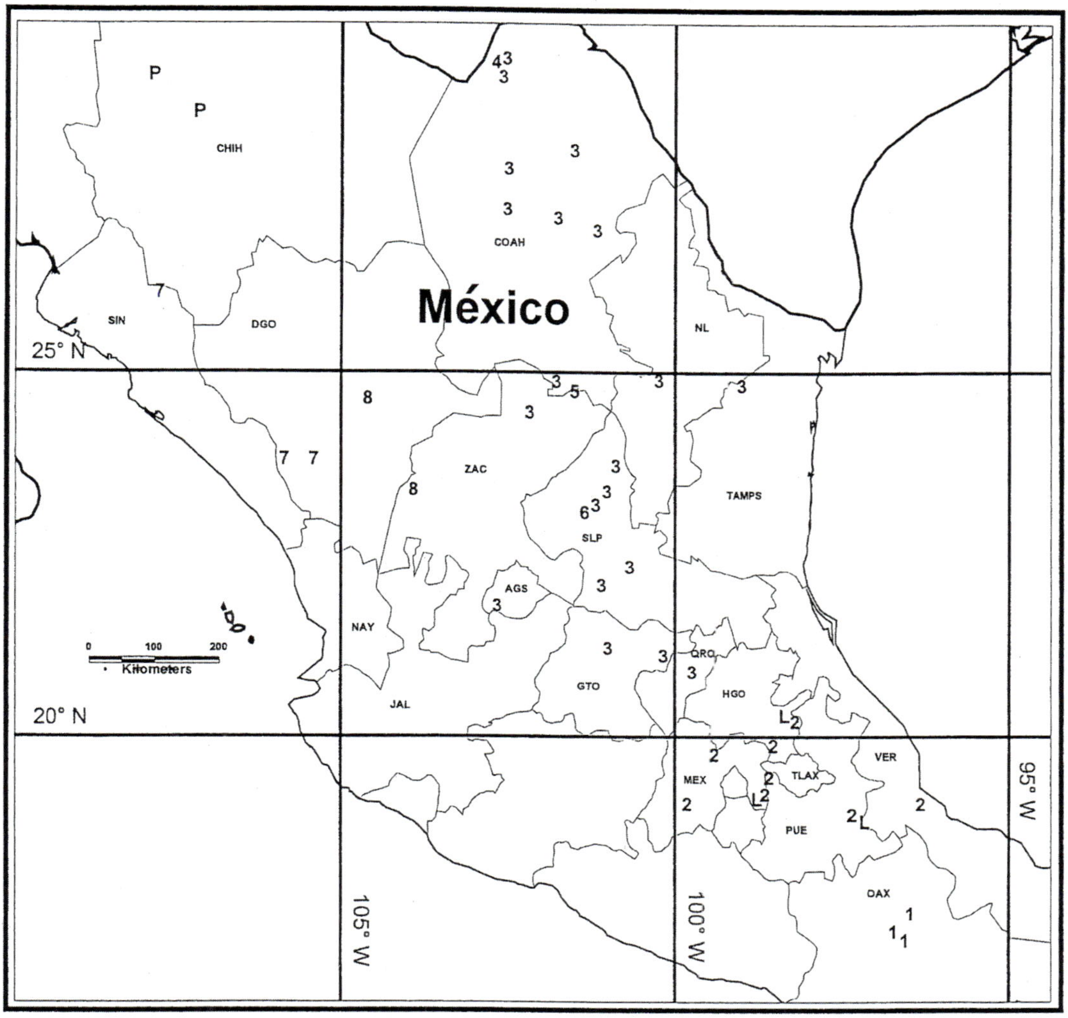

Fig. 92. Distribution of minor species of Section M. *Pedicellati*, as follows: **1** = *P. oaxacanus*; **2** = *P. esperanzae*; **3** = *P. polymorphus* var. *polymorphus*; **4** = *P. polymorphus* var. *albus*; **5** = *P. palmeri*; **6** = *P. purpusii*; **7** = *P. scabrellus*; **8** = *P. teulensis*; P = *P. pyramidalis*; **L** = *P. laxiflorus*.

hairs; internodes, maximum length 9.5–10.5 cm on upper mid-portions, with adventitious roots from sub-surface nodes, very nearly glabrous, few lateral branches, sometimes a witch's broom effect of many short (to 5–8 cm) fascicles of branches (virus or mycoplasma?). ***Stipules*** triangular, 3–5 mm long, 2 mm wide, foliaceous, distinctly 5-nerved, sessile, extended at 90°angle, sparsely covered by hirsute hairs, sparsely ciliate at margin. ***Leaves*** 12–18 cm long, indistinctly lobed; petiole 5–8.5 cm long, glabrous; petiolule 2.1 cm long, glabrous; stipels spatulate, 2 mm long, 0.75 mm wide, 1- to 2-nerved, purplish; basal pulvini 5 mm long, glabrous, lateral and terminal pulvini 2.5 mm long, glabrous; terminal leaflet elongate triangular-ovate, to 7 cm long, 6.3 cm wide at 1/3 from base, rounded or truncate at base, frequently obscurely lobed, slightly acuminate to about 4 mm long, apiculate, very thin membranous, nearly glabrous to very sparsely covered with hirsute hairs mostly on veins, the hairs to nearly 0.5 mm long; lateral leaflets similar but inequilateral and smaller. ***Inflorescence*** mostly small pseudoracemes but some larger ones become panicles with secondary branches of 1 to 6 nodes, often with the 2 lateral flowers on long pedicels and the third, centrally located flower on the short lateral raceme axis 2–6 mm long; peduncle 3.5–4–17.5 cm long, nearly glabrous to slightly covered with hirsute hairs, ciliate of very small hooked hairs; rachis 3–10 cm long with 2–10 flowering nodes and bearing from 20–25 flowers; primary bract oblong-ovate, 4–5 mm long, 1.5 mm wide, slightly 3- to 5-nerved, sparsely covered with hirsute hairs; pedicel delicate, 10–18 mm long at anthesis, nearly glabrous or covered with minute hooked hairs; pedicellar bracts narrowly lanceolate 1 mm long scale-like hyaline not nerved. ***Bracteoles*** ovate, sessile or sometimes alternate or sub-opposite on the pedicel and inserted 1.5–4 mm below calyx, 0.75 mm long, 0.5 mm wide, hyaline and ciliate, caducous shortly after anthesis, not nerved. ***Flower*** light purple, often relatively few on each plant; calyx campanulate, 4 mm long, the 2 upper teeth rounded and truncate at tip, 0.5–1.5 mm long, 3 mm wide, with a few scattered hairs and somewhat ciliate at margin, the 3 ventral teeth with the central 1.25 mm long, 2.5 mm wide, the lateral 1.25 mm long, 2 mm wide, rounded dentate, slightly more pubescent than upper teeth; standard light purple, reflexed at midpoint and hooded over keel, the upper portion mostly erect, broadly spreading laterally, the base somewhat restricted, the auricles inconspicuous; wings light purple, the blade nearly flat to somewhat cupped, 14 mm long, 11–12 mm wide, spreading, the basal claw 5–6 mm long, 0.5–0.75 mm wide, the spur orbicular 1 mm long, nearly 0.5–1 mm wide, quite firmly adhering to staminal tube; keel with 1 1/2 coils of 2 mm diam., the basal claws 5–6 mm long, the lateral knobs raised slightly less than 0.5 mm; vexillary stamen 10 mm long, the knob 0.75 mm from base, 0.5 mm high; stamen tube 9 mm long, reflexed at 7 mm, the knobs at 1.5 mm from base only slightly developed, the filaments 10+ mm long; anthers orbicular, 0.65 mm long, 0.35 mm wide, yellow; basal collar 0.75 mm long; ovary 5–6 mm long, 1.25 mm wide, laterally flattened, the thickened sutures heavily covered with hirsute hairs with somewhat yellowish hairs to 0.75 mm long, pubescence sparse on sides of young pod, 4 ovules; style, the thin basal portion 9 mm long to the thickened terminal coil of 1–1.5 mm in diameter; stigma lateral, straight, 1–1.25 mm long, narrow. ***Pod*** unknown. ***Seed*** unknown. ***Seedling*** unknown.

Specimens examined: **MÉXICO. México:** Mpio. Amecameca, 1 km E de San Antonio, (19°8'N, 98°45'W), 2550 m, 31 Oct 1968, *Rzedowski 26451* (in part) (MO). **Veracruz:** summit of pass "Paso del Viento," Km 30 Hwy 150 Orizaba-Tehuacán, 18°45'N, 97°10'W, 2120 m, 19 Sep 1978, *Freytag et al. 78-Mex-29* (ARIZ, BR, BYU, CSU, F, G, GH, IBUG, K, MEXU, MICH, MO, NA, TEX, UC, UC-R, US, WIS, WYAC).

Habitat.—This species is found in widely separated locations, in deep leaf mold, under hawthorn trees and other shrubs in dense shade or on weeds in open spaces, on steep nearly 30 degree slopes, and fairly frequent, spaced 2–20 m between plants.

Comments.—Honey is the station at the end of the railroad going NE of Tulancingo and from the best maps the senior author has available (Dirección General de Geografía 1982), this station seems to be on the Hidalgo side of the border between the states of Hidalgo and Puebla. However, Delgado (1985) has pointed out that the Trinidad Iron Works is located in the state of Puebla rather than Hidalgo as was indicated on the collector's label and as given by Piper (1926).

This species is very distinct in being a very delicate vine as compared with other species of this section and is unusual in having very delicate panicles rather than pseudoracemes, the pedicels are very long and delicate, and the bracteoles are often inserted one or two mm below the calyx. It also occupies a peculiar niche in the understory of the cloud forest environment subject to almost perennial rain and moisture of the eastern escarpment of the Sierra Madre Oriental of Mexico (see Color Plate II, photo 21).

Section N.—Chiapasana (Piper) Delgado (1985). TYPE SPECIES: *Phaseolus chiapasanus* Piper, Proc. Biol. Soc. Wash. 34:41. 1921.

Plant a climbing vine to 10 m long, all parts of which turn jet black on drying; root perennial; leaflets, large, broadly ovate; flowers large, lilac to purple; pods straight, the basal half (or 1/3) sterile, 2–6 large discoidal seeds.

Comments.—This section is named after Piper's species *P. chiapasanus* which is the only species in this section. It is also the only species in the genus which turns black on drying (also noted by Lackey 1983, and Delgado 1985), thus nearly all herbarium specimens of this species have nearly jet black stems, leaves, flowers and pods.

N.1.—Phaseolus chiapasanus Piper, Proc. Biol. Soc. Wash. 34:41. 1921. (**Figs. 48, 93**). TYPE: MÉXICO. CHIAPAS: Finca Mexiquito, (15°8'N, 92°18'W), Sep 1913, *Purpus 6881* (HOLOTYPE: US 567182; ISOTYPES: F, GH, UC).

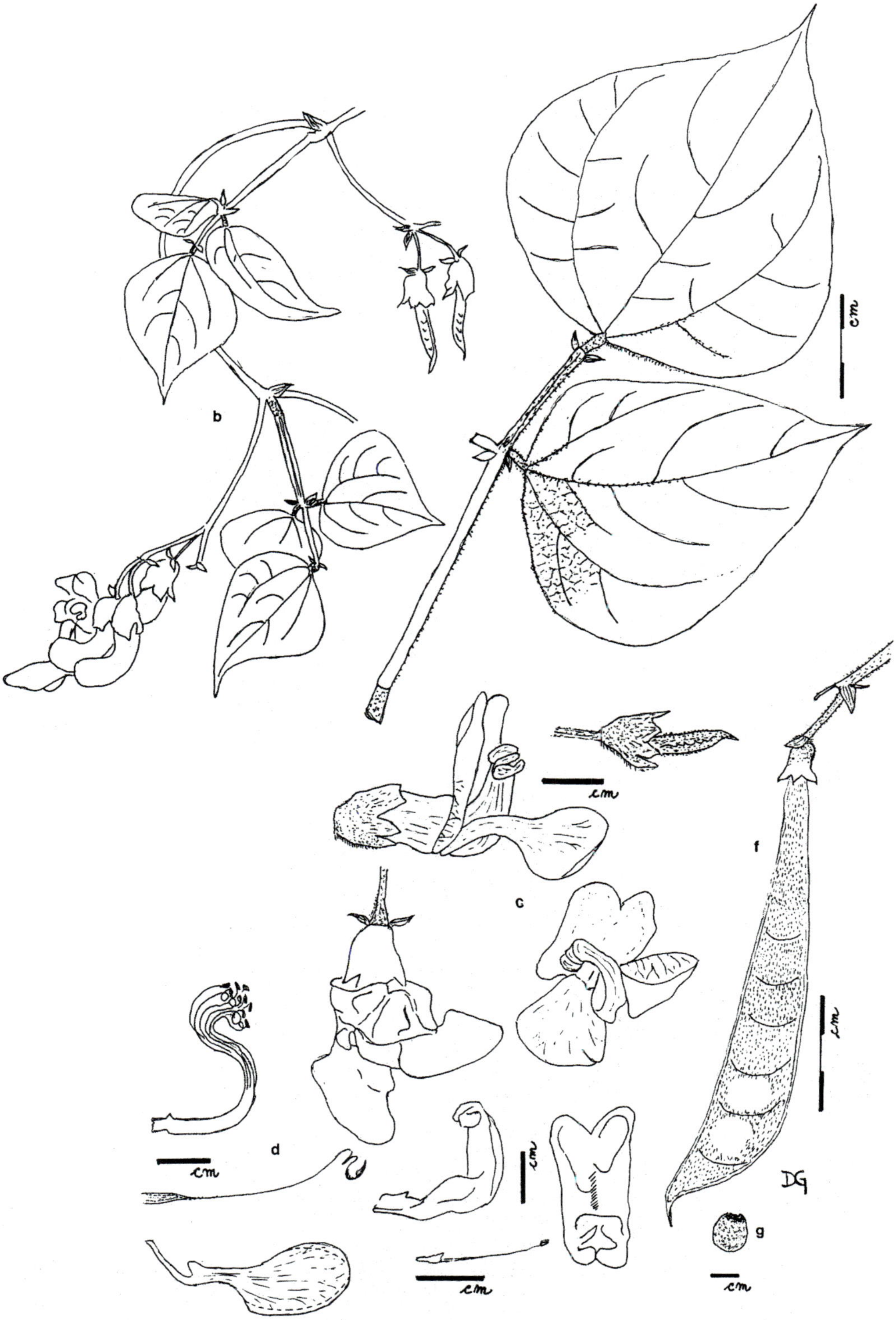

Fig. 93. Illustrations of *Phaseolus chiapasanus* Piper.—**b.** Flowering stem showing few-noded racemes.—**c.** Flower, side and front view.—**d.** Exploded view of flower showing all parts.—**f.** Pod, side view and dehisced.—**g.** Seeds. All drawings from living material grown in screenhouse in Popayán, Colombia, of seed of *Debouck et al. 2349* from near Pochutla, Oaxaca, México.

Aerial shoot a pluriannual, climbing, indeterminate vine, to 10 m long, all parts turning black on drying. ***Root*** perennial, globose, thickened, woody, to 10 cm long. ***Stems***, terete, densely covered with strigose hairs, somewhat yellowish pubescent especially on young vines, internodes 12–15–18 cm long. ***Stipules*** large, ovate, 9 mm long, 6 mm wide, multi-nerved, glabrous, appressed downward, margins heavily long strigose pubescent. ***Leaves*** 19.5–29.5 cm long, green, turning jet black on drying; petiole not canaliculate, 8–14 cm long, covered with long hooked hairs; petiolule 0.5–3.5 cm long slightly canaliculate; terminal leaflet broadly rounded-ovate to nearly triangular, 11–12 cm long, 12–15 cm wide, tip long acuminate, 7–10 mm long, apiculate, nearly glabrous except for scattered hispidulose hairs adaxially and covered with strigose pubescence mostly on veins abaxially and margins; lateral leaflet similar but slightly inequilateral and broadly ovate acuminate. ***Inflorescence*** to 18 cm long; peduncle 5–10 cm long; rachis 1–8 cm long of 2–3 nodes, 6- to 8-flowered; primary bract orbicular-acuminate, 8 mm long, multinerved, glabrous adaxially, covered with strigose hairs abaxially; pedicel 8–15 mm long, 20 mm long at pod, glabrous to densely covered with long strigose and hooked pubescence; pedicellar bracts elliptic, 7–9 mm long 3 mm wide, 5-nerved, covered by long strigose hairs, somewhat persistent. ***Bracteoles*** broadly ovate to elongate-oval, 6–7 mm long, 2.5–3 mm wide, obscurely 3- to 5-nerved, densely covered by long strigose hairs, often persistent to mid pod development. ***Flower*** pink-lilac to purple; calyx flaring-campanulate, 10 mm long, becoming larger in pod, densely covered with strigose hairs, the lobes all rounded acute, subequal, 3.5 mm long, 3.5 mm wide, the lower central somewhat larger to 5 mm long, densely covered with appressed-hispid hairs, covered with long strigose hairs mostly on abaxial surface, somewhat farinaceous turning black by mid pod development; standard lilac, the blade broadly rounded, 20–25 mm long, 25 mm wide, reflexed at midpoint, the terminal portion erect, the claw 4 mm long, 1.5 mm wide, the auricles well-developed, 1 mm high, prolonged into wing-like flaps 5 mm long on abaxial surface; wings, the blade obovate and broadly rounded, 20 mm long, 24 mm wide, spreading, the right one reflexed, the narrow basal claw 8–12 mm long, the spurs well-developed, 2 mm in diam., loosely adhering to keel; keel 5 mm broad, about 20 mm to bend and 11 more to base of the terminal 2 1/2 coils of 4 mm in diam.; vexillary stamen 3.5 cm long; stamen tube long, narrow, nearly linear-tubular, 20 mm to bend and about 18 mm more to divided filaments; anthers very large, 1.5–2 mm long, 1 mm wide; basal collar 2 mm long, minutely denticulate; ovary straight, linear, 12–14 mm long, 1.5 mm wide, densely covered by hispid hairs mostly on sutures, 5–7 ovules near apex of ovary; style about 36 mm long to the terminal thickened coil of 4 mm diam., the terminal coil densely covered with cannose hairs to base of stigma; stigma lateral, introrse, linear, 1.75 mm long. ***Pod*** straight, 11–12 cm long, 1.5–2 cm wide, thickened sutures, covered with yellowish-white, short hispid hairs, the basal half (or third) of pod sterile; beak stout, slightly curved, 2–5 mm long. ***Seed*** orbicular, 1–1.3 cm long, somewhat flattened, light brown mottled dark brown; hilum oblong-linear, 4.5 mm long; lens pronounced. ***Seedling*** from hypogeal germination; epicotyl glabrescent; stipules present, divided; primary leaves opposite, simple, ovate, truncate-cordate at base, acute, the petioles with basal and apical pulvini, stipels present.

Specimens examined: **MÉXICO. Chiapas:** Mpio. of Jitotol de Zaragoza, along Río Hondo 6.5 km N of Jitotol along road to Pichucalco, (17°5'N, 92°50'W), 1700 m, 26 Sep 1971, *Breedlove 19878* (CAS-DS, MO); Mpio. of Rayón, in the Selva Negra 10 km above Rayón Mezcalapa along road to Jitotol, (17°11'N, 92°58'W), 1700 m, 10 Sep 1981, *Breedlove 52677* (CAS); Toquitlán, Siltepec, (15°34'N, 92°18'W)?, 8 Aug 1937, *Matuda 1668* (K, MEXU, MICH, US); Monte Obando, Escuintla, 14 Nov 1945, *Matuda 6174* (MEXU); Cascada, Siltepec, 3 Aug 1937, *Matuda 219* (F, MICH); Finca Las Nubes, 2 Dec 1941, *Miranda 1678* (MEXU); Mpio. de Rayón, 3 Km N de Rayón, carr. a Pichucalco, 1150 m, 9 Dec 1980, *Sousa et al. 11410* (BM, CR, MO). **Oaxaca:** Pochutla, Candelaria Loxicha, 2 km NW de La Galera, 6 km S de La Soledad, 15°58'N, 96°30'W, 1260 m, 27 Oct 1987, *Debouck et al. 2349* (BR, CHAPA, COL, MICH, SI, US); Dist. de Pochutla, a 3 km al NE de La Galera, carr. Pochutla-Miahuatlán, 1210 m, 24 Oct 1976, *Sousa et al. 6488* (ARIZ, BM, DES, MEXU, MO); Dist. de Pochutla, a 2 km al S de La Soledad, 6 Km al NW de La Galera, 1450 m, 23 Jun 1977, *Sousa et al. 7574* (MEXU).

Habitat.—This species is reported growing on steep, moist, rocky slopes of undisturbed, dense montane rain forest with a short dry season which is termed "selva mediana subperennifolia" according to Campos Villanueva et al. (1992), "bosque tropical subcaducifolio" according to Rzedowski (1978), and "semi-evergreen seasonal forest" according to Breedlove (1973). It is found associated with *Pinus, Quercus, Liquidambar, Magnolia, Podocarpus, Caslatola, Platanus* and *Ardisia*, with vines of *Ipomoea* and *Cucurbita* and found on yellow, rocky, organic, clay soils derived from metamorphic schists.

Phaseolus chiapasanus has been reported from Veracruz (Delgado 1985; a specimen, apparently from the Huatusco area, that we have not seen), Oaxaca and Chiapas. Because of its distribution in the Soconusco area on the slopes of Volcán Tacaná close to the Guatemalan border, we suspect it might be present in the Department of San Marcos of Guatemala from where it has not been reported so far (Standley & Steyermark 1946). These original forests are being quickly replaced in México and Guatemala by cocoa and coffee (at lower and higher elevations, respectively) plantations.

Common name.—Frijol de Monte; Frijolillo.

Comments.—This species is not particularly outstanding in the wild though it is a very large vine with much larger flowers and pods in comparison to other species of the genus; however, on drying it is the only species which turns jet black. Another very unique trait is the discoid shape of the seed. It must be quite rare and of low population density in the forests since very few specimens have been collected and from only a few locations. This species will probably disappear together with the original forests as we have not seen it in secondary vegetation.

Placement of this taxon in the genus is beyond doubt, even though Piper, who first named it in 1921, inexplicably did not include it in his monograph of 1926! Lackey (1983) noticed the larger size of all parts of the plant, as well as the presence of hooked hairs on the inner face of the standard, but did not make a special section for it as did Delgado (Delgado et al. 1982; Delgado 1985), in part on the basis of pollen characteristics.

The pollen of *P. chiapasanus* shares some traits, namely that of infratectum, with American *Vigna* of the subgenus *Sigmoidotropis* (Delgado Salinas et al. 1982). Studies on cpDNA restriction site variation has shown some affinity between this taxon and *P. hintonii* (Delgado Salinas et al. 1993). ITS DNA sequencing data (Gaitán et al. 2000) show *P. chiapasanus* to be related to *P. xanthotrichus* and *P. pluriflorus*, confirming thus the finding by Delgado et al. (1999) who found it related to *P. zimapanensis*, *P. xanthotrichus* and *P. hintonii*. Pollen data as well as cpDNA information and the general characteristics of the plant justify, in our view, placing this taxon in its own separate section of *Phaseolus*. To our knowledge, it has not been included in widecrossing nor agronomic evaluation trials so far.

Section O.—Coriacei Freytag, sect. nov. TYPE SPECIES: *Phaseolus maculatus* Scheele, Linnaea 2:465–466. 1848.

Herba perennis prostrata non scandens crassa stricta indeterminata volubilis, radix diverse ramosa crassi carnosi globosi vel elongati, foliola rhomboidea coriacea cerecea, legumen brevem et latum, fibrosum.

Plant a prostrate, non-twining, thick and straight stemmed, annual, indeterminate vine; root perennial, variously branched, thick, fleshy, globose to elongate; leaflets rhombic, coriaceous, waxy; pod short and broad, fibrous.

Comments.—The similarity of species in this section was recognized by Piper (1926), but the flower size is large rather than "middle-sized" and the corollas, though brilliant purple, do not truly have a reddish cast. In reality a striking characteristic of the flowers is the sharply reflexed standard, to lie almost in a plane with the wings and almost parallel with the surface of the ground. Additionally, the group seems to have a series of self incompatibility genes which make it necessary to have a population and pollinators to produce good sets of pod and seed. It has been very difficult with only a few plants in the greenhouse to self or even cross pollinate flowers and obtain pods and seed.

Maréchal et al. (1978b) ended up placing most of the taxa of this group into a single species, perhaps an overly drastic treatment. After field work with Russ Buhrow (probably the person with the most field experience with this group), the senior author has come to the conclusion that there are well defined taxa in the complex which can be separated into species and subspecies at this time. However, a more definitive work by others, based on additional field collecting and genetic hybridizing and experimentation, will certainly help. Although most field collected specimens often lack some of the needed distinctive traits, there are nevertheless good distinguishing characteristics between the two subspecies *ritensis* and *maculatus* (see following Table).

KEY TO SPECIES AND SUBSPECIES

1. Leaflets mostly rhombic to nearly ovate; seed spherical to ovoid to oblongoid, never lenticular.

2. Leaflets not particularly strongly nor distinctly veined; stipules more than 2 mm long and ovate-elongate to lanceolate.
   3. Leaflets mostly rhombic to nearly ovate, usually obtuse, often quite large, the 1st leaf produced on new growth usually trifoliolate; stipules 5–10 mm long; pods broad and heavily sutured, usually without a stipe; seed large usually more than 7 mm diam.; common, in mountains of SW United States, NW México and S to Nayarit and Puebla; 480–2400 m ______ O.1.1. **P. maculatus** subsp. **maculatus**
   3. Leaflets mostly ovate, usually acute, usually small to medium size, the 1st leaf produced on new growth usually unifoliolate; stipules less than 5 mm long; pods falcate, stipitate, and not so heavily sutured; seed usually less than 7 mm diam.; common, in mountains of SW United States, NW México and S to Jalisco; 600–2600 m ______ O.1.2. **P. maculatus** subsp. **ritensis**
2. Leaflets strongly or distinctly veined; stipules usually less than 2 mm long and broadly ovate to orbicular; rare, in mountains of W central México, southern Chihuahua and S to Jalisco; 1650–2200 m ______ O.2. **P. venosus**

1. Leaflets mostly elongate, twice as long as wide, sometimes somewhat lobed at base, the veins distinct; seed lenticular or distinctly flattened; rare, found only in the mountains of W Durango; 2000–2400 m ______ O.3. **P. reticulatus**

O.1.1.—Phaseolus maculatus Scheele subsp. **maculatus**, Linnaea 2:465–466. 1848. (**Figs. 94, 97**). (not *P. maculatus* Mart. 1829—nomen nudum). *Phaseolus metcalfei* Woot. & Standl., Contr. U.S. Natl. Herb. 16:140. 1913. TYPE: UNITED STATES. TEXAS. Comal Co.: 15 mi W of New Braunfels, Cibolo, (29°40'N, 98°W), Aug-Sep 1846, *Lindheimer 263* (as Flora Texana exsiccata 367) (HOLOTYPE: B (lost or destroyed); neotype: MO 1731163 (designated by Delgado, 1985); ISOTYPES: CAN n.v., K, MO, US).

Phaseolus retusus Benth., Pl. Hartw. 11. 1839. (Not *P. retusus* Moench 1794, another species). TYPE: MÉXICO. AGUASCALIENTES: (no location), 1829, *Hartweg 59* (HOLOTYPE: K, designated by Delgado (1985); ISOTYPE: GH).

Phaseolus ovatifolius Piper, Contr. U.S. Natl. Herb. 22:688. 1926. TYPE: MÉXICO. NAYARIT: Sierra Madre between Santa Gertrudis and Santa Teresa, (22°25'N, 104°48'W), 8 Aug 1897, *Rose 2067* (HOLOTYPE: US; ISOTYPES: GH, MO).

Aerial shoot a long, prostrate, stout, indeterminate vine. ***Root*** a perennial, large, pyriform becoming globose with age, fleshy, woody, 20 cm or more in diam., with circular annular rings of a bright purplish red alternating with rings of a lighter orange color, with 2 or 3 major branches of some 3–4 cm diam., the crown 3–4 cm below soil surface. ***Stems*** terete, long to 5 m, 2–4 mm thick, prostrate and straight; internodes 5–6 cm long. ***Stipules*** lanceolate, 8 mm long, 2 mm wide. ***Leaves*** 10–18 cm long, leathery or coriaceous, covered with whitish pubescence; petioles about as long as leaflets, 4–5 cm long; petiolule 3–4 cm long; stipels ovate 1.3–3.5 mm long; terminal leaflet rhombic-ovate, 3–9 cm long, 3–8 cm wide, acute to obtuse or retuse, distinctly veined, ultimate veinlets distinct, glabrous except for ciliate hairs on margins and veins, scabrous on adaxial surface, smooth abaxially. ***Inflorescence*** long, erect; peduncle to 20 cm long; rachis 15 cm long of 5–10 flowering nodes; primary bracts lanceolate, 3.5 mm long, 0.75–1 mm wide, strongly 3- to 5-nerved, hispid hairs at apex, purple; pedicel 4–5 mm long, covered with minute hooked pubescence; pedicellar bracts aciculate 2 mm long 1-nerved ciliate margin. ***Bracteoles*** ligulate to lanceolate, 1–1.25 mm long, 0.25 mm wide, inserted somewhat below calyx (0.25 mm), acute, ciliate of minute hooked pubescence, purple. ***Flower*** purple; calyx campanulate, 5 mm long, the upper two lobes united, emarginate, 3.5 mm long, 6 mm wide, the 3 lower acute, 3.5 m long, 3.5 mm wide, the center lobe sparsely covered with strigose and hooked pubescence, scabrous, purple; standard purple, greenish on adaxial surface, emarginate, 11 mm long, 14 mm wide, deeply enrolled at edges, strongly reflexed at 4 mm from base, thickened at point of reflexion, the claw 1 mm long, 1 mm wide, the auricles not pronounced, less than 0.5 mm long; wings purple, the blade broadly rounded, rolled lengthwise, 16 mm long, 9 mm wide, the claw 3 mm long, the spur rounded, 1 mm in diam., loosely adhering to keel; vexillary stamen 16 mm long, the claw 1 mm long, the knob geniculate, 1.25 mm wide, 0.5 mm high; stamen tube 10 mm long, 6 mm long to bend, the ridges only slightly raised; stamens highly fertile producing copious pollen; keel short, purple, 5 mm to bend and 5 more to the green tip, the terminal coil of 2 mm in diam., the claws 2 mm long, 0.75 mm wide, the auricles well-developed; basal collar 1 mm long, denticulate; ovary straight, oval, 5 mm long, 1.5 mm wide, glabrous, 3–5 ovules; style 11–12 mm to the terminal thickened coil of 2.75 mm diam.; stigma lateral introrse, narrow linear, 0.75–1.25 mm long, 0.25 mm wide. ***Pod*** oblong, mostly straight and flattened and somewhat stipitate at base, 3–7.4 cm long, 1–1.5 cm wide, very heavy sutures, nearly glabrous, often purplish streaked, slightly restricted between seeds; beak short slightly recurved. ***Seed*** spherical, orbicular, or somewhat angular from seed compression, 7.5–13 mm diam., variously solid colors tan, brown, black or mottled, often a brown or black ring around the

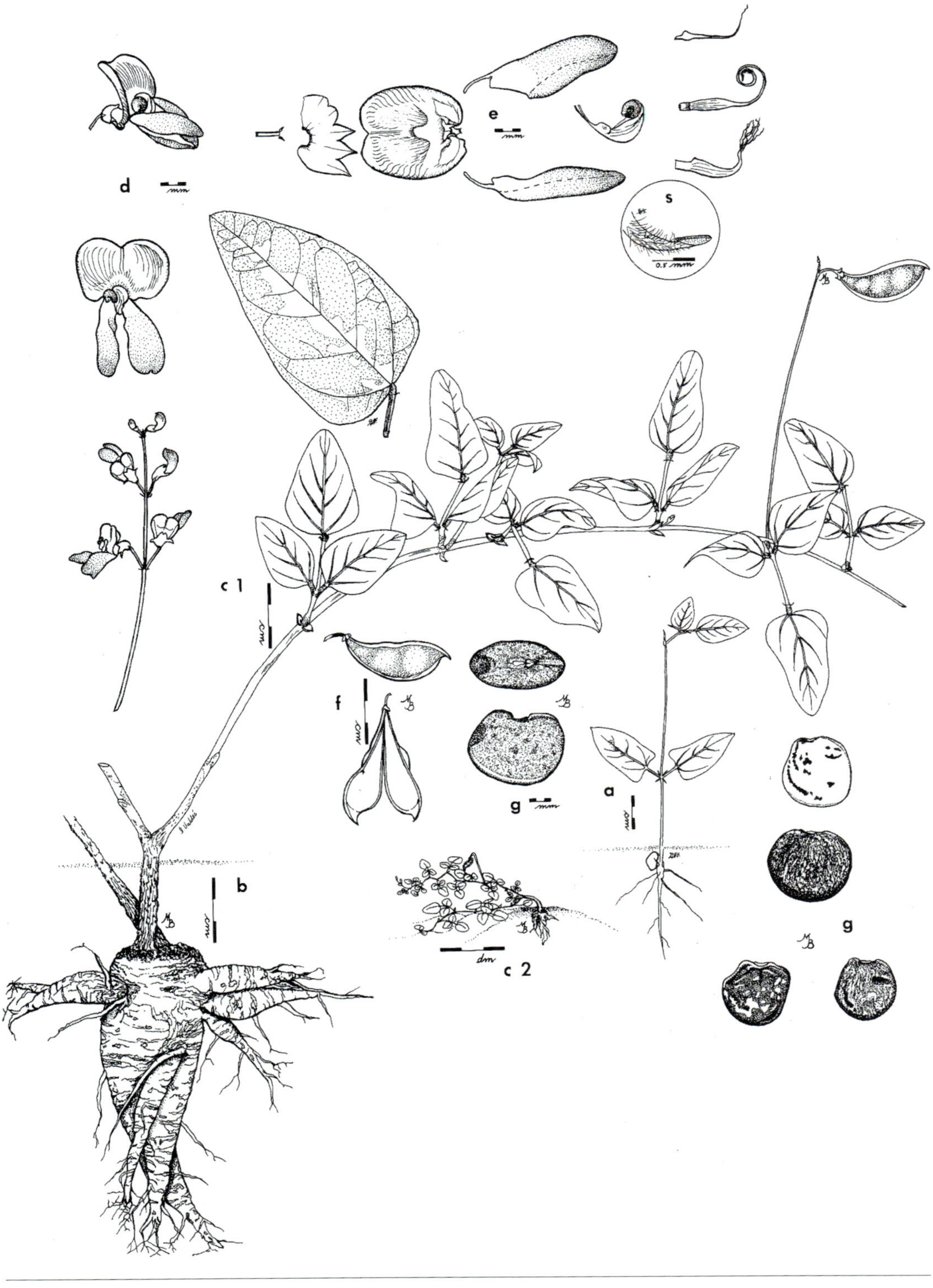

FIG. 94. Illustrations of *Phaseolus maculatus* Scheele subsp. *maculatus*.—**a.** Seedling a few weeks after germination.—**b.** Root after several years.—**c.1.** Lower portion of stem with basal trifoliolate leaves and inflorescence with pod and separately a flowering inflorescence and terminal leaflet; note large stipules, short petioles and long petiolules.—**c.2.** Trailing habit of young plant.—**d.** Flower, side view and front view; note sharply reflexed standard.—**e.** Exploded view of flower showing all parts including—**s.** Style tip and stigma as seen under the microscope.—**f.** Pods, side view and dehisced; note heavy sutures.—**g.** Seeds showing variation in size, shape and color patterns. All drawings made from living material grown in greenhouse at Mayagüez from seed of *Buhrow DRMT* (TARS #333) collected in the Dragoon Mts., Arizona, except **a.** seedling from seed of *Buhrow SLMT* (TARS #326) collected near San Lorenzo, New Mexico, USA.

hilum; hilum ovate to narrowly oblong, 1.3 mm long; lens not prominent. ***Seedling*** from hypogeal germination; epicotyl 6.5 cm long; stipules present; primary leaves opposite, simple, entire, the petiole 1.7 cm long, with basal and apical pulvini, stipels minute, the blade triangular-ovate, 6 cm long, 3 cm wide, truncate or very slightly cordate at base, obtuse.

Specimens examined: **MÉXICO. Aguascalientes:** 3 mi S of Aguascalientes, (21°45'N, 102°25"W), 2040 m, 25 Aug 1958, *Fearing 1173* (BRIT, TEX-LL); 19–20 mi W of Aguascalientes near Km 31 on road to Calvillo, (21°55'N, 102°35'W), 2200 m, 24 Aug 1960, *McVaugh 18251* (CAS). **Chihuahua:** Mpio. Guachochic, S side of Barranca del Cobre between Río Urique and Napuchi, 1800 m, 31 Jul 1973, *Bye 4428* (UCR); Mpio. Guachochic, NW of Norogachic, ca 3 road mi from airport near locality called To'achiachic, 2500 m, 23 Aug 1978, *Bye 8821* (UCR); 3 mi E of La Junta on Hwy 260 to Cuauhtemoc, 28°40'N, 107°25'W, 2310 m, 26 Jul 1949, *Freytag et al. 26* (GH, MO); Santa Eulalia, pass between Velardena and Cristo mines, 28°37'N, 105°21'W, 2134 m, 27 Jul 1948, *Hewitt 304* (GH); Culebra Mts., 17 Sep 1936, *Le Sueur 730* (F, GH, MO, UC, US); Mexican boundary line near White Water = Agua Blanca, 11 Sep 1893, *Mearns 2272* (CAS-DS, US); Mpio. de Temosachic, Cañon Huahuatan, 10 mi SE of Madera, (29°N, 107°55'W), 23 Sep 1939, *Muller 3455* (F, GH, MICH, NA, TEX-LL, UC); between Colonia García and Pratt's Ranch below Pacheco, (29°25'N, 104°55'W), 22–24 Aug 1899, *Nelson 6238* (GH, K, US); Dist. of Cusihuiriachic, 17 mi N of Rubio, 2134 m, 23 Jun 1937, *Shreve 7976* (ARIZ, F, GH, US); Santa Clara Mts, Cañon de las Varas, 1840 m, 13 Aug 1939, *Shreve 9043* (ARIZ, GH); Colonia García in the Sierra Madre, 2250 m, 27 Jul 1899, *Townsend et al. 97* (BM, EAP, F, G, K, MO, US); grown in UCR greenhouse from seeds (collected on Route 28 about 200 mi from Chihuahua toward Nuevo Casas Grandes, (30°15'N, 107°45'W)?, 9 Sep. 1976, *Waines 302*), 20 May 1982, as *Sanders et al. 2451* (TEX, UCR); E slope of the Sierra de la Breña, ca 10 km W of Estación Mata Ortiz, N Fork of the "Temporal" canyon, Benigno Mora Ranch, 30°9'N, 108°8'W, 1900–2100 m, 29 Jul. 1972, *Wilson et al. 8494* (TEX-LL); Prairies N of Chihuahua, (29°N, 106°20'W), 20 Aug 1846, *Wislizenus 109* (MO); Sierra del Nido complex, 7.5 mi W of Bella Vista on Mesa La Boquilla, 29°4'30"N, 106°29'30"S, 2020 m, 16 Jul 1981, *Worthington 7345* (ARIZ); Km 122 carr. Cuauhtémoc–La Junta, 2000 m, 25 Jul 1994, *Yen et al. 2666* (BRIT). **Coahuila:** W slope of Sierra de San Vicente, Cañon Espantosa, 20 km ESE of Cuatro Cienegas, (26°55'N, 101°56'W)?, 24 Apr 1941, *Schroeder 148* (GH). **Durango:** Durango, 12 Km N Castillo Najera, volcan El Jaguey, 24°22'N, 104°28'W, 1990 m, 29 Sep 1978, *Debouck et al. 272* (CHAPA, G, K); Peñon Blanco, Los Charcos, arroyo Las Minas, 24°45'N, 104°12'W, 1750 m, 5 Oct 1978, *Debouck et al. 299* (BR, CHAPA, COL, G, K, UC); Fco. I. Madero, arroyo entronque Mex 40 y camino microondas La Campaña, 24°26'N, 104°10'W, 1940 m, 5 Oct 1978, *Debouck et al. 305* (CHAPA, M); Mpio. Francisco Madero, La Galera, (grown at Campbell Ave Farm from seed of NI X160), *Leroy 17* (ARIZ). **Guanajuato:** Dolores Hidalgo, 24 km W de Dolores Hidalgo, 3 km NE de Ojo de Agua de Calvillo, 21°7'N, 101°6'W, 2040 m, 15 Nov 1987, *Debouck et al. 2394* (CHAPA, COL, MICH, US); Dolores Hidalgo, 8 km W de Mesa de San José, 21°4'N, 101°11'W, 2520 m, 15 Nov 1987, *Debouck et al. 2396* (CHAPA, US); 10 mi E of Guanajuato, (21°N, 100°40'W), 2286 m, 22 Jun 1963, *Harris 25671* (GH); 14.5 mi from Guanajuato on the road to Dolores Hidalgo, (21°10'N, 101°W), 2499 m, 5 Jul 1955, *Johnston 2625B* (TEX(2)). **Hidalgo:** near Tula, (20°5'N, 99°30'W), 2240 m, 18 Jul 1896, *Pringle 6360* (BM, CAS, CAS-DS, ENCB, F, G, GH, K, MEXU, MICH, MO, UC, US). **Puebla:** San Marcos, (18°40'N, 97°35'W), 2440 m, 4 Sep 1938, *Balls 5425* (BM, UC, US); entre Grajales y San José Chimapa, al oriente de La Malinche, (19°14'N, 97°45'W), 20 Jun 1977, *Martínez et al. MA1663* (CAS, MEXU); between Tepeaca and Santa Rosa, 27 Jun 1899, *Rose et al. 4738* (US). **Queretaro:** Cadereyta, (20°45'N, 99°50'W), 27 Jul 1952, *Kelly 709* (UC). **San Luís Potosí:** Mpio. Mezquitic de Carmona, 1 km S Milpillas, 8 Oct 1974, *Banda et al. 156a* (MEXU); 22°N, (101°W)?, 1829–2438 m, 1878, *Parry et al. 185* (GH, K, US). **Sonora:** Los Pinitos, 1829 m, 12 Oct 1890, *Hartman 117a* (GH); San José Mts., 1829 m, 5 Aug 1893, *Mearns 1637* (CAS-DS, US); base of San Luís Mts., (31°15'N, 108°45'W)?, 1829 m, 5 Sep 1893, *Mearns 2145* (CAS-DS, US); Quipur, 23 Jul 1970, *Pennington 84* (TEX); Maicoba, 1524 m, Jul 1968, *Pennington 180* (TEX). **Tlaxcala:** Cuapiaxtla, 2405 m, 24 Jul 1974, *Hernández et al. R-77* (MEXU). **Zacatecas:** Fresnillo, El Centro, La Mesa, 23°23'N, 103°19'W, 2150 m, 15 Oct 1978, *Debouck 344* (CHAPA, COL, M); ranch near Arroyo del Medio (Calera), NW of Zacatecas, (22°55'N, 102°40'W)?, 22 Aug 1948, *Dressler 321* (MO); Mpio. de Pinos, 7 km sobre camino de terraceria La Pendencia–Pinos, 2200 m, 20 Aug 1977, *García s.n.* (CAS, MEXU); Calera Victor Rosales, 20 km NE Zacatecas, 2200 m, 1 Aug 1974, *Rodríguez 1157* (ENCB, MEXU).

UNITED STATES. Arizona: Apache Co.: Fort Apache Indian Reservation, along N fork of White River between Indian Pine and Whiteriver, about 6 mi S of Indian Pine, (34°2'N, 109°54'W), 30 Aug 1963, *Schmidt 196* (ARIZ); White Mts, White Mtn Indian Reservation, 28 Jul 1905, *Thornber s.n.* (ARIZ). **Cochise Co.:** Coronado National Monument, Huachuca Mt., (31°30'N, 110°15'W)?, 2 Aug 1958, *Goodding 250-68* (ARIZ, CAS); near Croton Spring, 9 Sep 1858, *Hayes 149* (GH(2?)); Cave Creek Canyon, Chiricahua Mts., 1829–2438 m, 26 Jul–9 Aug 1927, *Kusche s.n.* (CAS); Herb Martyr Forest Camp, 152 m, 26 Aug 1965, *Lehto 5347A* (ASU(2), MSC); Sulphur Plains between Fort Bowie and Rucker Valley, (32°15'N, 109°30'W), Sep 1881, *Lemmon 54b* (GH, UC); Tanner's Cañon, Huachuca Mts., 2 Aug 1893, *Mearns 1574* (CAS-DS); campground, Chiricahua Nat'l Mon., 20 Aug 1977, *Nabhan GN694* (ARIZ). **Coconino Co.:** W Clear Creek, along Maxwell Trail, 2011 m, 24 Sep 1985, *Hodgson et al. 3954* (DES). **Gila Co.:** in Apache Mts., 23 mi E of Rice, 6 Jul 1918, *Ferris 1287* (CAS-DS); Sitgreaves Nat'l Forest, Chevelon Creek Canyon, end of FSR 91, 1950 m, 23 Aug 1985, *Hodgson et al. 3823* (DES). **Greenlee:** Blue Range Primitive Area, along FSR 567 just above Blue Crossing, 1935 m, 11 Aug 1984, *Hodgson 3175* (DES); Gila National Forest, 2.5 mi W of NM along road to Blue, just below Blue Vista, (33°40'N, 109°5'W), 14 Aug 1974, *Pinkava et al. P12440* (US). **Navajo Co.:** between Amos Lake and Dry Valley road, 2134 m, 25 Aug 1970, *Granfelt 70-135* (ARIZ). **Pima Co.:** Foothills of Santa Rita Mts., 8 Sep 1884, *Pringle 15857* (G, MO). **Yavapai Co.:** Fort Whipple, 1864, *Coury s.n.* (MO); Prescott, 26–31 Aug 1916, *Thornber 8648* (ARIZ). **County Unknown:** White Mts., 6–15 Aug 1903, *Griffiths 5360b* (US); Ryan Ranch, E of White River, Apache Reservation, 3 Oct 1927, *Harrison 4871* (ARIZ, US); Burro Mts., 1 Aug 1880, *Rusby 101* (F); Pine, (34°30'N, 111°15'W), 22 Jul 1892, *Toumey 566* (US). **New México: Catron Co.:** White Water, New Mexican side of boundary line, 18 Jun 1892, *Mearns 334* (US); Mogollon Mts., W fork of the Gila River, (33°15'N, 108°25'W), 2540 m, 5 Aug 1903, *Metcalfe 382* (ARIZ, CAS, CAS-DS, G, K, MO, UC, US); Mangas, (34°N, 108°15'W), 22 Aug 1897, *Metcalfe s.n.* (US(3)); Apache Forest, Dry Blue Creek, 2225 m, 12 Jul 1976, *Pase 2097* (ASU); Gila Cliff Ruins Nat'l Mon., 2 mi N of Gila Hot Springs, 1646 m, 8 Oct 1977, *Reichhardt et al.*

735 (ARIZ). **Doña Ana Co.:** valley of the Río Grande, below Doña Ana, (32°20'N, 106°50'W), 1878(?), *Parry et al. 205* (K, US). **Grant Co.:** Meerschaum Canyon near Lake Roberts, NE of Silver City, 2 Sep 1972, *Dale et al. 2069* (ARIZ); Gila National Forest on Hwy 25, 9 mi NE of Silver City, near Mill Creek, (32°50'N, 108°15'W), 2073 m, 3 Aug 1967, *Hess 1327* (BRIT, US); 1 mi S of Mimbres Post Office on E side of the river, Oct 1974, *Minnis s.n.* (MICH); above Cold Spring Canyon near Sherman, (32°40'N, 107°55'W), 488 m, 16 Aug 1976, *Minnis s.n.* (MICH); 400 m across river from Mimbres Foundation, Mimbres, 1554 m, 17 Aug 1977, *Nabhan GN669* (ARIZ); 2 mi NE of Pinos Altos on Inner Loop Drive, N of Silver City, 17 Aug 1977, *Nabhan GH674* (ARIZ). **Hidalgo Co.:** Peloncillo Mts., Tank Mt., 10 air mi SSW of Animas on S end of Tank Mt., 1524 m, 21 Sep 1986, *Worthington 15170* (ARIZ, DES, UCR). **Sierra Co.:** Kingston, S end of the Black Range, (32°55'N, 107°30'W), 2170 m, 10 Jul 1904, *Metcalfe 1104* (ARIZ, CAS, ENCB, GH, MO). **Socorro Co.:** Gila Forest, White Creek Ranger Station, 1 Aug 1920, *Eggleston 16892* (US). **County Unknown:** West slope Mts., W of Gray, 18 Sep 1937, *Goodding A9159* (ARIZ); Hick's Ranch, 20 Jul 1895, *Mulford 343* (MO); Burro Mts., 2 Sep 1880, *Rusby 107* (MO, US). **Texas: Brewster Co.:** Alpine, (30°15'N, 103°35'W), 22 Jul 1928, *Cory 1816* (GH); 17 mi S of Alpine, (30°5'N, 103°30'W), 11 Aug 1934, *Cory 9253* (GH). **Jeff Davis Co.:** 11 mi from Fort Davis, "Point of Rocks" park, (30°30'N, 103°55'W), 19 Sep 1966, *Correll 33676* (ENCB, GH, TEX-LL, UC); 21 1/4 mi NW of Fort Davis, Upper Madera, 12 Aug 1934, *Cory 9450* (GH); Davis Mts., Colleen Canyon, 1.9 mi above Tomahawk Trail, about 3 mi SE of the summit of Mt. Livermore, 1981 m, 7 Oct 1982, *Sanders et al. 3182* (UCR). **Kimble Co.:** Junction, (30°30'N, 100°W), 1900, *Armstrong s.n.* (US). **Presidio Co.:** Mt. Livermore, S ridge, Upper Goat Canyon, 21 Aug 1935, *Hinckley 365* (BRIT, F); Sierra Vieja Mts., Trans-Pecos, head Knox Canyon, (30°20'N, 104°15'W)?, 22 Jun 1941, *Hinckley 1796* (TEX, US). **County Unknown:** Valley of Limpio Mts., to Peña Colorada, Jul 188, *Havard s.n.* (US); dry bed of Cibolo River, 1846, *Lindheimer 367* (BRIT, G, GH, K, UC, US); Limpio, western Texas & Coppermines, Jun–Aug 1851, *Wright 947* (G, GH, MO, US).

Habitat.—This subspecies is found growing along streams and on grassy slopes in open oak woodland and Ponderosa pine forests, and in grassland with *Opuntia*. Dominant vegetation is *Agave, Aloysia, Nolina, Juniperus monosperma, Pinus edulis* and *Quercus griseus*, and grass is mostly *Bouteloua*.

Diseases and pests.—Leaves are often damaged by *Epilachna varivestis*.

Common name.—The large fleshy root is called "Cocolmeca" in Chihuahua (see Color Plate III, photo 36). Also reported as "frijolillo." Other names are reported by Nabhan et al. (1980).

Ethnobotany.—The root is used in preparing a purgative used by Pima indians in the high country. Uses of pods and seed in the Greater Southwest have been documented by Nabhan et al. (1980).

Comments.—This subspecies is easily recognized by its prostrate growth habit, large stipules, and large rhombic coriaceous leaves. The flowers are also unique with a greatly reflexed standard, nearly in the same plane as the extended wings. The 'Metcalfe' bean typically colonizes the grassland vegetations 'pastizal de grama', and savannahs with oak and juniper (see Gentry 1957). Although Piper's species *P. ovatifolius* does have some differences from typical *P. maculatus* such as smaller, variegated leaflets, long peduncles, and slightly smaller flowers, these are not enough for maintaining a species rank. The stipules and primary bracts are quite large so the taxon is placed as a synonym of subsp. *maculatus*.

O.1.2.—Phaseolus maculatus Scheele subsp. **ritensis** (Jones) Freytag, comb. & stat. nov. (**Figs. 95, 97**). *Phaseolus ritensis* Jones, Contr. West. Bot. 12:14. 1908. TYPE: UNITED STATES. ARIZONA: Santa Rita Mts., 1372 m, 24 Aug 1903, *Jones s.n.* (HOLOTYPE: POM 28786 designated by Delgado (1985) n.v., MICH (photo); ISOTYPES: GH, POM n.v.).

Aerial part an annual, prostrate, indeterminate vine, to 3–4 m long, the whole plant smooth except for the peduncles, petioles, and stems covered with minute pubescence. ***Root*** a perennial, large, elongate, woody, to 40 cm long, 2–3 cm wide, with several branches, the crown at 2–3 cm below soil surface with multiple stems. ***Stems*** terete striate sparsely covered with retrorse-strigillose hairs. ***Stipules*** triangular, 2.5 mm long, acute, glabrous, 5-nerved. ***Leaves*** 9–12 cm long, the first 1–2 produced at beginnning of each year's growth or axis are simple and subsequent leaves are bi- or trifoliolate; petioles about as long or longer or shorter than the leaflets, 0.4–4 cm long; petiolule 1.1–2 cm long; stipels linear, minute; terminal leaflets rhombic-ovate or ovate with a very broadly rounded or deltoid base, 3–6 cm long, nearly as broad, acute and slightly to strongly apiculate tip, rarely obtuse to retuse, reticulate, green on both sides, variegated on mid-vein, scaberulose and puberulent on adaxial surface, nearly glabrous abaxially except for a few appressed hairs on the veins and ciliate edges, ultimate veinlets indistinct; lateral leaflets similar but more ovate to orbicular and slightly inequilateral. ***Inflorescence*** a long raceme, dense in anthesis, usually about 20–26 flowering nodes; peduncle slender, 12–30–80 cm long; rachis 10–25–50 cm long, many flowered, often bearing 2–3 at a node, rarely with 2 lateral flowers inserted between calyx and bracteoles forming a cyme (see Color Plate 11-22), axis of secondary racemes often outgrown; primary bract minute, lanceolate, 1.5–3 mm long, acute,

FIG. 95. Illustrations of *Phaseolus maculatus* subsp. *ritensis* (Jones) Freytag.–**a.** Seedling several weeks after germination; note the long epicotyl, minute petiole and near linear eophyll.–**b.** Root after several years; note branches, corky and fissured cortex and large, elongate crown.–**c.** Short stem of plant with leaves and inflorescences; note simple lower leaves and small stipules.–**d.** Flower, side view and front view; note sharply reflexed standard.–**e.** Exploded view of flower showing all parts including–**s.** Style tip and stigma as seen under the microscope.–**f.** Pods, side view and dehisced; note stipitate base and sharply falcate shape.–**g.** Seeds, side view and view of hilum. All drawings made from living material grown in greenhouse at Mayagüez from seed of *Freytag 84-1* (TARS #328) collected on Mt. Lemmon, Arizona, except seed and pod from seed of *Buhrow s.n.?* (TARS #98) collected in Arizona, USA.

striate, glabrous, persisting; pedicel 10 mm long, very sparsely covered with strigose hairs, ciliate on margins of minute uncinate hairs; pedicellar bracts 1 mm long narrowly lanceolate 1-nerved ciliate margins often purplish. ***Bracteoles*** ovate, minute, scale-like, 0.5–1 mm long, 0.7 mm wide, indistinctly 1-nerved, puberulent and ciliate of minute hooked hairs. ***Flower*** purple to violet; calyx campanulate, 3 mm long, the two upper teeth united into a single short emarginate lip, 0.75 mm long, 10 mm wide, the lower lip with 2 broadly rounded triangular, obtuse teeth 1.25 mm long, 2 mm wide, the center lobe narrower, acute, 0.75 mm wide, mostly glabrous to sparsely hispid; standard purple, squarish, emarginate, with rounded lobes, somewhat greenish on adaxial surface and thickened at base, sharply reflexed at 4 mm from base, 10–11 mm long, the auricles pronounced, inflexed, 1 mm long, the stipe short and thick, 0.75 mm long; wings purple, rounded spatulate-oblong, 10–15 mm long, enrolled and cupped lengthwise, spreading from base, the claws 3.5 mm long, the spur pronounced, 1–1.25 mm wide, loosely adhering to keel; keel white at base with green tip, 12 mm long, the claws separated, 3 mm long, 4 mm more to bend and 7 mm more to the terminal 1 1/2 coils of 2.5 mm diam.; vexillary stamen with a pronounced geniculate knob at 0.75 mm from base, 1 mm wide, 0.5 mm high; stamen tube 7.5 mm long, 4 mm from base to bend, the lateral ridges very slightly raised, less than 0.25 mm; basal collar 1 mm long, minutely denticulate; ovary straight, 6 mm long, 1 mm wide, glabrous, 4–6 ovules; style 7 mm long to base of the thickened terminal coil of 2 mm diam.; stigma lateral introrse, linear, 1 mm long. ***Pod*** falcate and with a triangular, rather long-stipitate base, often curved, 2.5–5.3 cm long, 0.5–1.1 cm wide; valves coriaceus, nearly glabrous, the sutures somewhat pronounced, loosely twisted 1 turn at dehiscence; tip strongly beaked. ***Seed*** oblongoid, flattened, 4–7.5 mm long, 4–6 mm wide, 3–5 mm thick; variously mottled or streaked with black on a brownish background, a black circle around hilum; hilum ovate to lanceolate, 1.2 mm long; lens pronounced and divided. ***Seedling*** from hypogeal germination; epicotyl 9 cm long; stipules entire, bifid or divided; eophyll entire, simple, the petiole 2–3 mm long, the blade ovate-lanceolate to linear, rounded to truncate at base; subsequent 1–2 leaves usually unifoliolate then trifoliolate.

Specimens examined: **MÉXICO. Chihuahua:** Sierra Madre, Continental divide, México and NWRR, ridge between Río Chico and Río Caballo, 30 Sep 1911, *Barlow s.n.* (F); Canelas, Río Mayo, 8 Oct 1935, *Gentry 2006* (ARIZ, DES, F, GH, MO); Sierra Canelo, Río Mayo, 30 Aug 1936, *Gentry 2523* (ARIZ, F, GH, K, MO, UC); barranca, 19 mi by road W of Guachochic, (26°50'N, 106°45'W), 2286 m, 28 Oct 1966, *Gentry 22056* (NA, US); Barrancas del Divisadero, 2290 m, 28 Sep 1978, *Hernández 3248* (CAS(2)); Mojarachic, SW of San Juanito, 27°52'N, 108°W, 2140 m, 11 Aug 1954, *Knobloch 1213* (MICH, MSC); Areponapuchic, Barranca de Urique, 27°80'N, 107°50'W, 1000–1800 m, 23 Aug 1954, *Knobloch 1347* (MICH); 3 km N of town of Basaseachic, 28°14'N, 108°12'W, 2050 m, 15 Oct 1985, *Martin et al. s.n.* (ARIZ); Sierra Madre Occidental, NE of Colonia Pacheco, Strawberry Creek, (29°25'N, 104°55'W)?, 1900–2000 m, 22–24 Sep 1934, *Pennell 19189* (US); Mpio. Madera, 8 km al SE de Madera, Cuesta de la Borrega, (29°N, 107°55'W), 2100 m, 29 Sep 1982, *Tenorio et al. 1855* (CAS, MO); grown in a UCR greenhouse from seed (collected at Rt 28, 204.1 mi from the Motel Las Escobas, Chihuahua to Nuevo Casas Grandes, (30°15'N, 107°45'W)?, 9 Sep 1976, *Waines L302*), 20 May 1982, *Sanders et al. 2451* (ARIZ); Sierra Madre Occidental, 82 mi W of Vieja Casas Grandes, 2140 m, 25 Aug 1952, *Tucker 2525* (ARIZ, CAS, CAS-DS, UC); Mpio. de Ocampo, Parque Nal. Cascada de Basaseachic, 1950 m, 5 Aug 1994, *Yen et al. 2886* (BRIT). **Durango:** 32 mi W of Durango, along Hwy 40, 7.2 mi W of Río Mimbres, (23°50'N, 105°W), 2740 m, 2 Aug 1977, *Bennett et al. 793* (MO); 34 mi N of railroad at Coyotes, Laguna del Progreso, (24°10'N, 103°5'W), 2500–2600 m, 21 Jul 1955, *Maysilles 7909* (CAS, MICH); Río Chico, 3 mi NE of Otinapa, 2300–2400 m, 8 Jul 1950, *Maysilles 7330A* (MICH); Sierra Madre Occidental, 80 km W of Cd. Durango, N of Coyotes RR Station, 23 mi NW of Coyotes, Quebrada de San Juan, (28°50'N, 105°W)?, 2100 m, 30 Sep 1962, *McVaugh 21709* (MICH). **Jalisco:** Guadalajara, (20°40'N, 103°30'W), 29 Sep 1903, *Rose et al. 7430* (US). **Nayarit:** Mpio. El Nayar, at Mesa del Nayar, 1300 m, 11 Aug 1980, *Breedlove et al. 45621* (CAS); valley of the Río Jesús María near the village of Jesús María, (22°15'N, 104°30'W), 600–700 m, Sep 1960, *Feddema 1463* (MICH); Sierra Madre, near Santa Teresa, 9 Aug 1897, *Rose 2153* (MICH, US); Mpio. Nayar, NE of Villa de Guadalupe, Cerro Cangrejo, 22°16N, 104°38'W, 1200–1400 m, 18 Sep 1989, *Tenorio et al. 16222* (MO). **Sinaloa:** Culiacán, Cerro Colorado, (24°50'N, 107°30'W)?, 3 Nov 1904, *Brandegee s.n.* (UC); Ocurahui, Sierra Surotato, (26°N, 107°40'W)?, 1829 m, 27–30 Aug 1941, *Gentry 6173* (ARIZ); Quebrado de Mansana, Sierra Surotato, 1219 m, 10–14 Sep 1941, *Gentry 6559* (ARIZ, MICH, MO). **Sonora:** Arroyo Tejas, Sierra Charuco, 1372 m, 16 Oct 1966, *Gentry 22007* (US); 12 km W de El Talayote, Mpio. Yecora, 1750 m, 30 Sep 1985, *Tenorio et al. 10090* (MO); 3 km N de La Lobera, Ejido Zahuarivo, Mpio. Alamos, 27°12'N, 108°53'W, 1500 m, 25 Aug 1986, *Tenorio et al. 11946* (MO); loop of the Río de Bavispe, Cerro del Capulin, NW of Aribabi, (30°8'N, 109°10'W), 1859 m, 4 Sep 1939, *White 2707* (ARIZ, GH, MEXU, MICH); region of the Río de Bavispe, between Las Tierritas and El Tigre, 22 Aug 1940, *White 3453* (GH, MICH); region of the Río de Bavispe, NE of El Tigre, El Rancho del Roble, (30°45N, 109°10'W)?, 1829 m, 2–13 Sep 1941, *White 4369* (GH, MICH). **Zacatecas:** Sain Alto, sierra Chapultepec, camino a Valparaiso, 23°27'N, 103°08'W, 2420 m, 12 Oct 1978, *Debouck 329* (BR, CHAPA, COL, K, M); Fresnillo, 6 km W El Carizalillo a Menonita, 23°22'N, 103°23'W, 2290 m, 17 Oct 1978, *Debouck 362* (BR, CHAPA, K, M); Jiménez de Teul, terraceria a Refugio Los Pozos, Cienega San Pedro, 23°20'N, 103°54'W, 2200 m, 19 Oct 1978, *Debouck 379* (CHAPA, K); Jiménez de Teul, Arroyo de Milpillas, 23°07'N, 103°580'W, 2280 m, 19 Oct 1978, *Debouck 384* (CHAPA, COL, K, M).

UNITED STATES. Arizona: Cochise Co.: Whetstone Mts., 15 Aug 1979, *Buhrow 139* (ARIZ); Huachuca Mts., Carr Canyon, near waterfall below Reef Mine, 1981 m, 9 Sep 1944, *Darrow et al. 1421* (ARIZ); Huachuca Mts., near Bisbee, Muller's Canyon, (31°30'N, 110°W), 18 Aug 1909, *Goodding 425* (ARIZ); Chiricahua Mts., Cave Creek, in Herb Martyr Dam Area, 1767 m, 10 Aug 1960, *Throne s.n.* (WIS); Huachuca Mts., Ramsey Canyon above Bledsoe cabins, 1829 m, 23 Aug 1981, *Toolin 1676* (ARIZ); Huachuca Mts., Coronado National Monument, crest of Montezuma Pass, 2012 m, 25 Aug 1990, *Van Devender et al. 90-98* (ARIZ); near Fort Huachuca, Aug 1894, *Wilcox 307* (US). **Graham Co.:** Graham Mts., Arcadia Ranger Station, 20 Aug 1929, *Crider s.n.* (ARIZ); Pinaleno Mts., Goudy Canyon, 2060 m, 23 Sep 1914, *Shreve 4407* (ARIZ). **Pima Co.:** Santa Catalina Mts., Mt. Lemmon Road between Control Stations, (32°30'N, 110°35'W), 1981 m, 2 Sep 1938, *Benson 9084* (ARIZ, UC); Rincon Mts., Manning Trail, 2100 m, 15 Sep 1909, *Blumer 3360* (ARIZ, CAS-DS, F, GH, MO, UC); Rincon Mts., along the Miller Creek trail, 1402 m, 27 Aug 1982, *Bowers R511* (ARIZ, ASU); Santa Rita Mts., Mt. Hopkins Road, 2210 m, 19 Aug 1986, *Bowers et al. 3163* (ARIZ, ASU); Santa Rita Mts., Mt. Hopkins Road, 1.3 mi above gate near high end of old closed road, 1890 m, *Buhrow 25* (ARIZ); Quinlan Mts. on Hwy. to Kitt Peak, (32°N, 111°25'W), 1524 m, 21 Sep 1982, *Buhrow 182* (ARIZ); Santa Catalina Mts., Sabino Canyon on terrace of Sabino Creek, 8 Sep 1967, *Gillis 6706* (ARIZ); Baboquivari Canyon, (31°45'N, 111°35'W)?, 28 Sep 1931, *Gilman 229* (ARIZ); Mt. Lemmon, near Stratton mine, 6 Sep 1926, *Harrison et al. 3065* (ARIZ); Kitt Peak Observatory Road, 15 Sep 1979, *Hodgson 551* (DES); Rincon Mts., along the Miller Creek Trail, 1554 m, 6 Sep 1982, *McLaughlin et al. R640* (ARIZ) Greaterville Road 2.1 mi W of Hidden Springs Ranch Road, 25 Aug 1971, *Pinkava et al. 699* (ASU). **Santa Cruz Co.:** Patagonia Mts., 2.3 mi W of Duquesne on the road to Nogales, 1737 m, 22 Sep 1978, *Buhrow s.n.* (ASU); Patagonia Mts., 4.1 mi W of Duquesne on road to Nogales, (31°30'N, 110°45'W)?, 1737 m, 22 Sep 1978, *Buhrow 26* (ARIZ(2)); Huachuca Mts., near Sonoita Road, (31°45'N, 110°25'W)?, 2310 m, 4 Sep 1933, *Carter s.n.* (ARIZ); Sycamore Canyon, (31°25'N, 111°15'W)?, 22 Sep 1937, *Goodding A9171* (ARIZ); Atascosa Lookout Trail, 4.2 mi E of Sycamore Canyon turnoff, along Ruby-Arivaca Road, ca 3/4 mi from lookout, 11 Sep 1985, *Hodgson et al. 3885* (DES, UCR); Patagonia Mts., at intersection of FSR 61 & FSR 49, T23S R16E Sec 34, 1737 m, 5 Sep 1987, *Hodgson et al. 4692* (DES); Patagonia Mts., 1.7 mi NW of FSR 49-812 intersection, just east of FSR 812 (Flux Canyon Road), T22S R16E center Sec 32, 1493 m, 6 Sep 1987, *Hodgson et al. 4724* (DES); Patagonia Mts., 11.25 mi E of Hwy 82 along FSR 61 (Duquesne Road), 1676 m, 8 Oct 1988, *Hodgson et al. 5207* (DES, UCR); road to Ruby 150 m from entrance to Sycamore Canyon, 1150 m, 31 Jul 1977, *Nabhan et al. GN661* (ARIZ, ASU); Patagonia Mts., 18 Aug 1928, *Peebles et al. 5589* (ARIZ, US); Santa Rita Mts., Madera Canyon, (31°45'N, 110°45'W), 1981 m, 21 Aug 1974, *Reeves R1024* (in part) (ASU); Peña Blanca Canyon, 1219 m, 18 Aug 1966, *Tate 318* (ASU); Santa Catalina Mts., Sobrio Cañon, Box Springs, 10 Aug, *Thornber et al. s.n.* (ARIZ); Sonoita Creek, 1113 m, 10 Aug 1980, *Van Devender s.n.* (ARIZ). **Texas: El Paso Co.:** Franklin Mts., at Canutillo, (31°50'N, 106°20'W), (no date), *Barlow s.n.* (F). **Jeff Davis Co.:** Davis Mts., about 3 mi SE of the summit of Mt. Livermore, 1.9 mi above Tomahawk Trail, Colleen Canyon, 30°37'N, 104°7'W, 7 Oct 1982, *Sanders et al. 3182* (ARIZ).

Habitat.—This subspecies is quite common and is found growing in mixed forests of pine, oak, *Cupressus*, manzanita, and nogal with the dominant species being: *Arbutus sp., Juniperus deppeana, Pinus discolor, P. chihuahuana, P. engelmannii, Quercus arizonica, Q. emoryi, Q. hypoleucoides*, and *Q. viminea*. It is associated with *Acacia, Arctostaphylos, Bursera, Centrosema, Crataegus, Desmodiae, Dalea, Dasylirion, Epilobium, Erigeron, Eysenhardtia, Hymenothrix wrightii, Lupinus garcianus, Opuntia* and other cacti, *Rhynchosia, Ricinus, Silene, Tagetes, Tilandsia, Yucca schottii*, composites, Mimosoideae, and grasses. It is common along roadsides, steep rocky hillsides or along moist banks of streams, and in full sun or shade (see Color Plate V, photo 57). Soils are rocky and sandy, and derived from volcanic ash, rhyolitic, lava, granite, schist or limestone.

Common name.—Frijol de Monte; eggleaf stringbean (Niehaus et al. 1984).

Ethnobotany.—The root is reportedly used as a fermenting agent in making Warihio liquor called "batari."

Genetics.—Some hybrids have been made by Russ Buhrow who made voucher specimens as follows: *P. venosus* × *P. ritensis*, F_1 hybrid; NI696 × M36, greenhouse grown plant, Campbell Ave. Farm, Tucson, 17 Aug 1984, *Buhrow s.n.* (ARIZ). This plant has bracts without nectaries, is male sterile, and has secondary inflorescences at the base of the primaries. *P. sempervirens* × *P. ritensis*, F_1 hybrid; M4 × M36; grown in greenhouse at Campbell Ave. Farm, Tucson, 13 Sep. 1984, *Buhrow s.n.* (ARIZ).

Comments.—Placement at subspecies rank has been suggested on the basis of cluster analysis of electrophoretic patterns of seed proteins, which also shows that *P. polystachyus* is a close relative (Sullivan & Freytag 1986). Subspecies *ritensis* is much more common than the type of the species and is also much more variable and found at lower elevations (see also Nabhan 1990). This subspecies also has unifoliolate leaves on the seedling plantlet and on new growth produced at the beginning of each year (see Color Plate V, photo 60) and will sometimes produce a cymose type inflorescence (see Color Plate II, photo 22). Typically subspecies *ritensis* colonizes rocky slopes in oak forests and oak-pine forests (see Estrada & Martínez 2000).

The following table gives a comparison of characteristics of the two subspecies.

O.2.—Phaseolus venosus Piper, Contr. U.S. Natl. Herb. 22:687–688. 1926. (**Fig. 97**). TYPE: MÉXICO. JALISCO: near Colotlán, on the road to Plateado, (22°N, 103°30'W), 31 Aug 1897, *Rose 2688* (HOLOTYPE: US 301614).

Distinguishing characteristics for subspecies *maculatus* and *ritensis* (after Buhrow 1983; Nabhan et al. 1980).

Trait	*maculatus*	*ritensis*
seedling eophyll	wide, hastate	narrow
eophyll petiole	present	reduced to pulvinus
1st leaves	trifoliolate	simple
stipules	large	small
primary bracts	large	small
stipe on pod	absent	present
seed size	very large	medium

Plant a small, herbaceous, trailing and climbing, indeterminate vine. ***Root*** unknown. ***Stems*** terete, striate, covered with retrorse-strigillose, hirsute and minute uncinate hairs. ***Stipules*** lanceolate, 3 mm long, broadest at base, acute, 3- to 4-nerved, hyaline margins, puberulent. ***Leaves*** 4.5–7 cm long; petioles mostly shorter than the leaflets, 1.5–2 cm long, covered with short hispid and hooked pubescence; petiolules 1 cm long; stipels oblong, 1 mm long, 0.5 mm wide, acute, smooth and scale-like, glabrous; all leaflets rhombic-ovate, 2–4 cm long, 1–2 cm wide, cuneate at the base, strongly 3-nerved, coriaceous, acute, strongly apiculate, reticulate, sparsely covered with minute uncinate hairs on both surfaces, especially on the veins, somewhat farinose above, light green. ***Inflorescence*** a short, stout, pseudoraceme, longer than the leaves; peduncle 4–6–9 cm long, pubescent with minute hispid and strigose spreading hairs; rachis 7–9 cm long with short internodes of 5–10 mm and 20–30 flowers; primary bract orbicular to ovate-acuminate, 3 mm long, 1.5 mm wide, obscurely 5-nerved, covered by hispid pubescence, the secondary bracts oblong and much smaller; pedicel 5–7 mm long, covered by hispid pubescence. ***Bracteoles*** scale-like, ovate, 1–1.5 mm long, 0.5–0.75 mm wide, acute or rounded, ciliate. ***Flower*** purple; calyx campanulate, 4 mm long, nearly glabrous and scabrous, the 2 upper lobes rounded, 0.75 mm long, 1.5–2 mm wide, the 3 lower lobes subequal, broadly rounded, dentate, 1.25 mm long, 1.5 mm wide; standard purple, the claw 1 mm long, the blade broadly rounded, reflexed at 3–4 mm from base and 4–5 mm more to emarginate apex, 10 mm wide, the auricles very small or absent; wings purple, the blade broadly rounded, 8–15 mm long, inrolled at the margins, angled at base, the claw 4 mm long, the spur 1 mm wide, 2 mm long; keel, the claws divided 3 mm long, 3 mm more to upward bend and 4 mm more to base of the terminal 1 1/2 coils of 2.75 mm diam., the auricles pronounced, 2 mm in diam.; vexillary stamen, the claw 1.5 mm long, the geniculate knob about 0.5 mm diam, and the thickened portion only about 0.75 mm more; stamen tube 6 mm to upward bend and 4 mm more to end of united stamens, the ridges hardly developed; ovary straight, 5 mm long, 1 mm wide, flattened, pronounced sutures, glabrous, 5–6 ovules; style 7–8 mm to the terminal thickened coil of 2.25 mm diam.; stigma lateral, introrse and oblique, straight linear pointed, 1 mm long. ***Pod*** straight, 3.5–4 cm long, 6–8 mm wide, 7.5 mm thick, fibrous, brittle, tightly twisting at dehiscence, glabrous, thickened sutures; beak short, straight, 1–2 mm long; 2–6 seed. ***Seed*** oblongoid to trapezoid, flattened ends and sides, 5.3–7.8 mm long, 4.1–7.2 mm wide, 2.9–4.8 mm thick, solid black or dark brown, shiny; hilum oblong, 1.5 mm long, 0.75 mm wide; lens knobby. ***Seedling*** unknown.

Specimens examined: **MÉXICO. Aguascalientes:** 19–20 mi W of Aguascalientes, Km 31 road to Calvillo, (21°55'N, 102°35'W), 2200 m, 24 Aug 1960, *McVaugh 18251* (in part) (MICH). **Durango:** San Ramón, 21 Apr–18 May 1906, *Palmer 142* (K, US). **San Luis Potosí:** Mpio. de Mezquitic de Carmona, 1 km al S de Milpillas, 8 Oct 1974, *Banda et al. 156a* (CAS). **Zacatecas:** 4.6 mi W of Jalpa on road to Tlaltenango, 9 Aug 1983, *Buhrow et al. M19* (UCR); near Monte Escobedo, 27 Aug 1897, *Rose 2648* (US); Zacatecas, 22°20'N, 103°40'W, 4 Sep 1897, *Rose 2802* (US).

Habitat.—This species is found growing in lower pine-oak forests in pastures and on top of small hills. Soil is sandy and from riolite.

Comments.—Piper's description of his type does not adequately stress several characteristics of this species, viz. the leaflets are strongly 3nerved, primary bracts are broadly orbicular, pedicels are twice as long as the calyx, the stigma is terminal and lateral, and the pods have 5–6 seed.

Delgado places this species as a synonym of subsp. *maculatus*, with which the senior author does not agree since it looks more like a small-seeded *P. maculatus* subsp. *ritensis*.

O.3.—Phaseolus reticulatus Freytag & Debouck, sp. nov. (**Figs. 96, 97**). TYPE: MÉXICO. DURANGO: carr. a Mazatlán, Km 30 entre Ojo de Agua y Los Mimbres, 23°56'N, 104°54'W, 2320 m, 2 Nov 1978, *Debouck & Muruaga 409* (HOLOTYPE: US; ISOTYPES: CHAPA, K, M, MICH).

Persimilis *Phaseolo maculato* subsp. *ritensis*, sed foliolis terminalibus aliquantum late lobatis ad basas et argenteis infra, inflorescentiis longis erectis, bracteolis oblongis ad lanceolatis, leguminibus reticulatis. Crecit modo in locis regionis meridionali occidentali Durangensis rarus.

Aerial shoot a prostrate, indeterminate vine, to 2 m long. ***Root*** unknown. ***Stems*** terete, slightly striate, 2.5 mm diam., covered with minute hirsute hairs on lower portions, puberulent of minute uncinate hairs in upper portions of plant; internodes 5–9 cm long. ***Stipules*** triangular, 2.5–3 mm long, 1 mm wide, acuminate, strongly 3-nerved, sparsely covered with hirsute hairs, ciliate margins. ***Leaves*** 9.6–12.8 cm long; petioles 3–4 cm long, slender, glabrate; petiolules 10–12 mm long; pulvini, basal 4 mm long, glabrate, upper 2–2.5 mm long, hirsute; stipels 1–2 mm long, ciliate; terminal leaflet narrow ovate, those on upper parts of plant nearly linear, 5–7 cm long, 1.5–3 cm wide at base, somewhat broadly lobed at base, acute to obtuse, apiculate, glabrous, sparsely appressed-ciliate on margins, variegated and dark green adaxially, silvery abaxially; lateral leaflet similar but inequilateral and broader. ***Inflorescence*** long, erect, 25–30 cm long; peduncle 18–20 cm long, glabrous to puberulent; rachis 8–10 cm long of 8–15 nodes, covered with minute uncinate hairs; primary bract oblong-lanceolate, 3 mm long, 1–1.25 mm wide, heavily 3- to 5-nerved, covered with hirsute hairs, especially near apex; pedicels 4–8 mm long, delicate, scarcely covered with minute uncinate hairs, purplish. ***Bracteoles*** oblong to lanceolate, acute, 1 mm long, 0.3 mm wide, covered with hirsute hairs, ciliate. ***Flower*** purple striped; calyx campanulate, purple, covered with strigose hairs abaxially, covered with hirsute hairs adaxially, tube 2 mm long, the 2 upper lobes barely elongate, 1.25 mm wide, united into 1 emarginate, ciliate on margins, the 2 lateral lower lobes dentate, retuse, 1 mm long, 1.25 mm wide, ciliate, on margins, the center lobe, acute, covered with white strigose hairs; standard 3.5 mm to flexure and 4–5 mm more to tip, emarginate, 8 mm wide; wings, the blade broad spatulate, 7 mm long, 3.5 mm wide, the claw 6 mm long,; keel 6 mm to upward bend, 5 more to tip. ***Pod*** broad, falcate, 3 cm long, 1 cm wide, ligulate above, broad at tip; valves reticulate, laterally white streaked; sutures thickened; beak 1 mm long, minute; 4 seed. ***Seed*** round, lenticular, flattened, 6–7 mm diam., dull brownish tan or red; hilum orbicular, 1 mm long, 0.5 mm wide; lens prominent. ***Seedling*** from hypogeal germination; epicotyl 9 cm long; stipules united, divided at apex; eophyll petiole 1 mm long, the blade narrow triangular, 3.2 cm long, 1.4 cm wide, nearly truncate at base, acute; first true leaf fully trifoliolate.

PARATYPES. **MÉXICO. Durango:** paraje de Potrillos, entre el Alto de Caballos y el Alto de Potrillos, 2500 m, 6 Sep 1989, *Bénitez 818* (MEXU); Mpio. of Durango, 35–40 km SW of Durango City on road to La Flor, 2000 m, 15 Sep 1979, *Breedlove 44033* (CAS); 30 mi W of Durango, Hwy 40, 24 Jul 1958, *Correll et al. 20113* (TEX); 34 mi W of Cd. Durango, on Hwy. to El Salto, plateau W of Arroyo Mimbres, 2300–2400 m, 29 Aug 1951, *Maysilles 7559* (MICH).

Habitat.—It is found growing in openings of pine-oak forests with manzanilla, *Arbutus*, juniper and grasses. Soils are well drained, rocky and with organic matter, and limestone derived.

Diseases.—Reported to be attacked by *Septoria*.

Comments.—This species is very distinct from the others in this section which all have rather broad ovate leaflets while those of this species are quite narrow and lobed at base. The pod is also distinctly reticulate providing the base for the epithet. It has been shown using PCR-RFLPs on cpDNA to be close to subsp. *ritensis* (Fofana et al. 1999), and to *P. salicifolius* using ITS DNA sequencing (Gaitán et al. 2000).

FINAL DISCUSSION AND CONCLUDING REMARKS

From the previous taxonomic treatment, five points deserve further discussion: i) the discriminant value of characters and the finding of new species, ii) the organization of the genus into sections, iii) the ongoing evolution of the genus in certain parts of its range, iv) the pecularities of distribution, and v) the priorities for conservation. **First,** about the characteristics helping to the identification of new taxa, as expected in this group of Neotropical Phaseolinae (see Maréchal et al. 1981), the flower morphology specially from living specimens has been a powerful tool. The growing out of specimens or field work also allowed the observation of flower spatial configurations *in vivo*, and these

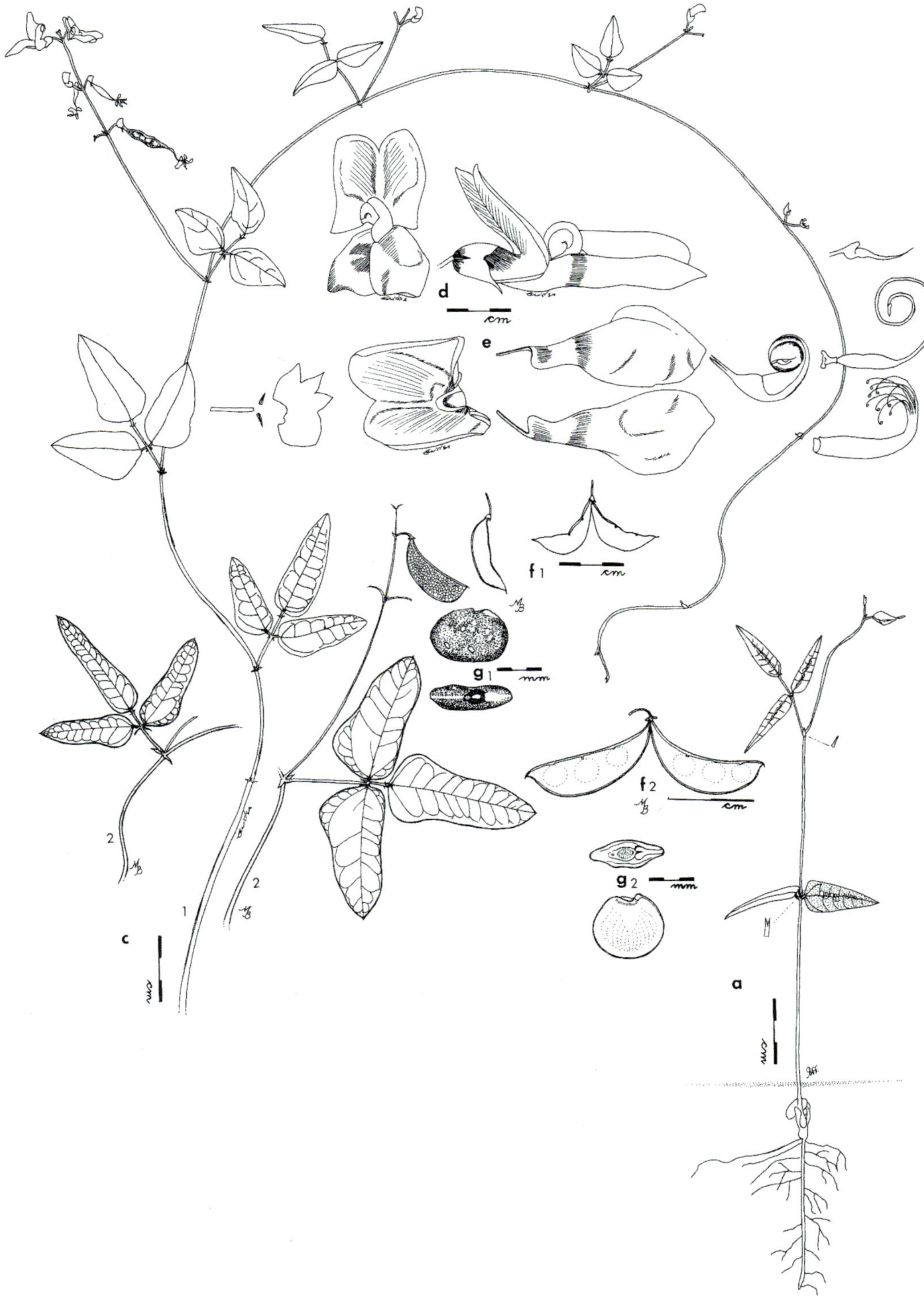

Fig. 96. Illustrations of *Phaseolus reticulatus* Freytag & Debouck.—**a.** Seedling several weeks after germination; note minute petiole and long triangular eophyll, and first true leaf.—**c.1, 2, & 3.** Portions of plant with mature leaves, inflorescences with flowers and pod and vine tip.—**d.** Flower, side view and view from front; note sharply reflexed standard.—**e.** Exploded view of flower showing all parts.—**f.1 & 2.** Pods, side view and dehisced.—**g.1 & 2.** Seeds, side views; note flattened shape and variation in shape and coloration. All drawings made from living material grown in greenhouse at Mayagüez from seed of *Debouck et al. 409* (TARS #125) collected near Km 30 carr. Durango–Mazatlán, Durango, México.

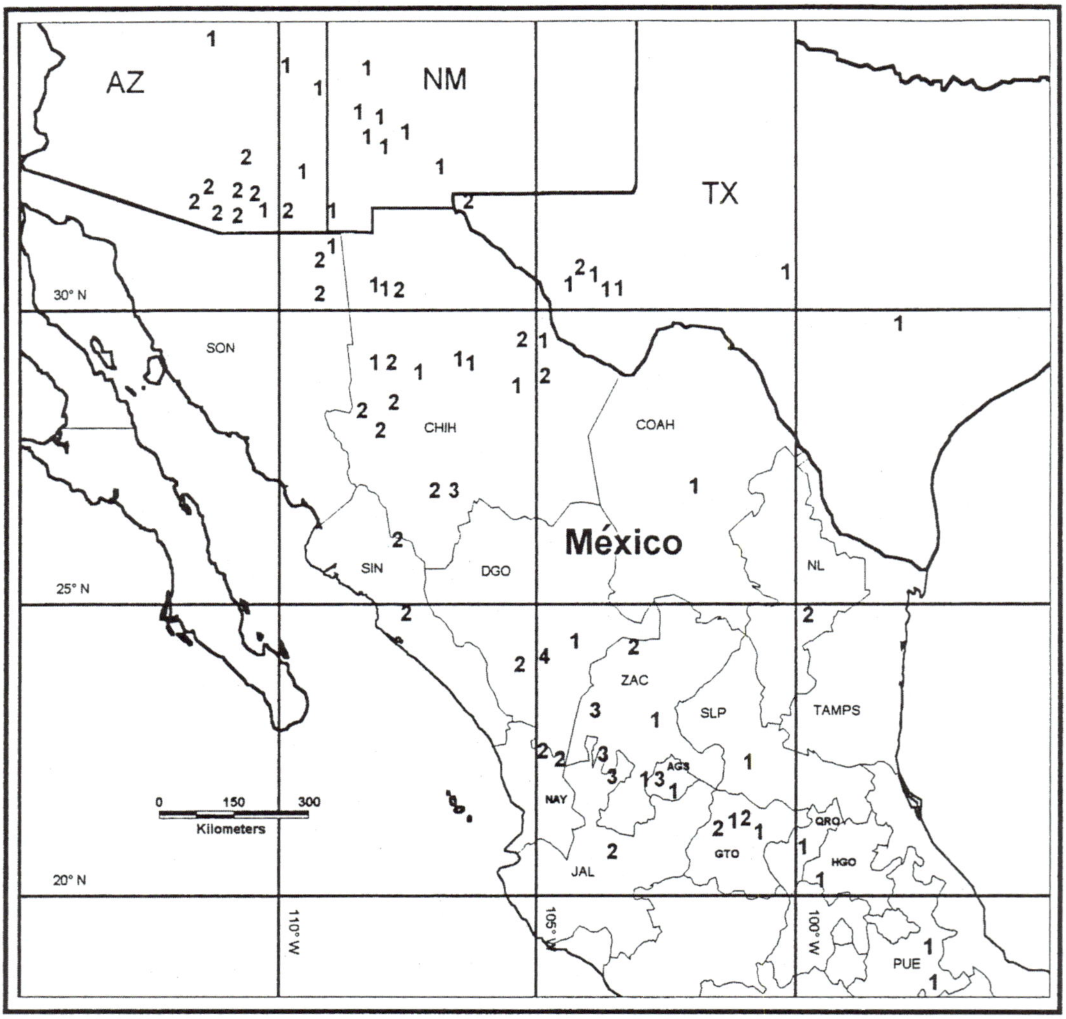

FIG. 97. Distribution of species of Section O. *Coriacei*, as follows: **1** = *P. maculatus* Scheele subsp. *maculatus*; **2** = *P. maculatus* subsp. *ritensis* (Jones) Freytag; **3** = *P. venosus* Piper; **4** = *P. reticulatus* Freytag & Debouck.

data have been of critical importance for the systematics of e.g. the *Minkelersia* group or the *Digitati*. Similarly, they allowed the observation of root systems, most often absent in herbaria. Root characteristics were particularly helpful in the biosystematics of section *Minkelersia*. Blastogeny, amplifying the observations of Baudet (1974), also has proven to be critical, namely to differentiate *P. maculatus* from *P. ritensis*, or *P. carteri* from *P. filiformis*. As a result, several new species have been described (e.g. *P. altimontanus*, *P. costaricensis*, *P. rotundatus*) or re-assessed (e.g. in the section M. *Pedicellati*: *P. esperanzae*, *P. oaxacanus*, *P. purpusii*). The original assumption by the co-authors following Maréchal and co-workers (1978b, 1981) that the kind of taxonomic characters had to be expanded beyond what can result from the direct examination of dried specimens has proven successful. The number of new species described here in addition to those recently described by Delgado (2000), Freytag & Debouck (1996), and Torres *et al.* (2001) may indicate that the exact content of the genus is still unknown, and that field work continues to be relevant and rewarding from a taxonomic perspective.

Second, the organization of the genus into sections raises three points: their identity, relationships, and genesis. Many of the sections recognized here (many of them as 'phylums' in Debouck 2000b or 'subclades' in Delgado et al. 1999) correspond to natural entities on the basis of evidences coming from disciplines as different as morphology, ecology, study of interspecific hybrids, and molecular genetics. And one could see the convergence of these evidences as an indication of robustness of the sections recognized herein. The *Phaseoli* section includes the common bean, and a few species shown to be of the same lineage because they share several cpDNA polymorphisms (Schmit et al. 1993; Llaca et al. 1994). If mtDNA polymorphisms are considered (Hervieu et al. 1993, 1994), then the evidence is becoming stronger, and goes together with results of ITS DNA sequencing (Delgado et al. 1999; Gaitán et al. 2000), showing that obviously section *Coccinei* (without *P. glabellus*) relates to the *Phaseoli*. Interestingly, Linnaeus in his 2nd edition of *Species Plantarum* in 1763 wrote about a *Phaseolus vulgaris* var. *coccineus*! This is the largest gene pool for three important cultigens, viz. the common bean, the scarlet runner, and the year-bean. Remotely sections *Acutifolii* and *Rugosi* (without *P. microcarpus*) are related to the *Phaseoli*, and the former is still phylogenetically related (Delgado et al. 1999; Gaitán et al. 2000). Another good example of natural entity is provided by the Rugosi (excepting *P. microcarpus*; Debouck 2000b). While the rugose testa is an easy character to identify them in the field, allozyme evidence (Jaaska 1996) and molecular data (Delgado et al. 1999; Fofana et al. 1999; Gaitán et al. 2000) have indicated that they can be grouped together. Similarly, these molecular data have also confirmed the grouping of the *Falcati*, viz. *P. leptostachyus*, *P. macvaughii* and *P. micranthus*, initially clustered together thanks to karyotype data (Mercado Ruaro & Delgado Salinas 1996, 1998). As anticipated by Maréchal et al. 1978b, the common bean may represent one end of an orthogenic sequence, while *P. lunatus* would be at the other extreme. Section *Coriacei* is related to the *Paniculati*, and as shown by our results (see also Debouck 2000a; Fofana et al. 1999; Maquet et al. 1999), the gene pool of Lima bean seems thus to be an ample one. The relationships between and within sections is perhaps the least academic question, as breeding programmes have not always embarked on the right species for widecrossing during decades!

If some sections are numerous, others continue to appear with few species, and one wonders whether the current figures are definitive. Over the past few years field work has proven to be useful in disclosing new taxa for sections with single species. *P. vulgaris* was long seen as isolated in the genus, thus with limited possibilities for widecrossing, while its section has now four species, with germplasm available thanks to recent explorations and specific relationships established through crossing experiments and molecular genetics. A similar situation exists with *P. neglectus*, forgotten for almost half century, and now with several sister species. This could be a strong indication that sections with one or few species may eventually include more taxa. However, in the case of *P. chiapasanus* or *P. glabellus* or *P. leptophyllus* or *P. microcarpus*, all herbarium work indicates that they may remain unique, although we have seen that current representativity in herbaria is a weak argument. That uniqueness raises the question of the genesis of the *Phaseolus* taxa, either as products of recent speciation or stable evolutionary dead-ends.

The genesis of sections, and largely that of species as well, is a page still to be written. Quite tentatively, one could think that a larger section of well differentiated and established species may mean a longer period of evolution or early differentiation. From that perspective, the phylum of Lima bean might be an ancient group with its tertiary gene pool in northern Mexico and the southern and southeastern USA, and its secondary gene pool in the Andes. This may also mean that eastern Central America may have been a doorway for a couple of *Phaseolus* species, and perhaps more than once. In that sense, the genus might be recent in the Andean region (see also Sousa & Delgado 1993). On the other hand, Gepts and co-workers (2000) have indicated that the separation of *P. coccineus* from *P. vulgaris* might have occurred two millions years ago, while in the latter the separation of gene pools between Mesoamerica and the Andes might have occurred half a million years ago. Although tentative, these indications are most useful, as they show the progressive establishment of the sections *Phaseoli* and *Coccinei* as recognized here, and the progressive differentiation occurring in a widely distributed species such as *P. vulgaris*.

Third, differentiation of new populations seems to take place actively for several taxa of the sections including *P. coccineus* and *P. pedicellatus*, contrasting with the apparent stability of the groups including *P. macrolepis* or *P. tuerckheimii*. While one could think about an expansion of the former (although now jeopardized by human activities) with the formation of new variants, the later seem to occupy the same range with little variation. Perhaps the high level of outcrossing noted in the *Coccinei*, facilitated by the extrorse stigma (Lackey 1983), is in part responsible for the huge variation observed in that section. The present treatment of the *Coccinei* is an attempt to classify that huge variation. For those who have seen var. *splendens*, var. *purpurascens* or var. *rigidicaulis* in the field (or in herbaria), these taxa are obviously something more than wild *coccineus*. Perhaps the level of outcrossing might not be as high in the *Pedicellati*; however under the selective pressure of the dry habitats in which they thrive they seem under active speciation. Beyond the need for additional collections, there is also a need to understand how microecological factors affect floral ontogenesis, and from there the genetic make-up of bean populations.

Fourth, although the geographic range of some species has been better sampled namely over the past few years (e.g. *P. dumosus, P. vulgaris*) or in the past (e.g. *P. filiformis, P. polystachyus*), the exact distribution of most species is still unknown. Entire regions (e.g. Coahuila, Zacatecas in México, or El Quiché in Guatemala) are relatively poorly sampled. But exploration has proven to be rewarding from this perspective too: the group of the *Digitati* has been better known only thanks to recent explorations in the northern part of the Sierra Madre Oriental. It is thus uneasy to conclude now about the rarity of certain taxa. Pending thus on additional explorations, it seems that some species are endemic, that is occupying quite a small geographic range; it would be the case of *P. amblyosepalus, P. macrolepis*, and *P. plagiocylix*. On the other hand, several taxa may have escaped the attention of botanists and bean germplasm collectors, simply because they bloom and set seeds at a time that is unusual for most species (namely if plant explorers are only after germplasm of wild common bean!). This seems to be the case of *P. parvifolius, P. sonorensis*, or *P. talamancensis*. There are however species which definitively are widespread: *P. leptostachyus, P. microcarpus*, and these species also colonize a wide range in altitude and habitats. Yet, and precisely about the later species where the range of distribution is relatively well explored (our present results), few populations are known from the SW USA about the former, and from Guatemala about the latter. Finally, some species colonize special ecological niches (*P. filiformis, P. macvaughii*), as the coastal areas along the Gulf of California, and may have some special attributes such as tolerance to salinity (Bayuelo et al. 2002a,b). Some species (*P. angustissimus, P. filiformis*) because they thrive in desert habitats may also be tolerant to extreme temperatures (Balasubramanian et al. 2002). Similarly, for tolerance to low temperatures, it would be worth exploring the extreme range of *P. polystachyus* in northeastern USA or that of wild *P. vulgaris* in southern Argentina (although beyond the scope of our study).

Fifth, the location of the populations of different *Phaseolus* species is an important step towards the conservation of these genetic resources. The conservation status of several species has been documented only recently and partially (e.g. Araya et al. 2001), and there are mixed signals. Some species (e.g. *P. lunatus*, and *P. vulgaris*) may have expanded their range due to deforestation and the conversion of many natural vegetations into anthropic savannahs. Not surprisingly, several of these species have been domesticated due to their ecological capability to withstand some periodic ecological disturbance. On the other hand, some species (e.g. *P. macrolepis, P. tuerckheimii*, or *P. xolocotzii*) seem to disappear together with the original climax (forest) vegetation. There are no recent collections for *P. leptophyllus, P. purpusii, P. polystachyus* subsp. *smilacifolius*, almost none since the original explorations, and one could even fear their extinction. Priorities for collecting clearly are where the changes in land use are likely to happen in the very near future, namely in or around the urban areas of Latin America. The junior author has collected populations at sites converted in urban areas since. The fact that several populations of *Phaseolus* species already thrive in protected areas (e.g. the Sierra de Manantlán Biosphere Reserve: Benz et al. 1990) is a positive indication. But a precise inventory of protected areas for *Phaseolus* wild relatives, that will increase the economic relevance of such areas, since the range of several *Phaseolus* species crosses with several biosphere reserves (for their location, see Davis et al. 1997), is still missing. Surely, some existing protected areas can be expanded in order to include known populations (see Araya et al. 2001), others can be established, often without setting apart huge acreage, for instance by adapting roadsides, fallows, recreational areas, etc. But for several populations, given the rhythm at which forests and other original vegetations are cleared, the sole alternative is collecting for *ex situ* conservation. The recording of all populations through herbarium specimens and germplasm samples is thus a powerful tool to monitor the intensity and geographic location of genetic erosion. Although not all populations of *Phaseolus* seen during the course of this study (over 5,000) have been reported here, our data can help to design a comprehensive conservation strategy first, and then to implement its appropriate monitoring.

We hope this work will stimulate new efforts at collecting germplasm, biosystematic and ecological information all leading to a better understanding of bean biology and to increased preservation of valuable habitats and plant resources, for the benefit of human societies.

ACKNOWLEDGMENTS

The senior author thanks the following institutions for economic assistance received: the Tropical Agricultural Research Station (TARS), ARS, USDA; and the Bean/Cowpea CRSP/AID Program at the University of Puerto Rico. Special thanks are due to Instituto Nacional de Investigaciones Agrícolas (INIA), SARH, México, and to Instituto de Ciencias y Tecnologia Agrícola (ICTA), MAG, Guatemala for assistance with field collecting trips. The senior author also acknowledges the valuable assistance extended by the National Seed Storage Laboratory, USDA\ARS at Fort Collins, Colorado for making it possible to continue work on this manuscript since early 1990.

The senior author also wishes to thank the curators of the herbaria of the Royal Botanic Gardens, Kew, the British Museum, London, England, the Instituto Botánico, Universidad de Guadalajara, Jalisco, and the Instituto de Biología, Universidad Nacional Autónoma de México for permission to study their collections at their institutions. These collections were not annotated by placing our labels on the sheets since they were visited before the inception of the idea of attempting a written treatise, but copious notes were taken and much of this information is incorporated herein. The senior author also thank the curators of the following herbaria from which collections were received on loan and annotated: CAS, CAS-DS, EAP, F, GH, MICH, MO, NA, TEX, TEX-LL, UC, US, and WIS while at the Tropical Agriculture Research Station, USDA\ARS, at Mayagüez, Puerto Rico; and from ARIZ, ASU, DES, UC-R and WIS while at the Colorado State University Herbarium (CSU) at Fort Collins, Colorado.

The senior author wishes to express his sincere thanks to the following people with whom field collecting trips were made: George Baxter (Chihuahua, México-1948); Howard Scott Gentry, deceased (Honduras-1966 and México-1967); Nader Vakili (México and Guatemala-1978); Joe Sullivan (México-1981); and Russ Buhrow (Arizona-1984). The senior author also wishes to express special thanks in memoriam to: Louis Williams, deceased and Paul Standley, deceased for sharing their knowledge and experience on many collecting trips from El Zamorano, Honduras (1950–52); and Efrain Hernández Xolocotzi, deceased, the senior author's field companion for nearly two years (1953–55) of collecting bean cultivars throughout most of central and southern México, sponsored by Oficina de Estudios Especiales, SARH and the Rockefeller Foundation.

We appreciate the viable seed received from: A. Evans, University of Cambridge, England; R. Buhrow, Dept. of Plant Sciences, University of Arizona, Tucson, AZ; E. Drijfhout, IVT, Wageningen, The Netherlands; R. Hannan, RPIS/USDA, Pullman, WA; R. Maréchal, Faculté des Sciences Agronomiques de l'Etat, Gembloux, Belgium; G. Nabhan, Desert Botanical Garden, Phoenix, AZ; J. G. Waines, University of California-Riverside, CA; and O. Toro and D. Wood, Genetic Resources Unit, CIAT, Cali, Colombia. Most of these seed collections were grown out, studied and their seed increased at TARS, USDA\ARS, Mayagüez, Puerto Rico, and are listed in a section near the end of the treatise.

We are grateful to the following persons who have assisted with review of the manuscript and provided much valuable advice and criticism: Hubert Bannerot, Ted Dudley (deceased), Claire Thomas Federici, Hugh Iltis, Duane Kolterman, Julio Lopez-Rosa, Amy McPherson, Robert Maréchal, Nancy Morin, Peter Raven, Howard Schwartz, Amy Schuler, Giles Waines, Jun Wen, and an anonymous reviewer.

We are deeply indebted to the following artists who worked long hours for many months under the senior author's supervision in Mayagüez, PR to provide the outline drawings: Ivonne Valdés, student; Enidsa Vásquez, portraitist; and Mary Breckon, botanical illustrator.

Many thanks are due to Duane A. Kolterman for providing some of the Latin diagnoses, and to the junior author who provided the majority of the others, and to Stanley L. Welsh and his students who reviewed and revised them all.

Sincere thanks for secretarial and computer assistance are extended to Hilda Carrero and Osvaldo González, TARS, USDA\ARS at Mayagüez, PR; and to the many people at the National Seed Storage Laboratory (NSSL), USDA\ARS at Fort Collins, Colorado who have made it possible to complete this monograph. Special thanks are extended to Steve Eberhart and Henry Shands, Directors, and Eric Roos, Phillip Stanwood and Christina Walters, Research Leaders during different periods, for their encouragement and the appointment of the senior author as a Collaborator at NSSL while completing the manuscript; to Phillip C. Stanwood, Research Agronomist (retired), who has provided much valuable assistance with the programs and computers used in making the distribution maps and color illustrations, and to Terri Harceg and Peggy Mattei for secretarial assistance.

The junior author acknowledges the financial support for many years of the Centro Internacional de Agricultura Tropical (CIAT), the International Board for Plant Genetic Resources (IBPGR), and the Administration Générale de la Coopération au Développement of Belgium. A total of 27 germplasm explorations throughout Latin America has been made possible thanks to grants from CIAT, IBPGR, the Word Conservation Union (IUCN), USDA, and the Central American network PROFRIJOL. The junior author thanks the Curators of the following herbaria for letting him to study their collections *in situ*: ARIZ, BAA, BM, BRIT, COL, CR, CUZ, DES, ENCB, G, HAO, HUT, K, LIL, LPB, MA, MEXU, MICH, MO, MOL, MSC, QCA, SGO, SI, US, USCG, USJ, USM (and partly those of BR, F, NY).

The junior author expresses special thanks to the following institutions for the support in the field work (for the area of study of this work): Instituto Nacional de Investigaciones Agrícolas de México (now INIFAP), the Universidad Autónoma de Nuevo León, the Instituto de Ciencia y Tecnología Agropecuaria de Guatemala, the Universidad de San Carlos de Guatemala, the Universidad de Costa Rica. The junior author would like to thank the following persons for different types of support, advice or lively discussions in 'phaseology' during the course of this work: Jorge Acosta G., Rodolfo Arraya V., César Azurdia, Rob Bertram, Hubert Bannerot, Francisco Cardenas, Paul Gepts, Guy Le Marchand (deceased), Rogelio Lépiz, Barney Lipscomb, Carmenza Llano, Robert Maréchal, José S. Muruaga M., John Nickel, Henry Shands, Juan José Soto, Joseph Tohme, Orlando Toro, Franciso J. Vásquez, and Trevor Williams.

REFERENCES

Al-Yasiri, S.A. and D.P. Coyne. 1966. Interspecific hybridization in the genus *Phaseolus*. Crop Sci. 6:59–60.

Alvarez, M.N., P.D. Ascher, and D.W. Davis. 1981. Interspecific hybridization in *Euphaseolus* through embryo rescue. HortSci. 16:541–543.

Andrade Aguilar, J.A. and M.T. Jackson. 1988. Attempts at interspecific hybridization between *Phaseolus vulgaris* L. and *P. acutifolius* A. Gray using embryo rescue. Plant Breed. 101:173–180.

Araya Villalobos, R., W.G. González Ugalde, F. Camacho Chacón, P. Sánchez Trejos, and D.G. Debouck. 2001. Observations on the geographic distribution, ecology and conservation status of several *Phaseolus* bean species in Costa Rica. Genet. Resources & Crop Evol. 48:221–232.

Baggett, J.R. 1956. The inheritance of resistance to strains of bean yellow mosaic virus in the interspecific cross *Phaseolus vulgaris* × *P. coccineus*. Plant Dis. Reporter 40:702–707.

Balasubramanian, P.M., A. Vandenberg, P.J. Hucl, and L.V. Gusta. 2000. Relative freezing resistance in the genus *Phaseolus*. Annual Rep. Bean Improvement Coop. (USA) 43:66–67.

Balasubramanian, P., A. Vandenberg, P. Hucl, and L. Gusta. 2002. Physiology of freezing resistance in the genus *Phaseolus*. Annual Rep. Bean Improvement Coop. (USA) 45:6–7.

Bannerot, H. and D.G. Debouck. 1992. L'importance de la double domestication pour l'amélioration du haricot commun (*Phaseolus vulgaris* L.). In: J.C. Mounolou et al. eds. Complexes d'espèces, flux de gènes et ressources génétiques des plantes. Diffusion Lavoisier, Cachan, France. Pp. 495–506.

Barbosa Fevereiro, V.P. 1986–87. *Macroptilium* (Bentham) Urban do Brasil (Leguminosae–Faboideae–Phaseoleae–Phaseolinae). Arq. Jard. Bot. Rio de Janeiro 28:109–180.

Bassett, M.J. 1988. Linkage mapping of marker genes in common bean. In: P. Gepts (ed), Genetic resources of

Phaseolus beans: their maintenance, domestication, evolution, and utilization. Kluwer Academic Publishers, Dordrecht, The Netherlands. Pp. 245–327.

Baudet, J.C. 1973. Intérêt taxonomique des caractères épidermiques dans le complexe *Phaseolus–Vigna*. Bull. Soc. Roy. Bot. Belg. 106:53–59.

Baudet, J.C. 1974. Signification taxonomique des caractères blastogéniques dans la tribu des *Papilionaceae–Phaseoleae*. Bull. Jard. Bot. Nat. Belg. 44:259–293.

Baudet, J.C. 1977. Origine et classification des espèces cultivées du genre *Phaseolus*. Bull. Soc. Roy. Bot. Belg. 110: 65–76.

Baudet, J.C. and R. Maréchal. 1976. Signification taxonomique de la présence de poils uncinulés chez certains genres de *Phaseoleae* et d'*Hedysareae* (*Papilionaceae*). Bull. Jard. Bot. Nat. Belg. 46:419–426.

Baudoin, J.P. 1990. L'amélioration de *Phaseolus lunatus* L. en zones tropicales. 6. Réservoirs génétiques, travaux de sélection, conclusions et perspectives. Bull. Rech. Agron. Gembloux 25:395–442.

Baudoin, J.P., V. Schmit, and B. Wathelet. 1991. Observations on seed protein electrophoretic patterns of some species within the genus *Phaseolus*. Annual Rep. Bean Improvement Coop. (USA) 34:85–86.

Bayuelo-Jímenez, J.S., R. Craig, and J.P. Lynch. 2002a. Salinity tolerance of *Phaseolus* species during germination and early seedling growth. Crop Sci. 42:1584–1594.

Bayuelo-Jímenez, J.S., D.G. Debouck, and J.P. Lynch. 2002b. Salinity tolerance in *Phaseolus* species during early vegetative growth. Crop Sci. 42:2184–2192.

Becerra Velásquez, V.L. and P. Gepts. 1994. RFLP diversity of common bean (*Phaseolus vulgaris*) in its centres of origin. Genome 37:256–263.

Belivanis, T. and C. Doré. 1986. Interspecific hybridization of *Phaseolus vulgaris* L. and *P. angustissimus* A. Gray using in vitro embryo culture. Plant Cell Rep. 5:329–331.

Bentham, G. 1837. Commentationes de leguminosarum generibus. Sollingeri, Vienna, Austria. Pp. 72–78.

Bentham, G. 1840. Commentationes de leguminosarum generibus. Annln. Wien Mus. Naturg. 2:61–142.

Bentham, G. 1844. Botany of the voyage of H.M.S. Sulphur. Botanical descriptions, London, United Kingdom. 195p.

Bentham, G. and J.D. Hooker. 1865. Genera plantarum, Vol. 1, Reeve & Co., London, United Kingdom, Pp. 434–600.

Benz, B.F., L.R. Sánchez Velásquez, and F.J. Santana Michel. 1990. Ecology and ethnobotany of *Zea diploperennis*: preliminary investigations. Maydica 35:85–98.

Berglund-Brücher, O. and H. Brücher. 1974. Murutungo, eine semi-domestizierte wildbohne (*Phaseolus flavescens* Piper) aus den tropischen Gebirgen Sudamerikas. Angew. Bot. 48:209–220.

Braak, J.P. and E. Kooistra. 1975. A successful cross between *Phaseolus vulgaris* L. and *P. ritensis* Jones with the aid of embryo culture. Euphytica 24:669–679.

Brandegee, T.S. 1893. Flora of the Cape region. Proc. Calif. Acad. Sci. Ser. 2, 3:130.

Brandegee, T.S. 1908. New species of Mexican plants. Zoe 5:248.

Brandegee, T.S. 1910. Plantae Mexicanae Purpusianae. II. Univ. Calif. Publ. Bot. 4:85–95.

Brandegee, T.S. 1912. Plantae Mexicanae Purpusianae. IV. Univ. Calif. Publ. Bot. 4:269–281.

Breedlove, D.E. 1973. The phytogeography and vegetation of Chiapas (México). In: "Vegetation and vegetational history of northern Latin America," A. Graham, ed. Elsevier Scientific Publishing Co., Amsterdam, The Netherlands. Pp. 149–165.

Breedlove, D.E. 1986. Listados florísticos de México. IV. Flora de Chiapas. Universidad Nacional Autónoma de México, Instituto de Biología, México, D.F., México.

Breedlove, D.E. and R.M. Laughlin. 1993. The flowering of man—A Tzotzil botany of Zinacantán. Smithsonian Contr. Anthropol. 35:1–706.

Britton, N.L. 1918. Flora of Bermuda. Charles Scribner's Sons, New York, USA. Pp. 183.

Brücher, H. 1988. The wild ancestor of *Phaseolus vulgaris* in South America. In: P. Gepts, ed. Genetic resources of *Phaseolus* beans: their maintenance, domestication, evolution, and utilization. Kluwer Academic Publishers, Dordrecht, The Netherlands. Pp. 185–214.

Bruneau, A., J.J. Doyle. 1990. A chloroplast DNA inversion as a subtribal character in the Phaseolae (Leguminosae). Syst. Bot. 15:378–386.

Bruneau, A., J.J. Doyle, and J.L. Doyle. 1995. Phylogenetic relationships in Phaseoleae: evidence from chloroplast DNA restriction site characters. In: M.D. Crisp and J.J. Doyle, eds. Advances in legume systematics 7: phylogeny. Royal Botanic Gardens, Kew, England. Pp. 309–330.

BUHROW, R.A. 1980. Frost tolerance of the Phaseolinae of the southwestern United States. Annual Rep. Bean Improvement Coop. (USA) 23:62–64.

BUHROW, R.A. 1983. The wild beans of southwestern North America. Desert Plants 5:67–69, 82–88.

BUKASOV, S.M. 1930. Cultivated plants of México, Guatemala, and Colombia. Bull. Appl. Gen. and Plant Breed. Supp. 47:151–176. Leningrad. Trans. H. J. Kidd.

BURKART, A. 1952. Las leguminosas argentinas silvestres y cultivadas. Acme Agency, 2nd ed., Buenos Aires, Argentina. 569p.

BURKART, A. and H. BRÜCHER. 1953. *Phaseolus aborigineus* Burkart, die mutmaßliche andine Stammform der Kulturbohne. Züchter 23:65–72.

CAMARENA, F. and J.P. BAUDOIN. 1987. Obtention des premiers hybrides interspécifiques entre *Phaseolus vulgaris* et *Phaseolus polyanthus* avec le cytoplasme de cette dernière forme. Bull. Rech. Agron. Gembloux 2:43–55.

CAMPOS VILLANUEVA, A., L. CORTÉS ARRIAGA, P. DÁVILA ARRANDA, A. GARCÍA MENDOZA, J. REYES SANTIAGO, G. TORIZ ACOSTA, L. TORRES COLÍN, and R. TORRES COLÍN. 1992. Plantas y flores de Oaxaca. Universidad Nacional Autónoma de México, Instituto de Biología, México, D.F., México. 62p.

CARDONA, C., C.E. POSSO, J. KORNEGAY, J. VALOR, and M. SERRANO. 1989. Antibiosis effects of wild dry bean accessions on the Mexican bean weevil and the bean weevil (Coleoptera: Bruchidae). J. Econ. Entomol. 82:310–315.

CHENG, S.S., M.J. BASSETT, and K.H. QUESENBERRY. 1981. Cytogenetic analysis of interspecific hybrids between common bean and scarlet runner bean. Crop Sci. 21:75–79.

CORRELL, D.S. and M.C. JOHNSTON. 1970. Manual of the vascular plants of Texas. Texas Research Foundation, Renner, Texas, USA. 1881p.

COYNE, D.P. 1964. Species hybridization in *Phaseolus*. J. Hered. 55:5–6.

DAVIS, S.D., V.H. HEYWOOD, O. HERRERA-MACBRYDE, J. VILLA-LOBOS, and A.C. HAMILTON, eds. 1997. Centres of plant diversity – A guide and strategy for their conservation. Volume 3 The Americas. World Wide Fund for Nature, Cambridge, United Kingdom. 562p.

DARWIN, C. 1858. On the agency of bees on the fertilization of papilionaceous flowers and on the crossing of kidney beans. Ann. Mag. Nat. Hist. 2:459–465.

DEBOUCK, D.G. 1983a. Contribution à l'étude des correlations périméristématiques intervenant dans les processus de ramification et de floraison chez les haricots (*Phaseolus*). Ph.D. Thesis, Faculté des Sciences Agronomiques, Gembloux, Belgium.

DEBOUCK, D.G. 1983b. Observations sur les processus de floraison et de ramification chez quelques *Phaseolus*. C. R. Acad. Sc. Paris. 296:425–428.

DEBOUCK, D.G. 1991. Systematics and morphology. In: A. van Schoonhoven and O. Voysest Voysest, eds. Common beans: research for crop improvement. Commonwealth Agricultural Bureaux International, Wallingford, United Kingdom. Pp. 55–118.

DEBOUCK, D.G. 1992. Frijoles, *Phaseolus* spp. In: E. Hernández Bermejo and J. León, eds. Cultivos marginados: otra perspectiva de 1492. Food and Agriculture Organization of the United Nations, Rome, Italy. Pp. 45–60.

DEBOUCK, D.G. 1999. Diversity in *Phaseolus* species in relation to the common bean. In: S.P. Singh, ed. Common bean improvement in the twenty-first century. Kluwer Academic Publishers, Dordrecht, The Netherlands. Pp. 25–52.

DEBOUCK, D.G. 2000a. Biodiversity, ecology and genetic resources of *Phaseolus* beans—Seven answered and unanswered questions. In: K. Oono, ed. Wild legumes. Ministry of Agriculture, Forestry and Fisheries, and National Institute of Agrobiological Resources, Tsukuba, Ibaraki, Japan. Pp. 95–123.

DEBOUCK, D.G. 2000b. Genetic resources of *Phaseolus* beans: patterns in time, space, and people. In: M.A. Fueyo, A.J. González, J.J. Ferreira, and R. Giraldez, eds. II Seminario de Judía de la Península Ibérica. Actas de la Asociación Española de Leguminosas, Villaviciosa, Spain. Pp. 17–39.

DEBOUCK, D.G., J.H. LIÑAN JARA, A. CAMPANA SIERRA, and J.H. DE la CRUZ ROJAS. 1987. Observations on the domestication of *Phaseolus lunatus* L. FAO/IBPGR Plant Genetic Resources Newsl. 70:26–32.

DEBOUCK, D.G., R. ARAYA VILLALOBOS, R.A. OCAMPO SÁNCHEZ, and W.G. GONZÁLEZ UGALDE. 1989a. Collecting *Phaseolus* in Costa Rica. FAO/IBPGR Pl. Genet. Resources Newsl. 78/79:44–46.

DEBOUCK, D.G., R. CASTILLO, and J. TOHME. 1989b. Observations on little-known *Phaseolus* germplasm of Ecuador. FAO/IBPGR Pl. Genet. Resources Newsl. 80:15–21.

DEBOUCK, D.G., A. MAQUET, and C.E. POSSO. 1989c. Biochemical evidence for two different gene pools in lima beans, *Phaseolus lunatus* L. Annual Rep. Bean Improvement Coop. (USA) 32:58–59.

Debouck, D.G. and J. Smartt. 1995. Beans, *Phaseolus* spp. (Leguminosae-Papilionoideae). In: J. Smartt and N.W. Simmonds, eds. Evolution of crop plants. Second Edition. Longman Scientific & Technical, London, United Kingdom. Pp. 287–294.

Debouck, D.G., O. Toro, O.M. Paredes, W.C. Johnson, and P. Gepts. 1993. Genetic diversity and ecological distribution of *Phaseolus vulgaris* (Fabaceae) in northwestern South America. Econ. Bot. 47:408–423.

de Candolle, A.P. 1825. Prodromus systematis naturalis regni vegetabilis 2. Paris, Strasbourg and London. 644 p.

de la Cruz S., J.R. and R. Sagastume L. 1983. Mapa de zonas de vida a nivel de reconocimiento. Escala 1:6000,000. Instituto Nacional Forestal, Instituto Geográfico Militar, Guatemala, Guatemala, C. America. 4 p.

Delgado Salinas, A.O. 1985. Systematics of the genus *Phaseolus* (Leguminosae) in North and Central America. Ph.D. Thesis, Univ. of Texas–Austin, Texas, USA.

Delgado Salinas, A.O. 1988. Variation, taxonomy, domestication, and germplasm potentialities in *Phaseolus coccineus*. In: P. Gepts, ed. Genetic resources of *Phaseolus* beans: their maintenance, domestication, evolution, and utilization. Kluwer Academic Publishers, Dordrecht, The Netherlands. Pp. 441–463.

Delgado Salinas, A.O. 2000. New species of Mexican *Phaseolus* (Fabaceae). Syst. Bot. 25:414–436.

Delgado Salinas, A.O. 2001. *Phaseolus* L. In: W.D. Stevens, C. Ulloa Ulloa, A. Pool, and O.M. Montiel, eds. Flora de Nicaragua—Angiospermas (Fabaceae–Oxalidaceae). Missouri Botanical Garden Press, St. Louis, Missouri, USA. Pp. 1042–1046.

Delgado Salinas, A.O., A. Bonet, and P. Gepts. 1988. The wild relative of *Phaseolus vulgaris* in Middle America. In: P. Gepts, ed. Genetic resources of *Phaseolus* beans: their maintenance, domestication, evolution, and utilization. Kluwer Academic Publishers, Dordrecht, The Netherlands. Pp. 163–184.

Delgado Salinas, A.O., A. Bruneau, and J.J. Doyle. 1993. Chloroplast DNA phylogenetic studies in New World Phaseolinae (Leguminosae: Papilionoideae: Phaseoleae). Syst. Bot. 18:6–17.

Delgado Salinas, A.O., E. Martínez Hernández, and P. Fernandez Ortuño. 1982. Estudio del polen de *Phaseolus chiapasanus* Piper (Leguminosae: Phaseolinae). Bol. Soc. Bot. México. 43:25–34.

Delgado Salinas, A.O., T. Turley, A. Richman, and M. Lavin. 1999. Phylogenetic analysis of the cultivated and wild species of *Phaseolus* (Fabaceae). Syst. Bot. 24:438–460.

Diels, L. 1937. Beiträge zur kenntnis der vegetation und flora von Ecuador. Bibliot. Bot. 29:98.

Dietrich, D. 1847. Synopsis plantarum seu enumeratio systematica. Sectio 4. Bernhard Friedrich Voigth. Weimar, Germany. Pp. 1188–1197.

Dirección General de Geografía. 1982. Carta de México 1:250,000. SPP, EUM, México, D.F. 255 p.

Don, G. 1832. General system of gardening and botany 2:349–356.

Donnell–Smith, J. 1913. Undescribed plants from Guatemala and other Central American republics XXXVII. Bot. Gaz. 56:51–62.

Doutt, M.T., 1932 Anatomy of *Phaseolus vulgaris* L. var. Black Valentine. Michigan Technical Bulletin 128:1–31.

Drijfhout, E. and W.J. Blok. 1987. Inheritance of resistance to *Xanthomonas campestris* pv. *phaseoli* in tepary bean (*Phaseolus acutifolius*). Euphytica 36:803–808.

Estrada C., A.E. and A. Martínez M. 2000. Legumes from the central part of the state of Chihuahua, México. SIDA 19:351–360.

Federici, C.T.& J.G. Waines. 1988. Interspecific hybrid compatibility of selected *Phaseolus vulgaris* L. lines with *P. acutifolius* A. Gray, *P. lunatus* L., and *P. filiformis* Bentham. Annual Rep. Bean Improvement Coop. (USA) 31:201–202.

Fermond, C. 1855. Recherches sur les fécondations réciproques de quelques végétaux. Bull. Soc. Bot. Fr. 2:748–754.

Fernald, M.L. 1942. The seventh century of additions to the flora of Virginia. Rhodora 44:416–452.

Fernald, M.L. and A. C. Kinsey. 1943. Edible wild plants of eastern North America. Idlewild Press, Cornwall-on-Hudson, New York, USA. 452p.

Fofana, B., J.P. Baudoin, X. Vekemans, D.G. Debouck, and P. du Jardin. 1999. Molecular evidence for an Andean origin and a secondary gene pool for the Lima bean (*Phaseolus lunatus* L.) using chloroplast DNA. Theor. Appl. Genet. 98:202–212.

Fozdar, B.S. 1963. Cytological investigation of parents, offspring and backcross derivatives involved in the interspecific cross *Phaseolus lunatus* L. × *P. polystachyus* (L.) B.S.P. Dissert. Abstr. 24:480–481.

Freeman, G.F. 1912. Southwestern beans and teparies. Univ. Ariz. Agric. Exp. Station Bull. 68:573–619.

Freeman, G.F. 1913 The tepary, a new cultivated legume from the Southwest. Bot. Gaz. 56:395–417.

Freytag, G.F. 1955. Variation of the common bean in Central America. Ph.D. Thesis, Washington University, St. Louis, Missouri, USA.

FREYTAG, G.F. 1965. Clasificación del frijol común (*Phaseolus vulgaris* L.) y especies afines. Ceiba 11:51–64.

FREYTAG, G.F. 1989. Inheritance of resistance to common bacterial blight in tepary bean. Annual Rep. Bean Improvement Coop. (USA) 32:101–102.

FREYTAG, G.F., R.O. CRAVIOTO, J. GUZMÁN, and G. MASSIEU. 1956. Estudio sobre las propriedades nutritivas del frijol. 1. El contenido total de proteina de los tipos de frijol. Tech. Bull. No. 19. O.E.E., S.A., México.

FREYTAG, G.F. and D.G. DEBOUCK. 1996. *Phaseolus costaricensis*, a new wild bean species (Phaseolinae, Leguminosae) from Costa Rica and Panama, Central America. Novon 6:157–163.

GAITÁN, E., O. TORO, J. TOHME, and D.G. DEBOUCK. 2000. Molecular taxonomy of the genus *Phaseolus* through ITS sequencing. Annual report. Centro Internacional de Agricultura Tropical, Cali, Colombia. Pp. 56–60.

GARVIN, D.F. and N.F. WEEDEN. 1994. Isozyme evidence supporting a single geographic origin for domesticated tepary bean. Crop Sci. 34:1390–1395.

GENTRY, H.S. 1942. Rio Mayo plants. Carnegie Institution, Publication No. 527, Washington, D.C., USA. 328p.

GENTRY, H.S. 1957. Los pastizales de Durango–Estudio ecológico, fisiográfico y florístico. Instituto Mexicano de Recursos Naturales Renovables, México, D.F., México. 361p.

GENTRY, H.S. 1969. Origin of the common bean, *Phaseolus vulgaris* L. Econ. Bot. 23:55–69.

GEPTS, P. 1993. The use of molecular and biochemical markers in crop evolution studies. In: M. K. Hecht, ed. Evolutionary Biology. Volume 27. Plenum Press, New York, USA. Pp. 51–94.

GEPTS, P. 1996. Origin and evolution of cultivated *Phaseolus* species. In: B. Pickersgill and J.M. Lock, eds. Advances in legume systematics. Part 8. Legumes of economic importance. Royal Botanic Gardens, Kew, United Kingdom. Pp. 65–74.

GEPTS, P. 1999. Development of an integrated linkage map. In: S.P. Singh, ed. Common bean improvement in the twenty-first century. Kluwer Academic Publishers, Dordrecht, The Netherlands. Pp. 53–91.

GEPTS, P. and F.A. BLISS. 1985. F_1 hybrid weakness in the common bean–Differential geographic origin suggests two gene pools in cultivated bean germplasm. J. Hered. 76:447–450.

GEPTS, P., T.C. OSBORN, K. RASHKA, and F.A. BLISS. 1986. Phaseolin protein variability in wild forms and landraces of the common bean (*Phaseolus vulgaris* L.): evidence for multiple centers of domestication. Econ. Bot. 40:451–468.

GEPTS, P., R. PAPA, S. COULIBALY, A. GONZÁLEZ MEJÍA, and R. PASQUET. 2000. Wild legume diversity and domestication: insights from molecular methods. In: K. Oono, ed. Wild legumes. Ministry of Agriculture, Forestry and Fisheries, and National Institute of Agrobiological Resources, Tsukuba, Ibaraki, Japan. Pp. 19–31.

GÓMEZ PIGNATARO, L.D. 1986. Vegetación de Costa Rica–Apuntes para una biogeografía costarricense. Vol. 1. Editorial Universidad Estatal a Distancia, San José, Costa Rica. 327p.

GRAY, A. 1850. Plantae wrightianae texano–neo-mexicanae: an account of a collection of plants made by Charles Wright, A.M., in an expedition from Texas to El Paso, New México, in the summer and autumn of 1849. Part I. Smithson. Contr. Knowledge 3(5):1–146.

GRAY, A. 1853. Plantae wrightianae texano–mexicanae. Part II. Smithson. Contr. Knowledge 5(6):1–119.

GREENE, E.L. 1881. New plants of New México and Arizona. Bot. Gaz. 6:217–219.

GREENMAN, J.M. 1907. New or noteworthy spermatophytes from México, Central America, and the West Indies. Field Columbian Mus. Publ. Bot. Ser. 2:247–287.

GUTIÉRREZ SALGADO, A., P. GEPTS, and D.G. DEBOUCK. 1995. Evidence for two gene pools of the Lima bean, *Phaseolus lunatus*, in the Americas. Genet. Resources & Crop Evol. 42:15–28.

HAMANN, A., D. ZINK, and W. NAGL. 1995. Microsatellite fingerprinting in the genus *Phaseolus*. Genome 38:507–515.

HARMS, H. 1921. Einige neue *Phaseolus*–Arten. Notizbl. Bot. Gart. Mus. Berlin–Dahlem 7:503–508.

HASSLER, E. 1923. Revisio specierum austro-americanorum generis Phaseoli L. Candollea 1:417–472.

HERMANN, F.J. 1948. Notes on North American Leguminosae. J. Washington Acad. Sci. 38:236–238.

HERNÁNDEZ XOLOCOTZI, E., H. CRUM, W.B. FOX, and A.J. SHARP. 1951. A unique vegetational area in Tamaulipas. Bull. Torrey Bot. Club 78:458–463.

HERNÁNDEZ XOLOCOTZI, E., S. MIRANDA Colín, and C. PRYWER. 1959. El origen de *Phaseolus coccineus* L. *darwinianus* Hdz. X. and Miranda C., subspecies nova. Rev. Soc. Mex. Hist. Nat. 20:99–121.

HERNÁNDEZ XOLOCOTZI, E., A. RAMOS RODRÍGUEZ, and M.A. MARTÍNEZ ALFARO. 1979. Etnobotánica. In: E.M. Engleman, ed. Contribuciones al conocimiento del fríjol (*Phaseolus*) en México. Colegio de Postgraduados, Chapingo, México, México. Pp. 113–138.

Hervieu, F., L. Charbonnier, H. Bannerot, and G. Pelletier. 1993. The cytoplasmic male-sterility (CMS) determinant of common bean is widespread in *Phaseolus coccineus* L. and *Phaseolus vulgaris* L. Curr. Genet. 23:149–155.

Hervieu, F., H. Bannerot, and G. Pelletier. 1994. A unique cytoplasmic male sterility (CMS) determinant is present in three *Phaseolus* species characterized by different mitochondrial genomes. Theor. Appl. Genet. 88:314–320.

Hickey, M. and C. King. 2000. The Cambridge illustrated glossary of botanical terms. Cambridge University Press, Cambridge, United Kingdom. 208p.

Hinton, J. and J. Rzedowski, 1972. George B. Hinton, collector of plants in southwestern México. J. Arnold Arbor. 53:141–181.

Honma, S. 1956. A bean interspecific hybrid. J. Hered. 47:217–220.

Honma, S. and O. Heeckt, 1959. Interspecific hybrid between *Phaseolus vulgaris* and *P. lunatus*. J. Hered. 50:233–237.

Hucl, P. and G.J. Scoles. 1985. Interspecific hybridization in the common bean: a review. HortScience 20:352–357.

Ibrahim, A.M. and D.P. Coyne. 1975. Genetics of stigma shape, cotyledon position, and flower color in reciprocal crosses between *Phaseolus vulgaris* L. and *Phaseolus coccineus* (Lam.) and implications in breeding. J. Amer. Soc. Hort. Sci. 100:622–626.

Isely, D. 1993. *Phaseolus*. In: J.C. Hickman, ed. The Jepson manual–Higher plants of California. University of California Press, Berkeley, California, USA. 641p.

Isely, D. and R. Polhill. 1980. Leguminosae subfamily Papilionoideae. Taxon 29:105–119.

Ivanov, N.R. 1928. Peculiarities in the formation of *Phaseolus* L. species in the Old and the New World. New. Bull. Applied Bot. Pl. Breed. 2(19); 185–208. Trans. by INSDOC, Delhi-110012.

Jaaska, V. 1996. Isoenzyme diversity and phylogenetic affinities among the *Phaseolus* beans (Fabaceae). Pl. Syst. Evol. 200:233–252.

Jacob, M., D. Zink, and W. Nagl. 1995. RFLPs of the rRNA genes in the genus *Phaseolus*. Genet. Resources & Crop Evol. 42:97–106.

Jones, M.E. 1908. New species and notes. Contr. West. Bot. 12:1–81.

Jones, M.E. 1929. Notes and new species of Mexican plants. Contr. West. Bot. 15:123–157.

Kaiser, W.J. 1981. Diseases of chickpea, lentil, pigeon pea and tepary bean in continental United States and Puerto Rico. Econ. Bot. 35:300–320.

Kami, J., V. Becerra Velásquez, D.G. Debouck, and P. Gepts. 1995. Identification of presumed ancestral DNA sequences of phaseolin in *Phaseolus vulgaris*. Proc. Natl. Acad. Sci. 92:1101–1104.

Kaplan, L. 1956. The cultivated beans of the prehistoric Southwest. Ann. Mo. Bot. Gard. 43:189–251.

Kaplan, L. and T. Lynch. 1999. *Phaseolus* (Fabaceae) in archaeology: AMS radiocarbon dates and their significance for pre-Colombian agriculture. Econ. Bot. 53:261–272.

Katanga, K. and J.P. Baudoin. 1987a. Obtention de nouveaux hybrides interspécifiques avec *Phaseolus lunatus* L. Bull. Rech. Agron. Gembloux 22:33–42.

Katanga, K. and J.P. Baudoin. 1987b. Obtention et observations d'un nouvel hybride interspécifique entre deux espèces: *Phaseolus filiformis* Benth. × *Phaseolus angustissimus* A.Gray. Bull. Rech. Agron. Gembloux 22:153–160.

Katanga, K. and J.P. Baudoin. 1990. Analyses méiotiques des hybrides F_1 et étude des descendances F2 chez quatre combinaisons interspécifiques avec *Phaseolus lunatus* L. Bull. Rech. Agron. Gembloux 25:237–250.

Kearney, T.H. and R.H. Peebles. 1939. Arizona plants. J. Wash. Acad. Sci. 29:485.

Kearney, T.H. and R.H. Peebles. 1960. Arizona flora. Second ed. University of California Press, Berkeley, California, USA. 1085p.

Kedar, N. and W.P. Bemis. 1960. Hybridization between two species of *Phaseolus* separated by physiological and morphological blocks. Am. Soc. Hort. Sci. 76:397–402.

Kelly, J.D., K.A. Schneider, J.M. Kolkman. 1999. Breeding to improve yield. In: S.P. Singh, ed. Common bean improvement in the twenty-first century. Kluwer Academic Publishers, Dordrecht, The Netherlands. Pp. 185–222.

Khairallah, M.M., B.B. Sears, and M.W. Adams. 1992. Mitochondrial restriction fragment length polymorphisms in wild *Phaseolus vulgaris* L.: insights on the domestication of the common bean. Theor. Appl. Genet. 84:915–922.

Koenig, R. and P. Gepts. 1989. Allozyme diversity in wild *Phaseolus vulgaris*: further evidence for two major centers of genetic diversity. Theor. Appl. Genet. 78:809–817.

Koenig, R., S.P. Singh, and P. Gepts. 1990. Novel phaseolin types in wild and cultivated common bean (*Phaseolus vulgaris*, Fabaceae). Econ. Bot. 44:50–60.

KOINANGE, E.M.K. and P. GEPTS. 1992. Hybrid weakness in wild *Phaseolus vulgaris* L. J. Hered. 83:135–139.

KOINANGE, E.M.K., S.P. SINGH, and P. GEPTS. 1996. Genetic control of he domestication syndrome in common bean. Crop Sci. 36:1037–1045.

KORNEGAY, J., C. CARDONA, and C.E. POSSO. 1993. Inheritance of resistance to Mexican bean weevil in common bean, determined by bioassay and biochemical tests. Crop Sci. 33:589–594.

KUBOYAMA, T., Y. SHINTAKU, and G. TAKEDA. 1991. Hybrid plant of *Phaseolus vulgaris* L. and *P. lunatus* L. obtained by means of embryo rescue and confirmed by restriction endonuclease analysis of rDNA. Euphytica 54:177–182.

LACKEY, J.A. 1978. Leaflet anatomy of Phaseoleae (Leguminosae–Papilionoideae) and its relation to taxonomy. Bot. Gaz. 139:436–446.

LACKEY, J.A. 1979. A chromosome atlas of the Phaseoleae (Leguminosae–Papilionoideae). Iselya 1:87–114.

LACKEY, J.A. 1980. Chromosome numbers in the Phaseoleae (Fabaceae: Faboideae) and their relation to taxonomy. Amer. J. Bot. 67:595–602.

LACKEY, J.A. 1981. Tribe 10. Phaseoleae DC. In: R.M. Polhill, and P.H. Raven, eds. Advances in legume systematics, Royal Botanic Gardens, Kew, England. Pp. 301–327.

LACKEY, J.A. 1983. A review of generic concepts in American Phaseolinae (Fabaceae, Faboideae). Iselya 2:21–64.

LACKEY, J.A. and W.G. D'ARCY. 1980. *Phaseolus*. Family 83. Leguminosae. Flora of Panama. Ann. Missouri Bot. Gard. 67:746–751.

LEAKEY, C.L.A. 1988. Genotypic and phenotypic markers in common bean. In: P. Gepts, ed. Genetic Resources of *Phaseolus* beans: their maintenance, domestication, evolution, and utilization. Kluwer Academic Publishers, Dordrecht, The Netherlands. Pp. 245–327.

Le MARCHAND, G., R. MARÉCHAL, and J.C. BAUDET. 1976. Observations sur quelques hybrides dans le genre *Phaseolus*. 3. *P. lunatus*: nouveaux hybrides et considérations sur les affinités interspécifiques. Bull. Rech. Agron. Gembloux 22:183–200.

Le MARCHAND, G. and R. MARÉCHAL. 1977a. Wide crossing in Lima beans. Trop. Grain Legumes Bull. 8:7–11.

Le MARCHAND, G. and R. MARÉCHAL. 1977b. Chromosome pairing in interspecific hybrids reveal the value of pollen morphology for deducting phylogenetic affinities in the genus *Phaseolus*. Proceedings 8th Congr. Eucarpia, Interspecific hybridization in plant breeding. Madrid, Spain. Pp. 335–337.

LEONARD, M.F., L.C. STEPHENS, and W.L. SUMMERS. 1987. Effect of maternal genotype on development of *Phaseolus vulgaris* L. × *P. lunatus* L. interspecific hybrid embryos. Euphytica 36:327–332.

LEPOIVRE, P. and J.P. BAUDOIN. 1994. Caractérisation de l'agent responsable de l'ascochytose de *Phaseolus vulgaris* en Afrique et en Amérique du Sud. Bull. Séances Acad. Roy. Sci. Outre-Mer (Belgique) 39:557–566.

LINNAEUS, C. 1753. Species Plantarum. A facsimile (1957–59) of the first edition, 1753. Ray Society, London, England. Pp. 723–725.

LIOGIER, H.A. 1988. Descriptive flora of Puerto Rico and adjacents islands. Editorial de la Universidad de Puerto Rico, Río Piedras, Puerto Rico. 481p.

LLACA, V., A. DELGADO SALINAS, and P. GEPTS. 1994. Chloroplast DNA as an evolutionary marker in the *Phaseolus vulgaris* complex. Theor. Appl. Genet. 88:646–652.

LORZ, A.P. 1952. An interspecific cross involving lima bean *Phaseolus lunatus* L. Science 115:702–703.

LYNCH, J., A. GONZÁLEZ, J.M. TOHME, and J. GARCÍA. 1992. Variation in characters related to leaf photosynthesis in wild bean populations. Crop Sci. 32:633–640.

MACFADYEN, J. 1837. The flora of Jamaica: description of the plants of that island. Vol. 1. Ranunculaceae–Leguminosae. Longman, Orme, Brown, Greene, and Longmans. London, United Kingdom. 351p.

MACMILLAN, C. 1892. The Metaspermae of the Minnesota valley. Harrison & Smith, Minneapolis, Minnesota, USA. 826p.

MANSHARDT, R.M. and J.G. WAINES. 1983. Isozyme variation and the origin of domesticated tepary beans (*Phaseolus acutifolius* Gray). Annual Rep. Bean Improvement Coop. (USA) 26:18–19.

MANSHARDT, R.M. and M.J. BASSETT. 1984. Inheritance of stigma position in *Phaseolus vulgaris* × *P. coccineus* hybrid populations. J. Hered. 75:45–50.

MAQUET, A., A. GUTIÉRREZ, and D.G. DEBOUCK. 1990. Further biochemical evidence for the existence of two gene pools in lima beans. Annual Rep. Bean Improvement Coop. (USA) 33:128–129.

MAQUET, A. and J.P. BAUDOIN. 1996. New lights on phyletic relations in the genus *Phaseolus*. Annual Rep. Bean Improvement Coop. (USA) 39:203–204.

Maquet, A., X. Vekemans, and J.P. Baudoin. 1999. Phylogenetic study on wild allies of Lima bean, *Phaseolus lunatus (Fabaceae)*, and implications on its origin. Plant Syst. Evol. 218:43–54.

Maréchal, R. 1970. Données cytologiques sur les espèces de la sous-tribu des *Papilionaceae-Phaseoleae-Phaseolinae*. Deuxième série. Bull. Jard. Bot. Nat. Belg. 40:307–348.

Maréchal, R. 1971. Observations sur quelques hybrides dans le genre *Phaseolus*. II. Les phénomènes méiotiques. Bull. Rech. Agron. Gembloux, N.S. 6:461–489.

Maréchal, R. and J.P. Baudoin. 1978. Observations sur quelques hybrides dans le genre *Phaseolus*. IV. L'hybride *Phaseolus vulgaris* × *P. filiformis*. Bull. Rech. Agron. Gembloux 13:233–240.

Maréchal, R., J. Mascherpa, and F. Stainier. 1978a. Combinaisons et noms nouveaux dans les genres *Phaseolus, Minkelersia, Macroptilium, Ramirezella* et *Vigna*. Taxon 27:199–202.

Maréchal, R., J. Mascherpa, and F. Stainier. 1978b. Etude taxonomique d'un groupe complexe d'espèces des genres *Phaseolus* et *Vigna* (Papilionaceae) sur la base de données morphologiques et polliniques, traitées par l'analyse informatique. Boissiera 28:1–273.

Maréchal, R., J. Mascherpa, and F. Stainier. 1981. Taxonometric study of the *Phaseolus–Vigna* complex and related genera. In: R.M. Polhill and P.H. Raven, eds. Advances in Legume Systematics, Royal Botanic Gardens, Kew, England. Pp. 329–335.

Martens, M. 1831. Auswahl Merkwürdiger Pflanzen:18–19 (+Pl. 12).

Martens, M. and H. Galeotti. 1843. Enumeratio synoptica plantarum phanerogamicarum ab Henrico Galeotti, in regionibus mexicanis collectarum. Bull. Acad. Roy. Brux. 10:178–200.

Martínez, M. 1979. Catálogo de nombres vulgares y científicos de plantas mexicanas. Fondo de Cultura Económica, México, D.F., México. 1247p.

McBryde, F.W. 1947. Cultural and historical geography of southwest Guatemala. Smithsonian Inst. Publ. 4:1–184.

McVaugh, R. 1977. Botanical results of the Sessé & Mociño expedition (1787–1803). 1. Summary of excursions and travels. Contr. Univ. Michigan Herb. 11:97–195.

McVaugh, R. 1987. Flora Novo-Galiciana—A descriptive account of the vascular plants of western México—Volume 5 Leguminosae. The University of Michigan Press, Ann Arbor, Michigan. 786p.

Mejía Jiménez, A., C. Muñoz, H.J. Jacobsen, W.M. Roca, and S.P. Singh. 1994. Interspecific hybridization between common and tepary beans: increased hybrid embryo growth, fertility, and efficiency of hybridization through recurrent and congruity backcrossing. Theor. Appl. Genet. 88:324–331.

Mendel, G. 1865. Versuche über pflanzenhybriden. Verh. Naturf. Vereins Brünn. 4:3–47 (verlag von Wilhelm Engelmann, 1901, Leipzig, Germany).

Mercado Ruaro, P. and A. Delgado Salinas. 1996. Karyological studies in several Mexican species of *Phaseolus* L. and *Vigna* Savi (Phaseolinae, Fabaceae). In: B. Pickersgill and J.M. Lock, eds. Advances in legume systematics part 8, Legumes of economic importance, Royal Botanic Gardens, Kew, England. Pp. 83–87.

Mercado Ruaro, P. and A. Delgado Salinas. 1998. Karyotypic studies on species of *Phaseolus* (Fabaceae: Phaseolinae). Amer. J. Bot. 85:1–9.

Micheli, M. 1903. Leguminosae Langlasseanae–Légumineuses récoltées dans les Etats Mexicains de Michoacán et de Guerrero pendant les années 1898 et 1899 par Eugène Langlassé. Mém. Soc. Phys. Hist. Natur. Genève. 34:261–262.

Miranda Colín, S. 1967. Origen de *Phaseolus vulgaris* L. (Frijol común). Agrociencia 1:99–109.

Miranda Colín, S. and A.M. Evans. 1973. Exploring the genetic isolating mechanisms between *Phaseolus vulgaris* L. and *P. coccineus* Lam. Annual Rep. Bean Improvement Coop. (USA) 16:39–41.

Mirkov, T.E., J.M. Wahlstrom, K. Hagiwara, F. Finardi Filho, S. Kjemtrup, and M.J. Chrispeels. 1994. Evolutionary relationships among proteins in the phytohemagglutinin-arcelin- a-amylase inhibitor family of the common bean and its relatives. Plant Mol. Biol. 26:1103–1113.

Mok, D.W.S., M.C. Mok, and A. Rabakoarihanta. 1978. Interspecific hybridization of *Phaseolus vulgaris* with *P. lunatus* and *P. acutifolius*. Theor. Appl. Genet. 52:209–215.

Moritz, C. and D.M. Hillis. 1996. Molecular systematics: context and controversies. In: D.M. Hillis, C. Moritz, and B.K. Mable, eds. Molecular systematics. Second edition. Sinauer Associates, Inc., Sunderland, Massachusetts, USA. Pp. 1–13.

Morton, C.V. 1944. Taxonomic studies of tropical American plants. Contr. U.S. Natl. Herb. 29:1–86.

Muñoz, L.C., M.W. Blair, and D.G. Debouck. 2002. Genetic diversity of the CIAT tepary bean (*Phaseolus acutifolius* A. Gray) collection measured with amplified fragment length polymorphism markers. Annual Rep. Bean Improvement Coop. (USA) 45:234–235.

Nabhan, G.P. 1979. Tepary beans: The effects of domestication on adaptations to arid environments. Arid Lands Newsletter 10:11–16.

Nabhan, G.P. 1985. Native crop diversity in Aridoamerica: conservation of regional gene pools. Econ. Bot. 39:387–399.

Nabhan, G.P. 1990. Wild *Phaseolus* ecogeography in the Sierra Madre Occidental, México: areographic techniques for targeting and conserving species diversity. International Board for Plant Genetic Resources, Rome, Italy. 35p.

Nabhan, G.P., J.W. Berry, and C.W. Weber. 1980. Wild beans of the greater southwest: *Phaseolus metcalfei* and *P. ritensis*. Econ. Bot. 34:68–85.

Nabhan, G.P. and R.S. Felger. 1978. Teparies in southwestern North America–A biogeographical and ethnohistorical study of *Phaseolus acutifolius*. Econ. Bot. 32:2–19.

National Research Council. 1979. Tropical legumes: resources for the future. National Academy of Sciences, Washington, D.C. 332p.

Niehaus, T.F., C.L. Ripper, and V. Savage. 1984. A field guide to Southwestern and Texas wildflowers. Houghton Mifflin Company, New York, New York, USA. 449p.

Osborn, T.C. 1988. Genetic control of bean seed protein. CRC Crit. Rev. Plant Sci. 7:93–116.

Osborn, T.C., T. Blake, P. Gepts, and F.A. Bliss. 1986. Bean arcelin. 2. Genetic variation, inheritance and linkage relationships of a novel seed protein of *Phaseolus vulgaris* L. Theor. Appl. Genet. 71:847–855.

Petzoldt, R. and M.H. Dickson. 1987. Interspecific hybridization of *Phaseolus vulgaris* with *P. angustissimus*, *P. filiformis* and *P. ritensis*. Annual Rep. Bean Improvement Coop. (USA) 30:94–95.

Piñero, D. and L. Eguiarte. 1988. The origin and biosystematic status of *Phaseolus coccineus* subsp. *polyanthus*: electrophoretic evidence. Euphytica 37:199–203.

Piper, C.V. 1921. Two new legumes from México and Costa Rica. Proc. Biol. Soc. Wash. 34:41–42.

Piper, C.V. 1926. Studies in American Phaseolinae. Contr. U.S. Natl. Herb. 22:663–701.

Pollard, C.L. 1896. Some new or rare plants. Bot. Gaz. 21:233.

Pratt, R.C. 1983. Gene transfer between tepary and common beans. Desert Plants 5:57–63.

Pratt, R.C. and G.P. Nabhan. 1988. Evolution and diversity of *Phaseolus acutifolius* genetic resources. In: P. Gepts, ed. Genetic Resources of *Phaseolus* beans: their maintenance, domestication, evolution, and utilization. Kluwer Academic Publishers, Dordrecht, The Netherlands. Pp. 409–440.

Prendota, K., J.P. Baudoin, and R. Maréchal. 1982. Fertile allopolyploids from the cross *Phaseolus acutifolius* × *Phaseolus vulgaris*. Bull. Rech. Agron. Gembloux 17:177–190.

Pueyo, J.J. and A. Delgado Salinas. 1997. Presence of a-amylase inhibitor in some members of the subtribe Phaseolinae (Phaseoleae: Fabaceae). Amer. J. Bot. 84:79–84.

Rabakoarihanta, A., C.T. Shii, M.C. Mok, and D.W.S. Mok. 1980. Meiosis and fertility of interspecific hybrids between *Phaseolus vulgaris* L. and *P. acutifolius* A. Gray. Theor. Appl. Genet. 57:59–63.

Ramírez Delgadillo, R. and A. Delgado Salinas. 1999. A new species of *Phaseolus* (Fabaceae) from west-central México. SIDA 18:637–646.

Robinson, B.L. 1892. Descriptions of new plants collected in México by C.G. Pringle in 1890 and 1891, with notes upon a few other species. Proc. Amer. Acad. Arts Sci. 27:165–185.

Robinson, B.L. and J.M. Greenman. 1894. Mexican plants. Proc. Amer. Acad. Arts Sci. 29:385–386.

Rodríguez Cabrera, M., S. Rodríguez Tijerina, M. Aguilar Sanmiguel, and D.G. Debouck. 1987. *Phaseolus* germplasm collection in México. FAO/IBPGR Pl. Genet. Resources Newsl. 69:30–38.

Romero Andreas, J. 1984. Genetic variability in the seed phaseolin of non domesticated bean (*Phaseolus vulgaris* var. *aborigineus*) and the inheritance and physiological effects of arcelin, a novel seed protein. Ph.D. Thesis, Univ. of Wisconsin, Madison, USA.

Rose, J.N. 1897. Studies of Mexican and Central American plants. II. Contr. US Nat. Herb. 5:142.

Rose, J.N. 1903. Studies of Mexican and Central American plants. III. Contr. US Nat. Herb. 8:1–55.

Roxburgh, W. 1832. Flora indica and description of Indian plants. Serampore, India. Second edition. Volume 3: 387–317.

Rydberg, P.A. 1965. Flora of the prairies and plains of Central North America. Facsimile of 1932 edition. Hafner Publishing Co., New York, USA. 969p.

Rzedowski, J. 1978. Vegetación de México. Editorial Limusa, México, D.F., México. 432p.

Rzedowski, J. 1993. Diversity and origins of the phanerogamic flora of México. In: Ramamoorthy, T. P., R. Bye, A. Lot and J. Fa, eds. Biological diversity of México: origins and distribution. Oxford University Press, Inc., New York, USA. Pp. 129–144.

Scheele, A. 1848. Beiträge zur flor von Texas. Linnea 2:465–466.

Scheerens, J.C., A.M. Tinsley, I.R. Abbas, C.W. Weber, and J.W. Berry. 1983. The nutritional significance of tepary bean consumption. Desert Plants 5:11–14; 50–56.

Schinkel, C. and P. Gepts. 1988. Phaseolin diversity in the tepary bean, *Phaseolus acutifolius* A. Gray. Pl. Breed. 101: 292–301.

Schinkel, C. and P. Gepts. 1989. Allozyme variability in the tepary bean, *Phaseolus acutifolius* A. Gray. Pl. Breed. 102: 182–195.

Schmit, V. 1992. Etude de *Phaseolus polyanthus* Greenman et autres taxons du complexe *Phaseolus coccineus* L. Ph.D. Thesis, Faculté des Sciences Agronomiques, Gembloux, Belgium.

Schmit, V. and D.G. Debouck. 1990. *Phaseolus glabellus* Piper, a noteworthy variant of the *P. coccineus* complex?. Annual Rep. Bean Improvement Coop. (USA) 33:124–125.

Schmit, V. and D.G. Debouck. 1991. Observations on the origin of *Phaseolus polyanthus* Greenman. Econ. Bot. 45: 345–364.

Schmit, V., D.G. Debouck, and J.P. Baudoin. 1996. Biogeographical and molecular observations on *Phaseolus glabellus* (Fabaceae, Phaseolinae) and its taxonomic status. Taxon 45:493–501.

Schmit, V., J.P. Baudoin, and B. Wathelet. 1991. Considerations about the taxonomical position of various taxa of the *P. coccineus* complex on the basis of seed protein electrophoresis and experimental hybridizations. Annual Rep. Bean Improvement Coop. (USA) 34:87–88.

Schmit, V., J.P. Baudoin, and B. Wathelet. 1992. Contribution à l'étude des relations phylétiques au sein du complexe *Phaseolus vulgaris* L.–*Phaseolus coccineus* L.–*Phaseolus polyanthus* Greenman. Bull. Rech. Agron. Gembloux 27:199–207.

Schmit, V., P. du Jardin, J.P. Baudoin, and D.G. Debouck. 1993. Use of chloroplast DNA polymorphisms for the phylogenetic study of seven *Phaseolus* taxa including *P. vulgaris* and *P. coccineus*. Theor. Appl. Genet. 87:506–516.

Schumann, K. and W. Nagl. 1995. PCR amplification of the phaseolin-genes in 18 species of the genus *Phaseolus*. Annual Rep. Bean Improvement Coop. (USA) 38:107–108.

Seaton, H.E. 1893. New and little known plants collected on Mt. Orizaba in the summer of 1891. Proc. Amer. Acad. Arts & Sci. 28:116–123.

Shaw, E.A. 1987. Charles Wright on the boundary 1849–1852, or Plantae Wrightianae revisited. Meckler Publishing Corporation Chadwyck–Healey Ltd. Westport, Connecticut, USA. 44p.

Shii, C.T., M.C. Mok, S.R. Temple, D.W.S. Mok. 1980. Expression of developmental abnormalities in hybrids of *Phaseolus vulgaris* L. Interaction between temperature and allelic dosage. J. Hered. 71:218–222.

Shii, C.T., A. Rabakoarihanta, M.C. Mok, and D.W.S. Mok. 1982. Embryo development in reciprocal crosses of *Phaseolus vulgaris* L. and *P. coccineus* Lam. Theor. Appl. Genet. 62:59–64.

Singh, S.P. 2001. Broadening the genetic base of common bean cultivars: a review. Crop Sci. 41:1659–1675.

Singh, S.P., D.G. Debouck, and W.M. Roca. 1997. Successful interspecific hybridization between *Phaseolus vulgaris* L. and *P. costaricensis* Freytag & Debouck. Annual Rep. Bean Improvement Coop. (USA) 40:40–41.

Skroch, P.W., R.C. Dobert, E.W. Triplett, and J. Nienhuis. 1993. Polymorphism of the leghemoglobin gene in *Phaseolus* demonstrated by polymerase chain reaction amplification. Euphytica 69:177–183.

Small, J.K. 1903. Flora of the southeastern United States. Published by the Author, New York, USA. 700p.

Smartt, J. 1970. Interspecific hybridization between cultivated American species of the genus *Phaseolus*. Euphytica 19:480–489.

Smartt, J. 1973. The possible status of *Phaseolus coccineus* L. ssp. *darwinianus* Hdz. X. et Miranda C. as a distinct species and cultigen of the genus *Phaseolus*. Euphytica 22:424–426.

Smartt, J. 1990. Grain legumes: evolution and genetic resources. Cambridge University Press, Cambridge, United Kingdom, 379 p.

Smartt, J. and N. Haq. 1972. Fertility and segregation of the amphidiploid *Phaseolus vulgaris* L. × *P. coccineus* L. and its behavior in backcrosses. Euphytica 21:496–501.

Sousa Peña, M., A. Wong León, and A. Delgado Salinas. 1996. Pollination dynamics and evolution in the *Phaseolus coccineus* L. complex. In: B. Pickersgill and J.M. Lock, eds. Advances in legume systematics. Part 8. Legumes of economic importance. Royal Botanic Gardens, Kew, United Kingdom. Pp. 75–81.

Sousa Sánchez, M. and A. Delgado Salinas. 1981. La identificación y clarificación de *Phaseolus pauciflorus* Sessé & Mociño ex G. Don (Leguminosae–Phaseolinae). Iselya 2:3–10.

Sousa Sánchez, M. and A. Delgado Salinas. 1993. Mexican Leguminosae: phytogeography, endemism, and origins. In: T.P. Ramamoorthy, R. Bye, A. Lot and J. Fa, eds. Biological diversity of México: origins and distribution. Oxford University Press, Inc., New York, USA. Pp. 459–511.

Stainier, F. 1974. Contribution à l' étude palynologique des Papilionaceae-Phaseoleae-Phaseolinae. II. Etude de quelques espèces des genres *Phaseolus* L., *Vigna* Savi et *Physostigma* Balf. Bull. Jard. Bot. Nat. Belgique 44:1–15.

Standley, P.C. 1916. New Panamanian Fabaceae. Contr. US Nat. Herb. 18:106–109.

Standley, P.C. 1937. Flora of Costa Rica. Field Mus. Nat. Hist. Bot. Ser. 18:1–1616.

Standley, P.C. 1940. Studies of American plants–IX. Field Mus. Nat. Hist. Bot. Ser. 22:3–62.

Standley, P.C. and J.A. Steyermark. 1946. Flora of Guatemala–Part V. Fieldiana, Botany 24:1–502.

Stearn, W.T. 1993. Botanical Latin–History, grammar, syntax, terminology and vocabulary. Fourth edition. David & Charles Publishers, Newton Abbot, Devon, United Kingdom. 546p.

Sullivan, J.G. and G. Freytag. 1986. Predicting interspecific compatibilities in beans (*Phaseolus*) by seed protein electrophoresis. Euphytica 35:201–209.

Thomas, C.V., R.M. Manshardt, and J.G. Waines. 1983. Teparies as a source of useful traits for improving common beans. Desert Plants 5:43–48.

Thomas, C.V. and J.G. Waines. 1984. Fertile backcross and allotetraploid plants from crosses between tepary beans and common beans. J. Hered. 75:93–98.

Tinsley, A.M., J.C. Scheerens, J.O. Alegbejo, F.H. Adan, K.C. Krumhar, L.E. Butler, and M.J. Kopplin. 1985. Tepary beans (*P. acutifolius* var. *latifolius*): a potential food source for African and Middle Eastern cultures. Qual. Plant Foods Hum. Nutr. 35:87–101.

Tohme, J., D.O. González, S. Beebe, and M.C. Duque. 1996. AFLP analysis of gene pools of a wild bean core collection. Crop Sci. 36:1375–1384.

Toro Ch., O., J. Tohme, and D.G. Debouck. 1990. Wild bean (*Phaseolus vulgaris* L.): description and distribution. International Board for Plant Genetic Resources and Centro Internacional de Agriculture Tropical. Cali, Colombia. 106p.

Toro Ch., O., L. Lareo, and D.G. Debouck. 1993. Observations on a noteworthy wild lima bean, *Phaseolus lunatus* L., from Colombia. Annual Rep. Bean Improvement Coop. (USA) 36:53–54.

Torres González, A.M., O. Toro Chica, and D.G. Debouck. 2001. *Phaseolus talamancensis*, a new wild bean species (Leguminosae, Phaseolinae) from montane forests of eastern Costa Rica. Novon 11:280–286.

Tucker, S.C. 1987. Pseudoracemes in papilionoid legumes: their nature, development, and variation. Bot. J. Linn. Soc. 95:181–206.

Turner, B.L. 1959. The legumes of Texas. University of Texas Press, Austin, Texas, USA. 284p.

Urban, I. 1928. Plantae cubensis novae vel rariores a clo. Er. L. Ekkman lectae. IV. Symb. Ant. 9:433–543.

Usher, G. 1996. Dictionary of botany. Wordsworth Editions Ltd., Ware, Hertfordshire, United Kingdom. 404p.

Vavilov, N.I. 1931. México and Central America as the principal centre of origin of cultivated plants of the New World. Bull. Appl. Bot. Gen. Pl. Breed. 26:135–199.

Vekemans, X., B. Berken, B. Fofana, and J.P. Baudoin. 1997. Preliminary results on chloroplast DNA diversity estimation within *Phaseolus* species using PCR-RFLP. Annual Rep. Bean Improvement Coop. (USA) 40:11–12.

Verdcourt, B. 1970. Studies in the *Leguminosae–Papilionoideae* for the 'Flora of Tropical East Africa'. IV. Kew Bull. 24:507–569.

Verdcourt, B. 1971. *Phaseolus*. In: J.B. Gillett, R.M. Polhill, and B. Verdcourt, eds. Flora of Tropical East Africa. Leguminosae 4. Papilionoideae 2. Crown Agents, London, England. Pp. 503–1107 (613–617).

von Schlechtendal, D. F. L. 1838. Einige bemerkungen über die wurzeln der pflanzen von H. F. Link. Linnaea 12:260–343 (327–328).

Voss, E. G. 1985. Michigan flora –Part II Dicots (Saururaceae –Cornaceae). Cranbook Institute of Science Bulletin 59, and University of Michigan Herbarium, Ann Arbor, Michigan, USA. 724 pp.

Voysest, O. and M. Dessert. 1991. Bean cultivars: classes and commercial seed types. In: A. van Schoonhoven and O. Voysest Voysest, eds. Common beans: research for crop improvement. Commonwealth Agricultural Bureaux International, Wallingford, United Kingdom. Pp.119–162.

Waines, J.G. 1978. Protein contents, grain weights, and breeding potential of wild and domesticated tepary beans. Crop Sci. 18:587–589.

Waines, J.G., R.M. Manshardt, and W.C. Wells. 1988. Interspecific hybridization between *Phaseolus vulgaris* and *P. acutifolius*. In: P. Gepts, ed. Genetic resources of *Phaseolus* beans: their maintenance, domestication, evolution, and utilization. Kluwer Academic Publishers, Dordrecht, The Netherlands. Pp. 441–463.

Wall, R.J. and T.L. York. 1957. Inheritance of seedling cotyledon position in *Phaseolus* species. J. Hered. 48:71–74.

Watson, S. 1882. Contributions to American botany. XVIII—List of plants from southwestern Texas and northern México, collected chiefly by Dr E. Palmer in 1879–80—1. Polypetalae. Proc. Amer. Acad. Arts Sci. 9:316–382.

Weilenmann de Tau, E., J.P. Baudoin, and R. Maréchal. 1986. Obtention d'allopolyploïdes fertiles chez le croisement entre *Phaseolus vulgaris* et *Phaseolus filiformis*. Bull. Rech. Agron. Gembloux 21:35–46.

Westphal, E. 1974. Pulses in Ethiopia, their taxonomy and agricultural significance. Agric. Res. Rep. Wageningen, The Netherlands, 815. Pp. 99–101.

Wiersema, J.H., J.H. Kirkbide, and C.R. Gunn. 1990. Legume (Fabaceae) nomenclature in the USDA germplasm system. U.S. Department of Agriculture, Technical Bulletin No. 1757, Springfield, Virginia, USA. 572p.

Wooton, E.O. and P.C. Standley. 1913. Descriptions of new plants preliminary to a report upon the flora of New México. Contr. U.S. Nat. Herb. 16:109–196.

Wooton, E.O. and P.C. Standley. 1915. Flora of New México. Contr. U.S. Nat. Herb. 19:9–794.

Zapata, M., G.F. Freytag, and R.E. Wilkinson. 1985. Evaluation for bacterial blight resistance in beans. Phytopathology 75:1032–1039.

Zink, D. and W. Nagl. 1998. A taxon identified by microsatellite–primed PCR and Southern hybridization in the secondary gene pool of the tepary bean. Annual Rep. Bean Improvement Coop. (USA) 41:107–108.

LIST OF SOME PUBLISHED NAMES AND EQUIVALENTS

Accepted valid names are given in *italics*. New names and new combinations from this manuscript are given in **bold face**. All others are set in Roman type face. Most equivalents are following synonymy as given by Maréchal et al. (1978a,b), Barbosa-Fevereiro (1986-87), McVaugh (1987), Wiersema et al. (1990) or as given in this manuscript. Focus is on taxa primarily distributed in the New World, and validly published.

Alepidocalyx
amblyosepalus Piper = *P. amblyosepalus* (Piper) Morton
anisophyllus Piper = *P. anisophyllus* (Piper) Freytag & Debouck
parvulus (Greene) Piper = *P. parvulus* Greene

Dolichos
polystachios L. = *P. polystachyus* subsp. *polystachyus* (L.)Britt., Sterns & Pogg.
Lipusa
formosa (Kunth) Alefeld = *P. coccineus* L.

Minkelersia
biflora Hemsl. = *P. pauciflorus* Sessé & Mociño ex G. Don
galactioides Mart. & Gal. = *P. pauciflorus* Sessé & Mociño ex G. Don
multiflora Rose = *P. pluriflorus* Maréchal, Mascherpa & Stainier
pauciflora Rose = *P. nelsonii* Maréchal, Mascherpa & Stainier
vulcanica Piper = *P. pluriflorus* Maréchal, Mascherpa & Stainier

Phaseolus
aborigineus Burk. = *vulgaris* L.
acariaeanthus Harms = *Macroptilium sabaraense* (Hoehne) Barbosa-Fevereiro
acinaciformis Freytag & Debouck
acutifolius A. Gray
acutifolius var. *acutifolius* A. Gray
acutifolius var. *latifolius* Freeman
acutifolius var. *tenuifolius* A. Gray
adenanthus G.F. Meyer = *Vigna adenantha* (G. F. Meyer) Maréchal, Mascherpa & Stainier
albescens McVaugh ex Ramírez & Delgado
albiflorus Freytag & Debouck
albinervus Freytag & Debouck
albiviolaceus Freytag & Debouck
altimontanus Freytag & Debouck
amabilis Standl.
amblyosepalus (Piper) Morton
amplus Bentham = *Vigna candida* (Vellozo) Maréchal, Mascherpa & Stainier
angularis (Willd.) W.F. Wight = *Vigna angularis* (Willd.) Ohwi & Ohashi
angustissimus A. Gray
angustissimus var. *latus* Jones = *angustissimus* A. Gray
anisophyllus (Piper) Freytag & Debouck
anisotrichos Schlecht. = *leptostachyus* Benth. var. *leptostachyus*
anisotrichos subsp. incisus Piper = *pluriflorus* Maré., Masch. & Stain.
antillanus Urban = *Vigna antillana* (Urban) Fawcett & Rendle
appendiculatus Bentham = *Vigna appendiculata* (Bentham) A. Delgado
atropurpureus DC = *Macroptilium atropurpureum* (DC) Urban
atropurpureus var. sericeus A. Gray = *Macroptilium atropurpureum* (DC) Urban
aureus Roxburgh = *Vigna radiata* (L.) Wilczek
barrancae M.E. Jones = *Canavalia villosa* Benth.
benghalensis scandens striato Bergen = *lunatus* L.
bilobatus Engelmann = *Macroptilium gibbosifolium* (Ortega) Delgado
bipunctatus Jacq. = *lunatus* L.
brachycalix Hassler = *Macroptilium brachycalix* (Hassler) Maréchal & Baudet
bracteatus Nees & Mart. = *Macroptilium bracteatum* (Nees & Mart.) Maréchal & Baudet
brevicalyx Micheli = *micranthus* Hook. & Arn.
buseri Micheli = *Ramirezella strobilophora* var. *buseri* (Micheli) Maréchal, Mascherpa & Stainier
campanulatus Freytag & Debouck
campestris Bentham = *Macroptilium longepedunculatum* (Bentham) Urban
candidus Vellozo = *Vigna candida* (Vellozo) Maréchal, Mascherpa & Stainier
canescens Mart. & Galeotti = *Macroptilium atropurpureum* (DC) Urban
caracalla L. = *Vigna caracalla* (L.) Verdcourt
carteri Freytag & Debouck
chiapasanus Piper
chiriquinus Standl. = *tuerckheimii* Donn. Smith
chrysanthos Savi = *Vigna angularis* (Willd.) Ohwi & Ohashi
coccineus L.
coccineus subsp. **coccineus** var. **coccineus** L.
coccineus subsp. **coccineus** var. **condensatus** Freytag
coccineus subsp. **coccineus** var. **argenteus** Freytag
coccineus subsp. **coccineus** var. **griseus** (Piper) Freytag
coccineus subsp. **coccineus** var. **lineatibracteolatus** Freytag
coccineus subsp. **coccineus** var. **parvibracteolatus** Freytag
coccineus subsp. **coccineus** var. **pubescens** Freytag
coccineus subsp. **coccineus** var. **semperbracteolatus** Freytag
coccineus subsp. **coccineus** var. **splendens** Freytag
coccineus subsp. **coccineus** var. **strigillosus** (Piper) Freytag
coccineus subsp. **coccineus** var. **tridentatus** Freytag
coccineus subsp. **coccineus** var. **zongolicensis** Freytag
coccineus subsp. darwinianus Hernández & Miranda = *dumosus* Macfady.
coccineus subsp. formosus (HBK) Maré., Masch. & Stain. = *coccineus* L.
coccineus subsp. obvallatus (Schlect.) Maré., Masch. & Stain. = *coccineus* L.
coccineus subsp polyanthus (Greenm.) Maré., Masch. & Stain. = *dumosus*
Macfady.
coccineus subsp. **striatus** var. **striatus** (Brandeg.) Freytag
coccineus subsp. **striatus** var. **guatemalensis** Freytag
coccineus subsp. **striatus** var. **minuticicatricatus** Freytag
coccineus subsp. **striatus** var. **pringlei** Rose ex Freytag
coccineus subsp. **striatus** var. **purpurascens** Freytag
coccineus subsp. **striatus** var. **rigidicaulis** Freytag
coccineus subsp. **striatus** var. **timilpanensis** Freytag
costaricensis Freytag & Debouck
communis L. ex Pritzel = *vulgaris* L.

coriaceus Desv. = *Vigna linearis* var. *latifolia* (Bentham) Maréchal, Mascherpa & Stainier
cuernavacanus Rose = *Vigna adenantha* (G.F.Meyer) Maréchal, Mascherpa & Stainier
dasycarpus Freytag & Debouck
dilatatus Woot. & Standl. = *angustissimus* A. Gray
diversifolius Pittier = *Macroptilium gracile* (Poepp.) Urban
dumosus Macfady.
elegans Piper = *Vigna elegans* (Piper) Maréchal, Mascherpa & Stainier
elongatus Rose = *Vigna linearis* (HBK) Maréchal, Mascherpa & Stainier var. *linearis*
erythroloma Mart. ex Bentham = *Macroptilium erythroloma* (Mart.) Urban
esculentus Salisb. = *vulgaris* L.
esperanzae Seaton
esquincensis Freytag
falcatus Benth. ex Hemsl. = *lunatus* L.
falciformis Piper
filiformis Benth.
flavescens Piper = *dumosus* Macfady.
floribundus Piper = *polymorphus* S. Wats.
foliaceus Piper = *grayanus* Woot. & Standl.
formosus HBK = *coccineus* L.
fulvus Brandeg. = *leptostachyus* Benth. var. *leptostachyus*
galactioides (Mart. & Gal.) Maréchal, Mascherpa & Stainier = *pauciflorus* Sessé & Mociño ex G. Don
gentryi Standl. = *Vigna gentryi* (Standley) Stainier & Horvat
geophilus Burk. = *Macroptilium geophilum* (Burk.) Debouck, Juárez & Pérez
gibbosifolius Ortega = *Macroptilium gibbosifolium* (Ortega) A. Delgado
glabellus Piper
glaber Schlecht. = *glabellus* Piper
gladiolatus Freytag & Debouck
gracilis Poepp. ex Benth. = *Macroptilium gracile* (Poepp.) Urban
grayanus Woot. & Standl.
griseus Piper = **coccineus** subsp. **coccineus** var. **griseus** (Piper) Freytag
harmsianus Diels = *dumosus* Macfady.
heterophyllus Willd. = *Macroptilium heterophyllum* (Willd.) Maréchal & Baudet = *Macroptilium gibbosifolium* (Ortega) Delgado
hintonii Delgado
inamoenus L. = *lunatus* L.
intonsus Piper = **leptostachyus** var. **intonsus** (Piper) Freytag
jaliscanus Piper
juquilensis Delgado
lambertianus Dietr. = *pauciflorus* Sessé & Mociño ex G. Don
lathyroides L. = *Macroptilium lathyroides* (L.) Urban
laxiflorus Piper
leiosepalus Piper = *coccineus* L.
leptophyllus G. Don
leptostachyus Benth.
leptostachyus var. *leptostachyus* Benth.
leptostachyus var. **intonsus** (Piper) Freytag
leptostachyus var. **pinnatifolius** f. **purpureus** Freytag
leptostachyus var. **pinnatifolius** f. **albus** Freytag
leptostachyus var. **lobatifolius** Freytag
leptostachyus var. **nanus** Freytag
leucanthus Piper = *dumosus* Macfady.
limensis Macfady. = *lunatus* L.
linearis HBK = *Vigna linearis* (HBK) Maréchal, Mascherpa & Stainier
linearis Sessé & Mociño = *leptophyllus* G. Don
lobatus Hooker = *Vigna hookeri* Verdcourt
longepedunculatus Mart. ex Benth. = *Macroptilium longepedunculatum* (Mart.) Urban
longiplacentifer Freytag
longirostratus Drucke = *Vigna caracalla* (L.) Verdcourt
lozanii Rose = *Ramirezella lozanii* (Rose) Piper = *Vigna lozanii* (Rose) Lackey ex McVaugh
lunatus L.
lunatus var. macrocarpus (Moench.) Benth. = *lunatus* L.
lunatus var. silvester Baudet = *lunatus* L.
macrocarpus Moench. = *lunatus* L.
macrolepis Piper
macropoides A. Gray = *Macroptilium gibbosifolium* (Ortega) Delgado
macropus Benth. = *Macroptilium gibbosifolium* (Ortega) Delgado
maculatus subsp. *maculatus* Scheele
maculatus subsp. **ritensis** (Jones) Freytag
maculatifolius Freytag & Debouck
magnilobatus Freytag & Debouck
marechalii A. Delgado
martii Benth. = *Macroptilium martii* (Benth.) Maréchal & Baudet
macvaughii A. Delgado
metcalfei Woot. & Standl. = *maculatus* subsp. *maculatus* Scheele
micranthus Hook. & Arn.
micranthus Martens & Galeotti = *Macroptilium gibbosifolium* (Ortega) Delgado
microcarpus Mart.
monophyllus Bentham = *Macroptilium monophyllum* (Benth.) Maréchal & Baudet
monospermus Robins. & Greenm. = *microcarpus* Mart.
montanus Brand. = *acutifolius* var. *tenuifolius* A. Gray
multiflorus Lam. = *coccineus* L.
multiflorus Willd. = *coccineus* L.
mungo L. = *Vigna mungo* (L.) Hepper
neglectus Hermann
nelsonii Maréchal, Mascherpa & Stainier
nodosus Freytag & Debouck
oaxacanus Rose
obvallatus Schlecht. = *coccineus* L.
occidentalis Rose = *Vigna adenantha* (G. F. Meyer) Maréchal, Mascherpa & Stainier
oligospermus Piper
opacus Piper
ovatifolius Piper = *maculatus* Scheele subsp. *ritensis* (Jones) Freytag
pachycarpus Standl. = *Ramirezella strobilophora* (Robinson) Rose
palmeri Piper
panduratus Benth. = *Macroptilium panduratum* (Benth.) Maréchal & Baudet
paniculatus Michaux = *polystachyus* subsp. *polystachyus* (L.) Britt., Sterns & Pogg.
parviflorus Schlecht. = *Macroptilium gibbosifolium* (Ortega) Delgado
parvifolius Freytag
parvulus Greene
pauciflorus Sessé et Mociño ex G. Don

pauper Standl. = *acutifolius* var. *latifolius* Freeman
peduncularis Kunth = *Vigna peduncularis* var. *peduncularis* (Kunth) Fawcett & Rendle
pedatus Rose = *Macroptilium pedatum* (Rose) Maréchal & Baudet
pedicellatus Benth.
perennis Walt. = *polystachyus* subsp. *polystachyus* (L.) Britt., Sterns & Pogg.
perplexus A. Delgado
persistentus Freytag & Debouck
phanerophlebius Standl. = *Vigna* sp.
plagiocylix Harms
pluriflorus Maréchal, Mascherpa & Stainier
polyanthus Greenm. = *dumosus* Macfady.
polymorphus S. Wats.
polymorphus var. *polymorphus* S. Wats.
polymorphus var. **albus** Freytag
polystachios var. aquilonius Fernald = *polystachyus* subsp. *polystachyus* (L.) Britt., Sterns & Pogg.
polystachyus (L.) Britt., Sterns & Pogg. = -
polystachyus subsp. *polystachyus* (L.) Britt., Sterns & Pogg.
polystachyus subsp. **sinuatus** (Nuttall ex Torrey & Gray) Freytag
polystachyus subsp. **smilacifolius** (Pollard) Freytag
prostratus Benth. = *Macroptilium prostratum* (Benth.) Urban
proriferus Jones = *coccineus* L.
psammodes Lindman = *Macroptilium psammodes* (Lindman) Drewes & Palacios
puberulus HBK = *lunatus* L.
pubescens (Rose) Morton = *Ramirezella pubescens* Rose
pulchellus Piper = *Vigna venusta* (Piper) Maréchal, Mascherpa & Stainier
purpusii Brandeg.
pyramidalis Freytag
reticulatus Freytag & Debouck
retusus Benth. = *maculatus* subsp. *maculatus* Scheele
revolutus Piper = *leptophyllus* G. Don
rigidus Piper = *Vigna speciosa* ? (HBK) Verdcourt
ritensis Jones = **maculatus** subsp. **ritensis** (Jones) Freytag
robiniiflorus Standl. = probably a *Vigna speciosa* (HBK) Verdcourt
rosei Piper = *lunatus* L.
rotundatus Freytag & Debouck
rotundifolius A. Gray = *Macroptilium gibbosifolium* (Ortega) Delgado
rubescens Brandeg. = *Macroptilium longepedunculatum* (Benth.) Urban
sabaraensis Hoehne = *Macroptilium sabaraense* (Hoehne) Barbosa-Fevereiro
saccharatus Macfady. = *lunatus* L.
salicifolius Piper
sanctorum S. Wats. = *filiformis* Benth.
savannarum Britton & Wilson = *Macroptilium gracile* (Poepp. ex Bentham) Urban
scabrellus Benth. ex S. Wats.
schaffneri Piper = *polymorphus* S. Wats.
scolecocarpus Piper = *Macroptilium lathyroides* (L.) Urban
scrobiculatifolius Freytag
seleri Harms = *Macroptilium gibbosifolium* (Ortega) Delgado
semierectus L. = *Macroptilium lathyroides* var. *semierectum* (L.) Urban
sempervirens Piper = *jaliscanus* Piper
sinuatus Nutt. ex Torr. & Gray = **polystachyus** subsp. **sinuatus** (Nuttall ex Torrey & Gray) Freytag
smilacifolius Pollard = **polystachyus** subsp. **smilacifolius** (Pollard) Freytag
sonorensis Standl.
speciosus HBK = *Vigna speciosa* (HBK) Verdcourt
spectabilis Standley = cfr *Vigna* subg. *Sigmoidotropis* (Piper)Vercourt
stenolobus Standley = cfr *Vigna* subg. *Sigmoidotropis* (Piper)Vercourt
striatus Brandeg. = **coccineus** subsp. **striatus** (Brandeg.) Freytag
strigillosus Piper = **coccineus** subsp. **coccineus** var. **strigillosus** (Piper) Freytag
strobilophorus (Robinson) Morton = *Ramirezella strobilophora* (Robinson) Rose
superbus DC = *coccineus* L.
supinus Wiggins & Rollins = *Macroptilium supinum* (Wiggins & Rollins) Delgado-Salinas & Torres-Colín
sylvestris HBK = *coccineus* L.
talamancensis Debouck & Torres
tenellus Piper
tenuifolius (A. Gray) Woot. & Standl. = *acutifolius* var. *tenuifolius* A.Gray
teulensis Freytag
trichocarpus C. Wright = *Vigna trichocarpa* (C. Wright) A. Delgado
trifidus Freytag
truxillensis HBK = *Vigna adenantha* (G.F. Meyer) Maréchal, Mascherpa & Stainier
tuerckheimii Donnell Smith
uleanus Harms = *Misanthus uleanus* (Harms) Lewis & Delgado
venosus Piper
vexillatus L. = *Vigna vexillata* (L.) Rich.
vignoides Rusby = *Vigna vignoides* (Rusby) Maréchal, Mascherpa & Stainier
viridis Piper = *lunatus* L.
vulcanicus (Piper) Maréchal, Mascherpa & Stainier = *pluriflorus* Maréchal, Mascherpa & Stainier
vulgaris L.
vulgaris L. subsp. aborigineus Burk. ex Burk. = *vulgaris* L.
vulgaris L. var. aborigineus (Burk.) Baudet = *vulgaris* L.
wrightii A. Gray = *filiformis* Benth.
wrightii A. Gray var. grayanus (Woot. & Standl.) Kearney & Peebles = *grayanus* Woot & Standl.
xanthotrichus Piper
xolocotzii Delgado
xuaresii Zucc. ex DC = *lunatus* L.
zimapanensis Delgado

List of Living Germplasm Grown Out and Studied at Mayagüez Collection Number Equivalents*

SPECIES	COLLECTOR** & NO.	LOCATION	ALTITUDE	TARS	USDA	NI	CIAT	UC-R
Section A. ACUTIFOLII PI or W6								
acut-acut	Gentry 22020	Son.-Sin., Mex.	365	1	319445	596	G40055	–
acut-acut	DGD 420	La Higuera, Culiacán, Sin., Mex.	210	2	W6 21079	720	G40103	–
acut-acut	DGD 300	Peñon Blanco, Dur., Mex.	1750	4	535200	843	G40086	–
acut-acut	DGD 455	San Isidro, Zapotan, Jal., Mex.	1600	5	535201	847	G40106	–
acut-acut	Buhrow SYACL	Sycamore Canyon, AZ, USA	1200	6	535202	775	G40169	–
acut-acut	Lehto L19321a	87 mi E Hermosillo, Son., Mex.	823	190	535203	838	G40073	L-176
acut-acut	Gentry 22292	San Bernardo, Son., Mex.	300	210	319442	601	G40053	–
acut-acut	Lehto L19297	75 mi E Hermosillo, Son., Mex.	2700	3	535204	839	G40074	L-177
acut-acut	"	"	"	191	535211	"	"	"
acut-acut	"	"	"	378	535217	"	"	"
acut-acut	"	"	"	571	–	–	–	L-177-3
acut-acut	Nabhan GN950 (= Buhrow SBACL)	65 km N, San Bernardo, Son., Mex.	–	7	535205	787	G40203	–
acut-lati	INIA, Son. 31A	Son., Mex.	–	8	535206	562	G40038	–
acut-lati	DGD 517A	Hecelchakan, Camp., Mex.	50	9	535207	694	G40110	–
acut-lati	DGD 288	V. Ocampo, Dur., Mex.	1920	10	535208	721	G40084	–
acut-lati	Gentry 21460	Santa Ana, El Salvador	660	137	312122	518	G40017	–
acut-acut	Gentry 17834	Matape, Son., Mex.	701	142	535209	502	G40056	–
acut-acut	Gentry 17834	Matape, Son., Mex.	701	156	535210	502	G40056	–
acut-acut	Gentry 17834	Matape, Son., Mex.	700	209	263590	586	G40056	–
acut-acut	Waines 266	Río Nozas, Dur.-Par., Dur., Mex.	–	229	535212	–	G40206	L-266
acut-lati	Gentry 22009	Arroyo Tenasas, Son., Mex.	610	302	319439	602	G40050	–
acut-lati	Nabhan GN, X-525	mix col., Rosemont,	1470	305	535213	–	G40177	–
acut-lati	Nabhan GN, X-525	Pima Co., AZ, USA	1470	433	440813	–	G40177	–
acut-acut	Buhrow	Todos Santos, B.C., Mex.	–	327	535216	–	G40274	–
acut-acut	Rogers et al. 113	Wilcox, Cochise Co., AZ, USA	–	565	–	–	G40286	–
acut-tenu	Buhrow MBACT	Molino Basin, AZ, USA	1350	11	535236	779	G40114	–
acut-tenu	(= Nabhan AZ PL-2)	Molino Basin, AZ, USA	1350	506	–	–	G40114	–
acut-tenu	Nabhan AZ-5	Santa Rita Mts., AZ, USA	–	12	535237	–	G40204	–
acut-tenu	(= AZ PL-5)	Santa Rita Mts., AZ, USA	–	294	535261	–	G40220	–
acut-tenu	Buhrow SRAC	Sta Cruz, Sta Rita Mts., AZ, USA	2000	13	535238	692	G40115	–
acut-tenu	Cambridge NA 13	USA	–	141	535239	–	G40117	–
acut-tenu	?	?	–	152	535240	–	G40077	–
acut-tenu	?	?	–	153	535241	–	G40078	–
acut-tenu	Leroi NE-28	41 km S Dur., Dur., Mex.	1800	157	535242	717	G40202	–
acut-tenu	DGD 388	Mezquital, Dur., Mex.	2110	158	535243	831	G40087	–
acut-tenu	Buhrow MCAC	Sta. Catalina, Pima Co., AZ, USA	1200	159	535244	700	G40114	–
acut-tenu	Waines 124	Chiricahua Mts., Portal, AZ, USA	–	192	535245	–	G40071	L-124
acut-tenu	Waines 124	Chiricahua Mts., Portal, AZ, USA	–	573	–	–	G40071	–
acut-tenu	Waines 137	Tucson, AZ, USA	–	193	535246	–	G40072	L-137
acut-tenu	UC L-320	Peña Blanca, AZ, USA	–	194	535247	–	G40100	L-320
acut-tenu	Waines 114	Portal, Cochise Co., AZ, USA	–	230	535248	–	G40207	L-114
acut-tenu	Waines 152	Pima Co., AZ, USA	–	231	535249	–	G40208	L-152
acut-tenu	Waines 154	Pima Co., AZ, USA	–	232	535250	–	G40209	L-154
acut-tenu	Waines 158	Madera Canyon, Sta. Cruz Co., AZ, USA	–	233	535251	–	G40210	L-158
acut-tenu	Waines 163	Peña Blanca Lake, Sta Cruz Co.,	–	234	535252	–	G40211	L-163
		AZ, USA	–	569	–	–	–	–
acut-tenu	Waines 166	Hank & Yank Spring, Sta. Cruz Co., AZ, USA	–	235	535253	–	G40212	L-166
acut-tenu	Buhrow et al.	Graham Co., AZ, USA	–	236	535254	–	G40213	L-466
acut-tenu	Buhrow et al.	Graham Co., AZ, USA	–	237	535255	–	G40214	L-469
acut-tenu	Buhrow et al.	Graham Co., AZ, USA	–	238	535256	–	G40215	L-477
acut-tenu	Marina Hoy	Cochise Co., AZ, USA	–	239	535257	–	G40216	L-123

SPECIES	COLLECTOR** & NO.	LOCATION	ALTITUDE	TARS	USDA	NI	CIAT	UC-R
acut-tenu	Nabhan AZ-3, PL	Molino Basin, Pima Co., AZ, USA	–	291	535258	–	G40217	–
acut-tenu	Nabhan AZ-3, PL	Molino Basin, Pima Co., AZ, USA	–	561	–	–	G40212	–
acut-tenu	Nabhan AZ-4, PL	Molino Basin, Pima Co., AZ, USA	–	292	535259	–	G40218	–
acut-tenu	Nabhan AZ-4	Molino Basin, Pima Co., AZ, USA	–	293	535260	–	G40219	–
acut-tenu	Nabhan AZ-5, PL	Molino Basin, Pima Co., AZ, USA	–	294	535261	–	G40220	–
acut-tenu	Nabhan AZ-6, PL	Molino Basin, Pima Co., AZ, USA	–	295	535262	–	G40221	–
acut-tenu	Nabhan AZ-7, PL	Molino Basin, Pima Co., AZ, USA	–	296	535263	–	G40222	–
acut-tenu	Nabhan AZ-8	Molino Basin, Pima Co., AZ, USA	–	297	535264	–	G40223	–
acut-tenu	Nabhan AZ-8	Molino Basin, Pima Co., AZ, USA	–	298	535265	–	G40225	–
acut-tenu	Nabhan AZ-10	Molino Basin, Pima Co., AZ, USA	–	299	535266	–	G40226	–
acut-tenu	Nabhan AZ-11	Molino Basin, Pima Co., AZ, USA	–	300	535267	–	+	–
acut-tenu	Nabhan AZ-12	Molino Basin, Pima Co., AZ, USA	–	301	535268	–	G40227	–
acut-tenu	Nabhan GN, 743	Molino Basin, Pima Co., AZ, USA	–	304	535269	–	G40228	–
acut-tenu	Buhrow DRACT	Dragoon Mts., AZ, USA	1524	330	535270	–	G40231	–
acut-tenu	Buhrow 85-2	Cerro Colorado Mts., AZ, USA	–	383	–	–	G40233	–
acut-tenu	Nabhan GN, X525 (mix)	Sta. Rita Mts., Rosemont, AZ, USA	–	433	440813	–	G40177	–
acut-tenu	FV 78-M-16	35 km W Jalapa, VC, Mex.	–	505	–	–	G40285	–
acut-acut	DGD 265	Durango-Carlos Real, Dur., Mex.	–	566	–	–	G40091	–
acut-tenu	Bye 10059	Km 27, Batopilas Rd, Chih., Mex.	–	572	–	–	G40273	–
parvifol	DGD 422	El Tepozal, Nay., Mex.	1220	14	535373	713	G40109	L-556
parvifol	DGD 390	El Troncón, Mezquital, Dur., Mex.	1920	15	535374	812	G40090	–
parvifol	FS 81-13	Km 35, Zacatepec Mixes, Oax., Mex.	1680	52	535375	1068	G40182	L-557
parvifol	FS 81-40	Los Cuartos, Jal., Mex.	1420	53	535376	1045	G40170	–
parvifol	DGD 413	El Salto, Dur., Mex.	1970	160	535382	705	G40102	–
parvifol	DGD 413	El Salto, Dur., Mex.	1970	463	–	–	G40102	–
parvifol	DGD 2427	Jalapa, Guat.	–	471	–	–	G40185	–
Section B. PHASEOLI								
costaric	DGD 2128	2 km N Herradura, Costa Rica	1690	420	–	–	+	–
costaric	DGD 2122	StaMaría a Copey, Costa Rica	1660	423	–	–	+	–
costaric	DGD 2126	0.8 km S Pueblo Nuevo, Costa Rica	–	435	–	–	+	–
costaric	DGD 2132	I km Providencia, Costa Rica	1990	436	–	–	+	–
costaric	DGD 2135	3 km SE Copey, Costa Rica	2080	437	–	–	+	–
dumosus	FV 78-Guat-47	Patzicia-Panajachel, Guat.	1775	71	–	1123	G35951	–
dumosus	Co 78-Guat-6	Semetabaj-Panajachel, Guat.	–	309	535291	1112	G35941	–
dumosus	DGD 1622	San Miguel Dueñas, Guat.	1622	412	–	–	G35908	–
dumosus	DGD 2444	6 km W Sn M. Dueñas, Guat.	2160	475	–	–	G35946	–
dumosus	DGD 2460	3 km NE Calderas, Guat.	–	481	–	–	G36045	–
vulg-abor	Burns A-76 (INIA)	Abancay, Apurímac, Peru	2300	111	535411	622	G7225	–
vulg-abor	Drijfhout IVT-61005	Argentina	–	112	535392	29	G7469	–
vulg-abor	Bannerot	Argentina	–	113	535393	190	G10024	–
vulg-abor	I-165	Argentina	–	114	535412	621	G10025	–
vulg-abor	Cambridge AB-5	?	–	148	535398	–	G13033	–
vulg-abor	?	Argentina	–	151	535400	–	G996	–
vulg-abor	?	Argentina	–	503	226910	–	G 996	–
vulg-abor	Vieira 283 (Burkart)	Argentina	–	283	535423	–	G 6388	–
vulg-abor	DGD 2152	Lampa, Junín, Peru	2650	452	–	–	G23421	–
vulg-abor	DGD 2295	Kishuara, Apurímac, Peru	2440	454	–	–	G23426	–
vulg-abor	DGD 2307	Abancay, Apurímac, Peru	2580	455	–	–	G23427	–
vulg-abor	DGD 2585	Mollepata, Cuzco, Peru	2340	493	–	–	G23492	–
vulg-abor	DGD 2484	S Totora, Cochabamba, Bolivia	2270	545	–	–	G23442	–
vulg-abor	DGD 2491	SSW Patatejas, Chuquisaca, Bolivia	2380	546	–	–	G23443	–
vulg-abor	DGD 2497	WNW Padilla, Chuquisaca, Bolivia	2040	547	–	–	G23444	–
vulg-abor	DGD 2501	WSW Tomatitas, Tarija, Bolivia	2100	548	–	–	G23445	–
vulg-abor	DGD 2580	WSW Limatambo, Cuzco, Peru	2460	549	–	–	G23454	–
vulg-mex	FS 81-19A	Chiquilistlán, Jel, Mex.	1630	34A	W6 21082	–	G50414	–

SPECIES	COLLECTOR** & NO.	LOCATION	ALTITUDE	TARS	USDA	NI	CIAT	UC-R
vulg-mex	Miranda (INIA) Mor. 622	Progreso, Mor., Mex.	1430	102	535404	401	G10008	–
vulg-mex	Miranda (INIA) Mor. 635	Progreso, Mor., Mex.		103	535405	404	G13505	–
vulg-mex	Miranda (INIA) Guer. 924	Teloloapan, Guer, Mex.	1150	104	535406	406A	G10000	–
vulg-mex	Miranda (INIA) Mich. 1001	Km 50, Temazcal, Mich, Mex.	888	105	535407	544	G10018	–
vulg-mex	Miranda (INIA) Oax. 910	Etla, Oax., Mex.	1900	106	535408	578	G13021	–
vulg-mex	Norvell 3365	Ahuachapán, Salvador	714	107	201014	748	G13504	–
vulg-mex	FS 81-21A (mix)	La Calaverna, Jal., Mex.	2000	109	535409	1061	G23506	–
vulg-mex	FS 81-21A (tans)	La Calaverna, Jal., Mex.	2000	109A	W6 21092	1061	G23506A	–
vulg-mex	FS 81-21A (pinto)	La Calaverna, Jal., Mex.	2000	109B	–	–	G23506A	–
vulg-mex	FS 81-36	El Chante-Manantlán, Jal., Mex.	1300	110	535410	1066	G23511	–
vulg-mex	FS 81-36 (tans)	Km 9, Manantlán, Jal., Mex.	1300	110A	W6 21093	1066	G23511A	–
vulg-mex	FS 81-36 (pinto)	Km 9, Manantlán, Jal., Mex.	1300	110B	–	1066	G23511B	–
vulg-mex	D 810119	?	–	115	535394	–	G23592	–
vulg-mex	FS 81-39	Km 56, Los Cuartos, Jal., Mex.	1420	134	535413	1050	G23512	–
vulg-mex	Miranda (INIA) Mor. 673	Progreso, Mor., Mex.	1430	144	535395	575	G13020	–
vulg-mex	Miranda (INIA) Mor. 650	Progreso, Mor., Mex.	1430	145	535396	581	G06386	–
vulg-mex	Miranda (INIA) Mor. 649	Progreso, Mor., Mex.	1430	146	535397	–	G10013	–
vulg-mex	Miranda (INIA) Guer.962	Ixcateopan, Guer., Mex.	1380	150	535399	–	G10004	–
vulg-mex	Co 78-Guat-3A	Jalapa-San Pedro Pinula, Guat.	–	187	535414	1043	G23593	–
vulg-mex	FS 81-1	Tepoztlán, Mor., Mex.	1900	198	535415	1055	G23505	–
vulg-mex	FS 81-23	Km 10, Nevado Colima, Col., Mex.	1920	199	535416	1082	G23507	–
vulg-mex	FS 81-31	El Chante-Manantlán, Jal., Mex.	900	200	535417	1093	G23508	–
vulg-mex	FS 81-32	El Chante-Manantlán, Jal., Mex.	1040	201	535418	1076	G23509	–
vulg-mex	FS 81-34	El Chante-Manantlán, Jal., Mex.	1300	202	535419	1075	G23510	–
vulg-mex	Hernández X & T 3 (yel)	36 km S Durango, Dur., Mex.	1800	280	533542	1042	G23504	–
vulg-mex	Gentry (Silv. 3-5)	Mex.	–	281	535421	–	G23596	–
vulg-mex	Gentry (Silv. 3-5)	Mex.	–	281	346956	–	G12886	–
vulg-mex	Gentry (Silv. 7, 1 5, 53)	Mex.	–	282	535422	–	G23597	–
vulg-mex	Gentry (Silv. 7, 1 5, 53)	Mex.	–	282	346957	–	G12887	–
vulg-mex	Gentry (Silv. 11, 13, 55)	Mex.	–	284	535424	–	G23598	–
vulg-mex	Gentry (Silv. 10)	Mex.	–	308	535425	848	G23503	–
vulg-mex	Leroi Mex-55	Zochitlán, Mor., Mex.	1750	335	535426	1052	G23522	–
vulg-mex	Leroi Mex-54	Zochitlán, Mor., Mex.	1750	336	535427	1057	G23521	-
vulg-mex	Leroi Mex-81	Tlalnepantla, Mor., Mex.	1950	337	535428	1059	G23527	–
vulg-mex	Leroi Mex-78	Amatlán, Mor., Mex	1700	340	535429	1065	G23524	–
vulg-mex	Leroi Mex-142	Tepoxtlán, Mor., Mex	1725	341	535430	1067	G23539	–
vulg-mex	Leroi Mex-127	Tlacotepec, Mor., Mex	1800	342	535431	1070	G23536	–
vulg-mex	Leroi Mex-137	Tepoxtlán Cliff, Mor., Mex	1775	343	535432	1071	G23537	–
vulg-mex	Leroi Mex-123	Tepoxtlán-Yautepec Rd., Mor, Mex.	1600	344	535433	1072	G23544	–
vulg-mex	Leroi Mex-114	Puente Caporal, Mor., Mex.	1250	345	535434	1073	G23532	–
vulg-mex	Leroi Mex-92	Oacalco, Mor., Mex.	1250	346	535435	1074	G23464	–
vulg-mex	Leroi Mex-124	Tepoxtlán-Yautepec Rd., Mor, Mex	1600	347	535436	1078	G23534	–
vulg-mex	Leroi Mex-107	Pantitlan, Oaxtepec, Mor., Mex	1350	348	535437	1080	G23531	–
vulg-mex	Leroi Mex-46	Sn Andrés de la Cal, Mor., Mex	1400	349	535438	1081	G23514	–
vulg-mex	Leroi Mex-126	Tepoztlán-Yautepec Rd., Mor, Mex.	1450	350	535439	1083	G23535	–
vulg-mex	Leroi Mex-118	Palo Grande, Mor., Mex.	1550	351	535440	1084	G23533	–

SPECIES	COLLECTOR** & NO.	LOCATION	ALTITUDE	TARS	USDA	NI	CIAT	UC-R
vulg-mex	Leroi Mex-9	Casa Grande, Tepoztlán, Mor, Mex.	1725	352	535441	1085	G23513	–
vulg-mex	Leroi Mex-48	San Andrés de la Cal, Mor., Mex.	1400	353	535442	1086	G23516	–
vulg-mex	Leroi Mex-47	San Andrés de la Cal, Mor., Mex.	1400	354	535443	1087	G23515	–
vulg-mex	Leroi Mex-79	Amatlán, Mor., Mex.	1700	355	535444	1088	G23525	–
vulg-mex	Leroi Mex-80	Tlalnepantla, Mor., Mex.	1950	356	535445	1089	G23526	–
vulg-mex	Leroi Mex-51	Tlalnepantla, Mor., Mex.	1950	361	535446	1098	G23519	–
vulg-mex	Leroi Mex-52	Santo Domingo, Mor., Mex.	1900	362	535447	1100	G23520	–
vulg-mex	Leroi Mex-98	Malinalco, Mex., Mex.	1850	363	535448	1102	G23530	–
vulg-mex	Leroi Mex-70	Los Laureles, Mor., Mex.	1750	364	535449	1103	G23523	–
vulg-mex	Leroi Mex-138	Tepoxtlán Falaise, Mor., Mex.	1775	366	535450	1105	G23538	–
vulg-mex	Leroi Mex-49	San Andrés de la Cal, Mor., Mex.	1400	367	W6 21113	1106	G23517	–
vulg-mex	Leroi Mex-86	Amatlán, Mor., Mex.	1400	368	535451	1113	G23528	–
vulg-mex	Leroi Mex-83	Tlalnepantla, Mor., Mex.	1950	373	–	1130	G23465	–
vulg-vulg	?	?	–	497	430208	–	G17131	–
vulg-mex	UC L-13	?	–	531	–	–	G12865	L-13
vulg-mex	DGD 2325	4 km NE Atlixco, Pue., Mex.	1430	543	–	–	G23429	–
vulg-mex	DGD 2355	9 km E Chilpancingo, Guer., Mex.	1620	544	–	–	G23430	–
Section C. COCCINEI								
cocc-cocc	FS 81-19 (mix)	Chiquilistlán, Jal., Mex.	1630	34B	535276	–	G35948	–
cocc-cocc	FS 81-19B	Chiquilistlán, Jal., Mex.	1630	35B	–	–	G35948	–
cocc-cocc	FS 81-5	Texcoco, Mex.	2670	39D	535277	+	G35949	–
cocc-cocc	FS 81-9	Zongolica, V.C., Mex.	1800	46A	535278	+	G35950	–
cocc-cocc	FS 81-9	Zongolica, V.C., Mex.	1800	46B	–	–	G35950	–
cocc-cocc	Co 78-Guat-2	San Juan Acatempa, Guat.	1500	100	–	–	+	–
cocc-cocc	FS 81-17	Tapalpa, Jal., Mex.	1980	203	–	1096	G35830	–
cocc-cocc	Co 78-Guat-4	Xroads Yepocapa-Duenos, Guat.	1980	212	535280	1111	G35844	–
cocc-cocc	Co 78-Guat-5	Cima Parra-Yepocapa, Guat.	2025	213	535281	1119	+	–
cocc-cocc	Co 78-Guat-8A	Km 159, Sta. Cl. Laguna, Guat.	1910	215	535282	–	+	–
cocc-cocc	Co 78-Guat-9	Hill to Godínez, Guat.	2310	216	535283	–	G36040	–
cocc-cocc	Co 78-Guat-10	Hill to Godínez, Guat.	2310	217	535284	–	G35953	–
cocc-cocc	Co 78-Guat-14	Mataque-Sn. Juan Pinula, Guat.	1950	221	535286	–	+	–
cocc-cocc	Co 78-Guat-16	Volcan de Agua, Guat.	1940	223	–	–	–	–
cocc-cocc	Co 78-Guat-17	Km 7, Totonicapán, Guat.	2490	224	535288	–	+	–
cocc-cocc	INIA, Mor. 667	Km 67, Mex.-Cuer., Mor., Mex.	1800	285	535289	1117	G35904	–
cocc-cocc	INIA, Mor. 664	Progreso, Mor., Mex.	1430	286A	535290	–	+	–
cocc-cocc	INIA, Mor. 664 = PR 7886	Progreso, Mor., Mex.	1430	286B	W6-21103	–	+	–
cocc-cocc	INIA, Mor. 663 = PR 11	Progreso, Mor., Mex.	1430	287	W6-21104	–	G36041	–
cocc-cocc	INIA, Mor. 662 = PR 12	Progreso, Mor., Mex.	1430	288	W6-21105	–	G35955	–
cocc-cocc	INIA, Mor. 671 = PR 4	Progreso, Mor., Mex.	1430	290	W6-21106	1125	+	–
cocc-cocc	INIA, Mor. 658	Progreso, Mor., Mex.	1430	303	–	1149	+	–
cocc-cocc	Leroi Mex. 103	Amatlán, Mor., Mex.	1700	369	W6-21114	1122	–	–
cocc-cocc	Leroi Mex. 71	Huertas, Mor., Mex.	2100	370	–	1118	G35944	–
cocc-cocc	Leroi Mex. 102	Amatlán, Mor., Mex.	1700	372	–	1129	G36047	–
cocc-cocc	INIA, Tlax. 659	Tlaxcala, Mex.	2250	438	–	–	–	–
cocc-cond	DGD 2037	Zimapán, El Salto, Hgo., Mex.	2037	411	–	+	G35867	–
cocc-semp	FS 81-2	Cuernavaca, Mor., Mex.	2400	30	W6-21081	1108	G35827	–
cocc-semp	FS 81-24	Grullo, Jal., Mex.	2000	204	–	+	G35952	–
cocc-stri	Co 79-Guat-18	San Marcos-Cucho, Guat.	–	183	–	–	–	–
cocc-stri	Co 78-Guat-11	Castillo Farm, Pat.-Ptz., Guat.	–	188	535279	1110	–	–
cocc-stri	Co 78-Guat-7	Sta. L. Utat.-Sta. C. Lag., Guat.	2330	214	W6-21647	–	–	–
cocc-stri	Co 78-Guat-12	Jalapa-Miramundo, Guat.	2100	219	535285	–	+	–
cocc-stri	Co 78-Guat-13	Jalapa-Miramundo, Guat.	2100	220	–	–	+	–
cocc-stri	Co 78-Guat-15	Road Volcán de Agua, Guat.	1940	222	535287	–	+	–
cocc-stri	Co 78-Guat-19	Eden-Sibilia, Guat.	2580	225	W6-21099	1109	G35954	–
cocc-stri	FV 78-Guat-95	Patzakan, Sibilia, Guat.	2580	311	W6-21107	+	G35956	–

SPECIES	COLLECTOR** & NO.	LOCATION	ALTITUDE	TARS	USDA	NI	CIAT	UC-R
cocc-trid	FS 81-37	Manantlán, Jal., Mex.	1500	36	W6-21083	–	G36039	–
cocc-trid	FS 81-38	Manantlán, Jal., Mex.	1750	164	–	–	–	–
cocc-zong	FS 81-7	Paso del Viento, V.C., Mex.	2200	31	W6-21643	1133	G36046	–
cocc-zong	FS 81-8	Zongolica, V.C., Mex.	1800	32	535275	–	+	–
stri-purp	FS 81-4	Tres Marias, Mor., Mex.	2880	38	535274	1092	+	–
stri-purp	FV 78-Mex-8	Km 45, Tex.-Tlax., Mex., Mex.	2670	273	–	–	G40746	–
(outside the Coccinei; see text)								
glabellus	DGD et al. 2041	La Misión, Hgo., Mex.	1610	–	–	–	G40584	–
glabellus	DGD et al. 2043	Xilitla, SLP., Mex.	1030	–	–	–	G40585	–
glabellus	DGD et al. 2075	Cd. Maíz, SLP, Mex.	1150	413	535311	1304	G40594	–
glabellus	Leroi	Jalapa, V.C, Mex.	1350	56	–	820	–	–
glabellus	Norvell M176	?	–	–	–	–	G40656	–
glabellus	Norvell HM 669	?	–	–	W6 15688	–	G40673	
Section D. PANICULATI								
jalis	Buhrow M 21	5.3 mi S Tequila, Jal., Mex.	1950	320	535313	1090	–	–
lunatus	Smartt	?		63	535339	516	G25220	–
lunatus	Evans PS16	Veracruz, Mex.	20	64	535340	543	G25221	–
lunatus	?	Canas, Guanacaste, Costa Rica	100	65	264609	554	G25226	–
lunatus	Johnson 285-73	5 km SE Sayulita, Nayarit, Mex.	–	66	535341	675	G25229	–
lunatus	DGD 492	Hopelchen, Campeche, Mex.	100	67	535342	738	G25706	–
lunatus	DGD 453	San Isidoro, Zapotan, Jal., Mex.	1390	68	535343	745	G25704	–
lunatus	ME F 3917	Chaco, Argentina	–	69	535344	814	G26404	–
lunatus	Norvell 2622	Gualan, Zacapa, Guat.	–	70	195333	823	G25222	–
lunatus	Norvell 2622	Gualan, Zacapa, Guat.	–	267	195333	823	G25222	–
lunatus	LE 1	Czechoslovakia	–	147	535345	–	G25822	–
lunatus	Vakili	Km 9, Hwy 505, Puerto Rico	460	149	W6 21645	–	+	–
lunatus	Venezuela 29	Venezuela	–	189	535346	–	G26658	–
lunatus	?	El Salvador	–	195	200913	–	G25273	–
lunatus	?	France	–	196	264239	–	G25582	–
lunatus	?	Costa Rica	–	197	264606	–	G25586	–
lunatus	CIAT	?	–	266	535347	–	G25220	–
lunatus	Leroi Mex. 77	Oacalco, Mor., Mex.	1250	371	–	1126	G25851	–
lunatus	DGD 1944	Peru	–	418	–	–	G25913	–
lunatus	DGD 2108	Costa Rica	–	441	–	–	G25965	–
lunatus	DGD 2103	Costa Rica	–	443	535348	–	–	–
lunatus	?	Costa Rica	–	502	264607	–	G25224	–
marechalii	(INIA) Mor. 623	Morelos, Mex.	–	124	W6 21095	402	G40506	–
polys-pol	Evans NC 67	México ?	–	40	535370	552	–	–
polys-pol	Bassett	Hort. Farm, Gainesville, FL	–	127	535371	–	+	–
polys-pol	Wolff 1950	Coal Co., Oklahoma, USA	–	374	–	–	+	–
polys-smil	Freytag 90-1	Kerrville, TX, USA	–	550	W6-21126	–	–	–
salic	Buhrow M 2	50 mi E Villa Unión, Sin., Mex.	2000	316	W6 21109	–	–	–
Section E. BRACTEATI								
none								
Section F. MINKELERSIA								
parvu	DGD 280	Los Altares, Parral, Chih., Mex.	2000	87	–	701	–	–
parvu	Buhrow BWPR	Arizona, USA	2400	171	535366	–	+	–
parvu	?	Pima Co., AZ, USA	–	246	W8 21100	–	–	L-138
perplex	DGD 2366	Coatepec, Mex., Mex.	–	468	W6-21129	–	+	–
plurif	DGD 436	Mazamitla, Jal., Mex.	1950	90	–	715	G40532	–
plurif	DGD 320	Registro, Durango, Dur., Mex.	2370	91	535368	869	+	–
plurif	FS 81-21-C	La Calaverna, Jal., Mex.	2000	131	535369	+	+	–
Section G. XANTHOTRICHA								
gladio	DGD 2040	Jacala, Hidalgo, Mex.	1550	396	535386	1239	G40583	–
hintonii	DGD 391 (mixed)	El Troncón, Mezquital, Dgo., Mex.	1940	119B	–	707	G40518	–
hintonii	FS 81-22	Veludero, Jalisco, Mex.	1650	132B	W6-21127	+	G40667	–
hintonii	FS 81-26	Mazos, Jalisco, Mex.	1110	133	535381	1069	G40556	–
magnilob	DGD 314	San Pedro, Madero, Dgo., Mex.	2130	120	535378	716	G40523	–

SPECIES	COLLECTOR** & NO.	LOCATION	ALTITUDE	TARS	USDA	NI	CIAT	UC-R
magnilob	DGD 403	Las Huertas, Madero, Dgo., Mex.	1950	121	535379	806	G40525	–
magnilob	FS 81-22 (mixed)	Veludero, Jalisco, Mex.	1650	132A	535380	+	G40666	–
xanthotr	DGD 1604	Sta. Maria Jesús, Guat.	1900	375	535452	–	G40575	–
xanthotr	Kretchmer 738	Costa Rica	–	388	W6 21115	333	G40027	–
xanthotr	DGD 1630	Panajachel, Guat.	1640	390	W6 21116	–	G40694	–
xanthotr	DGD 1623	Acatenango, Guat.	1790	391	W6 21117	–	G40714	–
xanthotr	DGD 1620	Sn. Miguel Dueñas, Guat.	1820	392	–	1207	G40577	–
xanthotr	DGD 2096	Costa Rica	–	442	–	–	+	–
xanthotr	DGD 2105	Costa Rica	–	444	–	–	G40596	–
zimapan	DGD 2073	Cd. Maíz, SLP, Mex.	1170	394	535384	1240	G40593	–
zimapan	DGD 2045	Ahuacatlán, SLP, Mex.	1430	395	535385	+	G40586	–
zimapan	DGD 2040	Jacala, Hgo., Mex.	1550	396	535386	1239	G40583	–
zimapan	DGD 2049	El Carrizal, Que., Mex.	890	398	535387	+	G40588	–
zimapan	DGD 2047	El Madroño, Que., Mex.	1720	399	535388	–	G40587	–
Section H. REVOLUTI								
none								
Section I. DIGITATI								
albiflor	DGD 1522	Nuevo León, Mex.	–	426	535390	1242	G40567	–
albiviol	DGD 2063	Cd. Victoria, Tam., Mex.	900	400	535389	1243	G40591	–
albiviol	DGD 2059	Zaragoza, Nuevo León, Mex.	1430	401	–	–	–	–
altimon	DGD 1523	Nuevo León, Mex.	–	427	535391	–	G40568	–
Section J. RUGOSI								
angus	Buhrow SCAN	Sunset Crater, AZ, USA	2050	l6	535271	788	G40548	–
angus	Buhrow SAAN, =(157) (mix)	St. Augustine Pass, Organ Mts., NM, USA	1500	166	535272	878	G40550	–
angus	Freytag 84-2	Organ Mts., NM, USA	–	329	535273	–	G40685	–
angus	Buhrow 85-1	Pinaleno Mts., AZ, USA	5000	382	–	–	G40669	85-1
carteri	León 3751	12 km S La Paz, Baja Cal., Mex.	–	535	W6 21124	–	G40675	–
fili	Kooistra (IVT 60143)	?	–	47	535292	31	G40501	–
fili	Carter 5930 (5390?)	Baja Cal., Mex.	–	48	535293	600	G40507	–
fili	Buhrow RTWR	Silver Bell Mts., Pima Co., AZ, USA	–	274	535308	–	G40657	–
fili	Buhrow RTWR-A	Silver Bell Mts., Pima Co., AZ, USA	1000	.49	535294	689	G40600	–
fili	Buhrow (RTWR?)	Silver Bell Mts., Pima Co., AZ, USA	–	116	535299	–	G40657	–
fili	Buhrow ATWR	Apache Trail, Maricopa Co., AZ, USA	550	50	535295	690	G40601	–
fili	Buhrow ATWR 1			173	535302	690	G40601	–
	Buhrow ATWR 4	Canyon Lake, AZ, USA	–	117	535300	–	G40512	–
fili	" " (10461)	Canyon Lake, AZ, USA	–	275	535309	–	G40512	–
fili	Buhrow BCFI	Baja Cal., Mex.	–	51	535296	–	+	–
fili	Nabhan	2 km Miraflores, BC, Mex.	–	54	535297	859	G40549	–
fili	Buhrow OPWR	AZ, USA	–	55	535298	–	G40602	–
fili	Cambridge F_1	Czechoslovakia (!)	–	143	535301	–	G40505	–
fili	Buhrow SAAN, (mixed)	St. Augustine Pass, Organ Mts., NM, USA	1500	166	535272	878	G40550	–
fili	UC ?	Yuma Co., AZ, USA	–	253	535303	–	G40658	L-297
fili	UC ?	Yuma Co., AZ, USA	–	254	535304	–	G40659	L-298
fili	UC ?	Yuma Co., AZ, USA	–	255	535305	–	G40660	L-301
fili	UC ?	N Silver Bell, Pima Co., AZ, US	2600	256	535306	–	G40661	L-318
fili	Felger 18734	Papago Tanks, Son., Mex.	150	257	535307	–	G40662	L-319
fili	Worthington 6814	El Paso, TX, USA	–	380	535310	–	G40663	–
fili	Buhrow MP-206	?	–	556	–	–	G40688	–
fili	Patton & Soulé s.n.	Son., Mex.	–	557	–	–	G40689	–
fili	Villareal et al. 4418	Torreón, Coah., Mex.	–	560	–	–	G40690	–
(outside the Rugosi; see text)								
micro	DGD 294	El Resbalón, Rodeo, Dgo., Mex.	1390	78	535352	703	G40526	–
micro	DGD 470	Linda Vista, Tecolotlán, Jal, Mex.	1650	79	W6-21087	708	G40537	–
micro	DGD 449	Tonala, Jal., Mex.	900	80	535353	709	G40535	–
micro	DGD 478	Sta. Rosa, La Huerta, Jal., Mex.	380	81	535354	728	G40538	–

SPECIES	COLLECTOR** & NO.	LOCATION	ALTITUDE	TARS	USDA	NI	CIAT	UC-R
micro	DGD 424	Chamela, La Huerta, Jal., Mex.	8	82	535355	733	G40530	–
micro	DGD 392	El Salitre, Mezquital, Dur., Mex.	1420	83	535356	773	G40519	–
micro	DGD 389	Matamoros, Mezquital, Dur., Mex.	2060	84	535357	809	G40539	–
micro	DGD H389	Matamoros, Mezquital, Dur., Mex.	2060	264	535363	809	G40539	–
micro	VF 78-M-65	Km 82, RT 190, Oax., Mex.	890	85A	535358	1039	G40646	–
micro	VF 78-M-65	Km 82, RT 190, Oax., Mex.	890	85B	–		G40670	–
micro	VF 78-M-69	Km 163, RT 190, Chiapas, Mex.	600	86	535359 W6 21088	1049	G40554	–
micro	VF 78-M-70	Km 177, Oax–Tehuan., Oax., Mex.	600	181	535361	–	G40552	–
micro	DGD H478 (DGD 78/144)	Huerta, Jal., Mex.	380	263	535362	–	G40538	–
micro	DGD 2339	Oaxaca, Mex.	–	464	535364	–	G40614	–
micro	DGD 2350	Juquila, Oax., Mex.	–	536	W6 21125	–	G40617	–
micro	DGD 2399	Volcan Batea, Gto., Mex.	–	537	+	–	G40627	–
micro	DGD 2328	Amatitlan, Pue., Mex.	–	538	–	–	G40613	–
Section K. FALCATI								
lept-inton	DGD 369	Corrales, Zac., Mex.	2250	25	535322	714A	G40516	–
lept-inton	Buhrow M44 (mixed)	E of Navojoa, Chih-Son., Mex.	1460	324A	535336	+	G40551	–
lept-lept	FS 81-14	Km 7, Tehuacan-Oaxaca, Oax., Mex.	1600	17	535314	–	+	–
lept-lept	FS 81-25	Piedra Ancha, 20 km de Cd. Guzmán, Jal., Mex.	2000	18	W6 21080	–	G40664	–
lept-lept	FS 81-27	Los Mazos, Jal., Mex.	1110	19	535316	–	+	–
lept-lept	FS 81-33	El Chante, Jal., Mex.	1040	20	535317	–	+	–
lept-lept	Hernández 224	UNAM, Cd. Mex, DF, Mex.	–	21	535318	353	G40502	–
lept-lept	DGD 365	Corrales, Zac., Mex.	2165	22	535319	699	G40515	–
lept-lept	DGD 410	Durango, Dur., Mex.	2320	23	535320	710	G40544	–
lept-lept	DGD 437	Mich., Mex.	1940	24	535321	725	G40533	–
lept-lept	DGD 423	Nay., Mex.	1220	155	535323	732	G40529	–
lept-lept	DGD 423	Tepozal, Nay., Mex.	1220	162	535324	732	G40529	–
lept-lept	DGD 480	Cuautitlán, Jal., Mex.	660	163	535325	743	G40545	–
lept-lept	FV 78-M-39	Oaxaca, Oax., Mex.	1750	174	535326	–	G40553	–
lept-lept	FV 78-M-40	Km 154, Huahuapan, Oax., Mex.	1750	175	W6 21097	–	+	–
lept-lept	FV 78-M-62	Km 52, Oax-Tehuantepec, Oax, Mex.	1680	177	535327	–	+	–
lept-lept	FV 78-M-64	Km 63, Oax-Tehuantepec, Oax, Mex.	1680	178	535328	–	G40557	–
lept-lept	FV 78-G-29	Km 124, Jalapa-Sn Pedro, Guat.	1180	179	535329	–	G40558	–
lept-lept	FV 78-G-3B	Km 24, Jalapa-Sn Pedro, Guat.	1180	182	535330	–	G40603	–
lept-lept	FV 78-M-39	Km 6, Oax.-Tehuacan, Oax., Mex.	1750	206	535332	+	G40553	–
lept-lept	FV 78-M-45	Km 102, Huahuapan-Oax., Oax, Mex.	2080	207	–	–	–	–
lept-lept	DGD 437	Zacapu, Mich., Mex.	1940	265	–	725	G40533	–
lept-lept	LM (TARS Gen 76-1)	– ?		269	W6 21101	–	+	–
lept-lept	LM (TARS Gen 76-2)	Univ. Guadalajara, Jal., Mex.	–	270	535333	353	G40502	–
lept-lept	Buhrow M16	Km 2088, Cerritos, SLP, Mex.	–	319	535334	–	+	–
lept-lept	Buhrow M44 (mixed)	E of Navajoa, Chih-Son., Mex.	1460	324	535335	+	G40551	–
lept-lept	Téllez T7282A	Pedregal, DF, Mex.	2240	325	535337	–	G40683	–
lept-lept	DGD 1629	Panajachel, Sol., Guat.	–	439	–	–	+	–
lept-lept	DGD 2107	Naranjo, Alaj., Costa Rica	–	440	W6 21121	–	G40597	–
lept-lept	DGD 2109	Monterrey, SnJosé, Costa Rica	–	448	–	–	G40598	–
lept-lept	DGD 1627	SnASemetabaj, Sol., Guat.	–	534	+	–	+	–
lept-nanu	FS 81-25	Piedra Ancha, 20 km de Cd. Guzmán, Jal., Mex.	2000	18	535315	–	G40664	–
lept-pin-al	FV 78-M-45	Km 102, Oax, Mex.	2080	176	W6 21098	+	G40665	–

SPECIES	COLLECTOR** & NO.	LOCATION	ALTITUDE	TARS	USDA	NI	CIAT	UC-R
macva	Norvell M-6999	Mazatlán, Sin., Mex.	1650	532	W6 21119 & W6 21122	–	G40656	–
micran	Buhrow B28	4-8 mi E of Tuito, Jal., Mex.	1000	322	–	–	G40559	–
micran	Buhrow B28	4-8 mi E of Tuito, Jal., Mex.	1000	424	535351	+	G40559	–
Section L. BREVILEGUMENI								
olig	Co 78-Guat-1	Mw Tower, San José Acatempa, Guat.	1500	101	535365	1116	+	–
olig	FV 78-Guat-18	Mw Tower, San José Acatempa, Guat.	1500	184	–	1115	G40560	–
olig	Xolocotzi X16694	Mex.	–	315	W6 21108	896	G40542	–
olig	DGD 2472	Guat.	–	487	–	–	–	–
Section M. PEDICELLATI								
esperan	DGD 2027	Tepeapulco, Hid., Mex.	2650	406	W6 21119	+	G40582	–
esperan	DGD 2382	Pue., Mex.	2400	470	–	–	–	–
gray	Buhrow ORGR	Oracle Ridge, Sta. Cat., AZ, USA	2350	058	–	792	–	–
gray	Buhrow CMGR1	Chiricahua Mts, AZ, USA	1950	331	W6 21110	–	–	–
gray	Buhrow DRGR	Dragoon Mts, AZ, USA	–	332	535312	–	+	–
oaxac	FS 81-11	Km 15, Oax-Tuxtepec, Oax., Mex.	2250	261	–	+	+	–
pedic	DGD 2030	Nicolas Flores, Hid., Mex.	2180	407	535367	1269	+	–
pedic	DGD 2035	Zimapan, Hid., Mex.	2320	408	W6 21120	–	+	–
teulen	DGD 380	Jiménez de Teul, Zac., Mex.	2400	405	–	724	G40517	–
Section N. CHIAPASANA								
none								
Section O. CORIACEI								
macu-macu	Buhrow PAMT	Pinos Altos, NM, USA	–	77	W6 21086	–	–	–
	Nabhan 734		–	167	–	805	–	–
macu-macu	Buhrow CMMT2	Chiricahua Mts., AZ, USA	2300	168	–	–	–	–
macu-macu	Buhrow SLMT	San Lorenzo, NM, USA	–	326	–	–	–	–
macu-macu	Buhrow DRMT	Dragoon Mts., AZ, USA	–	333	535350	–	–	–
macu-rite	Buhrow SRRI	Sta. Rita Mts., AZ, USA	2000	93	–	693	–	–
macu-rite	Buhrow GVRI	Geology Vista, Sta. Catalina Mts., AZ, USA	2050	95	535372	796	–	–
macu-rite	Buhrow (?) =MITA 10469	AZ, USA	–	98	W6 21090	–	–	–
macu-rite	Buhrow M36	Santiago Caballeros, Sin., Mex.	1340	323	–	1056	–	–
macu-rite	Freytag 84-1	Mt. Lemmon, AZ, USA	–	328	–	–	–	–
macu-rite	Buhrow WHRI	?	–	381	–	–	–	–
reticu	DGD 409	Ojo de Agua, Durango, Dur., Mex.	2320	125	W6 21096	702	–	–

*Abbreviations used for column headings:

TARS = Tropical Agriculture Research Station, USDA/ARS, Box 70, Mayagüez, PR 00681. (Seed no longer available)
PI = Plant Introduction, RPIS, USDA/ARS, 59 Johnson Hall, Washington State Univ., Pullman, WA
NI = Phaseolus Germplasm, Faculté des Sciences Agronomiques de l'Etat, B5800 Gembloux, Belgium
CIAT = Centro Internacional de Agricultura Tropical, Genetic Resources Unit, Apartado 6713, Cali, Colombia
UC-R = Plant Sciences Dept., University of California-Riverside, CA 92521

**Abbreviations used for names of collectors:

DGD = D.G. Debouck
FV = Freytag & Vakili
FS = Freytag & Sullivan
Co = Cojulún
H & T = Hernández & Tapia

LIST OF ADDITIONAL COLLECTIONS EXAMINED

Collections are listed in alphabetical order by the last name of the first collector (if more than one) followed by collection number and species (in parentheses) to which specimen belongs. Type numbers are given in **boldface**. Further information on each herbarium sheet listed below through the first collector and his (her, their) collection number is available from the authors.

List of species

1. **acinaciformis** Freytag & Debouck
2. *acutifolius* var. *acutifolius* Gray
3. *acutifolius* var. *latifolius* Freeman
4. *acutifolius* var. *tenuifolius* Gray
5. *albescens* McVaugh ex Ramírez & Delgado
6. **albiflorus** Freytag & Debouck
7. **albinervus** Freytag & Debouck
8. **albiviolaceus** Freytag & Debouck
9. **altimontanus** Freytag & Debouck
10. *amabilis* Standl.
11. *amblyosepalus* (Piper) Morton
12. *angustissimus* A. Gray
13. *anisophyllus* (Piper) Freytag & Debouck
14. **campanulatus** Freytag & Debouck
15. carteri Freytag & Debouck
16. *chiapasanus* Piper
17. *coccineus* subsp. *coccineus* var. **argenteus** Freytag
18. *coccineus* subsp. *coccineus* var. *coccineus* L.
19. *coccineus* subsp. *coccineus* var. **condensatus** Freytag
20. *coccineus* subsp. *coccineus* var. ***griseus*** (Piper) Freytag
21. *coccineus* subsp. *coccineus* var. **lineatibracteolatus** Freytag
22. *coccineus* subsp. *coccineus* var. **parvibracteolatus** Freytag
23. *coccineus* subsp. *coccineus* var. **pubescens** Freytag
24. *coccineus* subsp. *coccineus* var. **semperbracteolatus** Freytag
25. *coccineus* subsp. *coccineus* var. **splendens** Freytag
26. *coccineus* subsp. *coccineus* var. ***strigillosus*** (Piper) Freytag
27. *coccineus* subsp. *coccineus* var. **tridentatus** Freytag
28. *coccineus* subsp. *coccineus* var. **zongolicensis** Freytag
29. *coccineus* subsp. *striatus* var. **guatemalensis** Freytag
30. *coccineus* subsp. *striatus* var. **minuticicatricatus** Freytag
31. *coccineus* subsp. *striatus* var. **pringlei** Freytag
32. *coccineus* subsp. *striatus* var. **purpurascens** Freytag
33. *coccineus* subsp. *striatus* var. **rigidicaulis** Freytag
34. *coccineus* subsp. *striatus* var. ***striatus*** (Brandeg.) Freytag
35. *coccineus* subsp. *striatus* var. **timilpanensis** Freytag
36. *costaricensis* Freytag & Debouck
37. **dasycarpus** Freytag & Debouck
38. *dumosus* Macfady.
39. *esperanzae* Seaton
40. **esquincensis** Freytag
41. *filiformis* Benth.
42. *glabellus* Piper
43. **gladiolatus** Freytag & Debouck
44. *grayanus* Woot. & Standl.
45. *hintonii* Delgado
46. *jaliscanus* Piper
47. *laxiflorus* Piper
48. *leptophyllus* G. Don
49. *leptostachyus* var. ***intonsus*** (Piper) Freytag
50. *leptostachyus* var. *leptostachyus* Benth.
51. *leptostachyus* var. **lobatifolius** Freytag
52. *leptostachyus* var. **nanus** Freytag
53. *leptostachyus* var. **pinnatifolius** f. **purpureus** Freytag
54. *leptostachyus* var. **pinnatifolius** f. **albus** Freytag
55. **longiplacentifer** Freytag
56. *lunatus* L.
57. *macrolepis* Piper
58. **maculatifolius** Freytag & Debouck
59. *maculatus* subsp. *maculatus* Scheele
60. *maculatus* subsp. ritensis (Jones) Freytag
61. *macvaughii* Delgado
62. **magnilobatus** Freytag & Debouck
63. *marechalii* Delgado
64. *micranthus* Hook. & Arn.
65. *microcarpus* Mart.
66. *neglectus* Hermann
67. *nelsonii* Maréchal, Mascherpa & Stainier
68. **nodosus** Freytag & Debouck
69. *oaxacanus* Rose
70. *oligospermus* Piper
71. *opacus* Piper
72. *palmeri* Piper
73. **parvifolius** Freytag
74. *parvulus* Greene
75. *pauciflorus* Sessé & Moc. ex G. Don
76. *pedicellatus* Benth.
77. *perplexus* Delgado
78. **persistentus** Freytag & Debouck
79. *plagiocylix* Harms
80. *pluriflorus* Maréchal, Mascherpa & Stainier
81. *polymorphus* var. *polymorphus* S. Wats.
82. *polymorphus* var. **albus** Freytag
83. *polystachyus* subsp. *polystachyus* (L.) Britt. Sterns & Pogg
84. *polystachyus* subsp. ***sinuatus*** (Nutt. ex Torrey & Gray) Freytag
85. *polystachyus* subsp. smilacifolius (Pollard) Freytag
86. *purpusii* Brandeg.
87. **pyramidalis** Freytag
88. **reticulatus** Freytag & Debouck
89. **rotundatus** Freytag & Debouck
90. *salicifolius* Piper
91. *scabrellus* Benth.
92. **scrobiculatifolius** Freytag
93. *sonorensis* Standl.
94. *talamancensis* Debouck & Torres
95. *tenellus* Piper
96. **teulensis** Freytag
97. **trifidus** Freytag
98. *tuerckheimii* Donn. Smith
99. *venosus* Piper
100. *vulgaris* L.
101. *xanthotrichus* Piper
102. *xolocotzii* Delgado
103. *zimapanensis* Delgado

ADDITIONAL COLLECTIONS BY COLLECTOR

Adams, K.R. 222-83 (4), 281 (74), 328-84 (4), Adams, W.P. 19361 (83); Alcantara 2354 (76); Anderson, E. *s.n.* (56); Anderson, L.C. 7609 (83); Arias 628 (18); Arsène 71 (75), 390 (19), 1367 (50), 2250 (50), 3972 (75), 5234 (27), 6176 (75), 6942 (80); Atwood 1749 (12), 9456 (12), 17171 (74);

Baker, C.F. 640 (56); Balbo 4268 (50); Barkelew 211 (56); Barneby 2640 (12), 12918 (12); Barnes *s.n.* (12); Barr 60-337 (12), 64-193 (41); Bartholomew 3431 (42); Basurto 92-A (42), 655 (42); Benson 9085 (44), 9699 (74); Bernouilli 1170 (56); Bertelsen *s.n.* (41), 89-563 (4); Bessey *s.n.* (84); Bissell *s.n.* (83); Blakley 1567 (74); Block *s.n.* (81); Blumer 162 (44), 1347 (44), 1351 (74), 1676 (4), 2113 (4); 2237 (2), U99 (4), U221 (2), V232 (59); Boege 271 (50); Bohrer 733 (12); Borell *s.n.* (44); Botteri 301 (50), 738 (42); Bottimer 292 (4); Bourgeau 487 (31), 581 (in part) (31), 1375 (100), 1376 (56), 2919 (42), 3184 (50), 3351 (19); Bowers 1298 (41), R627 (4), R696 (4); Brandegee *s.n.* (41); Bravo 332 (60), 333a (60), 2375 (76); Breedlove 12407 (18), 12768 (18), 13415 (18), 39911 (18), 43996 (75), 46259 (67), 48191 (70), 52956 (18), 54626 (50), 55795 (98); Brenes *s.n.* (50), 353 (50), 522 (50), 16900 (56), 19834 (50); Briggs 1 (59); Bright *s.n.* (83), 1411 (83), 9723 (83); Brown, D. *s.n.* (41); Brown, J. G. *s.n.* (74), 89 (44); Buhrow *s.n.* (13 Apr 1978) (41), *s.n.* (29 Jul 1978) (44), *s.n.* (3 Aug 1978) (60), *s.n.* (4 Sep 1978) (59), 2 (4), 3 (4), 4 (4), 5 (4), 6 (4), 7 (4), 8 (4), 9 (4), 10 (4), 14 (44), 15 (44), 19 (74), 21 (59), 22 (59), 23 (60), 24 (60), 27 (41), 29 (41), 31 (41), 145 (74), 157 (12), 177 (12), 180 (12), 183 (4); Burger 4028 (98); Burgess, T.L. 6241 (41), 6287 (41); Butterwick 3881 (41), 5433 (12), 6617 (12); Bye, R. 5414 (50);

Cabrera 5930 (100); Calzada 1961 (42); Canby *s.n.* (83), 1757 (83); Cancino 31 (41); Carter 2063 (41), 2121 (41); 3759 (41), 3848 (41), 5020 (41), 5930 (41); Case 295 (in part) (19); Cazier 240 (44), 634 (74), 679 (44), 712 (74); Chambers 825 (41); Chanek 93 (56); Chapman *s.n.* (84); Chavelas 2813 (56); Chiang 9709 (41); Churchill 775 (12), Clark 8377 (59), 8402 (44), 8432 (44), 10860 (4), 11952 (12), 12467 (4); Clarke 361 (50), 395 (50), 690901-59 (75); Clover 7069 (12); Cochrane 11772 (56), 12293 (56); Collom 3901 (12); Conrad, A. *s.n.* (12); Conzatti 1029 (50), 1245 (65), 3651 (50), 4186 (24); Coon 146 (18); Correll 13706 (41), 24452 (41), 27816 (12), 30861 (41), 33771 (12); Cory 1714 (12), 7165 (44), 9256 (59), 30401 (44), 31152 (12); Cottam 5375 (12), 8716 (12), 13262 (12); Coues *s.n.* (12), 107 (12); Croat 45875 (50); Crane 260 (56); Crocker *s.n.* (41); Crosswhite 473 (41); Cruden 1194 (50); Cuevas 1179 (56), 1524 (75), 2362 (56), 2446 (27); Curtiss *s.n.* (83); 4252 (83), 5210 (83);

D'Arcy 9895 (36); Dana *s.n.* (83); Daniel, T.F. 1923 (41), 2378 (41); Darrow *s.n.* (11/27/37) (41), 2261 (4); Davidse 10145 (98), 31578 (50); Davidson, A. 274a (12); Davidson, M.E. 927 (98), 948 (98); Dawson 1213 (41); Deaver 6544 (44); Debouck 275 (50), 285 (44), 295 (3), 302 (3), 318 (44), 320 (80), 322 (50), 327 (50), 388 (4), 428 (3), 431 (50), 457 (50), 1513 (103), 1514 (50), 1526 (50), 1600 (18), 1604 (101), 1674 (50), 1676 (50), 2046 (42), 2056 (50), 2096 (101), 2113 (56), 2362 (50), 2370 (76), 2412 (50), 2413 (56), 2444 (38), 2453 (50), 2455 (50), 2466 (56), 3064 (56), 3120 (36); Degener 26610 (50); Delgado 28 (50), 67 (50), 202 (50), 240 (50); 332 (50), 475 (56), 516 (26), 517 (26), 519 (26), 520 (18), 1015 (6), 1130 (18); Demarée 41145 (12); Diego 306 (50); Dillon 1962 (41); Dressler 2299 (42); Drouet 3875 (41), 4016 (41); Dumond 1327 (83); Dunn 12647 (12); Dwyer *s.n.* (56), 318 (56);

Eastwood 3792 (12), 6254 (in part) (12), 8755 (41), 15965 (41), 17612 (12), 17667 (41); Eaton, D.C. *s.n.* (83); Eaton, F. M. 3144 (12); Echeverria, J.A. 1221 (50); Eggert *s.n.* (83); Eggleston 10773 (44), 10821 (74); Elias 9030 (4); Elmore A2 (41); Engard 102 (12); Equihua *s.n.* (41); Estrada-Castillón 719 (50);

Fearing 1037B (50), 1039 (50), 1040 (50), 1074 (75), 1078 (75), 2068 (4); Felger 6464A (41), 10604 (41), 12207 (41), 12415 (41), 86-457 (41); Fernald 8736 (83); Ferris 2797 (44), 5957 (64); Fisher, H. L. 9229 (83), U15162 (83); Fleetwood 228 (12), 11863 (83); Flores-Franco 960 (50), 1137 (50), 1169 (75), 1298 (50), 1372a (50), 1415 (75), 1455 (50), 1531 (2), 1623 (50), 2446 (64), 2615 (60), 3566 (50), 3897 (73); Forbes 1631 (12), 1759 (12); Foster 85 (41); Fox 2882 (84); Freytag 78-Guat-41-7 (26), 78-Mex-1 (32), 81-11 (69), 81-12 (69), 91-1 (15), #SC-1 (4), #SC-2 (2), #SC-3 (3),#SC-5 (2), #SC-6 (2), #SC-7 (3), #SC-8 (3), #SC-9 (3), #SC-10 (3), #SC-11 (4), #SC-12 (4), #SC-13 (4), #SC-14 (73), #SC-15 (73), #SC-16 (12), #SC-17 (50), #SC-18A (50), #SC-19 (50), #SC-22 (50), #SC-23 (50), #SC-24 (50), #SC-25 (49), #SC-34A (100), #SC-34B (18), #SC-36 (27), #SC-40 (83), #SC-44 (38), #SC-46A (18), #SC-46B (18), #SC-48 (41), #SC-49 (41), #SC-50 (41), #SC-51 (41), #SC-52 (73), #SC-54 (41), #SC-55 (41), #SC-63 (56), #SC-65 (66), #SC-67 (56), #SC-68 (56), #SC-69 (56), #SC-70 (56), #SC-71 (38), #SC-77 (59), #SC-78 (65), #SC-79 (65), #SC-80 (65), #SC-81 (65), #SC-82 (65), #SC-83 (65), #SC-84 (65), #SC-85 (65), #SC-86 (65), #SC-90-1 (85), #SC-102 (100), #SC-103 (100), #SC-104 (100), #SC-105 (100), #SC-106 (100), #SC-107 (100), #SC-109 (100), #SC-110 (100), #SC-111 (100), #SC-112 (100), #SC-113 (100), #SC-114 (100), #SC-115 (100), #SC-116 (41), #SC-117 (41), #SC-119B (45), #SC-120A (45), #SC-120B (62), #SC-127 (83), #SC-132B (45), #SC-133 1&2 (45), #SC-133 (45), #SC-134 (100), #SC-136 (65), #SC-142 (3), #SC-158 (4), #SC-160 (73), #SC-168 (59), #SC-174 (50), #SC-182 (65), #SC-187 (100), #SC-191 (3), #SC-194 (4), #SC-198 (100), #SC-199 (100), #SC-200 (100), #SC-206 (50), #SC-234 (4), #SC-237 (4), #SC-239 (4), #SC-261 (69), #SC-283 (100), #SC-324A (49), #SC-380 (41), #SC-394 (103), #SC-395 (103), #SC-396 (103), #SC-398 (103), #SC-400 (8), #SC-406 (39), #SC-407 (76), #SC-413 (42), #SC-424 (64), #SC-427 (9), #SC-460 (7), #SC-532 (61), #SC-535 (15), #SC-550 (85); Fuentes 598 (75); Fulton 7172 (12);

Galeotti 3169 (50), 3374 (50); Gallegos 371 (76); Gardner 218 (12); Garrett *s.n.* (59); Garwood 817 (56); Gaumer *s.n.* (56), 498 (56), 565 (56), 923-A (56), 1059 (56); Gentle 412 (56), 7845 (56); Gentry, A. 6516 (38); Gentry, H. S. *s.n.* (? Oct 1937) (4), 339M (50), 3855 (12), 4092 (41), 5858 (93), 6214 (60), 6272 (60), 6307 (80), 6376 (60), 6377 (60), 8829 (41), 22032 (50), 22143 (50), 22260 (100), 22329 (100), 22492 (100), 22511 (100); Gilenan 412 (12); Gilman *s.n.* (4), 98 (60); Gold 255 (76); Goodding 60-337 (12), 275-61 (4), 347-61 (4), 393-61 (4), 429-45 (41), 750 (44), 835 (60), 857-49 (44), 2395 (44), 4126 (44), 9347 (4), A9166 (4), A9454 (60), A9470 (44), A9650 (41), A9762 (4), A9794 (4), A9797 (4); A9807 (4), A9811 (12); Gould 2807 (in part) (4), 2895 (4); 3851 (12), 4098 (4), 4578 (44), 12230 (41); Graham *s.n.* (24 Aug 1927) (74), *s.n.* (28 Aug 1927) (60), 160 (80); Green, T. *s.n.* (83); Greene *s.n.* (29 Aug 1880) (12), *s.n.*=(*Greene 12435*? ? Sep 1880) (59), 295 (4); Griffiths 135 (60), 293 (60), 317 (60), 5362 (59), 6071 (60), 7167 (60), 7266(44); Guzmán H., F. 1451 (50);

Hagopian FCM-104 (44), FCM-146 (74), FCM-229 (59); Haines *s.n.* (41); Hammerly 23 (41); Harper 555 (84); Harris 25912 (18); Harrison 3064 (44), 3100 (4), 5655 (12), 8083 (74), 8223 (12); Haskell *s.n.* (4); Hastings 63-274 (41); Havard *s.n.* (12), 45 (12); Hawkes 2053 (100); Hayes 150 (2); Helmkamp *s.n.* (74); Hendricks 2084 (12); Hernández, E. X-184 (42); X-192 (18); Hernández, H.

526 (50); Hernández, R. 2375 (56), 2673 (50), 3233 (59), 6788 (18); Hess 878 (44), 1280 (44); Hesselberg *s.n.* (12); Hevly *s.n.* (12); Heyde 3753 (50), 4167 (in part) (100), 4168 (56); Hillyard 6339 (41); Hinckley *s.n.* (59), 399 (44); Hinton 2266 (50), 6528 (95), 9485 (50), 9567 (50), 10827 (50), 11551 (100), 11684 (100), 11771 (56), 17245 (76); Hitchcock, A.S. *s.n.* (8/7/1915) (12); Hitchcock, C.L. 4498 (12), 7337 (12); Hodgson 3834 (4), 3887A (4), 4807 (4), H971 (74), H1790 (74); Hoke *s.n.* (41); Holway 3729 (50); Holzner 1574 (59), 1712 (44), 2061 (44), 2077 (60); Hough 26 (12), 118 (12); Howell 10681 (41), 45057 (12), 45186 (12);

Ibaña 398 (74), 456 (74); Iltis 53 (100), 61 (100), 433 (27), 3123 (100), 3124 (27), 7832 (19), 28932 (100), 31236 (100); Irish, F. M. *s.n.* (41); Isaak 40 (12); Isely 10830 (44), 10845 (74), 10852 (44);

Janzen 15 (56); Jiménez, A. 3505 (56); Johnson, F.W. *s.n.* (9/18/1906) (42); Johnson, M.B. 89-105 (44); Johnston, I.M. 3794 (41), 4225 (41), 8900 (44), 8900A (44); Jones, D.J. 57 (12); Jones, M.E. *s.n.* (9/7/1903) (12), *s.n.* (9/16/1903) (18), *s.n.* (8/24/1903) (44); 81 (12), 87 (4), 88 (4), 591 (12), 592 (4), 593 (4), 4052 (12), 4252 (in part) (4), 22498 (in part) (4), 24141 (41), 24279 (41), 24280 (41), 24780 (41), 24879 (41), 24980 (41); Joseph *s.n.* (41); Joya (50); Judziewicz 5113 (18);

Kaiser 49-149 (44); Kamb 2011 (41); Kearney 8083 (74), 9226 (12), 9745 (12), 9885 (44), 10087 (4), 10090 (60), 10380 (4), 15135 (41); Keeler 18 (84); Keil 1175 (41), 1254 (41), 1325 (41), 1741 (12), 3431 (12); Kenoyer A61 (75), A69 (18); Killip 32715 (41); Kimball *s.n.* (41); Kleeberger *s.n.* (83); Kniffen *s.n.* (41); Knobloch 2365 (41), 5249 (74), 5525 (60), 5871 (75); Knowlton 227 (12); Kruckeberg 3892 (12); Kusche *s.n.* (8/9/1927) (44);

Landrum 5594 (12); Langlassé 204 (56); Langman 2557A (18), 2888 (6), 3533 (49); Lasseigne 1142 (83); Laughlin 1902A (in part) (50), 1902B (in part) (101), 2226 (18), 2310 (18), 2322 (18), 2400 (18), 2881 (56); Leavenworth 953 (32); Lehr 398 (12), 1008 (12), 1033 (12), 1151 (41), 2302 (60); Lehto *s.n.* (12), 707 (12), 2164 (4), 3308 (44), 3314 (4), 3648 (41), 5359a (4), 7933 (12), 10020 (4), 10261 (4), 14930 (4), 15127 (4), 17135 (4), 17167 (4), L20038 (41), L20185 (12), L20655 (12); Leithliter 175 (44), 195 (74), 594 (74), 736 (44); Lellbb *s.n.* (41); Lemmon *s.n.* (44), *s.n.* (59), 19 (100), 137 (in part) (12), 547 (44), 2694 (44); León, J.L. *s.n.* (41), 2211 (41), 3705 (41), 3987 (41), 4201 (41), 4227 (41); Le Sueur 688 (59), 708 (75), 1340 (74); Lewis 2037 (41); Liebmann 5318 (50); Lincoln 10636 (41); Lindheimer 31 (59); Lindsay 2155 (41); Livingston *s.n.* (44); Long, F. *s.n.* (41); López 279 (41), 617 (41); Lorence 3539 (24), 4688 (50), 4857 (42); Lundell 5521 (in part) (81), 16602 (56), 16836 (56); Lyonnet, E. 695 (50); Lyonnet, P. 2988 (76), 3360 (50), 3361 (75);

MacDaniels 363 (42), 785 (50), 787 (75); MacDougal *s.n.* (12), 48 (41), 111 (12); Magallanes 300 (65), 386 (56), 423 (56), 1387 (56), 1423 (64), 3315 (56); Maguire 11789 (12); Mahler 5479 (41); Marquez, 75 (76); Marsh 182 (44); Martínez 2455 (16), 5001 (50), 14229 (67), 19752 (98); Martínez-Ojeda 345 (56); Mason, H.L. 1668 (56), 1930 (41), 1988 (41); Mason, C.T. 1668 (41), 2631 (12), 2808 (12); Matuda 18891 (76), 19458 (50), 21292 (50), 21400 (75), 26554 (50), 27052 (50), 28825 (50); Maya 608 (50), 711 (50); Mayerhoff 104 (59); Maysilles 7528 (75), 7707 (75); McClintock *s.n.* (41), 52-73 (41), 52-544 (12); McCormick 349 (44), 410 (44), 728 (44); McCroskey *s.n.* (74); McFarlin 6579 (84); McKee 11368 (56); McLaughlin 12 (74), 1352 (59), 1428 (4), 2021 (41), 3926 (4), 3992 (60), 4085 (60), 4128 (4), R672 (74), R799 (74); McManus 106 (74); McVaugh 11586 (90); 13051 (75), 25113 (56); Mearns 852 (59), 1712 (44); Metcalfe, O.B. *s.n.* (4); Metcalfe, J.K. *s.n.* (59); Mexía 1372 (64); Minkley *s.n.* (12); Miranda 2291 (65), 2312 (65); Mishler *s.n.* (12); Mittleman 522 (4); Moir 558 (44); Moldenke 27994 (74); Molina 13130 (50), 18680 (70), 25183 (56), 25201 (56); Montalvo 5203 (18); Montufar 143 (41); Moore 5476 (50); Morales 1128 (70); Moran 2959 (41), 5843 (56), 7399 (41), 9057 (41), 19648 (41); Morefield 2782 (12); Mueller 8037 (44); Mulford 743 (44), 867 (44), 1110 (12); Müller 1700 (76); Mullinex *s.n.* (41); Munz 13073 (12);

Nabhan 184 (4), 586 (in part) (60), 677 (12), 726 (3), 743 (4), GN688 (59), GN695 (4), GN696 (4), GN699 (4), GN699a (4), GN702 (4), GN735 (59), X453 = GN240z (4), X523 (4), X942 (4); Nash 9973 (41); Neally 241 (74), 246 (60); Nelson, A. 1771 (41); Nelson, E.W. 1162 (24), 1222 (50), 1308 (65), 1502 (50), 1771 (41), 3177 (18), 4319 (56); Nicolas *s.n.* (50); Niering *s.n.* (44), 190 (44), 380 (74); Niles 342 (41); Norvell H310 (44), H330 (44);

O'Kennon 9945 (44), 10771 (44); Ochoa *s.n.* (56); Oefinger 43 (2); Ohlendorf 748 (41); Orcutt *s.n.* (41), 3490 (18); Orozco 376 (56), 422 (50), 734 (56), 880 (56); Ortega 870 (4), 5429 (56);

Palmer, E. *s.n.* (12), *s.n.* (59), 82 (41), 224 (65), 250 (41), 332 (18), 466 (56), 524 (65), 624 (18), 645 (4), 727 (41), 733 (41), 863 (50), 867 (100), 876 (41), 1116 (56), 6615 (4); Palmer, E.J. 10577 (83), 34353 (59); Parfitt 4144 (12); Parker 7071 (4), 7747 (41), 7916 (41); Parry 207 (44), 208 (12); Pase 1154 (4), 2149 (12); Pearson 53 (12); Peebles 570 (4), 570a (4), 1042 (41), 2451 (60), 2541 (60), 2651 (12), 2762 (4), 4316 (12), 4469 (44), 5230 (41), 8888 (12), 13491 (12); Pennell 17816 (18), 18271 (75), 18331 (74), 18336 (75), 18351 (11), 18453 (52), 19265 (59), 20114 (75); Pérez-Saldaña 21 (56); Perrill 5015 (41); Phillips, E.A. 636 (44), 835 (12); Pilz 1360 (18); Pinkava 2711 (41), 4366 (12), 5869 (12), 9973 (41), P12942 (90); Pittier 4992 (56), 9075 (36); Porter *s.n.* (83); Preece 2635 (12); Price *s.n.* (8/14/1893) (44), *s.n.* (8/7/1894) (59); Pringle *s.n.* (59), 126 (50), 576 (50), 1128 (2), 1232 (75), 5661 (69), 5670 (24), 5814 (24), 7116 (50), 9769 (76); Purpus *s.n.* (3/?/1903) (32), *s.n.* (10/?/1906) (50), 102 (41), 2686 (39), 3639 (39), 3889a (50), 4384 (39), 7099 (50), 8009 (50), 8180 (12), 8186 (12);

Quirós 1534 (56);

Ramirez, C. 213 (56); Ramírez, R. 614 (50), 767 (50); Ravenel *s.n.* (84); Reed, J. *s.n.* (41); Reeves 938 (74), 950 (74), R889 (74), R994 (44), R1017 (74), R1024 (in part) (74), R1309 (44), R1492 (4), R1584A (4), R1843 (44), R3613 (44), R3661 (44), R3805 (59), R4019 (4), R4279 (44), R4375 (44), R4569 (4); Reichhardt 87-95 (44); Rempel 156 (41); Richards *s.n.* (41); Ricketson 2237 (41); Rico, L. 614 (65); Ripley 4887 (12); Rodriguez 2081 (56); Roe 2081 (50); Rose *s.n.* (9/16/1897) (75), 2067 (60), 2725 (24), 5357 (31), 5694 (18), 6105 (42), 6106 (50), 6197 (59), 6515 (31), 6912 (50), 6984 (80), 7228 (80), 7451 (4), 7459 (49), 8318 (59), 8493 (50), 8622 (18), 9441 (50), 10215 (50), 13841 (56), 14424 (56), 16166 (41); Rossbach 4914 (44); Rothrock 417 (44), 535 (44); Rudd 3362 (41); Rusby *s.n.* (12), 107 (59), 582 (12), 583 (in part) (12); Russell, P.G. 182 (76); Russo 508 (41); Rugel *s.n.* (83), 131 (83); Ruth 309 (83), 2475 (83); Rzedowski, G.C. 26497 (41); Rzedowski, J. 1388 (50); 1770 (80), 1850 (50), 25552 (41);

Salvin *s.n.* (10 Oct 1873) (100); Sanders 2447 (41), 2449 (44), 2876 (44), 3046 (4), 3078 (4), 3150 (44), 3152 (44), 3154 (44), 3186 (59), 3240 (12), 4911 (90), 5243 (4), 5692 (41), 11647 (50), 11704 (73); Santana 3349 (56), 3965 (27), 4009 (27), 4112 (18), 4149 (27), 4225 (100), 4252 (100), 5453 (51), 5509 (50), 5525 (64), 5526 (18), 7800 (56), 8789 (50); Sargent 7893 (83); Schaffner 629 (59),

825 (59); Schipp 855 (56); Schallert *s.n.* (12); Schmalzel 1854 (98); Schmidt 66 (41), 102 (41); Schroeder *s.n.* (12); Schuerman *s.n.* (12); Schultes 463 (24); Schulz 2129 (59); Seler 3154 (50), 3380 (56); Sessé & Mociño *s.n.* 3700 & 3702 (50), *s.n.* 3701 (75); Shannon 541 (50); Shapiro 296 (56); Sharp 44934 (50); Sherman 157 (59); Shreve 7100 (41); Skutch 587 (18); Smith, C.L. 563 (24), 564 (50); Smith, E. C. 11052 (41), 14163 (74); Smith, J.D. 113 (83), 2297 (56); Solbrig 4571 (18); Sorensen 7832 (26); Soto 379 (50), 1954 (50), 3350 (50), 3400 (75), 4518 (50), 5966 (65), 10391 (75); Soulé 17 (44); Sousa 2942 (56), 3915 (56), 5060 (50), 5121 (76), 6001 (50), 6023 (50), 6087 (50), 6263 (50), 6282 (50), 6302 (50), 6395 (65), 6497 (50), 6708 (50), 7816 (69), 8180 (65), 8247 (50), 8261 (50), 8275 (65), 8298 (50), 8314 (50), 8412 (65), 8551 (50), 8886 (42), 9026 (50), 9720 (65), 10641 (50), 11374 (70; one specimen at MO mixed), 12481 (50), 12914 (50); Spellenberg *s.n.* (41); Sperry 651 (41), 1928 (44), 21137 (12); Stalmach 281 (4); Standley 272 (56), 904 (50), 916 (50), 1128 (56), 2147 (50), 2159 (50), 12869 (50), 14476 (56), 14777 (50), 16158 (56), 17689 (56), 19297 (56), 21181 (50), 21859 (50), 22425 (56), 23899 (50), 26603 (50), 27342 (50), 27655 (18), 28977 (50), 47339 (56), 57979 (50), 59779 (50), 62387 (98), 63642 (56), 66755 (56), 76999 (50), 77274 (50), 77326 (18), 77348 (18), 77477 (50), 79912 (50), 79990 (50), 81294 (50), 81540 (50), 81901 (50), 81911 (50), 83073 (50), 83556 (18), 83621 (18), 88532 (56), 88671 (56); Starr 680 (41), 700 (41); Stearns 169 (41); Steele *s.n.* (83); Stepan *s.n.* (41); Stephenson 67115 (41); Stevens 4484 (65); Stevenson 77 (12); Steyermark 32506 (18), 34043 (57), 48111 (50), 48126 (50); Streets *s.n.* (41); Sundberg 1307 (4), 1319 (44), 2135 (44);

Tate 650 (4); Tatnali 663 (83); Tedford *s.n.* (44); Téllez 4114 (50), 9930 (73), 11154 (56), 12135 (50), 12193 (2), 12224 (60), 12261 (50), 12306 (73); Tenorio 4297 (75), 6788 (50), 7547 (50), 9113 (50), 11946 (50), 12412 (50), 12894 (41), 12914 (41), 12991 (41), 13072 (41), 15606 (64), 15968 (75), 15975 (50), 16020 (50), 16087 (50), 16392 (50), 16428 (50), 18452 (50); Thames *s.n.* (41); Thomas 3739 (83), 7723 (41); Thornber *s.n.* (8/5/1907) (12), *s.n.* (8/4/1906) (74), 788 (12), 2638 (4), 4068 (12), 7491 (4); 7588 (4), 7998 (4), 8016 (4); Throne *s.n.* (8/5 & 8/10/60) (44); Ton 13 (18), 88 (18), 2762 (50), 2862 (18), 3498 (70), 7299 (98), 8450 (50); Tonduz 817 (50), 866 (50), 1453 (56), 1565 (56), 7200 (50), 7239 (56), 8970 (50); Toolin 52 (4), 145 (4), 903 (44), 1058 (74); Torres, M.L. 993 (65); Torres, R. 1735 (50), 1782 (50), 1882 (50), 1898 (100), 3520bis (49), 3631 (75), 3699 (75), 5647 (50), 5648 (50), 6268 (65), 6285 (65), 14050 (65); Toumey *s.n.* (12), 567 (12); Tucker *s.n.* (56), 516 (56); Turner, B.L. 5722 (4), 5726A (74), 5732 (44), 15621 (6); Turner, R.M. 59-5 (41), 67-95 (41), 83-26 (41);

Ugent 1948 (50);

Valdes R. 343 (50); Valerio, J. 2485 (56), 3062 (56), 3745 (56); Valerio, M. 79 (56), 1139 (56); Van Devender *s.n.* (9/11/77 & 9/6/81) (4), *s.n.* (3/1/73) (41), *s.n.* (9/3/83) (44), *s.n.* (8/28/83) (60), 17 (4), 87-6 (41), 91-390 (41); Vasek 66091-9 (4), 660918-2A (4); Vasey *s.n.* (?/?/81) (59); Vasey, G.R. *s.n.* (?/?/1881) (12), 442 (12); Vasquez, L. 1277 (50), 1389 (65), 1398 (100); Ventura, A. 318 (76), 3087 (75), 3223 (80), 3289 (80), 3566 (76), 3605 (76), 9983 (50), 10249 (50), 11163 (56); Villarreal, J.A. 4418 (41); Villarreal, L.M. 5389 (75); Villaseñor 39 (76); Vogel 110 (24); Voorhies 72-31 (50);

Wagenbreth 253 (50); Waines L220 (12); Walker 3708 (83); Ward *s.n.* (6/2/1901) (12); Warnock, B. H. 441 (in part) (44), 4375 (12), 13179 (41), 46658 (12), W819 (59), W820 (59); Warren *s.n.* (41); Weber, J. *s.n.* (5/5/52) (44); Weber, W.A. 12857 (59); Webster, G.L. 11516 (18), 22432 (41); Weldy 1047 (83); White 136 (50), 3408 (60), 3438 (44), 3473 (44), 4613 (2); Whitehouse *s.n.* (41); Whiting 756/940b (12); Wiggins 164 (41), 239 (41), 5523 (41), 6995 (in part) (49), 15123 (41), 15490B (41), 15512A (41), 15573 (41), 15831 (41), 16898 (41), 17152 (41), 17354 (41), 18072 (41); Wilbur 10879 (98); Williams, W. *s.n.* (41); Williams, L.O. 10555 (50), 11105 (18), 11892 (56), 15113 (56), 15116 (56), 16852 (18), 17292 (50), 18636 (18), 21837 (50), 22258 (50), 22432 (50), 24362 (98); Wilson, N. C. 126 (12); Wislizemus 109 (59); Wofford 87-60 (83); Wolf *s.n.* (12), 3089 (12); Woodward *s.n.* (83); Wooton *s.n.* (8/10 & 8/19/02) (4), *s.n.* (8/29/1894, 9/10/1899 & 9/11/1904) (12), *s.n.* (7/11/1900 & 8/19/02) (59), 434 (12); Worthington 3522 (12), 6814 (41), 6816 (12), 9070 (74), 9164 (41), 9224 (41), 13732 (41); Wright 107 (59), 108 (41), 111(41), 932 (44), 937 (12), 948 (4), 949 (3), 950 (4); Wyatt 48 (41);

Yatskievych 94-183 (59); Yong 266 (56); Young *s.n.* (59);

Zimmerman 2069 (59); Zizumbo 365 (65); Zuill 261 (50), 344 (50), 433 (18), 754 (98).

INDEX

Accepted names are italicized; new taxa and combinations are in **boldface;** all others are in roman type. Page numbers of major descriptions and citations are in **boldface.**

BIOGRAPHIC NOTES ABOUT THE AUTHORS

The senior author's (**Dr. George F. Freytag**) interest in *Phaseolus* taxonomy and genetics began in 1950 with a two year travel-study grant to the Escuela Agrícola Panamericana (El Zamorano), Honduras, Central America to undertake field work leading to a Ph.D. degree and thesis entitled "Variation of the common bean in Central America" (Washington University, St Louis, Missouri, 1955). Subsequently a two year program of collecting and studying cultivated beans throughout Mexico was pursued with the Oficina de Estudios Especiales, SARH, and field collecting with the late Prof. Efraín Hernández Xolocotzi provided numerous seed lots for field planting at El Horno, Chapingo and other field locations in Mexico for study. Subsequently, from 1955–1972 there followed many years of work at the Escuela Agrícola Panamericana at El Zamorano, Honduras as teacher, agronomist and bean breeder, and continued, from 1974–1990, as a USDA/ARS research geneticist and bean breeder based in Mayagüez, Puerto Rico. During these years an interest developed in utilizing the secondary and tertiary gene pools of the common bean for cultivar improvement. The principal focus was the study of variability within the *P. coccineus* complex, and a reassessment of the usual placement of the species *P. dumosus* (then known as *P. polyanthus*) within the *P. coccineus* complex, which eventually proved to be incorrect. Not until after several additional years of preliminary study and two field collecting trips, the first in 1978 to Mexico and Guatemala and the second in 1981 to Mexico, did it become evident that there were many, as yet unknown, taxa in the field to be collected and described.

At about the same time (1977), the junior author (**Dr. Daniel G. Debouck**) concluded with Prof. Robert Maréchal of the department of tropical agronomy in Gembloux, Belgium, that further progress in understanding taxonomy of Neotropical Phaseolinae would be possible thanks to field work. He moved to the Genetic Resources Unit of CIAT (International Center for Tropical Agriculture), and began a project to collect native bean cultivars and wild ancestral forms in Mexico, and soon became interested in wild bean speciation. Shortly thereafter the authors decided to collaborate on a taxonomic study of the cultivated species of the genus, and the relationships with their respective wild ancestors and the other wild species not touched by human domestication.

The senior author has studied living material in greenhouse and environmental chamber in Mayagüez, and has examined thousands of herbarium voucher specimens from the major herbaria holding collections of Neotropical legumes. He has been responsible for most of the description of species, the development of the keys, and the taxonomical treatments, namely that of the *Coccinei*. The junior author has studied and collected material in the field during twenty-seven germplasm explorations carried out in both Central and South America, resulting in more than 3,000 new germplasm accessions. He has studied material in herbaria of Latin America, and living material at CIAT. Through collaborations with colleagues and students in many universities, he has collated the data of molecular genetics, and compiled most of the literature.

Using their complementary expertise and long experience with this fascinating group of legumes, the authors have come with this taxonomical treatment; they would like to share with you their data, fascination and enthusiasm!